STRESS-CORROSION CRACKING

Edited by

RUSSELL H. JONES

ASM INTERNATIONAL®
The Materials Information Society

First printing, July 1992

Stress-Corrosion Cracking is a collective effort involving hundreds of technical specialists. It brings together in one book a wealth of information from world-wide sources to help scientists, engineers, and technicians solve current and long-range problems.

Comments, criticisms, and suggestions are invited, and should be forwarded to ASM International.

Library of Congress Catalog Card Number: 92-071978
ISBN 0-87170-441-2
SAN 204-7586

Production Manager - Linda Kacprzak
Production Assistant - Ann-Marie O'Loughlin

ASM International®
Materials Park, Ohio 44073-0002

Printed in the United States of America

Preface

The earliest reports of environment-induced cracking or embrittlement of materials include that of "season cracking" of brass, first observed more than 80 years ago (Ref 1), and the suggestion by W.H. Johnson in 1873 that hydrogen was a major cause of embrittlement in iron and steels (Ref 2). The incident of season cracking was discovered to result from the practice of storing cartridge cases in horse barns, where contact with ammonia vapors occurred. Hydrogen embrittlement was probably suspected by users of iron and steel products long before Johnson presented important details of this phenomenon. Many more cases of stress-corrosion or environment-induced cracking have been observed since these early studies.

Unfortunately, discovery of a particular stress-corrosion mechanism has often occurred because of a component failure that resulted in physical hazard or great economic cost. Early examples include the intergranular failure of a ferritic steel boiler (Ref 3), which was explained as hydrogen embrittlement (Ref 4), the explosion of a boiler by caustic "embrittlement" (Ref 5), and the failure of a steel hook (Ref 6) that caused the side of a building to fall on the street below. More recent examples include stress corrosion of steam turbine blading and rotors, nuclear piping, aircraft structures, and gas pipelines.

Stress-corrosion cracking (SCC) undoubtedly occurred prior to the earliest explanations cited above; however, the incidence of SCC has clearly increased since the beginning of the 20th century. This can be attributed to the use of more corrosion-resistant materials (which leads to increased susceptibility to localized corrosion), the use of higher stresses, and the complications caused by welding and complex forming operations. Our phenomenological understanding of SCC has also increased substantially since the beginning of the 20th century, as witnessed by the comprehensive nature of this book. Much is understood about the environmental, microstructural, microchemical, and mechanistic aspects that control stress corrosion, and it is often possible to avoid SCC failure of a particular component if these parameters are sufficiently well known. Unfortunately, relatively small alterations in environment or composition can easily place a material within the "susceptible zone," resulting in failure rather than extended life. Boundaries between susceptibility and resistance tend to be very distinct, so the exclusive use of phenomenological data to avoid SCC is often insufficient.

A mechanistic understanding of SCC can be used in conjunction with phenomenological data to identify whether environmental or material changes will increase the probability of failure. However, the present level of mechanistic understanding, while greatly increased over the last several decades, is still inadequate for prediction of component life. To underscore this point, the mechanisms for ammonia-induced stress corrosion of brass and hydrogen-induced crack growth of iron and steel, two of the earliest observations of environment-induced cracking, continue to be debated. Recent work by Pugh and his coworkers (Ref 7-9) and by Newman and Sieradzki (Ref 10, 11) has done much to settle the issue for transgranular SCC of brasses. The number of advances made in our mechanistic understanding of hydrogen-induced cracking is far too extensive to list, but includes models that describe atomistics, energetics, surface adsorption/mobility, fracture mechanics, dislocation, and cohesive energy concepts. For some aspects, modeling of hydrogen-induced cracking is sufficiently quantitative that crack growth rate predictions are possible. Yet discussion continues as to whether hydrogen-induced crack growth occurs by a cleavage or a slip mechanism.

This book thoroughly covers the phenomenological aspects of stress-corrosion cracking for a wide range of materials. An overview of SCC mechanisms is presented in Chapter 1, followed by chapters describing SCC of ferritic and austenitic steels and of nickel, copper, aluminum, magnesium, titanium, zirconium, and uranium alloys. Also included are chapters on irradiation-assisted stress corrosion, SCC of glasses, ceramics, and weldments, detection and sizing of cracks in boiling water reactors, evaluation of SCC, and failure analysis methods.

Russell H. Jones,
Battelle Pacific Northwest Laboratories

References

1. E.S. Sperry, *Brass World*, Vol 2, 1906, p 39
2. W.H. Johnson, *Iron*, Vol 1, 1873, p 291, 452-453
3. C.E. Stromeyer, *J. Iron Steel Inst.*, Vol 79, 1909, p 404
4. J.H. Andrew, *Trans. Faraday Soc.*, Vol 9, 1913, p 316
5. F.G. Straub, Illinois Experimental Station Bull. No. 216, 1930, p 8, 12
6. H.L. Logan, *J. Res., Natl. Bur. Stand.*, Vol 66c, 1962, p 347
7. E.N. Pugh, in *Theory of Stress Corrosion Cracking in Alloys*, J.C. Scully, Ed., NATO Science Affairs Division, 1971, p 418
8. J.A. Beavers and E.N. Pugh, *Metall. Trans. A*, Vol 11, 1980, p 809
9. M.T. Hahn and E.N. Pugh, *Corrosion*, Vol 36, 1980, p 380
10. R.C. Newman and K. Sieradzki, *Scr. Metall.*, Vol 17, 1983, p 621
11. K. Sieradzki and R.C. Newman, *Philos. Mag. A*, Vol 51, 1985, p 95

Contents

CHAPTER 1

Mechanisms of Stress-Corrosion Cracking

R.H. Jones, Pacific Northwest Laboratories
R.E. Ricker, National Institute of Standards and Technology

Stress-corrosion cracking (SCC) is a term used to describe service failures in engineering materials that occur by slow, environmentally induced crack propagation. The observed crack propagation is the result of the combined and synergistic interaction of mechanical stress and corrosion reactions.

Before SCC can be discussed in detail, we must clearly define the type of loading involved, the types of materials involved, the types of environments that cause SCC, and the nature of the interactions that result in this phenomenon. The term "stress-corrosion cracking" is frequently used to describe any type of environmentally induced or assisted crack propagation. However, this discussion will focus on the normal usage of the term as defined below.

One common misconception is that SCC is the result of stress concentration at corrosion-generated surface flaws (as quantified by the stress-intensity factor, K); when a critical value of stress concentration, K_{crit}, is reached, mechanical fracture results. Although stress concentration does occur at such flaws, it does not exceed the critical value required to cause mechanical fracture of the material in an inert environment ($K_{SCC} < K_{crit}$). Precorrosion followed by loading in an inert environment will not result in any significant crack propagation, while simultaneous environmental exposure and application of stress will cause time-dependent subcritical crack propagation. The term "synergy" is used to describe this process because the combined simultaneous interaction of mechanical and chemical forces results in crack propagation, whereas neither factor acting independently or alternately would result in the same effect. The exact nature of this interaction is the subject of numerous scientific investigations and will be covered in the section "Crack-Propagation Mechanisms" in this chapter.

The stresses required to cause SCC are small, usually below the macroscopic yield stress, and are tensile in nature. The stresses can be externally applied, but residual stresses often cause SCC failures. However, compressive residual stresses can be used to prevent this phenomenon. Static loading is usually considered to be responsible for SCC, while environmentally induced crack propagation due to cyclic loading is defined as corrosion fatigue. The boundary between these two classes of phenomena is vague, and corrosion fatigue is often considered to be a subset of SCC. However, because the environments that cause corrosion fatigue and SCC are not always the same, these two should be considered separate phenomena.

The term "stress-corrosion cracking" is usually used to describe failures in metallic alloys. However, other classes of materials also exhibit delayed failure by environmentally induced crack propagation. Ceramics exhibit environmentally induced crack propagation (Ref 1) and polymeric materials frequently exhibit craze cracking as a result of the interaction of applied stress and envi-

Table 1 Alloy/Environment Systems Exhibiting SCC

Alloy	Environment
Carbon steel	Hot nitrate, hydroxide, and carbonate/bicarbonate solutions
High-strength steels	Aqueous electrolytes, particularly when containing H_2S
Austenitic stainless steels	Hot, concentrated chloride solutions; chloride-contaminated steam
High-nickel alloys	High-purity steam
α-brass	Ammoniacal solutions
Aluminum alloys	Aqueous Cl^-, Br^-, and I^- solutions
Titanium alloys	Aqueous Cl^-, Br^-, and I^- solutions; organic liquids; N_2O_4
Magnesium alloys	Aqueous Cl^- solutions
Zirconium alloys	Aqueous Cl^- solutions; organic liquids; I_2 at 350 °C (660 °F)

Table 2 Alloy/Environment Combinations and the Resulting Films That Form at the Crack Tip

Metal or alloy	Environment	Initiating layer
α-brass, copper-aluminum	Ammonia	Dealloyed layer (Cu)
Gold-copper	$FeCl_3$	Dealloyed layer (Au)
	Acid sulfate	Dealloyed layer (Au)
Iron-chromium nickel	Chloride	Dealloyed layer (Au)
	Hydroxide	Dealloyed layer or oxide
	High-temperature water	Dealloyed layer or oxide
α-brass	Nitrite	Oxide
Copper	Nitrite	Oxide
	Ammonia (cupric)	Porous dissolution zone
Ferritic steel	High-temperature water	Oxide
	Phosphate	Oxide (?)
	Anhydrated ammonia	Nitride
	$CO/CO_2/H_2O$	Carbide
	CS_2/H_2O	Carbide
Titanium alloys	Chloride	Hydride
Aluminum alloys, steels	Various media	Near-surface hydrogen

ronmental reactions (Ref 2-5). The mechanical properties of composites will degrade if exposure to the environment attacks the matrix, the reinforcing phase, or the matrix-to-reinforcement interface; if crack propagation results during static loading, this degradation is SCC. Until recently, it was thought that pure metals were immune to SCC, but it is now known that this is not true (Ref 6, 7). Because considerably more research has been conducted on the SCC behavior of metallic alloys, the discussion in this chapter will focus on SCC of metals and their alloys.

Environments that cause SCC are usually aqueous and can be either condensed layers of moisture or bulk solutions. SCC is alloy/environment specific; that is, it is frequently the result of a specific chemical species in the environment. For example, the SCC of copper alloys, traditionally referred to as season cracking, is usually due to the presence of ammonia in the environment, and chloride ions cause or exacerbate cracking in stainless steels and aluminum alloys. Also, an environment that causes SCC in one alloy may not cause it in another. Changing the temperature, the degree of aeration, and/or the concentration of ionic species may change an innocuous environment into one that causes SCC failure. Also, different heat treatments may make the same alloy either immune or susceptible. As a result, the list of all possible alloy/environment combinations that cause SCC is continually expanding, and the possibilities are virtually infinite. Some of the more commonly observed alloy/environment combinations that result in SCC are listed in Table 1.

In general, SCC is observed in alloy/environment combinations that result in the formation of a film on the metal surface. These films may be passivating layers, tarnish films, or dealloyed layers. In many cases, these films reduce the rate of general or uniform corrosion, making the alloy desirable for resistance to uniform corrosion in the environment. As a result, SCC is of greatest concern in corrosion-resistant alloys exposed to aggressive aqueous environments. The exact role of films in the SCC process is the subject of current research and will be covered in the section "Crack-Propagation Mechanisms" in this chapter. Table 2 lists several alloy/environment combinations and the films that may form at the crack tip.

The Phenomenon of SCC

Stress-corrosion cracking is a delayed failure process. That is, cracks initiate and propagate at a slow rate (for example, 10^{-9} to 10^{-6} m/s) until the stresses in the remaining ligament of metal exceed the fracture strength. The sequence of events involved in the SCC process is usually divided into three stages:

- Crack initiation and stage 1 propagation
- Stage 2 or steady-state crack propagation
- Stage 3 crack propagation or final failure

The characteristics of each of these stages will be discussed in greater detail below. First, however, the techniques used to measure SCC will be reviewed briefly. Stress-corrosion cracking experiments can be categorized as:

- Tests on statically loaded smooth samples
- Tests on statically loaded precracked samples
- Tests using slowly straining samples

More detailed information on these tests can be found in the chapter "Evaluation of Stress-Corrosion Cracking" and in Ref 8.

Tests on statically loaded smooth samples are usually conducted at various fixed stress levels, and the time to failure of the sample in the environment is measured. Figure 1 illustrates the typical results obtained from this type of test. In Fig. 1, the logarithm of the measured time to failure, t_f, is plotted against the applied stress, σ applied, and the time to failure can be seen to increase rapidly with decreasing stress; a threshold stress, σ_{th}, is determined where the time to failure approaches infinity. The total time to failure at a given stress consists of the time required for the formation of a crack (the incubation or initiation time, t_{in}, and the time of crack propagation, t_{cp}). These experiments can be used to determine the maximum stress that can be applied in service without SCC failure, to determine an inspection interval to confirm the absence of SCC

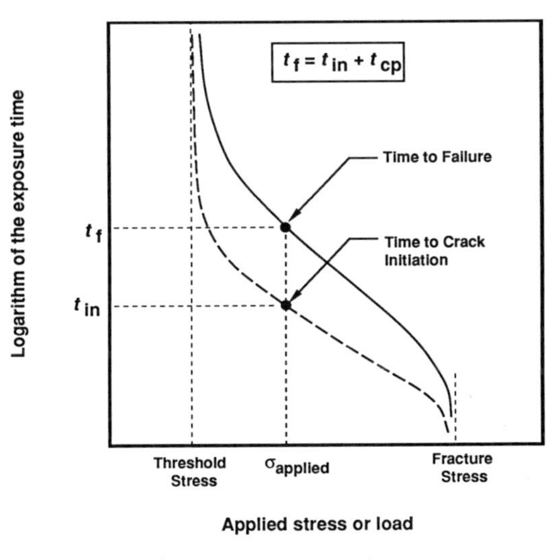

Fig. 1 Schematic of a typical time to failure as a function of initially applied stress for smooth-sample SCC tests

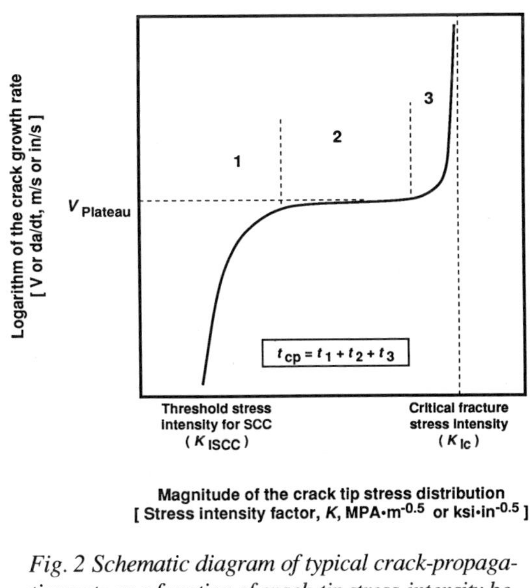

Fig. 2 Schematic diagram of typical crack-propagation rate as a function of crack-tip stress-intensity behavior illustrating the regions of stage 1, 2, and 3 crack propagation as well as identifying the plateau velocity and the threshold stress intensity

crack propagation, or to evaluate the influence of metallurgical and environmental changes on SCC. However, the time required for crack initiation is strongly dependent on a wide variety of parameters, such as surface finish. The presence of flaws that concentrate stress or crevices that alter the environment may dramatically change the threshold stress or the crack-initiation time (Ref 9, 10). The entire crack-initiation process is presently not well understood.

Tests on statically loaded precracked samples are usually conducted with either a constant applied load or with a fixed crack opening displacement, and the actual rate or velocity of crack propagation, da/dt, is measured (Ref 11). The magnitude of the stress distribution at the crack tip (the mechanical driving force for crack propagation) is quantified by the stress-intensity factor, K, for the specific crack and loading geometry. As a result, the crack-propagation rate, da/dt, is plotted versus K, as illustrated in Fig. 2. These tests can be configured such that K increases with crack length (constant applied load), decreases with increasing crack length (constant crack mouth opening displacement), or is approximately constant as the crack length changes (special tapered samples). Each type of test has its advantages and disadvantages. However, in service, most SCC failures occur under constant-load conditions, so that the stress intensity increases as the crack propagates. As a result, it is usually assumed in SCC discussions that the stress intensity is increasing with increasing crack length.

Typically, three regions of crack-propagation rate versus stress-intensity level are found during crack-propagation experiments. These are identified according to increasing stress-intensity factor as stage 1, 2, or 3 crack propagation (Fig. 2). No crack propagation is observed below some threshold stress-intensity level, K_{ISCC}. This threshold stress level is determined not only by the alloy but also by the environment and metallurgical condition of the alloy, and, presumably, this level corresponds to the minimum required stress level for synergistic interaction with the environment. At low stress-intensity levels (stage 1), the crack-propagation rate increases rapidly with the stress-intensity factor. At intermediate stress-intensity levels (stage 2), the crack-propagation rate approaches some constant velocity that is virtually independent of the mechanical driving force. This plateau velocity is characteristic of the alloy/environment combination and is the result of rate-limiting environmental processes such as mass transport of environmental species up the crack to the crack tip. In stage 3, the

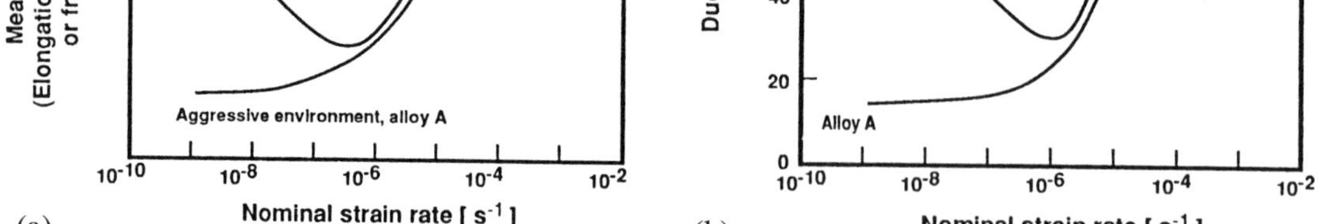

Fig. 3 Strain-to-failure plots resulting from slow-strain-rate testing. (a) Schematic of typical ductility versus strain-rate behavior of two different types of alloys tested by the slow-strain-rate technique. (b) Schematic of the ductility ratio versus strain-rate behavior of two different types of alloys. The ductility ratio is the ratio of a ductility measurement such as elongation, reduction in area, or fracture energy measured in the aggressive environment to that obtained in the inert reference environment.

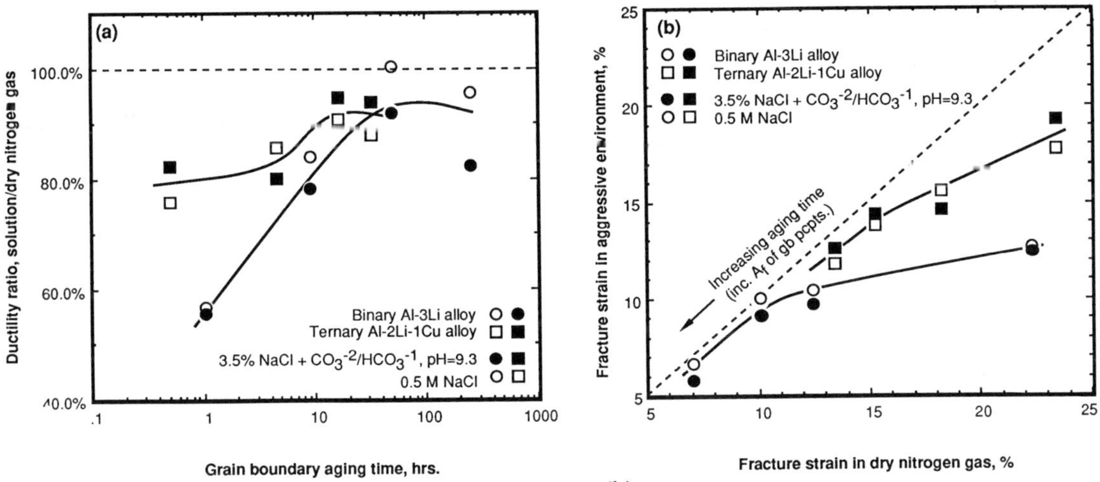

(a) (b)

Fig. 4 Comparison of (a) slow-strain-rate data plotted as a ductility ratio to (b) the same data plotted as an environment-dependent property versus the environment-independent value of the same property. Source: Ref 13

rate of crack propagation exceeds the plateau velocity as the stress-intensity level approaches the critical stress-intensity level for mechanical fracture in an inert environment, K_{Ic} (Ref 12).

Slow-Strain-Rate Testing. Stress-corrosion tests can also be conducted by slowly increasing the load or strain on either precracked or smooth samples. These tests are called constant-extension-rate tests, slow-strain-rate tests, or straining electrode tests. Usually, a tensile machine pulls a smooth sample that is exposed to the corrosive environment at a low crosshead speed (10^{-5} to 10^{-9} m/s). The strain to failure in the corrosive environment and the strain to failure in an inert environment can then be plotted against the strain rate, as shown in Fig. 3(a), or the ratio of these measurements can be plotted as shown in Fig. 3(b). The ratio of other tensile-property measurements, such as reduction in area and ultimate tensile strength, may be plotted.

Frequently, this type of test is used to evaluate the influence of metallurgical variables, such as heat treatment, on SCC resistance. This type of experiment yields rapid comparisons. However, since the mechanical properties of the samples also vary with the metallurgical condition, such evaluations can become difficult. As a result, it was proposed (Ref 13) that the environment-dependent property be plotted versus the inert-environment or environment-independent value of this property, as shown in Fig. 4. In this manner, the strength of the material in the environment, or the "situation-dependent strength" (Ref 10), and the extent of any environmental effect can be visualized simultaneously. However, the application of these data to the prediction of actual in-service lifetimes is difficult and unreliable.

Overview of SCC Mechanisms

Many different mechanisms have been proposed to explain the synergistic stress-corrosion interactions that occur at the crack tip, and there may be more than one process that causes SCC. The proposed mechanisms can be classed into two basic categories: anodic mechanisms and cathodic mechanisms. That is, during corrosion, both anodic and cathodic reactions must occur, and the phenomena that result in crack propagation may be associated with either type. The most obvious anodic mechanism is that of simple active dissolution and removal of material from the crack tip. The most obvious cathodic mechanism is hydrogen evolution, absorption, diffusion, and embrittlement. However, a specific mechanism must be able to explain the actual crack-propagation rates, the fractographic evidence, and the mechanism of formation or nucleation of cracks. Some of the more prominent of the proposed mechanisms will be covered in greater detail in the section "Crack-Propagation Mechanisms" in this chapter, but they usually assume that breaking of the interatomic bonds of the crack tip occurs either by chemical solvation and dissolution or by mechanical fracture (ductile or brittle). Mechanical fracture includes normal fracture processes that are assumed to be stimulated or induced by one of the following interactions between the material and the environment:

- Adsorption of environmental species
- Surface reactions
- Reactions in the metal ahead of the crack tip
- Surface films

All of the proposed mechanical fracture mechanisms contain one or more of these processes as an essential step in the SCC process. Specific mechanisms differ in the processes assumed to be responsible for crack propagation and the way that environmental reactions combine to result in the actual fracture process.

Controlling Parameters

The mechanisms that have been proposed for SCC require that certain processes or events occur in sequence for sustained crack propagation to be possible. These requirements explain the

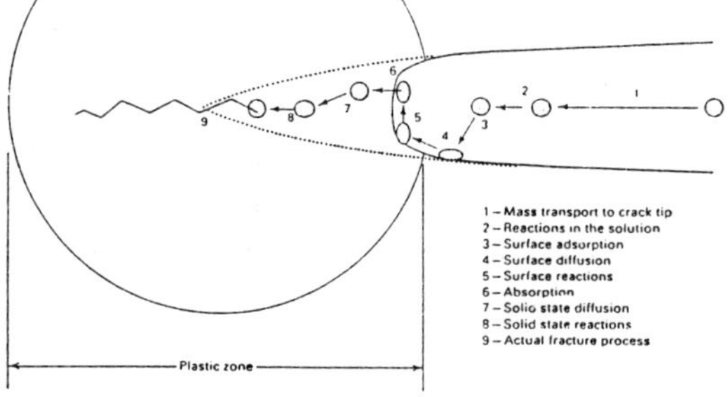

Fig. 5 Schematic of crack-tip processes that may be the rate-determining step in environmentally assisted crack propagation. For this illustration, an internal hydrogen embrittlement mechanism is assumed in order to maximize the number of possible rate-determining steps.

plateau region in which the rate of crack propagation is independent of the applied mechanical stress. That is, a sequence of chemical reactions and processes is required, and the rate-limiting step in this sequence of events determines the limiting rate or plateau velocity of crack propagation (until mechanical overload fracture starts contributing to the fracture process in stage 3). Figure 5 illustrates a crack tip in which crack propagation results from reactions in metal ahead of the propagating crack. This example was chosen because it maximizes the number of possible rate-limiting steps. Close examination of Fig. 5 reveals that potential rate-determining steps include:

- Mass transport along the crack to or away from the crack tip
- Reactions in the solution near the crack
- Surface adsorption at or near the crack tip
- Surface diffusion
- Surface reactions
- Absorption into the bulk
- Bulk diffusion to the plastic zone ahead of the advancing crack
- Chemical reactions in the bulk
- The rate of interatomic bond rupture

Changes in the environment that modify the rate-determining step will have a dramatic influence on the rate of crack propagation, while alterations to factors not involved in the rate-determining step or steps will have little influence, if any. However, significantly retarding the rate of any one of the required steps in the sequence could make that step the rate-determining one. In aqueous solutions, the rate of adsorption and surface reactions is usually very fast compared to the rate of mass transport along the crack to the crack tip. As a result, bulk transport into this region or reactions in this region are frequently believed to be responsible for determining the steady-state crack-propagation rate or plateau velocity. In gaseous environments, surface reactions, surface diffusion, and adsorption may be rate limiting, as well as the rate of bulk transport to the crack tip (Ref 14, 15).

Several different environmental parameters are known to influence the rate of crack growth in aqueous solutions. These include, but are not limited to:

- Temperature
- Pressure
- Solute species
- Solute concentration and activity
- pH
- Electrochemical potential
- Solution viscosity
- Stirring or mixing

Altering any of these parameters may affect the rate of the rate-controlling steps, either accelerating or reducing the rate of crack propagation. Also, it may be possible to arrest or stimulate crack propagation by altering the rate of an environmental reaction. It is well known and generally accepted that the environment at occluded sites, such as crack tips, can differ significantly from the bulk solution. If an alteration to the bulk environment allows the formation of a critical SCC environment at crack nuclei, then crack propagation will result. If the bulk environment cannot maintain this local crack-tip environment, then crack propagation will stop. As a result, slight changes to the environment may have a dramatic influence on crack propagation, while dramatic changes may have only a slight influence.

In addition to the environmental parameters listed above, stress-corrosion crack-propagation rates are influenced by:

- The magnitude of the applied stress or the stress-intensity factor
- The stress state, which includes (1) plane stress and (2) plane strain
- The loading mode at the crack tip (tension or torsion, for example)
- Alloy composition, which includes (1) nominal composition, (2) exact composition (all constituents), and (3) impurity or tramp element composition
- Metallurgical condition, which includes (1) strength level, (2) second phases present in the matrix and at the grain boundaries, (3) composition of phases, (4) grain size, (5) grain-boundary segregation, and (6) residual stresses
- Crack geometry, which includes (1) length, width, and aspect ratio, and (2) crack opening and crack-tip closure

Important Fracture Features

Stress-corrosion cracks can initiate and propagate with little outside evidence of corrosion and with no warning as catastrophic failure approaches. The cracks frequently initiate at surface flaws that either preexist or are formed during service by corrosion, wear, or other processes. The cracks then grow with little macroscopic evidence of mechanical deformation in metals and alloys that are normally quite ductile. Crack propagation can be either intergranular or trans-

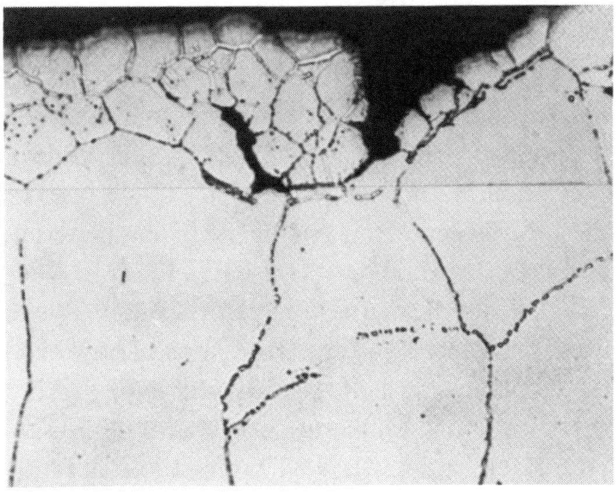

(a)

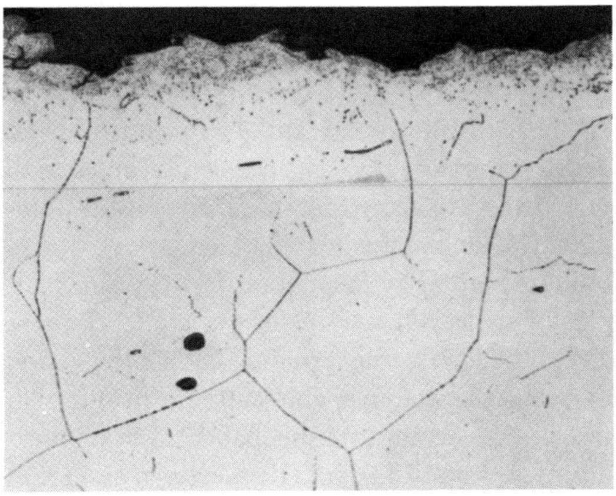

(b)

Fig. 6 Optical micrographs showing defects on the inner surface of type 304 stainless steel pipe near weld root (a) and near through-crack (b). Both 1000×

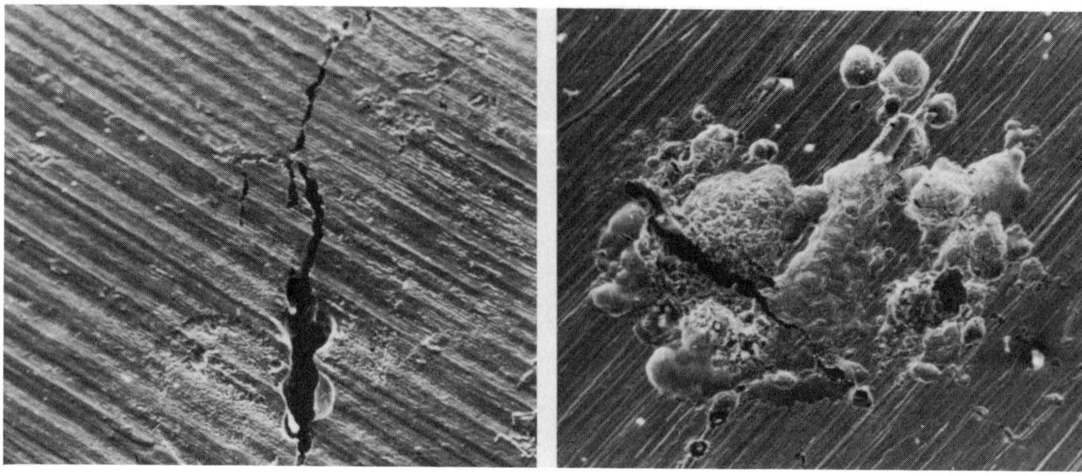

(a) (b)

Fig. 7 Stress-corrosion crack initiating from a corrosion pit in a quenched-and-tempered high-strength turbine disk steel (3.39Ni-1.56Cr-0.63Mo-0.11V) test coupon exposed to oxygenated, demineralized water for 800 h under a bending stress of 90% of the yield stress. (a) 275×. (b) 370×. Courtesy of S.J. Lennon, ESCOM, and F.P.A. Robinson, University of Witwatersrand

granular; sometimes, both types are observed on the same fracture surface. Crack openings and the deformation associated with crack propagation may be so small that the cracks are virtually invisible except in special nondestructive examinations. As the stress intensity increases, the plastic deformation associated with crack propagation increases and the crack opening increases. When the final fracture region is approached, plastic deformation can be appreciable, because corrosion-resistant alloys are frequently quite ductile.

Phenomenology of Crack-Initiation Processes

Crack Initiation at Surface Discontinuities. Stress-corrosion cracking frequently initiates at preexisting or corrosion-induced surface features. These features may include grooves, laps, or burrs caused by fabrication processes. Examples of such features are shown in Fig. 6; these were produced during grinding in the preparation of a joint for welding. The feature shown in Fig. 6(a) is a lap, which is subsequently recrystallized during welding and could then act as a crevice at which deleterious anions or cations concentrate. The highly sensitized recrystallized material could also more readily become the site of crack initiation by intergranular corrosion. A cold-worked layer and surface burrs, shown in Fig. 6(b), can also assist crack initiation.

Crack Initiation at Corrosion Pits. Stress-corrosion cracks can also initiate at pits that form during exposure to the service environment (Fig. 7) or during cleaning operations, such as pickling of type 304 stainless steel before fabrication. Pits can form at inclusions that intersect the free surface or by a breakdown in the protective film. In electrochemical terms, pits form when the potential exceeds the pitting potential. It has been shown that the SCC potential and pitting potential were identical for steel in nitrite solutions (Ref 16, 17).

The transition between pitting and cracking depends on the same parameters that control SCC, that is, the electrochemistry at the base of the pit, pit geometry, chemistry of the material, and stress or strain rate at the base of the pit. A detailed description of the relationship between these parameters and crack initiation has not been developed because of the difficulty in measuring crack initiation. Methods for measuring short surface cracks are under development, but are limited to detecting cracks that are beyond the initiation stage. The geometry of a pit is important in determining the stress and strain rate at its base.

Generally, the aspect ratio between the penetration and the lateral corrosion of a pit must be greater than about 10 before a pit acts as a crack-initiation site. A penetration to lateral corrosion ratio of 1 corresponds to uniform corrosion, and a ratio of about 1000 is generally observed for a growing stress-corrosion crack. As in the case of a growing crack, the pit walls must exhibit some passive-film-forming capability in order for the corrosion ratio to exceed 1.

A change in the corrosive environment and potential within a pit may also be necessary for the pit to act as a crack initiator. Pits can act as occluded cells similar to cracks and crevices, although in general their volume is not as restricted. There are a number of examples in which stress-corrosion cracks initiated at the base of a pit by intergranular corrosion. In these circumstances, the grain-boundary chemistry and the pit chemistry were such that intergranular corrosion was favored. Crack propagation was also by intergranular SCC in these cases.

Although the local stresses and strain rates at the base of the pit play a role in SCC initiation, there are examples of preexisting pits that did not initiate stress-corrosion cracks. This observation has led to the conclusion that the electrochemistry of the pit is more important than the local stress or strain rate (Ref 16). A preexisting pit may not develop the same local electrochemistry as one grown during service, because the development of a concentration cell depends on the presence of an actively corroding tip that establishes the anion and cation current flows. Similarly, an inability to reinitiate crack growth in samples in which active growth was occurring before the samples were removed from solution, rinsed, dried, and reinserted into solution also suggests that the local chemistry is very important.

Crack Initiation by Intergranular Corrosion or Slip Dissolution. Stress-corrosion crack initiation can also occur in the absence of pitting by intergranular or slip-dissolution processes. Intergranular corrosion-initiated SCC requires that the local grain-boundary chemistry differ from the bulk chemistry. This condition occurs in sensitized austenitic stainless steels or with the segregation of impurities such as phosphorus, sulfur, or silicon in a variety of materials. Slip-dissolu-

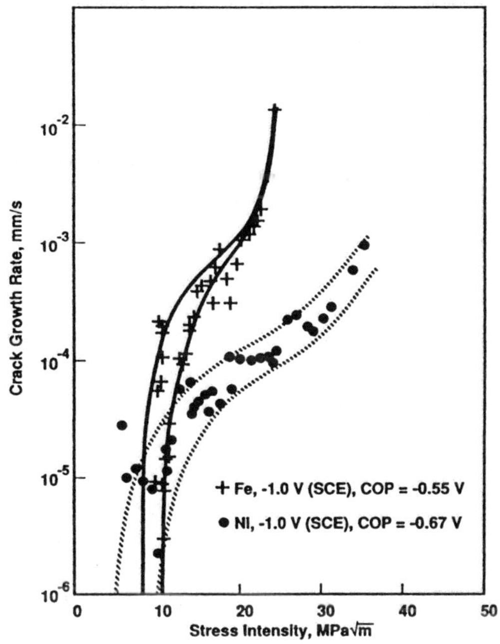

(a)

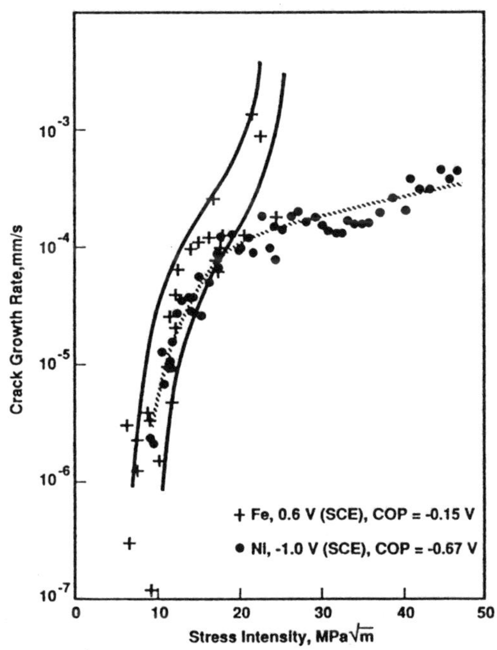

(b)

Fig. 8 Crack-growth rate versus elastic-plastic stress intensity for iron and nickel tested in 1 N H_2SO_4 at given cathodic overpotentials (COP). (a) 2 mm (0.08 in.) thick iron and nickel. (b) 10 mm (0.4 in.) thick iron and nickel tion-initiated SCC results from local corrosion at emerging slip planes and occurs primarily in low-stacking-fault materials. The processes of crack initiation and propagation by the slip-dissolution process are in fact very similar.

Crack initiation can be thought of as occurring in several stages, including the first few atomic layers that fail due to the transition from short-crack to long-crack behavior. Evidence obtained by electrochemical (Ref 18) and acoustic-emission monitoring (Ref 19) of crack initiation in austenitic stainless steel suggests that the process occurs by the initiation of multiple short cracks prior to the propagation of a single dominant crack. Examples of the electrochemical and acoustic-emission transients thought to be associated with the growth and retardation of multiple short cracks are given in Fig. 8. Retardation of the initiated cracks occurs because of variations in grain-boundary sensitization, crack angle, electrochemistry, and so on, with cracks being arrested after they grow about one grain diameter. Details of the conditions required for a single dominant crack to propagate have not been fully evaluated, but most likely it is a statistical process where there is a finite probability for a crack to find a path that does not retard its growth.

Another concept suggested by Parkins (Ref 20) for gas pipeline SCC initiation involves crack coalescence. It has been suggested that crack initiation in gas pipeline steels occurs by the following steps:

- Penetration of ground water to the pipeline surface and generation of a critical SCC environment
- Initially rapid formation of multiple cracks, the velocity decreasing with time to a constant value
- A period of constant crack-growth velocity
- Conventional stages 1, 2, and 3 as described by single-crack behavior, during which time crack coalescence leads to final fracture

Crack-Initiation Mechanisms

Although the features causing SCC initiation, such as pits, fabrication defects, and intergranular corrosion, are easily observed and identified, there are few well-developed models of SCC initiation, for several reasons. For example, the initiation of a crack is difficult to measure experimentally, even though it is not difficult to detect the location from which a growing crack has emanated. Furthermore, crack initiation has not been precisely defined, and it is difficult to determine at what point a pit is actually a crack and when intergranular corrosion becomes intergranular SCC. Also, the fracture-mechanics concept of design assumes preexisting flaws in materials, although these may not be surface flaws that can become stress-corrosion cracks. It has been demonstrated that the corrosion-fatigue threshold of 12% Cr and 2.0% NiCrMoV steels could be related to the minimum depth of surface pits, as shown in Fig. 9. Using a linear-elastic fracture-mechanics approach and relating a pit to a half-elliptical surface crack, one researcher has shown that the critical pit dimension could be expressed by the following relationship (Ref 21):

$$a_0 = \frac{1}{\pi}\left(\frac{\Delta K_{th}}{F\Delta\sigma_0}\right)^2 \qquad \text{(Eq 1)}$$

where ΔK_{th} is the corrosion fatigue threshold, F is a constant, and $\Delta\sigma_0$ is the alternating surface stress. A pit could be represented by a half-elliptical surface crack because it had intergranular corrosion at the base that caused the pit to have cracklike characteristics.

A model for stress-corrosion crack growth from pits in brass tested in a nontarnishing ammoniacal solution was developed (Ref 22). The researchers evaluated the corrosion and stress aspects of a pit to develop a model for crack initiation. They assumed that the corrosion conditions in the base of the pit were essentially the same as those on a flat surface and that initiation required that a critical crack-tip opening displacement be exceeded. Using linear-elastic fracture mechanics, the researchers developed the following relationship for the time to initiate a crack:

$$t_{inc} = \frac{(K_{ISCC})^2}{\pi B\,(\sigma^2 - \sigma_0^2)} \exp\left(\frac{-V_m}{V_0}\right) \qquad \text{(Eq 2)}$$

where K_{ISCC} is the stress-corrosion threshold for crack growth for brass in ammoniacal solution, σ is the applied stress, σ_0 is the stress needed to close the crack, B is a constant, $-V_m$ is the electrochemical potential of the sample, and V_0 is the reversible potential. However, the assumption that

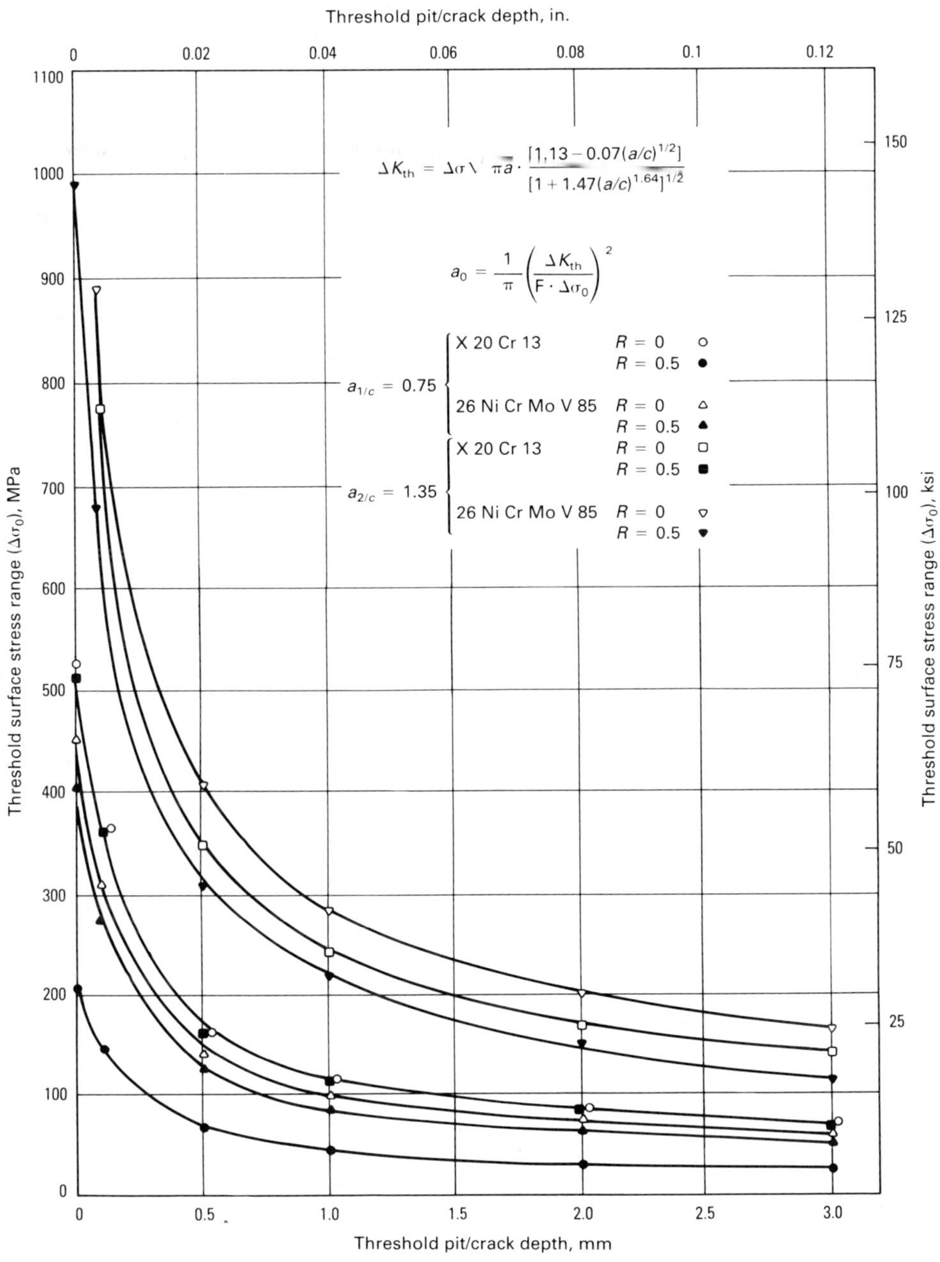

Fig. 9 Relationship between stress amplitude and minimum depth of surface defects for 12% Cr steel and a 2.0% NiCrMoV steel. Source: Ref 21

the pit corrosion conditions are equal to the flat surface conditions limits the applicability of this model, because it is generally accepted that the electrochemistry in a pit is considerably different from that of a flat surface. Also, this model does not explicitly describe the transition from a pit to a crack, but treats a pit as a small crack in which the crack-tip opening and crack depth are affected by corrosion.

The transition between intergranular corrosion and intergranular SCC was evaluated for nickel with segregated phosphorus and sulfur

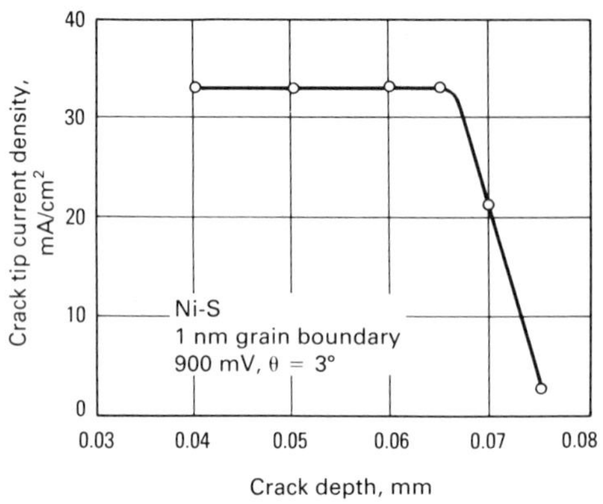

Fig. 10 Effect of crack depth on crack-tip corrosion rates for nickel with actively corroding crack walls

(Ref 23, 24). Because phosphorus and sulfur inhibit the formation of a passive film on nickel, it was expected that intergranular SCC would be similar in these two materials. It was observed, however, that only shallow intergranular corrosion to a depth of about 0.1 mm (0.004 in.) resulted in nickel in which sulfur was segregated to the grain boundary when tested at a strain rate of 10^{-6} in 1 *N* sulfuric acid (H_2SO_4). On the other hand, in a test under similar conditions, rapid intergranular SCC resulted in nickel in which phosphorus was segregated to the grain boundaries. Using crack-tip chemistry modeling, it was shown that intergranular corrosion did not convert to a stress-corrosion crack in Ni + S because the sulfur remained on the crack walls, causing the electrolyte in the solution to become saturated and the crack-tip corrosion rate to drop to zero. Starting from a flat surface, the corrosion current at the tip of the intergranular corrosion penetration was about 30 mA/cm^2 (195 mA/in.^{2}) and decreased to 0 at a depth of about 0.08 mm (0.003 in.), as shown in Fig. 10. A similar corrosion depth was measured for samples held at 900 mV for extended periods or for straining electrode samples, while for Ni + P a crack growth rate of 10^{-4} mm/s was measured. These results indicate the importance of electrochemistry in the base of a pit or an intergranular penetration and show that the occurrence of intergranular corrosion does not *a priori* indicate that intergranular SCC will follow.

Phenomenology of Crack-Propagation Processes

Crack initiation and propagation are related but different processes; however, by definition, if a crack initiates, it will propagate. As indicated in the section "Crack-Initiation Mechanisms" in this chapter, cracks may initiate at preexisting surface flaws if the necessary electrochemical, mechanical, and metallurgical conditions are met, or corrosion processes may create a surface flaw by pitting or localized corrosion. The conditions that create a pit or localized corrosion at chemical inhomogeneities such as grain boundaries, inclusions, second phases, and interphase boundaries are not necessarily the same as those needed for sustained crack growth. For example, the electrochemical conditions near the surface of a material are similar to the bulk electrolyte conditions; electrolyte conditions in a propagating crack generally differ from the bulk electrolyte. If the conditions at the tip of the growing pit or localized corrosion penetration cannot achieve the proper pH, potential, or chemistry, crack propagation may not proceed. Also, if the pit aspect ratio is not greater than a critical value, then the local stresses and strains resulting from a sustained sharp crack are not attained and crack growth cannot occur. Lastly, for stress-corrosion crack growth resulting from cathodic hydrogen, a short crack may have to initiate by an anodic dissolution process, while propagation depends on hydrogen activity.

Although there are conditions under which local corrosion or pitting may not result in crack propagation, more frequently they do lead to cracking. Three conditions frequently cited as requirements for sustained SCC are a susceptible material, a corrosive environment, and an adequate stress. The list of susceptible material conditions and corrosive environments that are known to cause SCC has been steadily growing. For example, pure metals are known to crack when impurities are segregated to grain boundaries, nonsensitized stainless steels crack in high-purity water, and ferritic steels crack in environments other than nitrate and caustic solutions. Similarly, the details of the microstructure and microchemistry of a susceptible material, specific ions and chemistry, pH, potential, the corrosion rates of the local crack tip associated with a corrosive environment, and the crack-tip stresses, strains, and strain rates associated with cracking are better known today than they were previously. Knowledge and understanding of the factors and mechanisms of stress corrosion are continuing to expand; however, a number of factors are well established. The phenomenological description of stress corrosion based on these known factors is presented in this section.

Environmental Factors

Environmental effects on SCC are frequently summarized simply by listing alloy/environment combinations in which SCC has been observed, such as those in Table 1. In recent years, the number of such combinations has increased. Added to this list are the transgranular SCC of copper, the intergranular SCC of pure metals such as iron and nickel, and the SCC of materials in high-purity water in the absence of specific anions. Stress-corrosion cracking of pure metals was thought to be impossible in early lists of material/environment combinations, but it is now realized that specific environment/metal reactions occur that allow crack advance. In the case of intergranular SCC of iron and nickel, this reaction occurs between the environment and impurities segregated to the grain boundaries (a condition that exists in many pure metals and engineering alloys).

This increase in known susceptible materials can be attributed to the use of new SCC tests, refined crack-propagation monitoring equipment, improved electrochemical control equipment, and, perhaps most importantly, the increased research activity in this field over the past 20 to 30 years. Expectations for material performance have also increased with time as materials are used in more aggressive environments under more demanding loading conditions. Lists such as that given in Table 1 can be useful in materials selection for a design involving corrosive environments, because they can lead the materials engineer to seek more specific information on the materials and environments in question. However, such lists can also be misleading; for example, service conditions may differ markedly from those under which the susceptibilities listed in Table 1 were determined. Because SCC is dependent on bulk alloy chemistry, microstructure, microchemistry, loading parameters, and specific environmental factors such as oxidizing potential and pH, lists of the type given in Table 1 should be used only for a general overview of SCC.

A complete description of SCC must treat both the thermodynamic requirements and kinetic aspects of cracking. A knowledge of the thermodynamic conditions will help determine whether cracking is feasible; kinetic information describes the rate at which cracks propagate. There are thermodynamic requirements both for anodically assisted SCC (where crack propagation depends either directly or indirectly on the oxidation of metal atoms from the crack tip and their dissolution in the electrolyte) and for hydrogen-assisted crack growth resulting from the reduction of hydrogen ions at the crack tip.

Thermodynamics of SCC. The thermodynamic conditions for anodically assisted SCC are that dissolution or oxidation of the metal and its dissolution in the electrolyte must be thermodynamically possible and that a protective film, such as an oxide or salt, must be thermodynamically stable. The first condition becomes a requirement because, without oxidation, crack advance by dissolution would not result. That a process is controlled by anodic dissolution does not *a priori* indicate that the total crack extension is the sum of the total number of coulombs of charge exchanged at the crack tip. There are crack-advance processes in which the crack advance is controlled by anodic dissolution but in which the total crack length is greater than can be accounted for by the total charge transfer. These processes are covered in the section "Crack-Propagation Mechanisms" in this chapter. However, it is important to note that if the brittle crack-advance process is initiated and controlled by anodic dissolution, the crack-growth rate will be zero if the anodic current density is zero and will increase with increasing current density. This also holds true for crack advance resulting from brittle crack jump.

The thermodynamic requirement of simultaneous film formation and oxidation for stress-corrosion crack growth can be understood from the diagram shown in Fig. 11, in which the ratio of

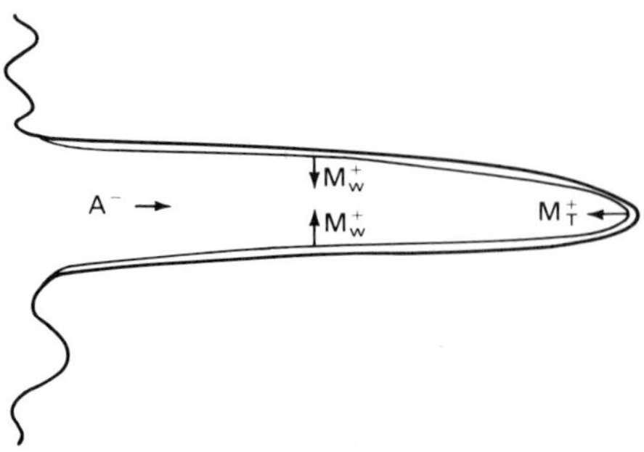

Fig. 11 Schematic of stress-corrosion crack showing important transport and corrosion reactions. A^- represents negatively charged anions migrating to the crack tip, M_w^+ represents metal ions entering the crack solution from the crack walls, and M_T^+ represents metal ions entering the crack solution from the crack tip.

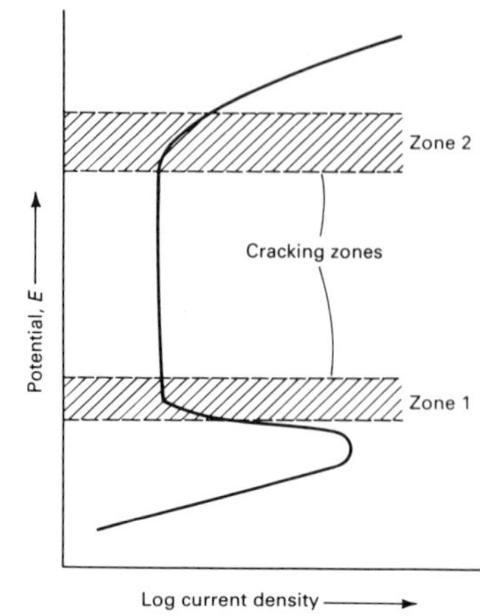

Fig. 12 Potentiokinetic polarization curve and electrode potential values at which SCC frequently appears

the corrosion currents from the walls relative to the crack tip is the critical parameter. This ratio must be substantially less than 1 for a crack to propagate; otherwise, the crack will blunt, or the crack-tip solution will saturate. Crack initiation can also be controlled by this ratio, because a pit with a high rate of wall corrosion will broaden as fast or faster than it will penetrate, resulting in general corrosion rather than crack growth. It is generally believed that the activity of the crack walls relative to the crack tip is a consequence of greater dynamic strain at the tip than along the walls.

A thermodynamic requirement of simultaneous film formation and oxidation of the underlying material led to the identification of critical potentials for the presence or absence of SCC. An example of these critical potentials is shown in Fig. 12 for a passive-film-forming material such as stainless steel. Zones 1 and 2 are those in which transgranular stress-corrosion crack growth is most likely to occur; intergranular stress-corrosion crack growth can occur over a wider range of potentials than these two zones. Transgranular SCC occurs in zone 1 because the material is in transition from active corrosion to passive-film formation such that the simultaneous conditions for film formation on the crack walls and corrosion at the crack tip are met. A similar condition exists in zone 2, with the added factor that these potentials are at or above the pitting potential so that cracks can initiate by pitting.

Intergranular SCC occurs over a wider range of potentials than those shown for zones 1 and 2 because chemical inhomogeneities at the grain boundary produce a different electrochemical response relative to the bulk material. Therefore, passive crack walls and active crack tips can result over the potential range from zone 1 to zone 2.

Examples of the critical potentials for SCC are shown in Fig. 13 for several materials. Identification of critical potentials for SCC has led to the use of electrochemical methods for assessing stress-corrosion susceptibility. Zones 1 and 2 are identified by determining the electrochemical potential versus current curves, as shown in Fig. 12 and 13. The shapes of these curves determined at high and low sweep rates are also used to indicate potentials at which the simultaneous conditions of film formation and metal oxidation occur.

Application of Potential-pH Diagrams. Critical potentials for SCC can also be related to potential-pH stability diagrams (Pourbaix diagrams), because these diagrams describe the conditions at which film formation and metal oxidation will occur. An example for carbon steel is given in Fig. 14, in which SCC is associated with potentials and pHs at which phosphate, carbonate, or magnetite films are thermodynamically stable while the species Fe^{2+} and $HFeO_2^-$ are

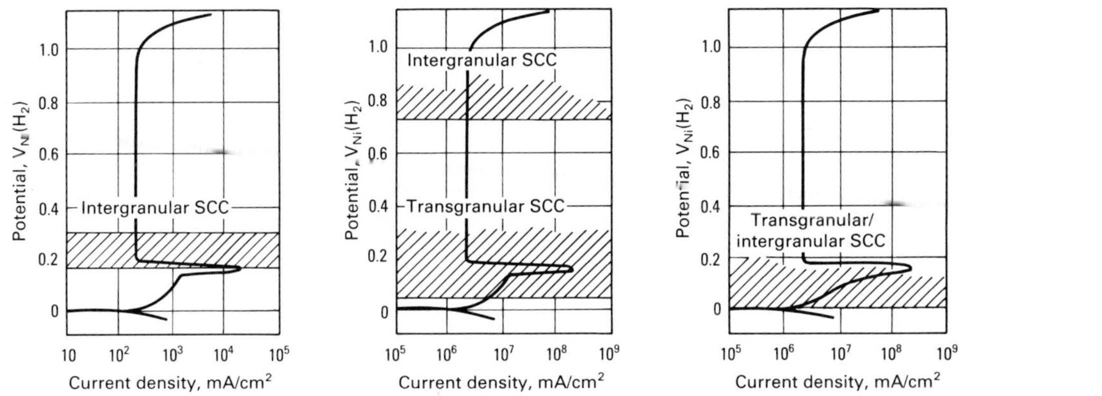

(a) (b) (c)

Fig. 13 Potentiokinetic polarization curve and electrode potential values at which intergranular and transgranular SCC appears in a 10% NaOH solution at 288 °C (550 °F). (a) Alloy 600. (b) Alloy 800. (c) AISI type 304 stainless steel

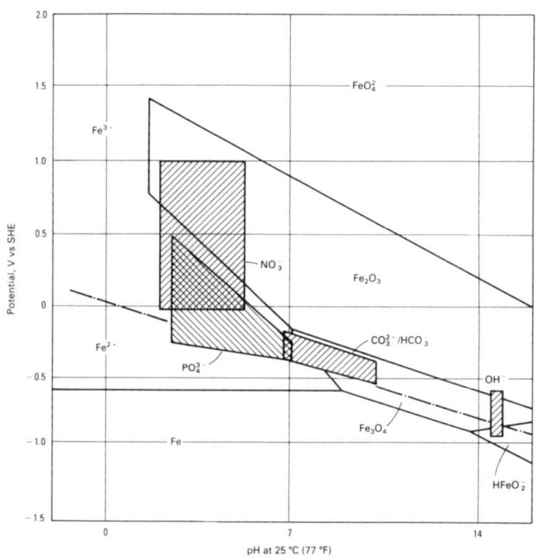

Fig. 14 Relationship between pH-potential conditions for severe cracking susceptibility of carbon steel in various environments and the stability regions for solid and dissolved species on the electrochemical equilibrium diagram. Note that severe susceptibility is encountered where a protective film (phosphate, carbonate, magnetite, etc.) is thermodynamically stable, but if ruptured, a soluble species (Fe^{2+}, $HFeO_2^-$) is metastable.

metastable. A second example of the potential-pH regimes in which SCC occurs is given in Fig. 15 for a 70Cu-30Zn brass in a variety of solutions.

The effect of many environmental parameters, such as pH, oxygen concentration, and temperature, on the thermodynamic conditions for SCC can be related to their effect on the potential-pH diagrams (as shown in Fig. 14 and 15) or on the material potential relative to the various stability regions. The potential-pH diagram for iron in water at 25 °C (77 °F) shown in Fig. 16 illustrates the effects of changing the pH and oxygen concentration. A decrease in the pH from 9 to 6 at a potential of –0.2 V versus standard hydrogen electrode (SHE) shifts iron from a region of stability to one of active corrosion. Based on the thermodynamic criteria for simultaneous stability and active corrosion, the critical pH would be expected to be 7, with decreasing susceptibility at higher pHs because of increased film stability and at lower pHs because of decreased film stability and the increase of general corrosion. Changes in oxygen concentration generally alter the electrode potential, with increasing oxygen concentration resulting in more oxidizing conditions. The effects of temperature on the potential-pH diagram must be determined for each temperature of interest as the regions of stability shift with temperature.

For materials in which SCC occurs by a hydrogen-induced subcritical crack-growth mechanism, the thermodynamic requirement for crack growth is governed by the hydrogen reduction line shown by dashed line *a* in Fig. 16. Hydrogen reduction of iron in water at 25 °C (77 °F) occurs at potentials below this line, but not above it.

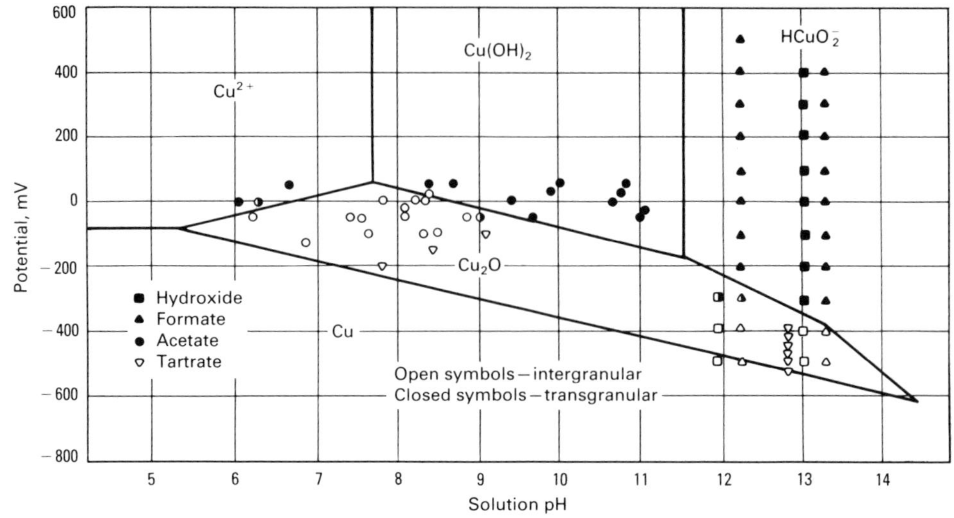

Fig. 15 Potential-pH diagrams showing the domains of failure mode for 70Cu-30Zn brass in various solutions, together with the calculated positions of various boundaries relating to the domains of stability of different chemical species

Therefore, the range of potentials at which hydrogen is available to cause crack growth increases and becomes more oxidizing with decreasing pH.

Applying potential-pH diagrams to identify specific conditions at which SCC will occur is limited by factors. These include the availability of these diagrams for complex solutions and for the temperatures of interest and the substantial deviation of the chemistry of a crevice or crack from the bulk solution chemistry. Also, the electrode potential at the crack tip can differ from that at the free surface of the material. These differences arise from the need for diffusion of metal ions away from the actively corroding crack tip, migration of anions into the crack, reactions along the crack walls, convection of the electrolyte in the crack, and, in some cases, gas bubble formation that causes potential drops. Efforts to measure the local crack-tip chemistry and potentials are restricted by crack sizes, which can be substantially smaller than 1 μm. Therefore, mathematical modeling of transport within cracks and the resulting crack-tip chemistry, reactions, and potentials has been actively pursued in recent years. The results of measurements in simulated crevices and modeling have clearly shown that the crack-tip chemistry can differ from the bulk conditions, but the specifics of these differences are not well known and depend on the material and environment being considered.

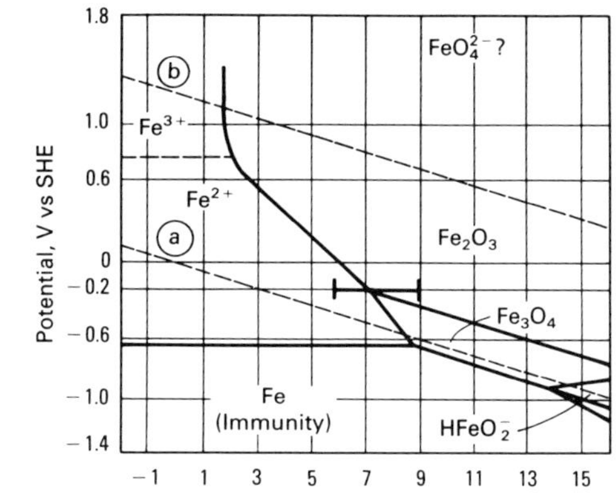

Fig. 16 Potential-pH Pourbaix diagram for iron in water at 25 °C (77 °F). A decrease in pH from 9 to 6 at potential of −0.2 V, which shifts iron from a region of stability to one of active corrosion, is indicated by the solid bar.

For example, based on evaluations by one researcher, the pH within cracks and crevices of structural steels in marine environments ranged from 4 to 11, of stainless steel from 0 to 3, of aluminum from 3 to 4, and of titanium from 1 to 2 (Ref 25). Shifts of up to 300 mV in the crack-tip potential in the anodic direction were observed for cathodically polarized steel, while other researchers found only a small shift in the crack-tip potential relative to the external potential under steady-state conditions (Ref 23) or a 300-mV shift in the cathodic direction when the crevice walls were active, reducing to a small shift when the walls passivated (Ref 26). Therefore, the critical potentials at which cracking may occur can differ from those given by the bulk equilibrium conditions used to determine the potential-pH diagrams. Although it is possible to determine the equilibrium conditions based on the crack-tip conditions, the current knowledge of crack-tip conditions is fragmentary at best. Other factors that limit the usefulness of the equilibrium potential-pH diagrams are the lack of available diagrams for specific materials and environments and the variable temperatures, chemistries, and potentials common in many industrial applications.

Kinetics of SCC. A knowledge of the thermodynamic conditions at which SCC can occur is insufficient without a corresponding understanding of the kinetics of crack growth, because the life of a component may be adequate if the crack-growth rate is sufficiently slow, even though SCC is thermodynamically possible. As in the thermodynamic conditions for SCC, environmental parameters such as potential, pH, oxygen concentration, temperature, and crack-tip chemistry have a strong effect on the crack-growth kinetics. The crack-tip reactions and the rate-determining steps controlling the crack-growth rate are specific to alloy/environment combinations; although a discussion of each system is beyond the scope of this chapter, some general observations will be made. Also, the crack-growth rate depends on the crack-advance process even though it is controlled by electrochemical reactions.

Detailed descriptions of various crack-growth mechanisms are given in the section "Crack-Propagation Mechanisms" in this chapter by categorizing these mechanisms as either anodic dissolution or mechanical fracture models. For the case of a crack growing by anodic dissolution alone, the total crack advance is a function of the total charge transfer at the crack tip, while the crack velocity is a function of the crack-tip current density. For a crack growing with mechanical fracture, the total crack advance exceeds the total charge transfer at the crack tip, but the crack velocity may still be controlled by the crack-tip density. A limiting velocity can be described for a crack advancing under pure anodic dissolution by the following Faradaic relationship:

$$\frac{da}{dt} = \frac{i_a M}{zF\rho} \qquad \text{(Eq 3)}$$

where i_a is the anodic current density of a bare surface, M is the atomic weight, z is the valence, F is Faraday's constant, and ρ is the material density. It has been shown (Ref 27) that this relationship (Eq 3) between the bare-surface current density and the crack-propagation rate is applicable to a wide variety of materials, as presented in Fig. 17. Equation 3 assumes that the crack tip is maintained in a bare condition, while the crack walls are relatively inactive. A bare crack tip and passive walls can result from the difference between the electrochemical conditions at the tip and other regions in the crack, a difference in the local chemistry of the material at the crack tip that causes the tip to be more active than the crack walls (such as at a sensitized or segregated grain boundary), or a crack-tip strain rate that is sufficiently high to prevent the formation of a protective film.

A number of factors can reduce the crack velocity below that given by Eq 3 and in Fig. 17. The most widely examined crack-growth retardation process is that resulting from the crack tip being covered by a film for some fraction of time. The process of crack growth in the presence of a film at the crack tip has been described by several mechanisms, such as slip dissolution and passive-film rupture, covered in the section "Crack-Propagation Mechanisms" in this chapter. In general, the crack-growth rate depends on the rate at which the film is ruptured and reformed. The amount of corrosion that occurs between these two events has been used to describe the crack-

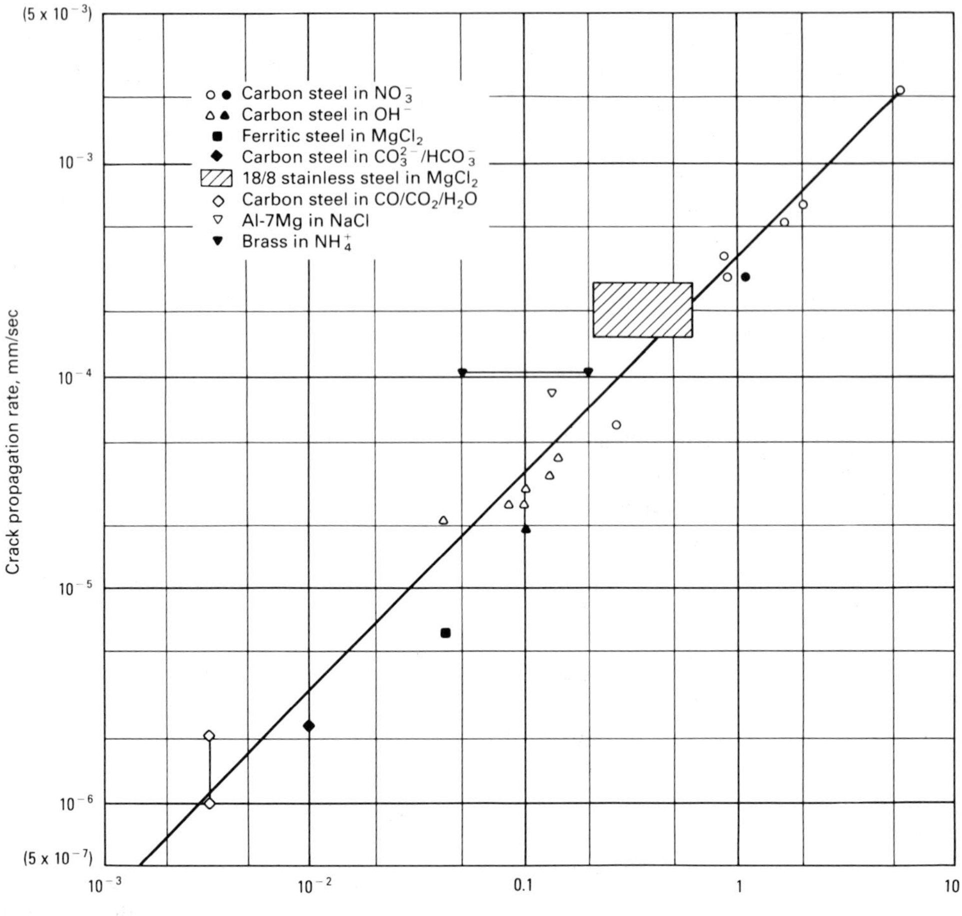

Fig. 17 Relationship between the average crack-propagation rate and the oxidation (i.e., dissolution and oxide growth) kinetics on a straining surface for several ductile alloy/aqueous environment systems

growth rate. This time period is determined by the crack-tip strain rate, film fracture strain, repassivation rate of the surface, maximum corrosion rate while the tip is bare, and the corrosion-rate decay with repassivation. Other factors that can reduce the crack-growth rate below that given by Eq 3 are limits in the diffusion rate of species into and out of the crack tip, crack deflection away from the principal stress, and changes in the local material chemistry. Transport of species into and out of cracks is considered a major limitation to attaining the crack-growth rate predicted by Eq 3. Factors such as the crack geometry or width, reactions or corrosion rate along the walls of the crack, diffusion rate of anions and cations, and metal salt solubility limits all contribute to transport-limited crack velocities.

A lack of clear knowledge about the specific conditions at the tip of the crack has limited the understanding of the role of specific species on crack-growth rates. At this time, it is possible only to describe the effect of bulk electrolyte conditions on crack growth rate; the local conditions may vary significantly. However, knowledge of the local crack-tip conditions is most important for understanding the mechanisms of cracking, although a knowledge of the external crack conditions is adequate for monitoring and controlling SCC. In the case of austenitic stainless steel, for example, the effects of electrochemical potential, oxygen content, and temperature are shown schematically in Fig. 18 to 20. Similar relationships exist for other anion concentrations, such as halides and sulfur species, but the data base is not as well developed as the cases shown. It is important to note that these are not independent effects, because the effects of oxygen concentration and temperature are probably related to the potential. The details of this relationship are complex and are beyond the scope of this chapter.

A mechanical fracture process in SCC can produce crack velocities exceeding those given by Eq 3 by some magnification factor, which can be as large as 100. The mechanism of brittle fracture induced by SCC, which is covered in the section "Crack-Propagation Mechanisms" in this chapter, is thought to involve the formation of a corrosion product at the crack tip in which a cleavage crack can initiate and propagate some depth into the ductile substrate. As mentioned earlier, this process would produce crack lengths that exceed those accounted for by the total charge transfer/metal oxidation process, but would be dependent on the crack-tip corrosion rate because the formation rate of the corrosion product would depend on the corrosion rate.

Early observations and proposals of this process have been made (Ref 28-32). A current list of materials in which SCC is thought to occur through a film-induced cleavage process as suggested in Ref 32 is given in Table 2, which shows a wide range of materials and environments. An attempt has been made to identify the corrosion product or layer that initiates the cleavage cracking process, but this identification must be considered speculative. The electrochemical conditions controlling the brittle SCC process have not been carefully catalogued; however, for many systems, the main issue is whether cracking can be completely described by anodic dissolution or whether a mechanical fracture process is involved. The greatest uncertainty involves the possibility of a transition between anodic and mechanical SCC processes that is affected by electrochemistry, material chemistry, or mechanics.

The electrochemical factors described in the previous paragraphs were generally drawn from examples of intergranular SCC, but many of the effects would be similar for brittle transgranular SCC. For example, the brittle transgranular crack-growth rate follows closely the dependence of the anodic current density on potential. Whether or not there are critical potentials at

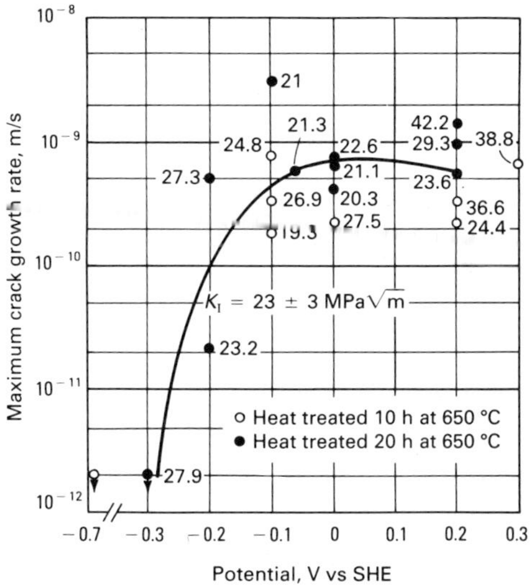

Fig. 18 Effect of potential on the maximum crack-growth rate in sensitized type 304 stainless steel in 0.1 M Na_2SO_4 at 250 °C (480 °F). Numbers denote K_I values.

which transgranular SCC may switch from a brittle mechanical fracture mode to a purely anodic dissolution mode is unknown. However, for many systems, the critical potentials describing SCC are valid control parameters, even though the process associated with these critical potentials may be uncertain. For an anodic dissolution-controlled process, the critical potentials were associated with the need for an active tip and passive crack walls. These conditions are still necessary for a brittle SCC process, because it is dependent on an anodic reaction to produce the corrosion reaction product. This product may be a film or a dealloyed layer, but the rate of formation, and therefore, the brittle crack growth rate are dependent on the crack-tip current density.

Material Chemistry and Microstructure

The relationship between material chemistry and microstructure and SCC is as complex as the relationship between the environment and SCC. Bulk alloy composition can affect passive-film stability and phase distribution (for example, chromium in stainless steel), minor alloying ele-

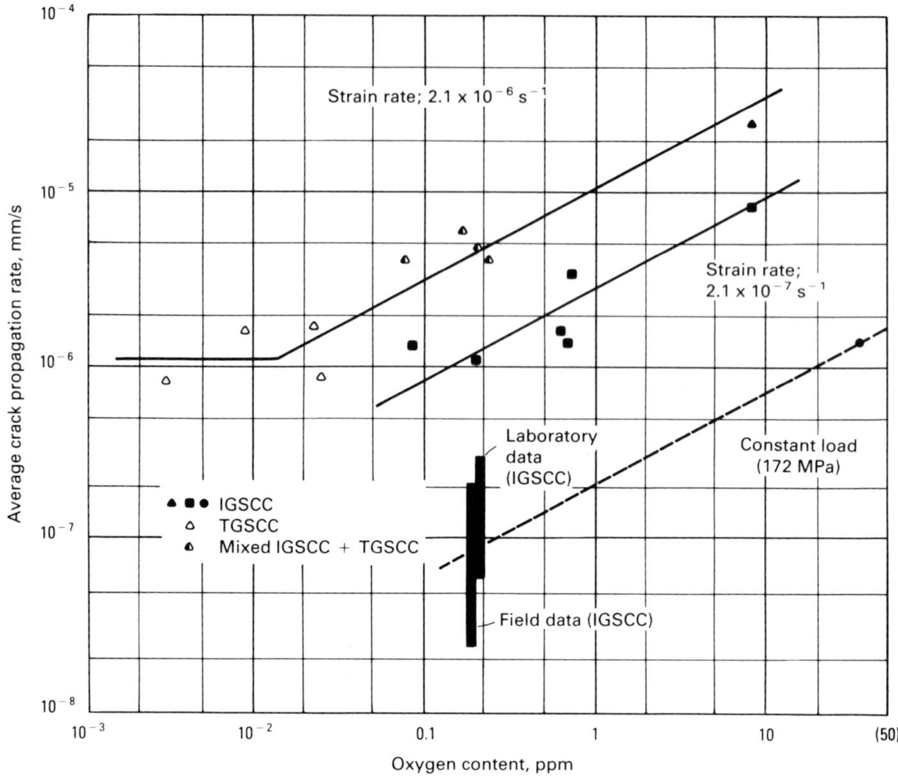

Fig. 19 Variation in the average crack-propagation rate in sensitized type 304 stainless steel in water at 288 °C (550 °F) with oxygen content. Data are from constant-extension-rate testing, constant-load testing, and field observations on boiling water reactor piping. IGSCC, intergranular stress-corrosion cracking. TGSCC, transgranular stress-corrosion cracking

ments can cause local changes in passive-film-forming elements (for example, carbon in stainless steel causing sensitization), impurity elements can segregate to grain boundaries and cause local differences in the corrosion rate (for example, phosphorus in nickel or nickel-base alloys), and inclusions can cause local crack-tip chemistry changes as the crack intersects them (for example, manganese sulfide in steel). Also, alloys can undergo dealloying, which is thought to be a primary method by which brittle SCC initiates. The following summary will be divided into intergranular and transgranular effects, because some of the primary factors can best be described in this way. It is important to recognize, though, that many of the material chemistry factors can affect both intergranular and transgranular cracking.

Intergranular Stress-Corrosion Cracking. Material chemistry and microstructural effects on intergranular SCC can generally be divided into two categories: grain-boundary precipitation and grain-boundary segregation. Grain-boundary precipitation effects include carbide precipitation in austenitic stainless steels and nickel-base alloys, which causes a depletion of chromium adjacent to the grain boundary and intermetallic precipitation in aluminum alloys, which are anodically active. Grain-boundary segregation of impurities such as phosphorus, sulfur, carbon, and silicon can produce a grain boundary that is up to 50% impurity within a region 1 to 2 nm thick. These impurities can alter the corrosion and mechanical properties of the grain boundary and can therefore cause cracking by anodic dissolution and perhaps mechanical fracture.

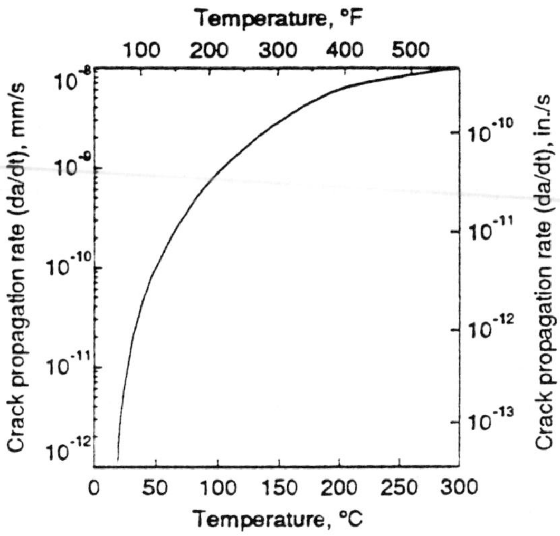

Fig. 20 Schematic of crack-growth rate versus temperature for intergranular SCC of type 304 stainless steel

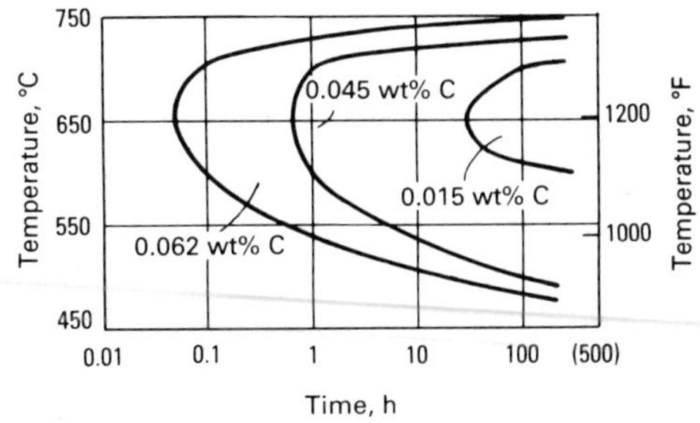

Fig. 21 Temperature/sensitization curves determined by EPR tests on type 304 stainless steel alloys with various carbon contents

Grain-Boundary Precipitation in Stainless Steels. Chromium carbide precipitation in stainless steels occurs in the temperature range of 500 to 850 °C (930 to 1560 °F), with the rate of precipitation controlled by chromium diffusion. For intermediate times, such as during heat treating and welding, chromium depletion occurs adjacent to the grain boundary during chromium carbide growth. This depletion can be described by the minimum chromium concentration adjacent to the carbide and the width of the depleted zone. Minimum chromium concentrations of 8 to 10 at.% have been measured by analytical electron microscopy, while the width of the depleted zone has been measured to range from 10 nm to hundreds of nanometers. After times long enough for carbide growth to reach completion, the chromium profile is eliminated, and the chromium concentration returns to the bulk value.

The intergranular SCC of austenitic stainless steel is primarily dependent on the nature of the chromium-depleted zone, which is generally explained by the depletion of a passive-film-forming element along a continuous path through the material. The stress-corrosion susceptibility and crack-growth rate of austenitic stainless steel can be described by the degree of sensitization (DOS) as measured by such corrosion tests as the Strauss or electrochemical potentiokinetic reactivation (EPR) test. Quantitative comparisons between susceptibility as measured by the presence or absence of intergranular SCC in an SCC test or the crack-growth rate and the DOS or chromium depletion parameters have been successful in cases in which sufficient data have been available, but these correlations are limited to specific alloys, environments, and stress conditions. Semiquantitative comparisons in which intergranular SCC is predicted for DOS values greater than some critical value are less costly to obtain and have been used more extensively.

Time/temperature/DOS curves at several carbon concentrations are shown in Fig. 21. The curves represent the conditions necessary for the development of a constant degree of sensitization, with the strong effect of bulk carbon concentration on the time needed to develop a sensitized microstructure. Therefore, the most common method for reducing the possibility of developing a sensitized microstructure is to reduce the carbon concentration or to control the thermal history of the material. Given that the material is sensitized, control of the environment and stress conditions can be used to reduce the crack-propagation rate.

Clearly, chromium and carbon are the two most important elements involved in the development of sensitization; however, other elements can play a role. Molybdenum has an effect similar to that of chromium — incorporation into the carbides and depletion adjacent to the grain boundaries — although it is generally present in smaller concentrations (for example, in type 316 stainless steel) than chromium. Nickel increases the activity of carbon in austenitic stainless steel; therefore, increases in nickel concentration can enhance sensitization for a given carbon concentration and thermal history. Manganese, silicon, and nitrogen have also been shown to affect sensitization, but comparatively sparse data are available on these elements. Silicon has been shown to segregate to grain boundaries when present in sufficient quantities or by the nonequilibrium processes that occur when the material is irradiated with neutrons. The presence of silicon in the grain boundaries has been found to be deleterious in oxidizing environments.

Grain-Boundary Precipitation in Nickel Alloys. The cause of intergranular SCC in nickel-base alloys is more complex than in austenitic stainless steel. Chromium carbide precipitation and chromium depletion occur in nickel-base alloys, such as alloy 600, as they do in austenitic stainless steels, but a clear connection between the presence of this microstructure and intergranular SCC has not been made. Carbon solubility is considerably lower and chromium diffusion is faster in nickel-base alloys than in austenitic stainless steels; therefore, the nose in the carbide precipitation or sensitization curves shown in Fig. 21 is shifted to shorter times and higher temperatures. Significant heat-to-heat variations in the chromium depletion of alloy 600 have been observed for a given heat treatment. This variation is attributed to the sensitivity of alloy 600 to carbon concentration and to the variability in mill-anneal heat treatment conditions. Chromium depletion is thought to accelerate cracking in oxygenated water, but has not been identified as a controlling factor in deaerated water or caustic environments at 300 °C (570 °F). The rapid carbide precipitation kinetics in nickel-base alloys allows the development of a healed microstructure in which carbide growth is complete, and the chromium profile is eliminated with relatively short heat treatment times.

Some significant differences between the intergranular SCC of nickel-base alloys and that of austenitic stainless steels include the positive effect of a semicontinuous distribution of carbides at the grain boundary in single-phase material and the galvanic couple between the γ' and γ phases in precipitation-hardened alloys. The beneficial effect of a semicontinuous distribution of carbides is thought to be related to dislocation generation from the carbides that results in a crack blunting or stress relaxation at the tip of a crack. The detrimental effect of the γ' phase is thought to result from a galvanic couple that polarizes the γ phase anodically relative to the γ' phase.

The conditions that promote SCC in nickel-base alloys are also different from those for austenitic stainless steels. In high-temperature water, SCC of nickel-base alloys is strongly temperature dependent and is accelerated by hydrogen concentration in the water and by cold work (Ref 33). Stress-corrosion cracking also occurs at potentials anodic and cathodic to the corrosion potential. The time to failure of U-bend samples, which is related to crack velocity, decreases with increasing temperature, with a single activation energy over the temperature range of 320 to 400 °C (610 to 750 °F) which includes the aqueous and steam phases of water.

Grain-Boundary Precipitation in Aluminum Alloys. Aluminum alloys are also noted for the occurrence of intergranular SCC in aqueous environments, including sodium chloride (NaCl) solutions. The details of the intergranular SCC process in aluminum alloys are very complex and vary with alloy composition, but some features can be summarized. Grain-boundary precipitation has been identified as a contributing factor in the intergranular SCC of aluminum alloys; galvanic effects between the precipitates and matrix are considered to be important (Ref 35). In some cases, the precipitates are anodic to the matrix; in others, they are cathodic.

In 7000 series alloys, peak-aged microstructures are the most susceptible to SCC; overaging usually reduces susceptibility. However, during normal aging treatments a large number of metallurgical factors, such as fracture toughness, yield strength, precipitate free zone width, and slip character, vary simultaneously with the precipitate size distribution. As a result, it is difficult to attribute the beneficial effect of overaging to any one factor without conducting experiments on samples heat treated in a manner designed to isolate these variables (Ref 13). Aluminum alloys are generally considered to be susceptible to cracking as a result of an anodic dissolution mechanism, but recent research has indicated that 7000 series alloys can absorb hydrogen from aqueous solutions and water vapor and that this absorbed hydrogen promotes intergranular fracture (Ref 36-38). In general, hydrogen evolution always accompanies the anodic reactions that occur on aluminum alloys even during anodic polarization (Ref 39), and, since the passivating layer on aluminum alloys is soluble in high pH solutions, cathodic polarization stimulates dissolution as well as hydrogen evolution (Ref 40). As a result, it is difficult to distinguish between anodic dissolution and hydrogen embrittlement for aluminum alloys exposed to aqueous solutions. Many investigators consider these alloys to be susceptible to both mechanisms, the dominant mechanism determined by environmental and metallurgical factors.

Grain-Boundary Segregation in Aluminum Alloys. Grain-boundary enrichment of magnesium without precipitation of a magnesium-rich phase has recently been identified as a factor in the intergranular SCC of aluminum alloys (Ref 41, 42). This enrichment is associated with increased corrosion activity at the grain boundary and the possible formation of magnesium hydride (MgH) or enhanced hydrogen entry along the grain boundaries. Crack advance is thought to occur by discontinuous crack jumps (possibly nucleated by hydrides) and crack arrest after about 150 to 350 nm of advance, followed by repetition of this process.

Grain-Boundary Segregation in Ferrous and Nickel Alloys. Grain-boundary enrichment of impurities can contribute to the intergranular SCC of iron-base alloys, austenitic stainless steels, and nickel-base alloys. The extent of their effect depends on the electrochemical potential, the presence of other stress-corrosion processes such as chromium depletion, and their concentration in the grain boundary. The enrichment of impurities to grain boundaries by equilibrium segregation can be described by the enrichment ratio, as shown in Fig. 22 for a variety of impurity/alloy combinations (Ref 43). The enrichment ratio defines the upper bound to the grain-boundary concentration that will be achieved under equilibrium conditions, which can be very long times at low temperatures. It can be seen from Fig. 22 that enrichment ratios as high as 10^5 are possible for some impurities and that ratios greater than 10^2 are not uncommon. For the bulk impurity concentrations present in most engineering alloys, grain-boundary impurity concentrations of 10 to 20 at.% are possible. Therefore, an intergranular stress-corrosion crack can propagate along a grain boundary that has a composition vastly different from that of the bulk alloy.

Ferritic steels exhibit intergranular SCC in hot nitrate, caustic, carbonate, and a variety of other environments. The presence of intergranular SCC is dependent on the electrochemical potential, because intergranular SCC is predominant at potentials in the active-passive transition. Early studies of this effect identified carbon segregation as the primary element of concern where the carbon atoms were said to provide suitable imperfection sites for adsorption of nitrate in an adsorption-in-

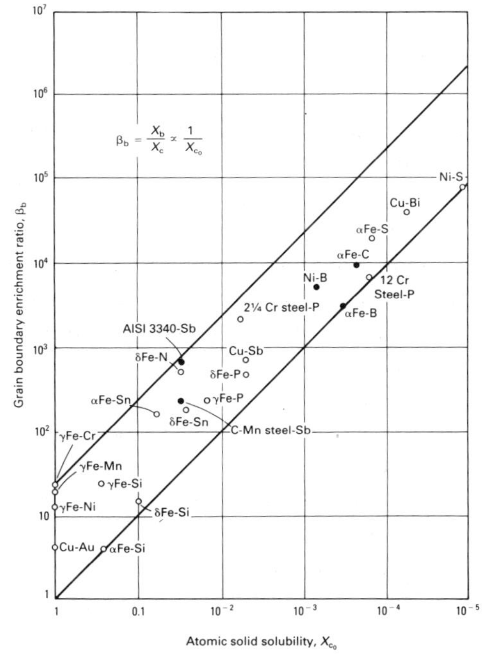

Fig. 22 Correlation between measured grain-boundary enrichment ratios, β, and the inverse of solid solubility, X_{C_0}. The symbols X_b and X_c represent the actual concentration of the species b and c in the bulk.

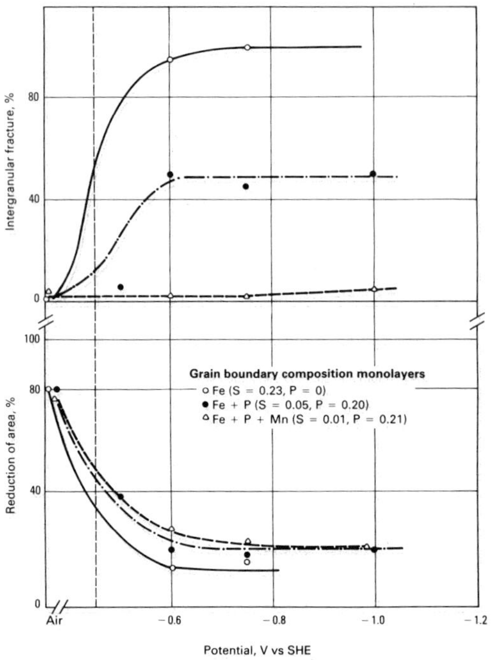

Fig. 23 Percent intergranular fracture, reduction in area, and strain to failure of iron, Fe + P, and Fe + P + Mn alloys tested at various cathodic potentials

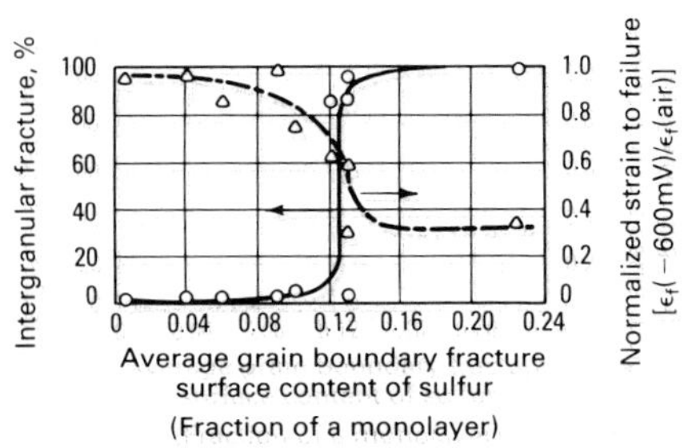

Fig. 24 Percent intergranular fracture and the normalized strain to failure plotted as a function of sulfur content at the grain boundary for straining electrode tests at a cathodic potential of −600 mV (SCE)

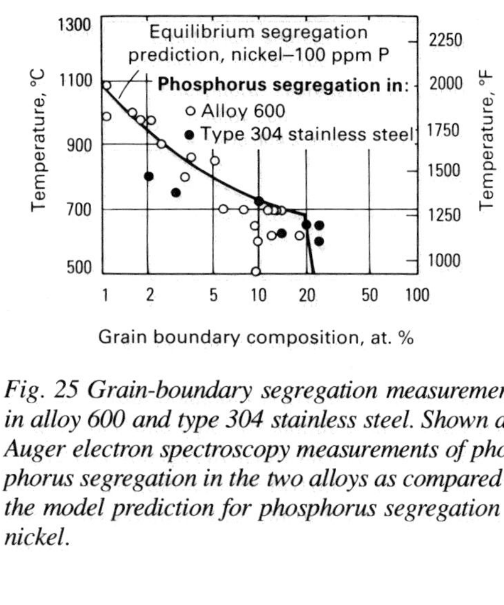

Fig. 25 Grain-boundary segregation measurements in alloy 600 and type 304 stainless steel. Shown are Auger electron spectroscopy measurements of phosphorus segregation in the two alloys as compared to the model prediction for phosphorus segregation in nickel.

duced crack growth mechanism (Ref 44). More recently, phosphorus segregation has been associated with the intergranular SCC of iron alloys in nitrate and caustic solutions (Ref 45, 46); grain-boundary concentrations as low as 2 to 3 at.% have been sufficient to alter the passivity of iron in hot nitrate solutions. A complicating factor in ferritic steels is their susceptibility to intergranular SCC and to hydrogen-induced subcritical crack growth. The temperatures and electrochemical potentials at which intergranular SCC and cathodic hydrogen-induced subcritical crack growth occur are generally not the same. Stress corrosion tends to dominate at temperatures above about 100 °C (212 °F) and at potentials in the active-passive transition; hydrogen effects are predominant at temperatures below 100 °C (212 °F), at more cathodic potentials, and at lower pH values. Examples of the effect of sulfur and phosphorus on the tendency toward intergranular fracture are given in Fig. 23 and 24, which show that sulfur was more effective than phosphorus in causing intergranular fracture of iron at cathodic potentials (Ref 47, 48). For the conditions of this test, a sulfur concentration of about 13 at.% was sufficient to change the fracture mode from transgranular to intergranular.

Grain-boundary impurity segregation of various impurities, including phosphorus, silicon, sulfur, and nitrogen, has been reported in austenitic stainless steels (Ref 49); however, no direct effect on intergranular SCC has been identified in high-temperature water. Phosphorus segregation has been shown to cause intergranular corrosion in highly oxidizing solutions, and impurity segregation of phosphorus and perhaps silicon has been suggested as a primary factor in irradiation-assisted SCC, which occurs in the oxidizing in-core environment of light water reactors (Ref 50). Phosphorus segregation can apparently contribute to the intergranular SCC of austenitic stainless steels in high-temperature water if the carbon concentration of the alloy is lower than 0.002%. At this low concentration, there is virtually no sensitization; therefore, the effect of phosphorus segregation is observed.

Phosphorus is also the primary grain-boundary segregant in alloy 600 and has been shown to segregate by an equilibrium process (Ref 51). Grain-boundary composition versus temperature curves for alloy 600 and type 304 stainless steel are shown in Fig. 25, along with a calculated curve for equilibrium segregation in nickel (Ref 52). It can be seen that grain-boundary phosphorus concentrations to 15 at.% are possible. Grain-boundary segregation of sulfur, boron, nitrogen, and titanium has also been observed in alloy 600; however, a clear connection between segregation and these impurities and intergranular SCC in nickel-base alloys has not been observed to date.

Shen and Shewmon (Ref 34) have proposed that the reaction of hydrogen with elemental carbon to form methane in the grain boundaries of alloy 600 promotes the formation of small, closely spaced bubbles along the boundaries, which in turn promotes intergranular failure. The activation energy for this process is equal to the grain-boundary self-diffusion in nickel, which is 115 kJ/mol. Shen and Shewmon (Ref 34) found an activation energy for the crack-growth rate of alloy 600 over the temperature range of 320 to 400 °C (610 to 750 °F) of 80 kJ/mol, while Bandy and Vanrooyen (Ref 33) found an activation energy of 138 kJ/mol from time to failure of U-bend specimens over the temperature range of 315 to 365 °C (600 to 690 °F). These experimental results span the value predicted by the Shen and Shewmon model and hence support its validity.

In contrast to the high-temperature water results for alloy 600, a clear effect of phosphorus segregation on the intergranular SCC of nickel was observed at oxidizing potentials in acidic solutions at 25 °C (77 °F) (Ref 53), as shown in Fig. 26. Intergranular SCC was observed at anodic potentials ranging from the active-passive transition to transpassive potentials in nickel with phosphorus-enriched grain boundaries but not sulfur-enriched grain boundaries. The role of phosphorus was identified with degradation of the passive

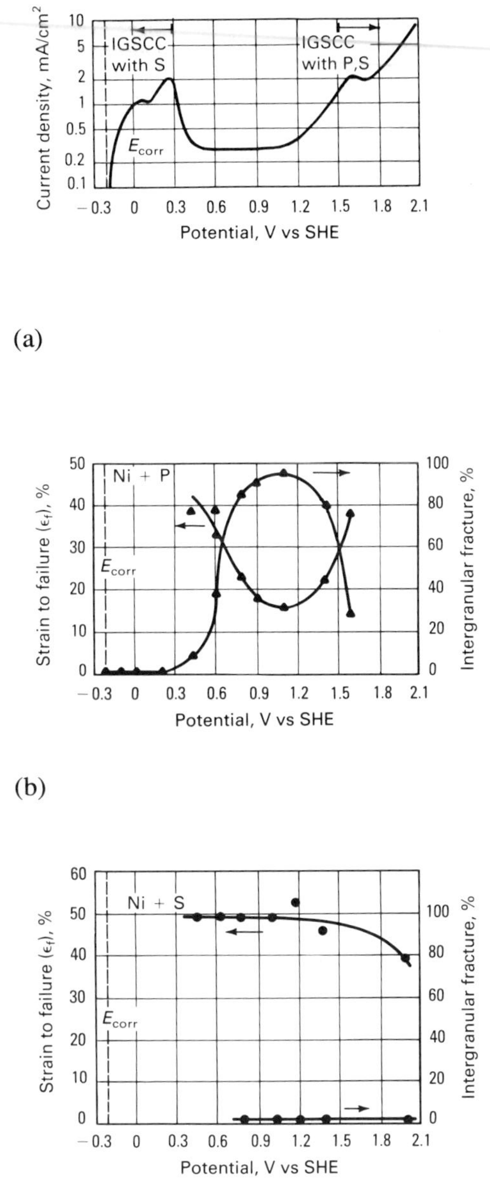

Fig. 26 Stress-corrosion cracking behavior of nickel with phosphorus and sulfur segregation. (a) Polarization curve for nickel in 1 N H_2SO_4 at 25 °C (77 °F). (b) Strain to failure and percent intergranular fracture for 26% phosphorus segregation of grain boundaries. (c) Strain to failure and percent intergranular fracture for 40% sulfur segregation at grain boundaries

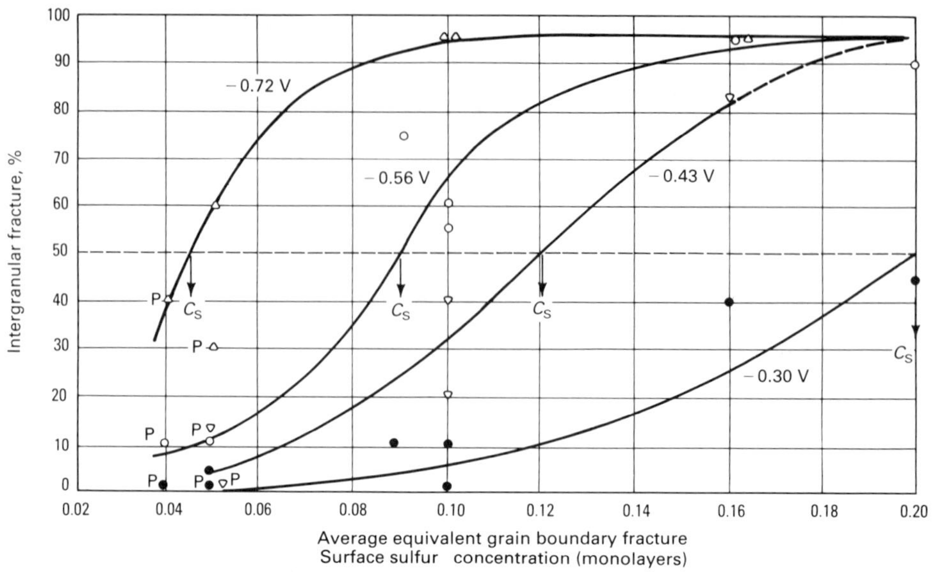

Fig. 27 Percent intergranular fracture and reduction in area versus grain-boundary composition of nickel for several cathodic test potentials. C_S is the critical sulfur concentration corresponding to 50% intergranular fracture. Points labeled P are equivalent sulfur concentrations for alloys with S + P at the grain boundaries. Source: Ref 53

film formed on nickel in acidic solutions. This example is a clear case of active-path corrosion; however, the cracking was clearly stress dependent, as evidenced by the alignment of cracks with the applied tensile stress and a threshold stress intensity below which SCC did not occur. Because there were no grain-boundary carbides or chromium depletion in the nickel, these results illustrate the potential effect of impurities when other grain-boundary processes are absent. Therefore, it appears, as in the case of type 304 stainless steel cited earlier, that impurity segregation may induce intergranular SCC in the absence of carbides and chromium depletion.

Like ferritic materials, nickel-base alloys and austenitic stainless steels are susceptible to hydrogen-induced intergranular subcritical crack growth. Impurity segregation is thought to play a key role in this hydrogen-induced fracture, as the results for nickel shown in Fig. 27 illustrate. These results show the combined effect of sulfur segregation and cathodic hydrogen on the amount of intergranular fracture in nickel. For the conditions of these tests, a given amount of intergranular fracture resulted for different combinations of grain-boundary sulfur and hydrogen reduction rate. A decrease in the amount of sulfur could be compensated for by an increased hydrogen activity to give the same amount of intergranular fracture.

It has also been demonstrated by Jones (Ref 54) that the hydrogen-induced subcritical crack-growth behavior of nickel can be remarkably similar to that of iron, given that their grain boundary chemistries are similar. For nickel and iron with about 0.3 monolayers of grain boundary sulfur, the K_{th} and stage 1 da/dt behavior are essentially identical for equivalent cathodic potentials (Fig. 8). This similarity can be understood in terms of hydrogen transport along sulfur-enriched grain boundaries being identical in both materials or a near-surface hydrogen-induced crack-growth process that does not require long-range diffusion.

Transgranular Stress-Corrosion Cracking. Numerous metallurgical factors affect transgranular SCC, for example, crystal structure, anisotropy, grain size and shape, dislocation density and geometry, yield strength, composition, stacking-fault energy, ordering, and phase composition. Some of these factors also affect intergranular SCC, as covered in the section "Intergranular Stress-Corrosion Cracking" in this chapter. Also, some of their effects on transgranular SCC are related to the corrosion behavior of the alloy, which can be understood from potential-pH diagrams or polarization curves. Because these effects are covered elsewhere (see the chapters on individual alloy systems in this book and in Ref 55), they will not be discussed here. Similarly, yield-strength effects will be considered in the section "Mechanical Factors" in this chapter. The effects of alloy composition on corrosion rate, hydrogen-exchange current densities, and hydrogen-absorption rates also influence the hydrogen-induced subcritical crack growth of materials in corrosive environments, but these are covered in the chapters dealing with specific metals and alloys and in Ref 55.

Alloying effects on slip planarity are a key metallurgical factor in transgranular SCC. Planar slip occurs in alloys with low stacking-fault energy, alloys containing ordered phases, or alloys exhibiting short- or long-range ordering. The consequences of planar slip on transgranular SCC have been explained by the slip-dissolution model discussed in Ref 56. In this model, the passive film is ruptured by the emergence of a slip step. In high-chloride environments, evidence was presented that preferential corrosion occurred along the high dislocation-density plane created by planar slip (Ref 57). A number of crack-growth processes were suggested based on the planar-slip-localized corrosion process — for example, control of crack advance solely by anodic dissolution of the slip plane, brittle fracture of the corrosion product or tarnish film formed along the localized corrosion path, and the tunneling process by which corrosion along the slip plane branches out into tunnels accompanied by mechanical fracture of the remaining ligaments. The planar slip concept of transgranular SCC appears consistent with a number of observations, such as alloying and strain-rate effects. Details of these mechanisms are presented in the section "Crack-Propagation Mechanisms" in this chapter.

A significant difficulty associated with the slip-dissolution mechanism is the nature of the transgranular SCC fracture surfaces, which are generally not on the slip planes and have cleavage-like features. For example, for admiralty brass tested in aqueous ammonia (NH_3) and Al-5.5Zn-2.5Mg tested in aqueous NaCl, the primary fracture facets were (110) planes, while for stable austenitic stainless steel tested in aqueous magnesium chloride ($MgCl_2$) at 155 °C (310 °F), the primary facets were on (100) planes. There is also evidence that crack advance in brass occurs in a discontinuous manner (Ref 31). Several mechanisms for the development of a cleavage crack in ductile face-centered cubic (fcc) alloys have been presented, but a definite correlation has not emerged.

One concept that has received considerable attention is that proposed in Ref 32, in which a rapid crack advance that begins in a brittle film at the crack tip induces cleavage fracture of the ductile material ahead of the crack tip. The brittle films that are regarded as the initiating layers are given in Table 2. These include a dealloyed layer, oxide, nitride, carbide, or hydride. A dealloyed layer acting as a cleavage-crack initiator is suggested for brass, copper-gold, and iron-chromium-nickel alloys. If this concept is supported by future research, the relative reactivity of the alloying elements will prove to be a key metallurgical factor in the transgranular SCC of materials. It has also been demonstrated that the dealloying rate of copper-aluminum and copper-zinc as a function of aluminum and zinc, respectively, corresponds to the crack-growth rate in ammoniacal solution (Ref 32).

Mechanical Factors

There is more in common between the mechanical aspects of intergranular SCC and transgranular SCC than there is between their environmental or metallurgical aspects. As indicated earlier, many of the environmental or metallurgical factors are specific to a given material/environment combination, so that a metallurgical factor may be significant in one environment but not in another. Threshold stress intensities and stresses, the presence of a stress-independent crack-growth regime, and a dependence of crack-growth rate on strain rate are features common to a variety of environmentally

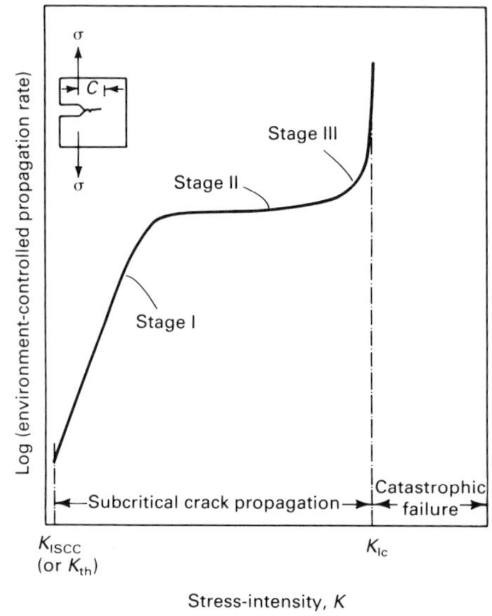

Fig. 28 Typical subcritical crack-propagation rate versus stress-intensity relationship. Stress intensity, K, is defined as $K = A\sigma\sqrt{\pi C}/B$*, where σ is the total tensile stress, C is the crack length, and A and B are geometrical constants.*

induced crack-growth processes and a variety of materials. Anodic- and cathodic-controlled processes show many of these common features.

The stress-intensity dependence of environmentally induced subcritical crack growth is shown schematically in Fig. 28, along with a relationship for the linear-elastic mode I stress intensity. The applied stress intensity is a function of the uniform stress and the crack length; therefore, for a constant-stress test, the stress intensity increases with increasing crack extension. For low-strength materials that form a significant plastic zone ahead of the crack, nonlinear fracture-mechanics relationships must be used to calculate K. The subcritical crack-growth behavior of many material/environment combinations is characterized by a threshold and by stages 1, 2, and 3 (see Ref 58 for a detailed discussion of modes of crack formation).

In Fig. 28, the threshold is defined by the minimum detectable or reliably measured crack-growth rate, while in some materials, stage 1 has a large K dependence such that a lower threshold does not result at lower crack velocities. There are a number of physical processes that can be associated with a threshold stress intensity, including a fracture strain for a passive-film-rupture mechanism, the critical resolved shear stress for the slip-dissolution mechanism, a fracture stress for a brittle-film-induced cleavage mechanism, or a critical crack-tip opening for transport of species in the crack. The threshold stress intensity is generally accompanied by the development of a plastic zone at the crack tip; however, calculations of crack-tip strains are not very accurate, and experimental correlations between plastic zone development and K_{ISCC} have been made in only a few cases.

Stage 1 subcritical crack growth is marked by a rapid increase in crack velocity, with a small increase in K. Two expressions used to describe the crack velocity-stress intensity relationship for this crack-growth regime are:

$$da/dt = AK^{m} \quad \text{(Eq 4)}$$

and

$$da/dt = B\exp(CK) \quad \text{(Eq 5)}$$

where da/dt is the crack velocity, K is the stress intensity, and A, B, C, and m are constants. Values for m in Eq 4 ranging from 2 to as high as 41 have been reported (Ref 59). A comprehensive explanation of stage 1 behavior has not been developed, but one explanation is that the crack-tip strain rate increases rapidly with K, and another is that the crack-tip corrosion rate and transport of species within the crack increase rapidly with increasing crack volume and hence with K. The film-rupture model of SCC developed by Ford (Ref 60) and Andresen and Ford (Ref 61) predicts an m value of between 2 and 4 for stainless steels; however, reported values for m range from 7 to 24. There is some correlation between the aggressiveness of the environment and the value of m, with the value of m increasing as the environment becomes more aggressive. Larger values of m reflect a greater increase in crack velocity for a given change in K resulting from the increased aggressiveness of the environment. Recently, Simonen and Jones (Ref 62) modeled the stage 1 behavior of Ni + P by considering that the dissolution of a salt film at the crack tip is controlled by the crack opening. This model provided an excellent correlation between experimental and theoretical results.

Rate-limiting processes in SCC include the transport of cations or anions in the crack electrolyte or the passive-film rupture and repassivation rate, depending on the material and environment. Environmental parameters such as crack-tip potential, corrosion rates, and pH also affect the stage 2 crack velocity, so that in many systems the limiting velocity is a function of crack-tip corrosion rate, strain rate, and transport rate. A steady-state crack transport calculation for intergranular SCC in Ni + P give crack-growth values within a factor of 5 of the measured stage 2 values (Ref 23). For transgranular SCC, there is increasing evidence that crack velocity is a function of corrosion rate, but that crack advance results from brittle crack jumps that exceed the depth of the corrosion reaction. In this case, the crack velocity may scale with the corrosion rate, but the crack length exceeds that determined from charge transfer.

Most stress-corrosion data have been obtained by using specimens loaded in a tensile mode in which the stress is perpendicular to the plane of the advancing crack; this condition is defined as mode 1 loading. In practice, a component may have a complex loading mode, with both torsion and shear components present. The effect of mixed loading modes on SCC is therefore an important consideration in component design; however, very few SCC data exist for other than mode 1 loading. Some examples of SCC in modes 1 and 3 for aluminum in 3.5% NaCl and type 304 stainless steel in boiling $MgCl_2$ are shown in Fig. 29, in which it can be seen that mode 1 loading gives the lowest threshold stress or stress intensity. Also shown in Fig. 29(a) is a schematic of the test specimen used to perform mixed-mode SCC tests. In both cases, the researchers concluded that the lower thresholds and higher crack velocities in mode 1 tests of these materials indicate a hydrogen-embrittlement component to SCC, because hydrogen effects are more prominent under tensile loading conditions (Ref 63, 64). However, anodic stress-corrosion processes are also aided

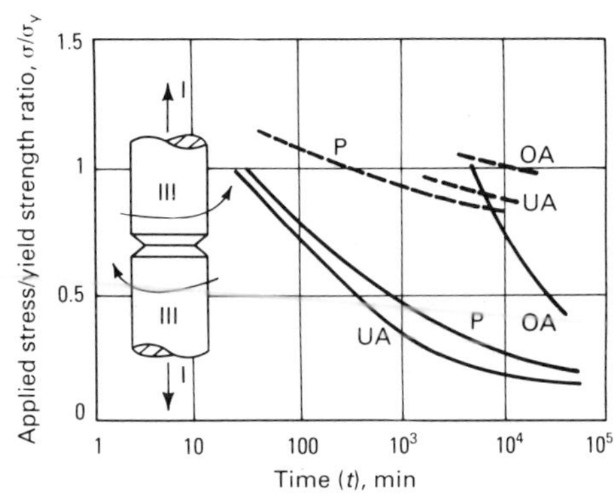

(a)

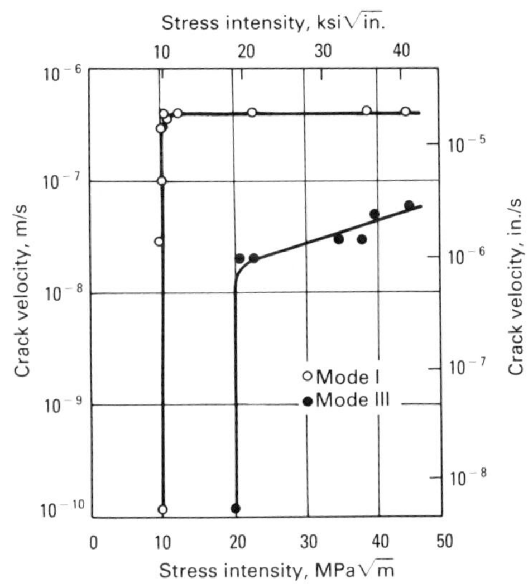

(b)

Fig. 29 Examples of SCC in mode 1 and mode 3. (a) Susceptibility of 7075 aluminum alloy to corrosion cracking in 3.5% NaCl solution. Mode 1, solid lines; mode 3, dashed lines. P, peak of aging; OA, overaged; UA, underaged. (b) Crack velocity versus stress-intensity curves for SCC of type 304 stainless steel in modes 1 and 3 in boiling $MgCl_2$

by the crack opening that results from mode 1 loading.

Stress-corrosion data are frequently given as applied stress versus time to failure, as shown in Fig. 30. These data are obtained using notched or unnotched specimens statically loaded in tension or bending. The threshold stress is sensitive to the environment, alloy composition, and structure and is not strictly a material property. The threshold stress is related to the yield strength of the material, with thresholds generally being greater than one-half the yield strength, and in many cases, the threshold is a significant fraction (about 0.8) of the yield strength of the material. Therefore, most SCC tests performed with this technique are performed at values of about 80% of the yield strength or greater.

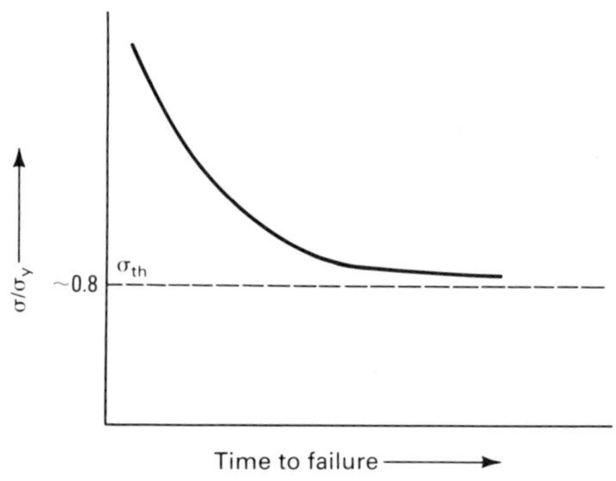

Fig. 30 Schematic diagram of time to failure versus applied stress, σ, normalized to the yield strength σ_y *for stress corrosion*

The data obtained with the constant-load method incorporate the stage 1, 2, and 3 regimes shown in Fig. 28 in addition to a crack-initiation stage. Crack growth begins with the development of a pit or intergranular corrosion groove that raises the stress intensity to the K_{ISCC}. Because the stress is constant, the stress intensity increases with increasing crack length, as given by the expression in Fig. 28, so that the crack progresses through stage 1 to stage 2 and ultimately to fracture in stage 3. Therefore, the failure time in a constant-stress test is dependent on the initiation rate, the value of the K_{ISCC}, the slope of the stage 1 regime, the crack-growth rate in stage 2, and the fracture toughness K_{Ic}. Constant-load test data are useful because they relate to the sequence that may occur in practice, but it is difficult to relate them to physical processes and rate-controlling steps. Crack-growth rates are frequently estimated from the crack depth divided by the total test time. Crack velocities obtained in this manner are minimum values. This is because the time for initiation is included; instantaneous velocities will always be greater.

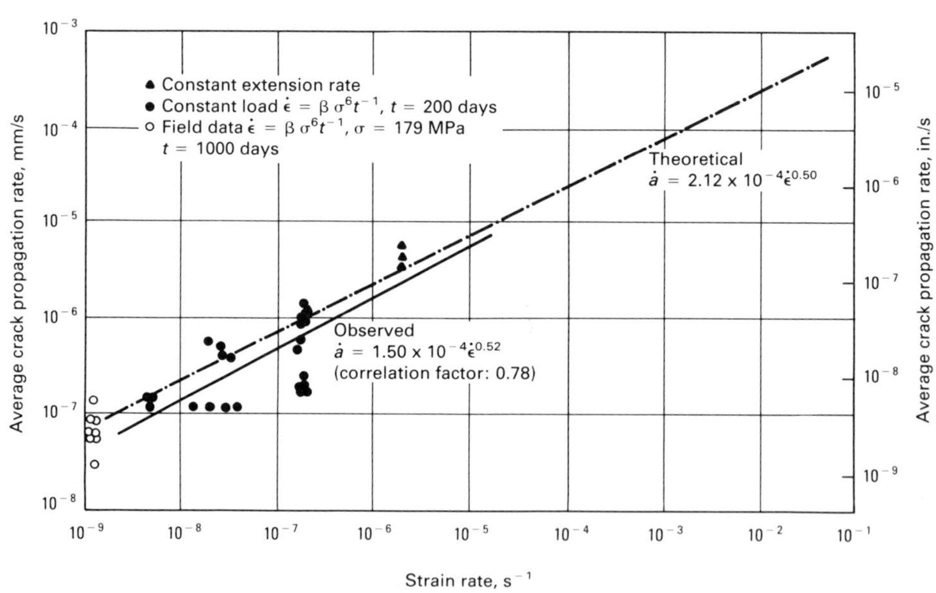

Fig. 31 Comparison between observed and theoretical crack-propagation/strain rate ($\dot{a}/\dot{\varepsilon}$) *relationships for furnace-sensitized type 304 stainless steel in water/0.2 ppm oxygen at 288 °C (550 °F)*

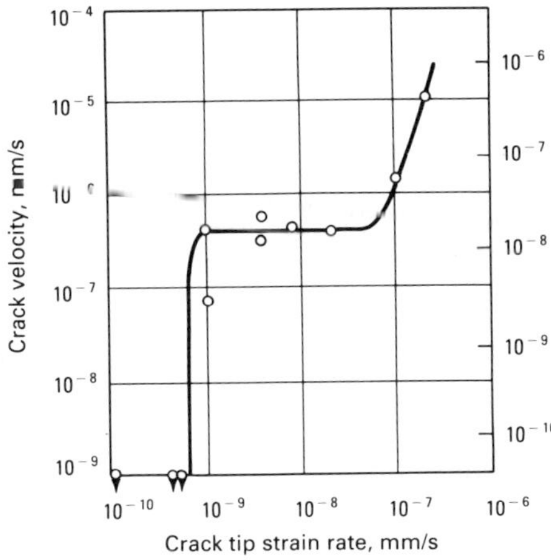

Fig. 32 Intergranular crack velocities for various applied crack-tip strain rates in a carbon-manganese steel immersed in 1 N Na_2CO_3 + 1 N $NaHCO_3$ at – 650 mV (SCE) and 75 °C (165 °F)

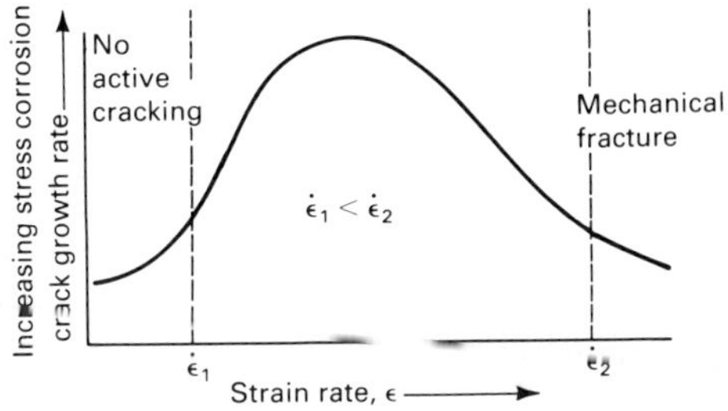

Fig. 33 Stress-corrosion crack growth as a function of the two strain-rate thresholds, $\dot{\varepsilon}_1$ and $\dot{\varepsilon}_2$

The threshold stress is perhaps the least complicated value obtained from constant-load tests. For film-forming materials, the threshold can be related to the film-formation and film-rupture rates. Below the threshold stress, the film-formation rate is sufficiently high to maintain the surface in a filmed condition; at higher stresses, the film-rupture rate exceeds the reformation rate and thus a pit or an intergranular corrosion groove can grow. For transgranular SCC initiated by a brittle corrosion film, the threshold stress could be related to the fracture stress of the film; however, this type of relationship has not been established for the brittle-film-induced cleavage process.

Constant-extension-rate for slow-strain-rate tests have essentially replaced the use of the constant-stress test for stress-corrosion testing. This has occurred because of the variability in times of failure with a constant-stress test and the long test times needed to define the stress threshold. Also, the concept that the crack-growth rate is a function of the strain rate, as shown in Fig. 29 for intergranular type 304 stainless steel, has led to the use of the constant-extension-rate test (CERT). The CERT specimens will fail within a given period of time, thus eliminating specimens that do not fail because they are loaded to a stress below the threshold. The crack-growth rates given in Fig. 31 are minimum values obtained from unnotched specimens, and the strain rate is the average applied rate. Clearly, once a crack develops, the crack-tip strain rate will differ from the average applied value. An example of crack velocity as a function of crack-tip strain rate is given in Fig. 32 for intergranular SCC of a carbon-manganese steel tested in a carbonate solution. A threshold strain rate is evident in these results that is the same as a threshold stress intensity. A second threshold strain rate exists at high strain rates because mechanical fracture occurs before a crack can initiate and propagate to a measurable length. The two strain-rate thresholds are illustrated in Fig. 33.

It is important to distinguish between static and dynamic effects. For example, a sample can be plastically strained 10% and the load dropped to below the threshold stress before being immersed in a corrosive environment and SCC may not occur. However, a sample can be slowly strained 10% in a corrosive environment, the result being the development of numerous cracks and perhaps even complete failure. Therefore, it is the dynamic strain and creation of fresh surface or dynamic fracture of a brittle film that is crucial to SCC. Corrosion-current measurements corroborate these observations; the current density of a prestrained sample is usually very similar to an unstrained sample, while the current density measured during dynamic strain is considerably greater than that in the unstrained sample.

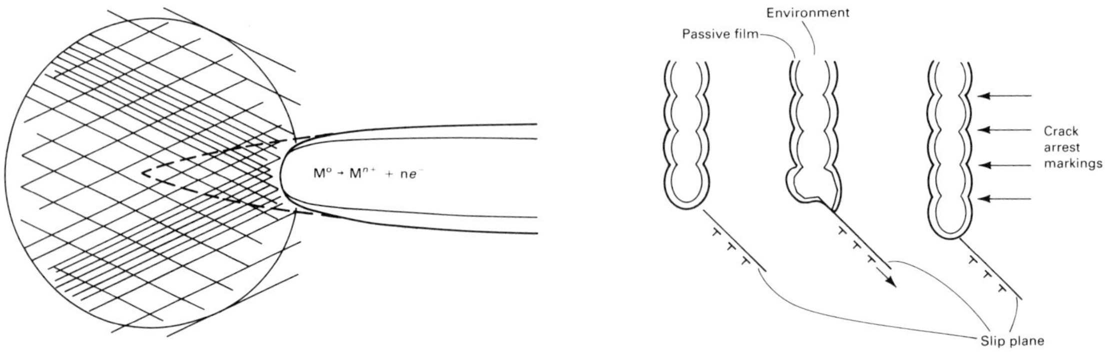

(a) (b)

Fig. 34 Schematic representation of crack propagation by the film-rupture model. (a) Crack tip stays bare as a result of continuous deformation (Ref 72). (b) Crack tip passivates and is ruptured repeatedly (Ref 78, 79).

Crack-Propagation Mechanisms

As indicated earlier, it is unlikely that a single mechanism of SCC exists; instead, there are probably two or three different operative mechanisms. Many models have been proposed, but only the more prominent will be discussed here. The proposed mechanisms for crack propagation fall into two basic classifications: those based on dissolution and those that involve mechanical fracture.

Dissolution Models

According to models of this type, a crack advances by preferential dissolution at its tip. A number of models have been proposed to account for this process. For example, preferential dissolution at the crack tip has been attributed to the formation of active paths in the material, stresses at the crack tip, and chemical-mechanical interactions (Ref 65-70). However, research has essentially eliminated all but the film-rupture model from serious consideration, and debate has recently centered on the details of this model.

The film-rupture model assumes that the stress acts to open the crack and rupture and the protective surface film. Two investigators, working independently, first postulated that localized plastic deformation at the crack tip ruptures the passivating film, exposing the bare metal (Ref 71, 72). This bare metal then dissolves rapidly, resulting in crack extension. Some investigators (Ref 73-75) assume that once propagation starts, the crack tip remains bare because the rate of film rupture at the crack tip is greater than the rate of repassivation (Fig. 34a). Others (Ref 76-79) assume that the crack tip repassivates completely and is periodically ruptured by the emergence of slip steps (Fig. 34b).

Considerable evidence has been found to support these mechanistic models, and intergranular corrosion may be considered the low-stress-limiting case of this mechanism (Ref 65-79). However, the observation of discontinuous cracking and crack-arrest markings (discussed previously) is an indication that crack propagation can be, and frequently is, discontinuous. Also, transgranular SCC fracture surfaces are flat, crystallographically oriented, and match precisely on opposite sides of the fracture surface (indicating very little dissolution during crack advance). As a result, film rupture and dissolution are accepted as viable mechanisms of intergranular SCC in some systems, but are not generally accepted as mechanisms of transgranular SCC (Ref 80).

Active-Path Intergranular SCC. Intergranular SCC can occur by an active-path process that results from a difference in the microchemistry of the material at the grain or interface bound-

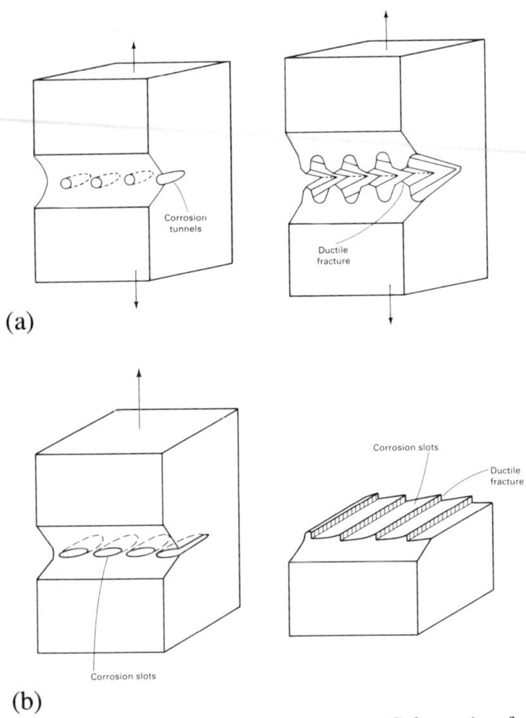

Fig. 35 Corrosion tunnel models. (a) Schematic of tunnel model showing the initiation of a crack by the formation of corrosion tunnels at slip steps and ductile deformation and fracture of the remaining ligaments. (b) Schematic diagram of the tunnel mechanism of SCC and flat slot formation as proposed in Ref 83

ary. This is in contrast to intergranular SCC in sensitized austenitic stainless steel, which also results from a different grain-boundary microchemistry, but which is considered to occur by a film-rupture process. The phenomenological characteristics of active-path intergranular SCC was described previously for grain-boundary segregation and precipitation. While there is no single, comprehensive mechanism to describe active-path intergranular SCC, the simplest description of the crack velocity is given by the Faradaic expression (Eq 3).

The key to the usefulness of Eq 3 for describing active-path intergranular SCC is knowledge of the grain-boundary microchemistry and the resulting corrosion rate, i_a. Methods for determining the grain-boundary microchemistry include Auger electron spectroscopy (AES) for monolayer grain-boundary segregation and scanning transmission electron microscopy/energy-dispersive x-ray analysis/electron energy loss spectroscopy (STEM/EDAX/EELS) for grain-boundary precipitation and multilayer enrichment or depletion. Specifics of the role that individual chemical species play when enriched in grain boundaries can be very complex, as noted by Jones *et al.* (Ref 24) for sulfur and phosphorus segregation in nickel and more recently for iron (Ref 81).

Active-path intergranular SCC crack velocity can exhibit positive or negative deviations from that predicted by Eq 3. Negative deviations result from the crack-deflection, branching, and bridging processes, which reduce the crack-tip stress intensity, and from crack-wall corrosion effects, which alter the crack-tip corrosion rate. Crack bridging results from ligaments left in the wake of the crack, such as particles or complete grains that the crack circumvented. Crack-wall corrosion effects can saturate the crack-tip region when a species adsorbs on the crack walls and inhibits passivation. Deviations in the crack-tip chemistry and electrochemical potential from those for flat-plate corrosion conditions, at which i_a is usually measured, can also cause the crack velocity to differ from that predicted by Eq 3; however, these deviations result from a lack of knowledge about the crack tip and are not really true deviations. Positive deviations in the crack velocity from Eq 3 result from crack jumps produced by mechanical fracture. Hydrogen-induced subcritical crack growth resulting from corrosion-induced hydrogen generation is the most common form of this type of mechanical fracture. Other processes such as adsorption and corrosion-induced enhancement in the local stress intensity have been postulated, but there is little direct evidence for these mechanisms.

Mechanical Fracture Models

Mechanical fracture models originally assumed that stress concentration at the base of corrosion slots or pits increased to the point of ductile deformation and fracture (Ref 70). These early proposals assumed that the crack essentially propagated by dissolution and that the remaining ligaments then failed by mechanical fracture (ductile or brittle). A refinement of this approach

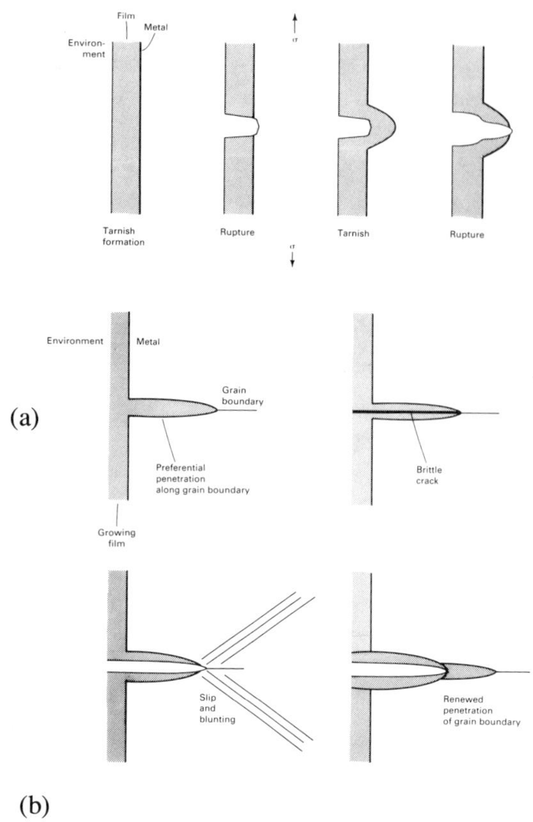

Fig. 36 Tarnish-rupture models. (a) Schematic of the tarnish-rupture model for SCC as proposed in Ref 87. (b) Modified tarnish-rupture model of SCC for systems with intergranular oxide-film penetration (Ref 89, 90)

has been proposed (Ref 57, 82); it is generally known as the corrosion tunnel model.

The corrosion tunnel model assumes that a fine array of small corrosion tunnels forms at emerging slip steps. These tunnels grow in diameter and length until the stress in the remaining ligaments causes ductile deformation and fracture. The crack thus propagates by alternating tunnel growth and ductile fracture (Fig. 35). Cracks propagating by this mechanism should result in grooved fracture surfaces with evidence of microvoid coalescence on the peaks, as illustrated in Fig. 35(a). This is not consistent with the common fractographic features discussed previously. As a result, it was suggested (Ref 83) that the application of a tensile stress results in a change in the morphology of the corrosion damage from tunnels to thin, flat shots, as shown in Fig. 35(b). The formation of slots was observed below the dealloyed sponge layer on the {110} -type planes in austenitic stainless steels, and it was found that the formation of these slots required the presence of a tensile stress. The width of the corrosion slots approached atomic dimensions, and, as a result, close correspondence of matching fracture, surfaces would be expected. It was concluded that transgranular SCC can be explained in terms of the formation and mechanical separation of corrosion slots (Ref 83).

Adsorption-Enhanced Plasticity. Studies of liquid-metal embrittlement (LME), hydrogen embrittlement, and SCC have led to the conclusion that similar fracture processes occur in each case (Ref 84-86). Because chemisorption is common to all three, this process was believed to be responsible for the environmentally induced crack propagation (Ref 84). Based on fractographic studies, it was concluded that cleavage fracture is not an atomically brittle process, but occurs by alternate slip at the crack tip in conjunction with formation of very small voids ahead of the crack. It was also proposed that chemisorption of environmental species facilitated the nucleation of dislocations at the crack tip, promoting the shear processes responsible for brittle, cleavagelike fracture (Ref 84-86). The origin of crack-arrest markings by this mechanism is, however, uncertain. Nevertheless, this mechanism promises to explain many similarities among SCC, LME, and hydrogen embrittlement.

The tarnish-rupture model was first proposed (Ref 87) to explain transgranular SCC, but was later modified by other researchers (Ref 88-90) to explain intergranular SCC. In the original model (Ref 87), a brittle surface film forms on the metal, fracturing under the applied stress. Fracture of the film exposes bare metal, which rapidly reacts with the environment to re-form the surface film. The crack propagates by alternating film growth and fracture, as shown by Fig. 36(a). This hypothesis was later modified to explain intergranular SCC based on the assumption that the oxide film penetrates along the grain boundary ahead of the crack tip, as shown in Fig. 36(b). Again, crack propagation consists of alternating periods of film growth and brittle-film fracture.

Film growth requires transport of species across the film, and, as a result, the thickness of the film is limited in the absence of stress (Ref 90). This model predicts crack-arrest markings on intergranular fracture surfaces and discontinuous acoustic emission during crack propagation that are not always observed during intergranular SCC. Also, it assumes penetration of the film into the grain boundary ahead of the crack tip, which may not be the case for all systems (Ref 91). At present, experimental results are insufficient to confirm or refute this model (Ref 89, 90).

Film-Induced Cleavage Model. In 1959, the hypothesis was presented that dealloying and/or vacancy injection could induce brittle fracture (Ref 92). However, the exact nature of the interactions and how they could induce the observed crack propagation was not thoroughly evaluated. More recently, a model was developed based on the hypothesis that a surface film could induce cleavage fracture (Ref 93, 94). This model assumes that:

- A thin surface film or layer forms on the surface
- A brittle crack initiates in this layer
- The brittle crack crosses the film/matrix interface with little loss in velocity
- Once in the ductile matrix, the brittle crack will continue to propagate
- This crack will eventually blunt and arrest, after which the entire process repeats itself

This model has the unique ability to explain the crack-arrest markings, the cleavage-like facets on the fracture surface, and the discontinuous nature of crack propagation.

The hypothesis that a brittle crack will continue to propagate after it has entered the normally ductile matrix is a critical point. This allows a thin surface layer to induce brittle crack propagation over distances much greater than the film thickness. Critical examination of this hypothesis led to the conclusion that a brittle crack can propagate in a ductile matrix if the crack is sharp and is propagating at high velocities before entering the ductile matrix (Ref 95). A computer model was developed for this process, and it was concluded that a surface layer can initiate brittle fracture even if the layer is ductile (depending on

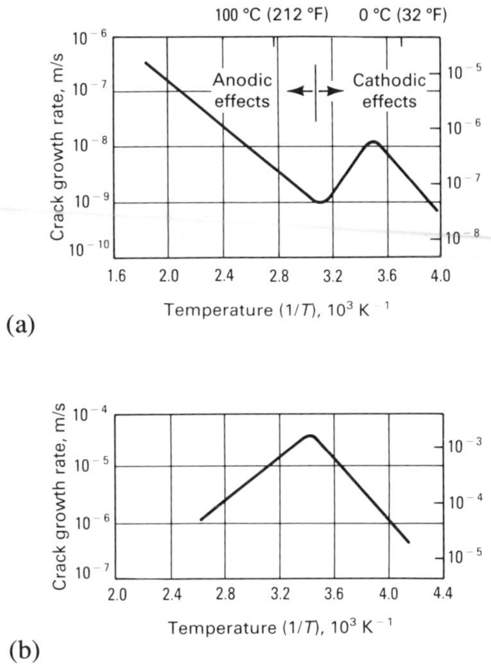

Fig. 37 Schematic of crack-growth rate versus temperature for (a) 3% Ni steel in water (Ref 98) and (b) 4340 steel in gaseous hydrogen (Ref 99)

lattice mismatch, and so on) (Ref 96). More research into surface films and brittle fracture is required before this model can be thoroughly evaluated.

Adsorption-Induced Brittle Fracture. This model, which was initially presented in Ref 68, is based on the hypothesis that adsorption of environmental species lowers the interatomic bond strength and the stress required for cleavage fracture. This model is frequently referred to as the stress-sorption model, and similar mechanisms have been proposed for hydrogen embrittlement and LME (Ref 97). This model predicts that cracks should propagate in a continuous manner at a rate determined by the arrival of the embrittling species at the crack tip. This model does not explain how the crack maintains an atomically sharp tip in a normally ductile material, because it does not include a provision for limiting deformation in the plastic zone. It also fails to explain the discontinuous nature of crack propagation.

Hydrogen Embrittlement. Stress-corrosion cracking in some material/environment combinations is the result of hydrogen-induced subcritical

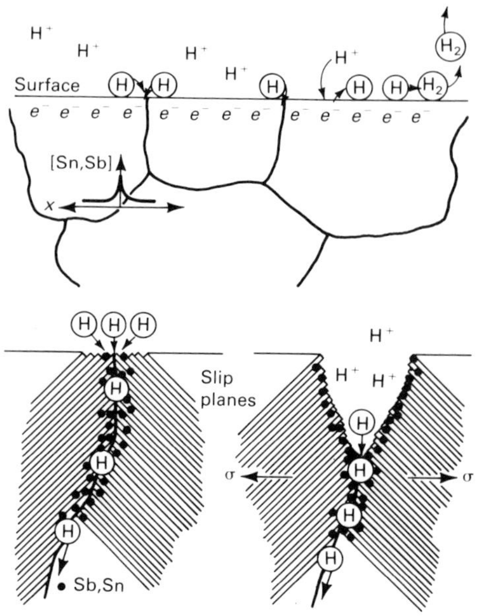

Fig. 38 Schematic showing effect of some impurities on mechanism by which intergranular embrittlement of nickel is presumed to occur at cathodic potentials

crack growth. Because the anodic reaction must have a corresponding cathodic reaction and because the reduction of hydrogen is frequently the cathodic reaction, hydrogen-induced subcritical crack growth can be responsible for subcritical crack growth in corrosive environments (SCC), and many features of hydrogen-induced subcritical crack growth from cathodic hydrogen are very similar to those produced by gaseous or internal hydrogen.

Once hydrogen has been absorbed by a material, its effect, whether from a gaseous or cathodic source, is essentially the same. This has been shown for a number of materials and a variety of properties. There are three primary differences between gaseous and cathodic hydrogen-absorption processes. First, cathodic hydrogen is present on the surface as atomic hydrogen (as reduced), while the gaseous hydrogen adsorbs in the molecular form and must dissociate to form atomic hydrogen. Desorption of loosely bound molecular hydrogen is relatively easy, while the dissociation of adsorbed molecular hydrogen can be the rate-limiting process in hydrogen absorption and embrittlement. Therefore, the desorption and absorption rates of gaseous and cathodic hydrogen may be substantially different for equal hydrogen fugacities. Second, the hydrogen activities produced by cathodic hydrogen can be quite large (thousands of psi) and are dependent on the electrode potential and reaction kinetics, while gaseous hydrogen pressures are generally much lower. Lastly, the surface of the material at a crack tip may be substantially different under electrochemical corrosion conditions from that in the presence of gaseous hydrogen containing substantial quantities of other gases, such as oxygen and carbon dioxide.

Hydrogen-induced crack growth has been suggested as the SCC mechanism for ferritic steels, nickel-base alloys, aluminum alloys, and intermetallic compounds. The effects of such factors as yield strength, impurity segregation, and temperature on the crack-growth behavior of ferritic materials in aqueous environments all follow the trends of gaseous hydrogen embrittlement. For example, the crack-growth rate of a 3% Ni steel as a function of temperature in water is shown in Fig. 37(a), while the crack-growth rate of 4340 steel in gaseous hydrogen is shown in Fig. 37(b). The maximum crack-growth rate occurs at about 20 °C (70 °F), with similar decreases at both higher and lower temperatures. However, anodic stress-corrosion processes become active at temperatures above 100 °C (212 °F) for steel tested in water. Similar trends have been shown for impurity segregation effects on hydrogen-induced crack growth of materials where cathodic and gaseous hydrogen produce essentially similar results.

Specific mechanisms of cathodic hydrogen-induced subcritical crack growth have not been developed, because it has generally been sufficient merely to identify hydrogen as the cause for cracking, in which case models for gaseous hydrogen-induced crack growth can be involved. However, a mechanism was presented in which grain-boundary impurities act as hydrogen recombinant poisons and enhance the uptake of cathodic hydrogen (Ref 100). A schematic of this process is shown in Fig. 38, in which the tin and antimony that segregated to the grain boundaries of nickel enhance the hydrogen-uptake kinetics. This mechanism does not propose a new mechanism by which hydrogen can cause cracking, but merely proposes a mechanism by which impurity segregation can enhance hydrogen uptake. There may be circumstances in which such a mechanism could tip the balance between an anodically driven process and a cathodically driven process or could accelerate cracking to a value that is measurable or of practical significance. However, a review of the combined effects of impurity segregation and hydrogen embrittlement concluded that grain-boundary impurities behave the same with cathodic and gaseous hydrogen in that they enhance crack growth by a combined grain-boundary embrittlement process but not by enhanced hydrogen uptake (Ref 101). It should be noted, however, that a material with impurities such as sulfur, phosphorus, antimony, and tin segregated to its grain boundaries is more susceptible to all forms of hydrogen — cathodic, gaseous, or internal. The reader is referred to Chapter 3 for a more detailed description of the micromechanics of hydrogen-induced subcritical crack growth.

Summary

Stress-corrosion cracking is a phenomenon in which time-dependent crack growth occurs when the necessary electrochemical, mechanical, and metallurgical conditions exist. Corrosion fatigue is a related process in which the load is cyclic rather than static as in stress corrosion. When hydrogen is generated as a product of the corrosion reaction, crack growth can occur by a hydrogen-embrittlement process in much the same way as if hydrogen were in the gaseous form. A common feature of each of these processes is subcritical crack growth, in which cracks grow from existing flaws or initiation sites and grow to a size at which catastrophic failure occurs. Such failure occurs because the combination of crack length and applied stress increases the stress intensity to a value equal to the fracture toughness of the material. A second common feature of stress corrosion, corrosion fatigue, and hydrogen-induced crack growth is that these mechanisms do not require the embrittlement of the entire component; the effect is localized to the crack-tip region.

This chapter was intended to familiarize the reader with the phenomenological and mechanistic aspects of stress corrosion in order to enhance understanding of other chapters in this book that deal with the occurrence and evaluation of stress corrosion in specific metals and alloy systems. The phenomenological description of crack initiation and propagation is based on well-established experimental evidence and observations of stress corrosion. Discussions of SCC mechanisms describe the physical processes involved in crack initiation and propagation. The physical processes related to crack growth have received more evaluation and are better understood than those responsible for crack initiation. There is some phenomenological understanding of stress-corrosion crack initiation, but detailed mechanistic information is lacking.

Stress-corrosion cracking occurs when certain critical conditions are achieved. These conditions include electrochemical, mechanical, and metallurgical factors that must exist simultaneously. A change in any one of these factors may eliminate SCC; therefore, a clear knowledge of these critical factors is important in system design. The important electrochemical parameters include oxidizing potential, pH, impurity concentration, and temperature. The important mechanical parameters include stress, stress intensity, and strain rate. The important metallurgical factors include localized microchemistry (such as the depletion of passive-film-forming elements or the enrichment of active corroding elements), bulk composition, deformation character, and yield strength.

There are a number of plausible mechanisms or physical processes that can account for SCC. No single mechanism is adequate to describe stress corrosion in the variety of materials in which it has been observed. For example, it is generally accepted that intergranular corrosion of sensitized stainless steel is the result of dissolution of the chromium-depleted region adjacent to chromium carbide precipitates and that in the presence of stress this is manifested as intergranular SCC. Ductile nickel aluminide (Ni_3Al + B), on the other hand, exhibits intergranular cracking in acidic environments where hydrogen evolution is the cathodic reaction supporting dissolution; suppression of hydrogen evolution prevents cracking (Ref 102). Transgranular cracking has been observed in brass under conditions where there is no detectable dissolution and hydrogen evolution is unlikely, and in copper-gold alloys where hydrogen absorption is not possible (Ref 103, 104). The obvious conclusion is that there is not a single mechanism of SCC and that there are at least three different mechanisms by which aggressive environments can cause cracking.

Stress-corrosion crack-propagation mechanisms can be divided into dissolution and mechanical fracture models, with the mechanical fracture models further subdivided into ductile and brittle crack-extension processes. Dissolution models include film-rupture and active-path processes, while ductile mechanical models include corrosion tunneling and adsorption-enhanced plasticity models. Brittle mechanical models include the tarnish-rupture and film-induced cleavage models.

References

1. T.A. Michalske and S.W. Freiman, *Nature*, Vol 295, 11 Feb 1982, p 511
2. J.F. Fellers and B.F. Kee, *J. Appl. Polymer Sci.*, Vol 18, 1974, p 2355
3. E.H. Andrews, *Developments in Polymer Science*, Applied Science, 1979
4. H.T. Sumison and D.P. Williams, in *Fatigue of Composite Materials*, STP 569, American Society for Testing and Materials, 1975, p 226
5. C. Zeben, in *Analysis of the Test Methods for High Modulus Fibers and Composites*, STP 521, American Society for Testing and Materials, 1973, p 65
6. K. Sieradski and R.C. Newman, *Philos. Mag. A*, Vol 5 (No. 1), 1985, p 95
7. T. Cassaigne, E.N. Pugh, and J. Kruger, National Bureau of Standards and Johns Hopkins University, unpublished research, 1987
8. A.J. Sedriks, *Stress Corrosion Test Methods*, National Association of Corrosion Engineers, 1990
9. C.M. Chen, M.H. Froning, and E.D. Verink, in *Stress Corrosion — New Approaches*, STP 610, American Society for Testing and Materials, 1976, p 289
10. R.W. Staehle, in *Environment-Induced Cracking of Metals*, R.P. Gangloff and M.B. Ives, Ed., National Association of Corrosion Engineers, 1990, p 561
11. C.J. Beevers, Ed., *The Measurement of Crack Length and Shape During Fracture and Fatigue*, Chameleon Press, 1980
12. E.N. Pugh, U. Bertocci, and R.E. Ricker, National Bureau of Standards, unpublished research, 1987
13. R.E. Ricker, J.L. Fink, and A.K. Vasudévon, *Metall. Trans. A*, Vol 22 (No. 1), 1991, p 264
14. D.P. Williams and H.G. Nelson, *Metall. Trans.*, Vol 1 (No. 1), 1970, p 63
15. R.E. Ricker and D.J. Duquette, The Role of Environment on Time Dependent Crack Growth, in *Micro and Macro Mechanics of Crack Growth*, K. Sadananda, B.B. Rath, and D.J. Michel, Ed., American Institute of Mining, Metallurgical, and Petroleum Engineers, 1982, p 29
16. R.M. Parkins, *Mater. Sci. Technol.*, Vol 1, 1985, p 480
17. K. Tamaki, S. Tsujikawa, and Y. Hisamatsu, in *Advances in Localized Corrosion*, H.S. Isaacs and V. Bertocci, Ed., National Association of Corrosion Engineers, 1991, p 207
18. J. Stewart, P.M. Scott, and D.E. Williams, Paper No. 285, presented at Corrosion/88, National Association of Corrosion Engineers, 1988
19. R.H. Jones, M.A. Friesel, and R. Pathania, Paper No. 499, presented at Corrosion/90, National Association of Corrosion Engineers, 1990
20. R.N. Parkins, *Mater. Sci. Eng.*, Vol A103, 1988, p 143-156
21. L. Hagn, Lifetime Prediction for Parts in Corrosion Environments, in *Corrosion in Power Generating Equipments*, Plenum Press, 1983
22. O. Buck and R. Ranjan, Evaluation of a Crack-Tip-Opening Displacement Model Under Stress-Corrosion Conditions, in *Modeling Environmental Effects on Crack Growth Processes*, R.H. Jones and W.W. Gerberich, Ed., The Metallurgical Society, 1986, p 209
23. M.J. Danielson, C. Oster, and R.H. Jones, *Corros. Sci.*, Vol 32 (No. 1), 1991, p 1
24. R.H. Jones, M.J. Danielson, and D.R. Baer, *J. Mater. Energy Syst.*, Vol 8, 1986, p 185
25. A. Turnbull, Progress in the Understanding of the Electrochemistry in Cracks, in *Embrittlement by the Local Crack Environment*, R.P. Gangloff, Ed., The Metallurgical Society, 1984, p 3
26. R.N. Parkins, Prevention of Environment Sensitive Fracture by Inhibition, in *Embrittlement by the Local Crack Environment*, R.P. Gangloff, Ed., The Metallurgical Society, 1984, p 385
27. R.N. Parkins, *Br. Corros. J.*, Vol 14, 1979, p 5
28. C. Edelneau and A.J. Forty, *Philos. Mag.*, Vol 46, 1960, p 521
29. J.A. Beavers and E.N. Pugh, *Metall. Trans. A*, Vol 11, 1980, p 809
30. M.T. Hahn and E.N. Pugh, *Corrosion*, Vol 36, 1980, p 380
31. E.N. Pugh, On the Propagation of Transgranular Stress Corrosion Cracks, in *Atomistics of Fracture*, R.M. Latanision and J.R. Pickens, Ed., Plenum Press, 1983, p 997

32. R.C. Newman and K. Sieradzki, "Film-Induced Cleavage During Stress-Corrosion Cracking of Ductile Metals and Alloys," NATO Advanced Research Workshop on Chemistry and Physics of Fracture, June 1986
33. R. Bandy and D. Vanrooyen, Mechanisms of Stress Corrosion Cracking and Intergranular Attack in Alloy 600 in High-Temperature Caustic and Pure Water, in *Corrosion of Nickel-Base Alloys*, American Society for Metals, 1985
34. C.H. Shen and P.G. Shewmon, A Mechanism for Hydrogen Induced Intergranular Stress Corrosion Cracking in Alloy 600, *Metall. Trans.*, 1990
35. G.M. Vgiansky, Ph.D. dissertation, Ohio State University, 1976
36. V.L. Ratke and W. Gruhl, *Werk. Korros.*, Vol 31, 1980, p 768-773
37. G.M. Scamans and C.D.S. Tuck, *Environment-Sensitive Fracture of Engineering Materials*, A.Z. Foroulis, Ed., TMS/AIME, 1979, p 464
38. R.E. Ricker and D.J. Duquette, *Metall. Trans. A*, Vol 19A, 1988, p 1775-1783
39. C.B. Bargeron and R.C. Benson, *J. Electrochem. Soc.*, Vol 127 (No. 11), 1980, p 2528-2530
40. R.E. Ricker, Ph.D. dissertation, Rensselaer Polytechnic Institute, 1983
41. N.J.H. Holroyd and G.M. Scamans, *Scr. Metall.*, Vol 19, 1985, p 915
42. E.C. Pow, W.W. Gerberich, and L.E. Toth, *Scr. Metall.*, Vol 15, 1981, p 55
43. E.D. Hondros and M.P. Seah, *Int. Met. Rev.*, Dec 1977, p 262
44. L. Long and H. Uhlig, *J. Electrochem. Soc.*, Vol 112, 1965
45. J. Kuppa, H. Erhart, and H. Grabke, *Corros. Sci.*, Vol 21, 1981, p 227
46. N. Bandyopadhyay, R.C. Newman, and K. Sieradzki, in *Proc. 9th Int. Congr. Metallic Corrosion* (Toronto, Canada), National Association of Corrosion Engineers, 1984
47. R.H. Jones, S.M. Bruemmer, M.T. Thomas, and D.R. Baer, Comparison of Segregated Phosphorus and Sulfur Effects on the Fracture Mode and Ductility of Iron Tested at Cathodic Potentials, *Scr. Metall.*, Vol 16, 1982, p 615
48. S.M. Bruemmer, R.H. Jones, M.T. Thomas, and D.R. Baer, Fracture Mode Transition of Iron in Hydrogen as a Function of Grain Boundary Sulfur, *Scr. Metall.*, Vol 14, 1980, p 137
49. A. Joshi and D.J. Stein, *Corrosion*, Vol 28 (No. 9), 1972, p 321
50. R.H. Jones, Some Radiation Damage-Stress Corrosion Synergisms in Austenitic Stainless Steels, in *Proc. 2nd Int. Symp. Environmental Degradation of Materials in Nuclear Power Systems — Water Reactors*, American Nuclear Society, Sept 1985, p 173
51. M. Guttman, P. Dumoulin, Nguyen Tan Tai, and P. Fontaine, *Corrosion*, Vol 37, 1981, p 416
52. S.M. Bruemmer, L.A. Charlot, and C.H. Henager, Jr., "Microstructural Effects on Microdeformation and Primary-Side Stress Corrosion Cracking of Alloy 600 Tubing," EPRI Final Report, Project 2163-4, Electric Power Research Institute, Aug 1986
53. R.H. Jones, S.M. Bruemmer, M.T. Thomas, and D.R. Baer, Influence of Sulfur, Phosphorus, and Antimony Segregation on the Intergranular Hydrogen Embrittlement of Nickel, *Metall. Trans. A*, Vol 14, 1983, p 223
54. R.H. Jones, Analysis of Hydrogen-Induced Subcritical Intergranular Crack Growth of Iron and Nickel, *Acta Metall.*, Vol 38, 1990, p 1703
55. *Metals Handbook*, Vol 13, 9th ed., *Corrosion*, ASM International, 1987
56. R.W. Staehle, *et al.*, *Corrosion*, Vol 26 (No. 11), 1970, p 451
57. H.W. Pickering and P.R. Swann, *Corrosion*, Vol 19 (No. 3), 1963, p 373
58. Fracture Mechanics, in *Metals Handbook*, Vol 8, 9th ed., *Mechanical Testing*, ASM International, 1985, p 439
59. R.H. Jones, M.J. Danielson, and D.R. Baer, Microchemistry and Mechanics Issues in Stress Corrosion Cracking, in *Proc. 20th Nat. Symp. Fracture Mechanics* (Bethlehem, PA), American Society for Testing and Materials, 1987, p 209
60. F.P. Ford, EPRI NP-2589, Electric Power Research Institute, 1982
61. P.L. Andresen and F.P. Ford, Life Prediction by Mechanistic Modeling and System Monitoring of Environment Cracking of Fe and Ni Alloys in Aqueous Systems, *Mater. Sci. Eng.*, Vol A103, 1988, p 167
62. E.P. Simonen and R.H. Jones, *Acta Metall.*, in press
63. A.W. Thompson and I.M. Bernstein, *Advances in Corrosion Sciences and Technology*, Vol 7, M.G. Fontana and R.W. Staehle, Ed., Plenum Press, 1980, p 53
64. R.M. Riecke, A. Athens, and I.O. Smith, *Mater. Sci. Technol.*, Vol 2, 1986, p 1066
65. E.H. Dix, *Trans. AIME*, Vol 137 (No. 11), 1940
66. R.B. Mears, R.H. Brown, and E.H. Dix, *Symp. Stress-Corrosion Cracking of Metals*, ASTM/AIME, 1944, p 323
67. E.H. Dix, Jr., *Trans. ASM*, Vol 42, 1950, p 1057
68. H.H. Uhlig, in *Physical Metallurgy of Stress Corrosion Fracture*, T.N. Rhodin, Ed., Interscience, 1959, p 1
69. L. Yang, G.T. Horne, and G.M. Pound, in *Physical Metallurgy of Stress Corrosion Fracture*, T.N. Rhodin, Ed., Interscience, 1950, p 29
70. J.J. Harwood, in *Stress Corrosion Cracking and Embrittlement*, W.D. Robertson, Ed., John Wiley & Sons, 1956, p 1

71. F.A. Champion, in *Symp. Internal Stresses in Metals and Alloys*, Institute of Metals, 1948, p 468
72. H.L. Logan, *J. Res. Natl. Bur. Stand.*, Vol 48, 1952, p 99
73. E.W. Hart, *Surfaces and Interfaces II*, Syracuse University Press, 1968, p 210
74. H.J. Engle, in *The Theory of Stress Corrosion Cracking in Alloys*, North Atlantic Treaty Organization, 1971
75. J.C. Scully, *Corros. Sci.*, Vol 15, 1975, p 207
76. D.A. Vermilyea, *J. Electrochem. Soc.*, Vol 119, 1972, p 405
77. D.A. Vermilyea, *Stress Corrosion Cracking and Hydrogen Embrittlement of Iron Based Alloys*, National Association of Corrosion Engineers, 1977, p 208
78. R.W. Staehle, in *the Theory of Stress Corrosion Cracking in Alloys*, North Atlantic Treaty Organization, 1971, p 223
79. R.W. Staehle, *Stress Corrosion Cracking and Hydrogen Embrittlement of Iron Based Alloys*, National Association of Corrosion Engineers, 1977, p 180
80. E.N. Pugh, *Corrosion*, Vol 41 (No. 9), 1985, p 517
81. R.H. Jones, B.W. Arey, D.R. Baer, and M.A. Friesel, Grain-Boundary Chemistry and Intergranular Stress Corrosion of Iron Alloys in Calcium Nitrate, *Corrosion*, Vol 45, 1989, p 494
82. P.R. Swann and J.D. Embury, *High Strength Materials*, John Wiley & Sons, 1965, p 327
83. J.M. Silcock and P.R. Swann, *Environment-Sensitive Fracture of Engineering Materials*, Z.A. Foroulis, Ed., The Metallurgical Society, 1979, p 133
84. S.P. Lynch, *Hydrogen Effects in Metals*, A.W. Thompson and I.M. Bernstein, Ed., The Metallurgical Society, 1981, p 863
85. S.P. Lynch, *Met. Sci.*, Vol 15 (No. 10), 1981, p 463
86. S.P. Lynch, *J. Mater. Sci.*, Vol 20, 1985, p 3329
87. A.J. Forty and P. Humble, *Philos. Mag.*, Vol 8, 1963, p 247
88. A.J. McEvily and P.A. Bond, *J. Electrochem. Soc.*, Vol 112, 1965, p 141
89. E.N. Pugh, in *Stress Corrosion Cracking and Hydrogen Embrittlement of Iron Based Alloys*, National Association of Corrosion Engineers, 1977, p 37
90. J.A. Beavers, J.C. Rosenberg, and E.N. Pugh, in *Proc. 1972 Tri-Service Conf. Corrosion*, MCIC-73-19, Metals and Ceramics Information Center, 1972, p 57
91. T.R. Pinchback, S.P. Clough, and L.A. Heldt, *Metall. Trans. A*, Vol 7, 1976, p 1241; *Metall. Trans. A*, Vol 6, 1975, p 1479
92. A.J. Forty, in *Physical Metallurgy of Stress Corrosion Fracture*, T.N. Rhodin, Ed., Interscience, 1959, p 99
93. K. Sieradzki and R.C. Newman, *Philos. Mag. A*, Vol 51 (No. 1), 1985, p 95
94. K. Sieradzki and R.C. Newman, *J. Phys. Chem. Sol.*, Vol 48 (No. 11), 1987, p 1101-1113
95. I.-H. Lin and R.M. Thomson, *J. Mater. Res.*, Vol 1 (No. 1), 1986
96. G.J. Dienes, K. Sieradzki, A. Paskin, and B. Massoumzadeh, *Surf. Sci.*, Vol 144, 1984, p 273
97. N.S. Stoloff, in *Environment-Sensitive Fracture of Engineering Materials*, Z.A. Foroulis, Ed., The Metallurgical Society, 1979, p 486
98. M.O. Spiedel and R.M. Magdowski, Stress Corrosion Cracking of Steam Turbine Steels — An Overview, in *Proc. 2nd Int. Symp. Environmental Degradation of Materials in Nuclear Power Systems—Water Reactors*, American Nuclear Society, 1986, p 267
99. D.P. Williams and H.G. Nelson, *Metall. Trans.*, Vol 1, 1970, p 63
100. R.M. Latanision and H. Opperhauser, *Metall. Trans.*, Vol 5, 1974, p 483
101. R.H. Jones, A Review of Combined Impurity Segregation-Hydrogen Embrittlement Processes, in *Advances in the Mechanics and Physics of Surfaces*, R.M. Latanision and T.E. Fischer, Ed., Scientific and Technical Book Service, 1986
102. R.E. Ricker, M.R. Stoudt, and J.L. Fink, in *Hydrogen Effects on Materials Behavior*, N.R. Moody and A.W. Thompson, Ed., TMS-AIME, 1990, p 499-506
103. U. Bertocci, R.I. Thomas, and E.N. Pugh, *Corrosion*, Vol 40 (No. 8), 1984, p 439-440
104. U. Bertocci, *J. Electrochem. Soc.*, Vol 136 (No. 7), 1989, p 1887-1892

CHAPTER 2

Stress-Corrosion Cracking of Carbon and Low-Alloy Steels (Yield Strengths Less Than 1241 MPa)

S.W. Ciaraldi, Amoco Norway Oil Company

Stress-corrosion cracking (SCC) of carbon and low-alloy steels is a significant problem for a variety of industries, such as those dealing with power generation (including nuclear), oil and gas production, refining/processing and pipeline transmission, chemical processing, paper and pulp manufacturing, and aircraft/aerospace. Stress-corrosion cracking is particularly insidious in that catastrophic failures can occur even at low applied stress levels (sometimes only residual in nature) and often with no warning. Environments causing SCC can be so mild that no evidence of corrosion can be detected (except possibly for SCC itself).

Stress-corrosion cracking is defined as environmentally induced subcritical crack growth under the influence of static or dynamic tensile stress (Ref 1). The effects of applied stress on SCC are briefly reviewed in the section "Test Methods" in this chapter. Environmental effects are discussed in some detail in the section "Environmental Aspects."

This chapter will review the SCC of steels containing less than approximately 5% total alloying elements and having yield strengths of less than about 1241 MPa (180 ksi). Table 1 lists some common carbon and low-alloy steels, along with their nominal compositions and typical mechanical properties. Two basic microstructural types of steels are included: those with ferritic-pearlitic structures and, for steels with yield strengths greater than about 620 MPa (90 ksi), those with tempered martensite structures. Discussion in this chapter will generally be limited to environments in which a water phase is present and thus will not include degradation such as high-temperature hydrogen attack or liquid-metal embrittlement.

A thorough review of SCC mechanisms is presented in Chapter 1, so they will not be discussed in detail here. It should be mentioned, however, that for carbon and low-alloy steels at least two mechanisms are thought to be operative: anodic (active-path) SCC and cathodic (hydrogen-embrittlement) SCC (Ref 4). These are sometimes differentiated by their electrochemical potential, pH, and temperature effects on SCC. When more negative potentials, lower pH values, and lower temperatures promote SCC, a hydrogen-embrittlement mechanism is generally thought to be active. The reverse trends are often believed to indicate an active-path mechanism. However, it has been suggested that SCC under both anodic and cathodic conditions is fundamentally the same process (Ref 5).

Fracture Modes

Stress-corrosion cracking most commonly initiates at external surfaces where there is contact with a conducive environment. Multiple points of crack initiation are often associated with SCC (Ref 6). Sometimes, however, single-site SCC initiation is observed; this is generally associated

Table 1 Typical Compositions and Mechanical Properties of Common Carbon and Low-Alloy Steels

UNS	ASTM	Composition, wt%	Yield strength MPa	Yield strength ksi	Tensile strength MPa	Tensile strength ksi	Elongation, %
K01700	A285A	0.17 C, 0.9 Mn	165	24	310–380	45–55	30
K03005	A53B	0.3 C, 1.2 Mn	241	35	415	60	...
K03006	A106B	0.3 C, 0.3–1.1 Mn, 0.1 Si (min)	241	35	415	60	...
G41300	4130 (AISI)	0.3C, 1 Mn, 0.25 Si, 1 Cr, 0.2 Mo	615–1358	89–197	676–1613	98–234	12–28
G43400	4340 (AISI)	0.4 C, 0.7 MN, 0.25 Si, 0.8 Cr, 1.8 Ni, 0.25 Mo	896–1572	130–228	979–1958	142–284	11–21
K21590	A387 (22)	0.15 C, 0.45 Mn, 0.5 Si, 2.25 Cr, 1 Mo	205–310	30–45	415–690	60–100	18

Source: Ref 2, 3

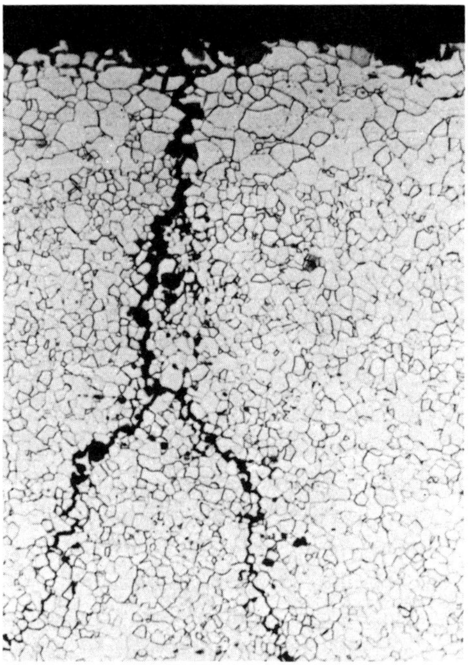

Fig. 1 Intergranular SCC of a carbon steel that occurred in a concentrated ammonium nitrate solution (Nital etch 100×). Source: Ref 13

with a notch or other stress concentrator. Initiation from corrosion pits is also common.

There are exceptions to surface initiation. Hydrogen blistering, also called hydrogen-induced cracking (HIC), can cause internal SCC initiation (Ref 7). Initiation is generally at nonmetallic inclusions and results from corrosion-generated hydrogen diffusing into and accumulating in the steel. This type of SCC initiation is most commonly observed in lower-strength steels in aqueous hydrogen sulfide or other hydrogen-charging environments (Ref 7).

Intergranular cracking (along prior-austenite grain boundaries for quenched and tempered steels) is commonly observed in neutral or acidic environments—for example, aqueous chloride solutions (Ref 1, 5). Intergranular SCC has also been reported for carbon steels in hydroxide-containing environments, aqueous nitrates, ammoniacal solutions, carbonates, and alkanolamine solutions (Ref 8-12). Typical microstructural appearance is shown in Fig. 1. A view by scanning electron microscopy is shown in Fig. 2.

Transgranular fracture, sometimes described as "cleavagelike" because of its brittle appearance, is often observed for SCC in environments that contain hydrogen sulfide, but intergranular cracking is also common (Ref 6, 15-17). An example of transgranular SCC is shown in Fig. 3. Similar observations have been reported for cracking in hydrogen gas (Ref 19, 20).

In lower-strength pressure vessel and pipeline steels in aqueous hydrogen sulfide, HIC arrays may begin to link together to form through-wall cracks (Ref 7, 21). Such cracking is often referred to as "step cracking" or "stress-oriented hydrogen-induced cracking" (SOHIC).

Propagation of stress-corrosion cracking most commonly occurs perpendicular to the principal tensile stress direction. Crack branching is generally thought to be a positive indication of SCC; however, lack of branching is not evidence

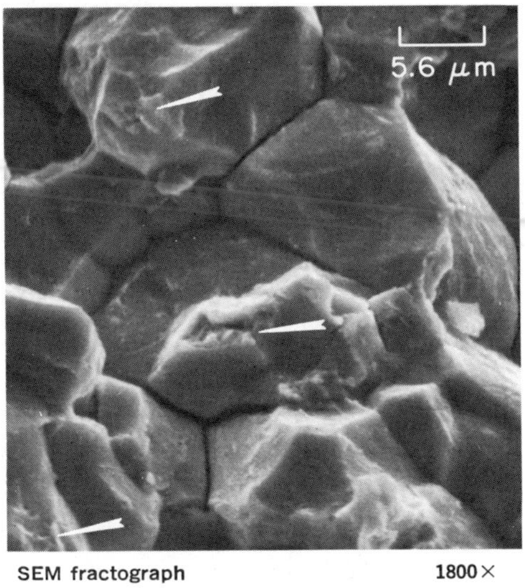

Fig. 2 Scanning electron fractograph of intergranular SCC in a modified AISI 4340 steel exposed to a salt-water environment (1800×). Source: Ref 14

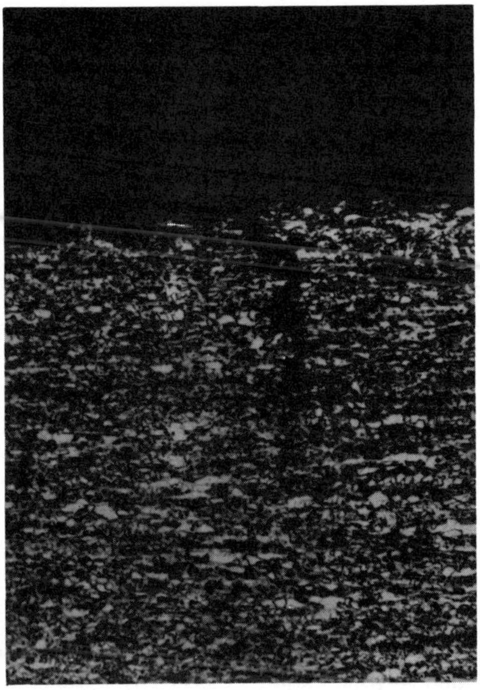

Fig. 3 Transgranular hydrogen sulfide SCC of a low-alloy steel (100×). Source: Ref 18

that SCC was not involved (Ref 6). As will be seen, branching is a consequence of constant crack-growth rate over a range of applied stress intensities, so that deviation in many directions from perpendicular to the principal tensile stress direction can more readily occur.

Microstructure can influence the tendency toward intergranular or transgranular cleavagelike cracking (Ref 22). For example, intergranular SCC is common for ferritic-pearlitic steels in ammonia-containing environments; for martensitic steels, both intergranular SCC and transgranular SCC are observed (Ref 10).

In hydrogen sulfides, steels with higher strength levels have a greater propensity for intergranular than for transgranular cracking; the opposite behavior is observed in lower-strength steels (Ref 23). Cathodic polarization of steel by coupling to a magnesium anode also reportedly increases the tendency for intergranular cracking in quenched and tempered steels in aqueous chlorides (Ref 22).

At high stress intensities, cracking can occur transgranularly by microvoid coalescence in steels in which intergranular or transgranular cleavagelike cracking predominates at lower stress intensities (Ref 22). High stress intensities are also thought to promote a lack of crack branching (Ref 15).

Test Methods

Evaluation methods for SCC are discussed in greater detail in Chapter 17, and will be reviewed only briefly here. Testing for SCC requires the application of tensile stress (applied or residual) and exposure to a conducive environment. The simplest test specimens include U-bends, bent beams, C-rings, and smooth tensile bars. In addition to being cost effective, these types of specimens (particularly U-bends) are the most practical for field SCC testing, as their compactness allows them to be easily inserted into equipment and environments of interest.

Resistance to SCC for these types of specimens is generally measured by time to failure or, if the maximum test duration is limited, the "apparent threshold stress" at and below which SCC does not occur. Figure 4 is an example of SCC measurement by this latter method.

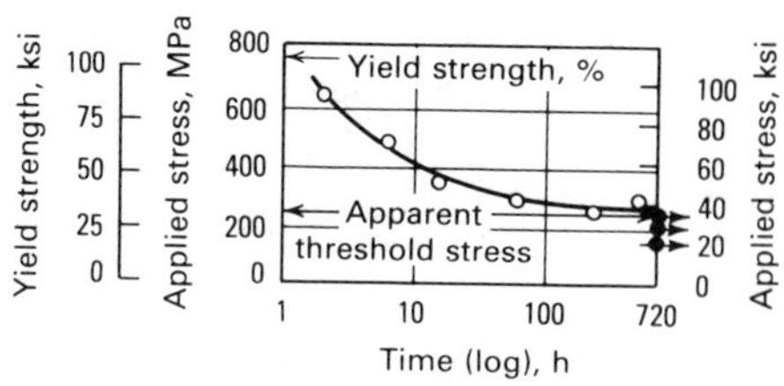

Fig. 4 Definition of "threshold stress" for SCC from applied stress versus time-to-failure plot. Source: Ref 24

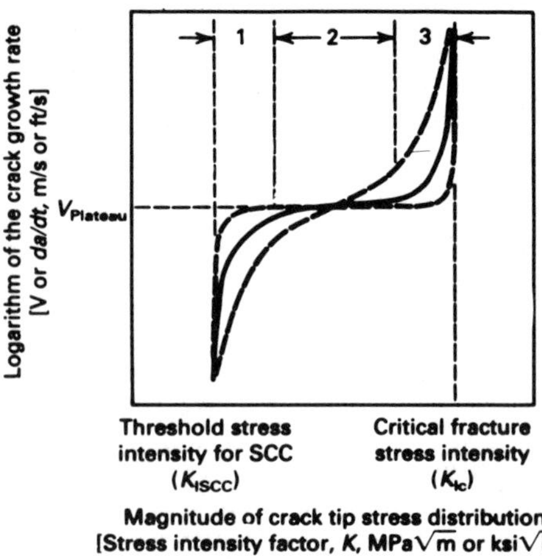

Fig. 5 Schematic of stress intensity versus crack velocity obtained using fracture mechanics methods. Stage II (plateau) velocity and K_{ISCC} are identified. Source: Ref 27

Tests using smooth specimens generally permit less than optimum reproducibility due to variability in SCC initiation times (Ref 5). Improved reproducibility is possible with notched or precracked specimens, for which SCC initiation time is reduced or eliminated. This is often accomplished by using fracture-mechanics specimens of standardized geometry (Ref 25). Fatigue precracking of such specimens further improves reproducibility.

For fracture-mechanics SCC testing of steels, stress state at the crack tip is well defined and described by stress-intensity relationships. Furthermore, conditions promoting plane strain (triaxiality) are believed to be the most conducive to SCC, provided the test specimen is of sufficient thickness (Ref 26).

Increasing stress or stress intensity increases SCC propagation rates and may also enhance crack initiation (Ref 5). Measurement of steel SCC crack-growth rates as a function of applied stress intensity will generally produce a curve of the type shown in Fig. 5. Note that a plateau region over which crack-growth rate is independent of stress intensity is often observed, as mentioned previously (Ref 28-30). Evaluations using fracture-mechanics specimens also employ the concept of "threshold stress intensity," the value at and below which no significant crack growth occurs, as a measure of SCC resistance.

Constant-extension-rate tests (CERT), often called slow-strain-rate tests, are also used for SCC evaluation of steels. In these tests, specimens are slowly strained in tension while simultaneously exposed to an SCC-conducive environment. Susceptibility is generally expressed by comparing SCC failure times and maximum stress, strain, and/or reduction in area to values obtained in tests conducted in inert environments. Initiation of multiple cracks is also an indicator of SCC.

SCC test specimens can be exposed to at least three different types of environments: actual service (field), simulated service, and accelerated, the latter two of which are generally performed in a laboratory. An accelerated environment is often produced by modifying a simulated service environment to make it more aggressive—for example, by changing the temperature, pH, and/or the concentration of the species promoting SCC. Note that electrochemical potentials are also applied during some SCC programs.

Evaluation of the resistance of steels to SCC in hydrogen sulfide environments (sour service) has been standardized by the National Association of Corrosion Engineers (NACE) in TM-01-77 (Ref 31). Current revision of the standard provides techniques for use of smooth tensile, bent-beam, C-ring, and double-cantilever-beam specimens.

The SOHIC resistance of low-yield-strength steels is generally performed using unstressed rectangular bar specimens. Such testing has also been standardized by NACE in TM-02-84 (Ref 32). After exposure to the test media, specimens are sectioned and examined metallographically for hydrogen-induced fissures.

Table 2 Effect of Chloride Solution Type and Yield Strength on SCC of AISI 4340 Steel

Yield strength		Threshold stress intensity: Water/3.5% NaCl		Seawater	
MPa	ksi	MPa$\sqrt{m}$	ksi$\sqrt{in.}$	MPa$\sqrt{m}$	ksi$\sqrt{in.}$
862	125	55	60	64	70
1034	150	48	53	54	59
1207	175	25	28	35	39
1379	200	9	10	12	13

Source: Ref 42

As mentioned previously, higher applied stresses strongly promote SCC, as evidenced by reduced failure times in laboratory tests or in service. Dynamic stress/strain can also promote SCC, particularly in environments where rupture of a protective surface film is involved in the SCC mechanism. Recent studies have demonstrated that correlations exist among SCC propagation rates from constant-load, fracture-mechanics, and CERT studies when crack-tip strain rates are used as the basis for comparison (Ref 33). The phenomenon of SCC threshold stress for statically loaded specimens can thus be explained by relaxation of applied stresses through creep and the reduction of crack-tip strain.

Simple load cycling stimulates crack-tip strain activity and hence may produce SCC at stresses below those produced by static loading, that is, below the threshold stress (Ref 34). This has been clearly demonstrated for the SCC of steels in carbonate-bicarbonate solutions, where laboratory results correlated well to an operating pipeline failure by SCC under repeated pressure cycling (Ref 34, 35). However, these effects should be differentiated from corrosion fatigue, in which enhanced fatigue cracking is much less environmentally specific.

Environmental Aspects

Aqueous Chlorides

Of the various forms of stress-corrosion cracking of carbon and low-alloy steels, SCC in aqueous chlorides is perhaps the most common. This is due largely to the variety of industries and environments in which aqueous chlorides are encountered, including marine, oil/gas production and refining, aerospace, construction, and power generation. As discussed below, SCC in aqueous chlorides is observed less frequently in steels with lower strength levels (Ref 36).

Stress-corrosion cracking of steels is most often associated with an aqueous phase. SCC has been reported in water vapor, but crack growth rates increase dramatically with relative humidity (Ref 37). At about 60% relative humidity, crack-growth rates are comparable to those observed in water, suggesting the presence of a condensed phase at the crack tip.

For steels with yield strengths of less than 1241 MPa (180 ksi), increasing chloride content of SCC-conducive environments apparently increases crack-growth rates, as indicated by reduced failure times (Ref 38-40). This effect is more pronounced at lower chloride levels. However, chloride concentration has been reported to have little influence on the threshold stress level of precracked specimens (Ref 41). As shown in Table 2, SCC in seawater produces threshold stress-intensity values comparable to those observed in laboratory-prepared sodium chloride solutions.

The pH of aqueous chloride solutions has a greater influence on the results of tests performed using smooth specimens than on those using precracked specimens; this is probably due to corrosion-related crack initiation—for example, pitting of smooth specimens (Ref 43). In general, however, pH values below 1 to 3 enhance SCC, and pH values above 9 retard SCC and at 13.6 can even stop crack propagation (Ref 42, 44). No change in SCC susceptibility is generally reported in the range of about 1-3 to 9-10, depending on exact test conditions (Ref 43, 45).

Temperature can have a significant effect on SCC in aqueous chlorides and is often used to demonstrate hydrogen embrittlement, which is more pronounced near room temperature. In general, SCC can be prevented by an increase to some temperature threshold (Ref 37, 46). However, depending on the environment, SCC can occur at quite elevated temperatures, for example, in mild steel in ferric chloride solutions at 316 °C (600 °F). Within the cracking range, threshold stress intensities of precracked specimens have been shown to be relatively constant with temperature (Ref 37, 47). Crack-growth rates may, however, actually increase with increased temperature (Ref 37, 47).

Cathodic polarization such as that obtained by coupling to zinc (*e.g.*, galvanized steels) can reduce SCC failure times, but apparently has little effect on threshold stress levels (Ref 42). These observations apply to high-strength steels and are not commonly seen in lower-strength structural steels (Ref 48-51). Anodic polarization has also been reported to promote SCC in laboratory tests (Ref 52, 53). This effect may have been due to the occurrence of pitting in the steel specimens, resulting in reduced SCC initiation times (Ref 5).

The effects of environmental additives on SCC can generally be rationalized in terms of a hydrogen-embrittlement mechanism, because additives known to promote hydrogen entry into steels accelerate SCC (Ref 49, 54, 55). These additives include sulfur-containing species (perhaps the most commonly encountered), as well as phosphorus, arsenic, selenium, and antimony (Ref 56). Such species have been shown to greatly increase hydrogen-permeation kinetics (Ref 57). By contrast, the presence of high levels of oxygen has been reported to inhibit SCC (Ref 58, 59).

Hydrogen Sulfide

Stress-corrosion cracking of carbon and low-alloy steels in environments that contain hydrogen sulfide is encountered most commonly in the production, transmission, and refining of oil and gas. Failures have been reported in gas/oil well tubulars, wellhead equipment, pipelines, process piping, and pressure vessels.

SCC problems in hydrogen sulfide are more prevalent for steels with higher strength levels and are often associated with hard zones caused by cold working or welding (Ref 60-62). As discussed below, influential environmental factors are consistent with a hydrogen-embrittlement SCC mechanism, although the term "sulfide stress-cracking" is often used to describe such failures.

The majority of evidence suggests that a free water phase must be present for SCC in environments that contain hydrogen sulfide (Ref 60, 63). Increased hydrogen sulfide concentrations result in increased likelihood of SCC, as shown in Fig. 6 (Ref 65, 66). Failures of high-strength steels can occur at very low hydrogen sulfide levels (for example, in the ppm range), but failure times generally increase compared to test results for higher concentrations (Ref 66, 67). SCC threshold stress values also decrease with increased hydrogen sulfide content (Ref 60, 68). Note that hydrogen sulfide concentration is often more conveniently referenced with respect to partial pressure in an associated gaseous phase.

Reduced pH levels significantly enhance the SCC of steels in solutions that contain hydrogen sulfide (Ref 61, 66, 69). Stress-corrosion cracking can occur at pH levels up to about 9 (to about 12 for materials with very high strengths) (Ref 49, 69, 70). In addition to reduced failure times, lower-pH solutions have been found to decrease SCC threshold stress levels (Ref 71-73).

Increased temperatures have been observed to decrease the propensity for SCC in hydrogen sulfide-containing environments (Ref 15, 68, 72, 74). This behavior is shown in Fig. 7. Maximum susceptibility to SCC is near room temperature (Ref 74).

Cathodic polarization of steels in environments that contain hydrogen sulfide has been reported to promote SCC (Ref 49, 75). Anodic

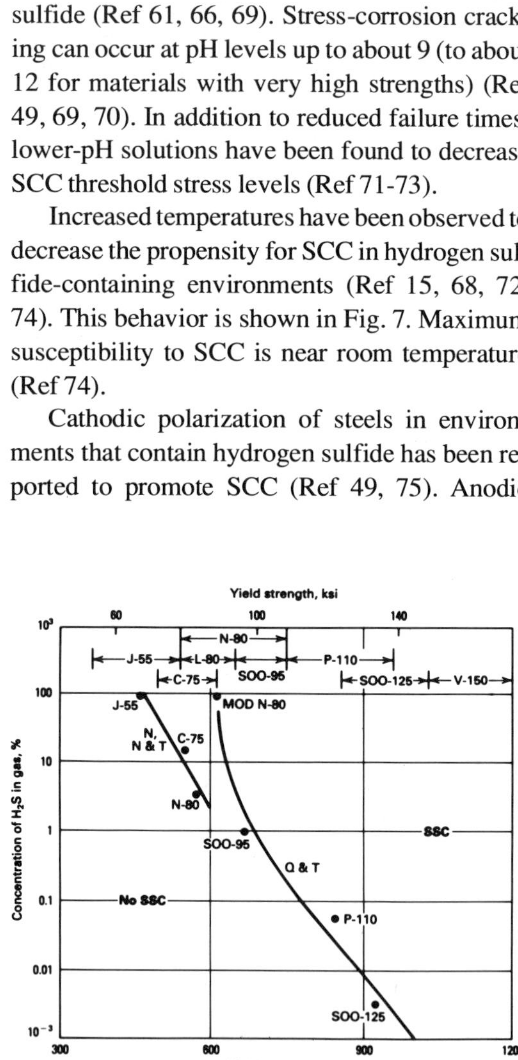

Fig. 6 Effect of hydrogen sulfide concentration on SCC of steels. Note also the effect of yield strength. Source: Ref 64

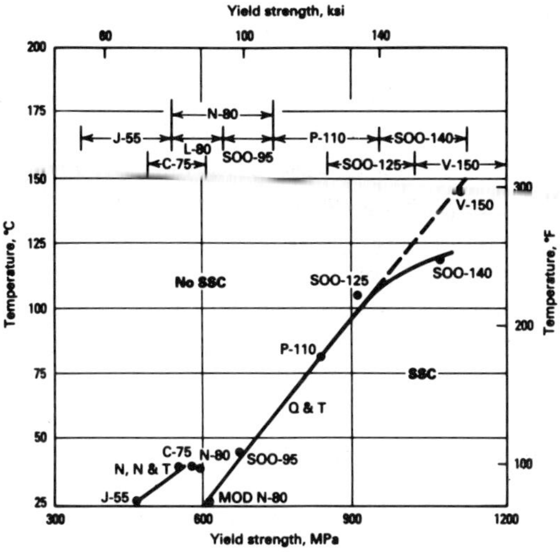

Fig. 7 Effect of temperature on the hydrogen sulfide SCC of steels. Note also the effect of yield strength. Source: Ref 64

polarization has been observed to retard SCC, but an accelerating effect, probably caused by increased electrochemical reaction kinetics, has also been reported (Ref 73, 75).

Additions of acetic acid to hydrogen sulfide-containing solutions promote SCC, probably by both reducing pH and removing protective sulfide films (Ref 76). This latter effect has also been suggested for cyanides (Ref 77). Carbon dioxide has generally been observed to promote SCC both in the laboratory and in service environments (Ref 15, 78). Increasing amounts of chlorides have also been reported to enhance SCC (Ref 79); however, at fixed hydrogen sulfide and pH levels, the effects have been minimal (Ref 72).

Most of the preceding discussion has concerned hydrogen sulfide SCC in moderate- and high-strength steels, a process often termed "sulfide stress-cracking." Recent cracking in low-strength refinery pressure vessels has been attributed to a process called "stress-oriented hydrogen-induced cracking," or SOHIC (Ref 7). Although most often associated with welds that have not been stress relieved, cracking in other regions of concentrated stresses has occurred. First observed in low-strength sour crude oil pipelines, cracking is believed to result when small hydrogen fissures in a steel link together under the action of hydrogen pressure within the fissures to form through-wall cracks (Ref 7, 21).

Sulfuric Acid

Stress-corrosion cracking of steel in sulfuric acid solutions is generally associated with high strength levels, although some cases of hydrogen blistering have been reported in mild steels (Ref 80, 81). Stress-corrosion cracking in sulfuric acid solutions has been suggested to be analogous to SCC in aqueous chlorides, producing many similar effects (Ref 44).

A wide temperature range for SCC—*i.e.*, room temperature to 204 °C (400 °F)—has been reported (Ref 9). Both anodic and cathodic polarization have enhanced the SCC of steels in sulfuric acid solutions (Ref 49, 82). Increased sulfuric acid concentrations also promote SCC, at least at levels below 5% (Ref 49, 83). Additions of hydrochloric acid or sodium chloride increase the SCC susceptibility of steels (Ref 1, 83).

Hydrogen Gas

A number of steel pressure vessel failures by high-pressure hydrogen gas have been reported (Ref 19, 20). Such cracking has been associated with welds and nozzle forgings of moderate-strength steels. However, laboratory studies have shown that SCC by gaseous hydrogen is much more likely to occur in higher-strength steels (Ref 84, 85). Note that for these latter studies and the work referenced below, nominally dry hydrogen gas conditions were employed (that is, the presence of moisture or an aqueous phase was not reported).

In general, increased pressure will promote SCC by gaseous hydrogen (Ref 85, 86). This effect is evident for precracked steels and is sometimes not observed in tests using smooth specimens (Ref 86). For high-strength steels, increased partial pressure of hydrogen lowers SCC resistance, even at pressures well below atmospheric (Ref 87). At higher pressures, a limited effect of hydrogen partial pressure on SCC is reported.

Maximum susceptibility of steels to SCC in gaseous hydrogen occurs at or near room temperature (Ref 87, 88). However, cracking has been reported over a wide temperature range (– 46 to 120 °C, or –50 to 250 °F) and appears to depend on steel hardness and loading conditions (Ref 85, 88). Above a threshold value, constant load or low loading rates promote SCC, whereas SCC is reduced under high-strain-rate conditions (Ref 89, 90).

The presence of even small amounts of oxygen or SO_2 can inhibit SCC in gaseous hydrogen (Ref 91, 92). By contrast, the presence of inert gases has no significant effect on SCC (Ref 93). Dissociation of hydrogen gas through use of a hot tungsten filament reportedly enhances SCC in gaseous hydrogen (Ref 89).

Nitrates

Reported SCC service failures by nitrates are generally associated with welded steel equipment and/or high-strength steels. To study such failures, laboratory investigations have sometimes used concentrated nitrate solutions. Increased nitrate concentrations reduce SCC failure times and threshold stress levels (Ref 94, 95). However, SCC can occur at quite low nitrate concentrations in boiling solutions. In these solutions, SCC may be insensitive to concentration increases above a certain level (Ref 49).

Increased temperatures result in increased crack-growth rates (Ref 82, 96). For this reason, most SCC testing of steels in nitrate solutions has been performed at boiling temperatures. However, available evidence shows that initiation of SCC and slow crack growth can occur over a wide temperature range given sufficient time (Ref 82, 96-98).

Lower pH values of nitrate solutions promote SCC, whether the lower pH results from a more aggressive cation or is obtained by acid additions (Ref 94). Although addition of NaOH can suppress SCC in nitrate solutions, SCC up to a pH of 12 has been reported for high-strength steel wires (Ref 49, 94).

A well-defined electrochemical potential threshold exists for the SCC of steels in nitrate solutions, below which SCC does not occur (Ref 96). This potential decreases as temperature is increased. Cathodic polarization generally results in decreased failure times or no SCC, whereas anodic polarization enhances cracking, as shown in Table 3 (Ref 49). These results suggest an active-path mechanism of SCC.

Table 3 Effect of Strength Level and Polarization (±0.5 V) on SCC of Quenched and Tempered Steels in Boiling Ammonium Nitrate

Yield strength		Time to failure, min		
MPa	ksi	Cathodic	Zero	Anodic
1075	156	>2500	>3000	>6000
1179	171	>3700	>8600	832
1282	186	>2600	6534	651

Source: Ref 49

Hydroxides

Stress-corrosion cracking of steels due to sodium hydroxide, generally called "caustic cracking" or "embrittlement," has been a phenomenon most commonly associated with steam boilers. Recent SCC problems with continuous digesters used in the manufacture of paper and pulp have also been attributed to a form of caustic cracking (Ref 99). Cracking often occurs in regions where hydroxides can concentrate and is generally associated with highly stressed areas. Plastic deformation has been demonstrated to be a prerequisite for SCC (Ref 100, 101).

Intermediate concentration ranges are most conducive to caustic cracking of steels — generally 5% to about 70% NaOH, depending on temperature (Ref 102, 103). Stress-corrosion cracking can occur over a wide temperature range, i.e., 100 to 349 °C (212 to 660 °F), with 200 to 250 °C (392 to 482 °F) being the most commonly reported range for boiler service failures (Ref 100).

Stress-corrosion cracking of steels in hydroxides has been shown to occur within a narrow electrochemical-potential range near the active current peak on potential-current diagrams (Ref 104, 105). An example from studies of SCC in continuous digesters is shown in Fig. 8. The potential range is thought to correspond to the region at which steel surfaces are partially covered with a protective magnetite film (Ref 106). Additives can influence whether the steel's potential falls within this range and hence whether SCC occurs. Small additions of oxygen and/or chlorides promote SCC, but larger additions can have a reverse effect (Ref 107, 108). In particular, large ad-

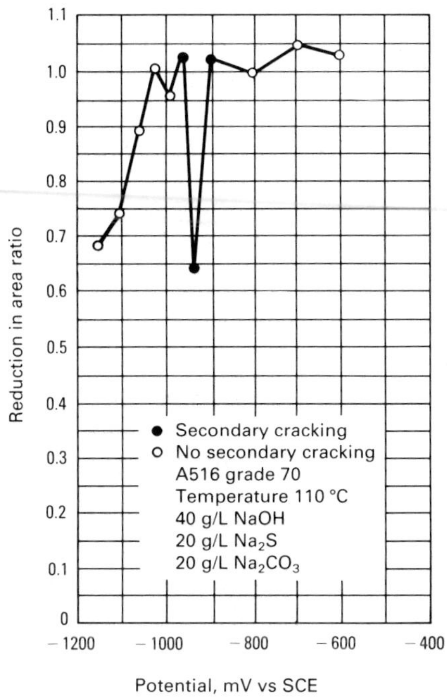

Fig. 8 Identification of SCC electrochemical-potential range for steels exposed to pulp manufacture digester liquor. Source: Ref 99

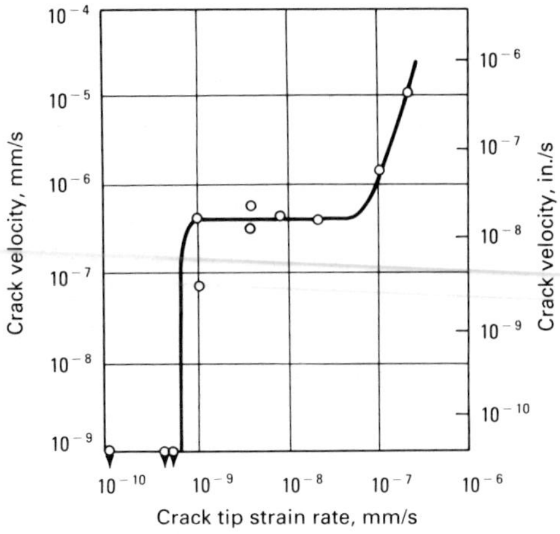

Fig. 9 Effect of strain rate on SCC of carbon-manganese steel in carbonate solution. Source: Ref 115

ditions of species having strong oxidizing effects can promote passivation and hence inhibit SCC (Ref 109).

Ammonia

Although steels have been widely used for transportation and storage of ammonia, recent surveys indicate that a significant percentage of these containers have leaked due to SCC (Ref 110). Stress-corrosion cracking is generally associated with cold-formed or welded steels and is not observed in stress-relieved equipment (Ref 110, 111). It has been shown that oxygen or the combined presence of oxygen and carbon dioxide is required for SCC (Ref 112). Laboratory tests in general have been performed in moisture-free environments. Some studies have suggested that small additions of water have an inhibitive effect, but service failures have been reported in the presence of water at similar low levels (Ref 10, 111).

Carbonate Solutions

Carbonate SCC has been reported for steels over a wide strength range; however, for lower-strength materials, it occurs only within a narrow electrochemical-potential range (Ref 66, 112, 113). This range corresponds to the transition between active corrosion and strong passivity and is therefore similar to steel behavior observed in nitrates and hydroxides (Ref 114). Cracking is also sensitive to strain rate, as shown in Fig. 9, possibly due to strain rate/repassivation interactions. Temperature and pH ranges reported for SCC are about 22 to 88 °C (71 to 190 °F) and 8 to 10.5, respectively (Ref 112).

Carbonate cracking has recently been blamed for piping failures at a refinery catalytic cracking unit, but was first reported as responsible for several failures of natural gas pipelines (Ref 11, 116, 117). For these failures, cracking was observed to initiate externally and propagate intergranularly (Ref 118). The presence of magnetite suggested that the cracking environment was essentially free of oxygen. Also, extensive siderite was observed in cracks, further evidence of a carbonate influence on SCC. Recent work indicates that the cracking media were actually concentrated carbonate/bicarbonate solutions (Ref 118).

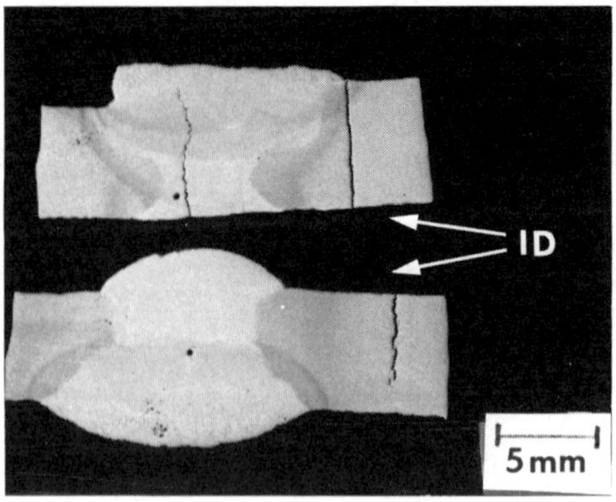

Fig. 10 Cross section of piping welds showing service-induced monoethanolamine SCC. Source: Ref 127

Other Environments

Stress-corrosion cracking has been a rare problem in low-alloy steels used in nuclear power water reactors (Ref 119-121). The environment of interest in this case is high-purity water with a very low oxygen content. Although susceptibility appears greatest at about 200 to 250 °C (392 to 482 °F), SCC at temperatures up to 320 °C (608 °F) has been reported (Ref 122-125). Oxygen and dynamic strain are considered important factors controlling crack-growth rates (Ref 126). Stress-corrosion cracking is thought to occur through an anodic dissolution mechanism.

Stress-corrosion cracking of low- and medium-strength carbon steels has been recently reported in refinery alkanolamine sweetening (hydrogen sulfide/carbon dioxide removal) units. Such failures are generally associated with welds that were not stress relieved (Fig. 10) and with elevated-temperature service. Certain laboratory investigations have suggested that carbonates play a role in the SCC process (Ref 128). However, the failure mechanism is still somewhat unclear, as other studies have been unable to confirm such findings (Ref 129, 130).

Cracking failures of steels in coal gas and chemical plant services have been attributed to SCC by water, carbon monoxide, and carbon dioxide (Ref 131, 132). Stress-corrosion cracking occurs only within a narrow electrochemical-potential range, and oxygen reportedly inhibits cracking (Ref 131, 133).

Ammonia, hydrogen sulfide, hydrogen cyanide, and carbon dioxide have been blamed for several SCC failures of welded and riveted steels used in gas and coke plants (Ref 134). Other failures in coke oven gas service have been attributed to hydrogen cyanide (Ref 135). Stress-corrosion cracking of lower-strength steels has also been reported in phosphorus trifluoride, hydrogen-antimony/aluminum chlorides, Na_3PO_4, $MgCl_2$-NaF, and hydrogen fluoride (Ref 75, 136-138).

Metallurgical Aspects

Effects of Composition

An overview of compositional effects on the SCC of two low-alloy steels in aqueous chlorides is given in Table 4. Although similar trends are sometimes observed for other steels and for other environments, the situation is much more complex in that composition can greatly influence heat-treatment response, strength, and welding characteristics, which also affect SCC. Further details are provided below.

As carbon content increases, the SCC resistance of steels in aqueous chlorides is reduced, with the most dramatic effects seen in the 0.2 to 0.4% range (Ref 139, 140). Increases in carbon content also increase the magnitude of the drop in SCC resistance with increasing yield strength, with lower-carbon steels showing a lesser effect (Ref 1).

For quenched and tempered steels in aqueous hydrogen sulfide, no significant influence of carbon on SCC resistance has been found. However, some researchers have reported that it causes a slight improvement; others have noted slightly detrimental effects (Ref 61, 141, 142).

Susceptibility of annealed and normalized carbon steels to SCC in nitrate solutions is determined primarily by carbon content (Ref 114). Steels are resistant at carbon contents below about 0.001% and at carbon contents above about 0.18%, but are susceptible between these values (Ref 114, 143). SCC susceptibility is thought to be largely due to carbide morphology and distribution. By contrast, SCC resistance of alloy steels appears to be independent of carbon content (Ref 144, 145). Although all plain carbon steels are susceptible to SCC in hydroxide solutions, sus-

Table 4 Effect of Alloying Within Normal Limits on the SCC Resistance of Martensitic Low-Alloy Steels to Chloride

Element	AISI 4120 (Yield strength = 1034 MPa, or 150 ksi)	AISI 4340 (Yield strength = 1172–1448 MPa, or 170–210 ksi)
Carbon	Decrease	Decrease
Manganese	No effect	Decrease
Nickel	Increase	No effect
Chromium	Increase	No effect
Molybdenum	Increase	No effect
Vanadium	Increase	...
Niobium	Increase	...
Titanium	Increase	...
Zirconium	Increase	...
Boron	No effect	...
Copper	No effect	...
Silicon	No effect	...
Sulfur	Beneficial	No effect
Phosphorus	Decrease	No effect
Oxygen	Decrease	...
Nitrogen	Decrease	No effect

Source: Ref 1, 38, 139

ceptibility to caustic cracking increases with carbon content; however, at levels above 0.2%, a beneficial effect is reported (Ref 100, 101, 146).

Manganese has an adverse effect on the chloride SCC of medium-carbon steels, but not on low-carbon materials (Ref 139). Increases in the manganese content of steels are reportedly detrimental to SCC resistance in hydrogen sulfide environments, possibly due to the segregation effects discussed below (Ref 147). The resistance of carbon steels to SCC in nitrate solutions is also adversely affected by increased manganese content, especially when steels are annealed at high temperatures for extended periods (Ref 148).

Nickel contents above 1% have been reported to reduce the SCC resistance of steels in hydrogen sulfide-containing environments (Ref 149). Other workers have not observed a significant effect (Ref 1). This contradiction may be related to heat-treatment specifics, as increasing nickel content increases the likelihood that untempered martensite may be present (see below). No effect of nickel on SCC in nitrate solutions has been reported (Ref 150).

Chromium contents below about 1.4% have been reported as beneficial to SCC in aqueous hydrogen sulfide; above this level, detrimental effects are generally observed (Ref 61, 147). In nitrate solutions, a similar beneficial effect of chromium on SCC resistance has been reported, at least to about 1.0% (Ref 150, 151).

Molybdenum is believed to increase SCC resistance to hydrogen sulfide by producing fine carbides during tempering (Ref 152, 153). Vanadium is thought to be beneficial by causing a similar carbide effect, as well as grain refinement (Ref 152). Titanium additions produce comparable results; however, at higher levels, SCC can be promoted through the presence of second-phase titanium carbonitride precipitation (Ref 153, 154). A beneficial effect of titanium on SCC resistance in nitrate solutions has been reported (Ref 151).

Copper has been observed to adversely affect the SCC resistance of steels in nitrate solutions (Ref 150). By contrast, silicon appears to be beneficial in such environments (Ref 143). Silicon contents of 1.58% or more have been reported to decrease SCC propagation rates in chloride solutions (Ref 155).

Aluminum in amounts of 0.3 to 0.6% reportedly increases SCC resistance to hydrogen sulfide (Ref 60). In nitrate solutions, aluminum does not affect SCC at typical deoxidizing levels, that is, less than about 0.01% (Ref 156). Improvements in SCC resistance are reported at levels above about 0.3% (Ref 150). A possible beneficial effect of aluminum on caustic cracking has also been reported (Ref 107).

High sulfur and phosphorus contents have reduced the SCC resistance of steels in hydrogen sulfide (see below) and can also affect SCC in nitrate solutions (Ref 157-159). No influence of oxygen and nitrogen has been reported for SCC in hydrogen sulfide solutions (Ref 143). Nitrogen levels below 0.01% promote SCC of steels in nitrate solutions, but higher levels appear to be beneficial (Ref 160, 161).

Effects of Microstructure and Heat Treatment

Steels of interest in this review can be divided into two main microstructural types: ferritic-pearlitic and quenched and tempered. Quench-and-temper heat treatments are generally used to obtain yield strengths above about 621 MPa (90 ksi).

Fine grain size has been shown to increase the SCC resistance of quenched and tempered steels in hydrogen sulfide solutions (Ref 149, 162). Similar effects have been observed for carbon steels in nitrates, where increased austenitizing temperature, resulting in grain growth, has adversely affected SCC resistance (Ref 163). Increased cooling rate after austenitizing has also been shown to decrease SCC resistance to nitrate environments (Ref 9). Prolonged aging at temperatures below about 593 °C (1100 °F) reportedly improves SCC resistance in nitrates, but aging at higher temperatures promotes SCC susceptibility (Ref 133, 164, 165).

For quenched and tempered steels, the presence of twinned martensite is thought to reduce SCC resistance as it does fracture toughness and hydrogen-embrittlement resistance (Ref 166-168). The presence of ε carbides and high dislocation densities also reduces SCC resistance (Ref 29, 169).

In aqueous hydrogen sulfide, quenched and tempered steels have been shown to have superior SCC resistance compared with normalized and tempered steels of similar strength levels (Ref 61). Martensitic microstructures generally give the best SCC resistance at a given strength level, especially if fine spheroidized carbides uniformly dispersed in ferrite (such as is obtained by high-temperature tempering) are present (Ref 158). Bainitic structures have approximately equivalent SCC resistance (Ref 170). The presence of untempered martensite is considered detrimental. Coarse or globular carbides that result from slow cooling and tempering produce an intermediate level of SCC resistance. In nitrate solutions, steels with pearlitic microstructures are superior in SCC resistance compared with steels having fine spheroidized carbides (Ref 171).

Manganese segregation in tempered steels can produce localized regions of increased hardness. This results in a preferential path for SCC in aqueous hydrogen sulfide and a corresponding drop in SCC resistance (Ref 62, 154).

Grain-boundary segregants can significantly influence the SCC of steels in environments in which intergranular cracking predominates. These effects have been systematically studied in elevated-temperature nitrate solutions (Ref 172-174). Following aging treatments to promote segregation, enhanced SCC susceptibility of steels to calcium nitrate was attributed to locally high phosphorus and sulfur levels at grain boundaries (Ref 172). Somewhat mixed results have been reported for silicon and tin in similar environments (Ref 173). In ammonium nitrate solutions, phosphorus, sulfur, arsenic, tin, and antimony (when present with silicon, germanium, selenium, tellurium, and bismuth) strongly promoted intergranular SCC (Ref 174).

Inclusions in steels act as SCC initiation sites and accelerate crack propagation in the major axis direction of elongated inclusions such as sulfides (Ref 175-176). Resultant anisotropy of hydrogen sulfide SCC has been reported in both plate and tubular products (see Table 5) (Ref 157, 177).

Effects of Mechanical Properties

Strength level is perhaps the greatest influence on the SCC resistance of steels under a variety of conditions, with resistance decreasing as strength is increased (Ref 178). Examples are shown in Table 2 and Fig. 6 and 7. Additional details are provided below.

Steels with yield strengths of less than 1241 MPa (180 ksi) are often considered resistant to SCC by aqueous chlorides, at least when not

Table 5 Effect of Sulfur Content and Sulfide Stringers on SCC of AISI 4130 (Modified) Steels in NACE TM-01-77 Solution
Yield strength = 691–760 MPa (98–110 ksi)

Sulfur, %	Sulfide stringer density, μm/mm^2	Threshold stress (% yield strength) Transverse	Parallel
0.004	nil	90	90
0.005	3.7	85	75
0.011	9.4	70	35
0.013	nil	95	55
0.018	4.3	85	50
0.028	5.2	90	55

Source: Ref 31, 157

highly stressed (Ref 36). By contrast, a value of 1034 MPa (150 ksi) was reported as the maximum yield strength for which good SCC resistance was observed using more highly stressed notched specimens (Ref 179). Stress-corrosion cracking has been reported in precracked steel specimens with yield strengths as low as 758 MPa (110 ksi) (Ref 1). In other work using ambient-temperature 15% HCl solutions, steels with yield strengths below 930 MPa (131 ksi) resisted SCC, even when stressed beyond yielding (Ref 180). By contrast, steels with yield strengths above 1282 MPa (186 ksi) exhibited SCC at applied stresses below yield.

Although a clearly defined yield-strength threshold for SCC in aqueous chlorides is not apparent, it can be generalized that steels with yield strengths of less than 689 MPa (100 ksi) are resistant (Ref 1). Above this value, SCC in aqueous chlorides can occur, with SCC resistance decreasing with increased yield strength. This strength level corresponds to a threshold stress intensity for SCC of about 110 MPa$\sqrt{m}$ (100 ksi$\sqrt{in.}$).

Resistance of steels to SCC in environments containing hydrogen sulfide also decreases with increasing strength level (Ref 61). Note that the threshold stress for cracking is often well below the yield strength of the steel. Stress-corrosion cracking in aqueous hydrogen sulfide has been reported at yield-strength levels down to 296 to 414 MPa (43 to 60 ksi) in laboratory tests (Ref 181).

Increases in steel strength result in decreases in cracking resistance in gaseous hydrogen as well (Ref 84). Similarly, steels with yield strengths below about 1034 MPa (150 ksi) are reportedly resistant to SCC in sulfuric acid solutions, with SCC observed within the 32 to 43 HRC range (Ref 180, 182, 183).

Steels with good fracture toughness also resist SCC. For example, a good correspondence between fracture toughness and SCC resistance in aqueous chlorides has been reported for 1034-MPa (150-ksi) steels (Ref 184). A similar correlation between Charpy fracture toughness and SCC in synthetic seawater was observed for steels with fracture toughnesses of 689 to 1379 MPa (100 to 200 ksi) (Ref 141). By contrast, no significant correlation between toughness and SCC resistance in aqueous hydrogen sulfide has been reported for AISI 4*xxx* series steels (Ref 184).

Welding Considerations

For steels to be welded, it is important that chemical composition be controlled or preheating be performed to prevent cold constraint cracking in the heat-affected zone (HAZ). If not detected, such cracks can be initiation sites for SCC (as well as other failure mechanisms). In practice, therefore, steels for welded applications will most often have maximum carbon contents of about 0.3% (Ref 1).

In moderate-strength steels, weld metal produced by a variety of techniques has been reported to be more susceptible to SCC in aqueous chlorides than base metal (Ref 185, 186). Similar observations of chloride SCC of HAZs were thought to be due to localized hardness increases

(Ref 187). Reduction of hardness through postweld stress relief reportedly increases SCC resistance, provided the temper-embrittlement range is avoided (Ref 188). Stress relief within this range, about 482 to 649 °C (900 to 1200 °F), is believed to adversely affect SCC resistance as well as toughness (Ref 1).

One study of SCC in hydrogen sulfide environments found that weldments of lower-strength steels had better resistance than their respective base metals; this was explained as a possible result of grain refinement (Ref 61). In most cases, however, reduced SCC resistance to hydrogen sulfide is associated with welds. This same study showed the SCC resistance of moderate-strength steels to be significantly lower in weld metal. Increased SCC susceptibility was also observed in the HAZs of these steels. For the most part, SCC behavior was directly related to increased localized hardness caused by welding. No consistent effects of preheating or of postweld heat treatment were reported, but it was nevertheless suggested that resultant reduction of residual stresses in actual equipment would probably improve SCC resistance (Ref 61). In another study, the beneficial effects of postweld stress relief on prevention of SCC in HAZs was clearly demonstrated (Ref 36).

Reduced SCC resistance of welded steels has also been reported for steels exposed to sour crude oils, with cracking occurring in or adjacent to HAZs (Ref 189). Similarly, tests conducted in a simulated pipeline environment containing only 112 to 130 ppm hydrogen sulfide resulted in SCC of HAZs with hardnesses above 28 HRC

Table 6 Effect of Cold Work on SCC of a Pipeline Steel Tested in 5% NaCl Saturated With Hydrogen Sulfide
Yield strength = 241 MPa (35 ksi)

Prestress, (% elongation)	Applied stress (% yield strength) 130	80
0	No SCC	No SCC
1	No SCC	No SCC
2	No SCC	No SCC
3	SCC	No SCC
5	SCC	No SCC

Source: Ref 66

(Ref 190). More recently, extensive SCC by hydrogen sulfide has been reported in welded refinery pressure vessels (Ref 191). Cracks were generally restricted to weld metal in which reported hardnesses were only 22 to 25 HRC.

Residual stresses due to welding alone are of sufficient magnitude to induce the SCC of steels in hot nitrate solutions (Ref 1). Preheating improves the SCC resistance of welded steels, as does postweld heat treatment (Ref 192). Although temperatures as low as 400 °C (750 °F) have an effect, optimum SCC resistance is restored by heating near about 650 °C (1200 °F), which is also much more effective in reducing residual stresses (Ref 9, 192). Postweld stress relief is important with respect to caustic cracking, as failures are generally reported only in welded steels that are not stress relieved (Ref 193). An exception is SCC in continuous digesters used in pulp manufacture, in which cracking was also observed in stress-relieved welds (Ref 99).

Effects of Cold Work

Cold working of steels can be performed deliberately during manufacture of equipment or can result from inadvertent mechanical damage. Resultant increases in strength, hardness, and residual stresses can significantly influence the SCC behavior of steels unless the materials are adequately stress relieved by thermal treatment.

In aqueous chlorides, plastic prestrain to about 5% has little effect on the SCC of low- and medium-strength steels, but as little as 1% prestrain can adversely affect higher-strength steels (Ref 141, 194). Plastic prestrain can, however, result in compressive residual stresses. In such cases, SCC resistance may increase due to inhibiting effects on SCC initiation (Ref 195). Thus, at similar strength levels, cold-drawn wires have been reported to have better SCC resistance than heat-treated wires (Ref 1).

In hydrogen sulfide environments, cold working as little as 1% has a strong detrimental effect on SCC resistance (Ref 62, 66, 71, 196). As shown in Table 6, cold working can impart SCC susceptibility to even low-strength steels that would not otherwise exhibit SCC. Such effects are believed to be due to increases both in available sites for SCC initiation and in hydrogen solubility (Ref 197, 198).

For low-carbon steels, cold working reportedly improves SCC resistance in nitrates (Ref 199, 200). At intermediate carbon levels (0.09 to 0.19%), SCC resistance initially decreases and then increases with increasing amounts of cold work (Ref 200). Cold working has also been reported to be more beneficial than heat treatment to a similar strength level for the SCC of steel wires in nitrate solutions (Ref 91). Introduction of residual compressive stresses by shot peening improves the nitrate SCC resistance of steels (Ref 201). Plastic straining appears to be a prerequisite for SCC in hydroxide environments (Ref 101).

Prevention and Control

Materials Selection

The most reliable method of preventing SCC of carbon and low-alloy steels is proper materials selection. Because of the predominant influence of strength level on SCC, limitations on steel strengths are the main selection criterion. For example, limiting the yield strength of steels used in aqueous chlorides to 689 MPa (100 ksi), as mentioned previously, is typical of this approach (Ref 1). Note that because of their simplicity and nondestructive nature, strength limitations are often accomplished in practice through hardness measurements and control.

Perhaps the most comprehensive SCC prevention and control guidelines have been developed for hydrogen sulfide service. These guidelines have been summarized in the document commonly referred to as MR-01-75, published by NACE (Ref 202). In general, the NACE guidelines have been supported by service experience, even though SCC can be obtained for some of the recommended materials in laboratory tests (Ref 62, 71, 189).

With respect to carbon and low-alloy steels, NACE MR-01-75 recommendations include limits on hardness levels to avoid hydrogen sulfide SCC. For most steels, hardness is restricted to a maximum of 22 HRC. Certain AISI 4*xxx* series steels are allowed at higher hardnesses in the quenched and tempered condition, but the user is cautioned that SCC testing of such materials is advisable. Additionally, MR-01-75 sets limits with respect to nickel content (1% maximum) and degree of cold work and specifies required stress-relief heat treatments.

Although not strictly applicable, the NACE MR-01-75 guidelines are often used for other environments — generally those which, like hydrogen sulfide, are believed to cause SCC by hydrogen embrittlement. Examples include gaseous hydrogen service and applications in which steel contact with an acidic aqueous phase could cause hydrogen adsorption.

Welding Considerations

NACE Standard MR-01-75 gives postweld heat-treatment guidelines for welded steels to be used in hydrogen sulfide service (Ref 202). For pressure vessel welds, a maximum hardness of 200 HB has been generally accepted (Ref 203). Similar guidelines are applied to welded steels in sulfuric acid service.

For refinery pressure vessels, it has now become accepted practice to stress relieve all welds in hydrogen sulfide service, not only to prevent SCC by lowering hardness but also to minimize chances of SOHIC (Ref 7). Stress-corrosion cracking in refinery carbonate service, alkonalamine gas sweetening units, and continuous digesters used for pulp manufacture are likewise minimized by stress relief of welds (Ref 11, 99, 127).

As-welded steels are considered acceptable for caustic service to about 60 °C (140 °F); above this temperature, however, stress relief is generally recommended (Ref 1). More detailed guidelines based on caustic strength are shown in Fig. 11. Selection of steels with minimum carbon contents of 0.2% also reduces the likelihood of SCC in hydroxide solutions (Ref 146).

Benefits of postweld heat treatment in the prevention of nitrate SCC have been reported, but are though to be due more to a reduction in residual stress levels than to microstructural or strength modifications (Ref 9). As mentioned earlier, higher-temperature stress relief near 649 °C (1200 °F) is considered most effective in this regard (Ref 192).

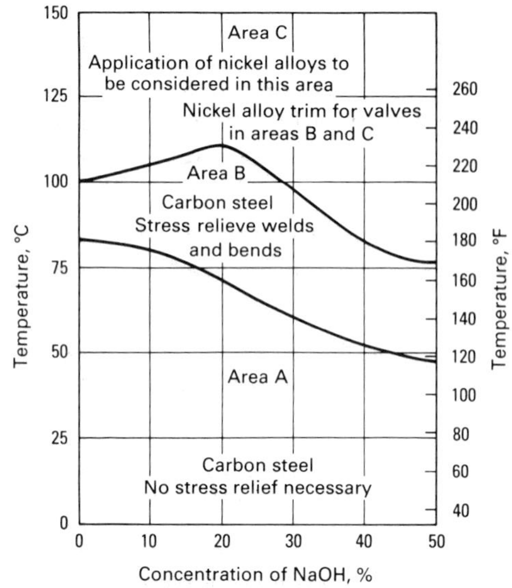

Fig. 11 Temperature and concentration limits for SCC susceptibility of carbon steels in caustic soda. Source: Ref 204

Other Methods

Since stress is required for SCC, reduced stresses help prevent cracking. This is one of the goals of stress relief following welding or cold working. Applied stresses can often be minimized by avoiding stress concentrators, employing suitable loading safety factors during design, and reducing fit-up and other installation and service loads. Note that because SCC can sometimes occur at very low applied stresses, reducing stresses alone may not be sufficient to prevent SCC.

Avoiding galvanic couples that could lead to hydrogen charging of steels is an important means of preventing unanticipated SCC in aqueous chlorides. By contrast, cathodic polarization can be useful in avoiding SCC of steels in nitrate-containing environments (Ref 205). Similarly, prevention of film formation by cathodic polarization can be utilized to deter caustic SCC (Ref 105). Anodic polarization has also been used to inhibit caustic SCC, but if insufficiently applied, may result in accelerated SCC (Ref 1). These methods have been reported to be somewhat successful in preventing SCC in continuous digesters (Ref 99).

Inorganic coatings are generally not used to prevent SCC, because holiday-free conditions would be required. However, metallic thermal-spray coatings have been used on a limited basis (Ref 99). Likewise, cladding of refinery pressure vessels with austenitic stainless steels is sometimes done to prevent SCC and SOHIC in wet hydrogen sulfide service (Ref 7). As mentioned previously, shot peening introduces residual stresses that inhibit SCC initiation.

Sodium hydroxide is an effective additive for preventing nitrate SCC of steels, even though at high concentrations it can cause SCC itself (Ref 94). Similarly, small nitrate additions are useful for inhibiting caustic SCC. A variety of other additives can inhibit nitrate SCC; halides, sulfates, and phosphonates are the most common (Ref 115, 206).

Substances that promote film formation are thought to be effective in preventing or minimizing caustic SCC (Ref 1). Examples include silicate, sulfate, nitrate, permanganate, phosphate, chromate, and carbonate. In typical boiler practice, a minimum ratio of inhibitor to sodium hydroxide content is usually maintained to prevent SCC (Ref 207, 208).

Conclusions

The factors that influence stress-corrosion cracking of carbon and low-alloy steels have been extensively studied, and they are now fairly well understood. Measures to avoid SCC must be taken at an early stage in the life of equipment, i.e., during materials selection. This selection process includes not only the steels themselves but also fabrication and installation practices.

Unfortunately, despite conscientious materials selection efforts, SCC is often unanticipated and thus can be a major threat to personnel, to the environment, and to the economic success of industrial endeavors. For this reason, inspection programs are indispensable safeguards. Given timely detection, proven corrective measures can be implemented to prevent SCC from reaching catastrophic proportions.

References

1. C.S. Carter and M.V. Hyatt, Review of Stress Corrosion Cracking in Low Alloy Steels with Yield Strengths Below 150 ksi, in *Stress Corrosion*

Cracking and Hydrogen Embrittlement in Iron Base Alloys, Vol 6, National Association of Corrosion Engineers, 1977, p 524-600

2. R.S. Treseder, Ed., *NACE Corrosion Engineers Handbook*, National Association of Corrosion Engineers, 1980, p 139
3. C.F. Lewis, Ed., *Materials Selector 1991*, Materials Engineering, 1990, p 37
4. B.F. Brown, Stress Corrosion Cracking and Corrosion Fatigue of High Strength Steels, in *Problems with the Load Carrying Capacity of High-Strength Steels*, DMIC Report 210, 1964, p 91
5. G.E. Kearns, M.T. Wang, and R.W. Staehle, Stress Corrosion Cracking and Hydrogen Embrittlement in High Strength Steels, in *Stress Corrosion Cracking and Hydrogen Embrittlement in Iron Base Alloys*, Vol 6, National Association of Corrosion Engineers, 1977, p 700-735
6. E.H. Phelps, Microscopic Identification of Stress Corrosion Cracking in Steels With High Yield Strength, in *Specialists Meeting on Stress Corrosion Testing Methods, AGARD Conf. Proc. No. 98*, AGARD CP-98, NATO, 1972
7. R.D. Merrick, *Mater. Perform.*, Vol 28 (No. 2), 1989, p 53-55
8. H.L. Logan, in *Physical Metallurgy of Stress Corrosion Fracture*, T.N. Rhonin, Ed., Interscience, 1959, p 295
9. R.N. Parkins, Stress Corrosion Cracking of Low Carbon Steels, in *Proc. Conf. Fundamental Aspects of Stress Corrosion Fracture*, National Association of Corrosion Engineers, 1967, p 361
10. A.W. Loginow and E.H. Phelps, *Corrosion*, Vol 18 (No. 8), 1962, p 299t
11. J.H. Kmetz and D.J. Truax, "Carbonate Stress Corrosion Cracking of Carbon Steel in Refinery FCC Main Fractionator Overhead Systems," Paper No. 206, presented at Corrosion/90, National Association of Corrosion Engineers, 1990
12. P.J. Teevens, "Toward a Better Understanding of the Cracking Behavior of Carbon Steel in Al-konolamine Sour Gas Sweetening Units: Its Detection, Monitoring and How to Avoid It," Paper No. 198, presented at Corrosion/90, National Association of Corrosion Engineers, 1990
13. Stress-Corrosion Cracking, in *Metals Handbook*, Vol 11, 9th ed., *Failure Analysis and Prevention*, American Society for Metals, 1986, p 211
14. Comparison of Scanning and Transmission Electron Microscopes for Fractography, in *Metals Handbook*, Vol 9, 8th ed., *Fractography and Atlas of Fractographs*, American Society for Metals, 1974, p 76
15. J.P. Frazer and R.S. Treseder, *Corrosion*, Vol 8, 1952, p 342
16. A.K. Farrel, *J. Iron Steel Inst.*, Vol 203, 1965, p 457
17. A.K. Farrel and A.G. Quarrel, *J. Iron Steel Inst.*, Vol 202, 1964, p 1002
18. T.G. Oakwood, Corrosion of Alloy Steels, in *Metals Handbook*, Vol 13, 9th ed., *Corrosion*, ASM International, 1987, p 532
19. W.T. Chandler and R.T. Walter, "Effects of High Pressure Hydrogen on Steels," paper presented to ASM and AWS, American Society for Metals, 1967
20. J.S. Laws *et al.*, "Hydrogen Gas Pressure Vessel Problems in the M-1 Facilities," NASA CR-1305, 1969
21. W. Bruckhoff *et al.*, Paper No. 389, presented at Corrosion/85, National Association of Corrosion Engineers, 1985
22. C.D. Beacham, *Metall. Trans.*, Vol 3, 1972, p 437
23. M. Watkins *et al.*, *Corrosion*, Vol 32 (No. 3), 1976, p 103
24. D.O. Sprowls, Evaluation of Stress Corrosion Cracking, in *Metals Handbook*, Vol 13, 9th ed., *Corrosion*, ASM International, 1987, p 271
25. B.F. Brown, *Met. Rev.*, Vol 13, 1968, p 171
26. W.F. Brown and J.E. Srawley, Plane Strain Crack Toughness Testing of High Strength Metallic Materials, in *STP 410*, American Society for Testing and Materials, 1966
27. R.H. Jones and R.E. Ricker, Stress-Corrosion Cracking, in *Metals Handbook*, Vol 13, 9th ed., *Corrosion*, ASM International, 1987, p 147
28. M.O. Speidel, Dynamic and Static Embrittlement of a High Strength Steel in Water, in *Proc. Hydrogen in Metals Int. Conf. (Paris)*, 1972, p 290
29. M.T. Wang and R.W. Staehle, Effect of Heat Treatment and Stress Intensity Parameters on Crack Velocity and Fractography of AISI 4340 Steel, in *Proc. Hydrogen in Metals Int. Conf. (Paris)*, 1972, p 342
30. S. Mostovoy *et al.*, *Eng. Fract. Mech.*, Vol 1, 1971, p 291
31. "Testing of Materials for Resistance to Sulfide Stress Cracking at Ambient Temperatures," NACE Standard TM-01-77-90, National Association of Corrosion Engineers, 1990
32. "Evaluation of Pipeline Steels for Resistance to Stepwise Cracking," NACE Standard TM-02-84-87, National Association of Corrosion Engineers, 1987
33. R.N. Parkins, *Corros. Sci.*, Vol 29, 1989, p 1019
34. R.N. Parkins, *Corrosion*, Vol 46 (No. 3), 1990, p 178
35. J.T. Evans and R.N. Parkins, *Acta Metall.*, Vol 25, 1976, p 511
36. E.H. Phelps, Stress Corrosion Behavior of High Yield Strength Steels, in *Proc. 7th World Petroleum Congr.*, Elsevier, 1967, p 27
37. H.H. Johnson and A.M. Willner, *Appl. Mater. Res.*, Vol 4, 1965, p 33

38. S. Yamamoto and T. Fujita, Delayed Failure Properties of High Strength Steels in Water, in *Fracture 1969, Proc. Int. Conf. Fracture*, P.L. Pratt, Ed., Chapman and Hall, 1969, p 425
39. V.J. Colangelo and M.S. Fergusen, "Susceptibility of Gun Steels to Stress Corrosion Cracking," WVT-7012, Watervliet Arsenal, U.S. Government Report AD 717 553, 1970
40. S. Fukui and A. Asada, *Trans. Iron Steel Inst. Jpn.*, Vol 9, 1969, p 448
41. A.H. Freedman, "Development of an Accelerated Stress Corrosion Test for Ferrous and Nickel Base Alloys," NOR 68-58, NASA Contract NAS-20333, Northrup Corp., 1968
42. G. Sandoz, High Strength Steels, in *Stress Corrosion Cracking in High Strength Steels and in Titanium and Aluminum Alloys*, B.F. Brown, Ed., Naval Research Laboratory, 1969
43. A.A. Scheinker, Ph.D. dissertation, Lehigh University, 1972
44. M.J. May and A.H. Priest, "The Influence of Solution pH on Stress Corrosion Cracking Resistance of a Low Alloy High Strength Steel," MG/A/45/68, BISRA, British Steel Corp., 1968
45. H. Yamaoko and G. Wranglen, *Corros. Sci.*, Vol 6, 1966, p 113
46. J.B. Greer *et al.*, *Corros. Sci.*, Vol 28, 1972, p 328
47. R.T. Ault, "High Strength Steels," Air Force Materials Symposium, AFML-TR-65-29, Wright-Patterson Air Force Base, 1969
48. J.T. Nieman, "Investigation of Mechanism of Delayed Failure in Prestressing Steel," U.S. Government Report PB180-194, 1968
49. J.D. Gilchrist and R. Narayan, *Corros. Sci.*, Vol 11, 1971, p 281
50. R.J. Scmitt and E.H. Phelps, *J. Met.*, 1970, p 47
51. NACE Unit Committee T-1H, *Mater. Protect.*, Vol 6 (No. 11), 1967, p 65
52. B.F. Brown, Stress Corrosion Cracking of High Strength Steels, in *The Theory of Stress Corrosion Cracking in Alloys, NATO Science Committee, Research Evaluation Conf.*, Brussels, 1971
53. C.F. Barth *et al.*, *Corrosion*, Vol 25, 1969, p 353
54. A. Tirman *et al.*, *Corrosion*, Vol 25, 1969, p 342
55. W. McCleod and R.R. Rodgers, *Corrosion*, Vol 22, 1966, p 143
56. C.G. Interrante, Interpretive Report on the Effect of Hydrogen in Vessels, Section I, Basic and Research Aspects, *Weld. Res. Coun. Bull.*, Vol 145, 1969
57. T.P. Rahakrishnan and L.L. Sheir, *Electrochim. Acta*, Vol 11, 1966, p 1007
58. C.G. Hancock and H.H. Johnson, *Trans. Met. Soc. AIME*, Vol 236, 1966, p 513
59. C.E. Barth and A.P. Troiano, *Corrosion*, Vol 28, 1972, p 259
60. H. Kihara *et al.*, Studies on Sulfide Corrosion Cracking of High Strength Steels Caused by H_2S, *Proc. 7th World Petroleum Congr.*, Elsevier, 1967
61. E. Snape, *Corrosion*, Vol 23, 1967, p 154
62. R.N. Tuttle, *Mater. Protect.*, Vol 9 (No. 4), 1970, p 11
63. W.P. Rees, Note on Stress Corrosion Cracking of Steels in the Presence of Hydrogen Sulfide, in *Symp. Metals and Alloys*, Institute of Metals, London, 1948
64. T.G. Oakwood, Corrosion of Alloy Steels, in *Metals Handbook*, Vol 13, 9th ed., *Corrosion*, ASM International, 1987, p 533
65. H. Ishizuki and K. Onishi, *J. Jpn. Inst. Met.*, Vol 30, 1966, p 846
66. C.M. Hudgins *et al.*, *Corrosion*, Vol 22, 1966, p 238
67. C.M. Hudgins Discussion to Ref 68
68. H. Ishizuki and K. Onishi, *Jpn. Chem. Q.*, Vol 3 (No. 2), 1967, p 30
69. D.G. Moore *et al.*, "Protection of Steel in Prestressed Concrete Bridges," National Cooperative Highway Research Program Report 90, Highway Research Board, National Academy of Sciences, 1970
70. D.R. Johnston and T.G. McCord, Stress Corrosion Cracking of Low Alloy Steels in Sodium Sulfide Solutions, in *Proc. NACE 26th Conf.*, National Association of Corrosion Engineers, 1970
71. R.S. Treseder and T.M. Swanson, *Corrosion*, Vol 24, 1968, p 31
72. L.M. Dvorecek, *Corrosion*, Vol 26, 1970, p 177
73. G.B. Kohut and W.J. McGuire, *Mater. Protect.*, Vol 7 (No. 6), 1968, p 17
74. H.E. Townsend, *Corrosion*, Vol 28, 1972, p 39
75. I. Class, Report on Investigations of Sulfide Stress Cracking of Steels, Particularly Steels of Comparatively Low Tensile Strength, in *Proc. 2nd Int. Congr. Metallic Corrosion*, National Association of Corrosion Engineers, 1963, p 342
76. J.P. Frazer *et al.*, *Corrosion*, Vol 14, 1958, p 517t
77. H.R. Copson, *Welding J.*, Vol 32, 1953, p 75s
78. K.H. Liebchen, *Mater. Protect.*, Vol 4, 1965, p 50
79. F.K. Bloom, *Corrosion*, Vol 11, 1955, p 351t
80. H.T. Effinger *et al.*, *Oil Gas J.*, No. 2, 1951, p 99
81. *Corros. Technol.*, Dec 1965, p 24
82. W.D. Everling, *Wire and Wire Prod.*, Vol 30, 1955, p 316
83. F.F. Azhogin, Stress Corrosion Cracking of High Strength Steels, in *3rd Int. Conf. Metallic Corrosion*, Moscow, Vol II, MIR, 1966
84. J.E. Campbell, "Effects of Hydrogen Gas on Metals at Ambient Temperature," DMIC Report S-31, April 1970
85. W. Haufman and W. Rouls, *Weld. J.*, Vol 44, 1965, p 225s
86. R.J. Walter and W.T. Chandler, *Mater. Sci. Eng.*, Vol 8, 1971, p 90

87. D.P. Williams and H.G. Nelson, *Metall. Trans.*, Vol 1, 1970, p 63
88. R.H. Cavet and H.C. Van Ness, *Weld. J.*, Vol 42, 1963, p 3165
89 H.G. Nelson *et al.*, *Metall. Trans.*, Vol 2, 1971, p 953
90. G. Sandoz, *Metall. Trans.*, Vol 3, 1972, p 1169
91. G.G. Hancock and H.H. Johnson, *Trans. Met. Soc. AIME*, Vol 326, 1966, p 513
92. H.W. Liu and F.J. Ficalora, *Int. J. Fract. Mech.*, Vol 8, 1972, p 223
93. R.J. Walter *et al.*, *Mater. Sci. Eng.*, Vol 5, 1969/70, p 98
94. R.N. Parkins and R. Usher, The Effect of Nitrate Solutions in Producing Stress Corrosion Cracking in Mild Steel, in *1st Int. Congr. Metallic Corrosion*, London, 1961, p 289
95. G.T. Spare, *Wire and Wire Prod.*, Vol 29, Dec 1954, p 1421
96. H. Mazille and H.H. Uhlig, *Corrosion*, Vol 28, 1972, p 427
97. R.E. Pollard, Stress Corrosion Tests of Bridge Cable Wire, in *Symp. Stress Corrosion Cracking of Metals*, ASTM/ASME, 1944, p 437
98. H.L. Logan, Stress Corrosion Cracking of Low Carbon Steels, in *The Stress Corrosion of Metals*, Wiley, 1966, p 43
99. R.A. Yeske, Corrosion by Kraft Pulping Liquors, in *Metals Handbook*, Vol 13, 9th ed., *Corrosion*, ASM International, 1987, p 1210
100. F.P.A. Robinson and L.G. Nel, A Contribution to the Study of Caustic Cracking, in *2nd Int. Congr. Metallic Corrosion*, National Association of Corrosion Engineers, 1963
101. K. Bohenkamp, Caustic Cracking of Mild Steel, in *Proc. Conf. Fundamental Aspects of Stress Corrosion Cracking*, R.W. Staehle, Ed., National Association of Corrosion Engineers, 1967, p 374
102. A.A. Berk and W.F. Waldeck, *Chem. Eng.*, Vol 57 (No. 6), 1950, p 235
103. H.W. Schmidt *et al.*, *Corrosion*, Vol 7, 1951, p 295
104. H. Grafen, *Corros. Sci.*, Vol 7, 1967, p 177
105. G. Herbsleb and W. Schwenk, *Stahl Eisen., Vol 90, 1970, p 903*
106. H. Grafen, *Werkst. Korros.*, Vol 20, 1969, p 305
107. W. Radeker and H. Grafen, *Stahl Eisen, Vol 76 (No. 11), 1956, p 1616*
108. J.E. Reinoehl and W.E. Berry, *Corrosion*, Vol 28, 1972, p 151
109. I.G. Podgornyi, *Chem. Abs.*, Vol 52, 1958, p 1102e
110. T.J. Dawson, *Weld. J.*, Vol 35, 1956, p 568
111. NACE Task Group T-5E-10 Minutes, 1972
112. J.M. Suitcliffe *et al.*, *Corrosion*, Vol 28, 1972, p 313
113. J.D. Gilchrist, *Metall. Club J.*, p 18
114. R.N. Parkins, Stress Corrosion Cracking of Low Strength Ferritic Steels, in *Conf. Theory of Stress Corrosion Cracking in Alloys*, J.C. Scully, Ed., NATO Science Committee Research Evaluation Conference, Portugal, 1971, p 167
115. B. Craig, Environmentally Induced Cracking, in *Metals Handbook*, Vol 13, 9th ed., *Corrosion*, ASM International, 1987, p 160
116. J.P. O'Donnell, *Oil Gas J.*, Vol 69, Jan 1970, p 77
117 H.E. Townsend, *Mater. Protect.*, Vol 11, Oct 1972, p 33
118. G.J. Ogundele and R.D. Venter, "On the External Stress Corrosion Cracking of Line-Pipe Steels," Paper No. 572, presented at Corrosion/89, National Association of Corrosion Engineers, 1989
119. B. Tomkins, Ed., International Cooperative Group on Cyclic Crack Growth: 10th Anniversary Report, *J. Pressure Vessels Piping*, 1990, submitted for publication
120. P.M. Scott, "A Review of Environmental Effects on Pressure Vessel Integrity," presented at 3rd Int. Symp. Environmental Degradation of Materials in Nuclear Power Systems, AERE Report 12780, Sept 1987
121. H. Hanninen *et al.*, Environment Sensitive Cracking in Pressure Boundary Materials of Light Water Reactors, CSNI Report 141, 1987
122. M.O. Speidel and R.M. Magdowski, "Stress Corrosion Cracking of Nuclear Reactor Pressure Vessel Steels in Hot Water," Paper No. 283, presented at Corrosion/88, National Association of Corrosion Engineers, 1988
123. M.E. Indig *et al.*, Environmental Aspects of Carbon Steels Stress Corrosion Cracking in High Purity Water, *Rev. Coatings Corros.*, Vol 5, 1982, p 173
124. J. Congelton *et al.*, *Corros. Sci.*, Vol 25, 1982, p 173
125. P. Hurst *et al.*, *Corros. Sci.*, Vol 25, 1985, p 651
126. P.M. Scott and D.R. Rice, *Nucl. Eng. Des.*, Vol 119, 1990, p 399
127. J. Gutzeit and J.M. Johnson, *Mater. Perform.*, Vol 25 (No. 7), 1986, p 18
128. R.N. Parkins *et al.*, *Mater. Perform.*, Vol 25 (No. 10), 1986, p 20
129. S.W. Ciaraldi and D.R. Grimes, "Corrosion and Stress Corrosion Cracking Behavior of Carbon Steels in Monoethanolamine Solutions," Paper No. 255, presented at Corrosion/89, National Association of Corrosion Engineers, 1989
130. R.D. Merrick, "Prevention of Cracking in Wet H_2S Environments," Paper No. 269, presented at Corrosion/89, National Association of Corrosion Engineers, 1989
131. A. Brown *et al.*, *Corros. Sci.*, Vol 10, 1970, p 547
132. M. Kowaka and S. Nagata, *Corrosion*, Vol 24, 1968, p 427
133. E. Houdremont *et al.*, *Stahl Eisen*, Vol 60, 1940, p 757
134. R.N. Parkins and R. Usher, *J. Appl. Chem.*, Vol 9, 1959, p 445

135. H. Buckholtz and R. Pusch, *Stahl Eisen*, Vol 62, 1942, p 445
136. R.L. Ritter, "Phosphorus Trifluoride Cylinder Failure," Report 1772, Union Carbide Corp., Nuclear Div., 1968
137. R.S. Treseder and A. Wachter, *Corrosion*, Vol 5, 1949, p 383
138. R.L. Schuyler III, *Mater. Perform.*, Vol 8, 1979, p 9
139. G. Sandoz, *Metall. Trans.*, Vol 2, 1971, p 1055
140. M.A. Eaglesham *et al.*, *Theoret. Appl. Frac. Mech.*, Vol 10, 1988, p 97
141. S.R. Novak, "Comprehensive Investigation of K_{ISCC} Behavior of Candidate HY180/210 Steel Weldments," 89.021-024(1) B63105, U.S. Steel Corp., Dec 31, 1970
142. P.G. Bastien, The Phenomena of Cracking and Fracture of Steel in the Presence of Hydrogen Corrosion Under Stress in the Presence of Moist Hydrogen Sulfide in *Physical Metallurgy of Stress Corrosion Fracture*, TMS-AIME, Interscience, 1959
143. L.M. Long and H.H. Uhlig, *J. Electrochem. Soc.*, Vol 112, 1965, p 964
144. G.J. Biefer, "Exploratory Stress Corrosion Cracking Tests on Some Low Alloy High Strength Steels," Tech. Bull. TB11, Dept. of Energy, Mines and Resources, Ottawa, Canada, May 1969
145. J. Flis and J.C. Scully, *Corros. Sci.*, Vol 8, 1968, p 235
146. M. Hugo *et al.*, Investigation of an Accelerated Stress Corrosion Cracking Method, in *Specialists Meeting on Stress Corrosion Testing Methods, AGARD Conf. Proc. No. 98*, AGARD CP-98, NATO, 1972
147. J.P. Frazer and G.G. Eldredge, *Corrosion*, Vol 14, 1958, p 524t
148. L.M. Long and N.A. Lockington, *Corros. Sci.*, Vol 7, 1967, p 447
149. N. Bailey, *Met. Constr. Brit. Weld. J.*, Vol 2, 1970, p 339
150. R.N. Parkins and A. Brown, *J. Iron Steel Inst.*, Sept 1959, p 45
151. L.M. Long and N.A. Lockington, *Corros. Sci.*, Vol 11, 1971, p 853
152. E. Herzog, *Ind. Eng. Chem.*, Vol 53, 1961, p 64A
153. C.N. Bowers *et al.*, *Corrosion*, Vol 8, 1952, p 333
154. S.W. Ciaraldi, *Corrosion*, Vol 40, 1984, p 77
155. C.S. Carter, *Corrosion*, Vol 25, 1969, p 423
156. E. Herzog, *Corros. Anticorros.*, Vol 2, 1954, p 3
157. S.W. Ciaraldi, Anisotropy of Sulfide Stress Cracking in C-90 Grade Oil Country Tubular Goods, in *Proc. 1981 Mechanical Working and Steel Processing Conf.*, AIME, 1981, p 605
158. E. Snape, *Corros. Anticorros.*, Vol 24, 1968, p 261
159. W.C. Shroeder *et al.*, *Met. Alloys*, Vol 8, 1937, p 320
160. A.E. Schuetz and W.D. Robertson, *Corrosion*, Vol 13, 1957, p 437t
161. J.T. Waber and H.J. McDonald, *Stress Corrosion Cracking of Mild Steels*, Corrosion Publishing, 1947
162. E.E. Hofmann *et al.*, *Arch. Eisenhüttenwes.*, Vol 39, 1968, p 677
163. M. Henthorne and R.N. Parkins, *Brit. Corros. J.*, Vol 5, 1967, p 186
164. M. Smialowski *et al.*, *Bull. Acad. Polon. Sci. Lett.*, A1 (No. 2 suppl), 1950, p 163
165. G. Athavale and W. Eilender, *Korros-Metalschutz*, Vol 16, 1940, p 127
166. S. Das and G. Thomas, *Trans ASM*, Vol 62, 1969, p 55
167. F. Watkinson *et al.*, *Brit. Weld. J.*, Vol 10, 1963, p 54
168. C.S. Carter, "Crack Extension in Several High Strength Steels Loaded in 3.5% Sodium Chloride Solution," D6-19770, Boeing Co., U.S. Government Report AD685377, 1967
169. D.A. Vaughan and D.I. Phalen, Reactions Contributing to the Formation of Susceptible Paths for Stress Corrosion Cracking in *STP 429*, American Society for Testing and Materials, 1967, p 209
170. W.M. Cain and A.R. Troiano, *Petrol. Eng.*, Vol 37, May 1965, p 78
171. R.L. McGlasson *et al.*, *Corrosion*, Vol 16, 1960, p 113
172. R.H. Jones *et al.*, *Corrosion*, Vol 45 (No. 6), 1989, p 494
173. K.L. Moloznik *et al.*, *Corrosion*, Vol 35 (No. 7), 1979, p 331
174. C. Lea and E.D. Hondros, *Proc. R. Soc. (London)*, Vol 377A, 1981, p 477
175. R.A. Davis, *Corrosion*, Vol 19, 1963, p 45t
176. N.A. Tiner and G.A. Gilpin, *Corrosion*, Vol 22, 1966, p 271
177. P.C. Hughes *et al.*, *J. Iron Steel Inst.*, Vol 203, 1965, p 154
178. E.H. Phelps, A Review of Stress Corrosion Cracking Behavior of Steels With High Yield Strength, in *Proc. Conf. Fundamental Aspects of Stress Corrosion Cracking*, Ohio State University, 1967
179. W.K. Boyd *et al.*, "Studies of Hydrogen Stress Cracking and Stress Corrosion Cracking of High Strength Bolts," Report 69-13, Battelle Memorial Institute, 1969
180. A.W. Coulter and T.S. Clairborne, *Mater. Protect.*, Vol 8, June 1968, p 23
181. H. Schenck *et al.*, *Stahl Eisen*, Vol 87, 1967, p 136
182. R.D. McDonald, *Weld. Res. Abroad*, June 1972, p 23
183. F.D. Sewell, "Hydrogen Embrittlement Challenges Tubular Goods Performance," Paper 65-PET-13, American Society of Mechanical Engineers, 1965
184. W.S. Pellini, "Advances in Fracture Toughness Characterization Procedures and in Quantitative Interpretations to Fracture Safe Design for Struc-

tural Steels," NRL 6713, Naval Research Laboratory, April 1968

185. J.H. Smith and J.A. Davis, "Stress Corrosion and Hydrogen Embrittlement Cracking Behavior of HY80 and HY130 Weld Metals," 89.021-024(4), U.S. Steel Corp., Jan 1971
186. J.H. Smith, "Final Report on the Development of an HY-110 Steel Weldment-Structural Evaluation," (39.018-019), U.S. Steel Corp., Jan 1971
187. J.G. Gouch, Stress Corrosion Testing of Welded Joints, in *Specialists Meeting on Stress Corrosion Testing Methods, AGARD Conf. Proc. No. 98*, AGARD CP-98, NATO, 1972
188. R.A. Swift and H.C. Rodgers, *Weld. J.*, Vol 50, 1971, p 357s
189. J.F. Bates, *Corrosion*, Vol 24, 1968, p 362
190. R.M. Vennet, "Safe Weld Hardness Levels to Insure Long Service Life in Offshore Pipelines," paper presented at 4th Annual Offshore Technology Conf., Houston, May 1972
191. W.J. Neill, *Mater. Protect.*, Vol 10, Aug 1971, p 33
192. R.N. Parkins, *Brit. Weld. J.*, Nov 1955, p 495
193. H.M. Wilton, *Met. Progr.*, Nov 1947, p 103
194. S.R. Novak, "Effect of Plastic Strain on the K_{ISCC} of High Strength Steels in Seawater," paper presented at Symp. Fracture and Fatigue, George Washington University, May 1972
195. C.S. Carter, *Metall. Trans.*, Vol 3, 1972, p 584
196. M.F. Baldy, *Corrosion*, Vol 17, 1961, p 509t
197. L.S. Darken and R.P. Smith, *Corrosion*, Vol 5, 1949, p 1
198. R.M. Hudson and G.L. Stragand, *Corrosion*, Vol 16, 1960, p 253t
199. H.H. Uhlig and J. Sava, *Trans. ASM*, Vol 56, 1963, p 361
200. M.J. Humphries and R.N. Parkins, Effects of Cold Work on Stress Corrosion Cracking of Mild Steel, in *Int. Congr. Corrosion*, Amsterdam, National Association of Corrosion Engineers, 1969, p 151
201. F.H. Cocks and J. Bradspies, *Corrosion*, Vol 28, 1972, p 192
202. "Sulfide Stress Cracking Resistant Material for Oil Field Equipment," NACE Standard MR-01-75, National Association of Corrosion Engineers, 1980
203. "Recommended Practice — Methods and Controls to Prevent In-Service Cracking of Carbon Steel (P1) Welds in Corrosion Petroleum Refining Environments," National Association of Corrosion Engineers, 1972
204. G. Kobrin, Materials Selection—Designing to Minimize Corrosion, in *Metals Handbook*, Vol 13, 9th ed., *Corrosion*, ASM International, 1987, p 328
205. R.N. Parkins, *J. Iron Steel Inst.*, Oct 1952, p 149
206. H.R. Baker and C.R. Singleterry, *Corrosion*, Vol 28, 1972, p 320
207. C.D. Weir and P. Hamer, *Chem. Ind.*, 1952, p 1040
208. R. Rath, *Chem. Ind.*, 1953, p 600

CHAPTER 3

Stress-Corrosion Cracking of High-Strength Steels (Yield Strengths Greater Than 1240 MPa)

P.G. Marsh and W.W. Gerberich, Department of Chemical Engineering and Materials Science, University of Minnesota

Steels with yield strengths greater than 1240 MPa (180 ksi), corresponding to hardnesses greater than 40 HRC, fall into a natural grouping in which hydrogen embrittlement is accepted as the predominant stress-corrosion cracking (SCC) mechanism. At this strength level, susceptibility to SCC is particularly acute. Environments that normally would cause minimal general corrosion, such as moist air, are found to promote the incidence of SCC in such steels.

The two commercially important types of steel with yield strengths* greater than 1240 MPa (180 ksi) are quenched and tempered low-alloy steels and maraging steels. Designations and compositions for these steels are given in Table 1. Precipitation-hardening stainless steels are included in Chapter 4. Other high-yield-strength steels of less commercial importance include ausformed steels and cold-worked fine pearlite steels. Ausformed steels have limited use as structural materials due to their high relative costs and to difficulties in controlling distortion (Ref 1). Cold-worked fine pearlite steels are normally limited to such applications as wire rope or reinforcing wire (Ref 1).

Typical applications for high-strength steels with yield strengths above 1240 MPa (180 ksi) include aircraft undercarriages, rocket motor casings, missile bodies, springs, high-strength fasteners—in general, applications where weight-saving is essential (Ref 1). The term "ultrahigh-strength steel" is often used in these applications, but this description is not clearly defined in terms of yield strength or heat treatment.

Owing to the inevitable increase in sensitivity to flaws that occurs with increase in yield strength, such steels are produced by using melting practices that minimize the negative effects of nonmetallic inclusions. Accordingly, melting practice has a strong influence on service performance not only in terms of SCC resistance but also in terms of fatigue strength and fracture toughness.

The quenched and tempered steels that have a tempered martensitic microstructure can be categorized as low-alloy and high-alloy grades, depending on total alloying elements. Alternative descriptions for low-alloy quenched and tempered steels include medium-carbon low-alloy steels and medium-alloy air-hardening steels. The latter type, commonly used for hot-working dies, has become less popular for structural applications, because superior fracture toughness is available in alternative grades at similar cost (Ref 2). For a comprehensive description of high-yield-strength quenched and tempered steels, see Ref 2.

As their name suggests, maraging steels develop strength by martensite formation followed by the growth of precipitates during aging. The

*That is, 0.2% proof stress; the term "yield strength" is used in the engineering sense.

Table 1 Nominal Composition of High-Strength Steels

Type or designation	Composition, % C	Cr	Ni	Mo	Mn	Co	Cu	Al	Ti	Si	V
Low-alloy martensitic											
4130	0.30	1.0	...	0.2	0.5	...	...	...	...	0.3	...
4340	0.40	0.8	1.9	0.25	0.7	...	...	...	...	0.6	...
D6-ac	0.45	1.15	0.55	1.0	0.8	...	...	...	...	0.25	0.05
H-11	0.40	5.0	...	1.3	0.3	...	...	...	...	0.90	0.50
High-alloy martensitic											
HP9-4-45	0.45	0.3	8.6	0.20	0.10	3.75	...	...	...	0.10	0.10
HP9-4-25	0.25	0.50	8.5	0.50	0.10	3.75	...	...	...	0.10	0.10
HP9-4-20	0.20	0.80	9.0	1.0	0.3	4.5	...	...	...	0.10	0.10
10Ni-2Cr-1Mo-8Co	0.11	2.20	10.0	1.0	0.06	8.0	...	...	...	0.1	...
Maraging, standard grades											
18Ni-200	0.03 max	...	18.0	3.2	...	8.0	...	0.1	0.2	...	...
18Ni-250	0.03 max	...	18.0	4.8	...	8.0	...	0.1	0.4	...	...
18Ni-300	0.03 max	...	18.0	5.0	...	9.0	...	0.1	0.6	...	...
Cobalt-free/low-cobalt-bearing grades											
Cobalt-free 18Ni(200)	0.03 max	...	18.5	3.0	...	...	...	0.1	0.7	...	...
Cobalt-free 18Ni(250)	0.03 max	...	18.5	3.0	...	...	...	0.1	1.4	...	...
Low-cobalt 18Ni(250)	0.03 max	...	18.5	2.6	...	2.0	...	0.1	1.2	...	...
Cobalt-free 18Ni(300)	0.03 max	...	18.5	4.0	...	...	...	0.1	1.85	...	...

Table 2 Variables that Influence SCC of High-Strength Steels

Variable	Influence on SCC behavior
Loading mode	Mode I most severe
State of stress	Plane-strain conditions more susceptible to crack growth
	Equibiaxial stresses most severe for initiation in uncracked members
Type of steel	Maraging steels less susceptible than other types of high-strength steels
Yield strength	Increasing yield strength for a given type of steel reduces K_{ISCC} and increases crack velocity
Composition	Cobalt-free grades of 18Ni maraging steel slightly more susceptible than standard grades
	Controlling residual elements to levels as low as possible decreases susceptibility; grain-boundary composition an important factor
Heat treatment	For quenched and tempered steels, small prior-austenite grain size improves resistance
	In maraging steels, crack velocity increases with decrease in aging temperature
Deformation	Cold working of maraging steels prior to aging decreases susceptibility
Electrochemical potential	In quenched and tempered steels, little influence on K_{ISCC}, but crack velocity is minimized at an impressed potential of ~–0.8 V (Ag/AgCl)
	Maraging steels, susceptible at all cathodic overpotentials greater than –0.6 V
pH	Little influence on susceptibility except at extremes
Temperature	Crack velocity increases exponentially with temperature up to the boiling point of water

most important precipitate is Ni_3Mo, and Ni_3Ti is also formed. Unlike the quenched and tempered steels, carbon is considered to be an impurity rather than an important alloying element (Ref 3). Carbon, when present, will form TiC, which reduces strength and fracture toughness. A typical heat treatment for an 18% Ni maraging steel is presented in Fig. 1. After annealing and air cooling, a relatively soft lath martensite is formed with a hardness of approximately 30 HRC. Subsequent aging increases hardness to levels typically above 50 HRC. The precise level of hardness largely depends on the aging temperature. As stated earlier, molybdenum is the major age-hardening addition. Cobalt, in the cobalt-bearing grades, functions to counteract the effects of other additions that lower the M_S temperature.

Maraging steels offer high strength and superior fracture toughness compared with competing types of steel at a higher material cost. Stainless maraging steels, in which some of the nickel is replaced by chromium, have been developed for improved corrosion resistance. However, their use appears to be limited. Reference 3 provides an excellent summary of the metallurgy, processing, properties, and weldability of maraging steels.

In the following two sections, SCC will be discussed with emphasis on external hydrogen embrittlement because of the weight of evidence indicating that hydrogen is the embrittling species. From Fig. 2 (Ref 4) it is apparent that the severity of SCC, including sulfide cracking, depends on the supply of atomic hydrogen generated by interaction with different environments.

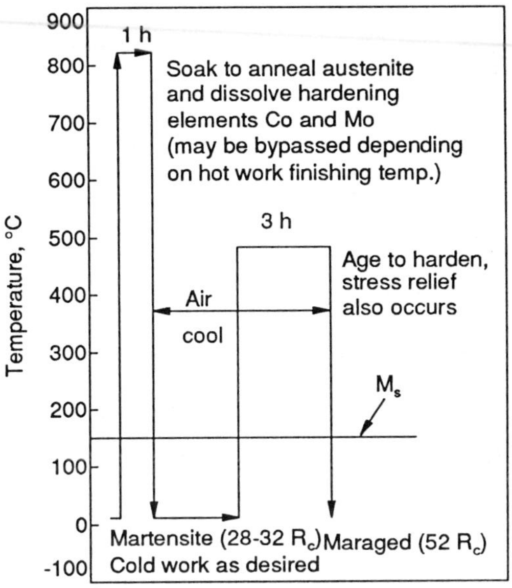

Fig. 1 Schematic diagram of a typical heat-treatment cycle for a 18Ni maraging steel. Source: Ref 1

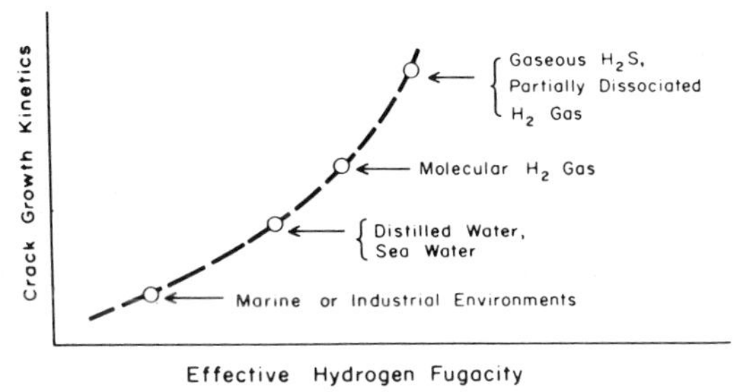

Fig. 2 Schematic illustration of severity of SCC environments. Source: Ref 4

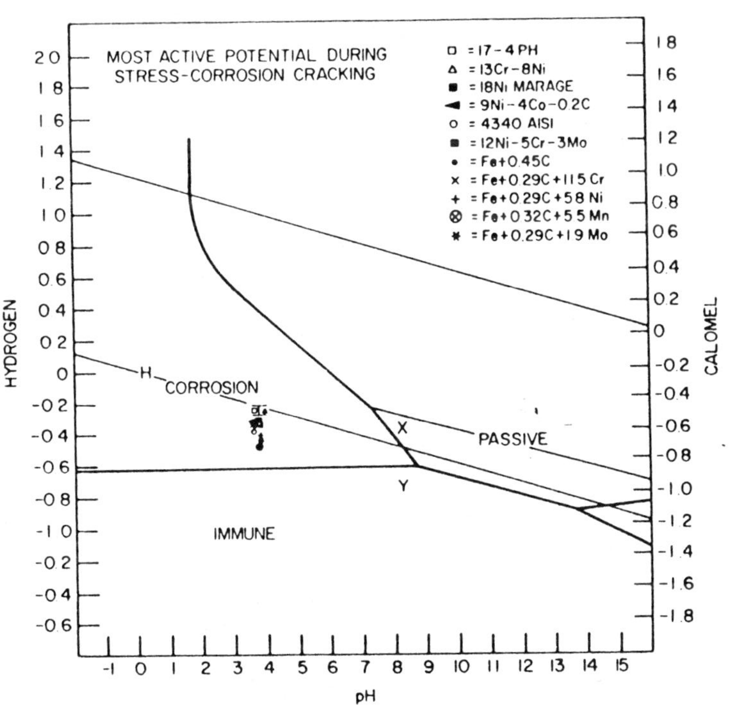

Fig. 3 Values of potential and pH at the crack tips of several steels during stress-corrosion crack propagation. Source: Ref 7

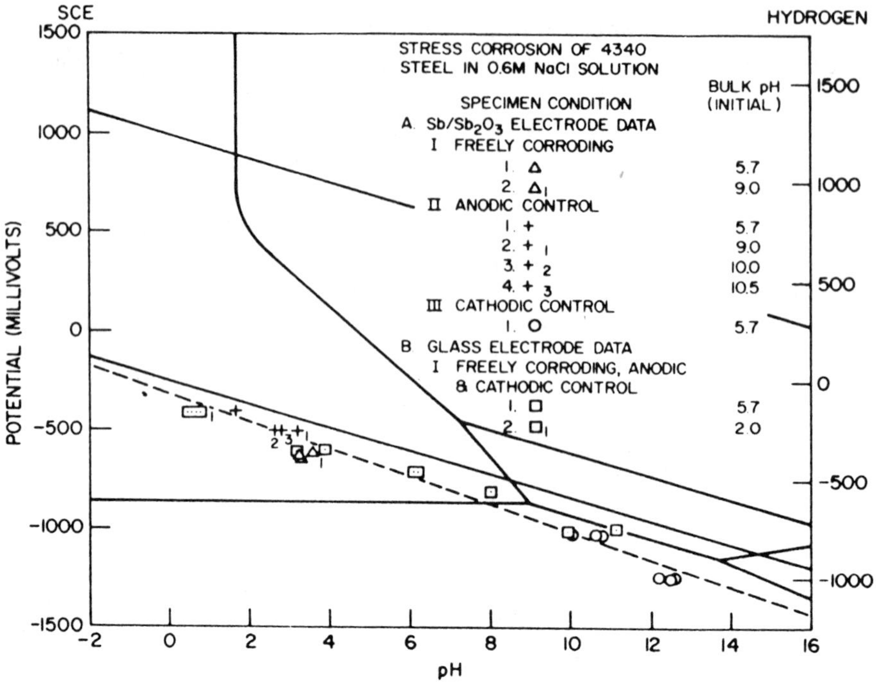

Fig. 4 Effects of anodic and cathodic (impressed) potential on the pH at the tip of propagating stress-corrosion cracks in AISI 4340 steel. Source: Ref 7

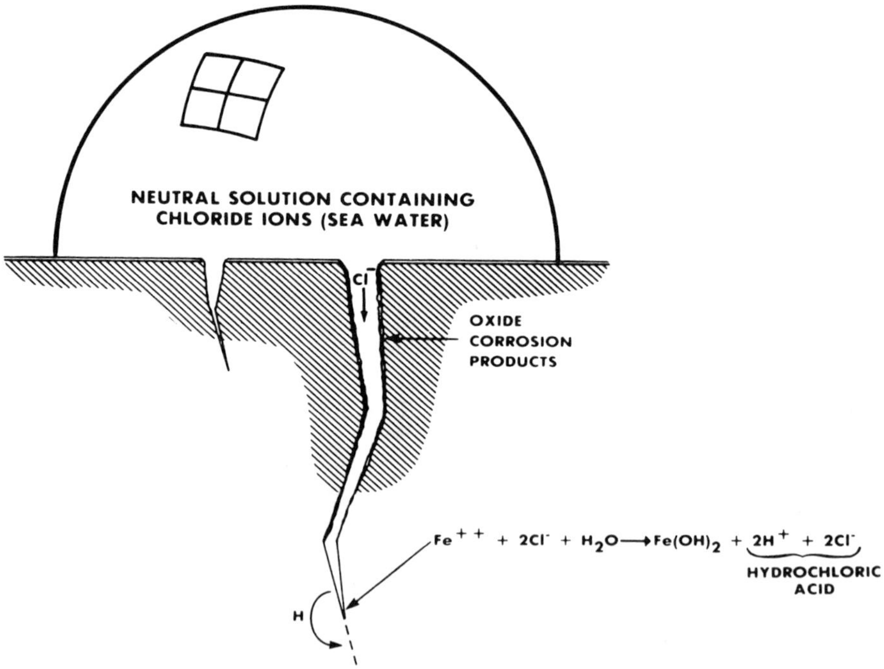

Fig. 5 Chemistry and pH changes in a crack growing in salt water. Note: Chloride ions have only a kinetic influence (see text). Source: Ref 6

Table 2 gives a brief overview of important variables that influence SCC behavior. These variables and others will be discussed in more detail in the following sections.

Overview of Theoretical Models

For hydrogen embrittlement to be the predominant mechanism for SCC of high-yield-strength steels, it is necessary to establish that the embrittling species can be present over a complete range of environmental conditions. If this is not so, more than one mechanism may be in operation. In the Pourbaix diagram for iron (Ref 5) (*i.e.*, the potential-pH equilibrium diagram for the iron-water system), the range of potentials and pH within which hydrogen can be released by reduction of water is defined. Below the sloping line marked "H" in Fig. 3 (the same line is unmarked in Fig. 4), it is possible for the following cathodic reaction to occur:

$$H_2O \rightarrow OH^- + \tfrac{1}{2}H_2 - e^-$$

This equation includes all the steps leading to molecular hydrogen evolution at the metal surface. The restriction on release of hydrogen defined by the Pourbaix diagram suggests that hydrogen embrittlement will be possible only under specific combinations of potential and pH. However, when the local conditions within the propagating stress-corrosion crack are evaluated, rather than conditions of the bulk environment, hydrogen embrittlement as the operating mechanism becomes more credible.

Figure 5 illustrates local chemistry and pH changes within a propagating crack in a neutral saline environment (Ref 6). The presence of chloride ions is not relevant to the thermodynamics of hydrogen reduction or crack-tip acidification. Rather, it is the metal ions in solution that cause hydrolysis and acidification. Accordingly, the equation in Fig. 5 is more correctly written as:

$$Fe^{2+} + H_2O \rightarrow Fe(OH)_2 + H^+$$

Because chloride-containing environments are so commonplace, it is worthwhile to digress and emphasize that the significance of the presence of chloride ions is in terms of reaction kinetics rather than thermodynamics. Chloride ions degrade passive films, thus increasing the rate of metal-ion dissolution. Returning to acidification, experimental evidence indicates that the pH of the local environment within a pit, crack, or crevice on a steel surface can be decreased to 3.5 (Ref 7).

Confirmation that the local environment dictates SCC behavior arose from the development of techniques for measuring local potential and pH at the tips of stress-corrosion cracks. When data points, so determined, are superimposed on the Pourbaix diagram for iron (Fig. 4), it is evident that local conditions are always favorable for release of hydrogen, irrespective of bulk pH and impressed potential. As can be seen, the dashed line in Fig. 4 lies below the solid water-reduction line. Similarly in Fig. 3, potential and pH are shown for free-corroding conditions. The local pH shown in Fig. 3 is consistent with the pH within pits, cracks, and crevices, discussed earlier.

Other supporting evidence for hydrogen embrittlement includes the similarity in fracture appearance of stress-corrosion cracks and cracks induced by gaseous hydrogen (Ref 4, 8). However, gaseous hydrogen evolved at the metal surface in aqueous environments does not constitute atomic hydrogen in the lattice. A number of critical steps, including absorption, must occur before hydrogen can diffuse to the embrittlement site. These critical steps include:

- Adsorption/reduction of hydrogen at the metal surface

  $$H^+ \rightarrow H_{ads} - e^-$$

- Recombination of hydrogen to its molecular form

  $$H^+ + H_{ads} \rightarrow H_2 - e^-$$

  or

  $$H_{ads} + H_{ads} \rightarrow H_2$$

- Absorption of hydrogen competing with recombination

As will be discussed later, cathodic poisons such as H_2S will act to promote hydrogen adsorption, while suppressing recombination.

A complete understanding of hydrogen embrittlement involves more than a knowledge of the basic mechanism of intrinsic embrittlement and the operation of microstructural influences. The kinetics of cracking, which consider stress state, alloy composition, microstructure, temperature, and hydrogen transport, must also be understood (Ref 9). Currently, hydrogen embrittlement is not fully understood.

Taking intrinsic embrittlement first, the two most widely accepted models are the decohesion theory (Ref 10, 11) and the enhanced-plasticity theory (Ref 12). Experimental evidence exists to support both models. Important earlier models include the pressure theory of Zapffe and Sims (Ref 13) and the surface-energy theory of Petch and Stables (Ref 14).

The pressure theory centers around the concept that molecular hydrogen precipitates in preexisting voids in the metal. Pressure is built up in the voids, adding to the effect of externally applied stress. Thus, the applied stress for fracture metal is reduced. The problem with the pressure theory appears to be that an uncharacteristically high hydrogen fugacity is required to build up local high pressure. Additionally, plastic deformation that might be expected to be associated with voids under high pressure is not observed (Ref 9).

The surface-energy theory (Ref 14) postulates that hydrogen adsorbed at the metal surface acts to lower surface energy. By applying a Griffith-like approach, it is apparent that a lower surface energy will give a lower required stress for fracture. However, a thermodynamics argument does not constitute an embrittlement mechanism (Ref 9). No description is available to explain how hydrogen is capable of causing crack advance.

The decohesion model has been developed from Troiano's original theory (Ref 10) that electrons donated from dissolved hydrogen atoms enter the incompletely filled *d*-bond of transition metals. The cohesive strength of the metal lattice is reduced by the increased electron density acting to increase the interatomic spacing of the lattice. At any locality of the lattice where hydrogen concentrates under the thermodynamic effect of hydrostatic tension, cohesive strength is reduced. Externally applied stress is magnified in the crack-tip region and overcomes the resistive-cohesive force between atoms. The resistive-cohesive force is diminished by the stress-induced local hydrogen concentration.

Recent work on iron aluminide (Ref 15) suggests that hydrogen embrittlement in the material results from a depletion of *d*-bonding charge from iron sites in the ordered lattice. Charge transfer from iron to dissolved hydrogen is observed. With a high hydrogen concentration, corresponding to all tetrahedral sites being occupied by a layer of hydrogen, calculated cleavage strength is reduced from 35 to 11 GPa (5076 to 1595 ksi), a reduction of 70%. Although iron aluminide and high-yield-strength steels are dissimilar, the nature of the iron-hydrogen interaction suggests that the underlying mechanism for hydrogen embrittlement in these materials may be the same.

The enhanced-plasticity model (Ref 12), which was proposed based on fractographic observation of low-strength steel, supposes that hydrogen reduces the shear stress required for dislocation motion at the crack tip. In this way, the local plastic-flow process associated with crack propagation is eased. Strong experimental support for enhanced plasticity has come from Birnbaum's experiments using an environmental cell in the transmission electron microscope (TEM) (Ref 16). Direct observation of foils has shown that the number of active dislocations suddenly increases when low-pressure hydrogen is introduced. The same behavior has been observed in iron and other materials.

The decohesion and the enhanced-plasticity models appear to be contradictory in that bond breaking is associated with cleavage fracture and plasticity should favor ductile fracture. However, enhanced plasticity, which has been observed to result from increased screw-dislocation mobility caused by the influence of hydrogen, is not general (Ref 17). Hardening effects due to hydrogen are also observed (Ref 17). Accordingly, no single model for embrittlement can account for observed behavior. Therefore, more than one mechanism may operate to cause embrittlement.

The issue of how microstructure influences cracking cannot be separated from that of the crack nucleation site. Consistent with the decohesion theory, hydrogen tends to diffuse to and collect at the highly dilated triaxial region ahead of the crack tip, as shown in Fig. 6(b). In addition to

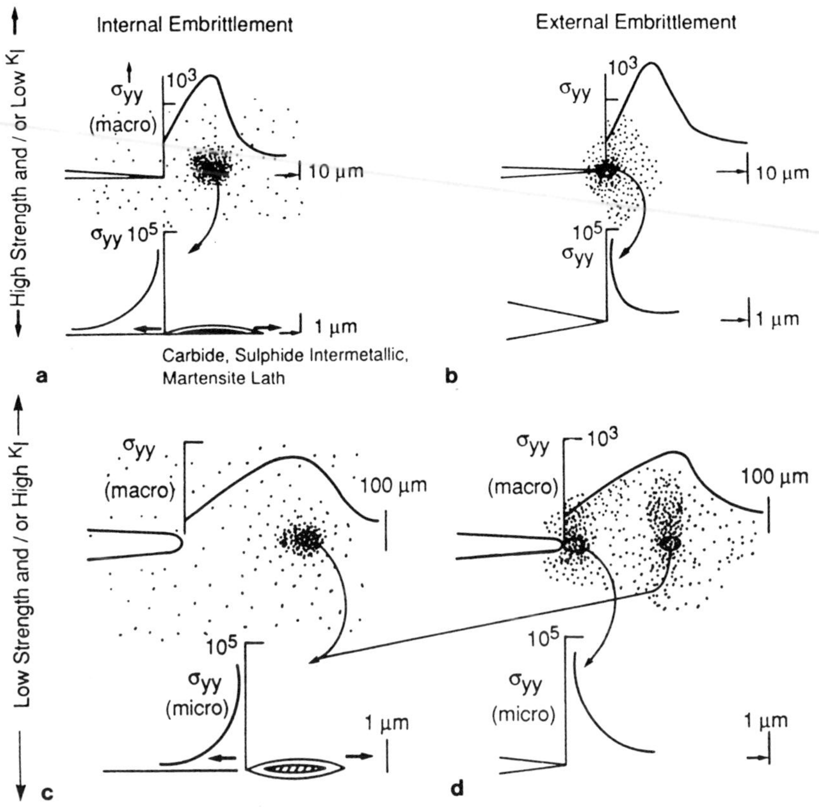

Fig. 6 Schematic of possible crack-initiation sites. (a) and (c) Internal environments. (b) and (d) External environments. Source: Ref 18

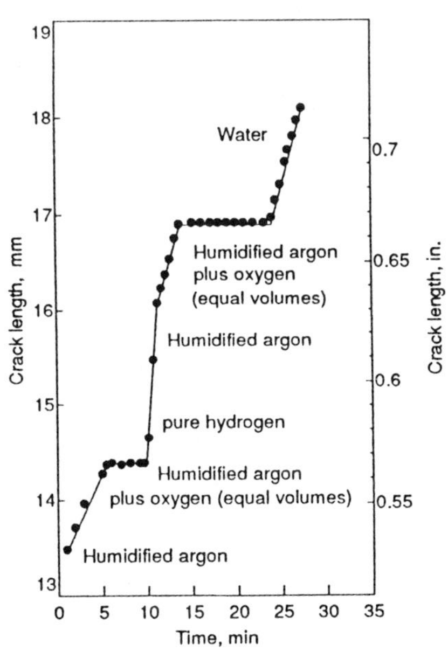

Fig. 7 Fast crack growth of H-11 steel in water and hydrogen. Source: Ref 19

local stress state, factors that influence the position of the crack nucleation site include local resistance and the time to reach local instability (Ref 18). As implied by Fig. 6(b) for a sharp crack, nucleation should occur at or very near the surface, because the combined local concentration and local stress is greatest there. The blunt crack schematic (Fig. 6d) is less applicable to a high-strength material. However, the nucleation site would be expected to be farther from the crack tip, because local concentration and local stress are highest in a subsurface location as shown. For a complete discussion of embrittlement site, see Ref 18.

As demonstrated by Hancock and Johnson's experiment (Ref 19), cracking is extremely rapid once an embrittling environment is introduced, as shown in Fig. 7. This might suggest that crack nucleation is at or very near the surface. However, 1 μm of diffusion path can be traversed by hydro-

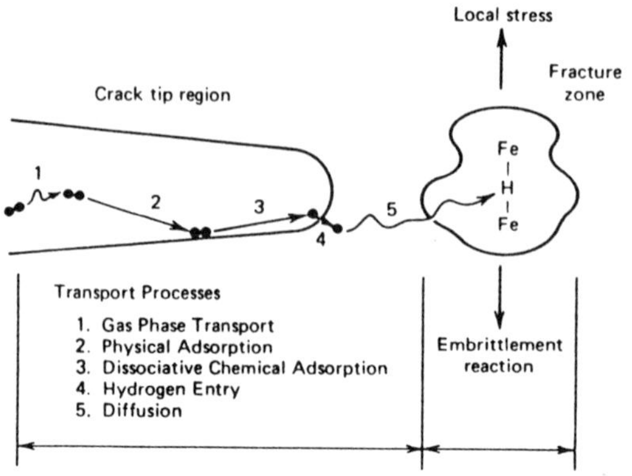

Fig. 8 Schematic illustration of various sequential processes involved in external hydrogen embrittlement. Source: Ref 21

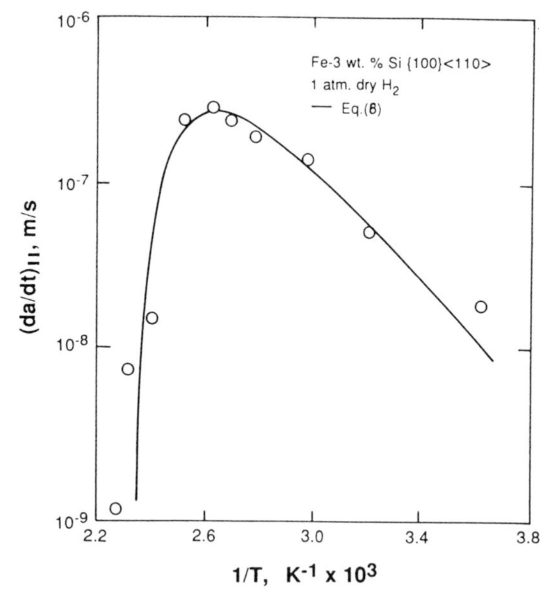

Fig. 10 Stage 2 (II) growth rate in 1 atm hydrogen as a function of temperature for {001} ⟨110⟩ orientated Fe-3Si single crystals. Source: Ref 22

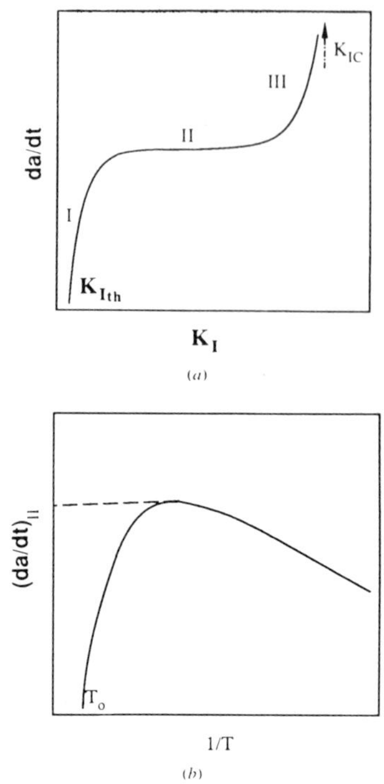

Fig. 9 Schematic of slow-crack-growth kinetics in steels due to hydrogen embrittlement. (a) Growth rate as a function of applied stress intensity. (b) Stage 2 (II) growth rate as a function of temperature. Source: Ref 22

gen in 1 s (Ref 17). Figure 7 also underlines the susceptibility of H-11 steel to SCC in moisture in the form of humidified argon. Also of interest is the ability of oxygen to almost instantaneously halt crack propagation.

In HY180 steel, discontinuous cleavage and secondary cracks have been found at least 50 μm from the crack surface (Ref 20). It is not presently clear whether crack nucleation can occur at a weak subsurface site (for example, an inclusion, carbide-particle interface, or prior-austenite grain boundary) to subsequently grow back toward the macrocrack by forming surfaces in the embrittled material. The evidence seems to suggest that subsurface nucleation is common in many materials, with the nucleation site associated with the region of highest stress beyond the crack tip.

Evaluation of crack nucleation site is important to understand the micromechanics and kinetics. Figure 8 (Ref 21) illustrates the various processes involved in external hydrogen embrittlement. For aqueous environments, the last three transport processes shown in the figure will be the same.

It is apparent from the work of many investigators that the kinetically controlling processes for a gaseous hydrogen environment at low temperatures differ from those at higher temperatures, as can be seen from the schematic in Fig.

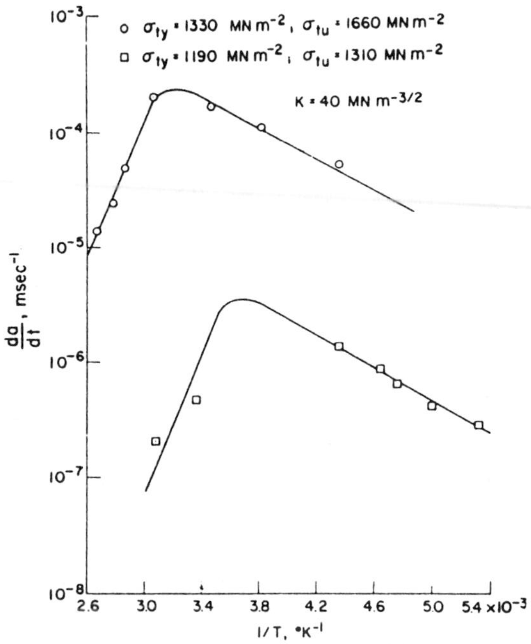

Fig. 11 Stage 2 (II) growth rates in 0.08 MPa (0.01 ksi) hydrogen as a function of temperature for 4130 steel with yield strengths of 1190 and 1330 MPa (173 and 193 ksi). Source: Ref 23

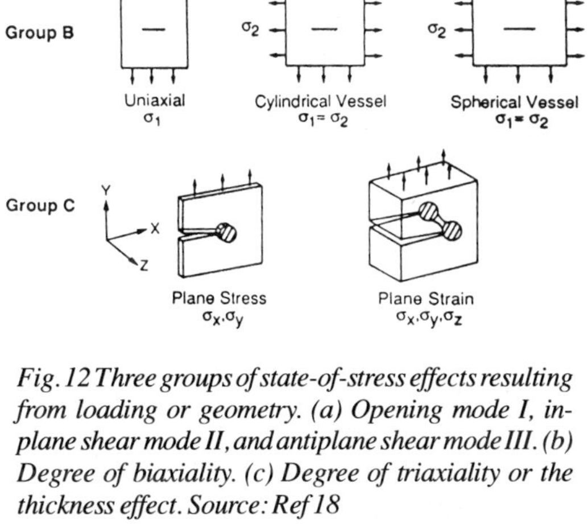

Fig. 12 Three groups of state-of-stress effects resulting from loading or geometry. (a) Opening mode I, in-plane shear mode II, and antiplane shear mode III. (b) Degree of biaxiality. (c) Degree of triaxiality or the thickness effect. Source: Ref 18

9(b) (Ref 22). It should be noted that the curve is representative of K-insensitive growth (*i.e.*, stage 2) as defined by Fig. 9(a). This type of behavior is observed in iron-base materials as diverse as Fe-3Si single crystals and AISI 4130 steel, as demonstrated by Fig. 10 (Ref 22) and Fig. 11 (Ref 23). Figure 11 shows that an increase in strength promotes higher crack velocities and causes a shift in maximum crack velocity to a higher temperature. For aqueous solutions, as will be discussed below, an inverted U-shape curve is not observed. For gaseous hydrogen at temperatures below that of maximum crack velocity, an Arrhenius relationship is observed, with the rate-controlling step believed to be that of hydrogen diffusion (Ref 24). At higher temperatures, the rate-controlling process is thought either to be associated with surface reactions or to be a binding energy associated with the collection of hydrogen at the nucleation site. At temperatures above that corresponding to the maximum growth rate for Fe-3Si single crystals, a sharp threshold is observed, while an Arrhenius-type relationship is observed for AISI 4130 steel.

Loading Mode and State-of-Stress Effects

State-of-stress effects resulting from loading or geometry are represented in Fig. 12. As illustrated, three groups can be considered.

Taking loading mode first, hydrogen embrittlement would be expected to be favored most under mode I loading. As originally suggested for the decohesion theory (Ref 10), embrittlement is most likely to occur in a dilatant region or in a region of elevated hydrostatic tension stress ($\sigma_{ii}/3$). For spherical strain centers in a dilute solution, Boltzmann statistics give the following (Ref 17):

$$C_H = C_o \exp\left(\frac{\sigma_{ii}\overline{V}_H}{3RT}\right)$$

where C_o is the initial concentration, C_H is the concentration in the hydrostatically stressed region, and $\overline{V}/RT$ is the partial molal volume term. When σ_{ii} is large, as in mode I, greater hydrogen is available. With $\sigma_{ii} = 0$, as in mode III, concentration should not occur, and thus embrittlement should not result.

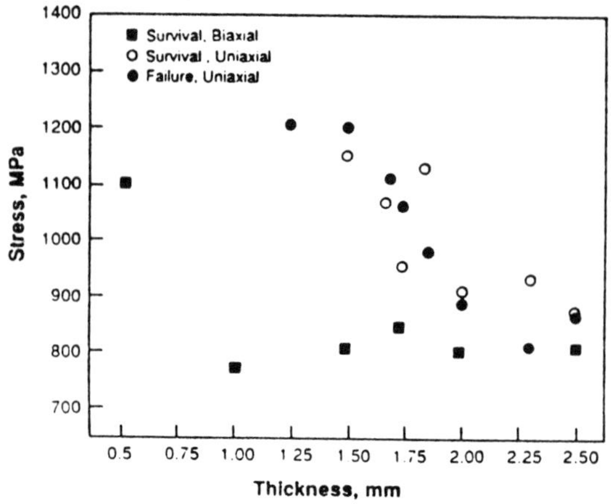

Fig. 13 Minimum failure stresses and maximum survival stresses for AISI 4340 steel as affected by degree of biaxiality at various thicknesses. Source: Ref 27

This is a simplistic approach that does not account for expected higher hydrogen concentration at grain boundaries. Therefore, it could be argued that mode I tensile stresses simply cause intergranular separation of grain boundaries greatly weakened by segregation of hydrogen that results from binding at trap sites. Such intergranular weakening would obviate the need for crack-tip dilation.

Mode II hydrogen embrittlement has been shown to occur under high fugacity (Ref 25). This result might indicate that hydrogen embrittlement does not require a significant bulk hydrostatic stress as inferred from the elasticity solutions. However, by using the Hutchinson, Rice, and Rosengren (HRR) stress-field approach for an elastic-plastic material, a dilational stress has been discussed as being possible for a mode II crack (Ref 26). Currently, it is not known whether hydrogen embrittlement of high-yield-strength steels depends on the bulk hydrostatic stress for mode II loading.

With regard to mode III loading, the generally accepted observation is that microscopic mode III cracking in the mode III direction is very difficult to achieve (Ref 18). When cracking does occur, the crack is actually propagating in the opening mode, since propagation is perpendicular to the maximum tensile axis, that is, at about ±45° to the load axis. This is analogous to brittle conchoidal fracture of chalk in torsion. This leaves the possibility that mode III and mode II can initiate cracking, but that mode I dominates crack propagation.

Biaxiality effects appear to have been studied through crack initiation in uncracked specimens (Ref 18) rather than precracked specimens as implied in Fig. 12. In 6 to 9 MPa (0.9 to 1.3 ksi) hydrogen, biaxial loading (punched test on clamped plate) was shown to be more detrimental than uniaxial loading (Ref 27) (actually, four-point bending). Holding untempered AISI 4340 steel with an ultimate tensile strength (UTS) of 2100 MPa (305 ksi) for 120 h gave the results shown in Fig. 13. Greater biaxial degradation was evident both in lower failure stresses and in more extensive intergranular cracking. During the long hold, it was assumed that dislocation activity becomes negligible in sustained-load tests. It was concluded that for hydrogen embrittlement of high-yield-strength steels, dislocation transport, internal pressure, strain softening, and hydrogen reduction of surface-energy effects do not play significant roles. Lattice dilation appeared to be the important factor. The drop-off in failure stress with increased thickness of the bend specimens was not addressed. This could be attributed to increased biaxiality effects in these uniaxial specimens (Ref 18).

Plane-strain versus plane-stress effects have been described by Brown (Ref 7). Unfortunately, no simple thickness relationship in terms of (K_{ISCC}/σ_{ys}) is available that would provide a basic index for K_{ISCC} in the same manner as for K_{Ic}. Increased resistance of thin-sheet (plane stress) compared with thick-plate (plane strain) conditions has been explained in terms of dilational $(\sigma_{ii}/3)$ and nominal stress (σ_{yy}^{max}) effects (Ref 28, 29). Crack tunneling that typically occurs in the central region of specimens has been explained in terms of embrittlement kinetics leading the tunneled crack. The plane-stress ligament that is holding on finally tears by a shearing mechanism. This explanation is consistent with decreased kinetics being associated with decreased thickness, *i.e.*, greater plane stress.

In summary, a strong state-of-stress influence exists in environmental degradation in general. For SCC of high-yield strength steels, for which brittle intergranular or cleavage modes are in-

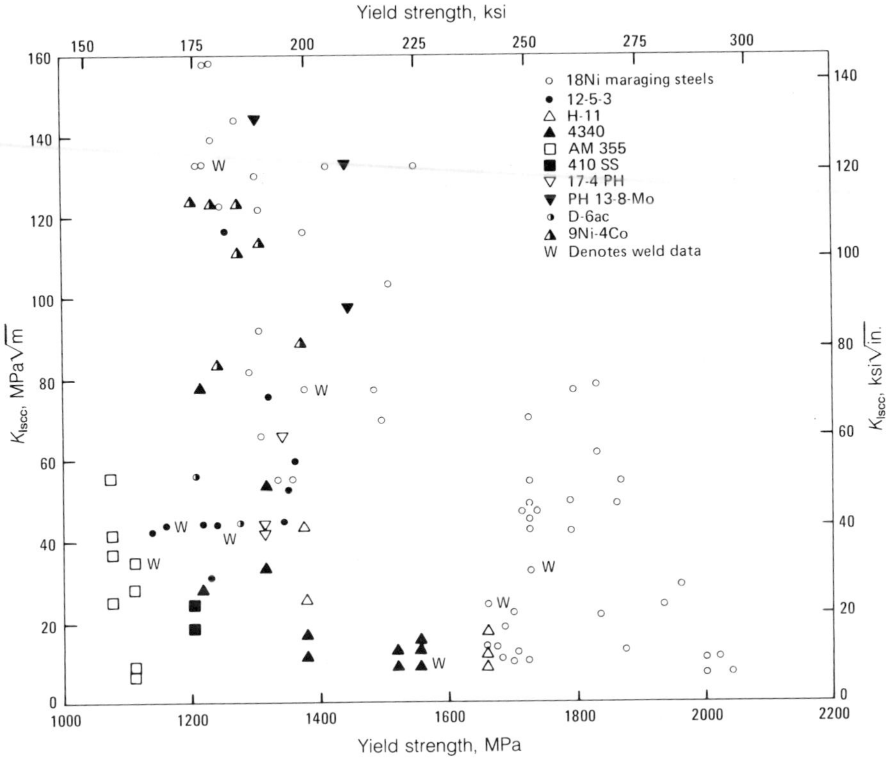

Fig. 14 Threshold stress intensity (K_{ISCC}) values for maraging steels and other high-strength steels as a function of yield strength. Source: Ref 3

volved, mode I plane-strain conditions are most severe for crack growth. In addition, equibiaxial stresses are most severe for initiation in uncracked members.

SCC Characteristics of Different High-Strength Steels

In general, maraging steels have better SCC resistance than other types of steels with comparable yield strengths (Ref 7, 8). The superiority of maraging steels in terms of K_{ISCC} and crack velocity for chloride-containing and nonchloride aqueous environments is evident from Fig. 14 and 15. Consistent results are found for smooth specimens; Fig. 16 shows data for bend specimens subjected to alternate immersion in 3.5% sodium chloride (NaCl) solution. Time to failure for the 18Ni-250 maraging steel is of the order of months for a yield-strength level of 1720 MPa (250 ksi), compared with hours or days for the D6-ac steel at a yield strength level of less than 1500 MPa (220 ksi).

Also apparent from Fig. 14 and 15 is the fact that yield strength has a strong influence on SCC resistance. An increase in strength decreases K_{ISCC} and increases crack velocity. These trends are not surprising, since decreasing the strength or hardness of a material to mitigate SCC is a common practice when possible. For items such as springs, where high strength is unavoidable, protective coatings are often necessary.

To gain perspective of the relative susceptibility of steels with yield strengths above 1240 MPa (180 ksi), Fig. 17 (Ref 30) shows that K_{ISCC} at this strength level, for AISI 4130 steel in seawater, is a fraction of that at 830 MPa (120 ksi). Also apparent is the inevitable decrease in K_{Ic} that occurs with increasing yield strength. An increase in yield strength decreases the applied stress intensity required for failure and increases the crack velocity. Note that although K_{Ic} is lower, the gap

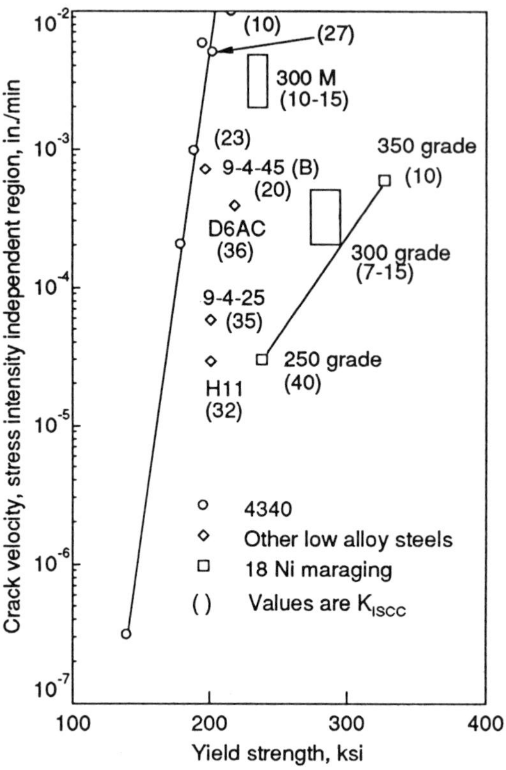

Fig. 15 Comparison of stress-corrosion crack velocities in maraging and low-alloy steels. Source: Ref 8

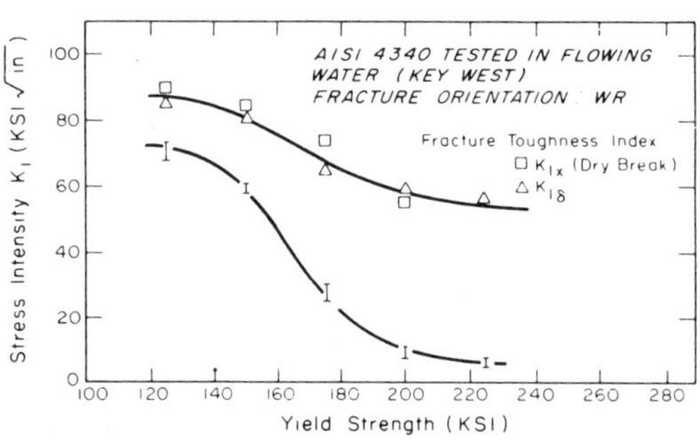

Fig. 17 Effect of strength level on K_{Ic} and K_{ISCC} for AISI 4340 steel. Source: Ref 30

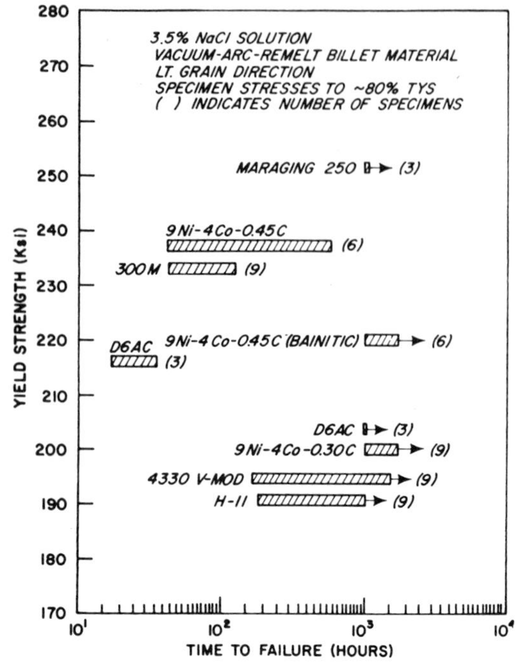

Fig. 16 Alternate-immersion stress-corrosion data from smooth bend specimens. Source: Ref 8

between K_{ISCC} and K_{Ic} is greater. From this discussion, it is apparent that the gap between K_{Ic} and K_{ISCC} is a measure of susceptibility. When K_{Ic} and K_{ISCC} converge, the material is not environmentally sensitive. From Fig. 17 it is obvious that, for AISI 4130 steel at least, the 1240 MPa (180 ksi) yield-strength level inevitably means SCC susceptibility in aqueous environments. K_{Ic} for this material is approximately 70 MPa$\sqrt{m}$ (64 ksi$\sqrt{in.}$) compared with a K_{ISCC} of 24 MPa$\sqrt{m}$ (22 ksi$\sqrt{in.}$). We will see later that chloride-ion content has little influence on K_{ISCC} for this type of steel.

Returning to Fig. 14, close examination reveals that for some grades of steel, the same form of reversed S-shape curve as shown in Fig. 17 is found. Separate curves are not plotted because of the large amount of scatter. The scatter reflects the fact that the data points come from different sources, and it is attributed to variations in testing procedure, melting practice, alloying content, processing, and so on (Ref 8). This perhaps illustrates the difficulty of ranking different grades of steel in terms of expected performance, because these variations will give a high degree of overlap between different grades with the same processing history. However, certain trends have been observed.

For a yield-strength level above 1380 MPa (200 ksi), quenched and tempered steels (*i.e.*,

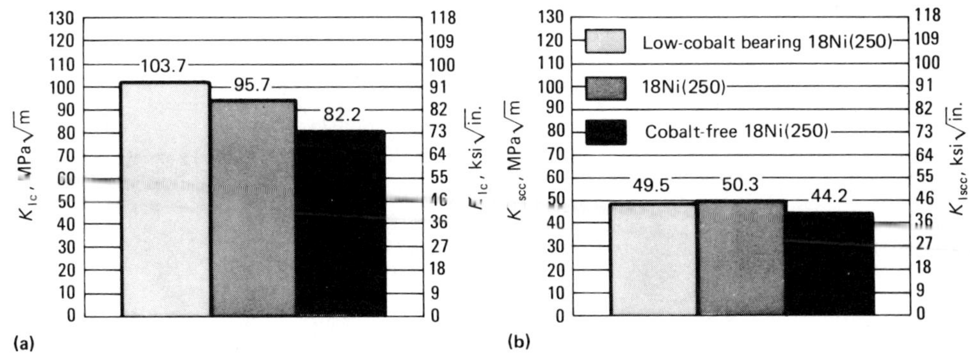

Fig. 18 Bar graphs comparing K_{Ic} (a) and K_{ISCC} (circumferential-radial specimen orientation) (b) of low-cobalt-bearing, standard, and cobalt-free 18Ni-250. K_{ISCC} testing was conducted in a marine atmosphere. Source: Ref 3

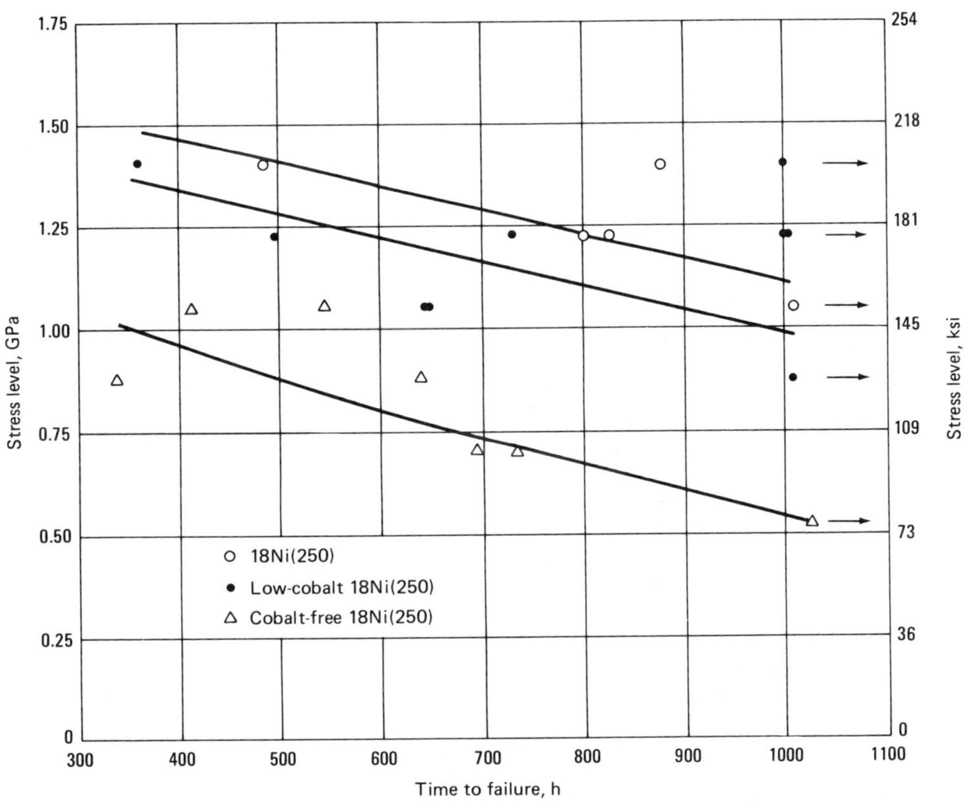

Fig. 19 Plot of stress level versus time to failure to compare behavior of low-cobalt-bearing, standard, and cobalt-free 18Ni-250 maraging steels tested in stagnant 3.5% NaCl solution for 1000 h using proof-ring tensile specimens. Source: Ref 3

martensitic microstructure) have similar levels of K_{ISCC} irrespective of alloy content. AISI 4130 material would be expected to show typical behavior (Ref 7). At yield strengths below 1380 MPa (200 ksi), alloying content has a greater influence, with high-alloy grades such as HP 9-4-20

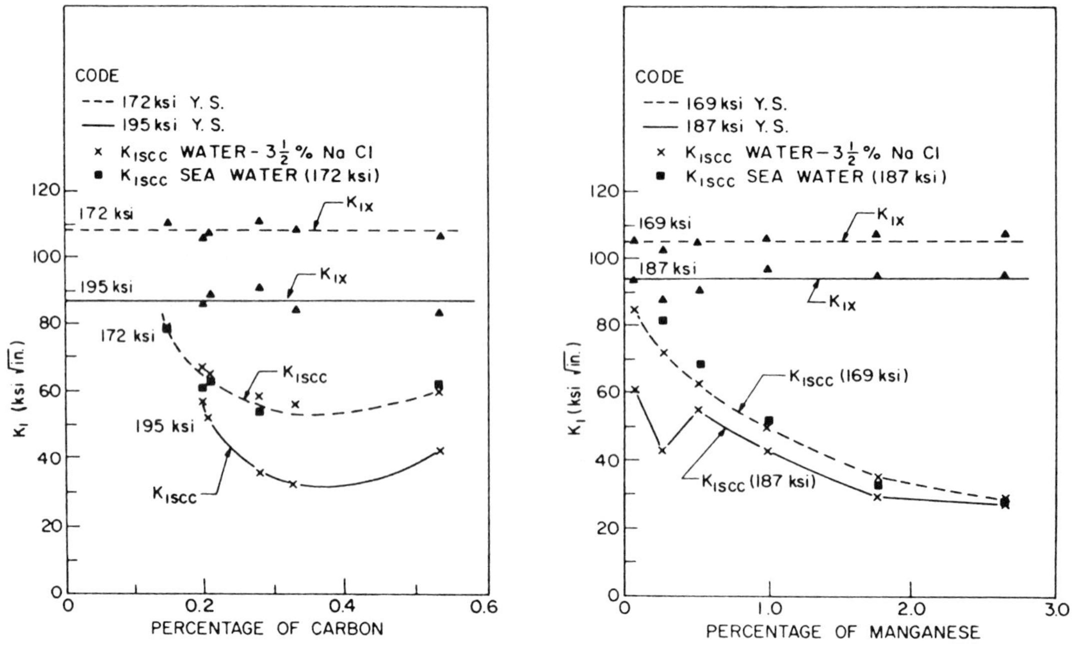

Fig. 20 Effects of carbon and manganese additions on the threshold stress intensity for crack growth in AISI 4340 steel. Source: Ref 7

exhibiting higher values of K_{ISCC}—approximately 104 to 137 MPa $\sqrt{m}$ (95 to 125 ksi$\sqrt{in.}$) compared with approximately 11 to 55 MPa$\sqrt{m}$ (10 to 50 ksi$\sqrt{in.}$) for AISI 4130 (Ref 7). At this strength level, the beneficial influence of nickel on the refinement of microstructure is probably not eclipsed by yield-strength effects.

For the 18Ni maraging steels, the prime competing factor for a given strength level is whether the material is a standard grade (8% Co, nominal) or a cobalt-free or low-cobalt (2% Co, nominal) bearing grade (Ref 3). At the 1720 MPa (250 ksi) yield-strength level, the influence of cobalt level can be seen in Fig. 18 for a marine atmosphere. Cobalt, which is added to lower the M_S temperature, thus producing a finer microstructure, exerts a mixed influence on K_{Ic} and K_{ISCC}. The highest value of K_{Ic} is obtained for the low-cobalt grade, while the two cobalt-bearing grades are not significantly different in terms of K_{ISCC}. The cobalt-free grade has only slightly lower resistance to SCC.

With regard to crack velocity, the yield-strength dependency is extremely well defined (Fig. 15). Maraging steels typically show crack velocities far below those of quenched and tempered steels for region II growth. Consistent with improved K_{ISCC}, cobalt additions will increase time to failure for maraging steels in stagnant 3.5% NaCl (Fig. 19). Consistent with Fig. 18 and 19, the data points in Fig. 15 show that a relatively low value of K_{ISCC} accompanies a relatively high value of crack velocity.

Influence of Composition and Steelmaking Practice

The influence of composition can be shrouded by yield-strength effects. Any alloy addition that increases yield strength is likely to increase susceptibility to SCC. However, several clear influences of composition have been found.

Sandoz (Ref 7, 31) investigated the influence of alloying elements on K_{ISCC} of AISI 4340 steel at different strength levels in seawater and 3.5% NaCl solution. Carbon and manganese additions were found to influence K_{ISCC} as shown in Fig. 20. The influence of carbon is interesting. Additions of up to 0.4% C lower K_{ISCC}, while additions above that level (*i.e.*, above that of the

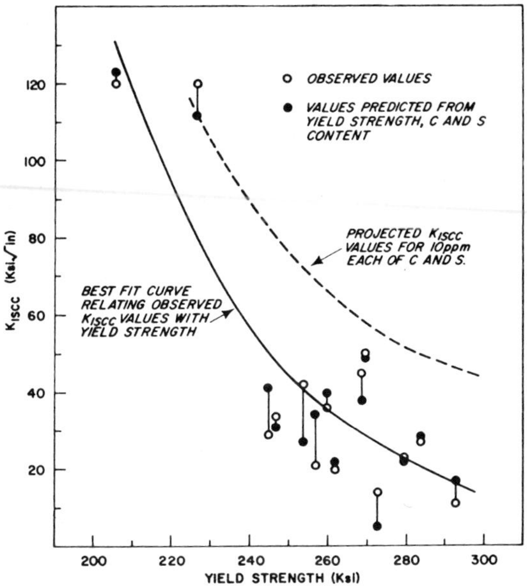

Fig. 21 Correlation of K_{ISCC} of 18Ni maraging steels with yield strength and carbon and silicon content. Source: Ref 8

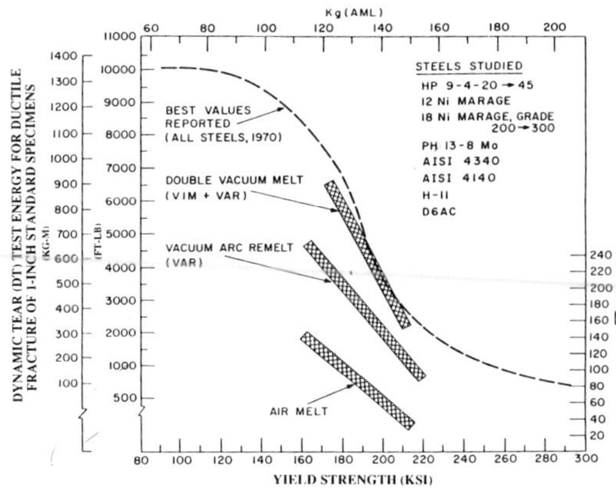

Fig. 22 Effects of melting practice and yield strength on the fracture toughness of high-strength steels. Source: Ref 35

nominal composition) improve resistance to some extent. The reason for the improvement is unclear. Manganese additions steadily decrease K_{ISCC}, as can be seen from Fig. 20. There is no explanation for the pronounced valley in the lower manganese curve. Other elements investigated include phosphorus (≤0.05%), sulfur (≤0.03%), chromium (≤2.5%), molybdenum (≤1.2%), and cobalt (≤3.5%), none of which caused any change in K_{ISCC}. Nickel in amounts up to 9% has been shown to have little influence (Ref 32).

The influence of silicon on the threshold stress-intensity factor is not clear. Some sources indicate little influence (Ref 7, 33), while others suggest that silicon and manganese can act together to promote phosphorus segregation during austenitization of AISI 4340 steel (Ref 34). The resultant phosphorus segregation reduced the threshold value in hydrogen. The complete absence of manganese and silicon resulted in a dramatic increase in K_{ISCC} for a high-purity vacuum induction melt, as will be seen in more detail below. For crack-velocity data, silicon appears to be the element most investigated in quenched and tempered steels. Contents above 1.5% have been found to decrease crack-growth rate and increase time to failure (Ref 33).

For maraging steels, the levels of the major alloying elements are kept within guidelines (Ref 3). Additionally, residual-element content is minimized to achieve optimum mechanical properties and fracture toughness. Accordingly, there appears to be limited scope for compositional influences within a given grade of material. The slightly superior SCC resistance exhibited by cobalt-bearing grades compared with cobalt-free grades has already been discussed.

The small amount of experimental data suggests that controlling carbon and sulfur to lower limits will improve resistance in terms of K_{ISCC} (Ref 8). Control of sulfur has been shown to be more important. Based on observed values and regression analysis, Fig. 21 was produced for a series of 18Ni maraging steel heats. The solid curve on the left corresponds to a best-fit polynomial expression describing K_{ISCC} variation with yield strength for various heats, including high-purity and commercial-purity materials. The benefits of controlling both carbon and sulfur to parts per million are apparent, but controlling residuals to such low levels involves special remelting practices, such as consumable-electrode vacuum remelting (CEVR).

Different melting/remelting practices include vacuum degassing, vacuum arc remelting (VAR), vacuum induction melting (VIM), double vac-

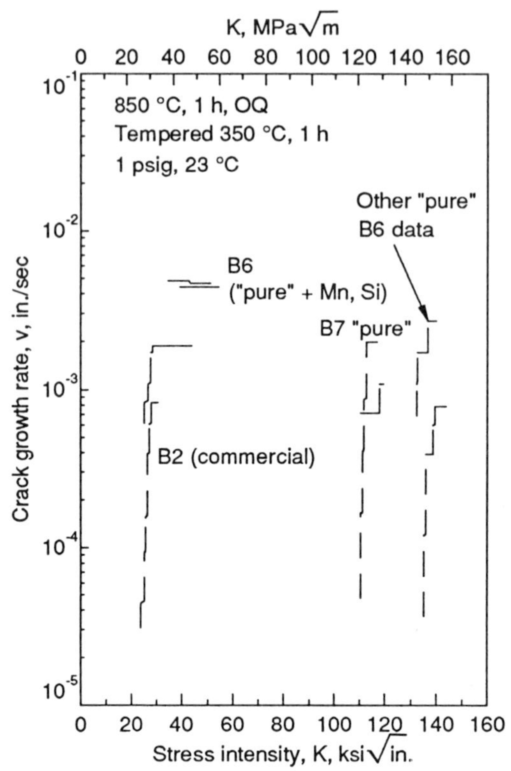

Fig. 23 Crack-growth rate as a function of stress intensity in steels B6 and B7 compared with that of steel B2. Source: Ref 34

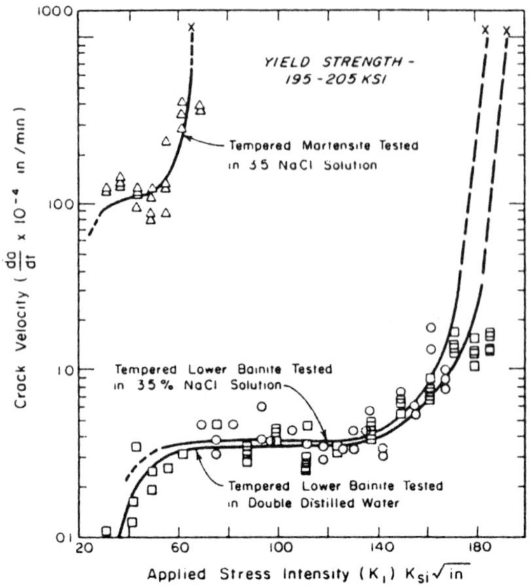

Fig. 24 Crack velocity versus stress intensity for AISI 4340 steel (martensitic and bainitic structures) in 3.5% NaCl solution (pH 6) and double distilled water. Source: Ref 36

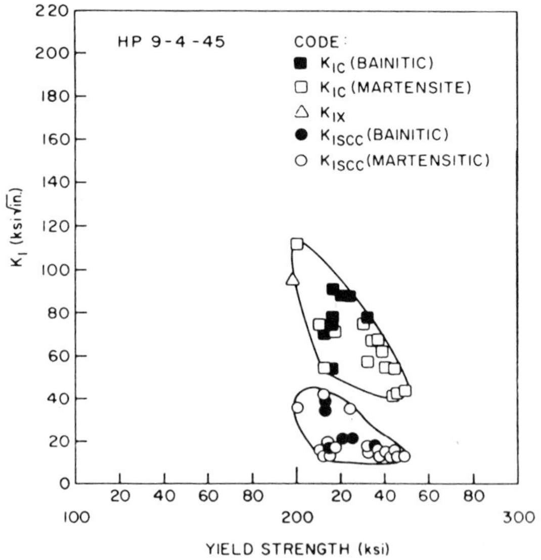

Fig. 25 Stress-corrosion cracking resistance and fracture toughness of HP 9-4-45 steel. Source: Ref 7

uum melting (VIM + VAR) and CEVR. These methods are given roughly in order of "cleanliness," which can be inferred from Fig. 22 (Ref 35). The shape of the "best values reported" curve is consistent with that seen earlier in Fig. 17, which reaffirms that there is an inevitable fall in fracture toughness/notch toughness as higher and higher yield strengths are approached. Although K_{ISCC} data are not included in Fig. 22, we have seen that K_{ISCC} tends to parallel K_{Ic} (Fig. 17). Thus, it appears reasonable to expect that melting practices that give optimum K_{Ic} will give optimum K_{ISCC}, purity being the consistent factor. Additionally, residual-element content is believed to be an important factor when scatter is observed in K_{ISCC} data for a given type of steel at a given yield strength (Ref 7), particularly for quenched and tempered steels.

At a strength level of 1240 to 1520 MPa (180 to 220 ksi), the benefits of vacuum remelting are obvious from Fig. 22. Above this strength level, to achieve useful fracture toughness, vacuum remelting would appear to be a necessity.

For maraging steels, there appears to be little opportunity for improvement in K_{Ic} or K_{ISCC} by

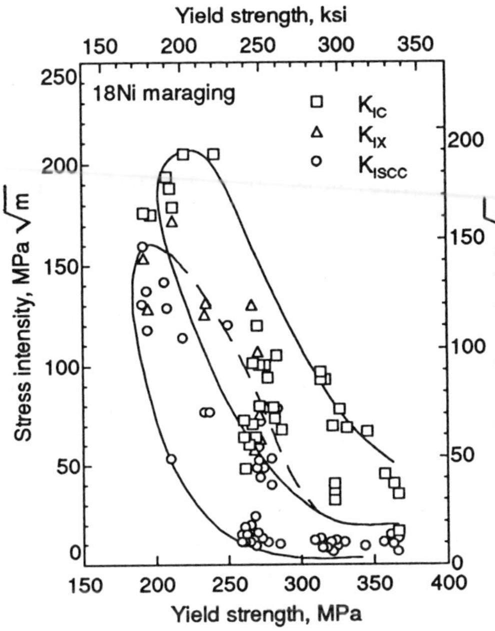

Fig. 26 Stress-corrosion cracking resistance and fracture toughness of 18Ni maraging steel. Source: Ref 7

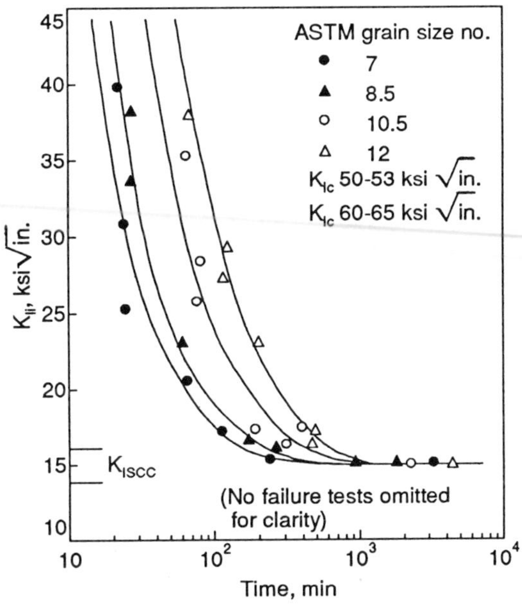

Fig. 27 Effects of grain size and initial stress intensity on delayed failure time for AISI 4340 steel. Source: Ref 36

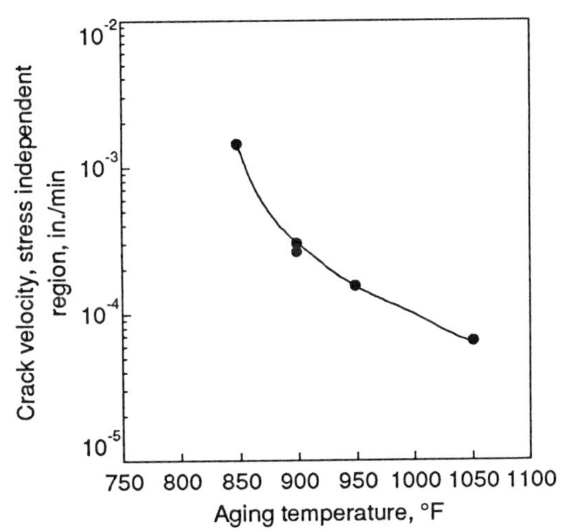

Fig. 28 Effect of aging temperature on crack-growth rates in modified 18Ni-300 maraging steel. Source: Ref 8

remelting practice. These materials require premium properties to justify their relatively high cost (arguably, easier fabrication in the annealed condition evens out cost differences), and these premium properties require high purity. For normal grades, melting practice usually consists either of air melting followed by VAR or of VIM + VAR (Ref 3). Triple remelting (air melting + VIM + VAR) is carried out for aerospace grades to produce very low levels of carbon, manganese, sulfur, and phosphorus (Ref 3).

Quenched and tempered steels are prone to temper embrittlement because of the migration of tramp elements such as arsenic, antimony, and phosphorus to grain boundaries. Because SCC is normally intergranular, it follows that the grain-boundary weaknesses should be minimized to achieve optimum SCC resistance, while at the same time improving K_{Ic}. For AISI 4340 steel, VAR significantly improves K_{Ic} compared with air melting at the 1650 MPa (240 ksi) strength level: *i.e.*, 66 MPa$\sqrt{m}$ (60.4 ksi$\sqrt{in.}$) compared with 45 MPa $\sqrt{m}$ (40.5 ksi$\sqrt{in.}$). As mentioned earlier, manganese and silicon appear to have an indirect effect on K_{ISCC} through phosphorus segregation that results in temper embrittlement. In the study described below (Ref 34), Auger electron spectroscopy (AES) was used to evaluate grain-boundary segregation of commercial grades of AISI 4340 steel. Phosphorus segregated at prior-austenite grain boundaries in VAR steel,

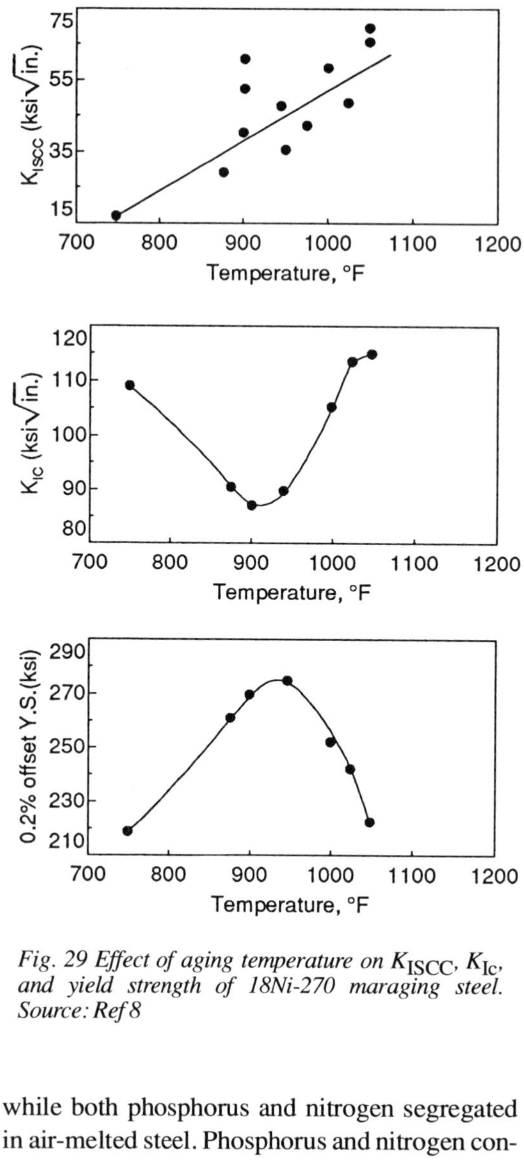

Fig. 29 Effect of aging temperature on K_{ISCC}, K_{Ic}, and yield strength of 18Ni-270 maraging steel. Source: Ref 8

while both phosphorus and nitrogen segregated in air-melted steel. Phosphorus and nitrogen concentrations did not appear to increase during tempering, but the observed segregation had been inherited during austenitizing.

In Fig. 23 (Ref 34), plots of crack velocity versus K_{ISCC} are given for different heats based on AISI 4340 composition at the 1380 MPa (200 ksi) strength level. Tempering was carried out at 350 °C (660 °F), within the 200 to 400 °C (390 to 750 °F) range that causes temper embrittlement. Heat B2 was air melted and vacuum degassed. B6 was a high-purity VIM heat, while B7 was a high-purity VIM heat produced from a NiCrMo-base alloy to give negligible manganese and silicon (0.007% Mn and 0.002% Si). From Fig. 23, it is apparent that freedom from manganese and silicon improves K_{ISCC} fivefold. The observed fracture mode for heat B7 involved rupture with few intergranular facets, confirming freedom from temper embrittlement. For heats B2 (air melted) and B6 (VAR), the slightly higher stress intensity necessary for crack growth exhibited by heat B6 is consistent with the AES results already described. Comparison of Fig. 23 with Fig. 17 for the 1380 MPa (200 ksi) strength level indicates that the absence of manganese and silicon results in K_{ISCC} for the B7 composition approaching K_{Ic} for the standard AISI 4340 grade. Unfortunately, using steels of the B7 composition is not practical, at least for normal steelmaking practice. K_{ISCC} for temper-embrittled AISI 4340 is in a similar range as nominally "unembrittled" material, suggesting that local grain-boundary composition usually exerts a controlling influence.

Effects of Heat Treatment, Microstructure, and Deformation

Maraging steels and quenched and tempered steels exhibit markedly different SCC susceptibilities. These differences cannot be explained in terms of general corrosion resistance. The acute susceptibility of quenched and tempered steels is illustrated in Fig. 24, in which martensitic and bainitic microstructures are compared. For AISI 4130 composition at the same yield strength, K_{ISCC} is lower and crack velocity higher for the martensitic microstructure (Ref 36). Similar behavior is observed for a higher-alloy quenched and tempered steel, HP 9-4-45, in Fig. 25. Even though alloy content is different, K_{ISCC} is similar for each microstructural type for the two steels, given the slight decreasing trend with increasing yield strength.

Explanations for the acute susceptibility of quenched and tempered steels include a combination of factors: presence of ε-carbide, high dislocation density, and coherency strains between twinned martensitic platelets (Ref 4). Tempering tends to promote the degradation of ε-carbide to cementite, reduce dislocation density, and reduce coherency strains, thus decreasing yield strength. Alternatively, and consistent with the assertion that hydrogen embrittlement is the operating

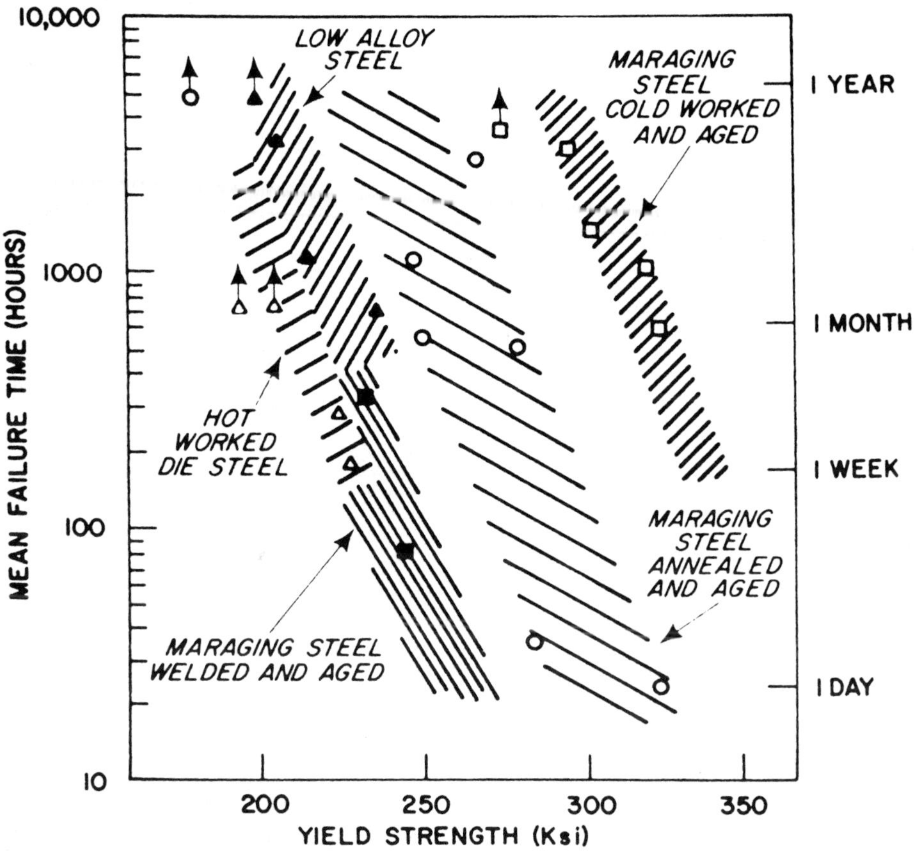

Fig. 30 Results of bent-beam test in aerated distilled water. Specimens were exposed at a stress of 75% yield. Source: Ref 8

mechanism, hydrogen is believed to be more effectively trapped at coherently twinned interfaces in the plate martensite microstructure, as opposed to incoherent bainite laths (Ref 36). Therefore, the almost-zero-carbon lath martensite in maraging steels must behave in a fashion similar to bainitic laths, being less able to trap hydrogen. Comparison of Fig. 25 (Ref 7) and Fig. 26 (Ref 7) suggests that K_{ISCC} for a bainitic microstructure is comparable to that for a maraging steel, and that therefore hydrogen trapping is probably the key factor. Decreased trapping leads to a decrease in hydrogen buildup for embrittlement at lath interfaces, compared with platelet interfaces.

For the higher crack velocities exhibited by quenched and tempered steels (Fig. 15 and 24), crack propagation is believed to be enhanced by the presence of transformation twins and preferential carbide precipitation on martensite platelet boundaries (Ref 7). Figure 24 illustrates that these microstructural features are responsible for increasing crack velocity by more than an order of magnitude.

Generally, SCC in high-yield-strength steels is intergranular, which implies that compositional variations and reactions at grain boundaries are important (Ref 7). As discussed earlier, residual elements for all the types of steels exert their deleterious influence at the prior-austenite grain boundaries. Therefore, prior-austenite grain size that is as fine as possible should provide maximum SCC resistance, at least in terms of crack velocity.

For precracked single-cantilever-beam specimens of AISI 4130 steel in 3.5% NaCl solution, prior-austenite grain size has been found to influ-

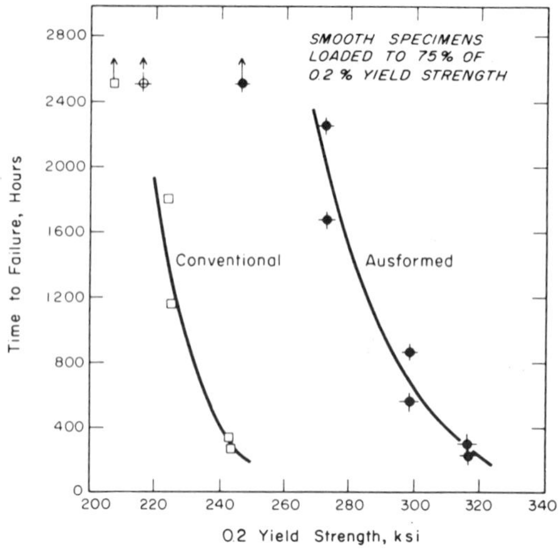

Fig. 31 Effect of prior ausforming process on delayed failure of D6ac steel in distilled water. Source: Ref 39

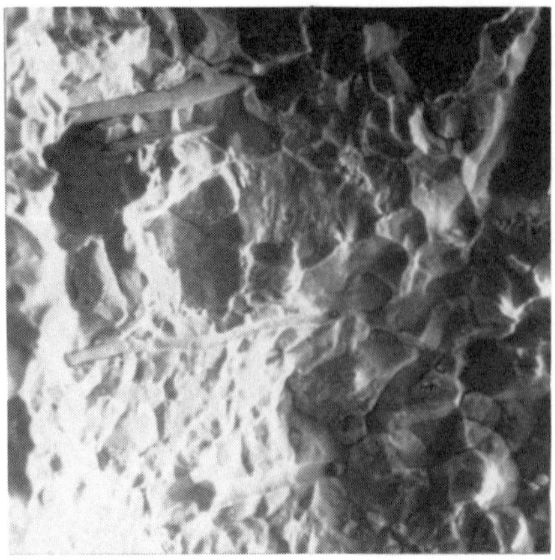

Fig. 32 Scanning electron micrograph showing typical fracture in AISI 4340 steel. Grain size, 20 μm

ence time to failure (Ref 37). Finer microstructure was found to be beneficial (see Fig. 27). For quenched and tempered steels, it appears that prior-austenite grain structure that is as fine as possible, along with tempering temperature that is as high as possible for the required level of strength, are desirable for minimizing susceptibility.

Likewise, for maraging steels, a lower annealing temperature, which produces a finer prior-austenite grain size, should be expected to improve SCC resistance, at least in terms of crack velocity. Experimental evidence suggests that this may be the case, but is inconsistent. The inconsistencies are probably caused by residual-element variation (Ref 8).

Different combinations of annealing and aging treatments that achieve the same level of yield strength for a 18Ni-300 maraging steel produced no significant change in K_{ISCC} (Ref 38). However, crack velocity increased when aging temperature was lowered (Fig. 28). When overaged to a yield strength of 1720 MPa (250 ksi), the 18Ni-300 steel exhibited crack velocities comparable to a 18Ni-250 grade aged at 480 °C (900 °F). K_{ISCC} was significantly lower for the overaged material, with intergranular cracking observed irrespective of aging treatment (Ref 26).

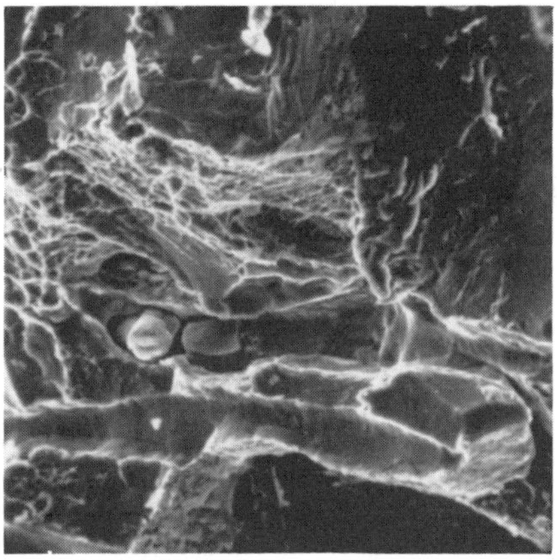

Fig. 33 Scanning electron micrograph of fracture area near large sulfide inclusions for AISI 4340 steel. Grain size, 90 μm

Figure 29 illustrates the effect of aging temperature on K_{ISCC}, K_{Ic}, and yield strength for an 18Ni-270 grade (Ref 8). Cracking was transgranular for the normal aging temperature, while

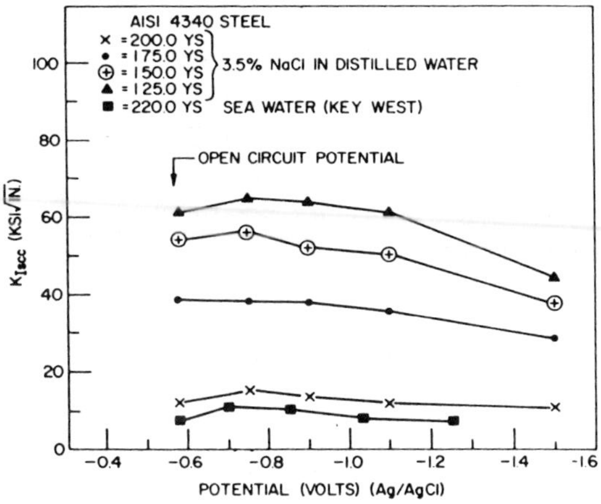

Fig. 34 Effect of impressed potential on cracking threshold stress intensity of AISI 4340 steel. Source: Ref 7

underaging resulted in a change to intergranular cracking, although no intergranular phase could be found. K_{Ic} varies consistently with yield strength, while K_{ISCC} increases with aging temperature. Taking Fig. 28 and 29 together, it is apparent that the higher the aging treatment, the lower the SCC susceptibility in terms of both K_{ISCC} and crack velocity (Ref 8).

Based on smooth-specimen data, cold working of 18Ni maraging steels prior to aging decreases susceptibility. This improvement is shown in Fig. 30, which also shows the ranking of various types and treatments of steels in terms of mean failure time. The strong dependency on yield strength is also evident. Welding prior to aging produces properties comparable to those of nonmaraging grades (Ref 8).

Ausforming and light tempering result in a very fine martensite plate structure, a finer car-

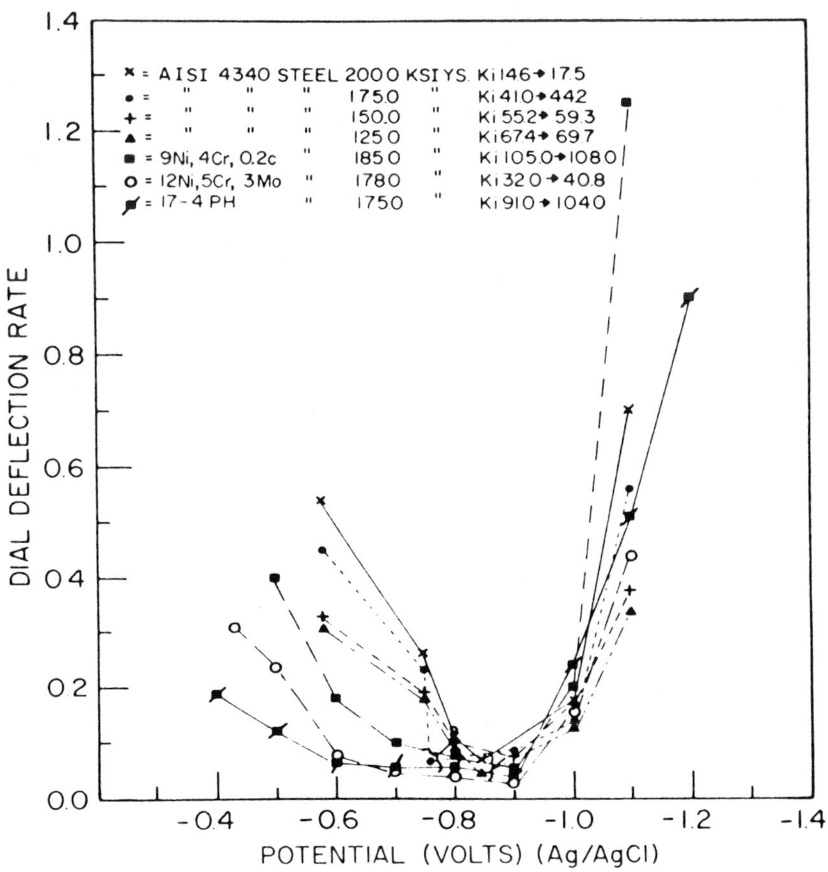

Fig. 35 Effect of applied potential on crack-growth rates in an aqueous chloride environment. Source: Ref 7

bide distribution, and an increased dislocation density compared with quenching and tempering (Ref 1). This heat treatment markedly improves SCC resistance, as Fig. 31 illustrates for a D6ac steel in distilled water (Ref 39).

Cold working of fine pearlite steels for use as high-strength bridge cable is necessary to achieve a useful service life. This fact has been attributed to the shifting of the critical potential to a potential more noble than the corrosion potential (Ref 40) and has been used as an argument against hydrogen embrittlement. However, cold work would be expected to make concentration of hydrogen at the crack tip more difficult by dislocation/hydrogen interaction.

Considering this along with the refined microstructures and/or increased dislocation densities associated with Fig. 30 and 31, it appears that significant gains may be made by providing more trap sites for hydrogen. While care must be taken to avoid interfaces that embrittle (twins, for example), other interfaces such as small carbides, grain boundaries and dislocations may in fact enhance resistance by trapping hydrogen so as to minimize larger concentrations at crack-initiation sites.

Illustrating the importance of microstructural features, Fig. 32 is part of the fracture surface of AISI 4340 steel subjected to hydrogen embrittlement. Fracture is clearly intergranular. Note the presence of large sulfide stringers, providing a fracture interface. The role of smaller inclusions can be seen in Fig. 33 for the same steel, heat treated to give a larger grain size. Fracture is predominantly intergranular, with dimpled areas that resulted from microvoid coalescence.

Effects of Electrochemical Potential and pH

For quenched and tempered steels, impressed cathodic potentials have been shown to have only a minor effect on K_{ISCC} (Ref 7). Test data for AISI 4340 steel in seawater and 3.5% NaCl solution are presented in Fig. 34. The familiar influence of a lower yield strength resulting in an improved K_{ISCC} is clearly apparent.

Impressed cathodic potential has a strong influence on crack velocity once K_{ISCC} has been exceeded. Figure 35 illustrates this for AISI 4340 steel of varying yield strengths, HP 9-4-20, and a 12Ni-5Cr-3Mo maraging steel (Ref 7). The potential corresponding to minimum crack-growth rate is about –0.8 V (Ag/AgCl).

The effect of pH on K_{ISCC} for quenched and tempered steels is not great except at extremes, *i.e.*, outside a range of 1 to 9. However, the stress required to fail-bend specimens in 4 weeks does have a pH dependency (Fig. 36) (Ref 41). At high pH, the prevention of pitting is believed to halt SCC (Ref 7).

It has been shown that the pH and potential conditions at the propagating crack tip promote local hydrogen reduction and hydrogen embrittlement. The effect of increasing yield strength is to provide a sharper crack, owing to reduced crack plasticity. Thus, stress relaxation due to yielding is reduced, and the dilational stresses necessary to concentrate hydrogen are higher. Therefore, a decrease in K_{ISCC} with increasing yield strength and a largely flat curve with change in potential are arguably consistent with hydrogen being the embrittling species. However, the minimum in crack velocity exhibited in Fig. 35 suggests that the supply of hydrogen at this potential has also been minimized. This behavior has been explained by relating the minimum in the rate curve to the transition from the region of corrosion to the region of immunity, as shown in Fig. 4 and 5 (Ref 7).

For maraging steels, applied potential appears to have a much greater influence on K_{ISCC} than for quenched and tempered steels (Fig. 37) (Ref 8). For cathodic overpotentials greater than –0.6 V, all strength levels are susceptible. Figure 37 was compiled from studies using both smooth and precracked specimens. The shaded areas indicate the range of applied potential in which SCC resistance is equivalent or better than for open-circuit conditions. Anodic potentials are seen to be harmful, as are cathodic potentials beyond the shaded areas. Safe potential ranges under cathodic conditions would be expected to be narrower than those presented for existing precracks.

With regard to pH, a minimum time to failure for foil specimens has been found to be within a range of 9 to 13 (see Fig. 38) for freely corroding

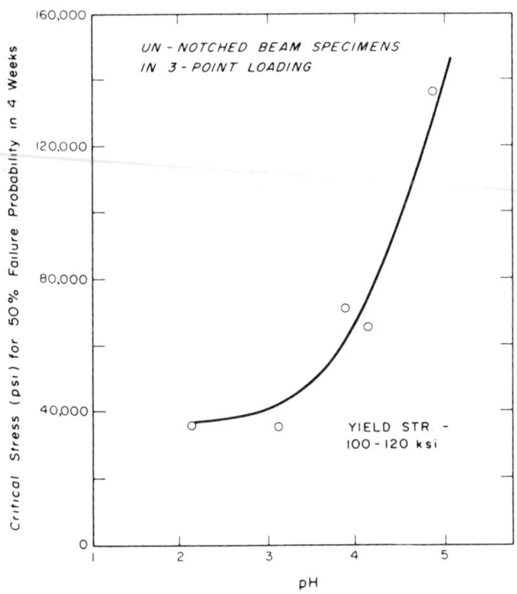

Fig. 36 Effect of pH on delayed failure stress for low-alloy steel in H_2S saturated acetic acid solution containing sodium acetate buffer. Source: Ref 41

conditions in various chloride solutions (Ref 8, 42). It is interesting to observe that chloride-ion content alters the pH value for minimum time to failure in the absence of precracks. This behavior is probably caused by the influence of pitting tendency on crack initiation (Ref 8). The fact that maraging steels exhibit behavior similar to that of other types of high-strength steel for precracked specimens is supported by Fig. 3, where the local crack-tip pH for the 18Ni maraging steel is 3.5, in common with the other steels shown. In addition, work on 18Ni-300 material has shown no systematic effect of pH on K_{ISCC} or crack velocity for a pH range of 0.5 to 11 (Ref 43). Consistent with quenched and tempered steels, no consistent differences were found for K_{ISCC} or crack velocity in distilled water or 3% NaCl solution.

SCC Behavior in Specific Environments

Generally, high-yield-strength steels are susceptible to SCC in most aqueous environments (Ref 4, 7). Figures 39 and 40 demonstrate that moisture (represented by moist argon) is alone capable of producing cracking in H-11 steel at the 1580 MPa (230 ksi) yield-strength level (Ref 44). Moisture is believed to cause condensation at the crack tip.

The most harmful environments are those that contain cathodic poisons (Ref 4, 7). These prevent the association of atomic hydrogen to molecular hydrogen and thus increase the supply of atomic hydrogen. Elements in groups V and VI of the periodic table, such as arsenic, selenium, tellurium, and sulfur, all act as poisons. These elements are listed in order of decreasing potency.

The most common cathodic poison is probably sulfur in the form of H_2S, as found naturally in "sour" oil wells. For exposure to these environments, hardness in steels is normally limited to 22 HRC (Ref 45). Because high-yield-strength steels usually have hardnesses in excess of 40 HRC, their use is inappropriate. Rapid failure would be expected, as implied by Fig. 1 and indicated by the low K_{ISCC} value shown in Fig. 41 (Ref 7).

Much experimental work has been carried out using chloride-containing solutions, reflecting the ubiquitousness of chlorides and the potential harm to aerospace components that marine environments pose. Interestingly, K_{ISCC} for AISI 4340 steel with yield strengths of 1380 and 1720 MPa (200 and 250 ksi) is very similar regardless of chloride-ion content (Ref 7). There is little variation in K_{ISCC} between distilled water and aqueous solutions containing varying amounts of chlorides. Similar behavior is found for the higher-alloyed HP 9-4-45 steel (Ref 7). For 18Ni-300 maraging steel, K_{ISCC} is the same in distilled water, 3% NaCl solution, and H_2SO_4 (Ref 7).

Other aqueous solutions that may promote SCC include those containing sulfate, phosphate, or nitrate ions (Ref 7). High-yield-strength steels are also susceptible to organic compounds. This behavior has been attributed to the effect of impurities, similar to the response of titanium alloys to methanol. Figure 42 shows the relative susceptibility of AISI 4130 steel to salt water and methanol (Ref 7).

The general pattern of behavior appears to be that K_{ISCC} is consistent in aqueous solutions, provided that a poison is not introduced. Relative humidity should be treated as if it were an aqueous solution. K_{ISCC} for a maraging steel will be con-

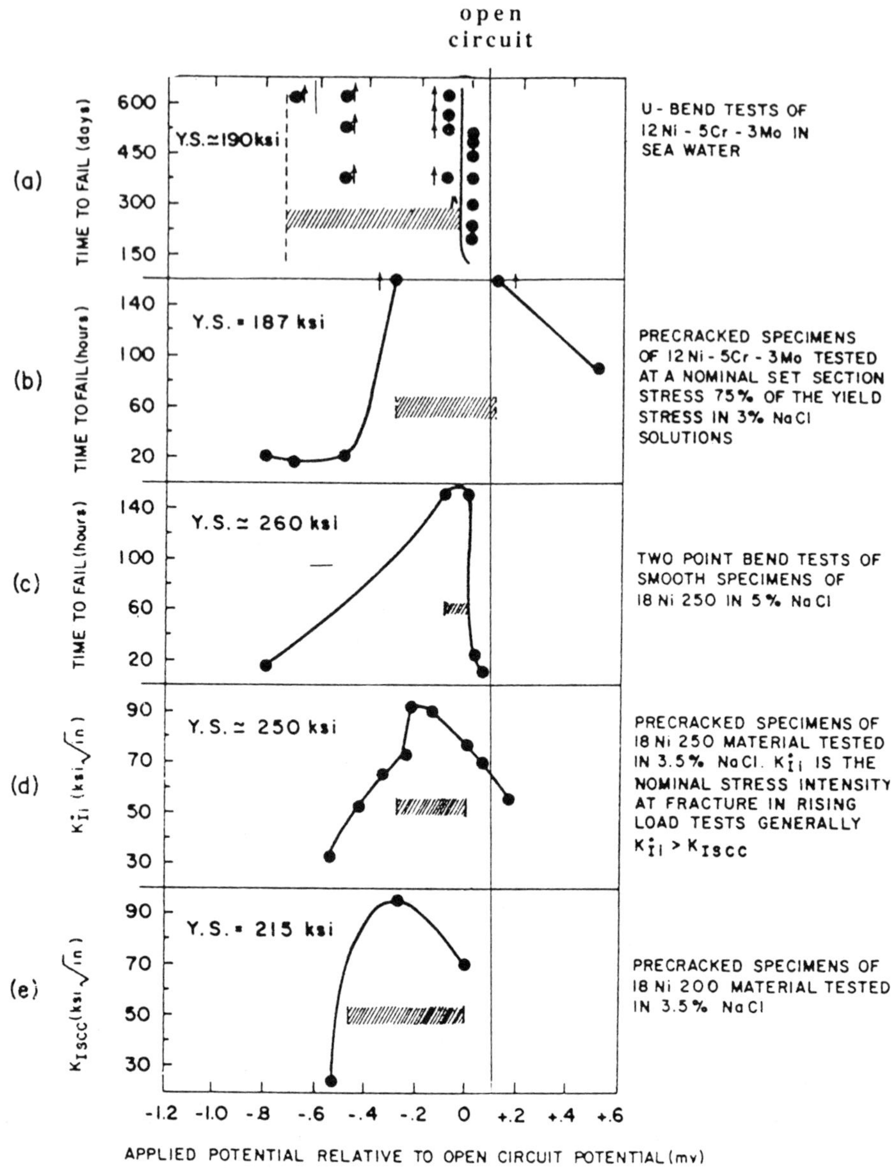

Fig. 37 Effect of applied potential on the SCC resistance of maraging steels. The cross-hatched regions represent potentials where the cracking resistance was as good or better than when freely corroding. Source: Ref 8

sistently higher than quenched and tempered steels at the same strength level. Additionally, any environment capable of producing hydrogen by chemical reaction, even from impurities, is suspect. Environments containing cathodic poisons, such as H_2S, promote rapid failure of high-yield-strength steels, even at low applied stress levels.

Effect of Temperature

Increases in temperature up to the boiling point of water will increase crack-growth rate. Earlier, for gaseous hydrogen environments, an inverted U-shape rate curve was found to be a common feature (see Fig. 9). It should be added that the temperature for maximum crack-growth rate is different for different materials. Addition-

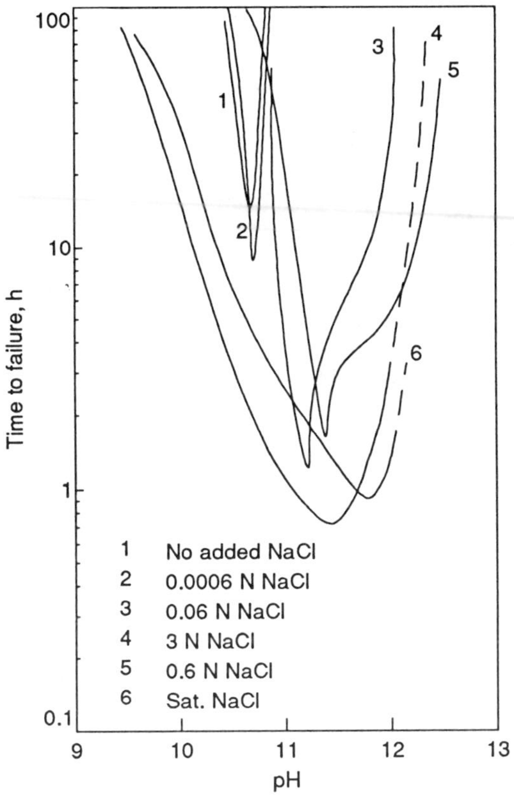

Fig. 38 Relation between time to failure of 18Ni maraging steel foils immersed in a series of environments, from deionized water saturated with NaCl to solutions containing no added NaCl, as a function of pH. Source: Ref 8

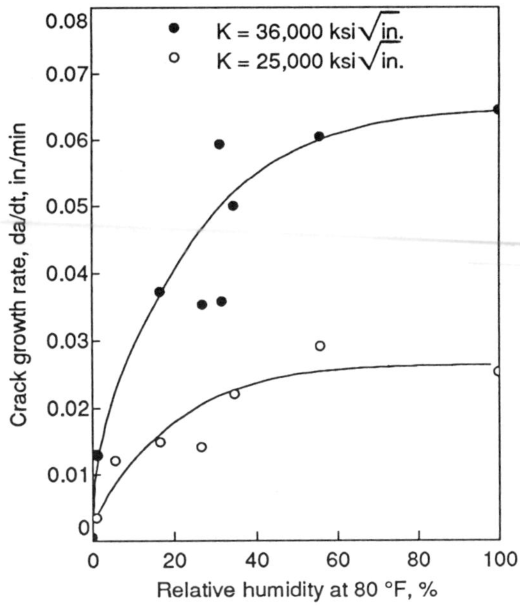

Fig. 39 Effect of relative humidity on crack velocity for H-11 steel with a yield strength of 1580 MPa (230 ksi) in moist argon environments. Source: Ref 44

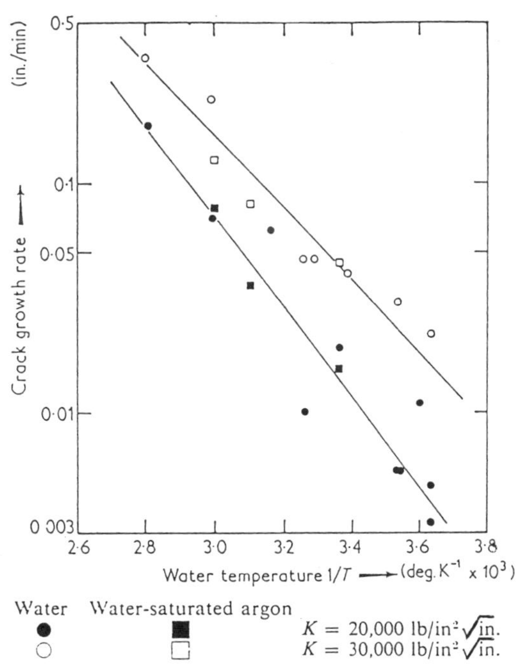

Fig. 40 Effects of temperature and stress intensity on crack-growth rate in water and water-saturated argon environments. Source: Ref 44

ally, crack velocity does not increase until a particular temperature is exceeded.

The rate-controlling step for crack growth at lower temperatures, *i.e.*, the Arrhenius region, has been attributed to diffusion of hydrogen into the bulk. From various temperature/crack-velocity curves for high-yield strength steels in aqueous environments, activation-energy values comparable to that of hydrogen in steel have been calculated (33 to 38 kJ/mol). In aqueous environments, a maximum crack-growth rate is not observed (Fig. 40) since the temperature is too low.

Figures 43 and 44 illustrate that threshold stress intensity for crack growth in gaseous hydrogen increases with an increase in temperature for both AISI 4130 and 4340 steels. Unfortunately, increasing temperature is not a useful means of avoiding SCC, as the maximum tem-

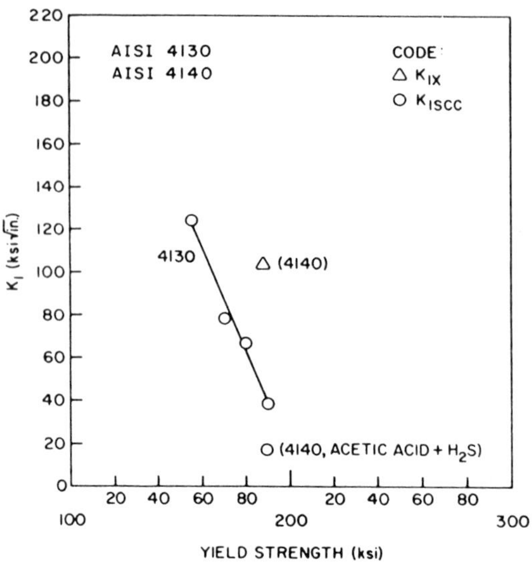

Fig. 41 Stress-corrosion resistance and fracture toughness of AISI 4130 and 4140 steels. Source: Ref 7

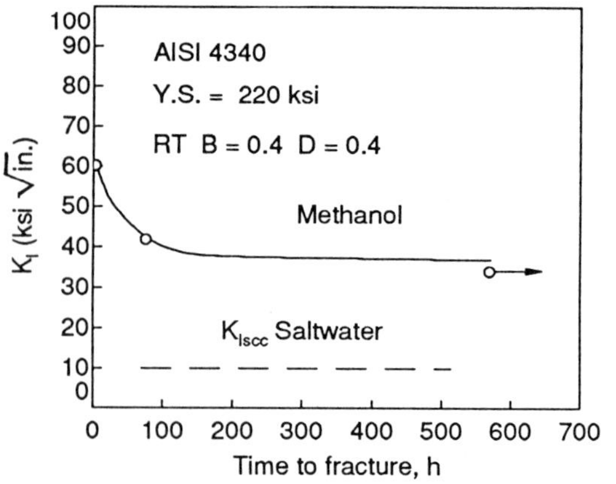

Fig. 42 Comparison of K_{ISCC} values in methanol and in salt water for AISI 4340 steel. Source: Ref 7

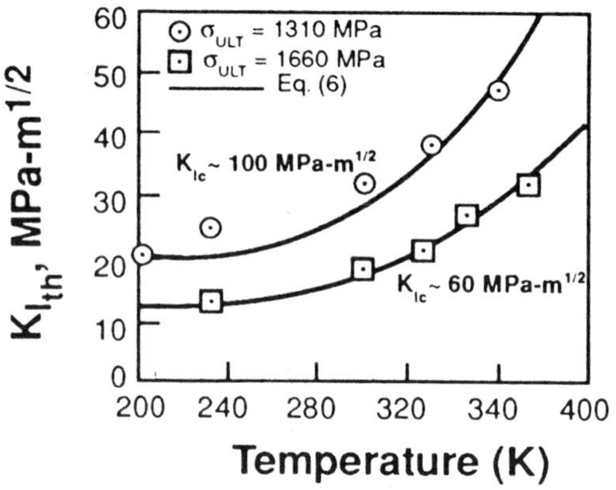

Fig. 43 Influence of temperature on threshold stress intensity for AISI 4130 steel in hydrogen. After Ref 46

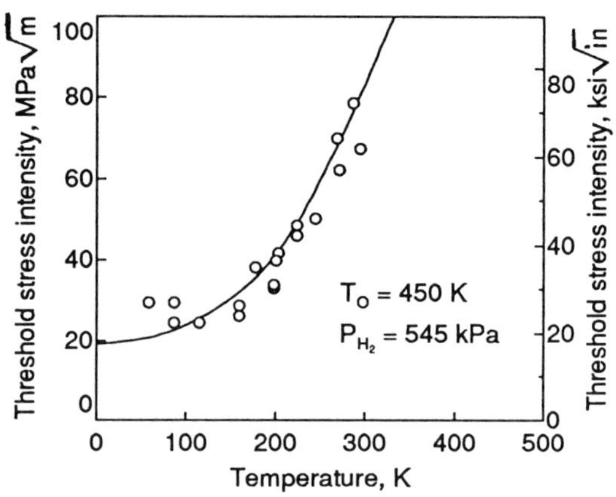

Fig. 44 Influence of temperature on threshold stress intensity for AISI 4340 steel in hydrogen. After Ref 46 (based on data of Ref 23 and 47)

perature for aqueous solutions is typically still in the Arrhenius-rate regime where crack velocity increases exponentially with temperature. That is, the kinetics illustrated in Fig. 9 dominate once the threshold is exceeded. A better strategy would be to reduce the temperature and/or applied stress if possible. Even if cracks are nucleated, lower crack velocity would better the chance of detection before unstable crack growth occurs.

Summary

High-yield-strength steels are inherently environmentally sensitive. Environments as apparently innocuous as moist air can promote cracking. It appears that there are limited options for avoiding SCC. Perhaps the most useful method for reducing the risk of SCC is to separate the material from the environment. When a choice of material is possible, maraging steels should give superior performance. Whichever type of steel is utilized, special attention should be paid to processing and heat treatment. Additionally, material should be specified with a yield strength not significantly greater than the minimum required.

Acknowledgment

The authors would like to acknowledge the support of the Office of Naval Research under Grant N00014-89-J-1726.

References

1. F.B. Pickering, *Physical Metallurgy and the Design of Steels*, Applied Science, 1978
2. T.V. Philip, Ultrahigh-Strength Steels, in *Metals Handbook*, Vol 1, 9th ed., *Properties and Selection: Irons, Steels, and High-Performance Alloys*, American Society for Metals, 1978
3. K. Rohrbach and M. Schmidt, Maraging Steels, in *Metals Handbook*, Vol 1, 10th ed., *Properties and Selection: Irons, Steels, and High-Performance Alloys*, ASM International, 1990
4. G.E. Kerns, M.T. Wang, and R.W. Staehle, Stress Corrosion Cracking and Hydrogen Embrittlement, in *Stress Corrosion Cracking and Hydrogen Embrittlement of Iron Base Alloys*, NACE-5, R.W. Staehle, J. Hochmann, R. McCright, and J. Slater, Ed., National Association of Corrosion Engineers, 1971, p 700-735
5. M. Pourbaix, *Atlas of Electrochemical Equilibria in Aqueous Solutions*, Pergamon Press, 1966
6. J.J. DeLuccia, Electrochemical Aspects of Hydrogen in Metals, in *Hydrogen Embrittlement: Prevention and Control*, STP 962, L. Raymond, Ed., American Society for Testing and Materials, 1988, p 17-34
7. B.F. Brown, Ed., *Stress Corrosion Cracking in High Strength Steels and in Titanium and Aluminum Alloys*, Naval Research Laboratory, 1972, p 80-145
8. D.P. Dautovich and S. Floreen, The Stress Corrosion and Hydrogen Embrittlement Behavior of Maraging Steels, in *Stress Corrosion Cracking and Hydrogen Embrittlement of Iron Base Alloys*, NACE-5, R.W. Staehle, J. Hochmann, R. McCright, and J. Slater, Ed., National Association of Corrosion Engineers, 1971, p 798-815
9. R.A. Oriani, Hydrogen Effects in High Strength Steels, in *Environment-Induced Cracking of Metals*, NACE-10, R.P. Gangloff and M.B. Ives, Ed., National Association of Corrosion Engineers, 1990, p 439-448
10. A.R. Troiano, *Trans. ASM*, Vol 52, 1960, p 54
11. R.A. Oriani, *Ber. Bunsenges. Phys. Chem.*, Vol 76, 1972, p 848
12. C.D. Beachem, *Metall. Trans.*, Vol 3, 1972, p 437
13. C.A. Zapffe and C.E. Sims, *Trans. AIME*, Vol 145, 1941, p 225
14. N.J. Petch and P. Stables, *Nature*, Vol 169, 1952, p 842
15. C.L. Fu and G.S. Painter, *J. Mater. Res.*, Vol 6 (No. 4), 1991, p 719-723
16. H.K. Birnbaum, in *Atomistics of Fracture*, R.M. Latanision and J.R. Pickens, Ed., Plenum Press, 1983, p 733
17. J.P. Hirth, Theories of Hydrogen Induced Cracking of Steels, in *Hydrogen Embrittlement and Stress Corrosion Cracking*, R. Gibala and R.F. Hehemann, Ed., American Society for Metals, 1984, p 29-59
18. W.W. Gerberich and S. Chen, Environment-Induced Cracking of Metals Fundamental Processes: Micromechanics, in *Environment-Induced Cracking of Metals*, NACE-10, R.P. Gangloff and M.B. Ives, Ed., National Association of Corrosion Engineers, 1990, p 167-187
19. G.G. Hancock and H.M. Johnson, *Trans. Met. Soc., AIME*, Vol 236, 1966, p 513
20. F.W. Frasar and E.A. Metzbower, in *Fractography and Materials Science*, STP 733, L.N. Gilbertson and R.D. Zipp, Ed., American Society for Testing and Materials, 1981, p 51
21. P.P. Wei and G. Shim, in *Corrosion Fatigue: Mechanics, Metallurgy, Electrochemistry and Engineering*, STP 801, T.W. Crookes and B.N. Leis, Ed., American Society for Testing and Materials, 1983, p 5
22. X. Chen and W.W. Gerberich, *Metall. Trans. A*, Vol 22, 1991, p 59-70
23. H.G. Nelson and D.P. Williams, Quantitative Observations of Hydrogen-Induced, Slow Crack Growth in a Low Alloy Steel, in *Stress Corrosion Cracking and Hydrogen Embrittlement of Iron Base Alloys*, NACE-5, R.W. Staehle, J. Hochmann, R. McCright, and J. Slater, Ed., National Association of Corrosion Engineers, 1971, p 390-404
24. R.P. Gangloff and R.P. Wei, *Metall. Trans. A*, Vol 8, 1977, p 1043
25. R.F. McGuinn and M. Aballe, *Br. Corros. J.*, Vol 17, 1982, p 18
26. R. Magdowski and W.W. Gerberich, *Scr. Metall.*, Vol 17, 1983, p 421
27. M.R. Louthan *et al.*, in *Hydrogen Effects in Metals*, J.M. Bernstein and A.W. Thompson, Ed., TMS-AIME, 1981, p 829
28. W.W. Gerberich and Y.T. Chen, *Metall. Trans A.*, Vol 13, 1982, p 305
29. W.W. Gerberich, Y.T. Chen, and C. St. John, *Metall. Trans. A*, Vol 6, 1975, p 1485
30. B.F. Brown, *Stress Corrosion Cracking of High Strength Steels, the Theory of Stress Corrosion Cracking in Alloys*, NATO Science Committee Research Evaluation Conference, 1971, p 186
31. G. Sandoz, *Metall. Trans.*, Vol 3, 1972, p 1169
32. G. Sandoz, *Metall. Trans.*, Vol 2, 1971, p 1055

33. C.S. Carter, *Corrosion*, Vol 25 (No. 10), 1969, p 423
34. S.K. Banerji, C.J. McMahon, Jr., and H.C. Feng, *Metall. Trans. A*, Vol 9, 1978, p 237-247
35. P.P. Puzak and E.A. Lange, "Fracture Toughness Characteristics of the New Weldable Steels of 180-210 ksi Yield Strengths," NRL Report 6951, Naval Research Laboratory, 1969
36. M.T. Wang and R.W. Staehle, in *Hydrogen in Metals*, Pergamon Press, 1972, p 342
37. R.P.M. Proctor and H.W. Paxton, *Trans. ASM*, Vol 62, 1969, p 989
38. C.S. Carter, *Metall. Trans.*, Vol 2, 1971, p 1621
39. R.T. Ault *et al.*, *Trans. ASM*, Vol 60, 1967, p 79
40. H.V. Uhlig and P.W. Rovie, *Corrosion and Corrosion Control*, Wiley, 1985
41. R.S. Treseder and T.M. Swanson, *Corrosion*, Vol 24, 1968, p 31
42. J.A.S. Green and E.G. Haney, *Corrosion*, Vol 23, 1967, p 5
43. A.J. Stavros and H.W. Paxton, *Metall. Trans.*, Vol 1, 1970, p 3049
44. H.H. Johnson and M. Willner, *Appl. Mater. Res.*, Vol 4, 1965, p 34
45. "Sulfide Stress Cracking Resistant Metallic Material for Oil Field Equipment," NACE Standard MR-01-75, National Association of Corrosion Engineers, 1980
46. W.W. Gerberich and T.J. Foecke, in *Hydrogen Effects on Material Behavior*, N.R. Moody and A.W. Thompson, Ed., Minerals, Metals, and Materials Society, 1990, p 687-702
47. W.G. Clark, *J. Mater. Energy Syst.*, Vol 1, 1979, p 33

CHAPTER 4

Stress-Corrosion Cracking of Stainless Steels

A. John Sedriks, Office of Naval Research

Stainless steels are iron alloys that contain a minimum of approximately 11% Cr, the amount needed to prevent the formation of rust in unpolluted atmospheres (hence the designation "stainless"). Today there are more than 180 different alloys that belong to the stainless steel group. The chromium content of some steels now approaches 30%, and many other alloying elements are added to provide specific properties or ease of fabrication (Ref 1). Figure 1 provides a useful summary of some of the compositional and property linkages in the stainless steel family. Because the susceptibility to stress-corrosion cracking (SCC) of stainless steels depends on alloy composition, structure, and thermal history (Ref 1), the alloys shown in Fig. 1 exhibit widely varying susceptibilities in different environments. For example, the austenitic (300 series) stainless steels can be susceptible to SCC in certain chloride environments, whereas the ferritic and martensitic grades (400 series) can be susceptible to hydrogen embrittlement. However, there is still lack of agreement about the mechanisms of cracking in many stainless steel/environment systems (Ref 2). In fact, the multiplicity of proposed mechanisms and the disagreement as to which mechanisms are operative (Ref 2), have led to the evolution of a simplistic nomenclature that describes SCC in terms of the environments that cause it. The terms "chloride SCC," "caustic SCC," and "polythionic acid SCC" are often used to describe SCC of stainless steels in those types of environments. The term "high-temperature water SCC" usually denotes SCC in simulated boiling water reactor (BWR) and pressurized water reactor (PWR) coolant environments. The term "irradiation-assisted SCC" refers to SCC in nuclear power station core components that are subjected to heavy doses of radiation within the core as well as to coolant water. The term "sulfide SCC" usually denotes hydrogen embrittlement, with the sulfide ion acting as a hydrogen ion recombination poison, promoting entry of hydrogen into the metal.

The extent to which these variously described forms of SCC are manifestations of the same process remains a matter of continuing debate that is not likely to be settled in the near future. Accordingly, in this chapter the various forms of SCC are discussed under the environment-related descriptors noted above. When considering the SCC of stainless steels, it is also useful to distinguish between cracking in sensitized and nonsensitized materials. Sensitization is particularly detrimental to the stress-corrosion resistance of austenitic stainless steels in a variety of environments, and in many cases the elimination of sensitization will eliminate SCC susceptibility. In view of the importance of sensitization to SCC, it is worthwhile to describe it here in some detail.

Sensitization

It is evident from Fig. 1 that in the development of the 300 series, considerable effort has gone into producing grades that resist sensitization. It is now widely accepted that this phenomenon is related to the precipitation of carbide at austenite grain boundaries. To understand this phenomenon in terms of microstructure, it is in-

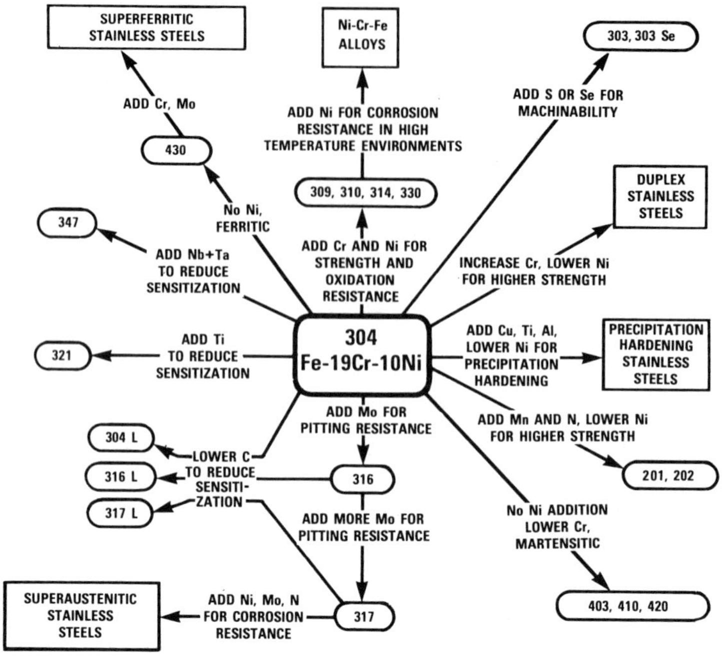

Fig. 1 Compositional and property linkages in the stainless steel family of alloys

structive to examine the equilibrium relationships and carbon solubility in the Fe-18Cr-8Ni alloy, illustrated in Fig. 2 (Ref 3). This figure shows that in alloys containing between about 0.03 and 0.7% C, the equilibrium structure should contain austenite, α-ferrite, and carbide ($M_{23}C_6$) at room temperature. In commercial alloys containing various austenite stabilizers, the reaction $\gamma + M_{23}C_6 \rightarrow \gamma + \alpha + M_{23}C_6$ (at line SK) is too sluggish to take place at practical rates of cooling from elevated temperatures. The same applies to the reaction $\gamma \rightarrow \alpha + M_{23}C_6$ at carbon contents below approximately 0.03%.

For commercial-purity materials, the transformation of austenite to α-ferrite is ignored in practice, and in considering carbon solubility in austenite, the simplified diagram (Ref 4) shown in Fig. 3 is often regarded as representative of real (*i.e.*, nonequilibrium) situations. In terms of this simplified diagram, up to 0.03% C should be soluble in austenite at temperatures up to 800 °C (1470 °F). Austenite that contains more than 0.03% C should precipitate $M_{23}C_6$ on cooling below the solubility line. However, at relatively rapid rates of cooling, this reaction is partially suppressed. This is the case in practice when type 304 stainless steel containing more than 0.03% C is heat treated at 1050 °C (1920 °F) to remove effects of cold working or hot working and cooled at a fairly rapid rate to room temperature. While some carbide may have precipitated on cooling, the room-temperature austenite is still largely supersaturated with respect to carbon.

If this supersaturated austenite is reheated to elevated temperatures within the $\gamma + M_{23}C_6$ field, further precipitation of the chromium-rich $M_{23}C_6$ will take place at the austenite grain boundaries. Certain time-temperature combinations will be sufficient to precipitate this chromium-rich carbide, but insufficient to rediffuse chromium back into the austenite near the carbide. This will result in the formation of envelopes of chromium-depleted austenite around the carbide (Ref 5, 6). Because the carbides precipitate along grain boundaries, the linking of the chromium-depleted envelopes provides a continuous path of lower corrosion resistance along the grain boundaries for the propagation of SCC. This type of structure is known as "sensitized," irrespective of whether the chromium depletion has been caused by slow

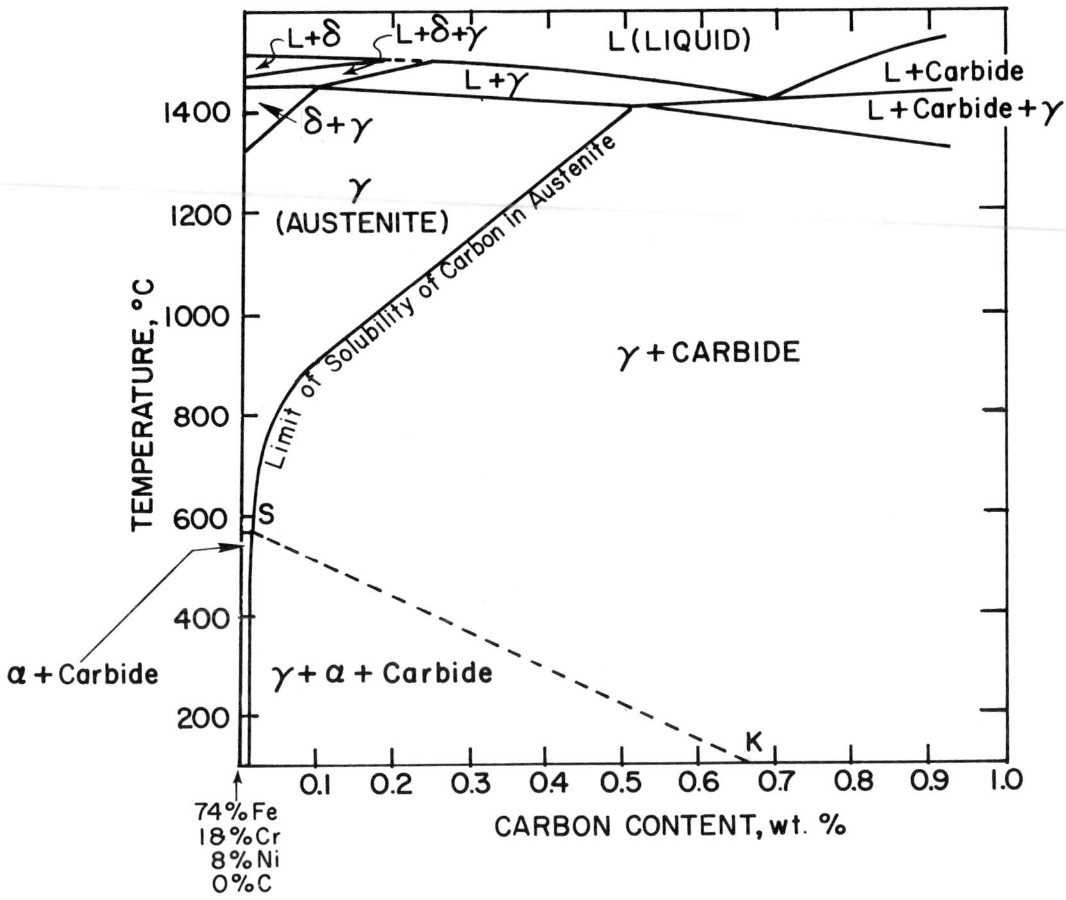

Fig. 2 Pseudo-binary phase diagram for an Fe-18Cr-8Ni alloy with varying carbon content. After Ref 3

cooling, heat treatment, elevated-temperature service, or welding. Sensitization also occurs in ferritic and martensitic stainless steels (Ref 1). The metallurgical remedies for reducing sensitization in austenitic stainless steels include (1) the use of low-carbon (0.03% maximum) grades of stainless steel (for example, types 304L, 316L, and 317L), (2) postweld heat treatment to rediffuse chromium back into the impoverished austenite, and (3) addition of titanium (type 321) or niobium plus tantalum (type 347) to precipitate the carbide at higher temperatures so that little carbon is left to precipitate as the chromium-rich grain-boundary carbide during cooling. Types 321 and 347 are sometimes given a stabilizing treatment at 900 to 925 °C (1650 to 1695 °F) to ensure maximum precipitation of carbon as titanium or niobium carbide. All of these remedies have certain advantages and disadvantages. Thus, postweld heat treatment is not always practicable in large structures, the stabilized grades can suffer another form of corrosive attack known as knife-line attack (Ref 1), and the low-carbon grades have lower strength.

To compensate for the lower strength of the low-carbon (L) grades, the nuclear power industry has supported the development of nitrogen-strengthened nuclear grade (NG) stainless steels (Ref 7). These are designated as types 304NG and 316NG and have the low carbon levels of the L-grades (that is, less than 0.03 wt% C), but contain added nitrogen at levels of $0.06 < N < 0.1$ wt% to increase strength. Sensitized types 304NG and 316NG have exhibited resistance to intergranular SCC when tested in high-purity water at elevated temperatures (Ref 7). However, other studies

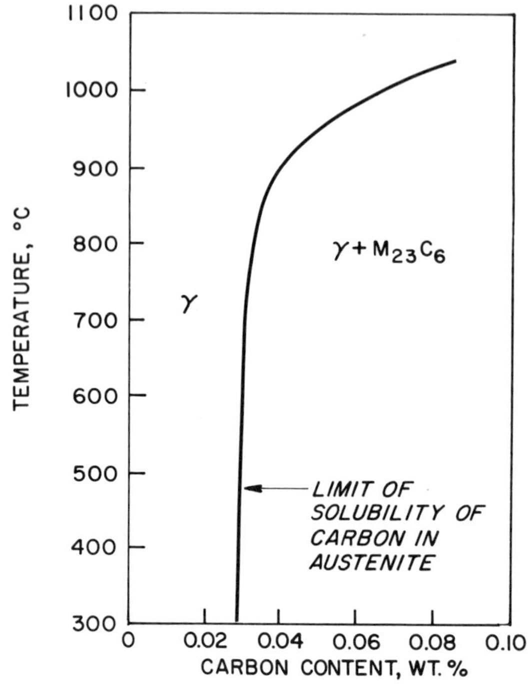

Fig. 3 Solid solubility of carbon in an Fe-18Cr-8Ni alloy. After Ref 4

(Ref 8) have shown that sensitized type 316NG exhibits transgranular SCC when tested in simulated BWR environments that contain sulfate ions as an impurity.

Other recent sensitization studies related to the nuclear power industry suggest that at BWR operating temperatures of about 288 °C (550 °F), radiation itself can cause chromium depletion at the grain boundaries in type 304 stainless steel and nickel-base alloys (Ref 9). In this radiation-induced phenomenon, described in greater detail in Chapter 6, there is no precipitation of chromium carbide and the chromium depletion is thought to occur by a process known as radiation-induced segregation. During the irradiation of an alloy, some constituents of the alloy migrate toward point defect sinks, such as grain boundaries or dislocation lines, and other constituents migrate away. This nonequilibrium segregation during irradiation was predicted in 1971 (Ref 10) and has subsequently received considerable attention. In austenitic steels, this segregation causes a depletion in the chromium levels and an enhancement in the nickel levels near grain boundaries during irradiation. Minor alloying elements are also redistributed; for example, silicon and phosphorus migrate toward the grain boundaries (Ref 9). These compositional changes occur in narrow regions close to grain boundaries. For example, the chromium depletion profiles suggest that the depletion widths are less than 20 nm (Ref 11). Cracking associated with this radiation-induced segregation is known as irradiation-assisted stress-corrosion cracking (IASCC) (Ref 12, 13) and has been shown to be dependent on temperature and radiation fluence. IASCC of solutionized stainless steel is reported (Ref 9) to occur in service at fluences above 0.3×10^{21} n/cm^2 and at temperatures above 150 °C (300 °F) (Ref 14). Maximum radiation-induced segregation is reported to occur in the temperature range of 400 to 500 °C (750 to 930 °F) (Ref 12).

Chloride SCC

Chloride SCC in stainless steels was first widely studied using austenitic stainless steels in boiling magnesium chloride solutions. Nickel-free ferritic stainless steels are highly resistant to SCC in boiling magnesium chloride and sodium chloride solutions. Hence, a misleading impression has emerged that all ferritic stainless steels are immune to all chloride environments. As will be shown in later discussion, there is no immunity associated with the body-centered cubic (bcc) ferrite lattice, and, as in the case of austenitics, susceptibility of ferritic and duplex stainless steels to chloride SCC depends on alloy composition and structure and on environmental parameters. Also, the danger of chloride SCC in austenitic stainless steels should not be minimized; it is reassuring to see the supplier literature (*e.g.*, Ref 15) using the caution "may be susceptible to chloride stress-corrosion cracking" when describing all variants of the austenitic types 304, 316, 317, 321, 347, and 348 and the austenitic high-manganese/high-nitrogen grades UNS S21904 and S20910.

Austenitic Stainless Steels

The Magnesium Chloride Test. Because chloride SCC was first noted in the more widely used austenitic stainless steels, these materials have received the greatest amount of study. Most early evaluations employed boiling magnesium chloride solutions, which are very severe environments. Nevertheless, the magnesium chloride test has continued to be used because of its severity and the belief among end-users that stainless alloys that can survive lengthy periods of exposure to concentrated boiling magnesium chloride are not likely to exhibit chloride cracking in service. It is also used by researchers because it is simple and rapid, and allows comparison among tests performed in different laboratories.

The boiling magnesium chloride test was first described in 1945 (Ref 16). It has now been standardized as ASTM G 36. The boiling point of magnesium chloride solution is strongly dependent on concentration, as shown in Fig. 4 (Ref 17), and ASTM G 36 recommends the use of a test solution that boils at 155 °C ± 1 °C. It is important to make alloy comparisons under the same conditions, since significant differences in the relative susceptibilities of types 304 and 316 are found at different temperatures or concentrations (Ref 18).

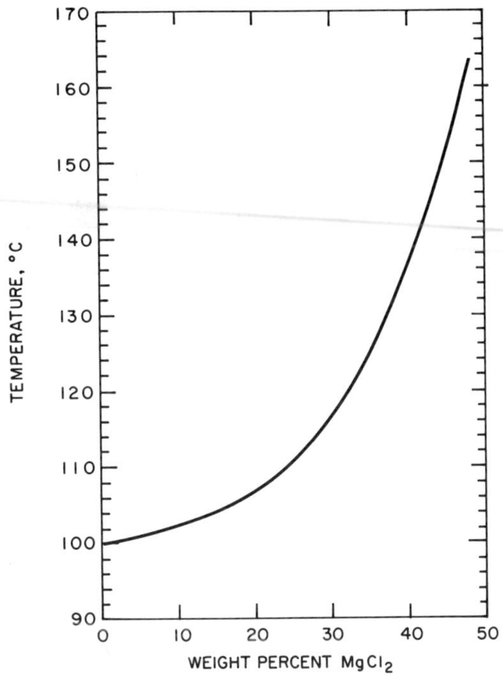

Fig. 4 Boiling points of aqueous magnesium chloride solutions at 1 atm as a function of concentration. After Ref 17

Other studies (Ref 19) using a 40% $MgCl_2$ solution have shown that keeping the solution at or near the boiling point for long periods of time causes it to become more alkaline. This increased alkalinity is reported to cause changes in the chloride SCC susceptibility of type 304, with different behaviors observed in freshly prepared and in aged (preboiled) solutions (Ref 19). The increased alkalinity reduces the transgranular crack-propagation rate and changes the potential dependence of the cracking process (Ref 19).

ASTM G 36 notes that any type of stress-corrosion test specimen can be used in the magnesium chloride test. For a comprehensive discussion of the various types of test specimens, consult Ref 20. One of the most popular is the U-bend specimen (described in ASTM G 30), which contains large elastic and plastic strains and provides one of the most severe configurations available for smooth (as opposed to notched or fatigue precracked) specimens. It is also simple and relatively inexpensive to make and use.

Effects of Composition. As noted before, there has been extensive research, often employing boiling magnesium chloride environments, on the effects of alloying additions on the chloride SCC resistance of stainless steels. The results, summarized in Fig. 5, have been taken largely from Ref 21 and 22. It is evident from Fig. 5 that many alloying additions appear to be detrimental to chloride SCC, but there are also those that can be categorized as "beneficial" and "variable." Those designated beneficial in Fig. 5 are nickel, cadmium, zinc, silicon, beryllium, and copper. It should be emphasized that the beneficial effect of nickel relates only to austenitic stainless steels. When present in relatively small quantities in ferritic stainless steels, nickel is detrimental to chloride SCC resistance. The beneficial effect of nickel on chloride SCC resistance of austenitic stainless steels and higher alloys has been extensively studied and is well documented (Ref 23-26).

Figure 6 (Ref 23) shows the effect of nickel content on the susceptibility to chloride SCC of stainless steel wire specimens tested in boiling magnesium chloride. The data of Fig. 6 suggest

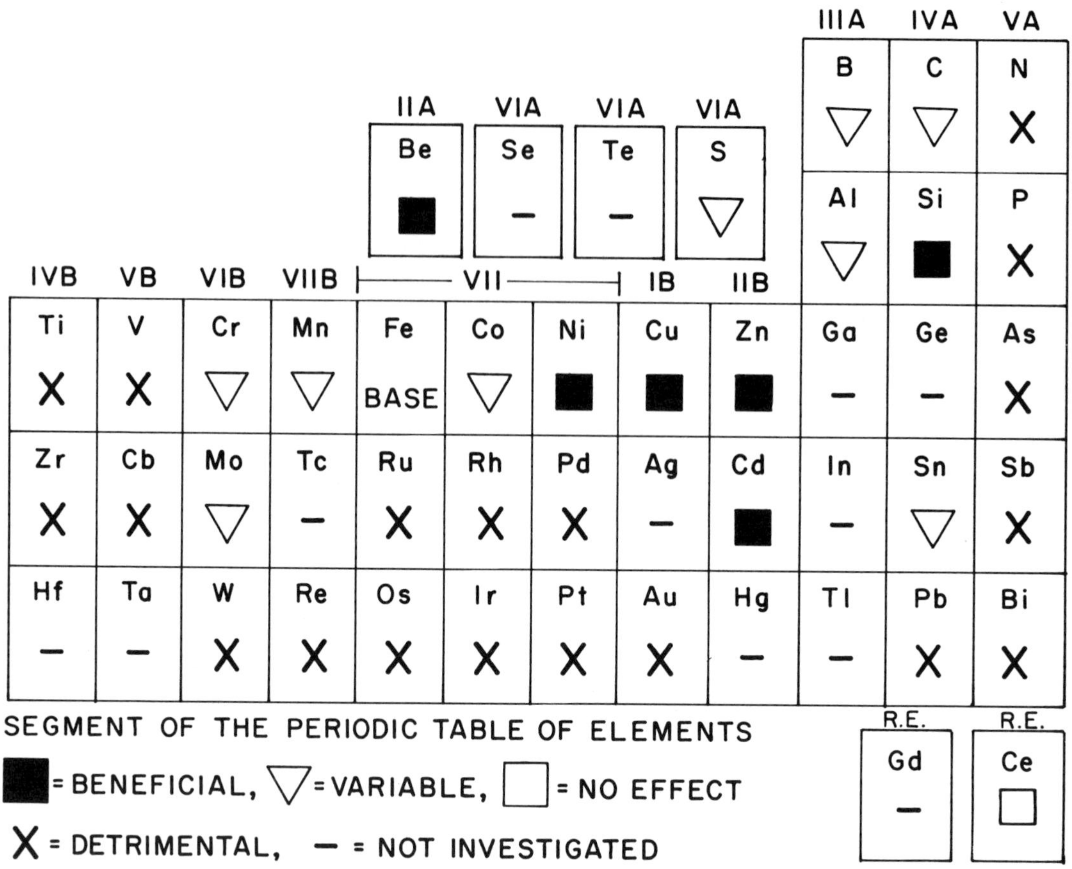

Fig. 5 Effect of various elements on resistance of austenitic stainless steels to SCC in chloride solutions

that very high nickel contents (about 50%) would be required for resistance to chloride SCC. This is at variance with general industrial experience, which is that chloride SCC is not usually observed at nickel contents at or above 42%. In fact, alloy 825 (42% Ni), Alloy G (44% Ni), and 20Cb3 (33% Ni) are often specified for applications requiring resistance to chloride SCC. Recent evaluations (Ref 27) using precracked fracture-mechanics specimens support the view that magnesium chloride is a much more severe environment than sodium chloride (Fig. 7), and suggest that the nickel level required for lack of susceptibility to SCC in sodium chloride may be between 30 and 40% (Fig. 8). In these tests, absence of crack propagation at a stress intensity of 60 MPa$\sqrt{m}$ (55 ksi$\sqrt{in.}$) is equated with resistance to SCC.

Additions of cadmium and zinc are also designated as beneficial (Fig. 5). Actually, there is very little information about the effects of these alloying elements. A 0.2% Cd addition to type 304 inhibited cracking in a vapor test that readily cracked type 304 (Ref 28), and a 1% Zn addition to an Fe-20Cr-15Ni alloy led to a fivefold improvement in SCC resistance (Ref 29). Therefore, the designation of these elements as beneficial must be regarded as tentative.

Alloying with silicon has been shown to be beneficial to chloride SCC resistance in boiling 42% $MgCl_2$ (Ref 30). However, other studies (Ref 31), summarized in Table 1, have shown that the beneficial effect of silicon noted in magnesium chloride tests is barely detectable, if at all, in sodium chloride solutions at higher temperatures. Beryllium and copper alloying additions have

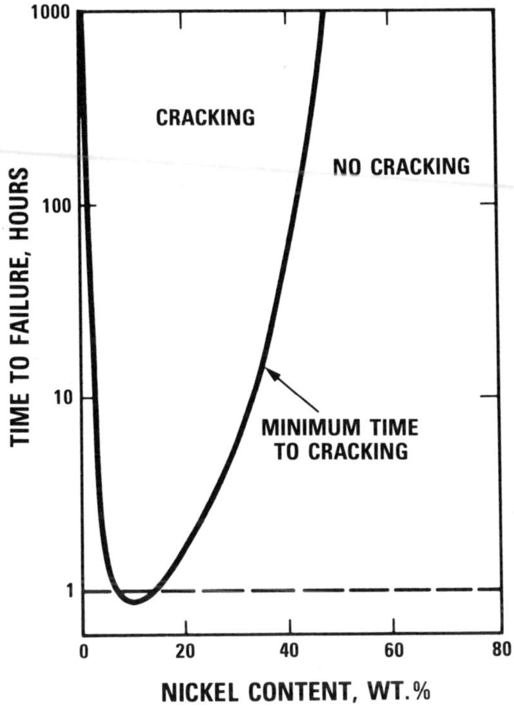

Fig. 6 Effect of nickel content on SCC susceptibility of stainless steel wires containing 18 to 20% Cr in a magnesium chloride solution boiling at 154 °C (309 °F). After Ref 23

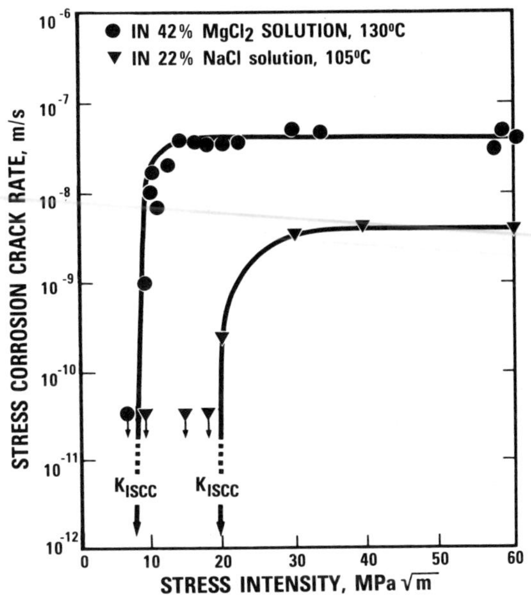

Fig. 7 Effect of stress intensity on the growth rate of stress corrosion cracks in type 304L stainless steel exposed to magnesium chloride and sodium chloride solutions. After Ref 27

been shown to be beneficial (Ref 32), although the effect of copper is slight.

The elements designated in Fig. 5 as having a variable effect on chloride SCC can be divided into the three categories shown in Table 2. Boron, aluminum, and cobalt appear to be detrimental in small quantities but beneficial in larger amounts (Ref 29, 32-36); tin and manganese appear to have no effect in certain ranges and beneficial or detrimental effects in other ranges (Ref 21, 36); carbon and chromium show minimal effect on chloride SCC resistance (Ref 21, 37). The effect of molybdenum appears to depend on the test environment. The magnesium chloride data suggest that molybdenum additions first decrease and then increase resistance to cracking, with a minimum resistance at about 1.5% Mo (Ref 37). No such pattern is in evidence in sodium chloride tests (Fig. 9) (Ref 27), which show cracking resistance increasing with increasing molybdenum content.

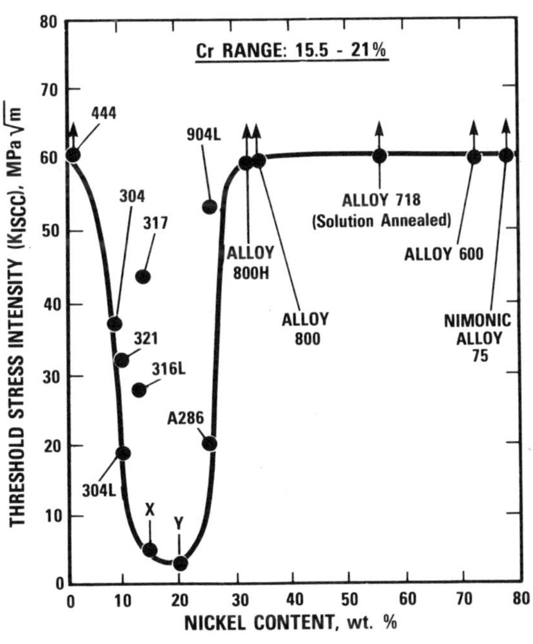

Fig. 8 Effect of nickel content on the stress-corrosion threshold stress intensity of various alloys in an aerated aqueous 22% NaCl solution at 105 °C (220 °F). Alloys X and Y are German heat-resistant grades. After Ref 27

Table 1 Effect of Silicon on SCC Resistance of Austenitic Stainless Steels in Various Chloride Media

Analysis, wt% (a)							Time to failure, h		
C	Si	Mn	Cr	Ni	N	Ferrite content, %	42% $MgCl_2$ at boiling point(b)	100 ppm NaCl at 250 °C (480 °F)	1000 ppm NaCl at 250 °C (480 °F)
0.020	0.46	0.83	18.28	15.10	0.07	0.5	219	28	10
0.022	0.80	0.77	18.54	15.20	0.08	0.5	292	24	5
0.031	2.48	0.83	18.58	15.23	0.07	1	1000 NF	92	32
0.031	3.69	0.88	18.46	15.20	0.07	1	1000 NF	81	15
0.031	4.53	0.86	18.42	15.28	0.08	2.5	1000 NF	54	50

(a) Balance iron. (b) 1000 NF, no failure in 1000 h
After Ref 31

Table 2 Effects of Various Elements on Resistance of Austenitic Stainless Steels to Chloride Cracking

Element	Beneficial	Detrimental
Boron	>0.1%	0.01%
Aluminum	0.1%	0.04%
Cobalt	>1.8%	1.50%
Tin	0.001–0.02%	No effect at 0.4%
Manganese	No effect at 0–2%	>2%
	Minimum in cracking resistance at:	
Carbon	0.06 – 0.1%	
Molybdenum	~1.5%	
Chromium	12 – 25%	

The foregoing survey of the effects of alloying additions on the chloride SCC resistance of austenitic stainless steels pertains to materials made by conventional melting practice, which contain many impurities. In this regard it has been shown that austenitic (as well as ferritic) stainless steels of very high purity are highly resistant to SCC in boiling magnesium chloride. Two examples are Fe-16Cr-20Ni (Ref 36) and Fe-18Cr-14Ni (Ref 38) high-purity alloys. In the latter case, the total metallic impurities were quoted at 1 ppm and the total nonmetal impurities (oxygen, nitrogen, sulfur, and phosphorus) at 10 ppm, the material having been prepared by plasma furnace melting. Other studies (Ref 39) have shown that decreasing the phosphorus content of an Fe-18Cr-10Ni stainless steel down to 0.003 wt% produces a material that is highly resistant to chloride SCC in a magnesium chloride solution

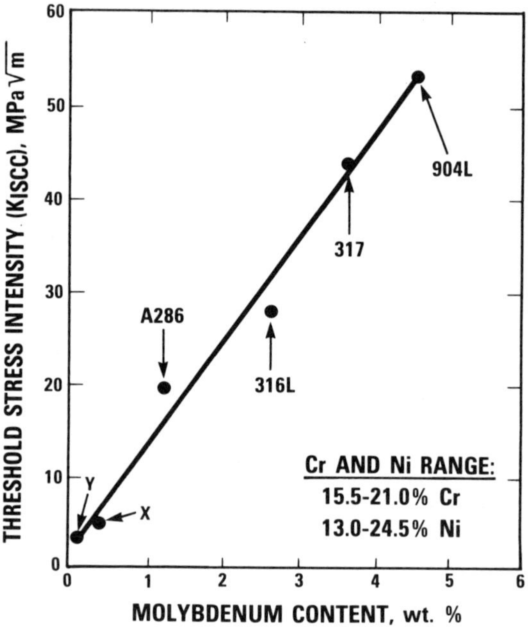

Fig. 9 Effect of molybdenum content on the stress-corrosion threshold stress intensity of Fe-Cr-Ni-Mo alloys in an aerated aqueous 22% NaCl solution at 105 °C (220 °F). Alloys X and Y are German heat-resistant grades. After Ref 27

boiling at 154 °C (309 °F) (Fig. 10). It has been suggested (Ref 40) that these high-purity alloys are resistant because their corrosion potentials are outside the potential range where SCC occurs.

Effects of Stress. As shown in Fig. 11 (Ref 41), decreasing the applied stress increases time to failure, and at low stresses a runout (threshold) stress is indicated. Careful studies (Ref 42) using

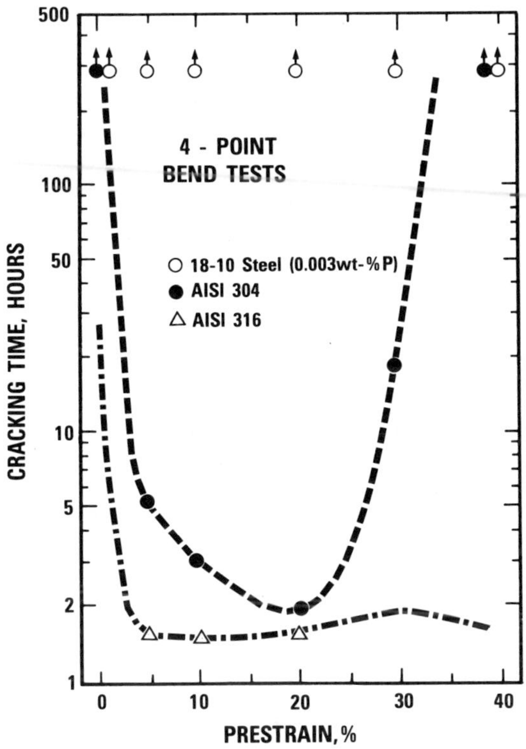

Fig. 10 Effect of cold work on the cracking time of a low-phosphorus (0.003% P) 18Cr-10Ni stainless steel and types 304 and 316 in magnesium chloride solutions boiling at 154 °C (309 °F) under an applied tensile stress of 196 MPa (28 ksi). After Ref 39

solution-annealed type 347 electropolished tensile specimens have established a threshold stress for this material at approximately 160 MPa (23 ksi), as indicated in Fig. 12. Threshold stress intensities (K_{ISCC}) obtained using precracked fracture-mechanics specimens in boiling magnesium chloride and sodium chloride solutions have been measured for various stainless steels, as shown in Fig. 7 to 9 (Ref 27). These threshold stress intensities should be regarded as stress-intensity factors below which the rate of cracking in a given environment is very low, rather than as absolute thresholds.

The method of increasing resistance to chloride SCC by reducing stress level has been studied. Stress, however, is a difficult parameter to control, and while it is obviously good practice to minimize applied (design) stresses, the presence of tensile residual stresses in the material can negate any benefit. The only sure way to control stresses is by stress-relief annealing the fully assembled structure. Full stress relief can be attained by annealing in the range of 800 to 900 °C (1470 to 1650 °F), since this causes a recrystallization of the deformed (internally stressed) grains. Unfortunately, this is often an option only for small components. For larger components, it is sometimes possible to achieve partial stress-relieving by heating to 400 to 600 °C (750 to 1110

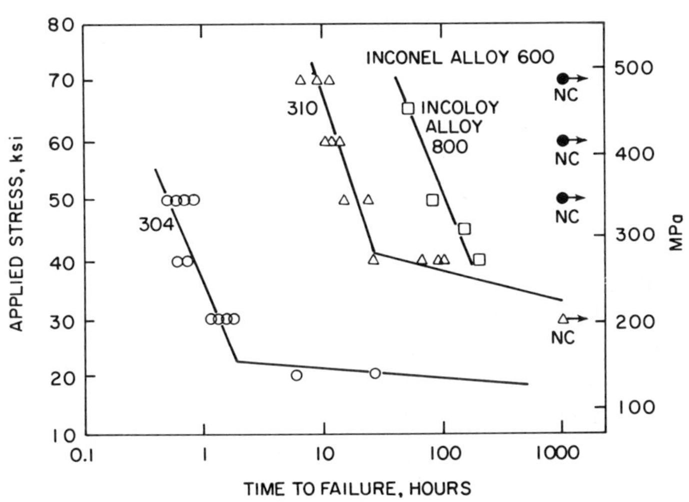

Fig. 11 Effect of applied stress on the times to failure of various alloys tested in a magnesium chloride solution boiling at 154 °C (309 °F). After Ref 41

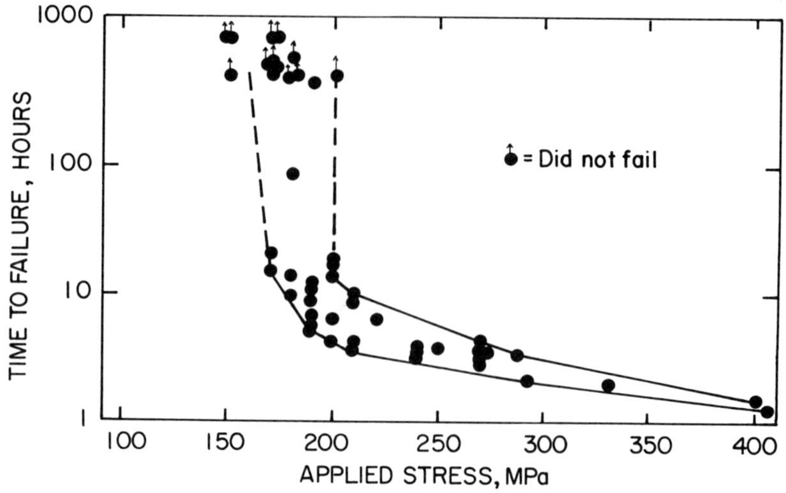

Fig. 12 Effect of applied stress on the time to failure of solution-annealed and electropolished type 347, tested in a magnesium chloride solution boiling at 145 °C (293 °F). After Ref 42

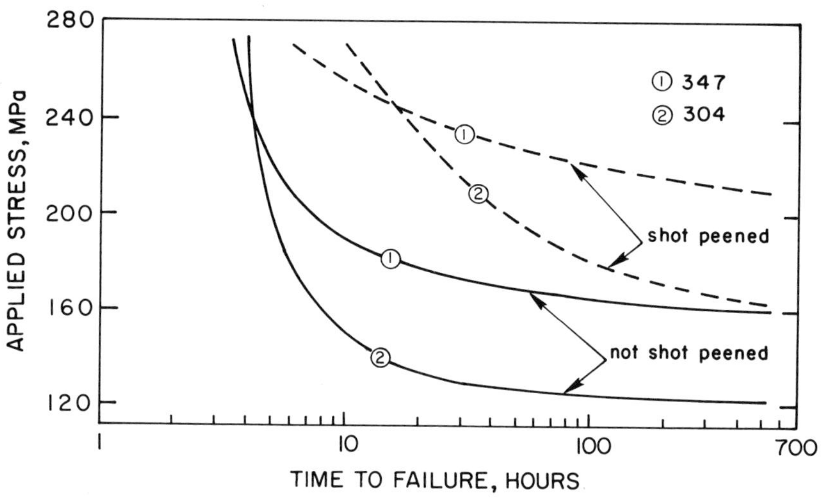

Fig. 13 Effect of peening with 40 to 80 μm glass shot on the times to failure of types 304 and 347 in a boiling 42% $MgCl_2$ solution. After Ref 43

°F), and it is believed, particularly in Europe, that even such partial stress-relieving can significantly improve resistance to chloride SCC. However, the benefits of any such procedure should be carefully weighed against potential sensitization problems. Compressive residual stresses, such as may be introduced by careful shot peening, have been shown to increase resistance to chloride SCC (Fig. 13) (Ref 43).

The effects of various surface finishes, which can produce stress variations due to local work hardening, martensitic transformation, residual stresses, and embedded material from abrasives and machining equipment, or provide stress raisers such as deep grooves or notches, are virtually impossible to quantify. All that can be said is that comparative evaluation of chloride SCC resistance should be carried out with nominally identical surface finishes. In particular, predictive laboratory evaluations should be performed using surface finishes identical to those encountered in service.

Effects of Microstructure. A schematic of the microstructural features found in stainless steels is shown in Fig. 14 (Ref 44), and many of these features have an effect on chloride SCC, as noted below.

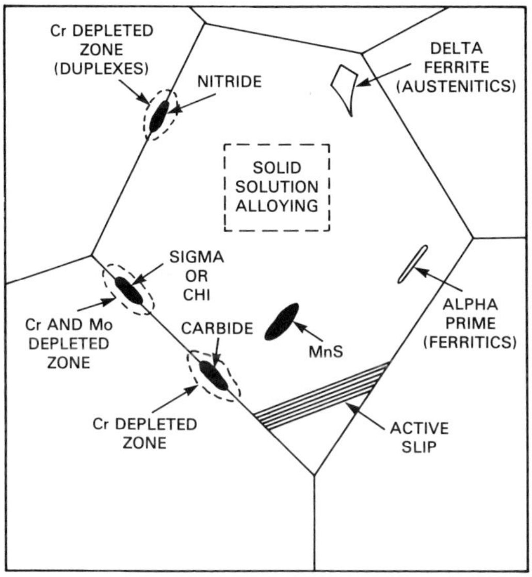

Fig. 14 Schematic of microstructural features found in stainless steels. After Ref 44

The presence of δ ferrite in austenitic stainless steels generally improves resistance to chloride SCC. An example of this for several cast austenitic stainless steels is shown in Fig. 15 (Ref 45). The beneficial effect of δ ferrite is generally attributed to its interference with the propagation of cracks across the austenite matrix. To obtain significantly improved resistance, however, considerable quantities of ferrite must be present, such as those found in duplex stainless steels. These are discussed in the section "Duplex Stainless Steels" in this chapter.

Regarding σ phase, few attempts have been made to evaluate its effect on chloride SCC resistance. A metallographic study (Ref 46) has shown that the transgranular cracks produced in boiling 42% $MgCl_2$ avoid or bypass σ phase produced in high-silicon types 316 and 317 by heating for 4 h at 871 °C (1600 °F).

Studies (Ref 46, 47) have failed to identify any relationship between sulfide stringers and the transgranular cracks produced in boiling magnesium chloride environments. The cracks appear to ignore the sulfide stringers. Other studies suggest

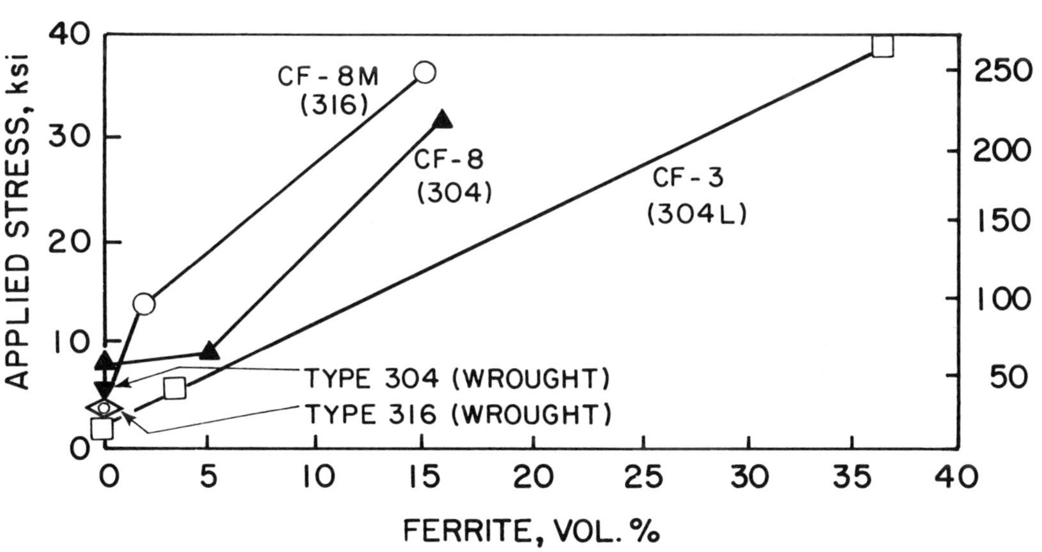

Fig. 15 Effect of ferrite content on the stress required to induce chloride SCC in various cast stainless steels. Materials exposed for 8 h in condensate from a 875 ppm chloride solution at 204 °C (400 °F). After Ref 45

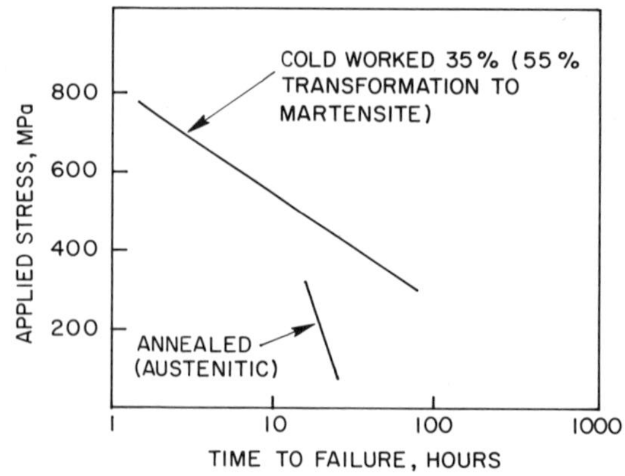

Fig. 16 Effect of cold work on the SCC susceptibility of type 301 in a boiling 40% $CaCl_2$ solution. After Ref 49

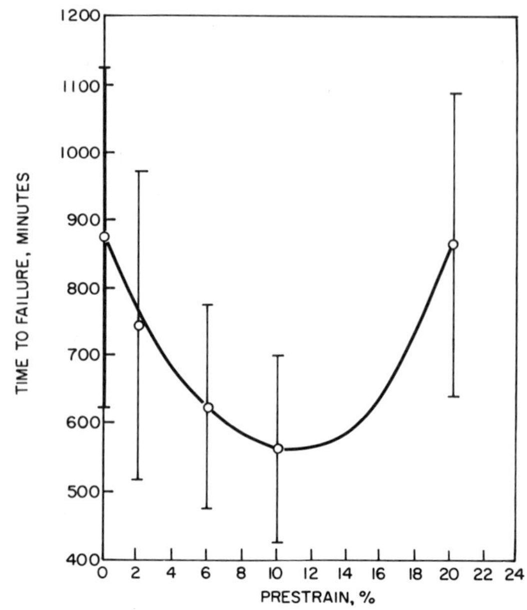

Fig. 17 Effect of prestrain on the time to failure of type 310 exposed to a magnesium chloride solution boiling at 154 °C (309 °F) and stressed at 90% of the yield stress. After Ref 50

that there may be an association between crack initiation and sulfides (Ref 48).

Plastic deformation of austenitic stainless steels results in both work hardening and partial transformation to α and ε martensites. Therefore, it is very difficult to separate effects caused by cold work (slip) from those caused by transformation to martensites. Studies (Ref 49) using type 301, which readily transforms to martensites on cold working, show that at comparable applied stresses (*e.g.*, 300 MPa, or 43.5 ksi) the cold-worked and partially transformed material exhibits longer times to failure than the untransformed material (Fig. 16). However, the deformed material has a higher yield stress. Studies (Ref 50) using prestrained wires of type 310, which shows little tendency for transformation to martensites, suggest a minimum at 10% prestrain (Fig. 17). Other reports of the observation of such minima have been compiled in Ref 21. An exception appears to be type 316 (Fig. 10), which shows a sharp decrease with strain but no increase at higher strains (Ref 39). Current understanding of this phenomenon can be summarized by stating that, in general, significant cold work (*e.g.*, 35%) increases time to failure, as shown by the data for type 321 in Fig. 18 (Ref 49).

Sensitization is very detrimental to chloride SCC resistance, and in many cases it appears to cause SCC. As shown in Table 3 (Ref 51), sensitization can cause many stainless steels to become susceptible to intergranular cracking in chloride environments such as marine atmosphere. Sensitization is also reported (Ref 52) to cause intergranular chloride SCC of types 304 and 316 in synthetic seawater and in 0.1 *M* NaCl solutions at 80 °C (175 °F). More quantitative tests using fracture-mechanics test specimens have shown that the threshold stress-intensity factor for SCC susceptibility, K_{ISCC}, of type 304 stainless steel tested in a 22% NaCl solution at 105 °C (220 °F) is decreased from a value above 35 MPa$\sqrt{m}$ (32 ksi$\sqrt{in.}$) to a value below 9 MPa$\sqrt{m}$ (8 ksi$\sqrt{in.}$) as a result of sensitization (Ref 27).

Stress-corrosion cracking in sensitized stainless steels generally propagates along an intergranular crack path, whereas nonsensitized materials exhibit predominantly transgranular cracking in environments such as boiling magnesium chloride. However, exceptions occur, and the factors determining the selection of a crack path require further research. Sensitization-induced SCC in the high-temperature water environments of nuclear light-water reactors is discussed later in this chapter.

Effects of Environmental Variables. Temperature is an important variable in determining

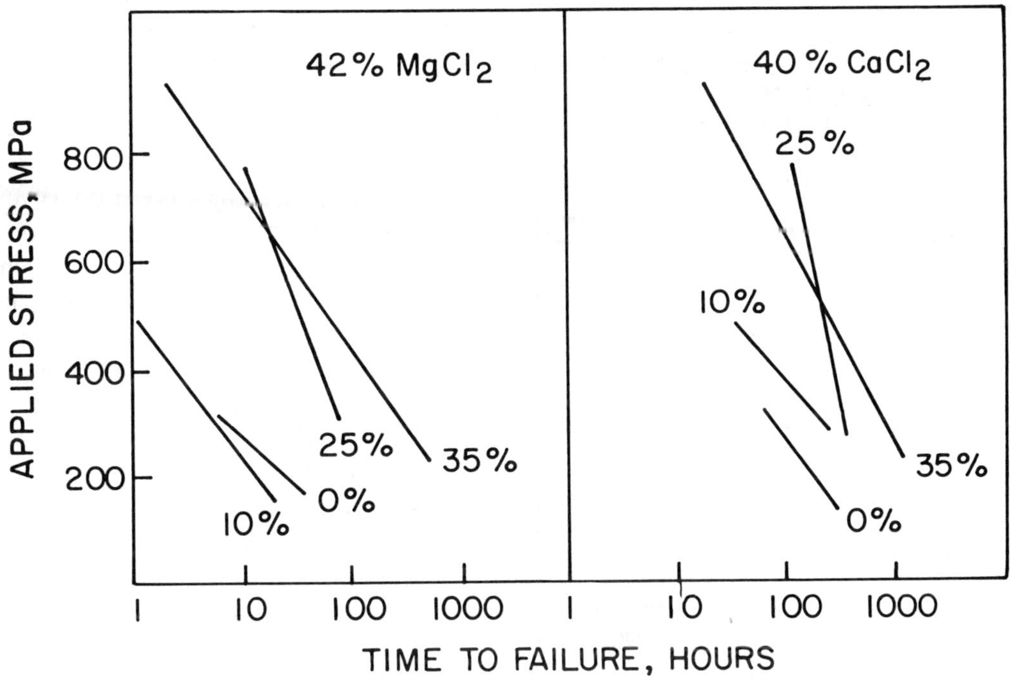

Fig. 18 Effect of cold work (%) on the SCC susceptibility of type 321 in boiling magnesium chloride and calcium chloride solutions. After Ref 49

Table 3 Results of SCC Tests in Which U-Bend Specimens Were Exposed for 5 Years to Marine Atmosphere in the 80-Ft Lot at Kure Beach

	Condition(a)			
Material	Annealed	Welded	Cold worked, 1/4 hard	Sensitized, 650 °C/1.5 h/FC
201	NF	NF	NF	IGA
301	NF	IGA	NF	IGA
302	NF	NF	NF	IGA
304	NF	NF	NF	IGA
304L	NF	NF	NF	NF
309	NF	NF	...	IGA
310	NF	NF	...	NF
316	NF	NF	NF	IGA
Carpenter 20Cb3	NF	NF	...	NF
Incoloy alloy 825	NF	NF	...	NF
Incoloy alloy 800	NF	NF	...	NF
Inconel alloy 600	NF	NF	...	NF

(a) NF, no failure in 5 years; IGA, intergranular attack; FC, furnace cooled
After Ref 51

whether chloride SCC will occur. The traditional engineering viewpoint, based on practical experience (Ref 53), has been that chloride SCC does not occur in nonsensitized austenitic stainless steels at temperatures below about 60 °C (140 °F) in near-neutral chloride solutions. Under the severe conditions encountered in a fracture-mechanics SCC test, this minimum temperature can

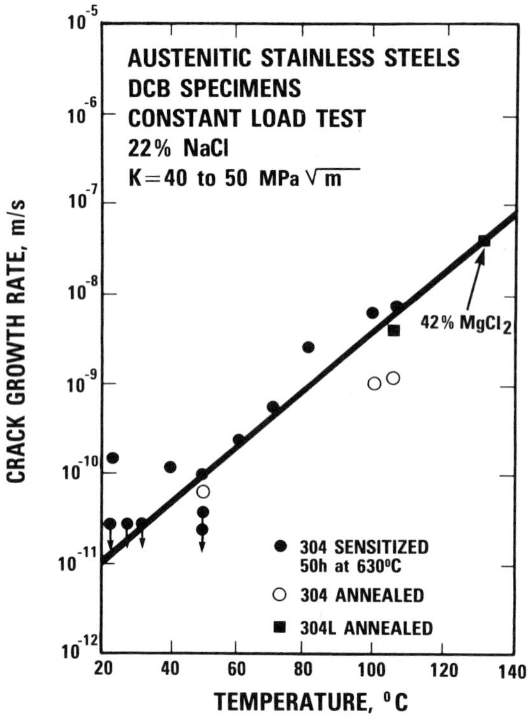

Fig. 19 Effect of temperature on SCC velocity for austenitic stainless steels in concentrated chloride solutions. After Ref 27

be lowered to 50 °C (120 °F) for annealed type 304 and to temperatures approaching ambient for sensitized type 304 (Fig. 19) (Ref 27).

Instances of transgranular chloride SCC in service in nonsensitized austenitic stainless steels at ambient temperature have been reported to occur in atmospheres above indoor swimming pools (Ref 54, 55) and in the marine atmosphere (Ref 56, 57). The swimming pool atmospheres have caused chloride SCC in ceiling support rods made from type 304, type 316, and a German titanium-stabilized type 316 (DIN 1.4571). The mechanism of these swimming pool atmosphere failures is not understood, and their occurrence was not predicted (Ref 54).

Marine atmospheres have caused transgranular chloride SCC in type 303 (free-machining grade) nuts after 14 years of service (Ref 56), in type 304L heater sheaths and dished ends (Ref 57), and in type 304 pipe for an emergency cooling system (Ref 57). Common factors in the marine atmosphere failures were the presence of a hot, humid coastal atmosphere, contamination of the stainless steel by iron that rusted, and residual stresses from fabrication and fit-up. Reference 58

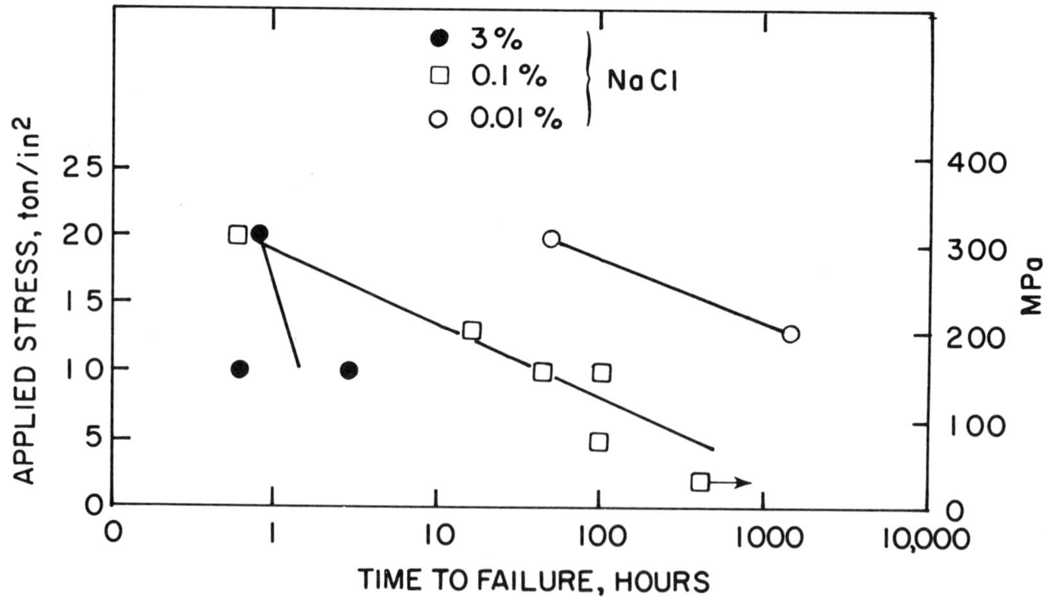

Fig. 20 Effect of chloride concentration on the SCC susceptibility of type 347 in oxygen-containing sodium chloride solutions at 250 °C (480 °F). After Ref 64

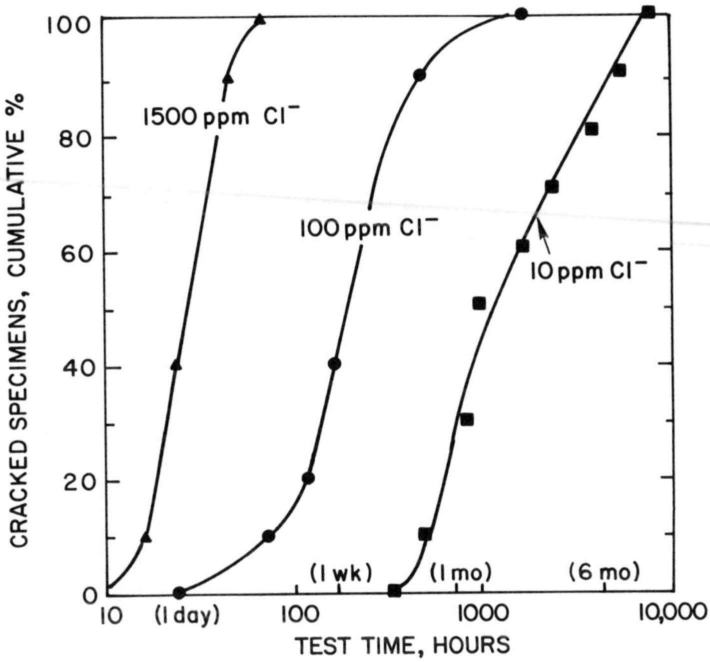

Fig. 21 Effect of chloride concentration on the SCC susceptibility of type 304 exposed at 100 °C (212 °F) under the concentrating conditions of the wick test. After Ref 65

outlines procedures for identifying and removing iron contaminants from stainless steel equipment.

Ambient-temperature chloride SCC of type 304 stainless steel has also been reported in solutions containing sulfuric acid and sodium chloride (Ref 59-61), and in hydrochloric acid solutions (Ref 59, 62). Type 304 U-bend specimens exposed to a 10 N H_2SO_4 + 0.5 N NaCl solution at 28 °C (82 °F) for 300 h exhibited primarily intergranular cracking, with only short transgranular cracks developing (Ref 59). However, type 316 tested under similar exposure conditions showed entirely transgranular cracking (Ref 59). Exposure of U-bend specimens of types 304 and 316 to a 5% HCl solution held at –5 °C (23 °F) for 742 h resulted in transgranular cracking (Ref 59). This form of acid-induced SCC occurs at potentials in the active region (Ref 62). Chloride SCC of stainless steels has also been reported (Ref 55) to occur on exposure to hydrogen chloride vapor at – 4 °C (25 °F).

Under immersion conditions, chloride SCC is facilitated by increased chloride concentrations, presence of oxygen, and lowered pH. The relationship between chloride concentration and SCC susceptibility is not simple, and in elevated-temperature environments it is linked with oxygen concentration. The presence of oxygen is not necessary to cause SCC in concentrated chloride solutions boiling at atmospheric pressure, since their oxygen contents should approach zero. However, the introduction of oxygen into such solutions by bubbling accelerates SCC (Ref 63).

There is some evidence to suggest that in oxygen-containing higher-temperature sodium chloride solutions, susceptibility to SCC decreases with decreasing chloride concentration, as indicated in Fig. 20 (Ref 64). However, the maintenance of low chloride levels cannot ensure freedom from SCC under conditions in which chloride can concentrate in crevices or shielded areas. The effect of temperature on chloride SCC susceptibility of type 304 under concentrating conditions is shown in Fig. 21 (Ref 65). In the wick test used to obtain the data of Fig. 21, a porous insulating material was used to draw the salt solution to the heated surface of a stressed speci-

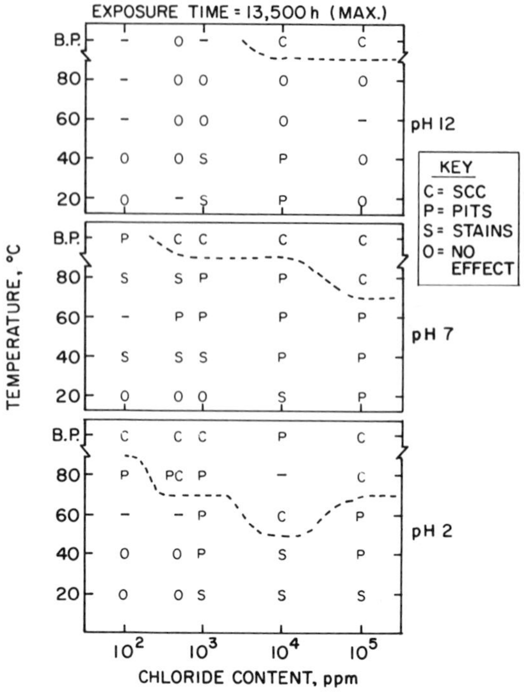

Fig. 22 Effect of pH on the chloride content and temperature required to produce SCC of type 304 in sodium chloride solutions. After Ref 69

men, where chloride can concentrate (Ref 20, 65, 66).

Regarding pH, it has been determined (Ref 67) that the pH of the solution at the tip of a crack in type 304 undergoing SCC in magnesium chloride is between 1.2 and 2.0. This suggests that, as in the case of pitting and crevice corrosion, acidification may be occurring by the hydrolysis of metal ions. In the case of pitting, increase in the bulk pH of the solution increases pitting resistance (Ref 68), and a similar improvement with increasing pH has been reported for chloride SCC resistance [see Fig. 22 (Ref 69) and Table 4 (Ref 70)]. However, as noted later, caustic SCC may occur in stronger caustic solutions. Acidity may also explain (Ref 55) why magnesium chloride and calcium chloride, which dissolve to form acidic solutions (Ref 19), cause chloride SCC.

Electrode potential has a major effect on susceptibility to chloride SCC. Detailed studies (Ref 71), summarized in Fig. 23, have established the relationships among applied potential, applied stress, and time to failure for type 304 stainless steel in boiling 42% $MgCl_2$. Increasing the potential in the noble direction facilitates cracking, whereas increasing the potential in the active direction reduces susceptibility until a cathodic protection potential is attained.

Regarding cathodic protection, it has been shown (Ref 72) that both impressed currents and sacrificial anodes can be used to prevent chloride SCC of type 304 in boiling 42% $MgCl_2$. Cathodic protection is also effective for preventing crack initiation in sensitized type 304 in NaCl solutions at 80 °C (175 °F) (Ref 73). Some researchers regard the cathodic protection potential as a critical cracking potential and equate it with the repassivation potential for crevice corrosion (Ref 74).

Inhibitors of chloride SCC have been extensively studied (Ref 72, 75-80). Silicates, nitrates, phosphates, carbonates, iodides, and sulfites have all proved to effectively inhibit chloride SCC if present in certain concentrations. Cadmium sulfate effectively inhibits chloride SCC of sensitized type 304 in 4 *N* NaCl solution acidified to a pH 2.3 with HCl at 100 °C (212 °F) (Ref 81).

Ferritic Stainless Steels

The high resistance to chloride SCC of types 405 and 430 ferritic stainless steels was first demonstrated in 1945 (Ref 16) and was confirmed in subsequent studies (Ref 72, 82). More recent studies have also demonstrated the SCC resistance of some of the newer ferritic grades tested as U-bend specimens (Ref 83) and precracked fracture-mechanics specimens (Ref 27). In the fracture-mechanics tests (Ref 27), the ferritic grades UNS S44400 (Remanit 4522) and S44800 (Remanit 4575) were fully resistant to chloride SCC at stress-intensity levels of 60 $MPa\sqrt{m}$ (55 $ksi\sqrt{in.}$) in 22% NaCl solutions at 105 °C (220 °F). The U-bend test data are shown in Table 5 (Ref 83), with types 409, 439, and 444 exhibiting resistance to chloride SCC in all three test environments.

The tests noted above have led to the misleading impression that all ferritic stainless steels are immune to SCC in all chloride environments. However, SCC of the ferritics has been reported in several instances: for type 434, type 430, and Fe-18Cr-2Mo in lithium chloride solutions (Ref 84), for sensitized type 446 in boiling magnesium

Table 4 Effects of pH and Heat Treatment on Time to Failure for Several Austenitic Stainless Steels in Boiling 26% NaCl Solution

pH	Material	Heat treatment(a)	No. of U-bend specimens tested	Time to failure, weeks(b) Minimum	Maximum
2	304	A	10	0.3	2.0
		A+L	10	1.0	1.0
	316	A	10	1.0	5.0
		A+L	10	0.7	2.3
	317	A	10	1.0	16.0
		A+L	10	1.0	8.0
4	304	A	4	NC	NC
		A+L	10	2.6	NC
	316	A	10	1.0	16.0
		A+L	10	1.0	11.0
	317	A	2	NC	NC
		A+L	2	NC	NC
6	304	A	2	NC	NC
		A+L	2	NC	NC
	316	A	2	NC	NC
		A+L	2	26(c)	NC
	317	A	2	NC	NC
		A+L	2	NC	NC
8	304	A	2	NC	NC
		A+L	2	NC	NC
	316	A	2	NC	NC
		A+L	2	NC	NC
	317	A	2	NC	NC
		A+L	2	NC	NC
10	304	A	2	NC	NC
		A+L	2	NC	NC
	316	A	2	NC	NC
		A+L	2	NC	NC
	317	A	2	NC	NC
		A+L	2	NC	NC
12	316	A	2	9.0(c)	NC
		A+L	2	NC	NC
	317	A	2	NC	NC
		A+L	2	NC	NC

(a) A = 27% CR + 1 h/1121 °C/WQ; L = 1 h/677 °C/WQ. (b) NC, no cracking in 30-week test period. (c) Crack in straight leg of U-bend away from bend
After Ref 70

chloride and sodium chloride solutions (Ref 85), for type 430F in the marine atmosphere (Ref 86), and for the nickel-containing superferritics AL 29-4-2 (UNS S44800), MONIT (UNS S44635), and SEA-CURE (UNS S44660) in boiling magnesium chloride (Ref 83). Data for the superferritics are shown in Table 5. The producers of AL 29-4C (UNS S44735), which is shown to be SCC-resistant in Table 5, caution that the addition of 1% Ni to the alloy composition will introduce susceptibility to chloride SCC in boiling magnesium chloride environments (Ref 87).

Factors that have so far been identified as detrimental to the chloride SCC resistance of ferritic stainless steels include the presence of certain alloying elements, sensitization (induced by heat treatment or welding), cold work, high-temperature embrittlement, and the precipitation of α' (475 °C, or 887 °F, embrittlement). Because of the metallurgical complexity of ferritic stainless

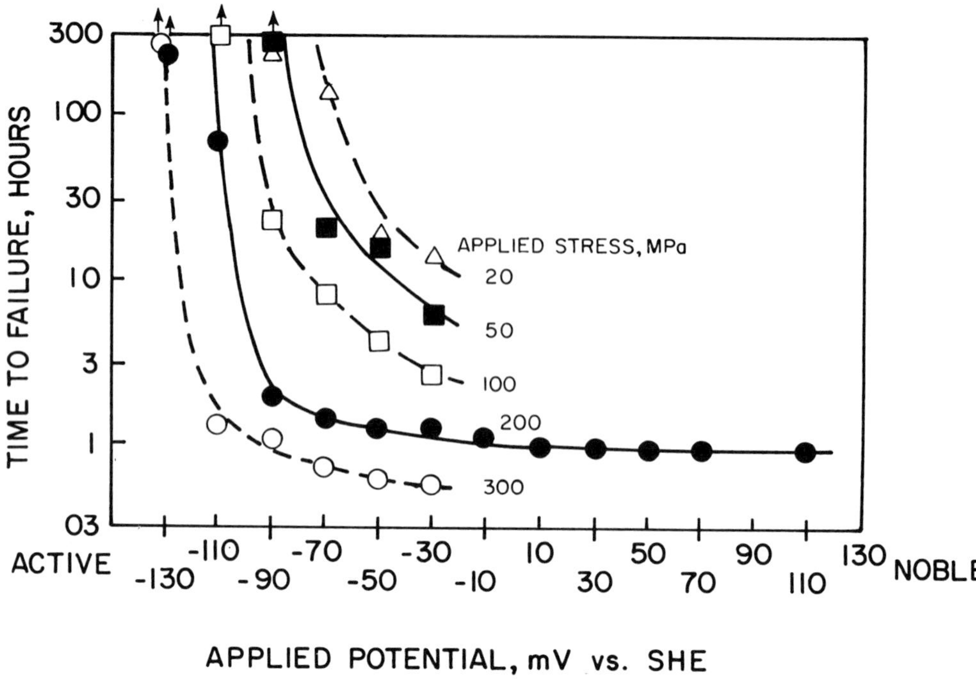

Fig. 23 Relationship among applied potential, applied stress, and time to failure of solution-annealed type 304 in a 42% $MgCl_2$ solution at 144 °C (291 °F). After Ref 71

Table 5 SCC Resistance of Stainless Steels

Material	SCC test(a)		
	Boiling 42% $MgCl_2$	Wick test	Boiling 25% NaCl
AISI type 304	F(b)	F	F
AISI type 316	F	F	F
AISI type 317	F	(P or F)(d)	(P or F)
Type 317LM	F	(P or F)	(P or F)
Alloy 904L	F	(P or F)	(P or F)
AL-6XN	F	P	P
254SMO	F	P	P
20Mo-6	F	P	P
AISI type 409	P(c)	P	P
Type 439	P	P	P
AISI type 444	P	P	P
E-Brite	P	P	P
Sea-Cure	F	P	P
Monit	F	P	P
AL 29-4	P	P	P
AL 29-4-2	F	P	P
AL 29-4C	P	P	P
3RE60	F	NT(e)	NT
2205	F	NT	(P or F)(f)
Ferralium	F	NT	(P or F)(f)

(a) U-bend tests, stressed beyond yielding. **(b)** Fails, cracking observed. **(c)** Passes, no cracking observed. **(d)** Susceptibility of grade to SCC determined by variation of composition within specified range. **(e)** Not tested. **(f)** Susceptibility of grade to SCC determined by variation of thermal history
After Ref 83

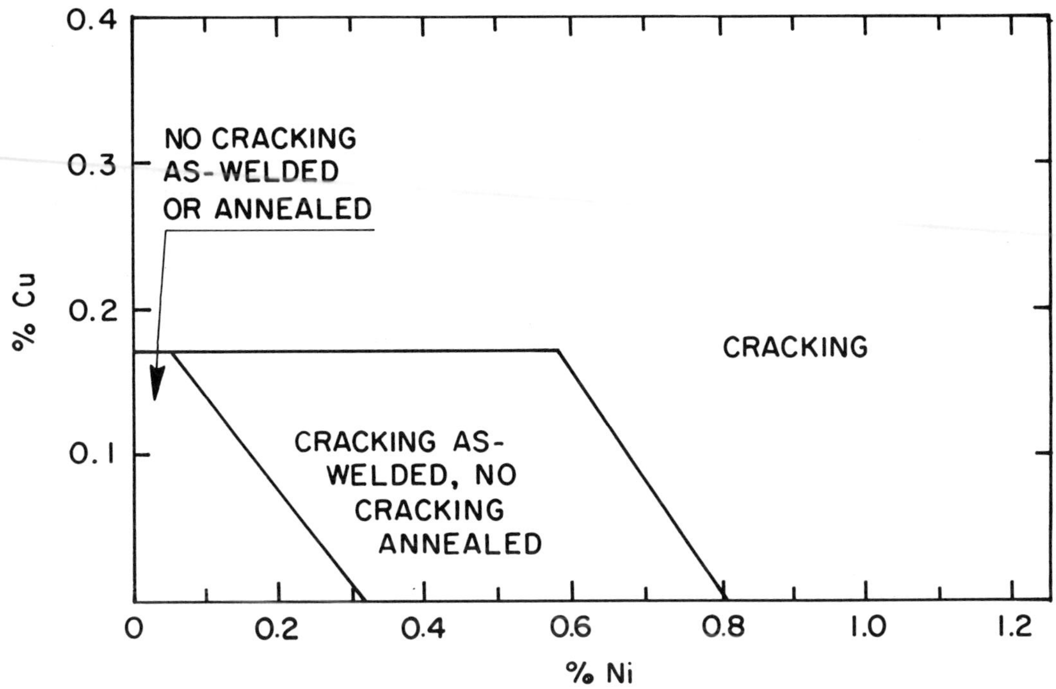

Fig. 24 Effect of copper and nickel contents on the SCC resistance of U-bend specimens of ferritic Fe-18Cr-2Mo-0.35Ti-0.015C-0.015N stainless steels exposed to a magnesium chloride solution boiling at 140 °C (284 °F). After Ref 88

steels, it is not known whether all of these factors are related to the phenomenon identified as chloride SCC in austenitic stainless steels or also represent manifestations of other phenomena, such as hydrogen embrittlement and stress-aided intergranular corrosion along sensitized grain boundaries. However, for the purpose of the present discussion, it is assumed that effects that can be detected in boiling magnesium chloride tests relate to chloride SCC.

Effects of Composition. Alloying elements that have been identified as detrimental to the chloride SCC resistance of ferritic stainless steels include copper, nickel, molybdenum (in the presence of nickel), cobalt (in the presence of molybdenum), ruthenium, and carbon. Sulfur, either in the alloy or as sulphur-containing gases in the environment, may also facilitate SCC in chloride environments.

The copper and nickel levels that introduce susceptibility to SCC in magnesium chloride environments in annealed and in welded Fe-18Cr-2Mo alloy are shown in Fig. 24 (Ref 88). The detrimental effect of molybdenum in the presence of nickel is shown in Fig. 25 (Ref 89). The detrimental effect of cobalt is noted in Ref 89. Since molybdenum can be an alloying addition, and copper, nickel, and cobalt can be present if scrap is used in the melt charge, these effects could become technologically significant. The reported detrimental effect of ruthenium (Ref 90) is of more academic interest, since significant amounts of this element are not generally found as an impurity in stainless steels. Regarding carbon, it has been shown (Ref 85) that, for a 28.5Cr-4.0Mo stainless steel, increasing the carbon content from 20 to 171 ppm leads to SCC in a magnesium chloride solution boiling at 155 °C (311 °F). The detrimental effect of sulfur was suggested by the report (Ref 86) that type 430F (0.15% S, minimum) can exhibit cracking in the marine atmosphere.

Effects of Microstructure. It is probably true that the presence of structural features that reduce

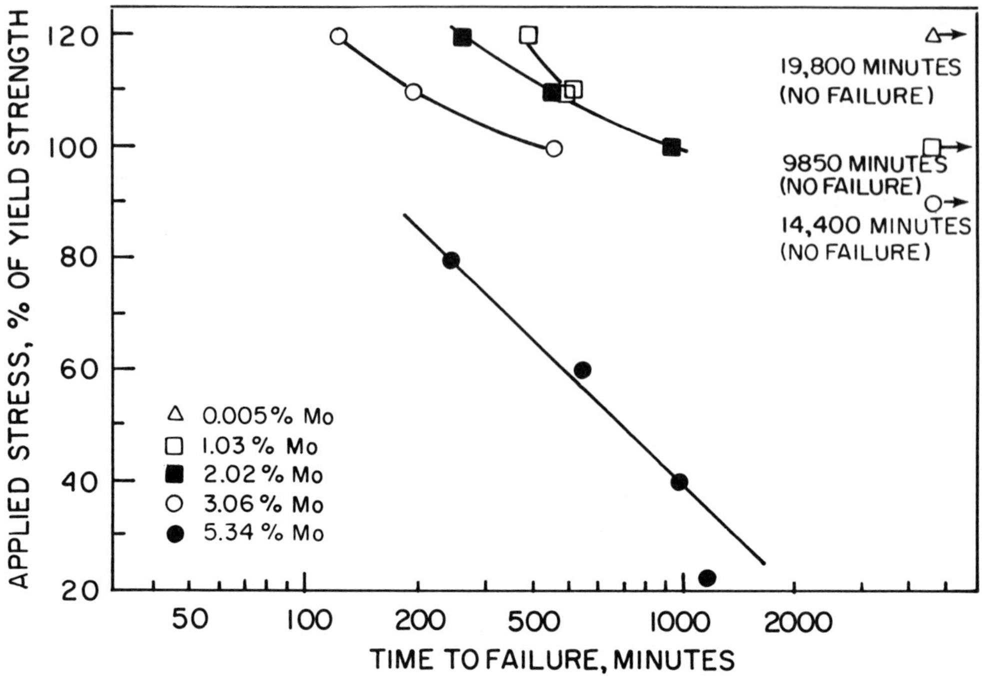

Fig. 25 Effect of applied stress on the times to failure of tensile specimens of experimental ferritic Fe-17Cr-1Ni stainless steels containing various amounts of molybdenum in a magnesium chloride solution boiling at 140 °C (284 °F). After Ref 89

the ductility of ferritic stainless steels increases their susceptibility to chloride SCC. Ductility is reduced by the precipitation of carbonitrides (high-temperature embrittlement), by α' (475 °C, or 887 °F, embrittlement), and, to some degree, by cold work. The detrimental effects on chloride SCC resistance of these three factors are noted in Ref 88, 91, and 88, respectively. There do not appear to have been any systematic investigations of the effect of σ phase or of sulfide inclusions on chloride SCC resistance. The detrimental effects of furnace sensitization of 28.5Cr-4.2Mo ferritic stainless steels that contain small amounts of copper and nickel have been reported and discussed in Ref 85 and 92.

Effects of Environmental Variables. The importance of the test environment in determining whether chloride SCC will occur is strikingly illustrated in Table 5. Several points emerge from these data. The first is that the boiling magnesium chloride test environment is much more severe than the boiling sodium chloride test environment. The second is that while the nickel alloying addition in the ferritic stainless steels (Sea-Cure, Monit, and AL 29-4-2 contain nickel) introduces SCC susceptibility in magnesium chloride solutions, it does not do so in the sodium chloride solution or in the wick test, which also uses sodium chloride. It is estimated (Ref 55) that 95% of the industrial and service occurrences of chloride SCC are due to sodium chloride. In a geothermal brine that contained alkali and alkaline earth chlorides, heavy-metal chlorides and sulfides, silica, and dissolved gases, both type 430 and E-Brite 26-1 ferritic stainless steels exhibited SCC (Ref 93).

Studies employing applied potentials show that, as in the case of austenitic stainless steels, chloride SCC of ferritic stainless steels can be prevented by cathodic protection. This is illustrated in Fig. 26 (Ref 94), which shows that a cathodic protection potential range (*i.e.*, –250 to – 650 mV, SHE) exists for an Fe-18Cr-1.5Ni alloy in a 42% $MgCl_2$ solution at 143 °C (289 °F).

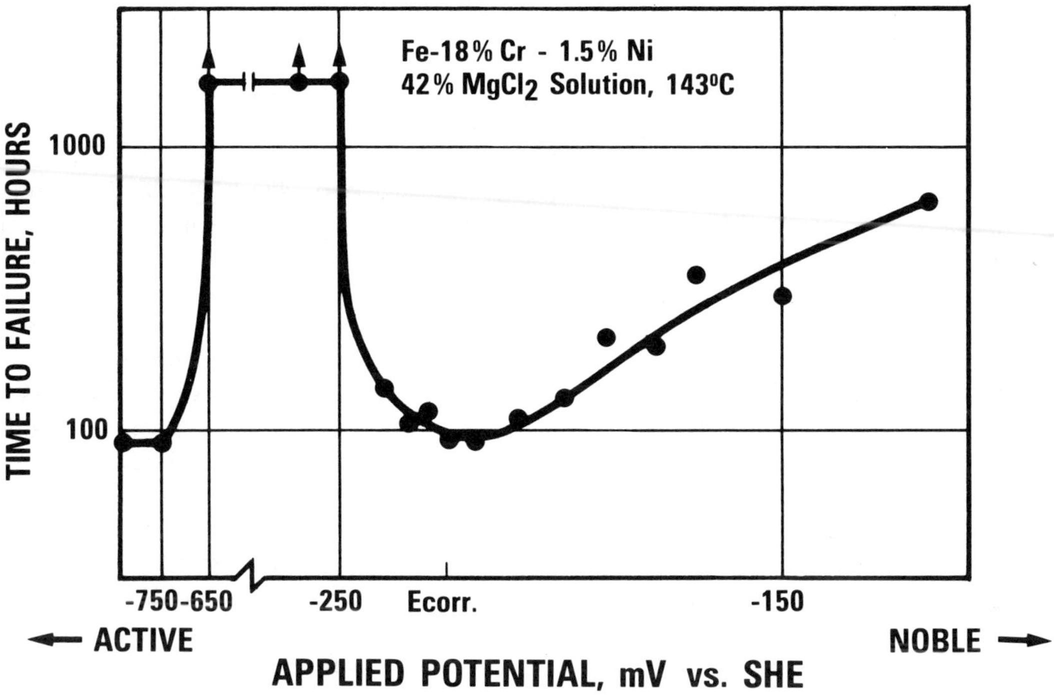

Fig. 26 Effect of applied potential on the times to failure of a ferritic stainless steel in a magnesium chloride solution. After Ref 94

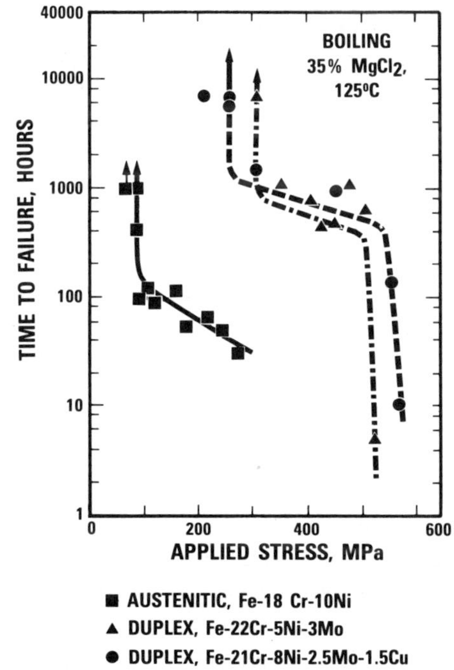

Fig. 27 Effect of applied stress on the times to failure of various stainless steels in a magnesium chloride solution. After Ref 95

Figure 26 also shows that at applied potentials active to – 650 mV (SHE), another cracking region exists, possibly caused by hydrogen embrittlement. Hydrogen embrittlement of ferritic stainless steels is discussed later in this chapter.

Duplex Stainless Steels

Stainless steels with 18 to 28% Cr and 4 to 8% Ni contain both austenite and ferrite and thus are known as duplex. Commercial practice is to process or heat treat these alloys at temperatures in the range from 1050 to 1150 °C (1920 to 2100 °F) and to water quench to maintain a structure of about 50% austenite and 50% ferrite. An attractive feature of the duplexes is their higher strength. For example, the 0.2% offset yield strengths of the austenitic type 304, the ferritic type 430, and the duplex type 329 are 241, 310, and 551 MPa (35, 45, and 80 ksi), respectively.

As shown in Table 5, the duplex grades 3RE60, 2205, and Ferralium are susceptible to chloride SCC in boiling magnesium chloride. However, the threshold stresses for chloride SCC

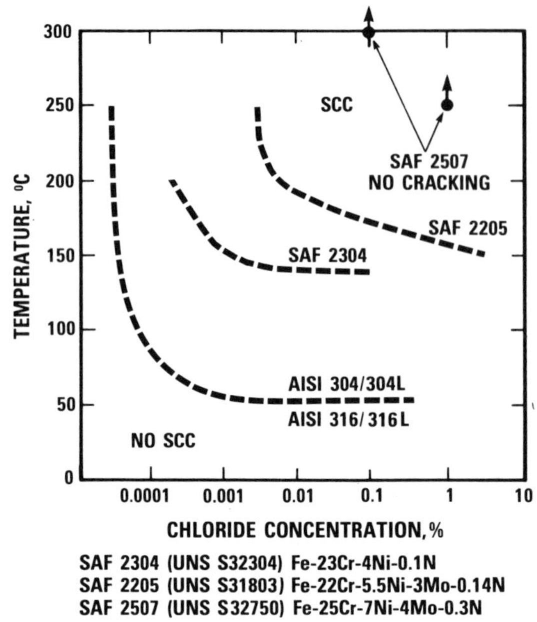

Fig. 28 SCC susceptibility of various stainless steels as a function of temperature and chloride concentration. Materials designated as 2304, 2205, and 2507 are duplex grades. After Ref 100

generally are higher for the duplex stainless steels than for the austenitics. An example of this is shown in Fig. 27 (Ref 95), where the threshold stresses for duplex stainless steels are about three times higher than that for an austenitic stainless steel. In fracture-mechanics tests (Ref 27) using precracked fatigue specimens in a 22% NaCl solution at 105 °C (220 °F), two variants of type 329 duplex stainless steel were found to be resistant to chloride SCC at stress-intensity levels of 60 $MPa\sqrt{m}$ (55 $ksi\sqrt{in.}$).

Among the microstructural parameters that have been examined in relation to chloride SCC susceptibility are the presence of σ phase and the ferrite content. The presence of σ phase has been shown (Ref 95) to be detrimental to chloride SCC resistance in boiling 35% $MgCl_2$ at 125 °C (257 °F). Ferrite content has also been shown to influence chloride SCC resistance in boiling 42% $MgCl_2$, with resistance reaching a maximum at 40% ferrite and decreasing on either side of this value (Ref 96-98). Ferrite formed by welding can also reduce resistance to chloride SCC (Ref 1, 95). From a metallurgical viewpoint, welding will destroy the duplex structure of an annealed material, creating continuous regions of ferrite in the heat-affected zone (HAZ), and, as discussed in the previous section, the precipitation of carbonitrides (high-temperature embrittlement), the precipitation of α′ (475 °C, or 887 °F, embrittlement), and sensitization can increase the susceptibility of ferrite to chloride SCC. Significantly, SCC has been reported in the ferrite portion of the HAZ of welds in a duplex stainless steel when welded U-bend specimens were exposed in the vapor phase above a 3% NaCl solution at 95 °C (203 °F) (Ref 1). Stress-corrosion cracking has also been observed in the HAZ of a welded duplex stainless steel when exposed as U-bend specimens to the marine atmosphere and a chemical plant atmosphere contaminated with traces of phosphoric and sulfuric acids (Ref 99). Tungsten-inert gas (TIG) welding is reported (Ref 95) to cause a much greater loss in chloride SCC resistance than manual arc welding. The latter is thought to provide larger heat input and longer cooling times, enabling some of the newly formed ferrite to transform back to austenite, thereby producing the more SCC-resistant duplex structure (Ref 95).

There have been no reports of systematic studies of the effects of single-alloying-element additions on chloride SCC for duplex stainless steels. There are, however, significant variations in chloride SCC resistance among the different commercial grades (Fig. 28) (Ref 100). The data of Fig. 28 suggest that increasing the total chromium, nickel, molybdenum, and nitrogen contents results in increased chloride SCC resistance. These data do not differentiate between effects due to individual alloying elements.

Studies of environmental variables have been somewhat more extensive. It has been established that, as in the case of austenitic and ferritic stainless steels, chloride SCC in duplex stainless steels can be prevented by cathodic protection.

Figure 29 (Ref 95) shows that the cathodic protection potentials of an austenitic and a duplex stainless steel are similar, despite the fact that the two steels were stressed at their widely differing ambient-temperature yield strengths during the SCC tests.

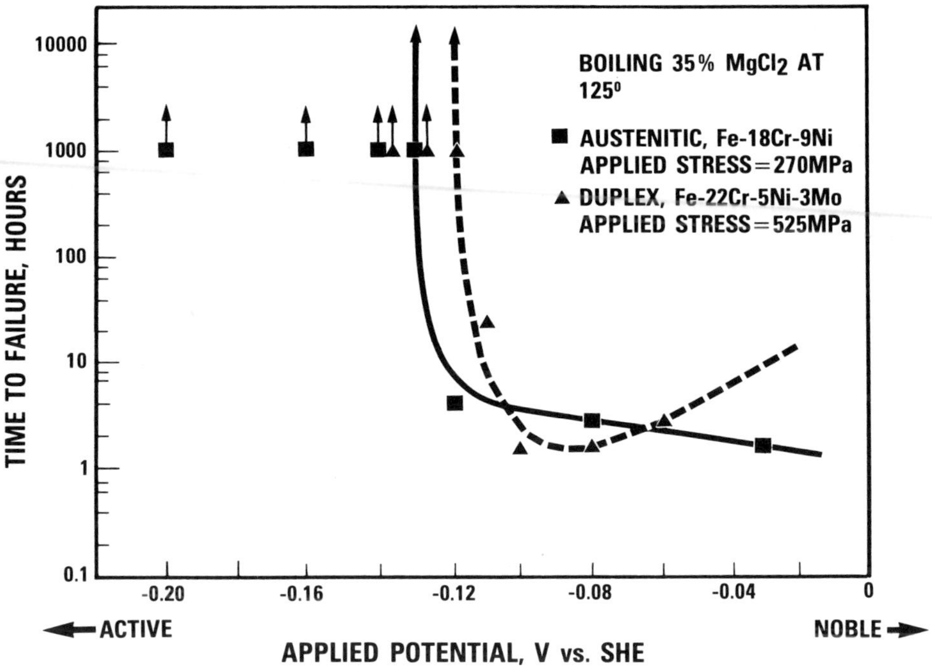

Fig. 29 Effect of applied potential on the times to failure of an austenitic and a duplex stainless steel in a magnesium chloride solution. After Ref 95

In addition to exhibiting SCC in concentrated chloride solutions such as boiling magnesium chloride, duplex stainless steels exhibit SCC in sour gas environments that contain chloride and hydrogen sulfide (Ref 101-104). Materials for use in sour-gas environments are often evaluated using NACE Standard Test TM-01-77 (Ref 105). This test employs a 5% NaCl + 0.5% CH_3COOH solution, saturated with hydrogen sulfide by continuous bubbling after nitrogen deaeration, which has a pH of 2.7 to 2.8. The test is sometimes modified by raising the temperature to 80 °C (175 °F) (Ref 104), bubbling with a gas mixture containing 90% CO_2 + 10% H_2S (Ref 103), or applying potentials (Ref 101). It has been found that in these environments increasing temperature (to 80 °C, or 175 °F) increases susceptibility to SCC (Ref 102) and that SCC is more severe in transverse specimens than in longitudinal specimens (Ref 102). Mechanistic assessments suggest that the environmentally induced cracking process of duplex stainless steels in sour-gas environments comprises an anodic chloride SCC step in which the crack is initiated in the austenite phase and a hydrogen-embrittlement propagation step in which the crack propagates through the ferrite phase (Ref 102, 104). Susceptibility to SCC can be reduced by lowering the hydrogen sulfide concentration to below 0.1 g/L or decreasing the applied stress level to below 60% of the yield stress (Ref 106). High-temperature annealing treatments designed to increase ferrite content do not affect SCC susceptibility in sulfide-containing environments (Ref 107).

Martensitic and Precipitation Hardening Stainless Steels

The martensitic and precipitation-hardening stainless steels are used in heat-treated conditions that give high strength. As noted by Brown (Ref 86), the modern view, arising from extensive studies, is that most of the service failures involving high-strength steels, including high-strength stainless steels, are caused by hydrogen embrittlement. This topic is discussed in the last major section of this chapter.

Caustic SCC

The handling and containment of caustic soda (sodium hydroxide) do not pose technological problems, because adequate materials are readily available. For example, the low-carbon grade of unalloyed nickel, Nickel 201 (0.02% C maximum), can handle caustic soda, without SCC, at all concentrations and temperatures of current technological interest (Ref 108). General corrosion resistance in caustic is almost directly proportional to the nickel content of a given alloy, and the nickel-containing austenitic stainless steels are used in some applications involving moderate-temperature caustic at concentrations of 50% or less (Ref 109, 110). However, it has been known for some time that at certain combinations of caustic concentration and temperature, stainless steels can exhibit caustic SCC (Ref 111, 112).

The 1970s witnessed increased interest in caustic SCC as a result of the use of austenitic alloys in heat-exchanger systems of water-cooled nuclear power plants. It is well known from conventional boiler technology that boiling and steam blanketing at heat-transfer surfaces can give rise to very high local concentrations of caustic. Apart from the power industry, caustic SCC resistance is also of interest in the chemical, petrochemical, and pulp and paper industries, although few publications attest to this fact.

Austenitic Stainless Steels and Nickel-Base Alloys

A summary (Ref 113) of available caustic SCC data for types 304, 347, 316, and 321 is shown in Fig. 30. While there is some uncertainty about the position of the conservative line designated "tentative safe SCC limit," the data show that there is an inherent danger of caustic SCC in strong caustic solutions when the temperature approaches 100 °C (212 °F). Sensitization appears to be detrimental at lower caustic concentrations. Figure 30 shows that at nuclear power plant steam generator temperatures of approximately 300 °C (570 °F), these steels could exhibit rapid cracking. This is a fact well recognized by steam generator designers, who utilize higher-nickel alloys

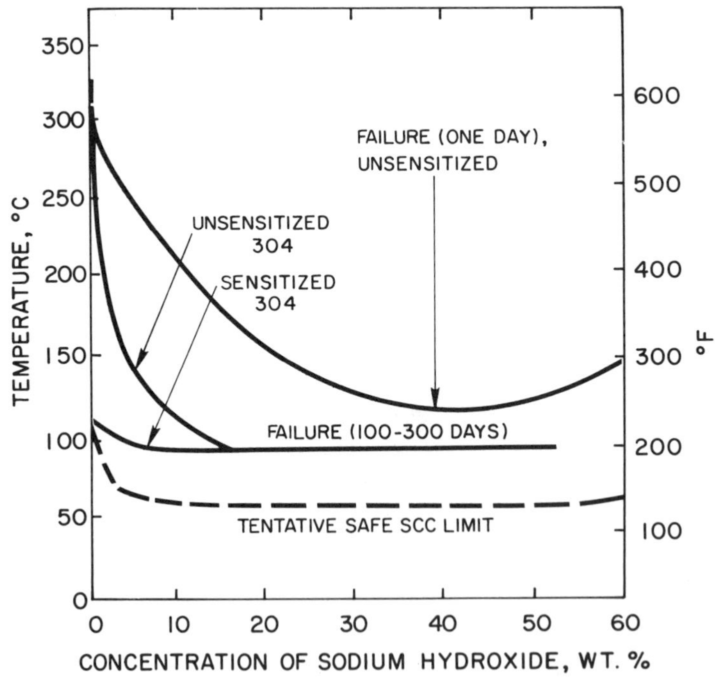

Fig. 30 Temperature and concentration limits for caustic SCC of types 304, 347, 316, and 321. After Ref 113

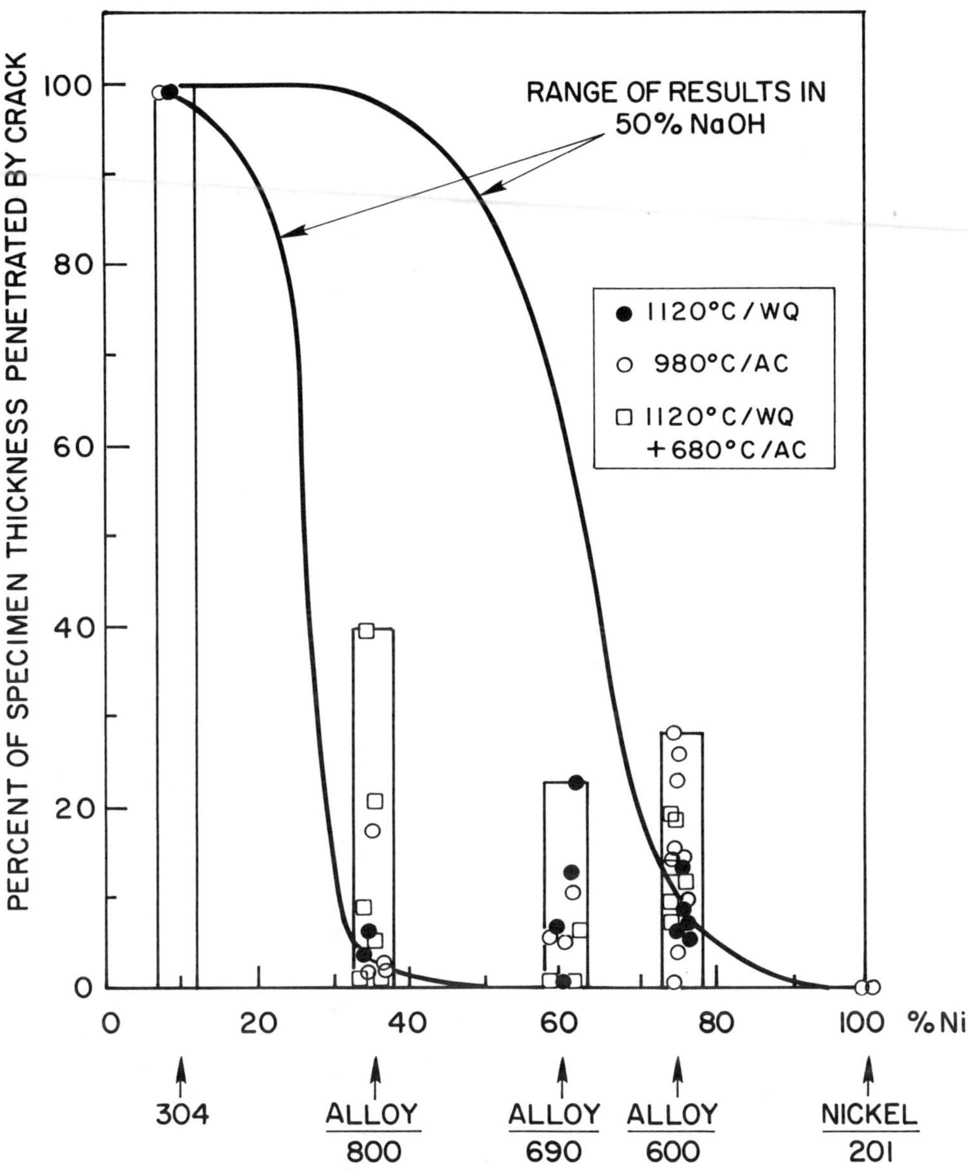

Fig. 31 Caustic SCC behavior of U-bend specimens of various alloys in deaerated 10% and 50% NaOH solutions at 316 °C (600 °F). Exposure time of 6 weeks. After Ref 116

with improved caustic and chloride SCC resistance for these applications.

When considering factors that affect caustic SCC resistance, careful attention must be paid to the oxygen level of the caustic solution. While there is no evidence that oxygen is necessary to cause caustic SCC, major changes in the ranking of alloys can occur depending on whether or not oxygen is present. Studies (Ref 114, 115) have shown that in deaerated concentrated (50%) sodium hydroxide solutions, cracking resistance increases with increasing nickel content of the alloy, with Nickel 201 exhibiting complete resistance (Fig. 31) (Ref 116). The data shown in Fig. 31 were obtained using U-bend specimens, which measure the effects of both crack initiation and propagation. Tests using precracked (wedge-opening-load) fracture-mechanics specimens

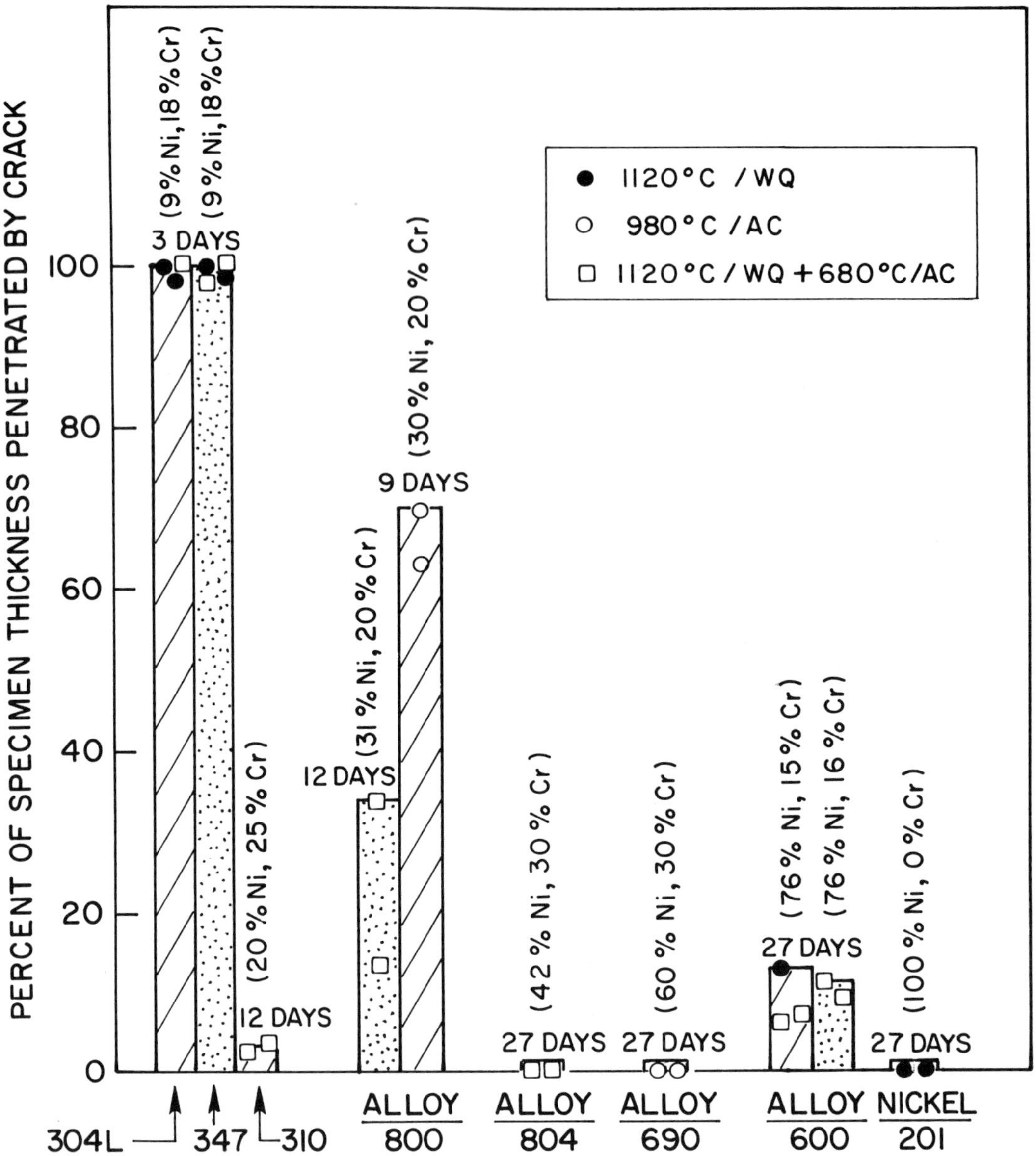

Fig. 32 Caustic SCC behavior of U-bend specimens of various alloys in an aerated 50% NaOH solution at 300 °C (570 °F). After Ref 115

have shown that the beneficial effect of nickel is also reflected in increasing threshold stress intensities (Ref 114). The presence of molybdenum as an alloying element does not improve SCC resistance in deaerated 50% NaOH (Ref 115). Figure 31 also shows that at lower caustic concentrations (*e.g.*, 10% NaOH), the beneficial effect of nickel is less pronounced, with all three high-nickel alloys (alloys 800, 690, and 600) exhibiting comparable SCC resistance.

In aerated (oxygen-containing) solutions of sodium hydroxide, high levels of both nickel and chromium appear to be necessary for resistance to caustic cracking, as indicated in Fig. 32 (Ref 115) and Fig. 33 (Ref 111). Figure 34 (Ref 64) illustrates the observation that in oxygenated caustic

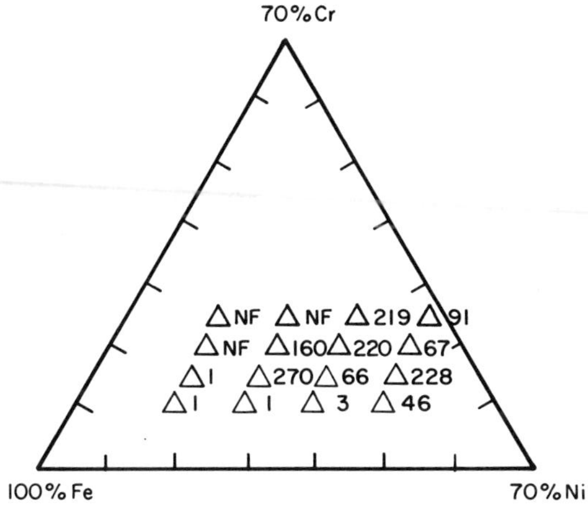

Fig. 33 Times to failure (hours) of various Fe-Cr-Ni alloys in a 50% NaOH solution at 300 °C (570 °F) with 14 MPa (2 ksi) oxygen. Specimens stressed at 140 MPa (20 ksi). NF, no failure. After Ref 111

solutions times to failure increase with decreasing applied stress. This suggests that stress-relief annealing could be a remedy. However, it has been reported (Ref 115) that annealing stressed U-bend specimens of type 304 stainless steel for 4 h at 593 °C (1100 °F) did not prevent cracking on subsequent exposure to deaerated 50% NaOH at 284 and 332 °C (543 and 630 °F). This, together with the observation that no meaningful threshold stress intensity could be determined for type 304 in this environment (Ref 114), poses questions both as to the effectiveness of stress-relief annealing for stainless steels in deaerated caustic solutions and as to the mechanism of cracking. The mitigation of caustic SCC by the addition of phosphate has received much more attention. Water treatments employing phosphate additions to prevent the development of free caustic have been known for several decades and have been used to prevent caustic SCC of stainless steels in boiler water (Ref 117-119). Chlorides (Ref 120) and chromates (Ref 121) have also been reported to inhibit the caustic SCC of austenitic stainless steels.

Ferritic, Duplex, and Martensitic Stainless Steels

The most extensive investigation of commercial stainless steels other than austenitics is probably that described in Ref 91. This investigation examined the caustic SCC behavior of ferritic, duplex, and martensitic stainless steels under conditions insufficiently severe to crack type 304 (*i.e.*, at 90% of the yield strength). The data are shown in Table 6 (Ref 91). Heat treatments designed to cause sensitization and 475 °C (887 °F) embrittlement were detrimental to the caustic SCC resistance of the higher-chromium ferritic stainless steels. Low-chromium type 405 exhibited heavy general corrosion. Heat treatments designed to produce 475 °C (887 °F) embrittlement in the duplex stainless steel were also detrimental. Martensitic type 410 exhibited SCC after tempering at 482 °C (900 °F), which produces a hardness of 50 HRC.

In other studies of ferritic stainless steels, it was reported (Ref 122) that types 430 and 434 exhibited uniform corrosion or intergranular corrosion on exposure to boiling 25% NaOH, rather than SCC. However, SCC of type 446 and E-Brite 26-1 heat treated at 871 °C (1600 °F) was reported (Ref 115) after exposure to deaerated 50% NaOH at 316 °C (600 °F).

In the case of duplex stainless steels, no SCC was reported in U-bend specimens of annealed IN-744 after a 15-day exposure to 90% NaOH at 300 °C (570 °F) (Ref 1). Other studies (Ref 123) have shown that introduction of the austenite phase into a ferritic matrix improves the caustic SCC resistance of experimental 25% Cr stainless steels (see Table 7), although general corrosion resistance is decreased.

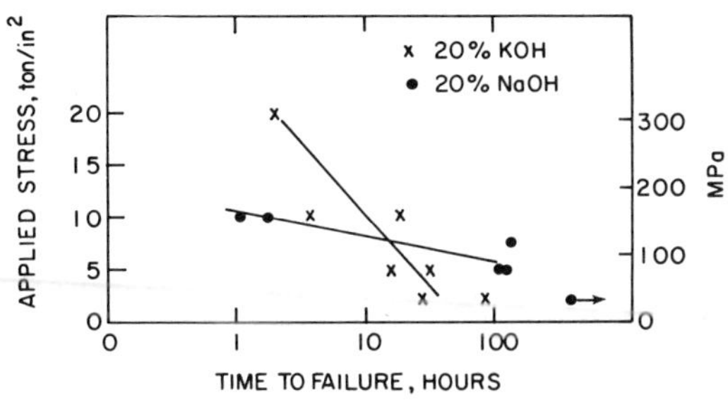

Fig. 34 Effect of applied stress on the times to failure of type 347 in oxygenated sodium hydroxide and potassium hydroxide solutions at 300 °C (570 °F). After Ref 64

Table 6 Cracking Behavior of Various Stainless Steels in Deaerated 10% NaOH Solution at 332 °C C-Ring Specimens Stressed at 90% of the Yield Strength

Material	Condition(a)	Observation(b)
Type 304 (austenitic)	Mill annealed	NC
	10 h/649 °C	NC
Type 405 (ferritic)	Mill annealed	NC, GC
	1 h/1010 °C/FC	NC, GC
Fe-18Cr-2Mo-0.5Ti (ferritic)	Mill annealed	NC, IGA
	1 h/1010 °C/FC	C
	3 h/475 °C	C
	20 h/475 °C	C
E-BRITE 26-1 (ferritic)	Mill annealed	NC
	1 h/1010 °C/FC	C
	100 h/475 °C	NC
	300 h/475 °C	C
3RE60 (duplex)	Mill annealed	NC, GC, IGA
	10 h/649 °C	NC, GC, IGA
	100 h/475 °C	C
	300 h/475 °C	C
Type 410 (martensitic)	Mill annealed	NC, GC
	Tempered 649 C (28 HRC)	NC, GC
	Tempered 565 C (41 HRC)	NC, GC
	Tempered 482 C (50 HRC)	C

(a) FC, furnace cooled. **(b)** NC, no cracking in 4800 h; C, cracking during 4800 h; GC, heavy general corrosion; IGA, intergranular attack
After Ref 91

SCC in Other Environments

High-Temperature Water SCC

As noted in the introduction to this chapter, "high-temperature water SCC" denotes SCC in the cooling water of nuclear reactors and has been used to describe the SCC of weld heat-affected (sensitized) zones in type 304 piping exposed to oxygenated BWR coolants. In the 1970s and early 1980s, more articles were written about the SCC of sensitized type 304 in BWR environments than about any other corrosion topic, and several detailed reviews are available (Ref 124-127). A compilation of test data from 17 test programs carried out in the 1970s is shown in Fig. 35 (Ref 113). The data clearly indicate that at dissolved oxygen contents of between 0.15 and 0.3 ppm and chloride contents of between 0.02 and 0.5 ppm, which span the BWR operating ranges (Ref 124), sensitized type 304 can undergo SCC. Resin intrusions can accelerate this cracking (Ref 126).

Considerable research has gone into understanding the mechanism of intergranular SCC of type 304 BWR piping, and a quantitative and predictive model has been developed (Ref 128-130). The model is based on the slip-dissolution/film-rupture model and relates crack advance to the oxidation reactions that are occurring at the crack tip, where a thermodynamically stable oxide (or protective film) is ruptured by an increase in the strain in the underlying matrix. The amount of subsequent crack propagation is related via Faraday's law to the oxidation charge density associated with both dissolution and oxide growth on the bare surface. Such an environmentally controlled mode of crack advance is sustained by a further increment in matrix strain, which reruptures the oxide after a certain time. This model successfully predicts to within a factor of 2 the environmentally controlled crack-propagation rates for stainless steel in water at 288 °C (550 °F). The model is used in the nuclear power industry as a crack-propagation prediction tool (Ref 131).

Several preventive measures have been developed for the intergranular SCC of BWR piping, including materials solutions such as the adoption of niobium-stabilized type 347L and of types 304NG and 316NG (described in the earlier

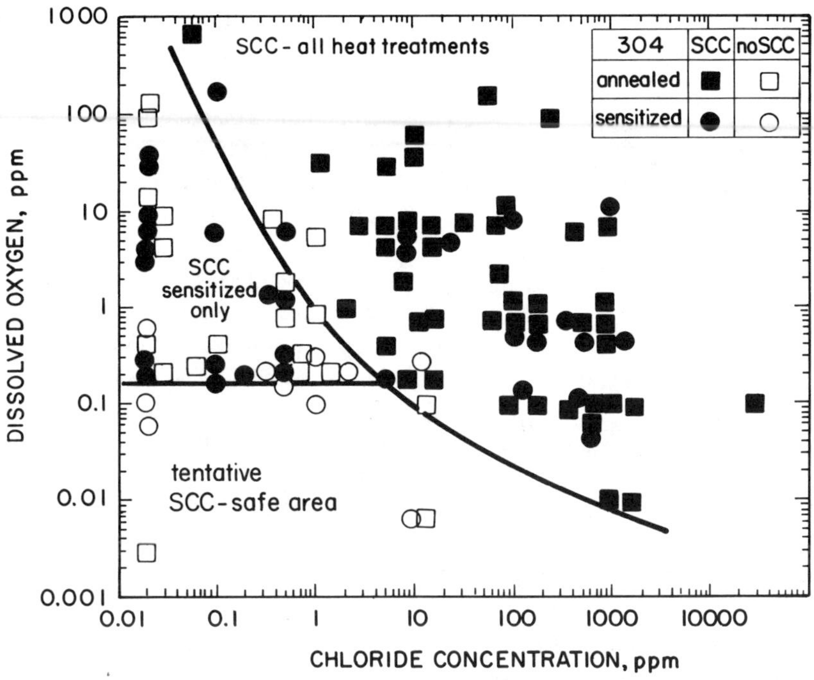

Fig. 35 Concentration ranges of dissolved oxygen and chloride that may lead to SCC of type 304 in water at temperatures ranging from 260 to 300 °C (500 to 570 °F). Applied stresses in excess of yield strength and test times in excess of 1000 h, or strain rates greater than 10^{-5}/s. After Ref 113

Table 7 Caustic Cracking Resistance of Various Experimental Stainless Steels Tested as Bent-Beam Specimens in Boiling 30% NaOH at 119 °C (246 °F) for 300 h

Alloy composition, wt%					0.5 h/900 °C/WQ		0.5 h/1100 °C/WQ		0.16 h/1300 °C/WQ	
Cr	Ni	Mo	C	N	Phases(a)	Result(b)	Phases(a)	Result(b)	Phases(a)	Result(b)
24.6	0.05	2.07	0.023	0.018	α	NF	α	C	α	C
24.9	0.50	2.11	0.029	0.020	α	NF	α	C	α	C
25.2	0.96	2.09	0.027	0.019	α	NF	α	C	α	C
24.9	1.95	2.09	0.030	0.018	α	NF	α	C	α	C
25.2	3.38	2.03	0.027	0.021	α+γ	NF	α	C	α	C
25.2	4.41	2.14	0.032	0.024	α+γ	NF	α+γ	C	α	C
24.9	5.92	2.09	0.027	0.017	...	...	α+γ	NF	α	C
24.8	7.63	2.10	0.024	0.014	...	...	α+γ	NF	α	NF
24.9	9.50	2.10	0.024	0.019	...	...	α+γ	NF	α	NF

(a)α, ferrite; γ, austenite. (b) NF, no failure; C, intergranular caustic cracking
After Ref 123

section "Sensitization"), solution treatment, cladding with types 308 and 308L, and weld overlays. Minimizing tensile stresses by procedures such as heat-sink welding and altering the water chemistry of the BWR coolant are also remedies. Reference 131 provides a comprehensive description of all these preventive measures.

Polythionic Acid SCC

In the 1950s a form of intergranular SCC became recognized as a technological problem in catalytic reformers used in the petroleum industry. Earlier studies (Ref 132, 133) showed that this form of attack occurred in the aqueous con-

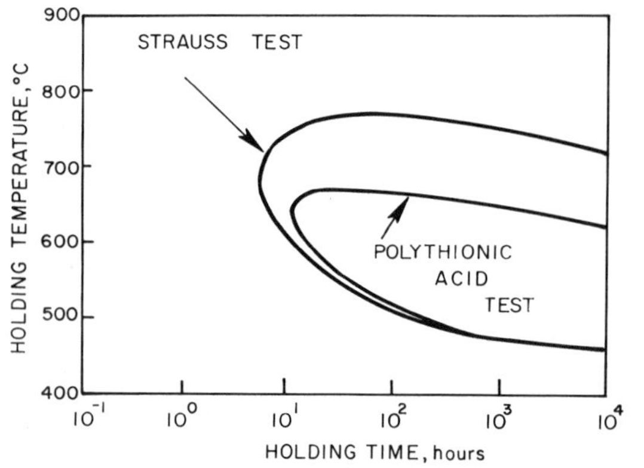

Fig. 36 Effect of heat treatment on the resistance of type 304 (0.04% C) in polythionic acid and Strauss tests. After Ref 133

densates of reforming furnaces when polythionic acids ($H_2S_xO_6$, where $x = 3, 4,$ or 5) were present. Subsequently, this form of attack became known as polythionic acid SCC. More recent studies (Ref 134) have shown that of all the polythionic acids, only tetrathionic acid ($H_2S_4O_6$) induces SCC in sensitized type 304. Tetrathionic acid has also been shown to be the critical species causing SCC of sensitized type 304 in tests employing sulfurous acid solutions (Ref 135).

A laboratory test for resistance to polythionic acid cracking has been standardized as ASTM Practice G 35. The solution is prepared by bubbling sulfur dioxide and then hydrogen sulfide through distilled water. Stressed U-bend specimens are often used in such tests, although any type of SCC test specimen can be used. Excellent correlation has been shown (Ref 133) between polythionic acid cracking resistance as determined by ASTM G 35 and intergranular corrosion resistance as determined by the Strauss test (ASTM A 393) for a dozen different austenitic stainless steels and higher alloys. An example of this correlation in terms of a TTS diagram is shown in Fig. 36 (Ref 133). As a result of this study, it is now generally accepted that polythionic acid cracking is a form of intergranular SCC linked to sensitization. Accordingly, attempts to mitigate this form of attack have included the metallurgical remedies used to minimize sensitization.

The technological problem encountered in the petroleum industry is essentially as follows. Stainless steel components are used at operating temperatures that may be high enough to sensitize the material. On subsequent shutdown of a refinery unit, polythionic acids may form by the interaction of sulfur compounds, moisture, and air at low temperatures and attack the chromium-depleted boundaries. A combination of operational procedures and metallurgical remedies has been used to minimize polythionic acid cracking during shutdowns. The operational procedures are described in NACE Recommended Practice RP-0170-85 and are based on preventing the condensation of water vapor or maintaining an alkaline environment by additions of ammonia or sodium carbonate.

Among the metallurgical remedies that have been tried are the use of low-carbon grades (*e.g.*, type 304L), the use of stabilized grades (type 321 or 347), and the use of higher alloys such as Incoloy alloy 801. However, the stabilized and the low-carbon grades of stainless steel can be sensitized at certain time-temperature combinations (Ref 1). Accordingly, it has been recommended that low-carbon grades be used for operating temperatures below 400 °C (750 °F), and that stabilized grades be thermally treated at temperatures of 900 to 925 °C (1650 to 1700 °F) to ensure maximum precipitation of carbon as titanium or niobium carbides prior to service (Ref 136). In cases where welding is necessary, a British practice has been to weld the thermally stabilized (920 °C, or 1690 °F) type 321 and to locally anneal the welds at 920 °C (1690 °F) (Ref 1). Higher alloys, such as the titanium-stabilized Incoloy alloy 801, have been used in the U.S. (Ref 137). Weld overlays, in the as-deposited state, of several austenitic stainless steels have been reported to be resistant to polythionic acid SCC (Ref 138).

SCC in the Presence of Other Sulfur Species

In addition to SCC in the presence of hydrogen sulfide and polythionic acids, SCC caused by thiosulfate, thiocyanate, and sulfate solutions has been investigated. Stress-corrosion cracking of sensitized type 304 in thiosulfate solutions at ambient temperatures is of interest to the nuclear industry. Borate solutions are stored in stainless steel tanks in containment buildings for use as sprays in case of nuclear accidents. In addition to borate, the solutions contain thiosulfate anions that would react with iodine fission products. Following earlier reports of SCC of sensitized austenitic stainless steels in these borate solutions (Ref 139), thiosulfate was identified as the critical species causing SCC (Ref 140). Laboratory tests showed that SCC could occur with very low concentrations (*e.g.*, 0.1 ppm) of $Na_2S_2O_3$, although in the presence of boric acid higher concentrations (*e.g.*, 1 ppm) were required (Ref 140). Cathodic protection was shown to prevent SCC (Ref 140).

Thiosulfate solutions have been used in recent work related to a percolation model of intergranular SCC in sensitized alloys (Ref 141). Essentially, this model states that a continuous path of susceptible grain boundaries must be present for a crack to propagate macroscopic distances, and the percentage of sensitized grain-boundary facets required to achieve this can be derived from a percolation model. A percolation threshold (p_c) of about 24% was derived for a model lattice of boundaries, and approximate correspondence to this was achieved in slow-strain-rate tests of material heat treated for various lengths of time. Below p_c, small cracks initiated but ran out of susceptible grain boundaries. A percolation model may explain (Ref 142) the extremely abrupt onset of SCC with increasing sensitization time.

Heavily sensitized type 304 stainless steel has been found (Ref 143) to be susceptible to SCC in thiocyanate solutions at concentrations greater than 10^{-4} mol/L.

Stress-corrosion cracking of tempered (sensitized) type 403 martensitic stainless steel is of interest to the power industry, since the martensitic grades are used as blade materials in steam turbines. The impurities found in the steam include sulfate. Tempered martensitic type 403 has been found (Ref 144) to suffer pitting corrosion and intergranular SCC in 0.01 *M* Na_2SO_4 at temperatures of 75 and 100 °C (167 and 212 °F), but not at the lower temperatures of 25 and 50 °C (77 and 122 °F). Significant sulfate contamination of the passive film from solution was found, but the

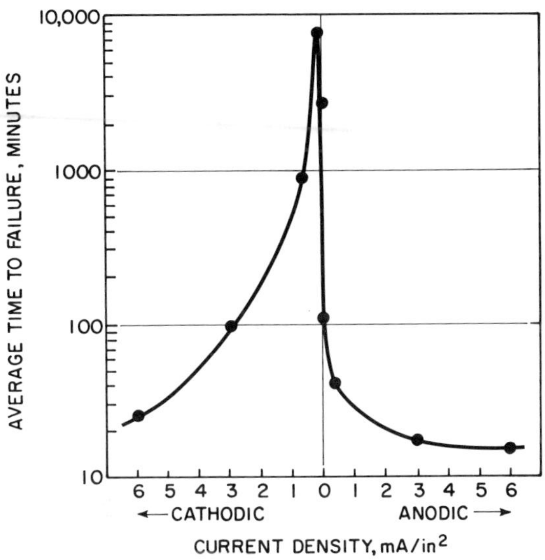

Fig. 37 Effect of applied current on the time to failure of an Fe-12Cr-1Mo-0.33V-0.25C martensitic stainless steel in an oxygenated 3% NaCl solution at pH 6.5. After Ref 146

level of contamination could not be correlated with the SCC susceptibility of the alloy. Nonmetallic inclusions (MnS) and carbide precipitates were found to act as nucleation sites for corrosion pits, which in turn gave rise to SCC (Ref 144).

Stress-corrosion cracking has been noted (Ref 53) in a 26Cr-1Mo low-interstitial ferritic stainless steel in 132 °C (270 °F) water containing chloride, hydrogen sulfide, ammonia, and traces of oil, thiocyanates, and organic acids. The possibility that this failure may be due to hydrogen embrittlement cannot be ruled out.

Hydrogen Embrittlement

Martensitic and Precipitation-Hardening Stainless Steels

For SCC in aqueous environments, such as saline solutions, two mechanisms have been discussed in the literature. The earlier discussions (Ref 145) distinguished between SCC that is accelerated by the application of anodic currents and SCC that is accelerated by cathodic currents. An example of this behavior is shown in Fig. 37 (Ref 146) for an Fe-12Cr-1Mo-0.33V-0.25C

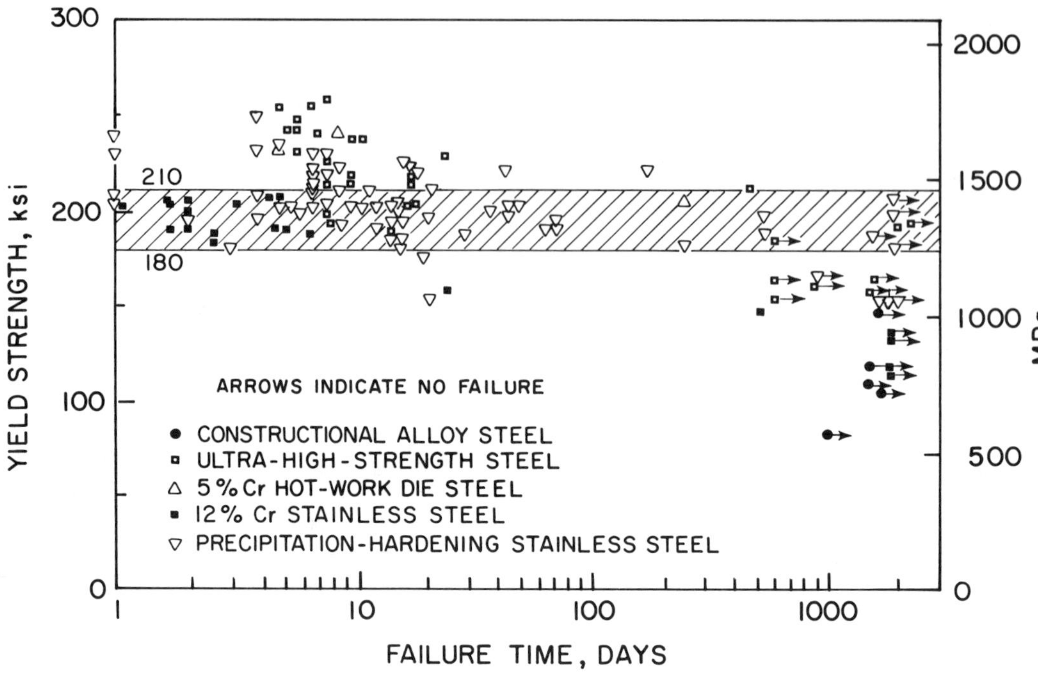

Fig. 38 Effect of yield strength on the times to failure of various steels exposed to a marine atmosphere at 75% of the yield stress. After Ref 148

martensitic stainless steel in a neutral salt solution. Subsequent studies (Ref 147) demonstrated that hydrogen enters steel when in contact with a neutral sodium chloride solution at both anodic and cathodic potentials. It is now widely accepted (Ref 95, 96) that in high-strength stainless steels, hydrogen embrittlement is the crack-propagation mechanism under both anodic and cathodic conditions. Anodic dissolution may, however, contribute to the initiation of the crack nucleus by processes such as pitting. It should also be noted that the presence of the chloride ion is not necessary to cause hydrogen embrittlement in the high-strength stainless steels (Ref 145).

When considering service environments, it has been customary to discuss the cracking of high-strength stainless steels in environments that contain sulfides and in environments that do not containing sulfides. Sulfide environments are of particular interest to the oil industry, and there is a large measure of agreement with the view that cracking in sulfide environments (known as sulfide SCC) is a form of hydrogen embrittlement. Sulfide is a hydrogen-ion recombination poison that promotes the entry of hydrogen into a metal.

Hydrogen Embrittlement in Sulfide-Free Environments. Most evaluations of the resistance of high-strength stainless steels in sulfide-free environments have been carried out either in natural marine atmospheres or in chloride solutions. Among the parameters that determine hydrogen-embrittlement resistance, yield strength has been identified as having a dominant effect. Hydrogen embrittlement can occur quite readily in most high-strength materials irrespective of composition or structure. This is illustrated by the data presented in Fig. 38 (Ref 148) for a variety of high-strength materials. It is evident that martensitic and precipitation-hardening stainless steels can exhibit hydrogen embrittlement in marine atmosphere at yield strengths above 1035 MPa (150 ksi). This applies to smooth specimens. In notched or precracked specimens, hydrogen embrittlement may occur at lower yield strengths.

Tempering, in the case of martensitic stainless steels, and overaging, in the case of precipitation-hardening stainless steels, can significantly lower

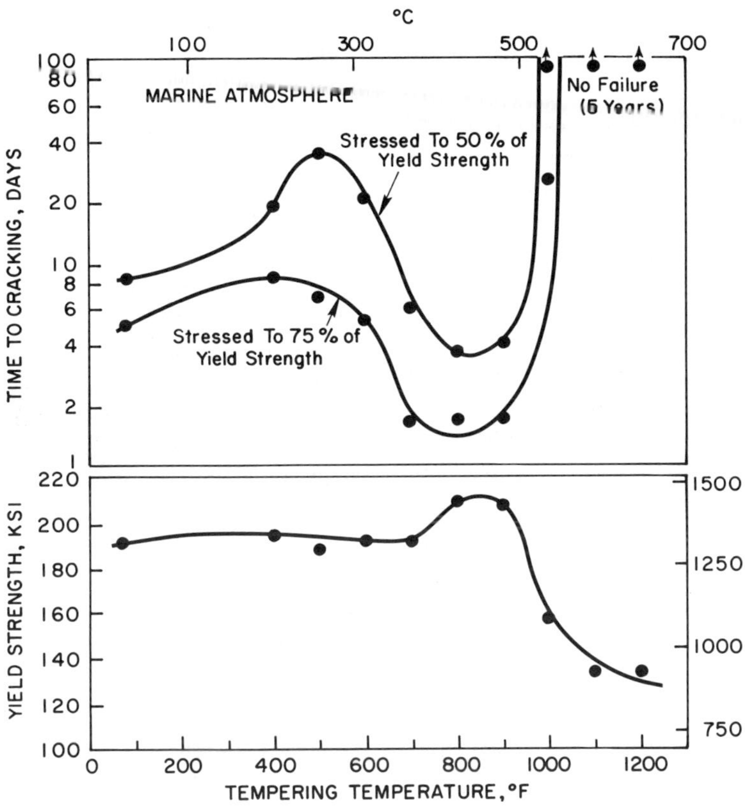

Fig. 39 Effect of tempering temperature on the cracking resistance and yield strength of an Fe-12Cr-1Mo-0.33V-0.25C martensitic stainless steel. After Ref 149

Table 8 K_{ISCC} Values for Several Precipitation-Hardening Stainless Steels Determined Using 3.5% NaCl Solution

Material	Yield strength		K_{Ic}		K_{ISCC}	
	MPa	ksi	MPa$\sqrt{m}$	ksi$\sqrt{in.}$	MPa$\sqrt{m}$	ksi$\sqrt{in.}$
17-4 PH	1215	176	56	51	56	51
17-4 PH	1090	158	131	119	131	119
15-5 PH	1210	175	106	96	85	77
15-5 PH	1090	158	132	120	132	120
PH 13-8Mo	1435	208	81	74	81	74
AM 362	1360	197	33	30	11	10
AM 362	1210	175	44	40	30	27
Custom 455	1695	246	79	72	79	72

After Ref 150

the yield strength and hence increase resistance to hydrogen embrittlement. An example of the effect of tempering on the yield strength and hydrogen embrittlement resistance of a martensitic stainless steel in marine atmosphere is presented in Fig. 39 (Ref 149). The relationship between yield strength and K_{ISCC} of a number of precipitation-hardening stainless steels in a 3.5% NaCl solution is shown in Table 8 (Ref 150). Other studies (Ref 27) employing fracture-mechanics specimens have shown that for type 431 martensitic stainless steel, the untempered (as-quenched) martensite exhibits much faster crack-growth rates and much lower K_{ISCC} values than the quenched and tempered material. Accordingly, it has become accepted practice to define resistance to hydrogen embrittlement in terms of yield strength, hardness, or heat-treatment conditions. For example, the Marshall Space Flight Center permits the use of precipitation-hardening stainless steels in space vehicles and associated equipment only in certain conditions of heat treatment known to improve resistance to hydrogen embrittlement, as outlined in Table 9 (Ref 151).

Regarding environmental factors, it has been shown (Ref 146) that pH can have an important effect on cracking resistance. For a martensitic stainless steel heat treated to give a yield strength of 1400 MPa (203 ksi), no hydrogen embrittlement was observed at a pH higher than 11.5, whereas rapid cracking was found at a low pH (Ref 146). However, the acidification of the solution within the crack could lead to hydrogen embrittlement not anticipated from tests in neutral solutions.

Resistance to hydrogen embrittlement can be improved by alloying with niobium (Ref 96) and by introducing δ ferrite into the structure (Ref 152). Detrimental features are manganese sulfide inclusions, which provide pits that can act as crack-initiation sites (Ref 153), and certain surface-machining procedures that generate localized subsurface cold-worked regions at which cracking can initiate (Ref 154).

Table 9 Precipitation-Hardening Stainless Steels With High Resistance to Hydrogen Embrittlement as Determined by the Marshall Space Flight Center

Material	Condition
A286	All
AM350	SCT 1000 and above
AM355	SCT 1000 and above
ALMAR 362	H1000 and above
CUSTOM 455	H1000 and above
15-5 PH	H1000 and above
PH 14-8 Mo	CH900 and SRH950 and above
PH 15-7 Mo	CH900
17-7 PH	CH900

After Ref 151

Hydrogen Embrittlement in Sulfide-Containing Environments. It is well known that the presence of hydrogen sulfide in various environments decreases the hydrogen embrittlement resistance of high-strength stainless steels. This problem became prominent in the 1950s as a result of failures in the oil industry and has received considerable attention. Most of the studies have dealt with steels other than high-strength stainless steels; however, there is sufficient information about the latter, because many evaluations have included some of these materials. The first extensive investigation (Ref 155), which included the precipitation-hardening stainless steels AM-350, AM-355, A-286, and 17-4PH, led to the conclusion that in sulfide-containing environments none of the steels tested completely resisted hydrogen embrittlement at yield strengths above 690 MPa (100 ksi). It was also established (Ref 155) that there exists a minimum hardness below which cracking does not occur for a given applied stress, and that this minimum hardness increases with decreasing hydrogen sulfide concentration. Numerous subsequent studies, covering a wide range of materials, have led to the issuance of NACE Standard MR-01-75 (Ref 156), which lists metallic material requirements for resistance to hydrogen embrittlement for petroleum production, drilling, gathering and flowline equipment, and field processing facilities to

Table 10 Stainless Steels Listed in NACE Standard MR-01-75(a) as Acceptable for Sulfide Environments

Stainless steel type	Acceptable grades
Austenitic(b)	302, 304, 304L, 310, 316, 316L 317, 321, 347, Carpenter 20Cb3
Ferritic(c)	405, 430
Martensitic(d)	410, CA15, CA15M
Precipitation-hardening	A286(e), 17-4PH(f)

(a) Revised annually. Reader should consult Technical Practices Committee of NACE for considered modifications. Some of the materials listed may be susceptible to chloride cracking in certain environments. **(b)** Annealed condition, not strengthened by cold work. **(c)** Annealed condition, hardness of 22 HRC (max). **(d)** Double tempered to hardness of 22 HRC (max). **(e)** Aged to hardness of 35 HRC (max). **(f)** Aged to hardness of 33 HRC (max)

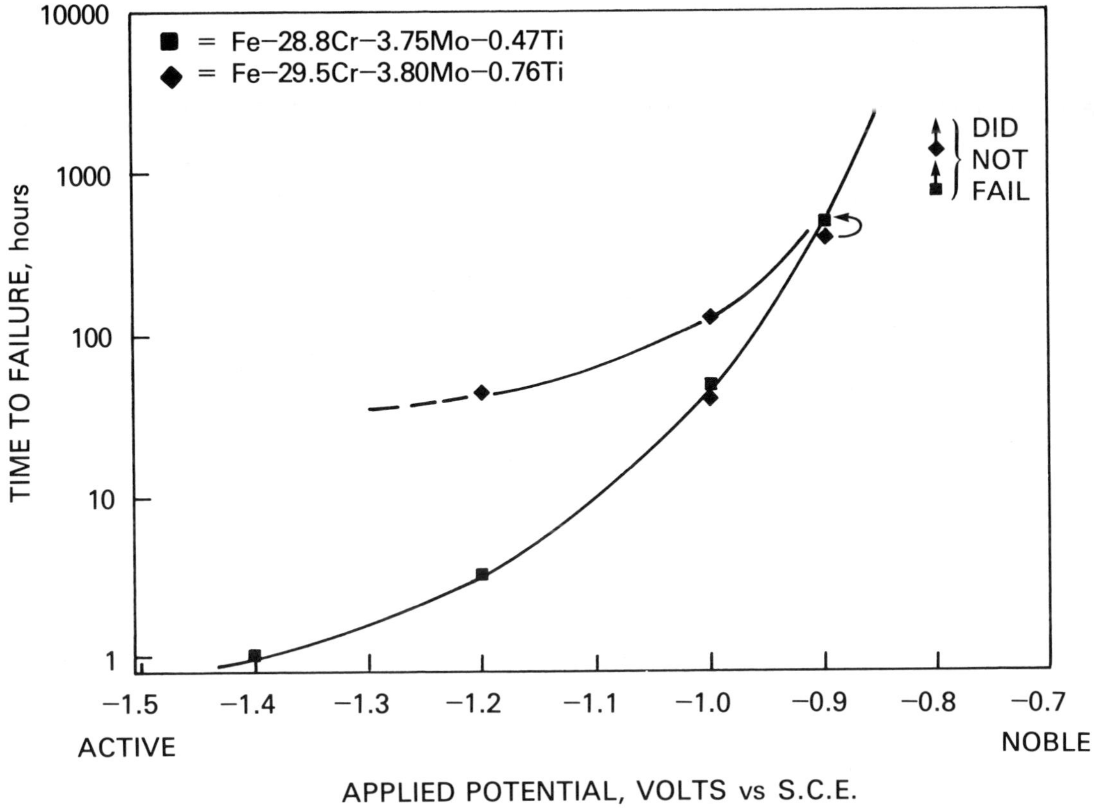

Fig. 40 Time to failure by hydrogen embrittlement of two heats of as-welded AL 29-4C stainless steel in ambient-temperature synthetic seawater as a function of applied potential. After Ref 158

be used in hydrogen sulfide service. This standard is revised annually. The part of the listing that relates to stainless steels is summarized in Table 10. Factors such as low pH, increase in temperature and hydrogen sulfide concentration, galvanic coupling to anodic materials, and cold work facilitate hydrogen embrittlement, and compliance with NACE Standard MR-01-75 may not be adequate to prevent failures under all conditions. In this regard, hydrogen-embrittlement failures have been reported (Ref 157) in type 410 (UNS S41000) and 17-4PH (UNS S17400) stainless steel tubing hangers in wellhead equipment, even though the hanger material met NACE Standard MR-01-75 hardness requirements.

Ferritic Stainless Steels

Hydrogen embrittlement was identified (Ref 158, 159) as a potential problem for superferritic stainless steel condenser tubes for seawater-cooled power plant condensers in the mid-1980s. In many instances, these tubes are used in conjunction with copper alloy tubesheets and cast iron or steel waterboxes, and the latter materials require cathodic protection by impressed-current systems. Cathodic protection at about – 0.8 V (SCE) is necessary to protect the tubesheets and waterboxes from corrosive attack. Both superferritics, 29-4C (UNS S44735) and Sea-Cure (UNS S44660), exhibit hydrogen embrittlement when they are cathodically polarized to potentials in the range of – 0.9 to –1.4 V (SCE). The data for 29-4C are shown in Fig. 40 (Ref 158), which demonstrates that overprotection at potentials more negative than – 0.8 V (SCE) must be carefully avoided. Since hydrogen can also be picked up from annealing in hydrogen atmospheres or acid pickling, these processes should be avoided for superferritic stainless steels. It is also claimed that the loss in ductility of the superferritics that results from hydrogen charging can be eliminated by outgassing the alloy at slightly elevated temperatures (Ref 159). Recently the UNS S44660 (Sea-Cure) alloy composition has been modified to contain %C + %N = 0.02 max, and the titanium stabilizing addition has been replaced by niobium (Ref 160). The new developmental alloy, designated by the trade name Sea-Cure Hy-Resist, reportedly can be cathodically protected to –2.0 V (SCE) without exhibiting hydrogen embrittlement (Ref 160).

Prior to the discovery of hydrogen embrittlement of superferritic stainless steels at cathodic potentials in seawater, most of the investigations of hydrogen embrittlement of ferritic stainless steels had been done under conditions of cathodic charging in sulfuric acid solutions containing arsenic (Ref 161-164). Applied stresses at or above the yield stress and the presence of a hydrogen-ion recombination poison (*e.g.*, an arsenic compound) in the sulfuric acid solutions are required for hydrogen embrittlement to occur (Ref 164). Increasing heat-treatment temperature and increasing the molybdenum content of the ferritic stainless steel reduce resistance to hydrogen embrittlement (Ref 164), as does applying test stresses perpendicular to the cold-rolling direction of the steel (Ref 162). The crack path of hydrogen embrittlement in the ferritic steels is transgranular (Ref 98, 164), with some evidence of intergranular cracking at the crack-initiation stage (Ref 98).

References

1. A.J. Sedriks, *Corrosion of Stainless Steels*, John Wiley and Sons, 1979
2. *Environment-Induced Cracking of Metals*, R.P. Gangloff and M.B. Ives, Ed., National Association of Corrosion Engineers, 1990
3. V.N. Krivobok, *The Book of Stainless Steels*, E.E. Thum, Ed., American Society for Steel Treating, 1933
4. C. Husen and C.H. Samans, *Chem. Eng.*, Jan 27, 1969, p 178
5. B. Strauss, H. Schottky, and J. Hinnuber, *Z. Anorg. Chem.*, Vol 188, 1930, p 309
6. E.C. Bain, R.H. Aborn, and J.B. Rutheford, *Trans. Am. Soc. Steel Treat.*, Vol 21, 1933, p 481
7. J.E. Alexander, "Alternative Alloys for BWR Piping Applications," Final Report, NP-2671-LD, General Electric Co., Oct 1982
8. P.S. Maiya and W.J. Shack, *Corrosion*, Vol 41, 1985, p 630
9. P.L. Andresen, F.P. Ford, S.M. Murphy, and J.M. Perks, State of Knowledge of Radiation Effects on Environmental Cracking in Light Water Reactor Core Materials, in *Proc. 4th Int. Conf. Environmental Degradation of Materials in Nuclear Power Systems—Water Reactors*, National Association of Corrosion Engineers, 1990
10. T.R. Anthony, Solute Segregation and Stresses Around Growing Voids in Metals, in *Proc. Radiation-Induced Voids in Metals and Alloys*, J.W. Corbet and L.C. Ianiello, Ed., U.S. Atomic Energy Commission Symp. Series 26, CONF 710601, 1971, p 630
11. S.M. Bruemmer, L.A. Charlot, and E.P. Simonen, "Grain Boundary Chemistry Effects on Irradiation-Assisted Stress Corrosion Cracking," Paper No. 506, presented at Corrosion/90, National Association of Corrosion Engineers, 1990
12. A.J. Jacobs and G.P. Wozadlo, *J. Mater. Eng.*, Vol 9 (No. 4), 1988, p 345
13. R.H. Jones, Some Radiation Damage-Stress Corrosion Synergisms in Austenitic Stainless Steels, in *Proc. 2nd Int. Symp. Environmental Degradation of Materials in Nuclear Power Systems—Water Reactors*, American Nuclear Society, 1985, p 173
14. G.R. Caskey, R.S. Ondrejcin, P. Aldred, R.B. Davis, and S.A. Wilson, "Effects of Irradiation on Intergranular Stress Corrosion Cracking of Type 304 Stainless Steel," Paper No. 504, presented at Corrosion/90, National Association of Corrosion Engineers, 1990
15. *Stainless Steel and High Alloy Tubing and Pipe*, Trent Tube, 1990
16. M.A. Scheil, in *Symposium on Stress Corrosion Cracking of Metals*, STP 64, American Society for Testing and Materials, 1945, p 395
17. I.B. Casale, *Corrosion*, Vol 23, 1967, p 314
18. M. Kowaka and T. Kudo, *Trans. Jpn. Inst. Met.*, Vol 16, 1975, p 385
19. I.A. Maier, C. Manfredi, and J.R. Galvele, *Corros. Sci.*, Vol 25 (No. 1), 1985, p 15
20. A.J. Sedriks, *Stress Corrosion Cracking Test Methods*, National Association of Corrosion Engineers, 1990
21. R.M. Latanision and R.W. Staehle, *Proc. Conf. Fundamental Aspects of Stress Corrosion Cracking*, National Association of Corrosion Engineers, 1969, p 214
22. G.J. Theus and R.W. Staehle, *Stress Corrosion Cracking and Hydrogen Embrittlement of Iron Base Alloys*, National Association of Corrosion Engineers, 1977, p 845
23. H.R. Copson, *Physical Metallurgy of Stress Corrosion Fracture*, Interscience, 1959, p 247

24. R.W. Staehle, *Trans. Inst. Chem. Eng.*, Vol 47, 1969, p T-227
25. A.J. Sedriks, *J. Inst. Met.*, Vol 101, 1973, p 225
26. A.J. Sedriks, *Corrosion*, Vol 31, 1975, p 339
27. M.O. Speidel, *Metall. Trans. A*, Vol 12, 1981, p 779
28. R.L. Beauchamp, M.S. thesis, Ohio State University, 1963
29. J.J. Royuela and R.W. Staehle, unpublished results quoted in Ref 21
30. J. Hochman and J. Bourrat, *Compt. Rend. Acad. Sci.*, Vol 255, Dec 17, 1962
31. J.E. Truman, Brown-Firth Research Laboratories, Sheffield, U.K., private communication
32. R.W. Staehle, J.J. Royuela, T.L. Raredon, E. Serrate, C.R. Morin, and R.W. Farrar, *Corrosion*, Vol 26, 1970, p 451
33. H.H. Uhlig and R.A. White, *Trans. ASM*, Vol 52, 1960, p 830
34. F.S. Lang, *Corrosion*, Vol 18, 1962, p 378t
35. D.G. Tufanov, *Metalloved. Term. Obrabotka Met.*, 15 April 1964
36. D. vanRooyen, in *Proc. 1st Int. Conf. Stress Corrosion Cracking*, Butterworths, 1961
37. J.G. Hines and E.R.W. Jones, *Corros. Sci.*, Vol 1, 1961, p 88
38. M. da Cunha Belo and J. Montuelle, in *Proc. NATO Conf. Theory of Stress Corrosion Cracking in Alloys*, Ericeiria, Portugal, March 1971
39. M. Kowaka and H. Fujikawa, *Sumitomo Search*, No. 7, 1972, p 10
40. H.E. Hanninen, *Int. Met. Rev.*, Vol 24 (No. 3), 1979, p 85
41. E.E. Denhard and R.R. Gaugh, *Stress Corrosion Testing*, STP 425, American Society for Testing and Materials, 1967, p 41
42. H. Spahn, G.H. Wagner, and U. Steinhoff, Methods to Prevent Stress Corrosion Cracking, *Techn. Ueberwachung*, Vol 14, 1973, p 292
43. K. Fassler in *Korrosionum*, H. Grafen, F. Kahl, and A. Rahmel, Ed., Vol 1, Verlag Chemie, 1974, p 136
44. A.J. Sedriks, *Corrosion*, Vol 42 (No. 7), 1986, p 376
45. M.G. Fontana, F.H. Beck, and J.W. Flowers, *Met. Prog.*, Vol 80 (No. 6), 1961, p 99
46. D. Warren, "Microstructure and Corrosion Resistance of Austenitic Stainless Steels," Sixth Annual Liberty Bell Corrosion Course, National Association of Corrosion Engineers, Sept 1968
47. F.S. Lang, *Corrosion*, Vol 18, 1962, p 378t
48. R.F. Overman, *Corrosion*, Vol 22, 1966, p 48
49. J.E. Truman, The Effects of Composition and Structure on the Resistance to Stress Corrosion Cracking of Stainless Steels, in *British Nuclear Energy Society Symposium on Effects of Environment on Material Properties in Nuclear Systems*, Paper No. 10, Institute of Civil Engineers, July 1971
50. R.W. Cochran and R.W. Staehle, *Corrosion*, Vol 11, 1968, p 369
51. K.L. Money and W.W. Kirk, *Mater. Perform.*, Vol 17 (No. 7), July 1978, p 28
52. F. Zucchi, G. Trabanelli, G. Rocchini, and G. Perboni, *Corros. Sci.*, Vol 29 (No. 4), 1989, p 417
53. E.C. Hoxie, Some Corrosion Considerations in the Selection of Stainless Steel for Pressure Vessels and Piping, in *Pressure Vessels and Piping: A Decade of Progress*, Vol 3, American Society of Mechanical Engineers, 1977
54. J.W. Oldfield and B. Todd, *Mater. Perform.*, Vol 29 (No. 12), 1990, p 57
55. C.P. Dillon, *Mater. Perform.*, Vol 29 (No. 12), 1990, p 66
56. R.M. Kain, *Mater. Perform.*, Vol 29 (No. 12), 1990, p 60
57. J.B. Gnanamoorthy, *Mater. Perform.*, Vol 29 (No. 12), 1990, p 63
58. J.Q. Lackey and M.A. Streicher, *J. Mater.*, Vol 3 (No. 4), 1968, p 983
59. N.A. Nielsen, *Corrosion*, Vol 27 (No. 5), 1971, p 173
60. J.P. Harston and J.C. Scully, *Corrosion*, Vol 25, 1969, p 493
61. S.J. Acello and N.D. Greene, *Corrosion*, Vol 18, 1962, p 286
62. S. Torchio, *Corros. Sci.*, Vol 20, 1980, p 555; A.I. Maier, C. Manfredi, and J.R. Galvele, *Corros. Sci.*, Vol 25, 1985, p 15
63. H. Grafen, *Corros. Sci.*, Vol 7, 1967, p 177
64. J.E. Truman, Methods Available for Avoiding SCC of Austenitic Stainless Steels in Potentially Dangerous Environments, in *Stainless Steels*, ISI Publication 117, Iron and Steel Institute, 1969, p 101
65. D. Warren, Chloride-Bearing Cooling Water and the Stress Corrosion Cracking of Austenitic Stainless Steel, in *Proc. 15th Annual Purdue Industrial Waste Conference*, Purdue University, May 1960, p 1
66. A.W. Dana and W.B. DeLong, *Corrosion*, Vol 12 (No. 7), 1956, p 309t
67. H.R. Baker, M.C. Bloom, R.N. Bolster, and C.R. Singleberry, *Corrosion*, Vol 26, 1970, p 420
68. Z. Szklarska-Smialowska, *Corrosion*, Vol 27, 1971, p 223
69. J.E. Truman, *Corros. Sci.*, Vol 17, 1977, p 737
70. J.L. Nelson, International Nickel Co., private communication quoted in Ref 1
71. E. Brauns and H. Ternes, *Werkst. Korros.*, Vol 19, 1968, p 1
72. A.S. Couper, *Mater. Protect.*, Vol 8, 1969, p 17
73. Y. Ogayu and T. Shoji, Cathodic Protection of Stress Corrosion Cracking of Austenitic Stainless Steels in Dilute NaCl Solutions, in *Proc. Life Prediction of Corrodible Structures*, Hawaii, 5-8 Nov 1991, National Association of Corrosion Engineers, to be published

74. R.L. Shamakian, A.R. Troiano, and R.F. Hehemann, *Corrosion*, Vol 36, 1980, p 279
75. J.H. Phillips and W.J. Singely, *Corrosion*, Vol 15, 1959, p 450t
76. S.P. Rideout, W.C. Rion, and J. Wade, *Trans. Am. Nucl. Soc.*, Vol 7, 1964, p 419
77. H.H. Uhlig, *J. Electrochem. Soc.*, Vol 116, 1969, p 173
78. M.L. Holzworth and A.E. Symonds, *Corrosion*, Vol 25, 1969, p 287
79. P.P. Snowden, *Nucl. Eng.*, Vol 6, 1961, p 409
80. S.A. Balezin and N.I. Podovaev, *J. Appl. Chem. USSR*, Vol 33, June 1990, p 1287
81. C.S. O'Dell, B.F. Brown and R.T. Foley, *Corrosion*, Vol 36 (No. 4), 1980, p 183
82. C. Edeleanu, *J. Iron Steel Inst.*, Vol 173, 1953, p 140
83. R.M. Davison, T. DeBold, and M.J. Johnson, in *Metals Handbook*, Vol 13, 9th ed., *Corrosion*, ASM International, 1987, p 547
84. L. Bednar, *Corrosion*, Vol 35, 1979, p 96
85. M.A. Streicher, "Stress Corrosion of Ferritic Stainless Steels," Paper No. 68, presented at Corrosion/75, National Association of Corrosion Engineers, 1975
86. B.F. Brown, *Stress Corrosion Cracking Control Measures*, NBS Monograph 156, 1977, p 55
87. *Stainless Steel: Allegheny Ludlum AL 29-4C*, Allegheny Ludlum Corp., 1990
88. R.F. Steigerwald, A.P. Bond, H.J. Dundas, and E.A. Lizlovs, *Corrosion*, Vol 33, 1977, p 279
89. A.P. Bond and H.J. Dundas, *Corrosion*, Vol 24, 1968, p 344
90. M.A. Streicher, *Platinum Met. Rev.*, Vol 21, 1977, p 51
91. I.L.W. Wilson, F.W. Pement, and R.G. Aspden, "Stress Corrosion Studies on Some Stainless Steels in Elevated Temperature Aqueous Environments," Paper No. 136, presented at Corrosion/77, National Association of Corrosion Engineers, 1977
92. M.A. Streicher, *Corrosion*, Vol 30, 1974, p 77
93. J.P. Carter and S.D. Cramer, *Mater. Perform.*, Sept 1980, p 13
94. S. Matsushima and T. Ishihara, *Trans. Natl. Res. Inst. Met.*, Vol 17, 1975, p 14
95. H. Spaehn, in *Environment-Induced Cracking of Metals*, R.P. Gangloff and M.B. Ives, Ed., National Association of Corrosion Engineers, 1990, p 449
96. J.E. Truman, *Int. Met. Rev.*, Vol 26 (No. 6), 1981, p 301
97. T. Suzuki, H. Hasegawa, and M. Watanabe, *Nippon Kinzaku Gakaishi*, Vol 32, 1968, p 1171
98. S. Shimodaira, M. Takano, Y. Takizawa, and H. Kamide, in *Stress Corrosion Cracking and Hydrogen Embrittlement of Iron Base Alloys*, R.W. Staehle, J. Hochmann, R.D. McCright, and J.E. Slater, Ed., National Association of Corrosion Engineers, 1977, p 1003
99. D.B. Anderson and C.J. Novak, "Stress Corrosion Cracking of High Strength Stainless and Low Alloy Steels in Chemical Plant Atmospheres," Paper No. 152, presented at Corrosion/76, National Association of Corrosion Engineers, 1976
100. *Sandvik SAF 2507: A High Performance Duplex Stainless Steel*, Sandvik Steel, Sweden, March 1990
101. R. Oltra, A. Desestret, E. Mirabal, and J.P. Bizouard, *Corros. Sci.*, Vol 27, 1987, p 1251
102. K. van Gelder, J.G. Erlings, J.W.M. Damen, and A. Visser, *Corros. Sci.*, Vol 27, 1987, p 1271
103. A. Cigada, T. Pastore, P. Pedeferri, and B. Vicentini, *Corros. Sci.*, Vol 27, 1987, p 1213
104. M. Barteri, F. Mancia, A. Tamba, and G. Montagna, *Corros. Sci.*, Vol 27, 1987, p 1239
105. "Testing of Metals for Resistance to Sulfide Stress Cracking at Ambient Temperatures," NACE Standard TM-01-77-86, National Association of Corrosion Engineers, 1986
106. G. Herbsleb and R.F. Poepperling, *Corrosion*, Vol 36 (No. 11), 1980, p 611
107. S.M. Wilhelm and R.D. Kane, *Corrosion*, Vol 40 (No. 8), 1984, p 431
108. "Corrosion Resistance of Nickel and Nickel-Containing Alloys in Caustic Soda and Other Alkalis," Corrosion Eng. Bull. CEB-2, International Nickel Co., 1976
109. M.G. Fontana and N.D. Greene, *Corrosion Engineering*, McGraw-Hill, 1978
110. J.K. Nelson, in *Metals Handbook*, Vol 13, 9th ed., *Corrosion*, ASM International, 1987, p 1174
111. J.E. Truman and R. Perry, *Br. Corros. J.*, Vol 1, 1966, p 60
112. H. Nathorst, Welding Research Council Bull. No. 6, Oct 1950, p 6
113. M.O. Speidel, "Stress Corrosion Cracking of Austenitic Stainless Steels," Ohio State University, Report to the Advanced Research Projects Agency, ARPA Order No. 2616, Contract No. N00014-75-C-0703, Aug 1977
114. A.J. Sedriks, S. Floreen, and A.R. McIlree, *Corrosion*, Vol 32, 1976, p 157
115. A.R. McIlree and H.T. Michels, *Corrosion*, Vol 33, 1977, p 60
116. A.J. Sedriks, J.W. Schultz, and M.A. Cordovi, *Corros. Eng. (Boshoku Gijutsu)*, Vol 28, 1979, p 82
117. M. Hecht, E.P. Partridge, W.C. Schraeder, and S.F. Hall, in *The Corrosion Handbook*, H.H. Uhlig, Ed., John Wiley & Sons, 1963, p 520
118. G.C. Wheeler and E. Howells, *Power*, Sept 1960, p 86
119. E. Howells, *Corros. Technol.*, 1960, p 368
120. A.V. Ryabchenkov, V.I. Gerasimov, and V.P. Sidorov, *Prot. Met. (USSR)*, Vol 2, May-June 1966, p 217

121. R.W. Staehle and A.K. Agrawal, "Corrosion, Stress Corrosion Cracking, and Electrochemistry of the Fe and Ni Base Alloys in Caustic Environments," Ohio State University, Report to ERDA, Contract E(11-1)-2421,, 1976
122. A.P. Bond, J.D. Marshall, and H.J. Dundas, in *Stress Corrosion Testing*, STP 425, American Society for Testing and Materials, 1967, p 116
123. M. Kowaka and T. Kudo, "Stress Corrosion Cracking Behavior of Ferritic and Duplex Stainless Steels in a Caustic Solution," paper presented at the Spring Meeting of the Japan Society of Corrosion Engineering, Tokyo, May 1977
124. B.M. Gordon, *Mater. Perform.*, Vol 19, 1980, p 29
125. S. Szklarska-Smialowska and G. Cragnolino, *Corrosion*, Vol 36, 1980, p 653
126. P.M. Scott, *Corros. Sci.*, Vol 25 (No. 8-9), 1985, p 583
127. J.C. Danko, in *Metals Handbook*, Vol 13, 9th ed., *Corrosion*, ASM International, 1987, p 927
128. F.P. Ford, D. Taylor, P.L. Andresen, and R.G. Ballinger, "Environmentally Controlled Cracking of Stainless and Low Alloy Steels in Light Water Reactors," Final Report, EPRI Contract RP 2006-6, Electric Power Research Institute, 1986
129. F.P. Ford and M. Silverman, *Corrosion*, Vol 36 (No. 10), 1980, p 558
130. F.P. Ford and P.L. Andresen, "The Theoretical Prediction of the Effect of System Variables on the Cracking of Stainless Steel and Its Use in Design," Paper No. 83, presented at Corrosion/87, National Association of Corrosion Engineers, 1987
131. B.M. Gordon and G.M. Gordon, in *Metals Handbook*, Vol 13, 9th ed., *Corrosion*, ASM International, 1987, p 927
132. A. Dravnieks and C. Samans, *Proc. Am. Pet. Inst.*, Vol 37 (No. 111), 1957, p 100
133. C. Samans, *Corrosion*, Vol 20 (No. 8), 1964, p 256t
134. S. Ahmad, M.L. Mehta, S.K. Saraf, and I.P. Saraswat, *Corrosion*, Vol 38 (No. 6), 1982, p 347
135. S. Ahmad, M.L. Mehta, S.K. Saraf, and I.P. Saraswat, *Corrosion*, Vol 37 (No. 7), 1981, p 412
136. C. Husen and C.H. Samans, *Chem. Eng.*, Jan 27 1969, p 178
137. C.D. Stephens and R.C. Scarberry, "Relationship of Polythionic Acid Cracking Susceptibility to Sensitization in Inconel Alloy 800 and 801," Paper No. 10, presented at Corrosion/69, National Association of Corrosion Engineers, 1969
138. E. Lendvai-Lintner, *Mater. Perform.*, March 1979, p 10
139. U.S. NRC I.E. Circular No. 76-06, Nov 26, 1976
140. H.S. Isaacs, B. Vyas, and M.W. Kendig, *Corrosion*, Vol 38 (No. 3), 1982, p 130
141. J. Stewart, P.M. Scott, D.B. Wells, and A.W. Herbert, "The Use of Percolation Theory to Predict the Probability of Failure of Sensitized Austenitic Stainless Steels by Intergranular Stress Corrosion Cracking," Paper No. 289, presented at Corrosion/88, National Association of Corrosion Engineers, 1988
142. R.C. Newman and A. Mehta, in *Environment-Induced Cracking of Metals*, R.P. Gangloff and M.B. Ives, Ed., National Association of Corrosion Engineers, 1990, p 489
143. P.M. Perillo and G.S. Duffo, *Corrosion*, Vol 46 (No. 7), 1990, p 545
144. B. Bavarian, Z. Szklarska-Smialowska, and D.D. Macdonald, *Corrosion*, Vol 38 (No. 12), 1982, p 604
145. E.H. Phelps, in *Fundamental Aspects of Stress Corrosion Cracking*, R.W. Staehle, A.J. Forty, and D. van Rooyen, Ed., National Association of Corrosion Engineers, 1969, p 398
146. H.J. Bhatt and E.H. Phelps, *Corrosion*, Vol 17, 1961, p 430t
147. B.E. Wilde, *Corrosion*, Vol 27, 1971, p 328
148. R.J. Schmitt and E.H. Phelps, *J. Met.*, March 1970, p 47
149. J.F. Bates and A.W. Loginow, *Corrosion*, Vol 20, 1964, p 189t
150. C.S. Carter *et al.*, *Corrosion*, Vol 27, 1971, p 190
151. D.B. Franklin, *Design Criteria for Controlling Stress Corrosion Cracking*, Marshall Space Flight Center Doc. MSFC-SPEC-522A, Nov 1977
152. P. Lillys and A.E. Nehrenberg, *Trans. ASM*, Vol 48, 1956, p 327
153. M.G. Fontana, "Stress Corrosion Cracking in Type 403 Stainless Steel", WADC Techn. Rep. 56-242, ASTIA Doc. No. AD-97215, Aug 1956
154. G.T. Murray, H.H. Honegger, and T. Mousel, *Corrosion*, Vol 40, 1984, p 146
155. C.M. Hudgins, R.L. McGlasson, P. Mehdizadeh, and W.R. Rosborough, *Corrosion*, Vol 22, 1966, p 238
156. "Sulfide Stress Cracking Resistant Metallic Materials for Oilfield Equipment," NACE Standard MR-01-75-91, National Association of Corrosion Engineers, 1991
157. R.M. Thompson, G.B. Kohut, D.R. Canfield, and W.R. Bass, *Corrosion*, Vol 47, 1991, p 216
158. J.F. Grubb and J.R. Maurer, "Use of Cathodic Protection With Superferritic Stainless Steels," Paper No. 28, presented at Corrosion/84, National Association of Corrosion Engineers, 1984
159. L.S. Redmerski, J. J. Eckenrod, K.E. Pinnow, and C.W. Kovach, "Cathodic Protection of Seawater Cooled Power Plant Condensers Operating With High Performance Ferritic Stainless Steel Tubing," Paper No. 208, presented at Corrosion/85, National Association of Corrosion Engineers, 1985
160. Crucible Materials Corporation, U.S. Patent No. 4942922

161. R.T. Newberg and H.H. Uhlig, *J. Electrochem. Soc.*, Vol 119, 1972, p 981
162. H.H. Uhlig and R.T. Newberg, *Corrosion*, Vol 28, 1972, p 337
163. J. Marquez, I. Matushima, and H.H. Uhlig, *Corrosion*, Vol 26, 1970, p 215
164. A.P. Bond and H.J. Dundas, in *Stress Corrosion Cracking and Hydrogen Embrittlement of Iron Base Alloys*, R.W. Staehle, J. Hochmann, R.D. McCright, and J.E. Slater, Ed., National Association of Corrosion Engineers, 1977, p 1136

CHAPTER 5

Stress-Corrosion Cracking of Nickel-Base Alloys

Narasi Sridhar and Gustavo Cragnolino, Center for Nuclear Waste Regulatory Analyses, Southwest Research Institute

Nickel and nickel-base alloys are specified for many critical service applications because of their fabricability, their fracture toughness, and their ability to be designed, by alloying, for a wide variety of environments. Additionally, unique intermetallic phases can form between nickel and some of its alloying elements, enabling formulation of alloys that exhibit high strength even at high temperatures. They are often used because of their resistance to stress-corrosion cracking (SCC). However, SCC of nickel-base alloys can occur under particular combinations of environment, microstructure, and stress. The number of environments that can cause SCC of nickel-base alloys (indeed of almost any alloy system) has grown with time (Table 1). This can be attributed to the increasing diversity of the application of nickel-base alloys, greater knowledge of service performance, and greater recognition of SCC problems. It is no longer necessary to think of stress-corrosion cracking as a manifestation of specific alloy/environment combinations, and attempts have been made to generalize various SCC systems within a potential-pH framework for a given metal such as nickel (Ref 1). Unfortunately, this approach is limited by:

- The paucity of systematic SCC data in terms of potential and pH
- The fact that SCC is not always controlled by the electrochemical stability of nickel, even in a single environment

Hence, for the present, we must be content to examine the various types of environments in which SCC has been observed in nickel-base alloys.

In the following sections, the physical metallurgy of nickel-base alloys is briefly discussed and then SCC is described in terms of various types of environments. The emphasis is on SCC at relatively moderate temperatures (less than 350 °C, or 660 °F). The information in this chapter is intended to provide the reader with an overview of the types of environment/alloy combinations that may lead to SCC. No attempt is made to synthesize the results in terms of various mechanisms. The reader is referred to other sources

Table 1 Evolution of Environments That Cause SCC of Nickel and Nickel-Base Alloys

Period	Environment
ca. 1950	H_2O-HF
	Fluosilicic acid
ca. 1978	H_2O-Cl^-
	H_2O-OH^-
	Polythionic acid
	Chromic acid
	Acetic acid
	Molten caustics
	Steam
	Organic liquids
	Liquid Li, Hg, Sn, Zn, Bi, Pb
	H_2O
	H_2
Late 1970s to present	H_2O-Cl^--CO_2-H_2S-S_8
	H_2O-Br^-
	H_2O-I^-
	Thiocyanates, tetrathionate, thiosulfate
	Sulfurous acid

Table 2 Nominal Chemical Compositions of Various Commercial Nickel-Base Alloys Used for Aqueous Corrosion Resistance

Common alloy designation	UNS No.	Chemical composition, wt% C(a)	Cr	Cu	Fe	Mo	Nb	Ni	Ti	Others
Nickel										
200	N02200	0.1	...	0.3(a)	0.4(a)	...	...	99.0(b)	0.1(a)	...
201	N02201	0.02	...	0.3(a)	0.4(a)	...	...	99.0(b)	0.1(a)	...
Ni-Cu										
400	N04400	0.15	...	31.5	1.25	...	...	66.6	...	...
Ni-Mo										
B-2	N10665	0.01	1.0(a)	...	2.0(a)	28	...	71	...	...
B	N10001	0.05	1.0(a)	...	5.0	28	...	66	...	V: 0.3
Ni-Cr-Fe										
600	N06600	0.08	16.0	0.5(a)	8.0	...	...	75	0.3(a)	...
800	N08800	0.1	21.0	0.8(a)	44.0	...	...	32.5	0.4(a)	...
Nimonic 75	...	0.15	19.5	0.5	5.0	...	...	73	0.4	...
690	N06690	0.02	29.0	...	10.0	...	...	61	0.3	...
800H	N08810	0.06-0.1	21.0	0.8(a)	44.0	...	...	32.5	0.4(a)	Al + Ti: 0.8(b)
Ni-Cr-Fe-Mo										
825	N00825	0.05	21.5	2.0	29.0	3.0	...	42	1.0	...
Sanicro 28	N08028	0.02	27	1.2	36	3	...	31.4	...	...
G-3	N06985	0.015	22.0	2.0	19.5	7.0	0.8(a)	44	...	W: 1.5(a)
2550	N06975	0.03	24.5	1.0	20.0	6.0	...	48	...	...
G-30	N06030	0.03	29.5	2.0	15.0	5.5	0.8	43	...	W: 2.5
G-50	N06950	0.02	20.0	0.5(a)	17.0	9.0	...	50(b)	...	...
Ni-Cr-Mo-W										
N	N10003	0.06	7.0	...	5.0(a)	16.5	...	71	...	...
W	N10004	0.12	5.0	...	6.0	24.0	...	63	...	...
625	N06625	0.1	21.5	...	5.0(a)	9.0	4.0	62	...	...
C-276	N10276	0.01	15.5	...	5.5	16.0	...	57	...	W: 4.0
C-4	N06455	0.01	16.0	...	3.0(a)	15.5	...	65	0.7(a)	...
C-22	N06022	0.015	22.0	...	3.0(a)	13.0	...	56	...	W: 3.0
ALLCORR	N06110	0.15	30.0	...	...	10.0	...	53	1.5(a)	W: 4(a)
Age-hardenable alloys										
K-500	N05500	0.25	...	29	2.0(a)	...	...	86	0.6	Al: 2.7
X-750	N07750	0.08	15.5	...	7.0	...	0.9	70(b)	2.5	Al: 0.7
718	N07718	0.05	18.0	...	19.0	3.0	5.0	52.5	0.4(a)	B: 0.006
Nimonic 80	N07080	0.10	19.5	...	3.0	...	...	73	2.25	Al: 1.4
Nimonic 105	...	0.20	14.5	...	2.0	5.0	...	56	1.2	Al: 4.7; Co: 20
R-41	N07041	0.09	19.0	...	5.0(a)	10.0	...	52	3.1	Al: 1.5; Co: 11
925	N09925	0.03	21.0	2.0	22(b)	3.0	...	43	2.1	...
625 Plus	N07716	0.03	21	...	5.0	8.5	3.5	61	1.3	Al: 0.35(a)
725	N07725	0.03	21	...	7.5	8.0	3.5	57	1.4	Al: 0.35(a)

(a) Maximum. **(b)** Minimum

(*e.g.*, Ref 2) for more in-depth discussion of the mechanistic aspects of SCC in nickel-base alloys. The test techniques that have typically been used to study stress-corrosion cracking and hydrogen embrittlement of nickel-base alloys are outlined in the last section. Approaches for the mitigation of SCC are intertwined with the discussion of the phenomenon in various environments.

Applications of Nickel-Base Alloys

Nickel-base alloys range from simple binary systems to multi-component systems containing seven intentional alloying elements. Additionally, various impurities, such as carbon, nitrogen, sulfur, phosphorus, tin, and silicon are present in varying concentrations. Table 2 lists representative commercial nickel-base alloys. Commercially pure nickel (UNS N02200) is used primarily in the handling of caustic and has been the subject of extensive mechanistic research in stress-corrosion cracking. The Ni-30Cu alloy, more commonly known as Alloy 400 (UNS N04400), is used in some seawater systems (condensers, offshore structures) and in the manufacture of hydrogen fluoride. The precipitation-hardenable Ni-Cu alloy designated K-500 (UNS N05500) is used in applications requiring higher strength, such as valve stems, tubing hangers, and bolts in offshore systems. The Ni-Mo alloy B-2 (N10665) is used in many chemical manufacturing operations that involve reducing acids or organic components. The Ni-Fe-Cr alloys, such as alloys 600 (UNS N06600) and 800 (N08800), are used widely in the power-generation industry as steam generator tubing. The Ni-Fe-Cr-Mo alloys, such as alloy 825 (N08825) and alloy G-3 (N06985), are used in a variety of industries, including oil and gas, pollution control, and chemical processing, and are being considered as candidate materials for high-level nuclear-waste containers. The Ni-Cr-Mo alloys, such as alloy C-276 (N10276), were used mainly in the chemical process industries, where their corrosion resistance to highly contaminated acidic systems was required. However, in the last decade, they have been used increasingly in more severe oil and gas, marine, and pollution-control applications. The precipitation-hardenable nickel-base alloys, such as alloy 718 (UNS N07718), were originally developed for aerospace applications, but are now being used in the nuclear and oil and gas industries. As the severity of downhole environmental conditions in oil and gas production increased, more corrosion-resistant, precipitation-hardenable grades, such as alloy 625 Plus, were developed.

Characteristics of Nickel-Base Alloys

The effects of alloying elements on the corrosion resistance of nickel-base alloys have been described in Volume 13 of the 9th edition of *Metals Handbook* (Ref 3). Copper and molybdenum are added to increase corrosion resistance to reducing environments, molybdenum being more effective. Chromium is added to provide resistance to oxidizing environments. Molybdenum and tungsten are added to chromium-containing alloys to increase resistance to oxidizing environments that contain halides. Silicon is typically present in small amounts and is considered detrimental. However, a Ni-9Si alloy has been used for its resistance to H_2SO_4 of all concentrations. Niobium and titanium are added to stabilize the chromium-containing alloys against precipitation of chromium carbides. In the precipitation-hardenable grades, niobium, aluminum, and titanium also combine with nickel to form coherent intermetallic phases that increase strength. Iron is typically used in nickel-base alloys to reduce cost. However, iron also increases the erosion resistance of Cu-Ni alloys, increases the solubility of carbon and nitrogen, and inhibits the formation of the intermetallic phase Ni_4Mo.

Carbides

Carbide precipitation is an important factor affecting SCC of nickel-base alloys. In nickel, carbon is known to form Ni_3C, but it decomposes to graphite, which weakens the grain boundaries (Ref 4, 5). The addition of copper reduces the tendency to graphitization. In the other nickel-base alloys, the type and composition of carbides depend on the composition of the alloy (Ref 4).

Carbides in these alloys can be classified as either primary or secondary. Primary carbides form during solidification in the interdendritic areas, which are generally rich in alloying elements. These carbides include the MC types, where M is niobium, tantalum, or titanium, and the M_6C types, where M is usually molybdenum or tungsten. Primary carbides are not easily dissolved during subsequent processing of the alloys and thus will be present as stringers along the rolling direction. A small amount of stringers must be tolerated in commercial alloys because they cannot be economically removed, but large amounts of stringers can be detrimental to fabricability and performance.

Secondary carbides precipitate as a result of thermal exposures during processing, heat treatment, fabrication, and service. They are usually intergranular, although in extreme cases precipitation inside the grains along slip lines and twin boundaries may occur. The amount of precipita-

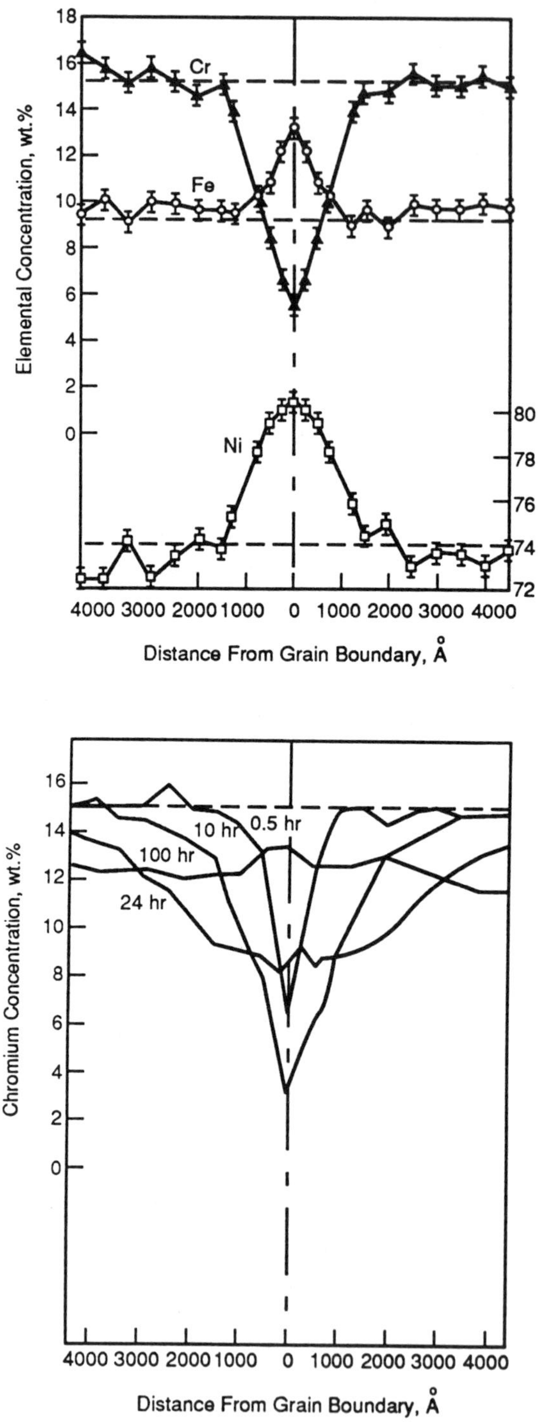

Fig. 1 (a) Grain-boundary chemistry of alloy 600, annealed 30 min at 1100 °C (2010 °F) and heated for 5 h at 700 °C (1290 °F). (b) Grain-boundary chromium depletion in alloy 600, annealed 30 min at 1100 °C (2010 °F) and heated at 700 °C (1290 °F) for various times. Source: Ref 6

tion depends on the concentration of carbon in solution, the stability of the alloy, time, temperature, presence of cold work, and grain size. Secondary carbides are often extremely detrimental to corrosion and SCC resistance because of their associated depletion zones (Fig. 1). For example, in alloy 600, precipitation of Cr_7C_3 carbide at the grain boundary results in a depletion of chromium adjoining the grain boundaries (Ref 6), the extent of which increases initially, then decreases at longer time periods due to back-diffusion of chromium from the bulk.

Carbides also can be classified as those that are rich in chromium and those that are rich in refractory elements (molybdenum, tungsten, niobium, titanium, and tantalum). One study (Ref 7) has shown that the chromium-rich carbides occur in alloys whose Cr/(Mo + 0.4W) ratio exceeds about 3.5. The chromium-rich carbides can be either Cr_7C_3 or $Cr_{23}C_6$, although the latter occurs in most alloys. In low-chromium alloys, such as alloy 600, that also contain low concentrations of other alloying elements, the predominant carbide that forms is Cr_7C_3, although $Cr_{23}C_6$ also has been reported (Ref 8). In more complex alloys, molybdenum, tungsten, iron, or nickel can substitute for chromium, and carbides of the form $Cr_{21}(Mo,W)_2C_6$ have been observed. The $M_{23}C_6$ carbides can occur at the grain boundaries as discrete globular particles, continuous films, or cellular precipitates. The temperature range of stability depends on alloy composition, with higher chromium, molybdenum, and tungsten contents increasing the stability to higher temperatures.

Time-temperature-sensitization diagrams for two alloys that form $M_{23}C_6$ carbides are shown in Fig. 2 and 3 (Ref 9, 10). Although sensitization diagrams indicate the kinetics of the effect of carbide precipitation on intergranular corrosion and not the kinetics of carbide precipitation (as do transformation diagrams), it can be seen that alloy 825, which contains 3% Mo, exhibits sensitization at higher temperatures than does alloy 800, which contains no molybdenum.

The refractory-metal carbides take on the form of MC, M_6C, or $M_{12}C$, depending on alloy composition. In the MC-type carbides, M is a mixture of niobium, titanium, and tantalum, but

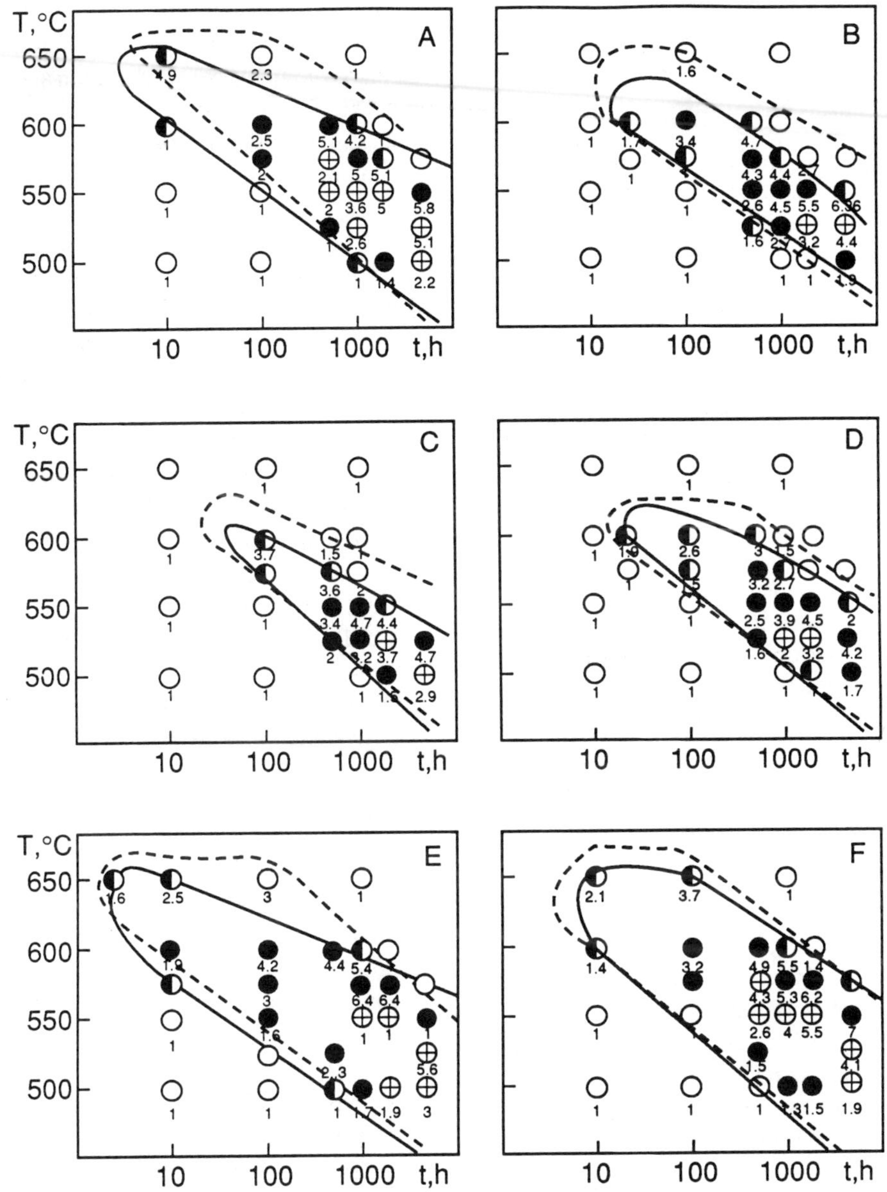

Fig. 2 Time-temperature-sensitization diagrams for various heats of alloy 800. The relative change of the magnetic susceptibility is reported below each point. The dotted lines define the conditions for chromium depletion. Source: Ref 9

may also contain some molybdenum or tungsten. If nitrogen is present in the alloy, it may substitute for some of the carbon, and the resulting carbonitrides may appear yellow or orange when examined metallographically, in contrast to the grayish color of the carbides. The MC-type carbides are extremely stable and generally do not break down. The M_6C-type carbides, also known as η-carbides, are formed by molybdenum and tungsten. However, major substitutional alloying elements, such as nickel, iron, and chromium, also are present in these carbides (Ref 11). Some typical compositions include Mo_6C, $(Ni,Co)_3Mo_3C_6$, and $(Mo,Ni,Cr,W)_6C$. Typically, these carbides dissolve at higher temperatures than the $M_{23}C_6$-type carbides. It must be

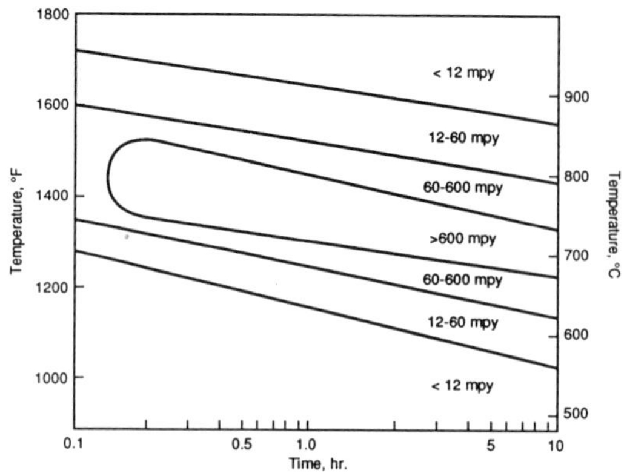

Fig. 3 Time-temperature-sensitization diagram for alloy 825, annealed at 1205 °C (2200 °F) for 1 h prior to sensitizing treatment. Source: Ref 10

noted that since these carbides are not usually rich in chromium, the depletion zones associated with them are molybdenum or tungsten depletion zones. "Sensitization" in these alloys means enhanced intergranular corrosion as a result of depletion of molybdenum or tungsten. Other types of refractory carbides are $M_{12}C$ and Mo_2C, which may form in a Ni-Mo system depending on the carbon and molybdenum contents (Ref 12).

Intermetallic Phases

The intermetallic phases in nickel-base alloys can be coherent phases such as γ' and γ'', which are often desirable for increased strength, or noncoherent phases, which are detrimental to toughness and corrosion resistance. The detrimental intermetallic phases are primarily the topologically close-packed (TCP) phases—σ, μ, and Laves—which form in the temperature range of 1100 to 800 °C (2010 to 1470 °F), depending on alloy composition. Sigma phase typically occurs in alloys that contain large amounts of chromium, and in Fe-Cr alloys it has the approximate composition of FeCr. Although σ-phase formation is not common in the Ni-Cr system, it can form in nickel-base alloys that contain moderate concentrations of molybdenum and iron. Other alloying elements that tend to stabilize σ phase include silicon and tungsten. Sigma phase has a complex, tetragonal, layered structure (Ref 13) and tends to nucleate from $M_{23}C_6$ carbides, to which it is structurally and compositionally related. Mu phase, which has a complex, rhombohedral structure, occurs in chromium-containing alloys with large amounts of molybdenum and tungsten and is stabilized by iron. In the Fe-Mo system, μ-phase has a composition similar to Fe_7Mo_6, but has not been observed in the Ni-Mo system. In Ni-Cr-Mo-Fe alloys, it has been found to have a composition similar to $(Ni_{0.36}Cr_{0.16}Fe_{0.04}Co_{0.02})$ $(Mo_{0.39}W_{0.03})$ (Ref 11). Mu phase tends to nucleate on M_6C-type carbides, which are structurally and compositionally similar to it. Laves phases have complex, hexagonal structures and are compounds of iron and molybdenum, tungsten, niobium, or tantalum. Other intermetallic compounds that are detrimental to corrosion or SCC include the ordered Ni_4Mo that forms in the commercial Ni-Mo alloy in the temperature range of 800 to 650 °C (1470 to 1200 °F) and the Ni_2Cr-type ordered particles that form in commercial Ni-Cr-Mo alloys in the temperature range of 550 to 200 °C (1020 to 390 °F).

Some intermetallic compounds can also be beneficial to properties such as strength. These are usually coherent precipitates with relatively low lattice mismatch, such as γ', which is a metastable form of $Ni_3(Al,Ti,Si)$, and γ'', which is a metastable version of Ni_3Nb. In both types of coherent precipitates, other alloying elements, such as chromium, molybdenum, and tungsten, tend to substitute for some of the nickel, although the γ' precipitates have been shown to be leaner in these alloying elements than the matrix phase (Ref 14).

A typical time-temperature-transformation diagram for alloy 625 (Ni-Cr-Mo) is shown in Fig. 4 (Ref 15). It can be seen that a variety of precipitates, including various types of carbides, metastable phases (γ''), and intermetallic phases, can form in a single alloy. In the absence of phase diagrams for the multicomponent alloys, semiempirical correlations have been used. For example, based on Pauling's analysis of the electronic structure of transition metals, Beck and others (Ref 16) developed a computerized calculation scheme known as PHACOMP. In this approach, the residual matrix alloying element concentrations are calculated after accounting for carbides, borides, and γ', which are then used to calculate an electron vacancy number. This is

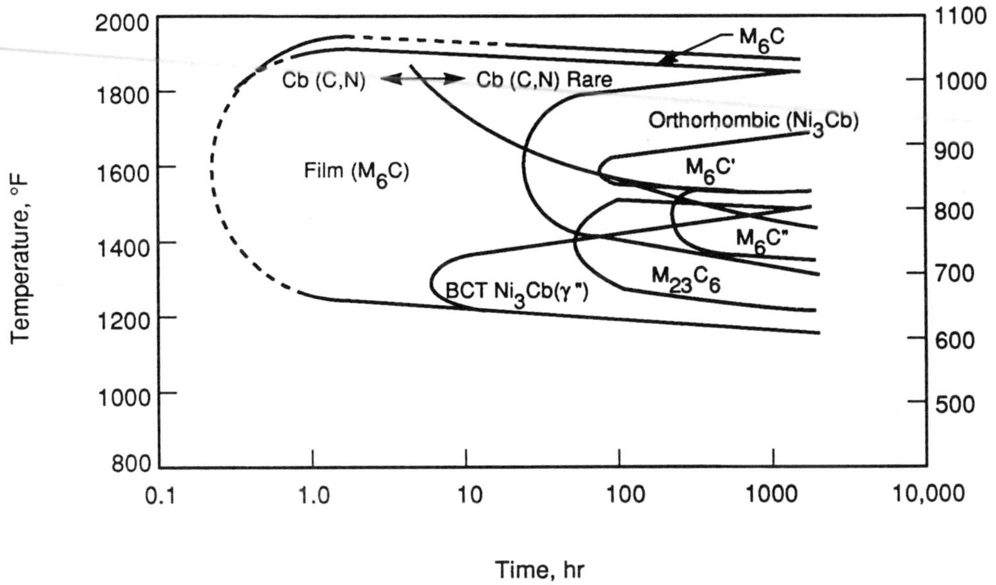

Fig. 4 Time-temperature-transformation diagram for solution-treated alloy 625. Lower γ″ limit determined by hardness measurement. Source: Ref 15

compared to a critical electron vacancy number for σ-phase precipitation. Similar approaches include SIGMA-SAFE (Ref 17) and New PHACOMP (Ref 18).

SCC in Halide Environments

Chlorides

The beneficial effect of nickel on the SCC resistance of Fe-Ni-Cr-Mo alloys to chloride solutions has been well established since the publication of the now-famous "Copson Curve" (Ref 19). This classic work was performed on wire samples in boiling 45% $MgCl_2$ solution and indicated that alloys containing above about 45% Ni did not crack in 30 days of testing. However, Staehle *et al.* (Ref 20) have shown that cracking can occur in the same nominal environment in nickel-base alloys that contain 32 to 52% Ni (Ref 20). They have also demonstrated the statistical nature of SCC.

The beneficial effect of nickel has been demonstrated in other chloride solutions using precracked samples (Fig. 5) (Ref 21) and, at higher temperatures, using two-point bend samples (Fig. 6) (Ref 22). At temperatures slightly above the boiling point (Fig. 5), essentially no cracking is observed in alloys with nickel contents above about 35% over a range of chromium contents. The corresponding stage 2 (stress-intensity independent) crack-growth rates are too low to be measurable. The lowest threshold stress intensity (highest susceptibility for SCC) in this study was observed for alloys containing about 20% Ni—unless the chromium content was higher than 25%, in which case the nickel content for the lowest threshold stress intensity was close to 4% (ferritic or duplex stainless steels). However, it must be noted that the stage 2 crack growth for the low-nickel ferritic/duplex stainless steels was low. It is likely that the presence of aeration enhanced SCC in these alloys. Stress-corrosion cracking can be observed even in high-nickel alloys, if the temperature is sufficiently high (Fig. 6). The critical temperature at and above which cracking was observed in this study (Ref 22) depended mainly on nickel content and not on chromium or molybdenum content. In the same study, similar response to temperature was noted in the same alloys in 25% NaCl and 23.7% $CaCl_2$, with the severity of the environment roughly following the sequence: $MgCl_2$ > NaCl > $CaCl_2$. All cracking was transgranular. It was also noted that the effect of cold

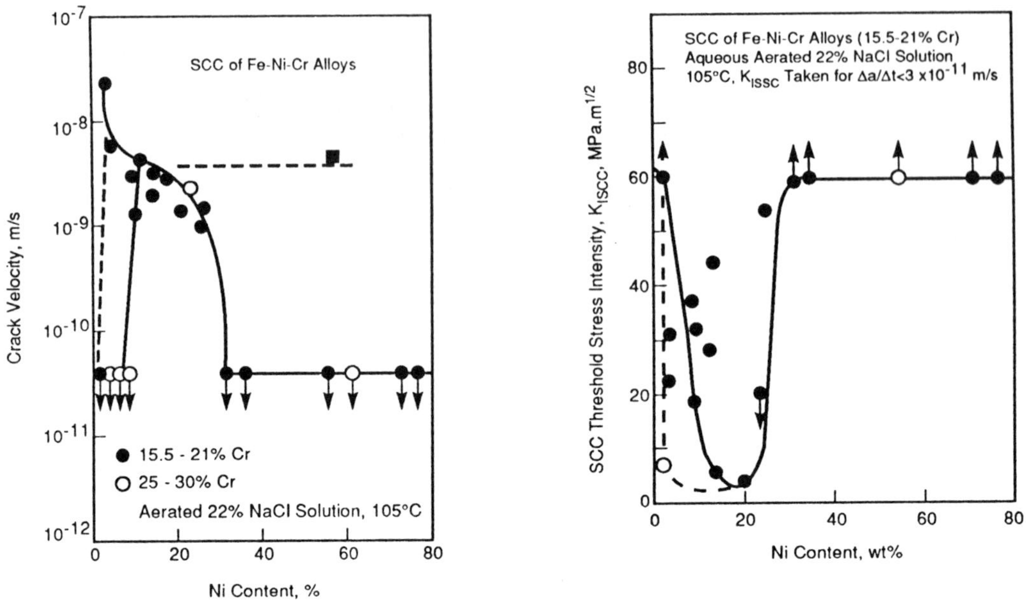

Fig. 5 Effect of nickel content on maximum crack-growth rate and threshold stress-intensity factor for Fe-Ni-Cr alloys in hot chloride solutions. Source: Ref 21

Table 3 SCC of Mill-Annealed Alloy 825 in Boiling $MgCl_2$ Solutions
Tests performed according to ASTM G 30 using two different types of condensers

$MgCl_2$, wt%	Boiling temperature °C	Boiling temperature °F	Time to failure, h: Finger condenser	Time to failure, h: Modified Allihn condenser, No purge	Time to failure, h: Modified Allihn condenser, Oxygen
42	145	293	155	1100	...
45	155	311	...	80	80
50	170	338	50(a)	1000(a)	...

(a) No cracking
Source: Ref 23

working was not as significant as that of temperature. Since the critical temperature for cracking depended only on nickel content, it is difficult to envision localized corrosion as a precursor to cracking in this environment.

Nickel may be beneficial in terms of either controlling the deformation mode or changing the corrosion potential away from the cracking potential. Further study using controlled-potential tests at various temperatures is needed. Chiang and Streicher (Ref 23) tested alloy 825 U-bend samples in boiling 42 and 45% $MgCl_2$ solutions (Table 3). Since finger-type condensers allow greater air intrusion into the flask, SCC occurred more readily. No cracking was found in a 50% $MgCl_2$ solution, suggesting that the lower oxygen solubility probably affected the test results. The effect of applied potential on SCC of sensitized nickel-base alloys in boiling dilute chloride and sulfate solutions was examined by Herbsleb (Ref 24). All cracking was intergranular and, as shown in Fig. 7 and 8, was observed only at high potentials well beyond the pitting potential.

Humphries and Nelson (Ref 25) observed that in a moist 3.5% NaCl environment, under alternating wet (10 min) and dry (50 min) conditions

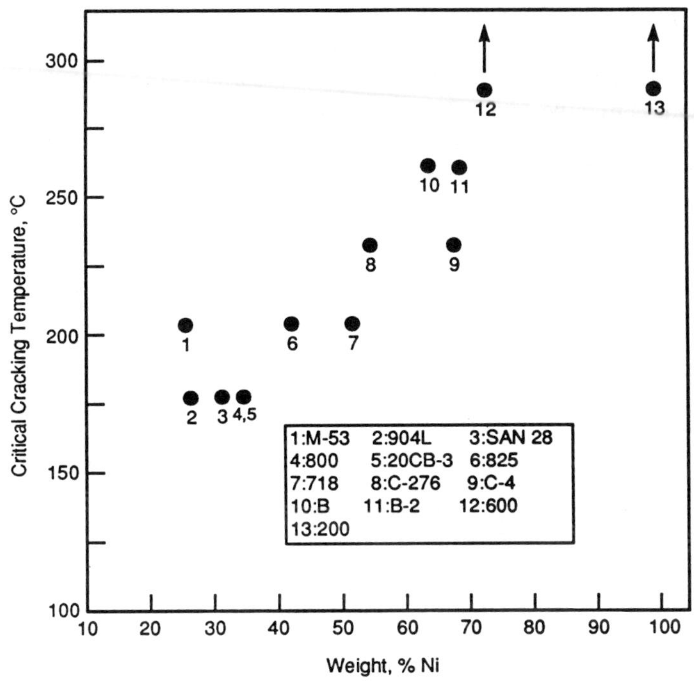

Fig. 6 Effect of nickel on SCC in 20.4% $MgCl_2$ deaerated with nitrogen. Source: Ref 22

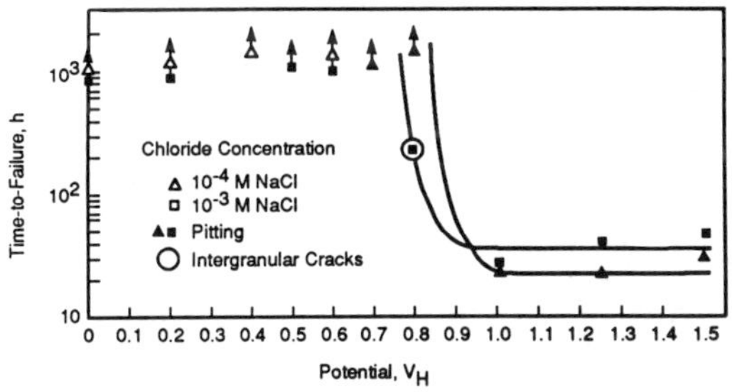

Fig. 7 Effect of chloride concentration on time-to-failure/potential curves of sensitized (1 h at 650 °C, or 1200 °F) alloy 800 tensile specimens ($\sigma_a = 1.75\ \sigma_{0.2}$) in boiling 10^{-2} M and 10^{-4} M NaCl. Source: Ref 24

at room temperature, no cracking was observed in alloy 718, alloy C (high-carbon C-276), and type 304 stainless steel. Cracking in concentrated HCl at room temperature has been observed in several nickel-base alloys (Fig. 9) (Ref 26). Cracking was observed only in precipitation-hardenable alloys and not in solid-solution-strengthened alloys (Ref 26). Stress-corrosion cracking can occur in alloys that do not exhibit any passive behavior.

Table 4 SCC of Nickel-Base Alloys in 1% HCl
Tests were conducted on two-point bend samples for 1000 h at 204 °C (400 °F); no deaeration. All alloys in mill-annealed condition except alloy 718, which was double aged (1 h at 968 °C/WQ/aged 8 h at 732 °C/cooled to 621 °C/aged 8 h/AC)

Alloy	Results of triplicate tests(a)
Type 316 stainless steel	C, C, C
Alloy 718	C, C, C
Alloy 825	C, C, C
Alloy G-30	C, C, C
Alloy 625	C, C, C
Alloy C-276	C, C, C
Alloy C-22	C, C, C
Titanium, grade 2	NC, NC, NC

(a) C, cracking; NC, no cracking
Source: Ref 27

This is shown in Table 4 for a variety of Ni-Cr-Mo-Fe alloys. In all the cases where cracking was observed, considerable corrosion was also observed. The addition of H_2S to 17% HCl has been shown to induce transgranular cracking in alloys 825, C-276, and C-22 even at room temperature (Ref 27).

The effect of precipitation hardening on SCC in boiling chloride solutions is shown in Table 5 (Ref 28). Both alloys 718 and R-41 suffered increased SCC upon aging. From the data on R-41, it can be seen that hardness is not necessarily the determining factor, since the cold-rolled material did not suffer from SCC. A combination of changes in distribution of alloying elements between the γ and γ' phases, chromium depletion at the grain boundaries due to carbide precipitation, and changes in the deformation mode from cross slip to a more planar mode may be responsible for the increased SCC susceptibility.

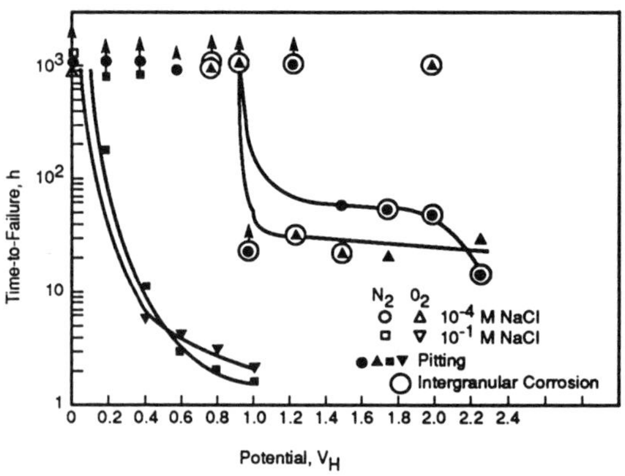

Fig. 8 Effect of chloride concentration on time to failure of sensitized (1 h at 600 °C, or 1110 °F) alloy 600 tensile specimens ($\sigma_a = 1.75\ \sigma_{0.2}$) in boiling NaCl ($H_2$ and O_2 atmosphere). Source: Ref 24

The effect of sensitization on SCC of Ni-Cr-Mo-W alloys with varying degrees of susceptibility to carbide and μ-phase precipitation was studied by Streicher (Ref 29). His results are summarized in Table 6. No cracking was observed in

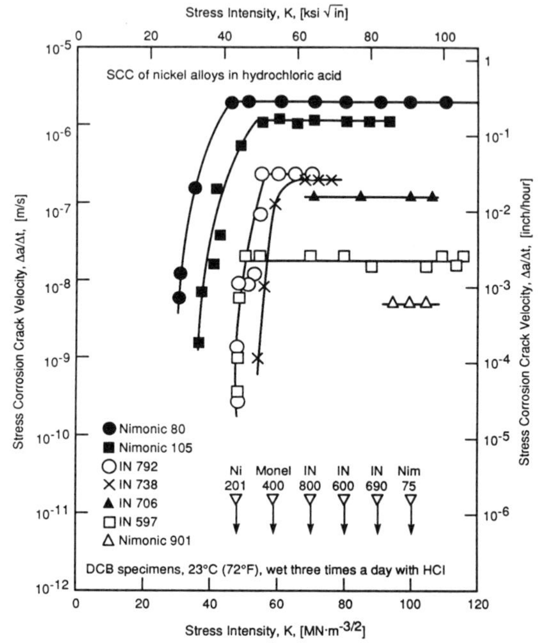

Fig. 9 Effect of stress intensity on SCC velocity of nickel alloys in concentrated aerated HCl at ambient temperature. Note that precipitation-hardened alloys crack, while solid-solution-strengthened alloys and pure nickel (Ni 201) resist cracking. Source: Ref 26

Table 5 Effect of Precipitation Hardening on SCC of Alloy 718 (UNS N07718) and Alloy R-41 (UNS N07041) in Chloride Environments

Two-point bend samples, tested for 1008 h

Alloy	Thermal treatment	Hardness	Time to failure, h: 20.4% $MgCl_2$, 177 °C (350 °F)	Time to failure, h: 25% NaCl, 204 °C (400 °F)
718	Mill annealed	80 HRB	408/408/408	48/48/504
	Aged 8 h at 663 °C	26 HRC	72/144/240	48/48/48
	MA + 8 h/718 °C + 8 h/621 °C	42 HRC	72/72/144	48/48/48
R-41	Mill annealed	100 HRB	...	NC/NC/NC(a)
	Cold rolled	45 HRC	...	NC/NC/NC
	1080 °C + 30 min	42 HRC	...	48/96/NC
	1066 °C/AC 16 h			
	760 °C			

(a) NC, no cracking
Source: Ref 28

Table 6 Effect of Grain-Boundary Precipitation on SCC of Ni-Cr-Mo-W Alloys in Boiling 45% $MgCl_2$

Alloy	Analysis, wt% Carbon	Silicon	Treatment	Type of precipitate	SCC in 45% $MgCl_2$ at 154 °C (310 °F) (a)
Alloy C	0.05	0.70	Solution annealed	None	NC, 2407 h
			1 h/704 °C	M_6C	IGC, 107-552 h
			1 h/871 °C	M_6C, μ	IGC, 90 h
			1 h/1038 °C	M_6C, μ	IGC, 552-625 h
C-276	0.01	0.01	1 h/704 °C	μ, small M_6C	NC, 6536 h
			1 h/871 °C	μ, M_6C	IGC,905-2100 h
			1 h/1038 °C	Small μ and M_6C	NC, 6536 h
	0.008	0.004	1 h/1149 °C	None	NC, 2847 h
			1 h/704 °C	μ, small M_6C	Cracking, 2252 h
			1 h/871 °C	μ, small M_6C	IGC, 957 h
			1 h/1038 °C	μ, small M_6C	IGC, 4359 h
	0.004	0.01	Mill annealed	None	NC, 2407 h
			1 h/704 °C	μ	NC, 2407 h
			1 h/871 °C	μ	NC, 2407 h
			1 h/1038 °C	Small μ	NC, 2407 h

(a) NC, no cracking in the stated time period; IGC, intergranular cracking
Source: Ref 29

Table 7 SCC in Simulated Packer Fluids

Solutions were deaerated and back-filled with nitrogen (1 atm); two-point bend testing of mill-annealed samples at 204 °C (400 °F)

Alloy	Time to failure, months(a) 4.7% $ZnBr_2$	43% $CaBr_2$
255 (Fe-25Cr-5.5Ni-3Mo-2Cu-0.2N)	1, 1, 2	4, 4, 4
825	1, 1, 2	>12, >12, >12
G-3	1, 2, 2	>12, >12, >12
C-276	1, 1, 2	>12, >12, >12
C-22	>12, >12, >12	>12, >12, >12

(a) Indicates no cracking within the stated time period
Source: Ref 30

the solution-annealed alloys. The precipitation of M_6C carbide, which was observed in alloys with a carbon content greater than 0.008 wt%, was much more detrimental to SCC resistance in $MgCl_2$ than the precipitation of μ phase, which was observed in low-carbon alloys but not low-tungsten alloys. Cracking in all cases was intergranular.

Bromides

Bromide solutions are encountered in chemical process industries (*e.g.*, flame-retardant chemicals) and the oil and gas industry (*e.g.*, packer and completion fluids). Due to the proprietary nature of both of these applications, extensive data are not available, especially for accurate comparison to SCC in chloride environments. In many plant failures related to SCC of stainless steels in bromide environments, some chloride is inevitably present. None of the Ni-Fe-Cr-Mo alloys suffered from SCC in two-point bend tests in 1% HBr. Among the Ni-Mo alloys, only alloy B cracked within 168 h in the same test. Stress-corrosion cracking has been observed in a number of nickel-base alloys in concentrated bromide solutions used as packer fluids. This is illustrated in Table 7 (Ref 30). Cracking in 4.7% $ZnBr_2$ is more severe than in 43% $CaBr_2$, probably because of the greater acidity of the former solution, especially in the presence of H_2S. Fricke

Table 8 SCC in Packer Fluids

	K_{ISCC}, MPa$\sqrt{m}$ (ksi$\sqrt{in.}$) (a)			
	1.5 g/mL NaBr 6.89 MPa H_2S + 4 MPa CO_2		1.49 g/mL $CaCl_2$ 6.89 MPa H_2 + 4.1 MPa CO_2	
Alloy	24 °C (75 °F)	232 °C (500 °F)	24 °C (75 °F)	232 °C (500 °F)
9Cr-1Mo steel	>81.7 (74.3)	>85.1 (77.4)	>68.2 (62)	>80.8 (73.5)
13Cr steel	33.1 (30.1), C	>119.7 (108.9)	25.8 (23.5), C	<86.3 (78.5)
22Cr duplex stainless steel	C	27.5 (25), C	<34.8 (31.7), C	28.6 (26), C
Sanicro 28	>114.7 (104.4)	<78.6 (71.5)	>124.3 (113.1)	>108.7 (98.9)
625	67.8* (61.7)	84.3 (76.7)	68.1* (62), C	51.1 (46.5)

Note: K_{ISCC} determined by DCB tests; 28-day C-ring exposure; stress = 90% longitudinal yield strength
(a) < indicates crack branching; > indicates no crack growth; * indicates that the specimen was coupled to steel; C indicate C-ring cracking
Source: Ref 31

Table 9 SCC in 1% HI
Two-point bend samples tested for 1000 h max

	Time to failure, h(a)		
Alloy	177 °C (350 °F)	204 °C (400 °F)	232 °C (450 °F)
B-2	168, 168, 366	48, 168, 168	48, 48, 48
200	NC, NC, NC	NC, NC, NC	NC, NC
400	NC, NC, NC	NC, NC, NC	NC, NC
600	NC, NC, NC	NC, NC	NC, NC, NC
825	NC, NC, NC	NC, NC, NC	NC, NC, NC
G-3	NC, NC, NC	NC, NC, NC	NC, NC, NC
C-276	NC, NC, NC	NC, NC, NC	NC, NC, NC

(a) NC, denotes no cracking in 1000 h

(Ref 31) investigated the SCC of a variety of materials in highly concentrated NaBr (1.5 g/mL) and $CaCl_2$ (1.49 g/mL) in the presence of H_2S using both C-ring and double-cantilever-beam (DCB) tests. His results (Table 8) indicate that the nickel-base alloy with low iron (alloy 625) suffered SCC when coupled to carbon steel at room temperature, which is probably related to hydrogen embrittlement rather than stress-corrosion cracking. In a NaBr + H_2S environment at 232 °C, all the highly alloyed materials underwent SCC, while the 9-Cr and 13-Cr alloys exhibited high rates of general and localized corrosion.

Iodides

Stress-corrosion cracking in iodide environments has been encountered in the manufacture of acetic acid (CH_3COOH), where potassium iodide is used as a promoter for the formation of acetic acid. The data to date indicate that cracking is specific to the Ni-Mo alloys (Table 9) and that cracking in mill-annealed alloys is transgranular.

Fluorides

Nickel-base alloys have been used quite successfully in fluoride-containing environments. Examples include alkylation in the refinery industry, semiconductor etching, and the manufacture of organic fluorides. However, intergranular cracking of cold-worked alloy 400 in fluosilicic acid was one of the earliest reported instances of SCC in nickel-base alloys (Ref 32). Since then, rather extensive data have been collected, mainly for Ni-Cu alloys—for example, by Copson and Chang (Ref 33), who showed that cracking occurred only in the vapor space in the presence of air, and by Graf and Wittich (Ref 34), who examined the SCC of Ni-Cu alloys in hydrogen fluoride containing CuF_2 as the oxidizer. Recently, Everhart and Price (Ref 35) have examined the ef-

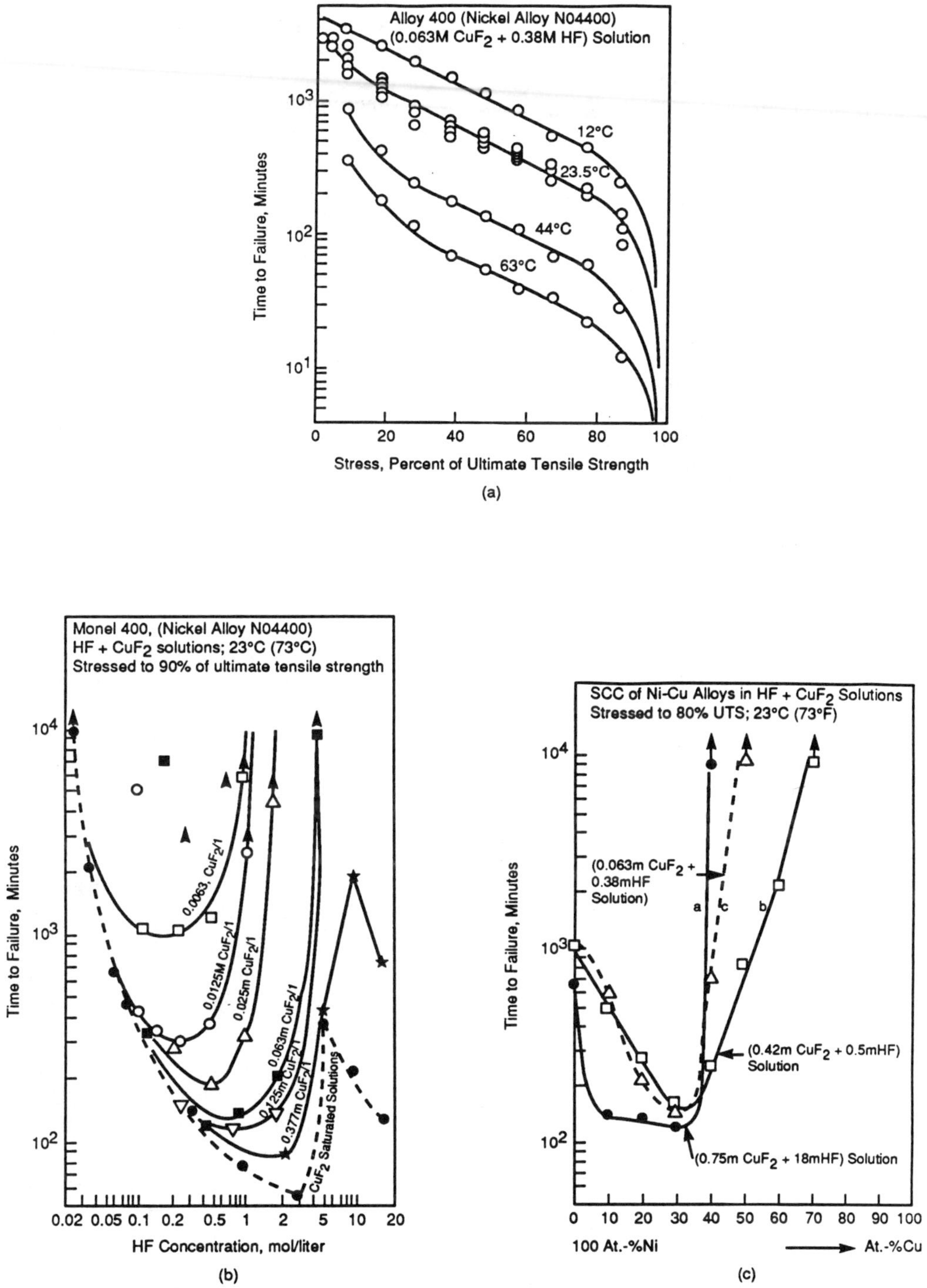

Fig. 10 SCC of Ni-Cu alloys in fluoride environments. Source: Ref 34

fect of strain rate, air, $CuCl_2$, and CuF_2 on SCC of alloys 400 and K-500. Stress-corrosion cracking of alloy 400 is found at the liquid-vapor interface or in the vapor phase because of the formation of a thin film of concentrated hydrogen fluoride and the easy access of oxygen to the surface, which increases the corrosion potential. Other oxidizers such as CuF_2 and $CuCl_2$ serve the same purpose. The effect of increasing applied stress and increasing concentrations of CuF_2 is shown in Fig. 10(a) and (b) (Ref 34). In slow-strain-rate tests (Ref 35), where the samples were partially immersed in HF + CuF_2 solutions, considerable interfacial corrosion and deposition of copper-rich product were noted along with cracking. However, the cracking was not associated with pitting. Intergranular cracking was observed at lower strains and strain rates, while transgranular cracking was observed at higher strains and strain rates. The effects of temperature and stress level are shown in Fig. 10(b). In a series of binary Ni-Cu alloys, Ni-30Cu exhibits the greatest sensitivity to cracking in these environments (Fig. 10b). Nickel-chromium alloys, such as alloy 600, have been used in components exposed to hydrogen fluoride vapors to minimize SCC.

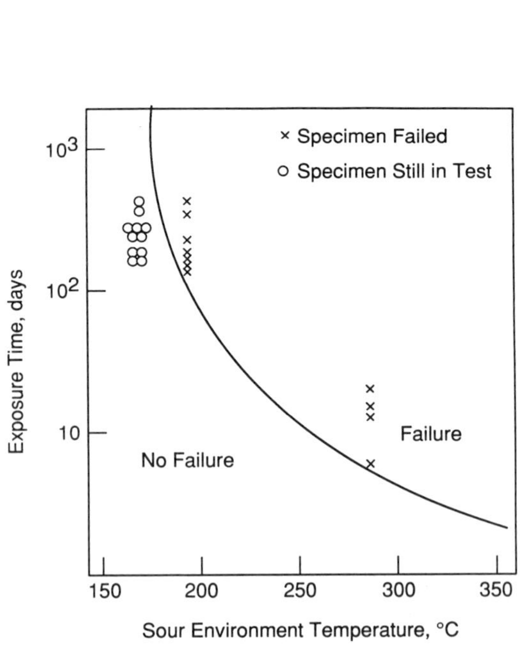

Fig. 11 SCC behavior of Hastelloy C-276 versus testing temperature. Heat treatment: 260 °C (500 °F) for 250 h; environment: 25% NaCl + 0.5% CH_3COOH + 1 g/L S_8 + H_2S; yield strength: 1050 to 1160 MPa (152 to 168 ksi); stress level: 70 and 90% of yield strength. Source: Ref 41

SCC in Environments Containing Sulfur Species

Environments that contain sulfur species are encountered in oil and gas production, both in downhole and surface components, and in refinery operations. Such environments can be divided into two major classifications:

- Cl^- + H_2S + CO_2 + elemental sulfur environments
- Other sulfur-containing environments (thiosulfate, tetrathionate, polythionic acid)

Environmentally induced cracking has been observed in the first type of environment, both at low and high temperatures. However, in low-temperature H_2S environments, cracking has been observed only when the nickel-base alloys have been coupled to carbon steel, thus cathodically polarizing them. Hence, this will be discussed under hydrogen-embrittlement phenomena.

Cl^- + H_2S + CO_2 Environments

Nickel-base alloys have come into prominence in the oil and gas industry because of the need to go to greater depths to find oil and natural gas. Because most of the nickel-base alloys used as downhole components (tubing, casing, valves) are solid-solution-strengthened alloys, the required strength levels (yield strengths ranging from 758 to 1103 MPa, or 110 to 160 ksi) can be achieved mainly through cold working. These are seldom welded, although some surface components such as flow lines can be joined by circumferential welding. Thick-wall components such as safety valves and tubing hangers, which cannot be cold worked economically or uniformly, are manufactured from precipitation-strengthened nickel-base alloys.

The behavior of solid-solution-strengthened alloys will be discussed first, followed by precipitation-strengthened alloys. The discussion is, of necessity, brief, and the interested reader is referred to several reviews (Ref 36-39). Investigation of the SCC of nickel-base alloys in H_2S + CO_2 + Cl^- environments began in the late 1970s (Ref 40). These early investigations emphasized low-temperature embrittlement under cathodic polarization conditions. However, initial data on anodic cracking in 5% NaCl + 0.1% CH_3COOH

Table 10 Effect of Test Technique on Observed SCC in a H_2S Environment
Solution: 25% NaCl + 2.4 MPa H_2S (RT) + 1.7 MPa CO_2 (RT)
Alloy: 50Ni-24Cr-15Fe-6Mo-1Cu-1Ti-1Mn

Test technique	Temperature °C	Temperature °F	No. cracked/Total
C-ring, constant deflection(a)	232	450	0/3
Slow strain rate(b)	232	450	3/6
	214	417	1/3
	204	400	0/3

(a) Stressed to 90% longitudinal yield strength; exposed for total of 6 months; same alloy bolt. (b) Extension rate, 4×10^{-6}/s; SCC based on secondary crack observation
Source: Ref 41

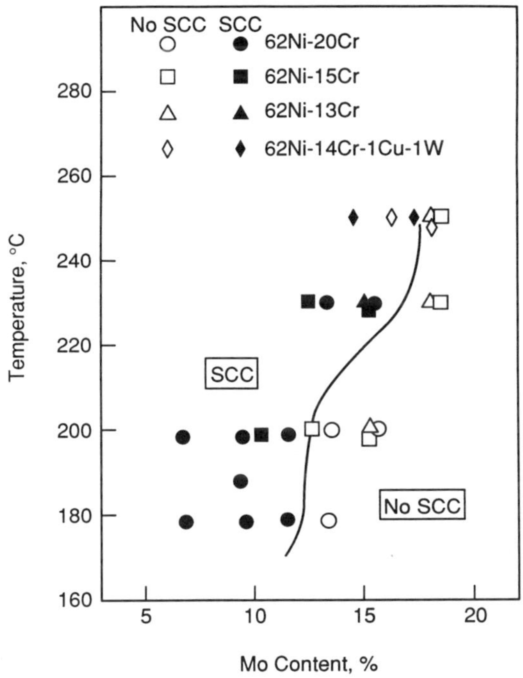

Fig. 12 Effect of molybdenum content on SCC resistance of Ni-Cr-Mo alloys in 20% NaCl + 0.5% CH_3COOH + 10 atm H_2S + 10 atm CO_2 + 1 g/L S_8. Source: Ref 42

+ H_2S (saturated at room temperature) at temperatures up to 204 °C (400 °F) showed that some of the high-carbon nickel-base alloys, precipitation-hardenable nickel-base alloys, and cobalt-base alloys were susceptible to cracking (Ref 40). The cracking was transgranular, in contrast to the intergranular low-temperature cracking under cathodic polarization. Watkins *et al.* (Ref 41) reported the results of SCC in alloy C-276 in a H_2S environment containing elemental sulfur (referred to in this chapter as S_8, although at the temperatures of interest in downhole applications, sulfur does not exist as a ring of eight atoms but as long chains of many lengths that contain biradicals of sulfur). Since then, many investigations of the effects of H_2S, S_8, Cl^-, CO_2, and pH have been carried out.

Effect of Temperature. In the absence of galvanic coupling with carbon steel, increasing the temperature increases the SCC susceptibility of nickel-base alloys. This is shown in Fig. 11 for alloy C-276 in a H_2S + S_8 environment. Watkins *et al.* (Ref 41) showed that the percentage of samples that cracked in the slow-strain-rate tests increased with temperature and that duplicate tests may not be sufficient to detect cracking. They also compared the results of slow-strain-rate tests to constant-deflection tests. Their results, shown in Table 10, indicate that slow-strain-rate tests are more severe than constant-deflection tests. This has not always been found to be the case, and cracking susceptibility may be controlled by a variety of factors, such as corrosion potential, surface effects, and sample orientation. In the same environment, Watkins *et al.* (Ref 41) found cracking at temperatures as low as 204 °C (400 °F). The detrimental effect of temperature has also been used to rank alloys in terms of their critical temperature for SCC in a given environment. This is illustrated for a variety of experimental alloys in Fig. 12 (Ref 42). It can be seen that increasing molybdenum content results in increased temperature above which cracking occurs in C-ring tests. In contrast, Ciaraldi (Ref 43) concluded, from DCB precracked specimens, that nickel exerts the dominant beneficial effect and chromium and molybdenum a lesser one. These may not be contradictory results, because the higher-nickel alloys (above 45% Ni) examined by Ciaraldi also contained higher molybdenum content.

Effect of H_2S. It has been well established that H_2S promotes cracking in nickel-base alloys. For example, Asphahani (Ref 44) showed that the addition of CO_2 to 5% NaCl did not cause cracking of nickel-base alloys, whereas H_2S additions resulted in SCC of 25% Ni at 177 °C (350 °F). No cracking was found in alloys containing higher amounts of nickel. However, unlike the cases of steels and duplex stainless steels, no systematic study to establish the minimum concentration of H_2S needed for cracking has been reported. Kolts (Ref 30) showed that for a 25Cr-5.5Ni-3Mo duplex stainless steel, 9 psia H_2S was sufficient to cause cracking (two-point bend tests) even at room temperature, whereas no cracking was reported in nickel-base alloys (31 to 56% Ni) in 100 psia H_2S + 25% NaCl at 204 °C (400 °F). Ikeda *et al.* (Ref 45) reported that a minimum of 10 atm H_2S (25% NaCl + 0.5% CH_3COOH + 10 atm CO_2) was needed to cause cracking under slow-strain-rate testing conditions (4×10^{-6}/s) in a 30Ni-22Cr-4Mo alloy at 150 °C (300 °F). Unlike the case of steels, cracking of nickel-base alloys in H_2S solutions at high temperatures is related to anodic dissolution mechanisms.

The effect of H_2S on localized corrosion has been studied more systematically than cracking. Tsujikawa (Ref 46) showed that the addition of 0.01 atm H_2S to a 20% NaCl solution at 80 °C (175 °F) and atmospheric pressure lowered the pitting potential from –25 to –200 mV (SCE). He also showed that the effect of H_2S on localized corrosion can be reproduced by the addition of $K_2S_2O_3$. The ranking of various alloys in terms of their localized corrosion resistance in NaCl + H_2S solution did not correspond to their ranking in an oxidizing chloride solution such as 6% $FeCl_3$. Similar results have been reported by Rhodes (Ref 39) and Kolts (Ref 30). As will be mentioned later, in environments containing S_8, H_2S does not seem to be necessary to cause localized corrosion or cracking, perhaps due to local generation of H_2S from electrolytic reduction of sulfur. The detrimental effect of H_2S may be related to adsorption of sulfur atoms in the passive film and eventual depassivation, as shown for nickel by Marcus (Ref 47).

Effect of CO_2. Carbon dioxide has been shown to have no significant effect on cracking of nickel-base alloys in chloride environments (Ref 30, 40, 44). In environments that contain H_2S and S_8, CO_2 may also inhibit cracking slightly (Ref 48). The role of CO_2 in inhibiting cracking is not clear. Bicarbonate has been shown to be a much greater inhibitor in a NaCl + H_2S + S_8 environment (Ref 48), perhaps due to its buffering action.

Effect of Elemental Sulfur (S_8). Most recent studies of nickel-base alloy in sour-gas environments have focused on the effect of S_8. It is clear from all the studies, dating back to the late 1970s, that S_8 is extremely detrimental to localized corrosion and cracking of nickel-base alloys in brine. What has not been clear until now is the mechanism by which S_8 affects cracking in these environments.

The structure and chemistry of sulfur as it pertains to sour-gas environments have been reviewed by Schmidt (Ref 49). Sulfur exists at room temperature in the α form, which has a nonplanar ring structure consisting of eight atoms. In this configuration it is quite stable. When heated above 95 °C (203 °F), the α phase transforms to a monoclinic β phase, which also has an eight-atom ring structure. The β phase melts at 119.5 °C (247 °F) to form the λ phase, which retains the ring structure. The easy displacement of the ring structure results in rather low viscosity. However, above 160 °C (320 °F), the λ phase transforms mostly to μ phase, with a small percentage of π phase. The μ phase has a zigzag chain structure with biradical ends. Because of formation of long polymeric structures, the viscosity increases above 160 °C (320 °F) rather dramatically. Above 160 °C (320 °F), H_2S and Cl^- can combine with the biradicals and reduce the viscosity (Ref 49). The melting point of sulfur can also be decreased by an increase in H_2S pressure, significant changes occurring only beyond 0.1 MPa (0.015 ksi). Ueda *et al.* (Ref 50) have shown that the melting point and viscosity changes are the same in aqueous brine solutions containing 0.1 MPa

(0.015 ksi) H_2S and 1 MPa (0.15 ksi) CO_2 as in pure sulfur. This agrees with the data given by Schmidt (Ref 49). It is well known that α, β, and λ sulfur are insoluble in water. Sulfur solubility has been shown to be increased by an increase in H_2S pressure in hydrocarbon gases due to the formation of polysulfide chains. However, very little solubility data in aqueous systems are available. Ueda *et al.* (Ref 50) have calculated the sulfur solubility in $H_2O + H_2S$ systems, based on thermodynamic self-oxidation and reduction reactions of sulfur:

$$S + 2H^+ + 2e^- = H_2S \qquad \text{(Eq 1)}$$

$$S + 4H_2O = SO_4^{2-} + 8H^+ + 6e^- \qquad \text{(Eq 2)}$$

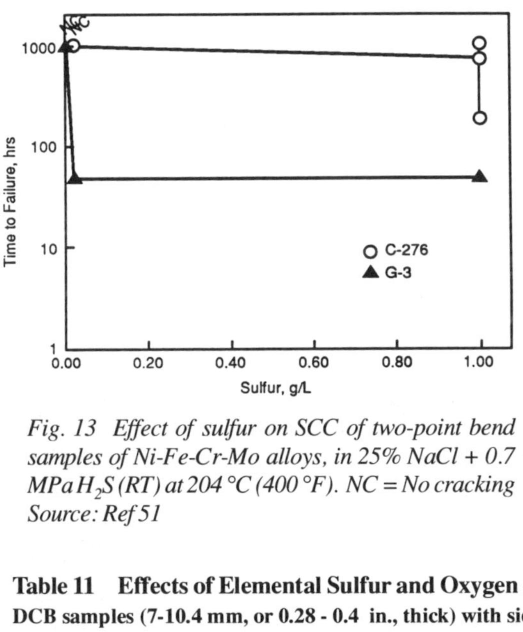

Fig. 13 Effect of sulfur on SCC of two-point bend samples of Ni-Fe-Cr-Mo alloys, in 25% NaCl + 0.7 MPa H_2S (RT) at 204 °C (400 °F). NC = No cracking Source: Ref 51

Based on these reactions, the sulfur solubility in H_2O at a pH of 3 and a temperature of 200 °C (390 °F) will be about 1 g/L and will be expected to be decreased by the presence of H_2S and increased by an increase in pH. Ueda *et al.* offer, as evidence for the validity of the above reactions, detection of H_2S in the gas phase and SO_4^{2-} in the aqueous phase after reaction of sulfur in a 5% NaCl solution in the presence of a platinum electrode at 250 °C (480 °F). Additionally, an increase in pH consistent with Eq 1 was noted upon exposure of a corroding Ni-Fe-Cr-Mo alloy in a Cl^--H_2S-S_8 environment. Schmidt (Ref 49) has proposed that chemisorption of sulfide molecules on molten sulfur particles serves to increase their electrical conductivity and thus facilitate electron-exchange reactions.

The effect of sulfur on SCC of two nickel-base alloys is illustrated in Fig. 13 (Ref 51). It can be seen that at 204 °C (400 °F) as little as 0.02 g/L of S_8 was sufficient to induce cracking of alloy G-3 in a 25% NaCl + 0.7 MPa (0.1 ksi) H_2S environment. Greater amounts of S_8 were required to crack alloy C-276. Cracking was transgranular in all the alloys and was invariably associated with deposits of dark sulfides, underneath which considerable corrosion also occurred. Miyasaka *et al.* (Ref 52) showed that H_2S is not necessary to cause cracking in the presence of S_8 because of the local H_2S created by the reduction of sulfur.

The mechanism of the effect of sulfur on cracking and the electrochemical behavior of nickel-base alloys is still being debated. Both Kolts (Ref 30) and Ciaraldi (Ref 43) have shown

Table 11 Effects of Elemental Sulfur and Oxygen on SCC of Nickel-Base Alloys

DCB samples (7-10.4 mm, or 0.28 - 0.4 in., thick) with side grooves tested at 204 °C (400 °F) in a solution of 20% NaCl + 0.1% CH_3COOH + 1.72 MPa H_2S + 1.72 MPa CO_2

Alloy, nominal composition	1 g/L S_8			0.07 MPa O_2		
	K_{ISCC}, MPa$\sqrt{m}$	K_{ISCC}, ksi$\sqrt{in.}$	SCC	K_{ISCC}, MPa$\sqrt{m}$	K_{ISCC}, ksi$\sqrt{in.}$	SCC
25Cr-7Ni	45.0	40.9	Yes	47.7	43.4	Yes
20Cr-26Ni	27.5	25.0	Yes	22.0	20.0	Yes
27Cr-31Ni	66.2	60.2	Yes	67.8	61.6	Yes
21Cr-36Ni	72.5	65.9	Yes	76.0	69.1	Yes
23Cr-42Ni	...	...	No	...	...	No
22Cr-45Ni	...	...	No	...	...	No

Note: Applied stress intensity ranged from 66 to 100 MPa$\sqrt{m}$ (60 to 91 ksi$\sqrt{in.}$)
Source: Ref 43

Table 12 Effect of Sulfur Solvents on SCC

Mill-annealed two-point bend samples tested (2.6-4% strain) at 232 °C (450 °F) in a solution of 25% NaCl + 0.5% CH_3COOH + 0.7 MPa H_2S + 1 g/L S_8

Alloy	No solvent	Time to failure, h(a) 5% diethylamine (pH 10-11)	5% diethylamine (pH 3.3-5.5)
825	960	>1440	>1440
G-3	1440	>1440	>1440
C-276	>8640	>8640	>8640

(a) > indicates no cracking
Source: Ref 51

that the effect of elemental sulfur is similar to that of oxygen (Table 11). In Ciaraldi's experiments, it is possible that oxygen converted H_2S to sulfur. However, no H_2S was used in the experiments reported by Kolts. Thus, the effect of sulfur may be viewed as that of an oxidizer or cathodic depolarizer. Rhodes (Ref 39) showed that sulfur increased the corrosion potential of alloy MP35N (35Co-35Ni-20Cr-10Mo). However, his potential measurements were carried out with reference to a platinum electrode immersed in the same solution. Since the platinum electrode also tends to be polarized by sulfur, the effect of sulfur on the corrosion potential of the alloy cannot be estimated accurately. Ueda *et al.* (Ref 50) showed that sulfur in a NaCl solution increased the open-circuit potential of platinum and a Ni-Fe-Cr-Mo alloy at 200 °C (390 °F) (measured against a pressure-balanced Ag/AgCl electrode at room temperature). However, in the presence of H_2S, no effect of sulfur on open-circuit potential was observed. They also found that sulfur increased the anodic currents and decreased the pitting potential even in the presence of H_2S. Miyasaka *et al.* (Ref 52) showed that sulfur increased the cathodic reaction kinetics but not the open-circuit potential or the anodic currents of Ni-Fe-Cr-Mo alloys and platinum. Wilhelm (Ref 48) found a slight decrease in the corrosion potential of alloy 825 with the addition of sulfur but an increase in cathodic reaction kinetics. He found a small decrease in anodic current density due to the addition of sulfur.

Sulfur may also affect cracking by increasing the anodic reaction or preventing repassivation of the alloy surface. Marcus (Ref 47) has examined the effect of adsorbed sulfur on the anodic reaction kinetics of nickel, Ni-Fe, and Fe-Ni-Cr-Mo alloys and has shown that adsorbed sulfur prevents passivation of the surface by excluding OH^- adsorption. However, his measurements have been carried out under conditions where atomic adsorption of sulfur at the surface was ensured and at low temperatures in nonchloride environments.

Effect of Inhibitors. Wilhelm (Ref 48) has shown that addition of bicarbonate inhibits cracking of alloy G-3 in a Cl^--H_2S-S_8 environment at 232 °C (450 °F). Similarly, addition of the sulfur solvents diethylamine and monoethanolamine has been shown (Ref 30, 51) to inhibit cracking of Ni-Cr-Mo alloys (Table 12).

Effect of Electrode Potential. It is possible that cracking in Cl^- + H_2S environments occurs in specific ranges of potential depending on the environmental and alloy compositions. The effect of applied potential on SCC of alloy 825 is shown in Fig. 14 (Ref 53). It is interesting to note that only a 10 mV increase in potential from the corrosion potential is needed to cause cracking in this environment. Such small differences can be easily caused by differences in deaeration and incidental galvanic contact of the sample with autoclaves and other metallic test accessories (Ref 53). These differences may account for the irreproducibility of slow-strain-rate data between various laboratories on identical materials and environments. For example, Ciaraldi (Ref 43) noted that no cracking occurred when the environment was completely deaerated, whereas even a slight exposure to air caused increased cracking. Kolts (Ref 30) also noted the adverse effect of

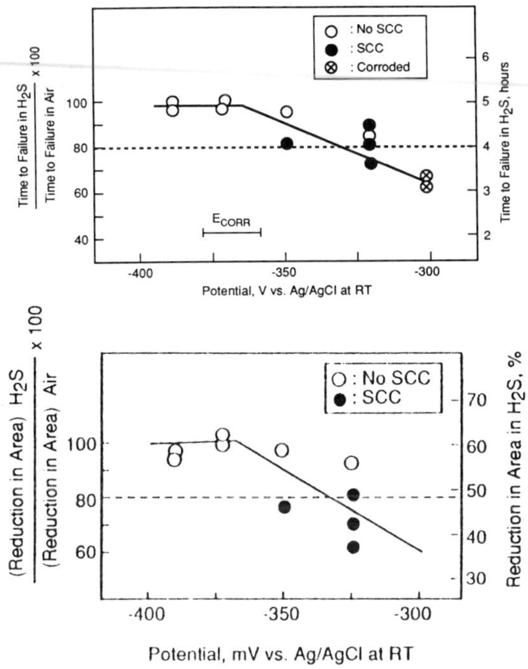

Fig. 14 Effect of potential on SCC resistance of alloy 825 in 25% NaCl + 0.5% CH_3COOH + 0.7 MPa H_2S, at 150 °C (300 °F); $\dot{\varepsilon} = 4 \times 10^{-6}$/s. Source: Ref 53

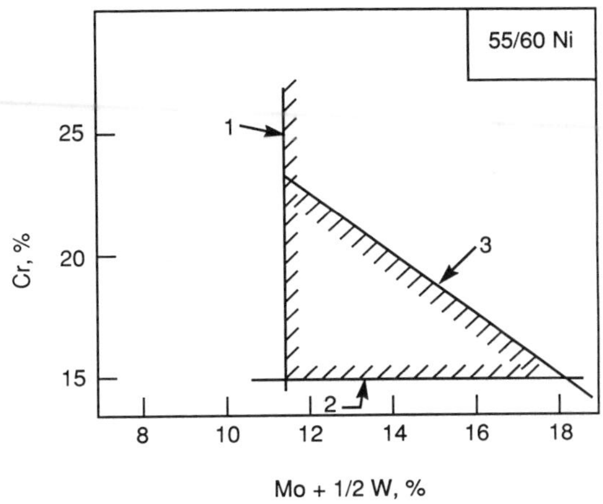

Fig. 15 Recommended region of chromium and molybdenum content of nickel-base alloy with approximately 55 to 60 wt% Ni in H_2S-CO_2-Cl^--S environment. Line 1: SCC; 230 °C (450 °F), 1 MPa H_2S + 1MPa CO_2 + 25 wt% NaCl + 1 g/L S_8, 336 h; four-point bent beam. Line 2: hydrogen embrittlement; 25 °C (77 °F), 5% NaCl + 0.5% CH_3COOH + 0.1 MPa H_2S, 720 h, $\sigma = \sigma_{0.2}$, iron coupling; four-point bent-beam. Line 3: phase stability. Source: Ref 55

aeration. Wilhelm (Ref 48) measured the corrosion potentials of alloy 825 in 15% Cl^- + 2.76 MPa (0.4 ksi) H_2S + 5.5 MPa (0.8 ksi) CO_2 environment at 204 °C (400 °F). The measured potentials were significantly higher than those measured by Ueda *et al.* (Ref 50) and cracking was noted in all the samples.

Effect of pH. Acidification by HCl or CH_3COOH has been shown to increase cracking susceptibility in both H_2S and H_2S + S_8 environments (Ref 30, 43, 48, 51). Conversely, increasing the pH by the addition of NaOH has been shown to prevent cracking in these environments (Ref 30, 48).

Effect of Chloride. Wilhelm (Ref 48) showed that chloride concentrations below about 9% did not cause cracking of alloys 825 and G-3 in a 2.76 MPa (0.4 ksi) H_2S + 5.5 MPa (0.8 ksi) CO_2 + 1 g/L S_8 environment at 204 °C (400 °F). Ikeda *et al.* (Ref 45) showed that cracking of a 31Ni-22Cr-4Mo alloy did not occur below 15% Cl^- when tested by the slow-strain-rate method in a 1 MPa (0.15 ksi) H_2S + 1 MPa (0.15 ksi) CO_2 environment at 150 °C (300 °F).

Effect of Alloy Composition. As shown in Fig. 12, the higher-molybdenum alloys exhibit superior SCC resistance, especially in environments containing elemental sulfur. This finding has been confirmed by many investigators (Ref 40-42, 45, 46, 48, 50, 51, 53, 54). In contrast, Ciaraldi (Ref 43) has indicated that nickel is the most beneficial element, with chromium less beneficial and molybdenum having no effect. However, since the high-nickel alloys (above 45% Ni) also contained significantly higher molybdenum than the lower-nickel alloys, the effect of molybdenum cannot be separated from that of nickel in his studies. It must be emphasized that increasing the contents of molybdenum and tungsten, as well as chromium and other refractory elements, increases the possibility of formation of intermetallic phases such as μ and Laves. Hence, metallurgical stability considerations limit the design of alloys. This is indicated in Fig. 15 (Ref 55), where the alloys of interest in H_2S + S_8 environments are plotted in terms of chromium versus molybdenum + 0.5 W. At Mo + 0.5

Table 13 Effect of Yield Strength on Sulfide Stress Cracking of Alloy C-276

C-ring samples (cold pilgered + aged at 204 °C, or 400 °F, for 200 h) tested in a solution of 5% NaCl + 0.1% CH_3COOH + 0.7 MPa H_2S (RT) + 1 g/L S_8

% Yield stress	Yield stress MPa	ksi	Time to failure, h(a) 177 °C (350 °F)	232 °C (450 °F)
70	1069	155	>8600	528
	1262	183	>8640	300
	1317	191	>8640	144
90	1069	155	>8688	528
	1262	183	>8688	312
	1317	191	>8688	168

(a) > indicates no cracking within the stated time period

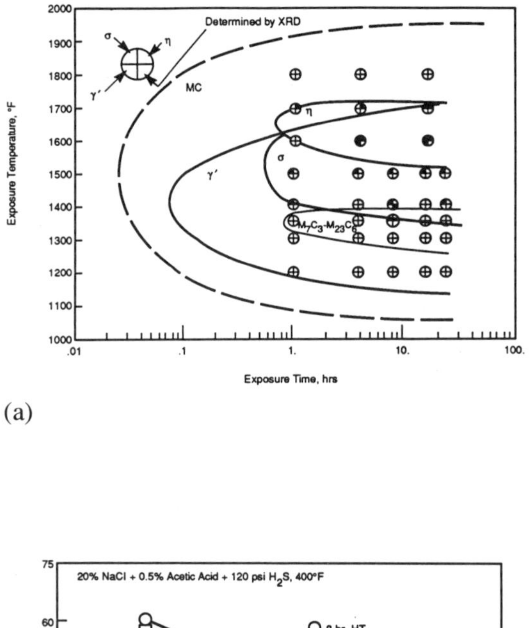

(a)

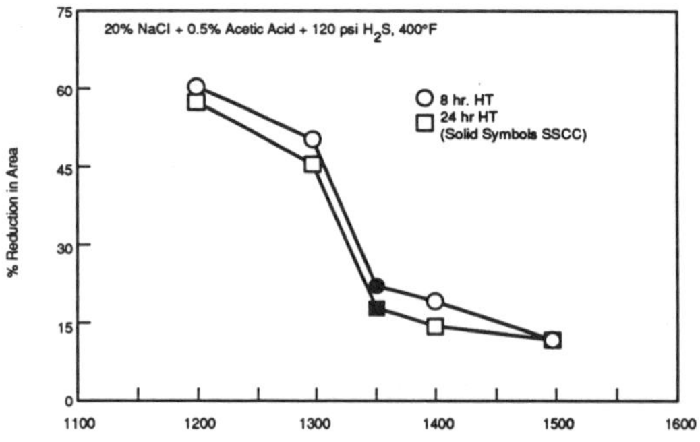

(b)

Fig. 16 (a) Time-temperature-transformation diagrams for annealed (1010 °C, or 1850 °F) alloy 925. (b) Slow-strain-rate test results for hot-rolled + annealed (1010 °C, or 1850 °F) + aged samples. Source: Ref 56

W levels (the factor 0.5 indicates a rough accounting for the relative atomic weight of tungsten with respect to molybdenum) below about 12% (line 1), the alloys do not possess adequate SCC resistance. Line 3, outside which the alloys are unstable, is calculated using the concept of M_d (Ref 18), which relates alloy composition to stability in terms of electron concentration. Line 2 refers to an area outside which the alloys will not be resistant to hydrogen embrittlement in H_2S environment. While this type of diagram is useful, its limitations are: (1) material stability is considered in terms of observed second-phase stringers after solution annealing rather than in terms of grain-boundary precipitation or sensitization, and (2) the authors did not present any data to justify the delineation of hydrogen embrittlement resistance by line 2.

Effect of Alloy Strength. As shown in Table 13, increasing the yield strength by cold work generally decreases the resistance to SCC (Ref 51). However, this relationship is not always monotonic and depends on the severity of the environment for a given alloy. Ikeda *et al.* (Ref 45) have shown that, in an environment of 25% NaCl + 0.5% CH_3COOH + 7 atm H_2S at 150 °C (300 °F), slow-strain-rate testing of alloys C-276 and 2550 showed no cracking at any strength level ranging from 758 to 1310 MPa (110 to 190 ksi). Asphahani (Ref 44) has shown that increased cold working increased the cracking tendency of 25 to 30% Ni stainless steels but not of the higher-Ni alloys in a 5% NaCl + 0.5% CH_3COOH + 1 atm H_2S (saturated at room temperature) at 177 °C (350 °F).

Table 14 Effect of Precipitation Hardening on SCC

Samples (solutionized at 1149 °C + 699 °C/20 h + AC) tested at a strain rate of 4×10^{-6}/s in a solution of 25% NaCl + 0.5% CH_3COOH + 0.7 MPa H_2S

Alloy	Room-temperature yield MPa	ksi	(Time to failure in H_2S)/(Time to failure in air), h 149 °C (300 °F)	177 °C (350 °F)	204 °C (400 °F)
23Cr-3.5Mo-4.7Nb (γ'' alloy)	896	130	16.7/16.8	16.3/16.9	4.0/17.0
19Cr-3Mo-0.9Ti-5.1Nb ($\gamma' + \gamma''$55 alloy)	931	135	6.3/17.7	6.9/17.2	...
22Cr-3Mo-2.67Ti (γ' alloy)	690	100	4.5/19.6	4.4/18.2	...

Source: Ref 59

Table 15 Effect of Grain Boundary Carbides on SCC of an Alloy Similar to Alloy 825

C-ring samples of 42Ni-22Cr-3Mo-1.7Cu-0.7Ti-0.016C tested at 100% of transverse yield strength in a solution of 20% NaCl + 1 MPa H_2S + 1 MPa CO_2

Annealing temperature °C	°F	Yield strength MPa	ksi	Test temperature(a) 200 °C (390 °F)	250 °C (480 °F)	275 °C (525 °F)	300 °C (570 °F)
950	1740	296	43	NC	NC	...	NC
		775	112	NC	NC	...	NC
		764	111	NC	NC	...	...
		859	125	NC	IGC	...	...
		924	134	NC	IGC	...	IGC
		937	136	NC	IGC	...	IGC
1100	2010	856	124	...	NC	IGC	...
		871	126	...	NC	IGC	...

(a) NC, no cracking in 30 days; IGC, intergranular cracking
Source: Ref 60

Effect of Precipitation. Incoherent, intermetallic precipitates such as μ, σ, and Laves phases are generally detrimental to the mechanical properties and corrosion resistance of nickel-base alloys. However, their effect on SCC in H_2S environments has not been clearly established. The detrimental effect of σ phase is illustrated in Fig. 16 (Ref 56). In this alloy, precipitation of both coherent phases (γ' and γ'') and incoherent phases (σ, η, carbides) occurs, depending on the time at temperature. Precipitation of σ phase occurred at temperatures above 704 °C (1300 °F) and caused SCC at 730 °C (1350 °F), but not at the higher temperatures. This may have been because of the low ductility in air that resulted in low times of exposure to the environment or due to "healing" of the sensitized regions. The precipitation of coherent phases at lower temperatures did not cause SCC. Precipitation of coherent or semicoherent γ'' phases in alloy 625 Plus (Ref 57) and alloy 725 (Ref 58) also did not result in loss of SCC resistance in 25% NaCl + 0.5% $CH_3COOH + S_8 + H_2S$ environments at temperatures up to 232 °C (450 °F). Igarashi *et al.* (Ref 59) have suggested that alloys that are strengthened by γ'' (by addition of niobium) are more resistant than alloys strengthened by γ' (by addition of titanium or aluminum) or by a mixture of $\gamma'' + \gamma'$. Their slow-strain-rate data in 25% NaCl + 0.5% CH_3COOH + 0.7 MPa (0.1 ksi) H_2S are presented in Table 14. They suggest that the γ' alloys showed a greater planarity of slip because of dislocations cutting the precipitates, which led to strain localization and cracking. On the other hand, heterogeneous slip was not observed in the γ'' alloy. They also showed that pitting susceptibility of the γ' alloy was greater, which is probably related to partitioning of alloying elements between the matrix and precipitates.

Carbide precipitation has been shown to be detrimental to SCC in sulfide environments. The results of Ishizawa *et al.* (Ref 60) are shown in Table 15. They annealed 42Ni-22Cr-3Mo-2Cu (similar to alloy 825) specimens at two temperatures and then cold worked them to various degrees. The lower annealing temperature produced grain-boundary $M_{23}C_6$-type carbides and chromium-depletion, as measured by scanning transmission electron microscopy (STEM). Although the investigators did not test multiple samples, the results in Cl^- + H_2S indicate that the presence of carbides resulted in cracking at a lower temperature. Cracking in both the sensitized and nonsensitized samples was intergranular. However, tests on the same alloy in a Cl^- + H_2S + S_8 environment did not show any effect of sensitization, and the cracking was all transgranular.

Sulfur-Oxyanions Solutions

Another class of aggressive sulfur-bearing species that can induce intergranular SCC of sensitized, single-phase Ni-Cr-Fe alloys is the family of metastable oxyanions, which includes thiosulfate ($S_2O_3^{2-}$) and polythionates ($S_xO_6^{2-}$, $x = 3, 4, 5, 6$).

Initial observations were related to the so-called "polythionic acid cracking" that affected sensitized stainless steels in catalytic reformers used in the petroleum industry in the early 1950s (Ref 61, 62). Samans (Ref 63) was the first to report the intergranular cracking in a polythionic acid solution at room temperature of alloys 600 and 800 after being sensitized at 650 °C (1200 °F) for 4 h. Samans observed no cracking in mill-annealed specimens.

Polythionic acid solutions are commonly prepared by bubbling H_2S into a cold solution of SO_2 in water (Wackenroder's solution). These solutions contain a mixture of the different polythionic acids. Solutions prepared according to ASTM Standard Practice G35-88 have a pH of about 1.15 and approximately 0.13 mol/L of polythionic acids, in addition to 1.62% H_2SO_4 and 3.02% of H_2SO_3 (Ref 64, 65). The predominant species is tetrathionic acid at a concentration of approximately 0.074 mol/L, which represents around 57% of the total content of polythionic acids. Ahmad *et al.* (Ref 65) also showed that of all the polythionic acids, only tetrathionic acid is a potent cracking agent for sensitized AISI 304 stainless steel.

Scarberry *et al.* (Ref 66) confirmed that annealed alloy 600 was not susceptible to cracking in polythionic acid solution after 1000 h of exposure under constant-load conditions over a wide range of nominal stress, whereas a sensitization treatment at 593 °C (1100 °F) for 4 h induced intergranular cracking in less than 10 h. As expected, the time to failure increased with decreasing stress, and an apparent threshold stress value as low as 69 MPa (10 ksi) was reported. Similar behavior was observed in alloy 800 U-bend specimens exposed to polythionic acid solution (Ref 67). In relatively short times (4 to 48 h), cracks were detected in specimens solution annealed at 955 °C (1750 °F) for 1 h and heat treated at 650 °C (1200 °F) for 1 and 2 h. Prolonged heat-treatment times did not lead to cracking in 1000 h of exposure. As suggested by all these authors, susceptibility to intergranular SCC is related to the existence of a chromium-depleted region along grain boundaries caused by the intergranular precipitation of chromium carbides. However, correlation with the results of boiling nitric acid tests (Ref 66, 67) was not entirely satisfactory since this test leads to relatively high corrosion rates even for microstructures that are not prone to cracking. This was more evident in the case of stabilized alloys, such as alloy 801 (a titanium-stabilized version of alloy 800), which was found to be resistant to cracking even though high corrosion rates (30.5 mm/yr, or 1200 mil/yr) were measured in the boiling nitric acid test.

Intergranular SCC of these sensitized Ni-Cr-Fe alloys is by no means confined to the very acidic polythionic acid solutions. Berge and Donati (Ref 68) reported the intergranular SCC of alloy 600, sensitized at 700 °C (1290 °F) for 1 h, in 0.1 mol/L sodium tetrathionate (pH 3) at room temperature. A threshold stress of about 100 MPa (14.5 ksi), which is almost identical to that obtained in the presence of polythionic acids, was determined in that solution. However, at higher stress levels the failure time was found to be shorter in polythionic acids.

The discovery of extensive intergranular cracking of alloy 600 tubing in the once-through steam generators at the Three Mile Island Unit 1 Nuclear Power Plant, following a prolonged shutdown, prompted an intensive investigation (Ref 69). As previously reported for sensitized AISI 304 stainless steel (Ref 62, 70) the thiosulfate ($S_2O_3^{2-}$) anion was identified as one of the primary causative species for the failure. Furthermore, very small concentrations (of the order of parts per million of equivalent sulfur) were sufficient to induce cracking. In a series of publications, Newman and coworkers (Ref 71-74) examined mechanical, material, and environmental factors relevant to the intergranular cracking of alloy 600 in dilute thiosulfate and tetrathionate solutions.

The threshold stress for intergranular SCC of sensitized alloy 600 in thiosulfate-containing solutions, as measured in constant-load tests, was found to be about 95% of the yield strength (Ref 73). However, this value was obtained at 22 °C (72 °F) in a relatively mild environment containing 1.3% boric acid, thiosulfate equivalent to 0.7 ppm sulfur, and 0.7 ppm lithium added as LiOH. In more aggressive solutions, such as polythionic acid (Ref 66) or concentrated tetrathionate (Ref 68), the apparent threshold stress for alloy 600 is close to 30% of the yield strength. It has been suggested (Ref 73) that the threshold stress may approach zero for conditions at which deep intergranular corrosion occurs (low-pH solutions or inside crevices). Although no fracture-mechanics tests have been conducted, it can be concluded that, under a wide range of environmental conditions in terms of anion concentration, pH, and temperature, crack growth occurs primarily in the stage 2 (stress-intensity independent) region.

The degree of sensitization of alloy 600 is a critical factor in the intergranular SCC of this alloy in thiosulfate and tetrathionate solutions. Whereas solution- or mill-annealed material is not susceptible to cracking even in relatively concentrated thiosulfate and tetrathionate solutions, sensitization as a result of heat treatment at 600 to 700 °C (1110 to 1290 °F) for a short period induces severe cracking (Ref 73, 75). On the other hand, prolonged heat treatment at those temperatures renders the alloy immune to cracking because of replenishment of the chromium-depleted regions along grain boundaries.

The initial microstructure of alloy 600 before sensitization heat treatment is extremely important in determining its SCC susceptibility in the presence of metastable sulfur oxyanions. Solution-annealed material contains more available carbon in solid solution than does mill-annealed material and will therefore experience more severe chromium depletion along grain boundaries after a sensitization heat treatment. This was confirmed (Ref 73) by comparing the behaviors of solution- and mill-annealed samples with a further heat treatment at 621 °C (1150 °F) for 18 h. The solution-annealed and sensitized (SAS) alloy was found to be susceptible to intergranular SCC in neutral thiosulfate solution, which was not the case for the mill-annealed and sensitized (MAS) alloy. However, in more aggressive environments (tetrathionate or acidified thiosulfate), both materials cracked. In order to differentiate the degree of sensitization of the MAS and SAS materials, electrochemical potentiokinetic reactivation (EPR) tests were performed in 0.5 *M* H_2SO_4 + 0.01 *M* KSCN solution. A semiquantitative correlation with the tendency to SCC was observed, since the reactivation charges per unit grain-boundary area for the mill-annealed, MAS, and SAS materials were found to be 111, 209, and 774 C/cm^2, respectively.

However, as noted by Was and Rajan (Ref 76), degree of sensitization is a rather ambiguous concept that arises from the fact that sensitization is defined in terms of the tendency to intergranular corrosion experienced by a material in a given environment, generally evaluated according to the EPR test or other standardized ASTM A262 test methods. Different responses are obtained according to the test environment. By conducting slow-strain-rate tests in $Na_2S_4O_6$ solutions using a high-purity Ni-16Cr-9Fe alloy heat treated under different time and temperature conditions, Was and Rajan (Ref 77) showed that the level of chromium depletion at the grain boundary (see Fig 1) is the determining factor in the propensity to intergranular SCC. They noted that, irrespective of heat-treatment conditions, a grain-boundary chromium level below 5 wt% resulted in 100% intergranular fracture, whereas levels above 8.3 wt% produced no intergranular fracture. In addition, the variation in the slope of the

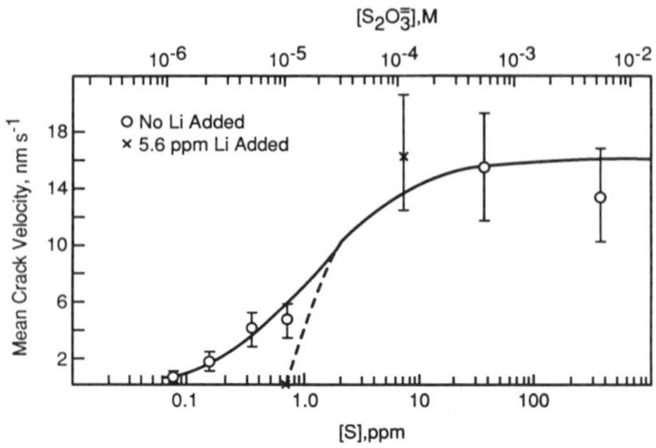

Fig. 17 Mean crack velocity of constant-extension-rate tests on sensitized alloy 600 as a function of $S_2O_3^{2-}$ concentration in air-saturated 1.3% H_3BO_3 at 40 °C (104 °F). Source: Ref 71

chromium concentration profile at the grain boundary had no effect on the propensity to intergranular SCC. On the other hand, the EPR test in H_2SO_4 + KSCN solutions is more sensitive to broad, shallow, chromium-depleted regions in which the chromium concentration can be as high as 13 wt%. From these observations, it can be concluded that the results of EPR tests cannot be used to quantitatively predict the SCC susceptibility of sensitized alloy 600.

From the previous discussion, it is apparent that intergranular SCC of Ni-Cr-Fe alloys in the presence of metastable sulfur oxyanions can be avoided by using prolonged heat treatments (*i.e.* 15 h at 700 °C, or 1290 °F, for alloy 600) that replenish the chromium-depleted regions. For nickel-base alloys, the decrease in the carbon content does not provide adequate improvement, as in the case of austenitic stainless steels (*i.e.*, types 304L and 316L), because of the low solubility of carbon in nickel. It is expected that stabilized alloys, such as alloys 801, 601, and 625, and those containing higher chromium levels, such as alloy 690, are resistant to cracking. However, no systematic investigation has been reported for the stabilized alloys. In the case of alloy 690, even though chromium depletion down to a concentration of 15 wt% has been measured at grain boundaries for certain heat treatments, no SCC was detected in 0.001 *M* $Na_2S_2O_3$ solution at pH 3 (Ref 78). It is obvious that the level of chromium depletion attained, even for prolonged heat treatments at 600 to 700 °C (1110 to 1290 °F), is not sufficient to induce intergranular SCC in this alloy. It is worthwhile to note that Berge and Donati (Ref 68) have reported that high-purity heats (carbon content below 0.003%) of Ni-xCr-8Fe alloys, where $7 < x < 16$, were immune to cracking in the solution-annealed condition, with the exception of the alloy containing 7% Cr, which exhibited transgranular cracking in polythionic acid solution.

Environmental factors are also extremely important in the intergranular SCC of these alloys in sulfur-oxyanions solutions. Very low concentrations of either thiosulfate or tetrathionate anions (about 10^{-5} mol/L) are sufficient to promote cracking in severely sensitized materials. Bandy *et al.* (Ref 73) have determined threshold concentrations of about 2×10^{-6} mol/L thiosulfate and 1×10^{-6} mol/L tetrathionate in air-saturated solutions at 22 °C (72 °F). Although the effects of both metastable sulfur oxyanions are similar, it appears that the propensity to cracking initiation in thiosulfate is lower than in tetrathionate. The presence of oxygen seems to be necessary to initiate intergranular SCC in thiosulfate solutions. Oxygen is not necessary in the case of tetrathionate, presumably because of the higher oxidizing ability of tetrathionate with respect to thiosulfate. In a related study (Ref 72) using reverse U-bend specimens of alloy 600, no cracking was observed in 0.1 *M* thiosulfate, whereas tetrathionate caused cracking down to 10^{-5} *M* concentration. However, rapid cracking occurred when the pH of the thiosulfate solution was reduced to 3.

Intergranular SCC of sensitized alloy 600 in 0.5 *M* $Na_2S_2O_3$ at 22 °C (72 °F) occurs in a potential range that extends from –300 to 300 mV (SCE) (Ref 71). Very rapid crack-growth rates (4×10^{-7} m/s) were measured at approximately 0 mV (SCE), whereas under open-circuit conditions, crack velocities up to 1.5×10^{-8} m/s were observed over a wide range of $S_2O_3^{2-}$ concentrations (10^{-2} to 10^{-4} mol/L) in air-saturated H_3BO_3 at 40 °C (104 °F), as shown in Fig. 17. Crack velocity also increased with temperature, but it leveled off at approximately 80 °C (175 °F), presumably due to the decrease in the concentration of oxygen dissolved in the solution and the concurrent decrease in the potential. An important observation in the studies of Newman *et al.*

(Ref 71, 74) is that cracking in solutions containing H_3BO_3 and a low concentration of $Na_2S_2O_3$ can be prevented by the addition of LiOH. Although it is not shown in Fig. 17, the addition of 20 ppm Li^+ to the boric acid solution containing 7 ppm sulfur as $S_2O_3^{2-}$ decreases the mean crack velocity to values lower than 1 nm/s. The same inhibiting effect was exhibited when 5.6 ppm Li^+ were added to a solution containing 0.7 ppm sulfur, as shown in Fig. 17. In the presence of LiOH, the concentration of $H_2BO_3^-$ increases, and cracking is prevented by the competitive electromigration of this anion with $S_2O_3^{2-}$.

It appears that the intergranular cracking of alloy 600 in tetrathionate and thiosulfate solutions is a true SCC phenomenon in the presence of low concentrations of these anions. However, with increasing concentrations, intergranular corrosion assisted by stress seems to be the dominant phenomenon. Under certain low-pH conditions or in crevices, intergranular corrosion rather than intergranular SCC predominates. Localized attack in the form of pitting near grain boundaries has been identified by Bandy *et al.* (Ref 72, 73) as the initial step in the nucleation of intergranular cracks.

Finally it should be noted that other sulfur-bearing compounds may be involved directly or indirectly in the intergranular SCC of sensitized Ni-Cr-Fe alloys. As in the case of sensitized austenitic stainless steels, the SCN^- anion may be a cracking agent under certain environmental conditions, and SO_2 and/or H_2S dissolved in water in the presence of air and a catalytic metal surface are known to generate polythionates as predominant ionic species. On the other hand, a thermodynamically stable sulfur-oxyanion such as sulfate does not promote intergranular SCC, even in heavily sensitized alloys. As reported by Herbsleb (Ref 24), sensitized alloys 600 and 800 were found to be resistant to intergranular cracking in boiling Na_2SO_4 solutions over a wide range of applied potentials.

SCC in High-Temperature Water and Dilute Aqueous Solutions

It is now well recognized that solid-solution-strengthened nickel alloys, such as alloy 600, and precipitation-hardened alloys, such as alloys X-750 and 718, are susceptible to intergranular SCC in deaerated high-purity water at temperatures of about 300 to 350 °C (570 to 660 °F). For alloy 600, very long initiation times (of the order of months) and relatively high stress levels (above macroscopic yielding) are required to develop intergranular cracks at the lower end of that temperature range. Under identical environmental and stress conditions, the same phenomenon has not been observed in alloys 800 or 690, which are used, as is alloy 600, as tubing materials in steam generators of pressurized nuclear water reactors.

Coriou and coworkers were the first to report the intergranular SCC of alloy 600 in deaerated high-temperature pure water (Ref 79). For nearly 20 years, the general validity of their observation was questioned, but the widespread occurrence of intergranular cracking originating from the primary side of recirculating steam generators in many nuclear power plants has eliminated any doubts about the generic character of the phenomenon (Ref 80, 81).

Most of the testing has employed reverse U-bend specimens cut from split tubing, since this is the main product form for alloy 600. Significant variations in failure times have been observed by comparing different heats, reflecting the important influence of relatively minor differences in thermomechanical treatments. Bandy and Van Rooyen (Ref 82) found that for mill-annealed material and cold-worked material tested under constant-load conditions in both pure water and primary water (H_3BO_3 + LiOH) at 365 °C (690 °F), time to failure is related to applied stress by the following expression:

$$t_f = k\,\sigma^{-4.0} \qquad \text{(Eq 3)}$$

indicating a strong dependence of failure time on applied stress.

Intergranular SCC has also been observed in slow-strain-rate tests at strain rates ranging from 3×10^{-8}/s to 1×10^{-6}/s over the temperature range of 290 to 365 °C (555 to 690 °F) (Ref 83, 84). Although crack-initiation times are usually shortened under dynamic straining conditions, it is difficult to initiate cracking of alloy 600 in high-temperature water within the duration of slow-strain-rate tests. It has been shown (Ref 85)

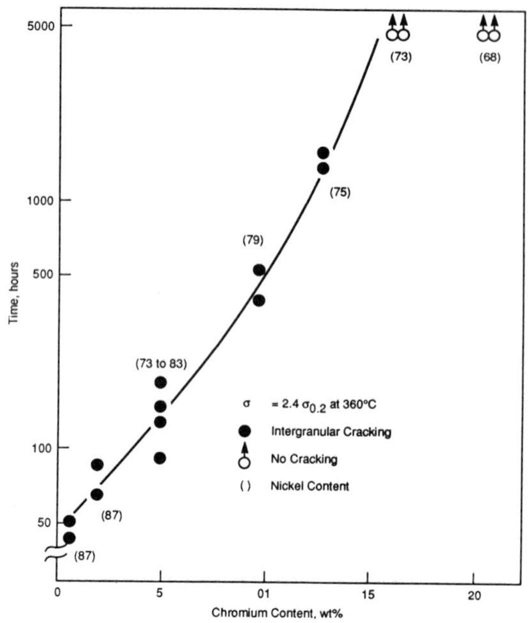

Fig. 18 Effect of chromium content on the time to failure of solution-annealed Ni-Cr-Fe alloys tested in H_3BO_3 + LiOH solutions containing hydrogen at 360 °C (680 °F). Source: Ref 86

that tensile specimens with a pressed hump, which introduces high levels of cold work locally, exhibit short initiation times.

A singular aspect of the intergranular SCC of alloy 600, in addition to the fact that it occurs in apparently innocuous environments such as deaerated high-purity water, is associated with alloy composition. As noted above, Ni-Cr-Fe alloys with lower nickel contents, such as alloys 800 and 690, are not susceptible to this form of cracking. However, results of constant-load tests in a hydrogenated H_3BO_3 + LiOH solution at 360 °C (680 °F) (Fig. 18) indicate that the resistance to cracking of solution-annealed Ni-Cr-Fe alloys increases with increasing chromium content (Ref 86). It can be inferred, therefore, that the higher chromium content is the factor that renders alloys 800 and 690 immune to intergranular SCC under those environmental conditions. The presence of carbon and other impurities, such as phosphorus and boron, does not necessarily play a detrimental role, since high-purity alloys of similar composition are also susceptible to cracking (Ref 87, 89).

One of the most important factors affecting intergranular SCC of alloy 600 is microstructure, which depends on the thermomechanical history and carbon content of the material. As noted by several authors (Ref 90, 91), alloy 600 tubing subjected to mill-annealing treatments at temperatures lower than 950 °C (1740 °F) exhibits a fine-grained microstructure with relatively heavy transgranular carbide precipitation and almost no carbide precipitation on grain boundaries. This microstructure is characteristic of a material highly susceptible to intergranular SCC in high-temperature water. On the other hand, a higher-temperature annealing (>1000 °C, or 1830 °F) increases the grain size, decreases the transgranular precipitation, and induces preferential grain-boundary carbide precipitation, leading to a microstructure that, although not immune, is more resistant to intergranular SCC. In this regard, a carbon content of at least 0.02 wt% in the commercial alloy seems necessary for improving SCC resistance in high-temperature water. By heat treatment at 700 °C (1290 °F) for several hours, a semicontinuous grain-boundary precipitation of Cr_7C_3 occurs. This treatment renders the alloy more resistant to intergranular SCC (Ref 87, 92, 93). The degree of improvement is extremely dependent on the prior microstructure of the mill-annealed material. This implies that mill annealing should be done at a temperature sufficiently high (>1000 °C, or 1830 °F) to maintain enough carbon in solid solution. In addition, the treatment at 700 °C (1290 °F) should be extended for a period long enough (>15 h) to replenish, at least partially, the chromium-depleted region along grain boundaries to avoid intergranular cracking in the presence of acidic contaminants or metastable sulfur oxyanions.

Environmental variables also have an important effect on the intergranular SCC of alloy 600 in high-temperature water. However, very little effort has been made to evaluate systematically the effect of compositional variables in the medium, including pH, concentration of dissolved gases (such as hydrogen and oxygen), and, above all, potential.

Van Rooyen and coworkers (Ref 82-84) did not find the environment to significantly affect crack velocity as measured in slow-strain-rate tests at 345 °C (653 °F) comparing deaerated water, water containing hydrazine (to scavenge oxygen) and morpholine (to adjust the pH to 9), and primary water, which contains H_3BO_3, LiOH, and an overpressure of hydrogen gas. A slightly accelerating effect was noted for primary water with respect to pure water. Using U-bend specimens, they found that crack-initiation time was shorter for the pH-adjusted waters than for pure water. It appears that variations of the pH_{25C} within the range of 7 to 9 did not affect the susceptibility to intergranular SCC.

Slow-strain-rates and reverse U-bend tests conducted in pure water and in water containing H_3BO_3 and LiOH showed a definite accelerating effect on SCC susceptibility or rate upon addition of hydrogen to the environment at partial pressures of the order of a few bars, followed by a decrease at higher pressures, as summarized by Smialowska (Ref 94). Tests for a 12-week period on U-bend specimens from nine heats of alloy 600 produced approximately 2% failures in pure deaerated water at 365 °C (690 °F) without hydrogen, but 83% failures in water containing hydrogen at concentrations typical of primary water in pressurized water reactors (PWRs) (Ref 82). Airey (Ref 93) found that the crack-initiation time decreases by a factor of 5 due to the presence of hydrogen in high-temperature (360 °C, or 680 °F) pure water. These initial observations were confirmed for pure water and also for steam at temperatures ranging from 360 to 400 °C (680 to 750 °F) (Ref 95, 96). In both environments, the percentage of failed U-bend specimens increased with the partial pressure of hydrogen up to 0.27 MPa (0.04 ksi), but decreased at higher pressures. On the other hand, no intergranular SCC has been observed in alloy 600 exposed to dry hydrogen, even at a pressure of 35 MPa (5 ksi) in the temperature range of 300 to 400 °C (570 to 750 °F) (Ref 94). As noted above, the detrimental effect of hydrogen has also been observed in 0.01 *M* H_3BO_3 + 0.001 *M* LiOH at 350 °C (660 °F) using slow-strain-rate tests. The percentage of intergranular SCC in the fracture surface increased from approximately 10 to 60% by increasing the hydrogen partial pressure from 0.005 to 0.1 MPa (0.0007 to 0.015 ksi) (Ref 96, 97). A high hydrogen content was measured in the most strained volume of the specimens (close to the fracture surface) at the highest hydrogen partial pressure, but absorption of hydrogen was also noted at the lowest pressure.

The effect of dissolved oxygen on the intergranular SCC susceptibility of mill-annealed alloy 600 in high-temperature water has been reviewed in detail (Ref 98, 99). The accelerating effect of oxygen is well documented, and it is even more significant in creviced specimens (Ref 100). However, the effect of heat treatment on cracking susceptibility is precisely the opposite of that found in deaerated water. Alloy 600, with a carbon content of approximately 0.05%, exhibited the greatest susceptibility when sensitized at 677 °C (1250 °F). The time required to nucleate cracks decreases with increasing oxygen concentration (from 5 to 100 ppm) in pH 10 ammoniated water at 315 °C (600 °F) (Ref 100). It appears that, depending on the oxygen concentration and, therefore, on the electrode potential, two different mechanisms are operative. At high oxygen levels, the behavior is similar to that observed in high-temperature acidic solutions. In high-temperature (290 °C, or 555 °F) dilute H_2SO_4 solutions, sensitized alloy 600 is extremely susceptible to intergranular SCC at potentials above 0 V (SHE), whereas annealed and annealed plus cold-drawn materials are slightly susceptible (Ref 101, 102). These observations have been confirmed and extended recently by several authors (Ref 103-105) to a range of environmental conditions in terms of pH and solution composition. Under the same testing conditions, alloy 690 is not susceptible to cracking (Ref 103).

In this context, it is worth noting that the variation of the corrosion potential of alloy 600 as a function of dissolved oxygen concentration and temperature has been studied in recent years (Ref 106, 107). Sigmoidal curves showing a potential difference of about 600 mV between deaerated (<10ppb O) and air-saturated (8 ppm O) solutions have been determined at 250 °C (480 °F). As in the case of other passive Fe-Cr-Ni alloys, as well as for platinum, alloy 600 attains a corrosion potential in high-temperature deoxygenated solutions that is close to the reversible value for the hydrogen evolution reaction. Over a wide range of dissolved hydrogen concentrations, the corro-

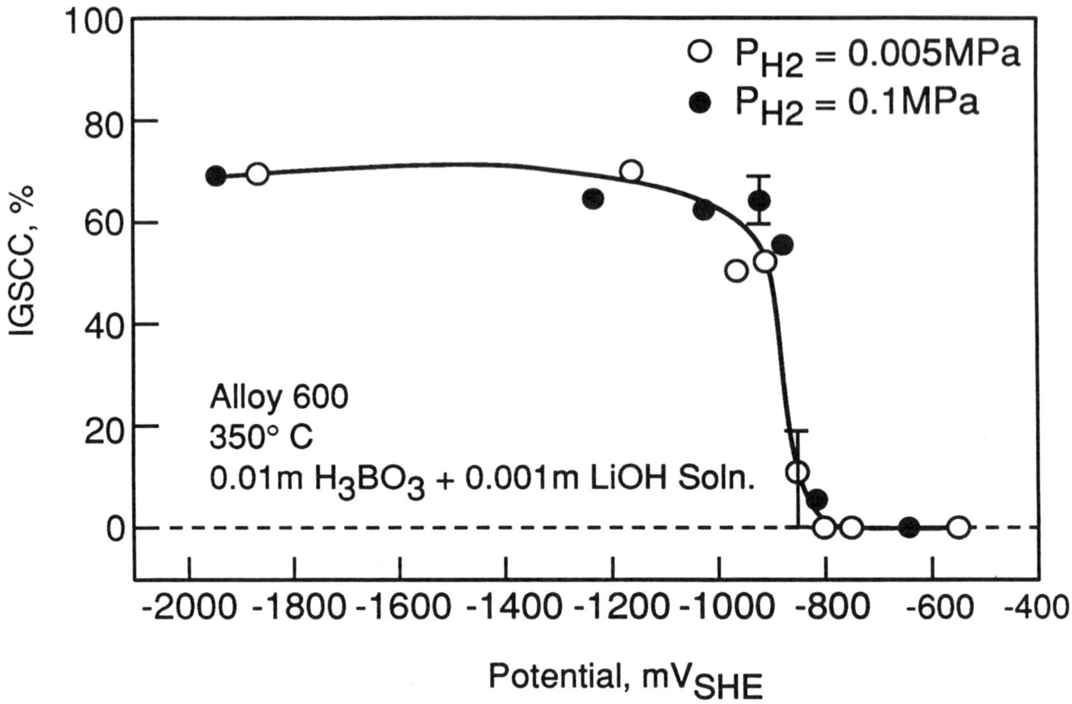

Fig. 19 Percentage of intergranular SCC on the fracture surface of alloy 600 as a function of applied potential in 0.01 M H_3BO_3 + 0.001 M LiOH solution at 350 °C (660 °F). Corrosion potentials for both hydrogen partial pressures are indicated by vertical bars. The extent of the bars shows the variation of percentage of intergranular SCC for three independent tests. Source: Ref 97

sion potential depends on the partial pressure of hydrogen, as expected for the reversible hydrogen electrode, and therefore will decrease (become more negative) with increasing hydrogen overpressure.

The effect of applied potential on the intergranular SCC of as-received alloy 600, as determined by slow-strain-rate tests conducted at 350 °C (660 °F) in solutions containing H_3BO_3 + LiOH at two different partial pressures of hydrogen, is shown in Fig. 19 (Ref 97). A small decrease of potential in the proximity of the corrosion potential measured in the solution with a partial pressure of hydrogen equal to 0.005 MPa (0.0007 ksi) produces a significant increase in the intergranular SCC susceptibility. A further decrease to more cathodic potentials has almost no effect. On the other hand, no intergranular SCC was detected at anodic potentials, at least up to potentials 400 mV higher than the corrosion potential. It is apparent from Fig. 19 that the major effect produced by the increase in the partial pressure of hydrogen is the displacement of the corrosion potential to lower potential values where susceptibility is higher.

Temperature is another environmental variable that strongly affects crack initiation and propagation. Temperature dependence has been represented according to an Arrhenius plot with an apparent activation energy that varies from 125 to 290 kJ/mol within the temperature range of 325 to 365 °C (615 to 690 °F), according to the results of different authors (Ref 82, 83, 95, 108). A lowest value of 74 kJ/mol has been reported by Totsuka *et al*. (Ref 109) for the temperature range of 310 to 350 °C (590 to 660 °F). Putting aside mechanistic interpretations and the wide variability in the data that make such interpretations difficult, it should be noted that the high value of the apparent activation energy indicates that both crack propagation and initiation become very

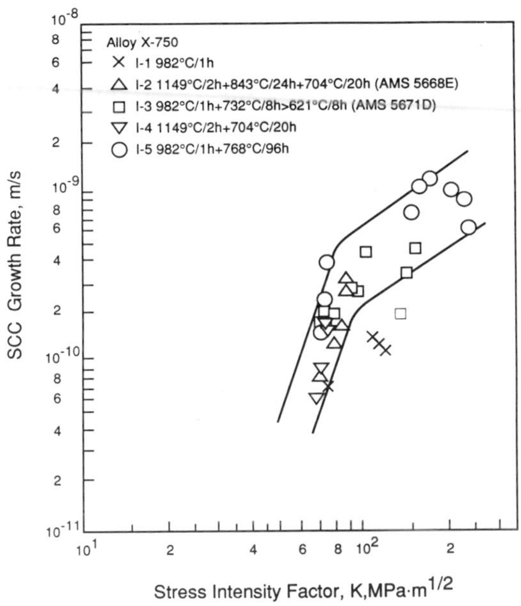

Fig. 20 Effect of stress intensity on crack-growth rates of alloy X-750 in oxygenated pure water at 300 °C (570 °F). Source: Ref 116

slow processes at temperatures below 300 °C (570 °F).

The principal remedial measures for mitigating and controlling intergranular SCC of alloy 600 in existing steam generators have been the application of local stress-relief heat treatments to reduce the level of residual stress and the introduction of compressive stresses in the inner tubing surface by shot peening and roto-peening.

Other nickel-base alloys that are susceptible to intergranular SCC in high-temperature deaerated water are precipitation-hardened alloys X-750 and 718 (Ref 110, 111). These alloys are used in water-cooled nuclear reactors in hold-down springs in fuel assemblies, high-strength pins and bolts in reactor cores, and other structural components. A larger number of failures in service have been reported for alloy X-750 than for alloy 718, probably because of the widespread use of alloy X-750 and the fact that it has been more extensively studied.

In several studies on the cracking of alloy X-750 (Ref 112-115), a pronounced effect of heat treatment on intergranular SCC susceptibility has been reported. A heat treatment that causes less

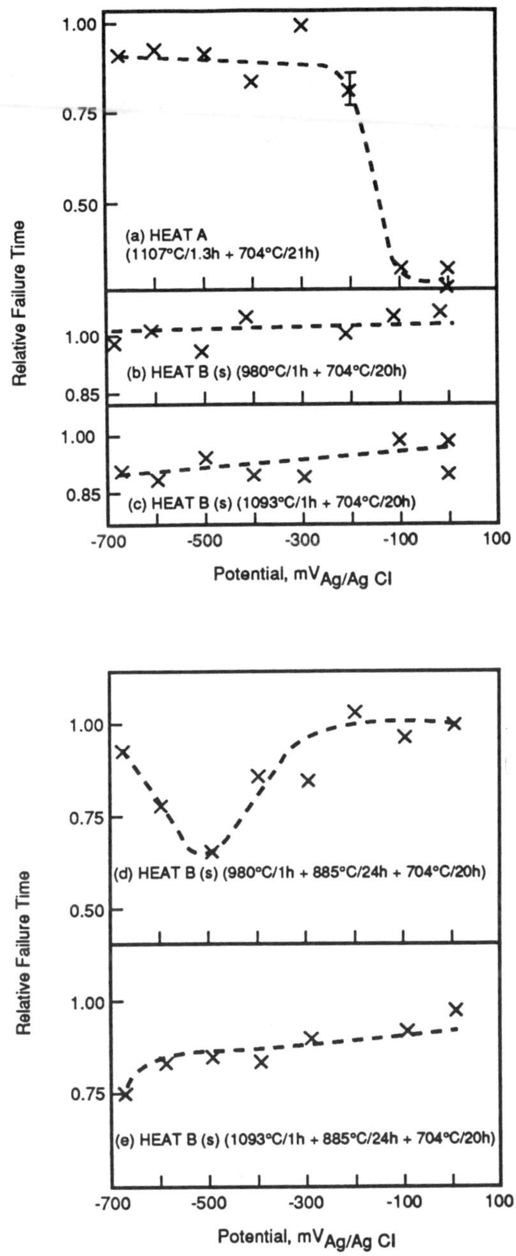

Fig. 21 Effect of potential on relative failure time for alloy X-750 (different heats and thermal treatments) in H_3BO_3 + LiOH solution at 340 °C (645 °F); $\dot{\varepsilon} = 1.5 \times 10^{-6}$/s. Source: Ref 115

susceptibility consists of solution annealing at a relatively high temperature (>1050 °C, or 1920 °F), followed by a single-step aging treatment of approximately 20 h at about 700 °C (1290 °F).

This heat treatment, although it improves cracking resistance in deaerated high-temperature water, produces a microstructure that is susceptible to intergranular SCC under oxidizing conditions, as a result of chromium depletion at the grain boundaries. On the other hand, a two-step aging heat treatment, originally adopted from previous applications in the aircraft industry, was found to promote intergranular SCC in both deaerated and aerated water—typical of PWR and boiling water reactor (BWR) environments, respectively.

As shown in Fig. 20, crack velocity of alloy X-750 in oxygenated water depends on the stress-intensity factor, as for other alloys, but a *K*-independent region is not discernible due to the limited number of data points at high *K* values (Ref 116). In this type of oxidizing environment, crack-growth rates are not affected significantly by heat treatment. Under deaerated conditions, even though the crack-growth rates are comparable to those shown in Fig. 20, a lower value of K_{ISCC} is observed (Ref 111). Also, K_{ISCC} is very sensitive to the effect of heat treatment. While work conducted with U-bend specimens suggests that alloy 718 is more resistant than alloy X-750 to intergranular SCC in deoxygenated water at 360 °C (680 °F) (Ref 113), fracture-mechanics tests using wedge-loaded compact tension specimens indicated that K_{ISCC} is lower for alloy 718 (Ref 117). Miglin *et al.* (Ref 118) reported that the presence of a heterogeneous, banded microstructure containing δ and Laves phases in a commercial heat of alloy 718 gives rise to a lower K_{ISCC} than a homogeneous microstructure.

The significant effect of potential on the intergranular SCC behavior of different heats and heat treatments of alloy X-750 is shown in Fig. 21 (Ref 115). While in the case of heat A the pronounced susceptibility at high potentials is noticeable, the beneficial effect of the single-step heat treatment for heat B is apparent over a wide range of potentials.

Chromium content is also important to the SCC resistance of alloy X-750 in high-purity water containing dissolved hydrogen. As clearly

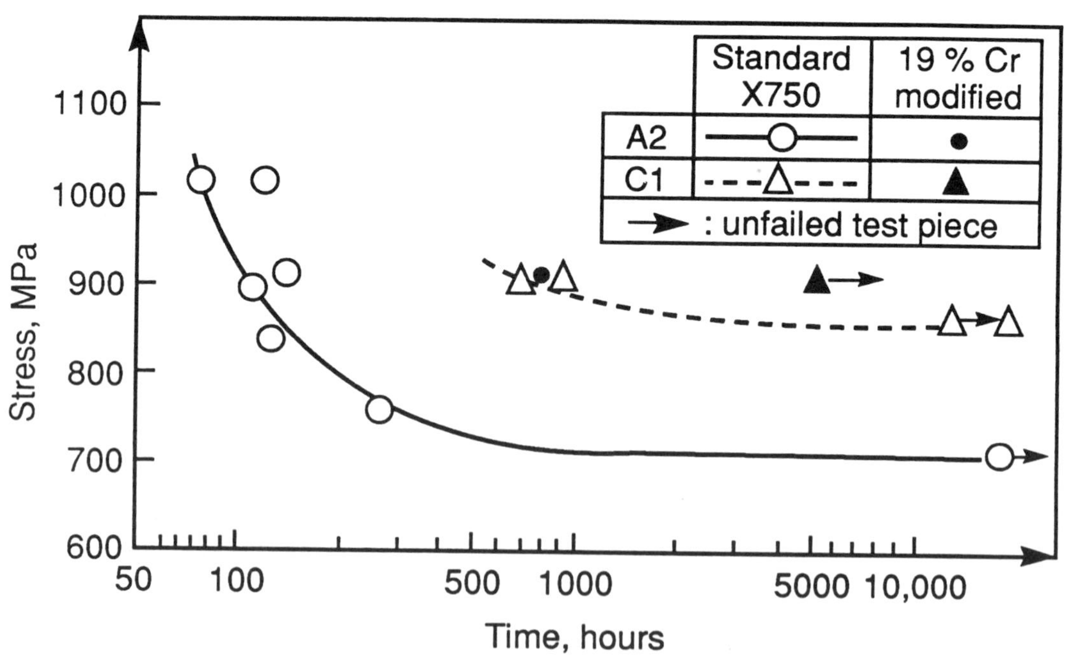

Fig. 22 SCC tests on tensile specimens of standard and modified alloy X-750 in pure water with 25 to 50 mL H_2 (STP)/kg H_2O at 350 °C (660 °F). A2: 885 °C/24 h, AC + 730 °C/8 h; furnace cooled at 14 °C/h down to 620 °C/8 h, AC. C1: 1093 °C/1 h, AC + 704 °C/20 h, AC. Source: Ref 119

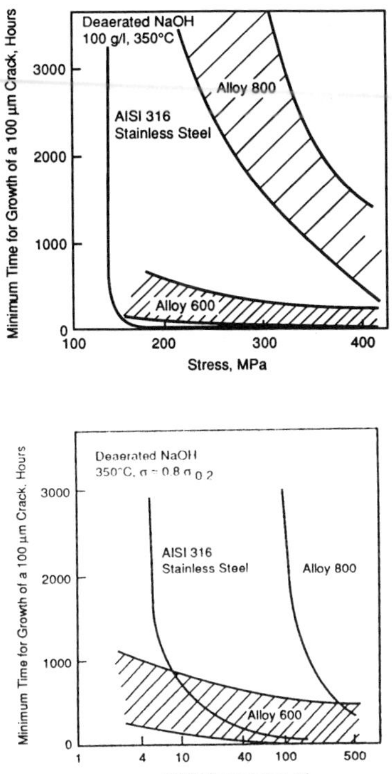

Fig. 23 Comparison of the SCC resistance of alloy 600, alloy 800, and AISI 316 stainless steel in deaerated caustic soda solutions at 350 °C (660 °F). (a) Effect of stress (NaOH = 100 g/L). (b) Effect of caustic soda concentration ($\sigma \approx 0.8\ \sigma_{0.2}$). Source: Ref 121, 123

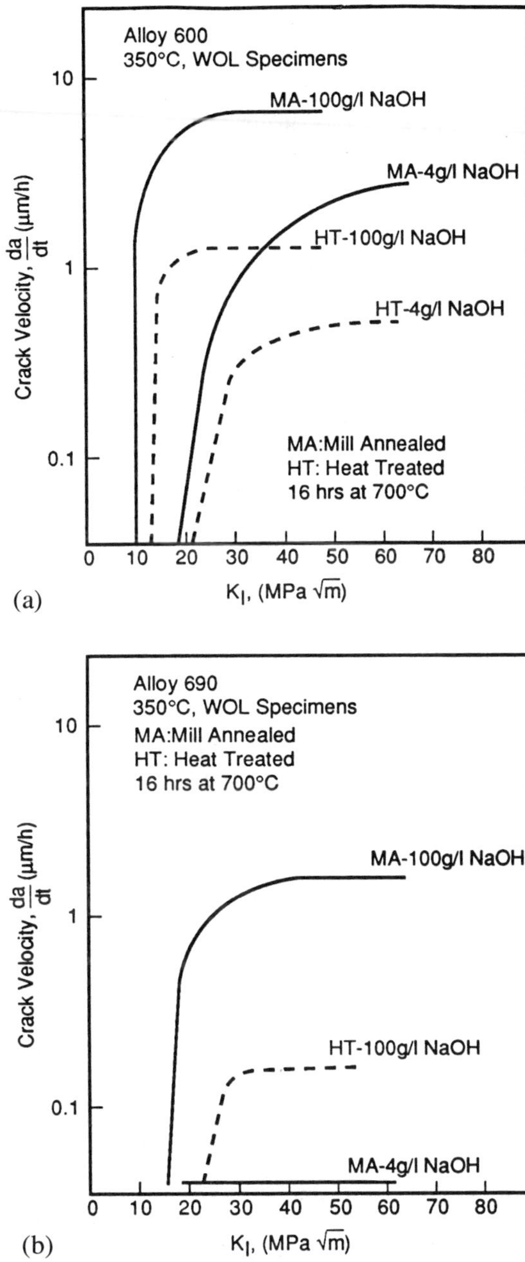

Fig. 24 Effect of stress intensity on crack velocity in deaerated NaOH at 350 °C (660 °F). WOL, wedge opening loaded. (a) Alloy 600. (b) Alloy 690. Source: Ref 68

illustrated in Fig. 22, a modified alloy with an increased chromium content of 19% is significantly more resistant to cracking (Ref 119). The effect of chromium, in this regard, is similar to that observed in single-phase Ni-Cr-Fe alloys. A Ni-20Cr-15Fe alloy containing 1.5% Ti, 3.7% Nb, and 3.1% Mo was found to be extremely resistant to intergranular SCC even in oxidizing environments (Ref 120).

SCC in Caustic Environments

Stress-corrosion cracking of solid-solution-strengthened nickel-base alloys in caustic environments has been extensively studied as a result of many service failures experienced in the nuclear power industry. Most of the studies concentrated on alloy 600 because of its extensive use as steam generator tubing in PWRs. Intergranular SCC of this alloy, initiated in the outer surface of the tubing, has been attributed to the formation of highly concentrated alkaline solutions in the annular gap between the tube and the tube-sheet and also within the sludge pile in the secondary side of recirculating steam generators.

The threshold stress for the SCC of nickel alloys in high-temperature NaOH solutions has not been determined precisely. However, as illustrated in Fig. 23(a), it appears that at a given NaOH concentration the threshold stress is lower for alloy 600 than it is for alloy 800 (Ref 121, 123). Additionally, the exposure time required to observe cracks of a given length (approximately 100 μm) is far less dependent on stress for alloy 600 than for alloy 800. As reported by Theus (Ref 122), cracking of as-received alloy 600 has been observed at stress levels as low as 20% of the room-temperature yield strength in deaerated 10% NaOH solution at 288 °C (550 °F). The threshold stress for cracking of alloy 690 is significantly higher than those for the other two alloys.

The effect of stress intensity on the crack-growth rate of alloy 600 in NaOH solutions at 350 °C (660 °F) is shown in Fig. 24(a) (Ref 68, 123). It is seen that K_{ISCC} increases slightly with decreasing NaOH concentration. Additionally, heat treatment at 700 °C (1290 °F) increases K_{ISCC} and decreases crack velocity in the plateau region. On the other hand, alloy 690 exhibited a higher K_{ISCC} in the most concentrated NaOH solution than alloy 600 (Fig. 24b) and no crack growth in 4 g/L NaOH. The beneficial effect of heat treatment is noticeable in Fig. 24(b).

As shown in Fig. 23 and 24, alloy composition is extremely important in determining the susceptibility of austenitic Ni-Cr-Fe alloys to intergranular SCC in caustic solutions. For deaerated, concentrated NaOH solutions at temperatures from 280 to 350 °C (535 to 660 °F), stainless steels and alloys 800 and 600 are far less resistant to cracking than alloy 690, while nickel 201 (low-carbon nickel) is immune. However, it is not nickel content alone that determines cracking susceptibility. A high chromium content, as in alloy 690, is required to reduce the susceptibility of Ni-Cr-Fe alloys to caustic cracking. In addition, molybdenum-containing alloys, such as alloys C-276 and 825, were found to be susceptible to cracking in concentrated caustic solutions (Ref 124). While transgranular cracking of both alloys was observed in 50% NaOH solution, a higher temperature (>150 °C, or 300 °F) was required for alloy 825, indicating that this alloy is slightly more resistant to cracking in this environment than alloy C-276.

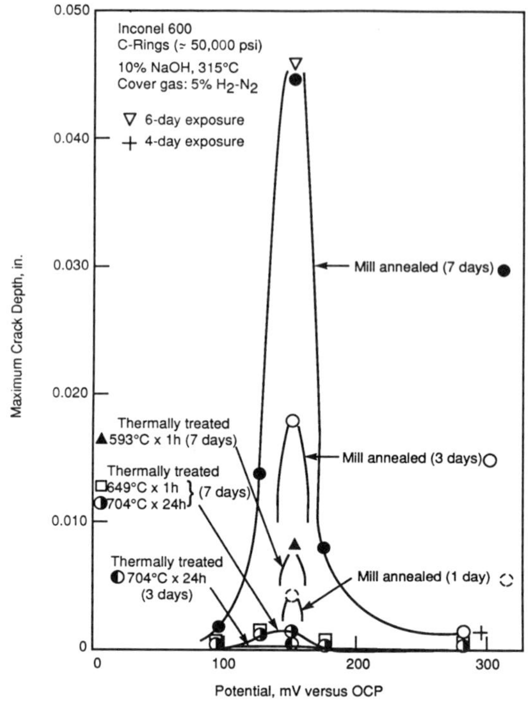

Fig. 25 Variation of maximum SCC depth with specimen potential for mill-annealed and thermally treated alloy 600 exposed to 10% NaOH at 315 °C (600 °F). OCP, open-circuit potential. Source: Ref 135

As shown in Fig. 24(a), prolonged heat treatment within the carbide-precipitation temperature range renders alloy 600 more resistant to intergranular SCC in hot caustic solutions (Ref 125-128). The same effect has been observed in the case of alloy 690 (Fig. 24b). However, as pointed out by Crum (Ref 129), the inherently greater SCC resistance of alloy 690 makes the effect of heat treatment less important than in the case of alloy 600. Although there are conflicting results, it seems (Ref 130) that heat treatment at a temperature ranging from 700 to 850 °C (1290 to 1560 °F) is beneficial for alloy 800. However, Mignone *et al.* (Ref 131) found that a sensitized material (heat treated at 575 °C, or 1065 °F, for 100 h) is more susceptible to intergranular SCC than the mill-annealed alloy.

In general, the severity of SCC in caustic solutions is a function of concentration and temperature. Nevertheless, according to Fig. 23(b), NaOH concentration has a minor influence on the cracking of alloy 600 in deaerated NaOH solutions at 350 °C (660 °F). On the other hand, the susceptibility to cracking of alloys 800 and 690 seems to be more dependent on NaOH concentration. The effect of temperature on the intergranular SCC of alloy 600 is apparently not as strong as it is in pure water. Some authors (Ref 121, 132) have observed only a minor temperature effect in deaerated, concentrated NaOH solutions at 284 to 350 °C (543 to 660 °F). However, by comparing results obtained at 316 and 343 °C (600 and 650 °F), Airey (Ref 93) found temperature to have an accelerating effect in 10% NaOH solutions for both mill-annealed and thermally treated alloy 600.

Potential is a dominant factor in the intergranular SCC susceptibility of alloy 600 in hot caustic solutions. Many authors (Ref 133-137) have conducted potentiostatic SCC tests in concentrated NaOH solutions to identify potential ranges over which cracking occurs. With minor variations, most authors have observed intergranular SCC above 300 °C (570 °F) within an anodic potential region that extends from 80 to 250 mV with respect to the open-circuit potential (corrosion-potential), as shown in Fig. 25 (Ref 135). This potential range corresponds to the active-passive transition observed in the anodic polarization curve. At even higher potentials, within the passive region, no cracking has been observed, whereas at potentials near the corrosion potential, extended intergranular attack has been detected (Ref 137, 138). Several anions, such as SO_4^{2-}, CO_3^{2-}, and PO_4^{3-}, increase SCC susceptibility, whereas SO_3^{2-} acts as an inhibitor (Ref 137).

Most workers who have conducted potentiostatic tests in solutions containing approximately 10% NaOH at about 320 °C (610 °F) (Ref 126, 133-137) have found that prolonged heat treatment at the sensitization temperature (*i.e.*, 620 °C for 18 h or 700 °C for 15 to 24 h) significantly improves cracking resistance at the potential of maximum susceptibility (approximately 90 to 150 mV anodic with respect to the open-circuit potential). At lower temperatures (140 °C, or 285 °F) in more concentrated NaOH solutions, cracking behavior is more complex (Ref 128, 139). Regions of intergranular SCC have been identified over most of the range of primary passivity, whereas transgranular SCC has been observed at the open-circuit potential and within the secondary passivation region. The beneficial effect of heat treatment on intergranular SCC was also confirmed at this high NaOH concentration.

The effect of potential on the cracking behavior of alloys 800 and 690 has been studied less extensively. Mill-annealed alloy 800 seems to be more resistant to intergranular SCC than alloy 600 at potentials close to the active-passive transition in 50% NaOH solution at 140 °C (285 °F) (Ref 139). However, it is more susceptible to SCC in the secondary passivation region. The addition of SiO_2 to 50% NaOH solution at 140 °C (285 °F) seems to inhibit cracking of alloy 600, but does not affect the tendency to intergranular cracking of alloy 800 (Ref 140). At higher temperatures (300 °C, or 570 °F) in a more dilute solution (deaerated 10% NaOH), no cracking of alloy 800 was found at the open-circuit potential, but transgranular SCC was observed at anodic potentials ranging from 70 to 300 mV with respect to the open-circuit potential. Maximum susceptibility was observed at 200 to 250 mV (Ref 134). However, intergranular SCC at a potential of 60 mV anodic with respect to the open-circuit potential has been reported (Ref 130) in a more concentrated solution (15% NaOH). Cracking was also observed at the open-circuit potential. For alloy 690, few studies on the effect of potential have been reported. Very short, intergranular cracks (>130 μm) were detected in annealed alloy 690 using the slow-strain-rate technique in 10% NaOH solution at 316 °C (600 °F) under an applied potential of 150 mV above the open-circuit potential (Ref 129). Under the same conditions, alloy 600 exhibited crack depths close to 900 μm. Sarver *et al.* (Ref 141) found that the cracking range for alloy 690 is narrower than that for alloy 600 in 10% NaOH solution at 288 °C (550 °F). Tests were conducted at potentials ranging from 150 to 250 mV above the open-circuit potential, but intergranular cracks were detected only at 170 mV.

In summary, all the austenitic Ni-Cr-Fe alloys are susceptible to SCC in hot, concentrated caustic solutions. Alloy 690 is probably the most resistant of the commercial alloys, but only pure nickel seems to be immune to cracking. Most of these alloys crack intergranularly; however, alloys with a lower nickel content, such as alloy 800, frequently exhibit transgranular cracking.

Cracking in Other Environments

Liquid-Metal Embrittlement. Although liquid-metal embrittlement does not fall under the restricted category of stress-corrosion cracking, a brief summary of the liquid metals that have been shown to embrittle nickel and its alloys is presented here. Shunk and Warke (Ref 142) summarized the combinations of metal-embrittler couples for various metals discussed in literature up to 1974. Nickel and its alloys were shown to be embrittled by mercury, lithium, lead, and tin. However, Johnson *et al.* (Ref 143) examined several commercial Ni-Cr-Mo-Fe alloys as well as a Co-Cr-Mo alloy and found that tin did not wet the alloys, while lead and zinc wetted but did not embrittle the alloys. Molten zinc reacted profusely with the nickel-base alloys and formed brittle corrosion products. Sadigh (Ref 144) also reported that molten zinc or its vapor did not cause cracking of commercial nickel-base alloys such as 825, C-276, 400, 600, 800, and 200, but did cause brittle Ni-Zn corrosion products. He also confirmed the well-known susceptibility of type 304 stainless steel to crack rapidly in molten zinc. An intermediate nickel-containing stainless steel (alloy 20Cb-3) also showed a slight susceptibility to cracking in molten zinc. Kelly (Ref 145) reported that molten calcium that occurs in the processing of ferrites for the electronic industry caused cracking of retorts built of alloy RA 330 (36Ni-19Cr-1.1Si-0.05C). He also reported cracking of some nickel-base alloys in molten lead, lithium, and probably samarium, neodymium, and promethium. Hemsworth *et al.* (Ref 146) have reported cracking of type 316 stainless steel in molten tellurium, particularly when tellurium is mixed with cerium. The most systematic investigation by far has been conducted in mercury cracking of alloy 400 (Ni-30Cu) (Ref 147, 148).

Organic Environments. Takizawa and Sekine (Ref 149) reported cracking of a Ni-Mo alloy (alloy B-2) in an organic environment of un-

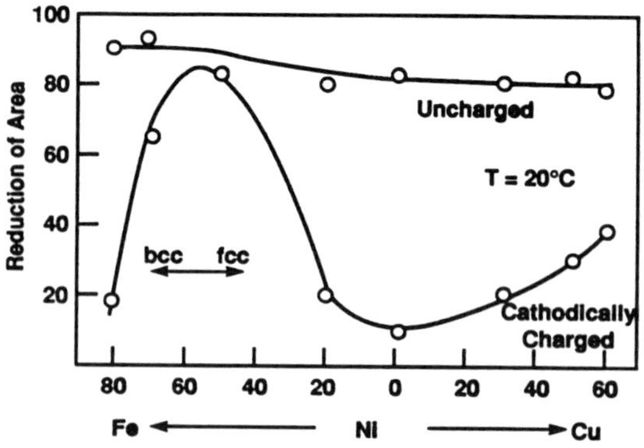

Fig. 26 Fracture behavior of cathodically charged nickel containing iron or copper. Source: Ref 150

known composition that contained small amounts of sulfuric acid and water. Intergranular cracking was found near welds on alloys whose compositions tended to form Ni_4Mo phase. Laboratory tests on the plant solution at 130 °C (265 °F) of samples that were aged at 750 °C (1380 °F) for various times ranging from 1 to 30 min also showed susceptibility to intergranular cracking.

Hydrogen Embrittlement

Embrittlement due to hydrogen in nickel-base alloys can manifest itself in a variety of ways, including embrittlement due to absorbed atomic hydrogen either from a gaseous or an aqueous source, hydrogen reaction with carbides or other reactive phases, and pressure buildup due to hydrogen gas formation inside defects. Of these, hydrogen embrittlement due to absorption of hydrogen from an aqueous phase will be discussed here, because it is closely allied to many SCC processes and has been postulated as a mechanism for SCC in some cases. Nickel-base alloys embrittle in hydrogen, although not to the same extent as iron-base alloys. This is illustrated in Fig. 26 (Ref 150) for a series of Ni-Cu and Ni-Fe alloys. As the iron content of the nickel alloys increases, the tendency toward hydrogen embrittlement decreases until sufficient iron is added to change the alloy from a face-centered cubic (fcc) to a body-centered cubic (bcc) crystal structure.

Hydrogen embrittlement in nickel-base alloys is predominantly intergranular, although some embrittlement has also been observed to be by

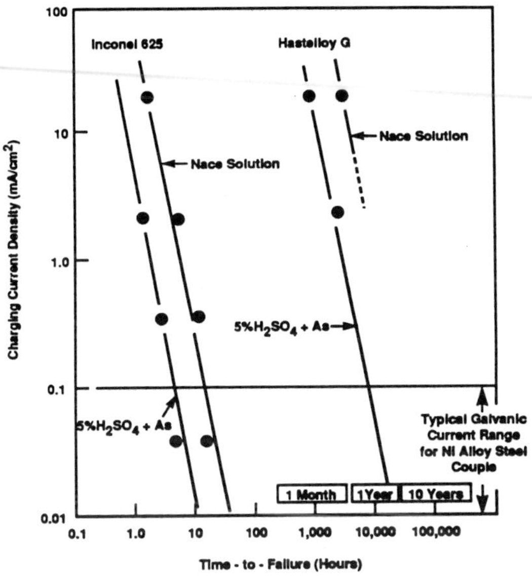

Fig. 27 Effect of test environment and charging current density on the failure time of stressed C-ring specimens of alloy 625 (59% cold rolled + 500 °C for 50 h) and Hastelloy G (59% cold rolled + 260 °C for 250 h). Room temperature; 100% yield stress. Source: Ref 152

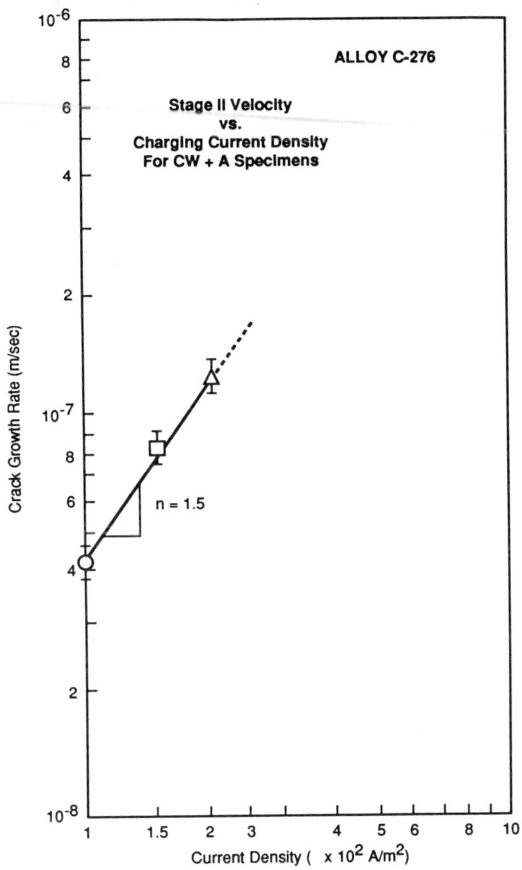

Fig. 28 Stage 2 (stress-intensity independent) crack velocity as a function of charging current density for cold-worked and aged specimens of alloy C-276 at room temperature. Source: Ref 153

dimpled rupture (Ref 151). In contrast, hydrogen embrittlement in iron-base alloys is predominantly transgranular, although in some cases cracking has occurred along prior-austenitic grain boundaries. The addition of copper to nickel also decreases the tendency toward hydrogen embrittlement. Pure copper has not been shown to be embrittled by hydrogen. The three most important environmental factors affecting hydrogen embrittlement are (1) hydrogen fugacity, as dictated by the cathodic current density in aqueous systems, (2) recombination poisons such as sulfur, and (3) temperature. The metallurgical factors that control embrittlement include (1) alloy composition (*e.g.*, iron content), (2) deformation mode, as controlled by stacking-fault energy and coherent precipitation, (3) segregation of metalloid elements to the grain boundaries, (4) grain size and orientation, and (5) strength level.

Effect of Hydrogen Fugacity

Hydrogen fugacity in aqueous systems is controlled by the cathodic current density, which in turn can be controlled by factors such as galvanic coupling between nickel-base alloy and carbon steel components and the potential-pH conditions prevailing in defects such as cracks, pits, or crevices. Galvanic effects often occur, for example, in oil and gas production operations when a steel liner comes in contact with high-nickel tubing and in offshore structures when nickel alloy bolts come in contact with steel structural elements. The effect of cathodic current density on time to failure is shown for two nickel-base alloys in Fig. 27 (Ref 152). Its effect on stage 2 (stress-intensity independent) crack-growth rate for alloy C-276 is shown in Fig. 28 (Ref 153).

Effect of Recombination Poisons

Minor concentrations of impurities in the environment or alloy can greatly increase the entry

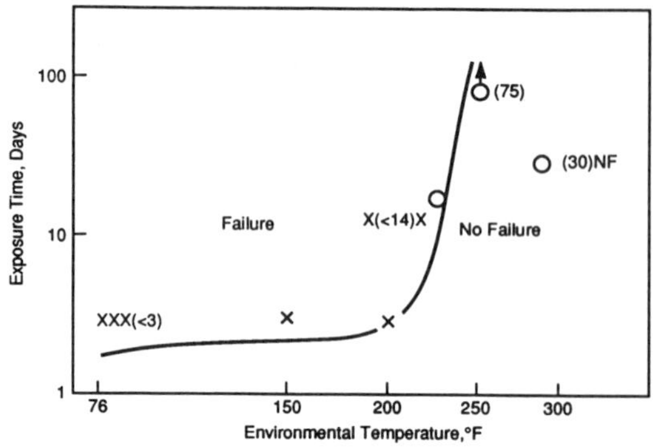

Fig. 29 Effect of environmental temperature on hydrogen embrittlement of alloy C-276. Environment: 4% NaCl + 0.5% CH_3COOH + H_2S (1 atm), coupled to carbon steel. Stressed to 90% yield. Alloy aged at 500 °C (930 °F) for 100 h. Source: Ref 41

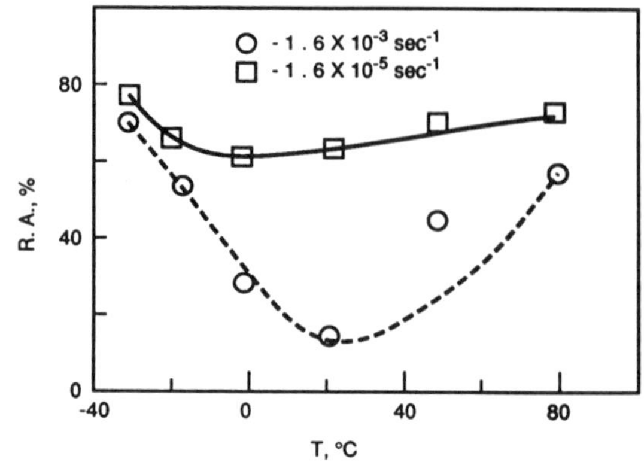

Fig. 30 Reduction in area at fracture of Monel 400 tested in the presence of hydrogen. Source: Ref 148

of hydrogen into the metal by preventing recombination of adsorbed hydrogen atoms, resulting in the evolution of hydrogen gas. A rather extensive review of the recombination poisons was performed by McCright (Ref 154). These include sulfur, phosphorus, arsenic, antimony, and tin. In sour-gas environments, H_2S has been shown (Ref 37) to increase hydrogen permeation (related to the concentration of adsorbed hydrogen) in steels. Some of these elements are also present as impurities in alloys and can influence hydrogen entry into the metal. However, these alloyed impurities may also influence embrittlement directly by reducing the grain-boundary cohesion synergistically with hydrogen. The effect of these internal impurity elements will be discussed in the section on metallurgical variables.

Effect of Temperature

The effect of temperature on hydrogen embrittlement depends on temperature range, alloy composition, and the environmental conditions under which hydrogen entry takes place. Wilcox and Smith (Ref 155) examined the delayed failure of prestrained, hydrogen-charged nickel between –50 and 0 °C (–58 to 32 °F) and found that increasing the temperature decreased the time to failure (increased embrittlement), with an activation energy of 6.8 kcal/mol. Chandler and Walter (Ref 156), however, showed that when electroformed nickel was tested in high-pressure gaseous hydrogen over a wider temperature range (–73 to 170 °C, or –100 to 340 °F), embrittlement initially increased with temperature up to 24 °C (75 °F), but decreased thereafter. Most investigations of hydrogen embrittlement at temperatures close to 0 °C (32 °F) and above have found that increasing the temperature decreases embrittlement. This is illustrated in Fig. 29 and 30. In the case of the Ni-Cr-Mo alloy (C-276), embrittlement in a NaCl + H_2S solution occurs only upon cathodic polarization, either by galvanic coupling to carbon steel (Fig. 29) or by applying a cathodic current. The embrittling tendency decreases at temperatures above approximately 120 °C (250 °F) (Ref 41). Similar results have been observed for other nickel-base alloys in this type of H_2S test (Ref 30, 40). In the case of alloy 400 (Fig. 30), the loss in ductility (percentage of reduction in area of a tensile sample) from air to hydrogen increases initially with temperature and then decreases at temperatures higher than 20 °C (68 °F). Some exceptions to the above trends include the delayed failure of alloy K-500 (Ni-30Cu-3Al), where embrittlement due to precharged hydrogen has been found to increase with temperature even at temperatures as high as 177 °C (350 °F) (Ref 157), and in hydrogen-purged high-purity water, where cracking of alloy 600 has been shown to increase with temperatures of up to 350 °C (660 °F) (Ref 158).

Effect of Metallurgical Variables

Impurity segregation and its associated effects of grain size have been examined for a num-

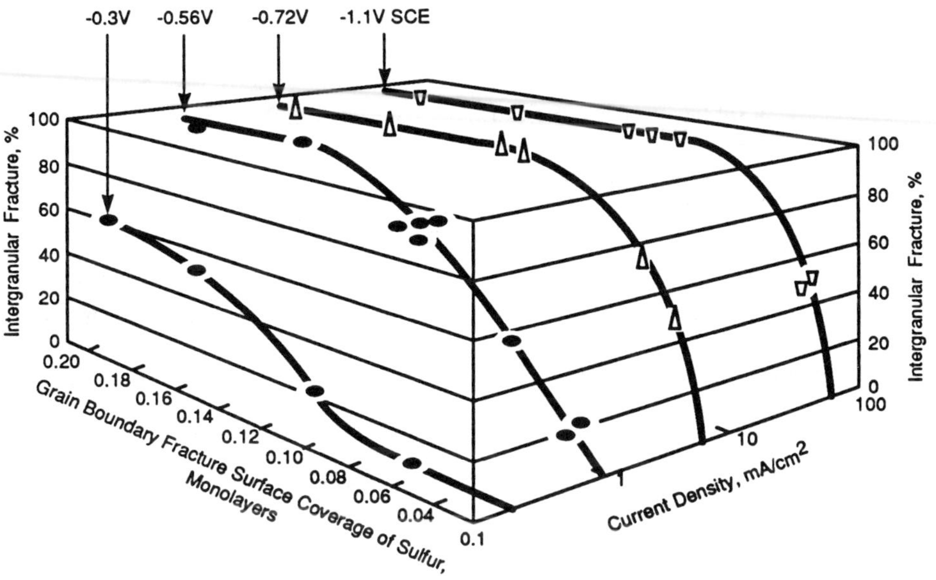

Fig. 31 Interrelationships among cathodic current density, grain-boundary sulfur composition, and fracture mode in straining electrode tests of nickel. Source: Ref 159

Table 16 Effect of Bulk Phosphorus Content on Hydrogen Embrittlement in a Chloride + H_2S Environment

Bent sheet samples coupled to carbon steel tested at room temperature in a solution of 5% NaCl + 0.5% CH_3COOH + 1 atm H_2S (saturated at RT)

Condition	Aging temperature °C	Aging temperature °F	Aging time, h	Time to failure, days(a) 0.023% P	Time to failure, days(a) 0.002% P
50% cold-rolled sheet	Unaged		...	>36	>36
	204	400	200	13	>36
	371	700	20	<3	9
	482	900	0.16	<3	>36
	482	900	900	<3	<3
60% cold-rolled sheet	Unaged		...	>36	>36
	204	400	200	<3	>36
	371	700	20	<3	>36
	482	900	0.16	<3	>36
	482	900	900	<3	<3

(a) > indicates no cracking in the stated time period; < indicates cracking
Source: Ref 160

ber of alloys and for pure nickel. Impurities include sulfur, phosphorus, antimony, tin, and carbon. In pure nickel, sulfur segregation to the grain boundaries is the dominant factor (Ref 2). This is illustrated in Fig. 31 (Ref 159). Intergranular fracture is observed when the grain-boundary concentration of sulfur exceeds about 0.1 monolayers at an applied cathodic potential of –0.3 V (SHE). However, sulfur can have an interactive effect with hydrogen fugacity; at more negative cathodic potentials (greater hydrogen charging current or fugacity), less sulfur is needed at grain boundaries in order for intergranular cracking to occur. Jones and Bruemmer (Ref 2) report a very high tendency for sulfur segregation to grain boundaries, such that even 1 ppm bulk concentration can result in 10% monolayer concentration.

Phosphorus segregation has been reported to be more of a problem in nickel-base alloys, although not all the reported results agree on this point (Ref 160-168). Unlike the case of pure nickel, however, increased embrittlement cannot be attributed solely to grain-boundary segregation; other metallurgical changes can occur simultaneously. For example, Berkowitz and Kane (Ref 160) showed that hydrogen embrittlement and grain-boundary phosphorus concentration of a Ni-Cr-Mo-W alloy (alloy C-276) increased with low-temperature aging at 500 °C (930 °F). A low-phosphorus alloy exhibited less sensitivity toward aging-induced increase in hydrogen embrittlement (Table 16).

In contrast, Asphahani (Ref 161) tested the same Ni-Cr-Mo-W alloys, varying in bulk phosphorus concentration from 0.003 to 0.086%, and showed no link between bulk phosphorus increase and the tendency toward hydrogen embrittlement. However, he used a much greater hydrogen fugacity than Berkowitz and Kane (cathodic current density of 40 mA/cm^2). It must be noted that an ordering reaction of the type Ni_2Cr occurs in the same temperature region in these alloys. Ladna and Kargol (Ref 162), using a Ni-22Cr-6Mo-20Fe alloy (alloy G), which does not undergo ordering, showed that only a small increase in crack growth (in a ductile mode) occurred due to aging at 500 °C (930 °F). Hence, increased crack growth due to low-temperature aging in these alloys cannot be attributed to phosphorus segregation alone. The effect of phosphorus versus ordering in this alloy system has been a matter of considerable controversy, because the prescription for aging-enhanced hydrogen embrittlement depends heavily on which factor is considered important.

Berkowitz *et al.* (Ref 163) and Kurkela *et al.* (Ref 169) further examined this issue by testing pure Ni_2Cr after various aging treatments to produce various degrees of long-range ordering. They showed that the alloy was most susceptible to hydrogen embrittlement in the completely disordered and completely ordered conditions and that susceptibility decreased as the percent ordering (calculated from aging time) increased. On the other hand, Lehman *et al.* (Ref 164) tested Ni_2Cr with various phosphorus contents and showed that high levels of phosphorus increased ordering kinetics significantly. From a practical perspective, it is difficult to reach bulk phosphorus levels below about 3 ppm economically, and most commercial alloys contain about 10 ppm P. Ordering reactions in the nickel-base alloys can be prevented by increasing the iron content above about 10%; however, some corrosion resistance will be sacrificed, because alloys containing that amount of iron cannot sustain as much molybdenum and tungsten.

Cornet *et al.* (Ref 165) showed that addition of phosphorus to a high-purity Ni-16Cr-8Fe alloy (residual phosphorus level was 0.004%) may actually lower hydrogen embrittlement susceptibility. Their investigation was carried out on alloys that were solution annealed at 1100 °C (2010 °F) but not aged. In a more recent investigation (Ref 166), Ni-20Cr alloy was doped with 0.015% bulk P and tested under cathodic hydrogen charging conditions. The authors contend that the effect of phosphorus is more complex than portrayed by previous reports. Phosphorus can be detrimental if its segregation to the grain boundaries occurs without coprecipitation of carbides; it can be beneficial at temperatures that allow carbide precipitation. Other species shown to be detrimental to hydrogen-embrittlement resistance of nickel-base alloys include antimony and tin (Ref 165, 168, 170).

Deformation mode in nickel-base alloys depends on stacking-fault energy and precipitation of second phases. Stacking faults are created in fcc crystal structures by alteration of the sequence of the stacked layers of atoms (Ref 171). Stacking faults are enclosed by partial dislocations—the type of dislocation depending on the type of stacking fault as well as the crystal structure. Because stacking faults are internal surfaces, energy is required to alter their area. The lower the stacking-fault energy, the larger the area of the stacking fault or the higher the stacking-fault probability. Because of energy constraints, dislocations that contain stacking faults must move along a given slip plane (the plane along which shear deformation occurs). This type of deformation, termed

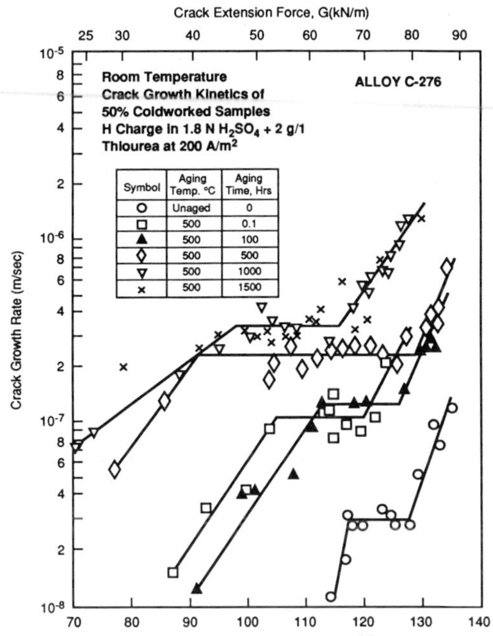

Fig. 32 Effect of stress intensity on crack-growth rate of alloy C-276 as a function of aging time at 500 °C (930 °F). Source: Ref 173

planar deformation, creates large inhomogeneities in slip as well as great stress concentrations at obstacles such as grain boundaries. It has been shown that lower stacking-fault energies can lead to greater hydrogen embrittlement and SCC (Ref 172). This relationship becomes especially obvious in nickel-base alloys as cobalt concentration becomes progressively larger. Cobalt reduces the stacking-fault energy of nickel, and the hydrogen embrittlement resistance of Ni-Co alloys decreases with increasing cobalt content. However, stacking-fault energy is not the only factor affecting hydrogen embrittlement; the addition of iron decreases stacking-fault energy, but increases resistance to hydrogen embrittlement.

As mentioned before, short- and long-range ordering reactions can also cause the planar deformation mode. In the case of short-range ordering, plastic deformation by dislocation motion tends to disorder the material, thus raising the energy. This in turn tends to force dislocation motion along certain planes. In the case of long-range ordering, several mechanisms can come into play to force planar deformation. For example, in alloy C-276 long-range ordering of the type Ni_2Cr forces the alloy to deform by twinning. The effect of low-temperature aging on hydrogen-induced crack-growth rate in alloy C-276 is shown in Fig. 32 (Ref 173). Two stages of increase in crack growth can be seen: one at time periods of a few minutes to 100 h, and the second after 500 h or more of aging. The first increase has been attributed to a combination of phosphorus segregation and short-range ordering and the second to long-range ordering.

Coherent precipitates such as γ' and γ'' also induce planar deformation, as mentioned earlier. The effects of coherent precipitation on hydrogen embrittlement of alloys K-500 and X-750 have been experienced in oil and gas production processes (Ref 157, 174-176). For example, in offshore platforms, alloy K-500 bolts, which were threaded after precipitation hardening, failed when cathodically coupled to carbon steel (Ref 175). Failure analysis suggested that hydrogen embrittlement was the failure mechanism and that threading after precipitation hardening caused the hardness at the thread roots to exceed 35 HRC. It was recommended that the bolts be solution annealed after threading and then hardened to a maximum of 35 HRC. However, even after being treated in accordance with the recommendations, and with a hardness of 25 HRC, the bolts still failed in service. This was shown (Ref 176) to be caused by polarization to a more negative potential, which resulted from the bolts being coupled to steel cathodically protected by aluminum anodes. Results of laboratory slow-strain-rate tests that reproduced this type of cracking are shown in Table 17. While these cracking occurrences have been at ambient temperatures, hydrogen embrittlement of alloy K-500 and other precipitation-hardenable alloys, such as alloy 925, has been shown to occur at temperatures up to 125 °C (257 °F) (Ref 175) in $MgCl_2$ brines under cathodic polarization.

Another example of failure in oil and gas production is cracking of tubing hangers made of alloy X-750, a γ' precipitation-hardened alloy, which was shown to be related to hydrogen embrittlement in combination with hardnesses above 35 HRC (Ref 177). Hydrogen embrittle-

Table 17 Results of Slow-Strain-Rate Tests on Samples From 50 mm (2 in.) Diameter Alloy K-500 Bolts
Hardness of precipitation-hardened samples: 25 HRC

Environment	Time to failure, h	Elongation, %	Reduction in in area, %	Fracture type
Air	36.3	24.6	36	Ductile
Seawater, no anode	39.8	27.4	42	Ductile
Seawater, steel anode, –620 mV (AgCl)	28.0	19.0	45	Ductile
Seawater, steel anode precoupled 8 days, –620 mV (AgCl)	30.0	21.0	46	Ductile
Seawater,aluminum anode, –900 mV (AgCl)	6.1	4.2	8	Intergranular
Seawater, aluminum anode precoupled 8 days, –900 mV (AgCl)	4.1	3.5	7	Intergranular
Seawater, potentiostat, –800 mV (AgCl)	5.8	4.0	...	Intergranular

Source: Ref 176

ment of alloy 718, a γ″-type precipitation-hardenable alloy has occurred in high-pressure hydrogen gas when the alloy was double-aged to produce hardnesses above 40 HRC (Ref 178). However, hydrogen embrittlement in H_2S-containing solutions has not generally been observed in this alloy, with the exception of a result from slow-strain-rate testing (Ref 179). Precipitation-hardenable alloys with higher molybdenum contents, such as alloys 625 Plus and 725, have been shown to be resistant to hydrogen embrittlement, even when subjected to a combination of precipitation-hardening and low-temperature aging treatments (Ref 180).

SCC Testing Methods

This section summarizes the test techniques that are used to study SCC and hydrogen embrittlement of nickel-base alloys. The reader is referred to Chapter 17 for a more detailed discussion of SCC test methods.

Environmental Conditions. While the environmental conditions are specific to the aims of a particular study, certain general considerations apply. In SCC tests, it is desirable to monitor the open-circuit or corrosion potential of the stressed sample and, if possible, test under a variety of applied potentials. Small variations in potential that may be produced by relatively small differences in deaeration or in experimental arrangements (*e.g.*, galvanic contacts with test apparatus) can substantially affect the measured cracking susceptibility of nickel-base alloys (Ref 53). However, when monitoring or control of the potential is not possible, concentrations of oxidizing species must be maintained at relatively constant levels. For example, oxygen is detrimental in many SCC environments, because its presence generally increases the corrosion potential. The oxygen concentration in a solution can be affected by a variety of factors, including condenser design, temperature, chloride concentration, and any unintentional deaeration performed prior to or during the test (Ref 23).

Constant-deflection tests are the most common type of SCC test and use a variety of configurations: U-bend, C-ring, two-point bend, four-point bend, and so on. Constant-deflection tests do not require elaborate loading apparatus and can be adapted to a wide range of product forms. The C-ring test has been used widely in the oil and gas industry for testing seamless tubular products. One of the most important variables in this test is surface condition. The usual procedure is to grind the outer surface, which will be stressed in tension to a given finish—typically 120-grit, wet ground. However, grinding or machining may remove any surface chromium depletion, thus affecting SCC evaluation of a given alloy product (Ref 181). Removing oxide films, altering the rest potential, or introducing surface stresses by grinding may affect crack initiation.

In some cases, a crevice is formed at the apex of the stressed sample by spot welding another strip of the same alloy onto the stressed sample. In elemental sulfur environments, it has been shown that cracking occurs under crevices formed either intentionally or unintentionally. Another important factor in constant-deflection tests is the assumed stress level. Typically, samples are stressed to a given percent of the yield strength. In the case of tubular products, yield strength in the transverse direction is difficult to measure and is usually assumed to be about 90% of the longitudinal yield strength. In the case of other types of products, it is important to specify the direction of stress relative to the rolling direction. It has been shown (Ref 182) that the hydrogen embrittlement tendency of highly cold-worked nickel-base alloys is more pronounced when a sample is stressed in the transverse direction, which is the usual stressing direction.

Constant-Load Tests. Some of the earliest testing of nickel-base alloys was performed on wire samples using constant load (Ref 19). Wire samples offer the advantage of simplicity in experimental design, but they cannot be used to examine the behavior of the products used in most applications (sheet, plate, tubing, and bar). A commonly used constant-load test is the NACE TM-01-77 tensile test (Ref 183). In this test, a cylindrical sample, usually 6.3 mm (0.25 in.) in gage diameter is spring loaded in a proof-ring apparatus and the time to failure monitored. Because samples are machined, the effect of surface condition cannot be tested. An advantage of the constant-load test over the constant-deflection test is that cracking is accelerated once it is initiated, because the cross section available to sustain the load decreases and the stress level increases.

Precracked samples offer the advantage of separating crack-initiation processes from crack growth, and they usually allow more accelerated testing than do smooth samples. Because of the well-defined stress conditions at the crack tip, precracked samples are more amenable to modeling than smooth samples, in which multiple cracking can occur. The types of samples that have been used in testing nickel-base alloys include compact tension, wedge-opening-loaded (WOL), modified WOL (bolt-loaded), double-cantilever-beam, and double torsion samples. Selection of a particular type of sample depends on the particular test apparatus, economics, the availability of suitably sized material, and the type of information sought.

Because most corrosion-resistant nickel-base alloys exhibit high fracture toughness (K_{Ic} > 1000 MPa$\sqrt{m}$, or 910 ksi$\sqrt{in.}$) and relatively low yield strengths (300 to 1000 MPa, or 43.5 to 145 ksi), the thickness requirements for plane-strain conditions over a wide range of crack-growth rates may not be met by the material sizes that are available. In many tests, significant crack branching occurs, which can sometimes be overcome by using side grooves. The double torsion test was initially developed to test ceramic materials (Ref 153, 173), but has also been used to study nickel-base alloys. Although the test is not well established, its advantages include constant stress intensity at constant applied load and ease of monitoring crack-growth rate through load changes.

Slow-strain-rate testing or constant-extension-rate testing involves slow straining of a tensile sample to failure while it is exposed to the environment of interest. The susceptibility to embrittlement depends on the strain rate, but the optimum strain rate for nickel-base alloys appears to be lower for SCC than for hydrogen embrittlement. For example, Ueda *et al.* (Ref 53) showed that the highest embrittlement for a Fe-21Cr-31Ni-4Mo-1.5Cu alloy in a 25% NaCl + 0.5% CH_3COOH + 1 MPa (0.15 ksi) H_2S + 1 MPa (0.15 ksi) CO_2 environment at 150 °C (300 °F) occurred at a nominal strain rate of 4×10^{-6}/s. Most testing in H_2S, chloride, and caustic environments has used strain rates ranging from 1×10^{-6}/s to 4×10^{-6}/s. In high-purity water, strain rates of 1×10^{-7}/s have been used (Ref 184), although some evidence indicates that even this strain rate is not low enough to cause cracking. Introduction of cold work in the gage section of tensile specimens overcomes this problem (Ref 85). For hydrogen embrittlement, much higher strain rates (10^{-4}/s to 10^{-2}/s) may be used (Ref 165, 185).

References

1. R.W. Staehle, Understanding "Situation Dependent Strength": A Fundamental Objective in Assessing the History of Stress Corrosion Cracking, in *Environment Induced Cracking of Metals*, R.P.

Gangloff and M.B. Ives, Ed., National Association of Corrosion Engineers, 1990, p 561

2. R.H. Jones and S.M. Bruemmer, Environment-Induced Crack Growth Processes in Ni-Base Alloys, in *Environment Induced Cracking of Metals*, R.P. Gangloff and M.B. Ives, Ed., National Association of Corrosion Engineers, 1990, p 287
3. A.I. Asphahani, Corrosion of Nickel-Base Alloys, in *Metals Handbook*, Vol 13, 9th ed., *Corrosion*, ASM International, 1987
4. W.Z. Freind, *Corrosion of Nickel and Ni-Base Alloys*, John Wiley and Sons, 1980
5. W. Betteridge, *Nickel and Its Alloys*, Ellis Horwood Ltd., 1984
6. G.S. Was, H.H. Tischner, and R.M. Latanision, The Influence of Thermal Treatment on the Chemistry and Structure of Grain Boundaries in Inconel 600, *Metall. Trans. A*, Vol 12, 1981, p 1397
7. R.L. Dreshfield, The Effect of Refractory Elements on the Stability of Complex Carbides in Ni-Base Superalloys, *Trans. ASM*, Vol 51, 1968, p 352
8. C.L. Briant, C.S. O'Toole, and E.L. Hall, The Effect of Microstructure on the Corrosion and Stress Corrosion Cracking of Alloy 600 in Acidic and Neutral Environments, *Corrosion*, Vol 42, 1986, p 15
9. A. Borello, S. Casadio, A. Saltelli, and G. Scibona, Susceptibility to the Intergranular Corrosion of Alloy 800, *Corrosion*, Vol 37, 1981, p 498
10. E.L. Raymond, Mechanisms of Sensitization and Stabilization of Incoloy Ni-Fe-Cr Alloy 825, *Corrosion*, Vol 24, 1968, p 180
11. M. Raghavan, B.J. Berkowitz, and J.C. Scanlon, Electron Microscopic Analysis of Heterogeneous Precipitates in Hastelloy Alloy C-276, *Metall. Trans. A*, Vol 13, 1982, p 979
12. C.R. Brooks, J.E. Spruiell, and E.E. Stansbury, Physical Metallurgy of Nickel-Molybdenum Alloys, *Int. Met. Rev.*, Vol 29, 1984, p 210
13. H.J. Wernik, Topologically Close Packed Structures, in *Intermetallic Compounds*, John Wiley and Sons, 1967
14. Y.C. Fayman, γ - γ' Partitioning Behavior in Waspaloy, *Mater. Sci. Eng.*, Vol 82, 1986, p 203
15. J.R. Crum, M.E. Adkins, and W.G. Lipscomb, Performance of High Nickel Alloys in Refinery and Petrochemical Environments, *Mater. Perform.*, Vol 25, 1986, p 27
16. C.T. Sims, Occurrence of Topologically Close-Packed Phases, in *The Superalloys*, C.T. Sims, Ed., John Wiley and Sons, 1972
17. E.S. Machlin and J. Shao, SIGMA-SAFE: A Phase Diagram Approach to the Sigma Problem in Superalloys, *Metall. Trans. A*, Vol 9, 1978, p 561
18. M. Morinaga, N. Yukawa, H. Adachi, and H. Ezaki, New PHACOMP and Its Application to Alloy Design, *Superalloys Source Book*, American Society for Metals, 1984
19. H.R. Copson, in *Physical Metallurgy of Stress Corrosion Fracture*, T.N. Rhodin, Ed., Interscience, 1959, p 247
20. R.W. Staehle, J.J. Royuela, T.L. Raredon, E. Serrate, C.R. Morin, and R.V. Farrar, Effect of Alloy Composition on SCC on Fe-Cr-Ni Base Alloys, *Corrosion*, Vol 26, 1970, p 451
21. M.O. Speidel, Stress Corrosion Cracking of Stainless Steels in NaCl Solutions, *Metall. Trans. A*, Vol 12, 1981, p 779
22. J. Kolts, "Temperature Limits for Stress Corrosion Cracking of Selected Stainless Steels and Ni-Base Alloys in Chloride-Containing Environments," Paper No. 241, presented at Corrosion/82, National Association of Corrosion Engineers, 1982
23. Y.L. Chiang and M.A. Streicher, "The Effect of Condenser Design on Stress Corrosion Cracking of Stainless Alloys in Boiling Chloride Solutions," Paper No. 353, presented at Corrosion/85, National Association of Corrosion Engineers, 1985
24. G. Herbsleb, The Stress Corrosion Cracking of Sensitized Austenitic Stainless Steels and Nickel-Base Alloys, *Corros. Sci.*, Vol 20, 1980, p 243
25. T.S. Humphries and E.E. Nelson, "Stress Corrosion Evaluation of Several Ferrous and Nickel Alloys," NASA Report TM X-64511, April 1970
26. M.O. Speidel and R.W. Staehle, Ed., *ARPA Handbook of Stress Corrosion Cracking and Corrosion Fatigue*, to be published
27. J. Kolts, C.C. Burnette, and M.W. Joosten, Stress Corrosion Cracking of Nickel-Base Alloys in Room-Temperature HCl Containing H_2S, in *Environment Induced Cracking of Metals, R.P.* Gangloff and M.B. Ives, Ed., National Association of Corrosion Engineers, 1990
28. J. Kolts, "Heat Treatment and Environmental Embrittlement of High Performance Alloys," Paper No. 407, presented at Corrosion/86, National Association of Corrosion Engineers, 1986
29. M.A. Streicher, Effect of Composition and Structure on Crevice, Intergranular, and Stress Corrosion of Some Wrought Ni-Cr-Mo Alloys, *Corrosion*, Vol 32, 1976, p 79
30. J. Kolts, "Laboratory Evaluation of Corrosion Resistant Alloys for the Oil and Gas Industry," Paper No. 323, presented at Corrosion/86, National Association of Corrosion Engineers, 1986
31. J. Fricke, "Corrosion in Packer and Completion Fluids," Paper No. 300, presented at Corrosion/87, National Association of Corrosion Engineers, 1987
32. W.K. Boyd and W.E. Berry, Stress Corrosion Cracking of Nickel and Nickel Alloys, in *Stress Corrosion Cracking of Metals—A State of the Art*, STP 518, American Society for Testing and Materials, 1972, p 58
33. H.R. Copson and C.F. Cheng, *Corrosion*, Vol 12, 1956, p 71t

34. L. Graf and W. Wittich, *Werkst. Korros.*, Vol 5, 1966, p 385
35. L.G. Everhart and C.E. Price, Stress Corrosion Cracking in Monel at Room Temperature, in *Corrosion Testing and Evaluation*, STP 1000, R. Baboian and S.W. Dean, Ed., American Society for Testing and Materials, 1990
36. R.N. Tuttle and R.D. Kane, Ed., *H_2S Corrosion in Oil and Gas Production—A Compilation of Classic Papers*, National Association of Corrosion Engineers, 1981
37. R.D. Kane, Roles of H_2S in Behavior of Engineering Alloys, *Int. Met. Rev.*, Vol 30, 1985, p 291
38. I. Matsushima, T. Shimada, Y. Ishizawa, K. Masamura, J. Sakai and M. Tanimura, "Effects of Alloy Composition and SCC Resistance of High Alloys OCTG in Sour Well Environments," Paper No. 233, presented at Corrosion/85, National Association of Corossion Engineers, 1985
39. P.R. Rhodes, "Stress Cracking Risks in Corrosive Oil and Gas Wells," Paper No. 322, presented at Corrosion/86, National Association of Corrosion Engineers, 1986
40. A.I. Asphahani, "High Performance Alloys for Deep Sour Gas Wells," Paper No. 42, presented at Corrosion/78, National Association of Corrosion Engineers, 1978; also published in Ref 36
41. M. Watkins, H.E. Chaung, and G.A. Vaughn, "Laboratory Testing of the SCC Resistance of Stainless Alloys," Paper No. 283, presented at Corrosion/87, National Association of Corrosion Engineers, 1987
42. Y. Ishizawa, T. Takaoka, H. Misao, K. Masamura, and J. Matsushima, Development of Ni-Base Corrosion Resistant Alloy OCTG, *NKK Tech. Rev.*, No. 55, 1989
43. S.W. Ciaraldi, "Stress Corrosion Cracking of Corrosion-Resistant Alloy Tubulars in Highly Sour Environments," Paper No. 284, presented at Corrosion/87, National Association of Corrosion Engineers, 1987
44. A.I. Asphahani, Evaluation of Highly Alloyed Stainless Materials for CO_2/H_2S Environments, *Corrosion*, Vol 37 (No. 6), 1981, p 327
45. A. Ikeda, M. Ueda, and H. Tsuge, "Environmental Cracking Evaluation of Corrosion Resistant Alloys in H_2S-CO_2-Cl^- Environment by Slow Strain Rate Test," Paper No. 7, presented at Corrosion/89, National Association of Corrosion Engineers, 1989
46. S. Tsujikawa, "A New Test Method for Predicting Pitting Corrosion Resistance of CRA's in Sour Environments," Paper No. 64, presented at Corrosion/88, National Association of Corrosion Engineers, 1988
47. P. Marcus, The Mechanisms of Sulfur-Induced Corrosion of Nickel and Nickel Alloys, in *Advances in Localized Corrosion*, H. Isaacs, U. Bertocci, J. Kruger, and S. Smialowska, Ed., National Association of Corrosion Engineers, 1990, p 289
48. S.M. Wilhelm, "Effect of Elemental Sulfur on Stress Corrosion Cracking of Nickel Base Alloys in Deep Sour Gas Well Production," Paper No. 77, presented at Corrosion/88, National Association of Corrosion Engineers, 1988
49. G. Schmidt, Effect of Elemental Sulfur on Corrosion in Sour Gas Systems, *Corrosion*, Vol 47, 1991, p 285
50. M. Ueda, H. Tsuge, and A. Ikeda, "Influence of Elemental S on Corrosion Behavior of CRA in H_2S-CO_2-Cl^- Environment," Paper No. 8, presented at Corrosion/89, National Association of Corrosion Engineers, 1989
51. N. Sridhar and S.M. Corey, "The Effect of Elemental Sulfur on Stress Cracking of Ni-Base Alloys," Paper No. 12, presented at Corrosion/89, National Association of Corrosion Engineers, 1989
52. A. Miyasaka, K. Denpo, and H. Ogawa, "Environmental Aspects of SCC of High Alloys in Sour Environments," Paper No. 70, presented at Corrosion/88, National Association of Corrosion Engineers, 1988
53. M. Ueda and T. Kudo, "Evaluation of SCC Resistance of CRA's in Sour Service," Paper No. 2, presented at Corrosion/91, National Association of Corrosion Engineers, 1991
54. T. Takaoka, Y. Ishizawa, K. Masamura, H. Misao, and T. Inazumi, "Effect of Alloying Elements on Environmental Cracking of Ni-Base Alloys in Sour Gas Environments," Paper No. 73, presented at Corrosion/88, National Association of Corrosion Engineers, 1988
55. A. Ikeda, M. Igerashi, M. Ueda, Y. Okada, and H. Tsuge, "On the Evaluation Methods of Ni-Base Corrosion Resistant Alloys for Sour Gas Exploration and Production," Paper No. 65, presented at Corrosion/88, National Association of Corrosion Engineers, 1988
56. P. Ganesan, E.F. Clatworthy, and J.A. Harris, "Development of a Time-Temperature-Transformation Diagram for Alloy 925," Paper No. 286, presented at Corrosion/87, National Association of Corrosion Engineers, 1987
57. R.B. Frank and T.A. DeBold, "Heat-Treatment of an Age-Hardenable Corrosion Resistant Alloy—UNS N07716," Paper No. 59, presented at Corrosion/90, National Association of Corrosion Engineers, 1990
58. E.L. Hibner, Corrosion Behavior of Age-Hardenable Alloy UNS N07725 for Oil Field and Other Applications, in *Environmental Effects on Advanced Materials*, R.D. Kane, Ed., National Association of Corrosion Engineers, 1991
59. M. Igarashi, S. Mukai, T. Kudo, Y. Okada, and A. Ikeda, "Precipitation Hardened Nickel-Base Al-

loys for Sour Gas Environments," Paper No. 287, presented at Corrosion/87, National Association of Corrosion Engineers, 1987

60. Y. Ishizawa, T. Shimada, T. Inazumi, T. Takaoka, and M. Tankmura, Effects of Heat Treatment and Microstructure on Stress Corrosion Cracking of High Nickel Austenitic Alloys in Sour Gas Environments," Paper No. 234, presented at Corrosion/85, National Association of Corrosion Engineers, 1985
61. A.J. Sedriks, *Corrosion of Stainless Steels*, John Wiley and Sons, 1979
62. G.A. Cragnolino and D.D. Macdonald, Intergranular Stress Corrosion Cracking of Austenitic Stainless Steel at Temperatures Below 100 °C—A Review, *Corrosion*, Vol 38, 1982, p 406-424
63. C.H. Samans, Stress Corrosion Cracking Susceptibility of Stainless Steels and Nickel-Base Alloys in Polythionic Acids and Acid Copper Sulfate Solution, *Corrosion*, Vol 20, 1964, p 256t-262t
64. S. Ahmad, M.L. Mehta, S.K. Saraf, and I. P. Saraswat, Stress Corrosion Cracking of Sensitized 304 Austenitic Stainless Steel in Petroleum Refinery Environment, *Corrosion*, Vol 38, 1982, p 347-353
65. S. Ahmad, M.L. Mehta, S.K. Saraf, and I.P. Saraswat, Effect of Polythionic Acid Concentration on Stress Corrosion Cracking of Sensitized Austenitic Stainless Steel, *Corrosion*, Vol 39, 1983, p 333-338
66. R.C. Scarberry, S.C. Pearman, and J.R. Crum, Precipitation Reactions on Inconel Alloy 600 and Their Effect on Corrosion Behavior, *Corrosion*, Vol 32, 1976, p 401-406
67. C.D. Stephens and R.C. Scarberry, The Relation of Sensitization to Polythionic Acid Cracking of Incoloy Alloys 800 and 801, *Proc. 25th Conf. NACE*, 1969, p 583-586
68. P. Berge and J.R. Donati, Materials Requirements for Pressurized Water Reactor Steam Generator Tubing, *Nucl. Technol.*, Vol 55, 1981, p 88-104
69. R.L. Jones, R.L. Long, and J.S. Olszewski, "The Origin of the Extensive Cracking of the Steam Generator Tubing at TMI Unit 1," Paper No. 141, presented at Corrosion/83, National Association of Corrosion Engineers, 1983
70. H.S. Isaacs, B. Vyas, and M.W. Kendig, The Stress Corrosion Cracking of Sensitized Stainless Steel in Thiosulfate Solutions, *Corrosion*, Vol 38, 1982, p 130-136
71. R.C. Newman, R. Roberge, and R. Bandy, Environmental Variables in the Low Temperature Stress Corrosion Cracking of Inconel 600, *Corrosion*, Vol 39, 1983, p 386-390
72. R. Bandy, R. Roberge, and R.C. Newman, Low Temperature Stress Corrosion Cracking of Sensitized Inconel 600 in Tetrathionate and Thiosulfate Solutions, *Corrosion*, Vol 39, 1983, p 391-398
73. R. Bandy, R. Roberge, and R.C. Newman, Low Temperature Stress Corrosion Cracking of Inconel 600 Under Two Different Conditions of Sensitization, *Corros. Sci.*, Vol 23, 1983, p 995-1006
74. R.C. Newman, R. Roberge, and R. Bandy, Evaluation of SCC Test Methods for Inconel 600 in Low-Temperature Aqueous Solutions, in *Environment-Sensitive Fracture: Evaluation and Comparison of Test Methods*, STP 821, American Society for Testing and Materials, 1984, p 310-322
75. K.H. Lee, G.A. Cragnolino, and D.D. Macdonald, Effect of Heat Treatment and Applied Potential on the Caustic Stress Corrosion Cracking of Inconel 600, *Corrosion*, Vol 41, 1985, p 540-553
76. G.S. Was and V.B. Rajan, On the Relationship Between the EPR Test, Sensitization, and IGSCC Susceptibility, *Corrosion*, Vol 43, 1987, p 576-579
77. G.S. Was and V.B. Rajan, The Mechanism of Intergranular Cracking of Ni-Cr-Fe Alloys in Sodium Tetrathionate, *Metall. Trans. A*, Vol 18, 1987, p 1313-1323
78. G-P. Yu and H-C. Yao, The Relation Between the Resistance of IGA and IGSCC and the Chromium Depletion of Alloy 690, *Corrosion*, Vol 46, 1990, p 391-402
79. H. Coriou, L. Grall, and S. Vettier, Corrosion sous contrainte de l'Inconel dans l'eau a haute temperature, in *Third Saclay Metallurgy Colloquium*, North Holland, 1960, p 161
80. P. Berge, Can PWR U-Tube Steam Generators Work Safely?, in *3rd Int. Symp. Environmental Degradation of Materials in Nuclear Power Systems—Water Reactors*, G.J. Theus and J.R. Weeks, Ed., The Metallurgical Society, 1988, p 49-54
81. C.S. Welty, Jr. and J.C. Blomgren, Steam Generator Issues, in Proc. 4th Int. Symp. Environmental *Degradation of Materials in Nuclear Power Systems—Water Reactors*, D. Cubicciotti, Ed., National Association of Corrosion Engineers, 1990, p 1.27-1.36
82. R. Bandy and D. Van Rooyen, Stress Corrosion Cracking of Inconel Alloy 600 in High Temperature Water—An Update, *Corrosion*, Vol 40, 1984, p 425-430
83. T.S. Bulischek and D. Van Rooyen, Stress Corrosion Cracking of Alloy 600 Using the Constant Strain Rate Test, *Corrosion*, Vol 37, 1981, p 597-607
84. R. Bandy and D. Van Rooyen, Mechanisms of Intergranular Failures in Alloy 600 in High Temperature Environments, in *Proc. 9th Int. Cong. Metallic Corrosion*, Vol 2, National Research Council of Canada, 1984, p 202-209
85. N. Totsuka, E. Lunarska, G. Cragnolino, and Z. Szklarska-Smialowska, A Sensitive Technique for

Evaluating Susceptibility to IGSCC of Alloy 600 in High Temperature Water, *Scr. Metall.*, Vol 20, 1986, p 1035-1040

86. T. Yonezawa and K. Onimura, "Effect of Chemical Composition and Microstructure on the Stress Corrosion Cracking Resistance of Nickel Base Alloys," presented at EPRI Meeting on Intergranular Stress Corrosion Cracking Mechanisms, Washington, DC, April 27-May 1, 1987
87. J. Blanchet, H. Coriou, L. Grall, C. Mahhiew, C. Otter, and G. Turler, Historical Review of the Principal Research Concerning the Phenomenon of Cracking of Nickel-Base Austenitic Alloys, in *Stress Corrosion Cracking and Hydrogen Embrittlement of Iron-Base Alloys*, R.W. Staehle, J. Hochmann, R.D. McCright, and J.E. Slater, Ed., National Association of Corrosion Engineers, 1977, p 1149-1160
88. V.B. Rajan, J.K. Sung, and G.S. Was, Stress Corrosion Cracking of Controlled Purity Inconel 600-Type Alloys in High Purity Water and Caustic Solutions, in *Proc. 3rd Int. Symp. Environmental Degradation of Materials in Nuclear Power Systems—Water Reactors*, G.J. Theus and J.R. Weeks, Ed., The Metallurgical Society, 1988, p 545-550
89. J.K. Sung and G.S. Was, The Role of Grain Boundary Chemistry in Pure Water Intergranular Stress Corrosion Cracking of Ni-16Cr-9Fe Alloys, in *Proc. 4th Int. Symp. Environmental Degradation of Materials in Nuclear Power Systems—Water Reactors*, D. Cubicciotti, Ed., National Association of Corrosion Engineers, 1990, p 6.25-6.37
90. A.A. Stein, A. Deleon, and A.R. McIlree, "Relationship of Annealing Temperature and Microstructure to Primary Side Cracking of Alloy 600 Steam Generator Tubing," EPRI Steam Generator Owners Group, SCC Contractors Workshop, San Diego, CA, March 1985
91. D. Garriga Majo, J.M. Gras, Y. Rouillon, F. Vaillant, J.C. Van Duysen, and G. Zacharie, "Prediction of the Stress Corrosion Cracking Resistance of Alloy 600 in Pure/Primary Water," Paper No. 56, presented at Int. Symp. Life Prediction of Corrodible Structures, Cambridge, UK, Sept 1991
92. H.A. Domian, R.H. Samuelson, L.W. Sarver, G.J. Theus, and L. Katz, Effect of Microstructure on Stress Corrosion Cracking of Alloy 600 in High Purity Water, *Corrosion*, Vol 33, 1977, p 26-37
93. G. Airey, "Optimization of Metallurgical Variables to Improve Corrosion Resistance on Inconel 600," EPRI NP-3051, Electric Power Research Institute, 1983
94. Z. Szklarska-Smialowska, Factors Influencing IGSCC of Alloy 600 in Primary and Secondary Waters of PWR Steam Generators, in *Proc. 4th Int. Symp. Environmental Degradation of Materials in Nuclear Power Systems—Water Reactors*, D. Cubicciotti, Ed., National Association of Corrosion Engineers, 1990, p 6.1-6.24
95. G. Economy, R.J. Jacko, and F.W. Pement, IGSCC Behavior of Alloy 600 Steam Generator Tubing in Water or Steam Tests Above 360°C, *Corrosion*, Vol 43, 1987, p 727-734
96. N. Totsuka, E. Lunarska, G. Cragnolino, and Z. Szklarska-Smialowska, Effect of Hydrogen on the Intergranular Stress Corrosion Cracking of Alloy 600 in High Temperature Aqueous Environments, *Corrosion*, Vol 43, 1987, p 505-514
97. N. Totsuka and Z. Szklarska-Smialowska, Effect of Electrode Potential on the Hydrogen-Induced IGSCC of Alloy 600 in an Aqueous Solution at 350 °C, *Corrosion*, Vol 43, 1987, p 734-738
98. R.L. Cowan and G.M. Gordon, Intergranular Stress Corrosion Cracking and Grain Boundary Composition in Fe-Ni-Cr Alloys, in *Stress Corrosion Cracking and Hydrogen Embrittlement of Iron Base Alloys*, R.W. Staehle, J. Hochmann, R.D. McCright, and J.E. Slater, Ed., National Association of Corrosion Engineers, 1977, p 1023-1070
99. D. Van Rooyen, Review of the Stress Corrosion Cracking of Inconel 600, *Corrosion*, Vol 31, 1975, p 327-337
100. H.R. Copson and G. Economy, Effect of Some Environment Conditions on Stress Corrosion Behavior of Ni-Cr-Fe Alloys in Pressurized Water, *Corrosion*, Vol 24, 1968, p 55-65
101. D.A. Vermilyea, Stress Corrosion Cracking of Iron- and Nickel-Base Alloys in Sulfate Solutions at 289°C, *Corrosion*, Vol 29, 1973, p 442-448
102. D.A. Vermilyea, Susceptibility of Iron- and Nickel-Base Alloys to SCC in pH 2.5 H_2SO_4 at 289°C, *Corrosion*, Vol 31, 1975, p 421-424
103. J.F. Newman, "Stress Corrosion of Alloys 600 and 690 in Acidic Sulfate Solutions at Elevated Temperatures," EPRI NP-3043, Electric Power Research Institute, 1983
104. P.L. Andresen, "The Effects of Sulfate Impurities in 288°C Water on IGSCC of Inconel 600 in Constant Load and SSRT Experiments," Paper No. 177, presented at Corrosion/84, National Association of Corrosion Engineers, 1984
105. C.A. Caraminhas and D.F. Taylor, Constant Extension Rate Tensile (CERT) Tests of Creviced Alloy 600 Specimens in Aqueous Environments at 288°C, *Corrosion*, Vol 40, 1984, p 382-385
106. R.W. Staehle, J.A. Begley, A.K. Agrawal, F.H. Beck, and J.B. Lumsden, "Corrosion and Corrosion Cracking of Materials for Water Cooled Reactors," EPRI NP-1741, Electric Power Research Institute, 1981
107. D.D. Macdonald, Z. Szklarska-Smialowska, G. Cragnolino, F.H. Beck, A. Boateng, A. Moccari, and S.P. Pednekar, "Corrosion and Corrosion Cracking of Materials for Water Cooled Reactors,"

EPRI Progress Reports FCC 7805 (Jan-June 1980), Electric Power Research Institute

108. P. Berge, "Modeling Corrosion for Life Prediction of Nuclear Reactor Components," Paper No. 1, presented at Int. Symp. Life Prediction of Corrodible Structures, Cambridge, UK, Sept 1991
109. N. Totsuka and Z. Szklarska-Smialowska, Hydrogen Induced IGSCC of Ni-Containing FCC Alloys in High Temperature Water, in *Proc. 3rd Int. Symp. Environmental Degradation of Materials in Nuclear Power Systems—Water Reactors*, G.J. Theus and J.R. Weeks, Ed., The Metallurgical Society, 1988, p 691-696
110. A.R. McIlree, Degradation of High Strength Austenitic Alloys X-750, 718, and A-285 in Nuclear Power Systems, in *Proc. Int. Symp. Environmental Degradation of Materials in Nuclear Power Systems—Water Reactors*, National Association of Corrosion Engineers, 1984, p 838-850
111. H. Hanninen and I. Aho-Mantila, Environment-Sensitive Cracking of Reactor Internals, in *Proc. 3rd Int. Symp. Environmental Degradation of Materials in Nuclear Power Systems—Water Reactors*, G.J. Theus and J.R. Weeks, Ed., The Metallurgical Society, 1988, p 77-92
112. K. Hosoi, S. Hattori, Y. Urayama, I. Masaoka, and R. Sasaki, Relation Between Susceptibility to Stress Corrosion Cracking in High Temperature Water and Microstructure of Inconel Alloy X-750, in *Proc. Int. Symp. Environmental Degradation of Materials in Nuclear Power Systems— Water Reactors*, National Association of Corrosion Engineers, 1984, p 334-344
113. T. Yonezawa, K. Onimura, N. Sakamoto, N. Sasaguri, H. Nakata, and H. Susukida, Effect of Heat Treatment on Stress Corrosion Cracking Resistance of High Nickel Alloys in High Temperature Water, in *Proc. Int. Symp. Environmental Degradation of Materials in Nuclear Power Systems—Water Reactors*, National Association of Corrosion Engineers, 1984, p 345-367
114. C.A. Grove and L.D. Petzold, Mechanisms of Stress Corrosion Cracking of Alloy X-750 in High Purity Water, in *Proc. Conf. Corrosion of Nickel-Base Alloys*, R.C. Scarberry, Ed., American Society for Metals, 1985, p 165-180
115. P. Skeldon and P. Hurst, Environmentally Assisted Cracking of Inconel X-750, in *Proc. 3rd Int. Symp. Environmental Degradation of Materials in Nuclear Power Systems—Water Reactors*, G.J. Theus and J.R. Weeks, Ed., The Metallurgical Society, 1988, p 703-710
116. T. Kawakubo and B. Rosborg, Stress Corrosion Cracking Threshold Testing—SCC Growth Behavior of Alloy X-750 and Type 304 Stainless Steel Under Slow Rising Load, *Proc. 2nd Int. Atomic Energy Agency Specialists Meeting on Subcritical Crack Growth*, Vol 1, NUREG/CP-0067, U.S. Nuclear Regulatory Commission, 1986, p 477-485
117. I.L. Wilson and T.R. Mager, Stress Corrosion of Age-Hardenable Ni-Fe-Cr Alloys, *Corrosion*, Vol 42, 1986, p 352-361
118. B.P. Miglin, M.T. Miglin, J.V. Monter, T. Sato, and K. Aoki, Stress Corrosion Cracking of Commercial-Grade Alloy 718 in Pressurized Water-Reactor Primary Water, in *Proc. 4th Int. Symp. Environmental Degradation of Materials in Nuclear Power Systems—Water Reactors*, D. Cubicciotti, Ed., National Association of Corrosion Engineers, 1990, p 11.32-11.44
119. J.R. Donati, M. Guttman, Y. Rouillon, P. Saint-Paul, and J.C. Van Duysen, Stress Corrosion Cracking Behavior of Nickel-Base Alloys With 19% Chromium in High Temperature Water, in *Proc. 3rd Int. Symp. Environmental Degradation of Materials In Nuclear Power Systems—Water Reactors*, G.J. Theus and J.R. Weeks, Ed., The Metallurgical Society, 1988, p 697-701
120. S. Hattori, K. Yamauchi, Y. Urayama, J. Kuniya, and H. Itow, Evaluation of Stress Corrosion Margin of Precipitation Hardenable Nickel Base Alloys in High Temperature Water, in *Proc. 4th Int. Symp. Environmental Degradation of Materials in Nuclear Power Systems—Water Reactors*, D. Cubicciotti, Ed., National Association of Corrosion Engineers, 1990, p 11.1-11.11
121. P. Berge and J.R. Donati, An Evaluation of PWR Steam Generator Tubing Alloys, *Nucl. Energy*, Vol 17, 1978, p 291-299
122. G.J. Theus, "Summary of the Babcock and Wilcox Company's Stress Corrosion Cracking Tests of Alloy 600," EPRI Workshop on Cracking of Alloy 600 U-Bend Tubes in Steam Generators, EPRI-WS-80-0136, Electric Power Research Institute, 1980
123. P.Berge, J.R. Donati, B. Prieux and D. Villard, Caustic Stress Corrosion of Fe-Cr-Ni Austenitic Alloys, *Corrosion*, vol 33, 1977, p 425-435
124. A.I. Asphahani, Slow-Strain-Rate Technique and Its Applications to the Environmental Stress Cracking of Nickel-Base and Cobalt-Base Alloys, in *Stress Corrosion Cracking—The Slow Strain Rate Technique*, STP 665, G.M. Ugianski and J. Payer, Ed., American Society for Testing and Materials, 1979, p 279-293
125. I.L. W. Wilson and R.G. Aspden, Caustic Stress Corrosion Cracking of Iron-Nickel-Chromium Alloys, in *Stress Corrosion Cracking and Hydrogen Embrittlement of Iron Base Alloys*, R.W. Staehle, J. Hochmann, R.D. McCright, and J.E. Slater, Ed., National Association of Corrosion Engineers, 1977, p 1189-1204

126. G. Theus, Caustic Stress Corrosion Cracking of Inconel 600, Incoloy 800 and Type 304 Stainless Steel, *Nucl. Technol.*, Vol 28, 1976, p 388-397
127. J.R. Crum, Effect of Composition and Heat Treatment on Stress Corrosion Cracking of Alloy 600 Steam Generator Tubing in Sodium Hydroxide, *Corrosion*, Vol 38, 1982, p 40-45
128. K.H. Lee, G. Cragnolino, and D.D. Macdonald, Effect of Heat Treatment and Applied Potential on the Caustic Stress Corrosion Cracking of Inconel 600, *Corrosion*, Vol 41, 1985, p 540-553
129. J.R. Crum, Stress Corrosion Cracking Testing of Inconel Alloys 600 and 690 in High Temperature Caustic, *Corrosion*, Vol 42, 1986, p 368-372
130. R.S. Pathania and R.D. Cleland, The Effect of Thermal Treatments on Stress Corrosion Cracking of Alloy 800 in a Caustic Environment, *Corrosion*, Vol 41, 1985, p 575-581
131. A. Mignone, M.F. Maday, A. Borello, and M. Vittori, Effect of Chemical Composition, Thermal Treatment and Caustic Concentration on the SCC Behavior of Alloy 800, *Corrosion*, Vol 46, 1990, p 57-65
132. A.R. McIlree and H.T. Michels, Stress Corrosion Behavior of Fe-Cr-Ni and Other Alloys in High Temperature Caustic Solutions, *Corrosion*, Vol 33, 1977, p 60-67
133. J.R. Cels, Stress Corrosion Cracking of Stainless Steels and Nickel Alloys at Controlled Potential in 10% Caustic Soda Solutions at 550°F, *J. Electrochem. Soc.*, Vol 12, 1976, p 1152-1156
134. J.R. Cels, Caustic Stress Corrosion Studies at 288 °C (550 °F) Using the Straining Electrode Potential Technique—Comparison of Alloy 600, Alloy 800 and Type 304 Stainless Steel, *Corrosion*, Vol 34, 1978, p 198-209
135. N. Pessal, G.P. Airey, and B.P. Lingenfelter, The Influence of Thermal Treatment on the Stress Corrosion Cracking Behavior of Inconel 600 at Controlled Potential in 10% Caustic Soda Solutions at 315 °C, *Corrosion*, Vol 35, 1979, p 100-107
136. R. Roberge, R. Bandy, D. Van Rooyen, "IGA of Alloy 600 in High Temperature Solutions of Sodium Hydroxide Contaminated With Carbonate," EPRI NP-3059, Electric Power Research Institute, 1983
137. G. Pinard Legry and G. Plante, "Intergranular Attack of Alloy 600: High Temperature Electrochemical Tests," EPRI NP-3062, Electric Power Research Institute, 1983
138. D. Van Rooyen and R. Bandy, "Mechanism of Intergranular Attack and Stress Corrosion Cracking of Alloy 600 by High Temperature Caustic Solutions Containing Impurities," EPRI NP-5129, Electric Power Research Institute, 1987
139. A. Kawashima and M. Takano, Stress Corrosion Cracking of Inconel 600 and Incoloy 800 in 50% NaOH Aqueous Solutions at 140 °C, *Boshuku Gijutsu*, Vol 27, 1978, p 340-347
140. H. Hickling and N. Wieling, Electrochemical Aspects of the Stress Corrosion Cracking of Fe-Ni-Cr Alloys in Caustic Solutions, *Corros. Sci.*, Vol 20, 1980, p 269-279
141. J.M. Sarver, J.V. Monter, and B.P. Miglin, The Effect of Thermal Treatment on the Microstructure and SCC Behavior of Alloy 690, in *Proc. 4th Int. Symp. Environmental Degradation of Materials in Nuclear Power Systems—Water Reactors*, D. Cubicciotti, Ed., National Association of Corrosion Engineers, 1990, p 5.47-5.63
142. F.A. Shunk and W.R. Warke, Specificity as an Aspect of Liquid Metal Embrittlement, *Scr. Metall.*, Vol 8, 1974, p 519
143. P.D. Johnson, A.I. Asphahani, and C.W. Allen, "Response of Several Nickel and Cobalt Base Alloys to Molten Pb, Sn, and Zn," Paper No. 128, presented at Corrosion/79, National Association of Corrosion Engineers, 1979
144. S. Sadigh, Stress Cracking of Stainless Steel and High Alloys in Molten Zinc at High Temperature, *Mater. Perform.*, July 1981, p 16
145. J. Kelly, Rolled Alloys Inc., private communication, Nov 1986
146. J.A. Hemsworth, M.G. Nicholas, and R.M. Crispin, Tellurium Embrittlement of Type 316 Steel, *J. Mater. Sci.*, Vol 25, 1990, p 5248
147. C.E. Price and R.S. Fredell, A Comparative Study of the Embrittlement of Monel 400 at Room Temperature by Hydrogen and by Mercury, *Metall. Trans. A*, Vol 17, 1986, p 889
148. C.E. Price and R.N. King, The Embrittlement of Monel 400 by Hydrogen and Mercury as a Function of Temperature, *Corrosion Cracking*, V.S. Goel, Ed., American Society for Metals, 1985
149. Y. Takizawa and I. Sekine, "Stress Corrosion Cracking Phenomena on Ni-Mo Alloys in High Temperature Nonaqueous Solution," Paper No. 355, presented at Corrosion/85, National Association of Corrosion Engineers, 1985
150. G.C. Smith, in *Hydrogen in Metals*, I.M. Bernstein and A.W. Thompson, Ed., American Society for Metals, 1974, p 485
151. A.W. Thompson, Hydrogen Assisted Fracture in Single-Phase Nickel Alloys, *Scr. Metall.*, Vol 16, 1982, p 1189
152. R.D. Kane, Accelerated Hydrogen Charging of Ni-Base and Co-Base Alloys, *Corrosion*, Vol 34, 1978, p 442
153. N. Sridhar, J.A. Kargol, and N.F. Fiore, Hydrogen-Induced Crack Growth in a Ni-Base Superalloy, *Scr. Metall.*, Vol 14, 1980, p 225
154. R.D. McCright, Effects of Environmental Species and Metallurgical Structure on the Hydrogen Entry Into Steel, in *Stress Corrosion Cracking and Hy-*

drogen Embrittlement of Iron-Base Alloys, R.W. Staehle, J. Hochmann, R.D. McCright, and J.E. Slater, Ed., National Association of Corrosion Engineers, 1977, p 306
155. B.A. Wilcox and G.C. Smith, *Acta Metall.*, Vol 13, 1964, p 331
156. W.T. Chandler and R.J. Walter, Testing to Determine the Effect of High Pressure Hydrogen Environments on the Mechanical Properties of Metals, in *Hydrogen Embrittlement Testing*, STP 543, American Society for Testing and Materials, 1974
157. J. Papp, R.F. Hehemann, and A.R. Troiano, Hydrogen Embrittlement of High-Strength FCC Alloys, in *Hydrogen in Metals*, A.W. Thompson and I.M. Bernstein, Ed., American Society for Metals, 1974, p 657
158. N. Totsuka and Z. Szklarska-Smialowska, Activation Energy for IGSCC of Alloy 600 in an Aqueous Solution Containing Dissolved H_2 at 300 to 350°C, *Scr. Metall.*, Vol 21, 1987, p 45-47
159. S.M. Bruemmer, "Grain Boundary Composition Effects on Environmentally Induced Cracking of Engineering Materials," Paper No. 185, presented at Corrosion/87, National Association of Corrosion Engineers, 1987
160. B.J. Berkowitz and R.D. Kane, The Effect of Impurity Segregation on the Hydrogen Embrittlement of a High-Strength Ni-Base Alloy in H_2S Environment, *Corrosion*, Vol 36, 1980, p 24
161. A.I. Asphahani, "Hydrogen Cracking of Ni-Base Alloys," 2nd Int. Cong. Hydrogen in Metals, Paris, 1977
162. J.A. Kargol and B. Ladna, The Roles of Ordering and Impurity Segregation on the Hydrogen-Assisted Propagation in Ni-Base Stainless Alloys, *Scr. Metall.*, Vol 16, 1982, p 191
163. B.J. Berkowitz, M. Kurkela, and R.M. Latanision, Effect of Ordering on the Hydrogen Permeation and Embrittlement of Ni_2Cr, in *Effect of Hydrogen on Behavior of Metals*, The Metallurgical Society, 1980, p 411
164. L.P. Lehman and T.H. Kosel, Influence of Phosphorus on the Kinetics of Ordering in Ni_2Cr, in *High Temperature Ordered Intermetallic Alloys*, C.C. Koch *et al.*, Ed., Materials Research Society Symposia Proceedings, Vol 39, 1985
165. M. Cornet, C. Bertrand, and M. Da Cunha Belo, Hydrogen Embrittlement of Ultra-Pure Alloys of the Inconel 600 Type: Influence of the Additions of Elements (C, P, Sn, Sb), *Metall. Trans. A*, Vol 13, 1982, p 141
166. J. Kupper-Feser and H.J. Grabke, Effect of Grain Boundary Composition on Tensile Strength and Hydrogen Induced Stress Corrosion Cracking of Ni-20Cr, *Mater. Sci. Technol.*, Vol 7, 1991, p 111
167. A.W. Funkenbusch, L.A. Heldt, and D.F. Stein, *Metall. Trans. A*, Vol 13, 1982, p 611
168. S. Hinotani, Y. Ohmori, and F. Terasaki, *Mater. Sci. Technol.*, Vol 1, 1985, p 297
169. M. Kurkela, R. Latanision, and B.J. Berkowitz, Effect of Ordering on the Hydrogen Permeation and Embrittlement of Ni_2Cr, in *Effect of Hydrogen on Behavior of Metals*, The Minerals, Metals, and Materials Society, 1980
170. R.M. Latanision and H. Opperhauser, The Intergranular Embrittlement of Nickel by Hydrogen: The Effect of Grain Boundary Segregation, *Metall. Trans.*, Vol 5, 1974, p 483
171. J.P. Hirth and J. Lothe, *Theory of Dislocations*, 2nd ed., Wiley-Interscience, 1982
172. A.W. Thompson and I.M. Bernstein, The Role of Metallurgical Variables in Hydrogen-Assisted Environmental Fracture, in *Advances in Corrosion Science and Technology*, Vol 7, R.W. Staehle and M.G. Fontana, Ed., Plenum Press, 1977
173. N. Sridhar, J.A. Kargol, and N.F. Fiore, Effect of Low Temperature Aging on Hydrogen-Induced Crack Growth in a Ni-Base Superalloy, *Scr. Metall.*, Vol 14, 1980, p 1257
174. J.G. Erlings, H.W. deGroot, and J.F.M. van Roy, Stress Corrosion Cracking and Hydrogen Embrittlement of High-Strength Nonmagnetic Alloys in Brines, *Mater. Perform.*, Oct 1986, p 28
175. K.D. Efird, Failure of Monel Ni-Cu-Al Alloy K-500 Bolts in Seawater, *Mater. Perform.*, April 1985, p 37
176. L.H. Wolfe and M.W. Joosten, Failures of Nickel/Copper Bolts in Subsea Applications, *SPE Prod. Eng.*, Aug 1988, p 382
177. J.W. Kochera, A.K. Dunlop, and J.P. Tralmer, Experience With Stress Corrosion Cracking of Nickel Base Alloys in Sour Hydrocarbon Production, in *H_2S Corrosion in Oil and Gas Production—A Compilation of Classic Papers*, R.N. Tuttle and R.D. Kane, Ed., National Association of Corrosion Engineers, 1981, p 413
178. J. Kolts, Alloy 718 for Oil and Gas Industry, in *Superalloy 718—Metallurgy and Applications*, E.A. Loria, Ed., The Minerals, Metals, and Materials Society, 1989, p 329
179. O.A. Onyewuenyi, Alloy 718—Alloy Optimization for Applications in Oil and Gas Production, in *Superalloy 718—Metallurgy and Applications*, E.A. Loria, Ed., The Minerals, Metals, and Materials Society, 1989, p 345
180. R.B. Frank and T.A. DeBold, "Heat Treatment of an Age-Hardenable Corrosion-Resistant Alloy—UNS N07716," Paper No. 59, presented at Corrosion/90, National Association of Corrosion Engineers, 1990
181. M.C. Place, P.R. Rhodes, and R.D. Mack, "Qualification of Corrosion Resistant Alloys for Sour Service," Paper No. 1, presented at Corrosion/91,

National Association of Corrosion Engineers, 1991

182. R.D. Kane, M. Watkins, D.F. Jacobs, and G.L. Hancock, Factors Influencing the Embrittlement of Cold Worked High Alloy Materials in H_2S Environments, *Corrosion*, Vol 33, 1977, p 309

183. "Testing of Metals for Resistance to Sulfide Stress Cracking at Ambient Temperatures," NACE Standard Test Method TM-01-77, National Association of Corrosion Engineers, 1986

184. T.S. Bulischeck and D. Van Rooyen, "Experimental Investigation of the SCC of Inconel 600 and a Predictive Model for Evaluating Service Performance," NUREG/CR-1675, Brookhaven National Laboratory, 1980

185. R.J. Coyle, J.A. Kargol, and N.F. Fiore, The Effect of Aging on Hydrogen Embrittlement of a Nickel Alloy, *Metall. Trans.A*, Vol 12, 1981, p 653

Selected References

- P. Gangloff and M.B. Ives, *Environment Induced Cracking of Metals*, National Association of Corrosion Engineers, 1990
- D.D. Macdonald and G.A. Cragnolino, Corrosion of Steam Cycle Materials, in *The ASME Handbook on Water Technology for Thermal Power Systems*, P. Cohen, Ed., American Society of Mechanical Engineers, 1989, Chap 9
- A.J. McEvily, Jr., Ed., *Atlas of Stress Corrosion and Corrosion Fatigue Curves*, ASM International, 1990
- R.W. Staehle, A.J. Forty, and D. Van Rooyen, Ed., *Fundamental Aspects of Stress Corrosion Cracking,* National Association of Corrosion Engineers, 1969
- R.W. Staehle, J. Hochmann, R.D. McCright, and J.E. Slater, Ed., National Association of Corrosion Engineers, 1977
- R.N. Tuttle and R.D. Kane, Ed., *H_2S Corrosion in Oil and Gas Production—A Compilation of Classic Papers*, National Association of Corrosion Engineers, 1981
- G.M. Ugianski and J.H. Payer, Ed., *Stress Corrosion Cracking—The Slow Strain Rate Technique*, STP 665, American Society for Testing and Materials, 1979

CHAPTER 6

Irradiation-Assisted Stress-Corrosion Cracking

Peter L. Andresen,
GE Corporate Research & Development Center

Irradiation-assisted stress-corrosion cracking (IASCC) describes premature cracking of material/environment systems exposed to ionizing irradiation (Ref 1-5). This broad interpretation attributes to IASCC all instances of environmental cracking accelerated by radiation, whether acting singly or conjointly in altering water chemistry, material microchemistry, material hardness, creep behavior and so on. While initially interpreted as inducing a form of cracking that would not occur in its absence, radiation is now more broadly viewed as an accelerate of the environmental cracking process, as shown schematically in Fig. 1. Several reviews of IASCC have been written (Ref 1, 3-5); this chapter discusses some of the primary radiation effects on IASCC and the our current understanding of and capability for predicting the individual phenomena and their collective effect, with an emphasis on stainless steels.

From a practical perspective, IASCC is principally associated with intergranular cracking of austenitic alloys in high-temperature water under neutron-irradiation exposure, which alters many material properties and produces radiolysis of water. Classic examples of the neutron-fluence dependence of IASCC of austenitic stainless steel are shown in Fig. 2(a) (Ref 6,7) for control-blade sheath cracking in high-conductivity boiling water reactors (BWRs) and in Fig. 2(b) (Ref 8) for laboratory tests on pre-irradiated material. In both instances, a characteristic rise in intergranular cracking occurs for fluencies above approximately 5×10^{20} n/cm^2 (energy > 1 MeV), which corresponds to nearly 0.7 lattice displacements per atom (dpa). The contribution of oxidizing water chemistries produced by radiolysis is also indicated by the strong dependence of cracking on dissolved oxygen at a fluence of approximately 3×10^{21} n/cm^2 in Fig. 2(b).

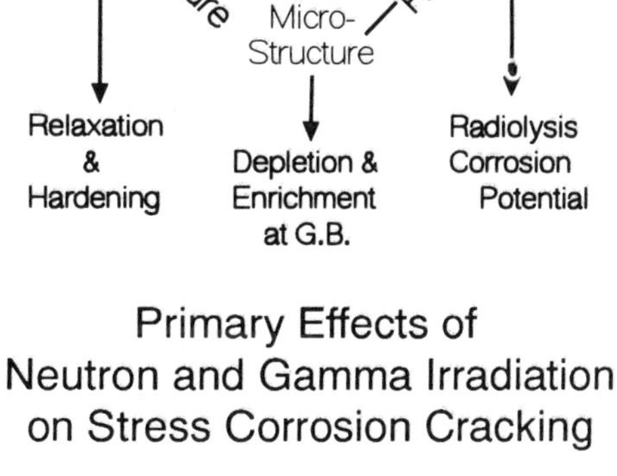

Fig. 1 Engineering factors (stress, environment, and microstructure), underlying fundamental phenomena (mass transport, oxide rupture, and repassivation), and primary effects of radiation on crack advance processes.

Initial concern for aggressive high radiation conditions was reflected in the specification of solution-annealed materials for most light water reactor (LWR) core components. Nevertheless,

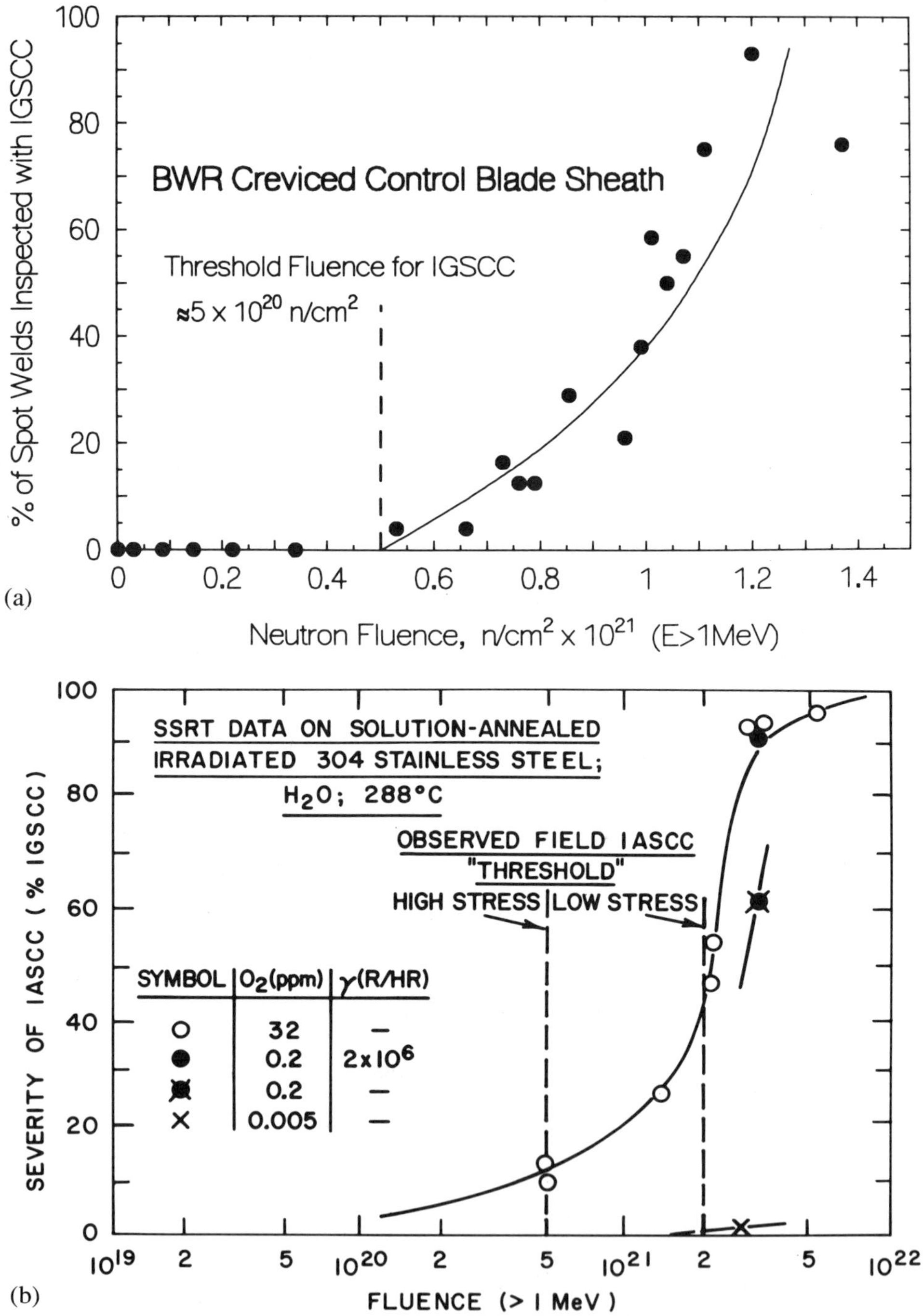

Fig. 2 Dependence of IASCC on fast neutron fluence for (a) creviced control-blade sheath in high conductivity BWRs (Ref 6) and (b) as measured in slow-strain-rate tests at 3.7×10^{-7}/s on preirradiated type 304 stainless steel in 288 °C (550 °F) water (Ref 8). The effect of corrosion potential via changes in dissolved oxygen is shown at a fluence of approximately 2×10^{21} n/cm^2

IASCC was first observed in the early 1960s in fuel elements (Ref 1, 9-17), where high stresses associated with fuel swelling were considered an essential and unusual ingredient. However, IASCC was later reported in a variety of high- and low-stress core components (Ref 1, 3-24, 30) and *in situ* test specimens (Ref 1, 22, 23, 25-29) in BWRs (Ref 1, 3-16), commercial pressurized water reactors (PWRs) (Ref 1, 4, 5, 10, 14-23), U.S. Navy test PWRs (Ref 16), and steam-generating heavy water reactors (SGHWRs) (Ref 24, 30), although there is not complete consensus on whether cracking is properly attributed to IASCC in all instances. While IASCC is of primary interest in LWR core materials, it is also of concern in applications such as fusion reactors and high-level radioactive waste containers.

To date, IASCC in LWR reactor internals has been discovered during routine inspections and thus has not greatly affected plant availability. Nonetheless, its implications are significant, both in terms of repair and outage costs as well as the potential for cracking in components that may be extremely difficult to repair. These concerns led to the formation of the International Cooperative Group on IASCC (Ref 2), whose primary objectives include the development of fundamental understanding of and life-prediction capabilities for IASCC, which are considered crucial to LWR plant management and life-extension decisions.

Overview of Service Experience

Incidents of IASCC were first reported in the early 1960s (Ref 1, 3-5, 9-21) and involved intergranular cracking of stainless steel fuel cladding. The conclusions drawn from those early investigations can be summarized as:

- Intergranular cracking morphology was predominant. Initiation occurred on the water side, and multiple cracks branching were observed.
- In contrast, only ductile, transgranular cracking was observed in post-irradiation mechanical tests performed in inert environments and at various temperatures and strain rates.
- Grain-boundary carbide precipitation was not observed by optical microscopy or transmission electron microscopy (TEM), nor inferred from acid copper sulfate tests.
- A correlation between time-to-failure and stress level was reported, with failure occurring first in thin-wall rods with small fuel-cladding gaps, where swelling strains were largest.
- The highest incidence of cracking occurred in peak heat flux regions, corresponding to the greatest fuel-cladding interaction and highest stresses and strains.
- Similar stainless steel cladding in PWR service exhibited fewer instances of intergranular failure. At that time, the PWR failures were attributed to off-chemistry conditions or stress rupture.

Subsequently, IASCC was observed in other stainless steel core components, such as neutron source holders in 1976 and control-rod absorber tubes in 1978. Instrument dry tubes and control-blade handles and sheaths, which are subject to very low stresses, have also cracked, although generally in creviced locations and at higher fluencies (Ref 6, 7).

Reported failures of reactor internal components are summarized in Table 1, suggesting that IASCC is not confined to a particular reactor design. For example, stainless steel fuel-cladding failures were reported in commercial PWRs and in PWR test reactors (Ref 1, 4, 5, 10, 14-21). At the West Milton PWR test loop, intergranular failure of vacuum-annealed type 304 stainless steel fuel cladding was observed (Ref 16) in 316 °C (600 °F) ammoniated water (pH 10) when the cladding was stressed above yield. Similarly, IASCC was observed in creviced stainless steel fuel-element ferrules in the Winfrith SGHWR (Ref 24, 30), a 100 MWe plant in which light water is boiled within pressure tubes, giving rise to a coolant chemistry similar to other BWR designs. The 20Cr-25Ni-Nb stainless steel differs from type 304 primarily in nickel and niobium contents, as well as in its lower sulfur (~0.006%) and phosphorus (~0.005%) contents. The ferrules were designed for 5-year exposure, during which the peak fast-neutron flux is 2 to 3×10^{13} $n/cm^2 \cdot s$ ($E > 1.5$ MeV), yielding a peak fluence over 5 years of 3 to 5×10^{21} n/cm^2.

Comparisons of reactor types were also made in swelling-tube tests performed in the cores of a BWR and a PWR (Ref 22, 23) on a variety of commercial- and high-purity heats of types 304,

316, and 348 stainless steel and alloys X-750, 718, and 625. Swelling was controlled by varying the mix of Al_2O_3 and B_4C within the tubes; the latter swells as neutrons are absorbed. Nominally identical strings of specimens were substituted for fuel rods. The distinction between the IASCC response of the two reactor types was small (Fig. 3).

Because the oxidizing potential in a PWR core is expected to be much lower than in BWRs, PWR cracking incidents have also been attributed to low-ductility stress rupture, although this cracking mechanism must be considered equally applicable to BWRs. In support of a mechanical cracking interpretation, there have been laboratory reports of intergranular cracking of irradiated stainless steels in slow-strain-rate tensile tests in approximately 300 °C (570 °F) inert environments (see, for example, Ref 31), although in related experiments (for example, Ref 8) no similar evidence of intergranular failure was found. Although intergranular cracking in inert tensile tests is not surprising, evidence against it as the sole basis for service failures includes the general observation that crack initiation occurs from the water side. Since there is a consistent trend toward increasing IASCC susceptibility with increasing corrosion potential in BWRs (see Fig. 2b), PWRs should be less susceptible to IASCC. Thus, other factors must be important in PWRs, including the borated-lithiated water chemistry additives, the generally higher temperatures and fluxes, and the possibility of localized boiling and thermal concentration cells in crevices from gamma heating, which, coupled with the borated-lithiated chemistry, could lead to aggressive local chemistries.

IASCC field-history details and dependencies were reported by Brown and Gordon (Ref 6, 7), who accumulated and analyzed data for cracking in alloy 600 shroud-head bolts (first observed in 1986), stainless steel safe ends (first observed in 1984), and in-core instrumentation tubes (first observed in 1984). They focused on components that were creviced, a factor known to exacerbate cracking (Ref 6, 7, 9). The highest radiation exposure occurred for the intermediate-range and source-range monitor (IRM/SRM) dry tubes, which contain flux monitors housed in thin-wall annealed stainless steel tubes. Cracking initiated in the crevice between the spring housing tube and the guide plug at fluences between 0.5 to 1.0

Table 1 IASCC Service Experience

Component	Material	Reactor type	Possible sources of stress
Fuel cladding	304 SS	BWR	Fuel swelling
Fuel cladding	304 SS	PWR	Fuel swelling
Fuel cladding(a)	20Cr-25Ni-Nb	AGR	Fuel swelling
Fuel-cladding ferrules	20Cr-25Ni-Nb	SGHWR	Fabrication
Nuetron source holders	304SS	BWR	Welding, Be swelling
Instrument dry tubes	304 SS	BWR	Fabrication
Control-rod absorber tubes	304 SS	BWR	B_4C swelling
Fuel-bundle cap screws	304 SS	BWR	Fabrication
Control-rod follower rivets	304 SS	BWR	Fabrication
Control-blade handle	304 SS	BWR	Low stress
Control-blade sheath	304 SS	BWR	Low stress
Control blades	304 SS	PWR	Low stress
Plate-type control blades	304 SS	BWR	Low stress
Various bolts(b)	A-286	PWR, BWR	Service
Steam separator dryer bolts(b)	A-286	BWR	Service
Shroud-head bolts(b)	600	BWR	Service
Various bolts	X-750	BWR, PWR	Service
Guide-tube support pins	X-750	PWR	Service
Jet-pump beams	X-750	BWR	Service
Various springs	X-750	BWR, PWR	Service
Various springs	718	PWR	Service

(**a**) Cracking in AGR fuel occurred during storage in spent fuel pond. (**b**) Cracking of core internals occurs away from high neutron and gamma fluxes.

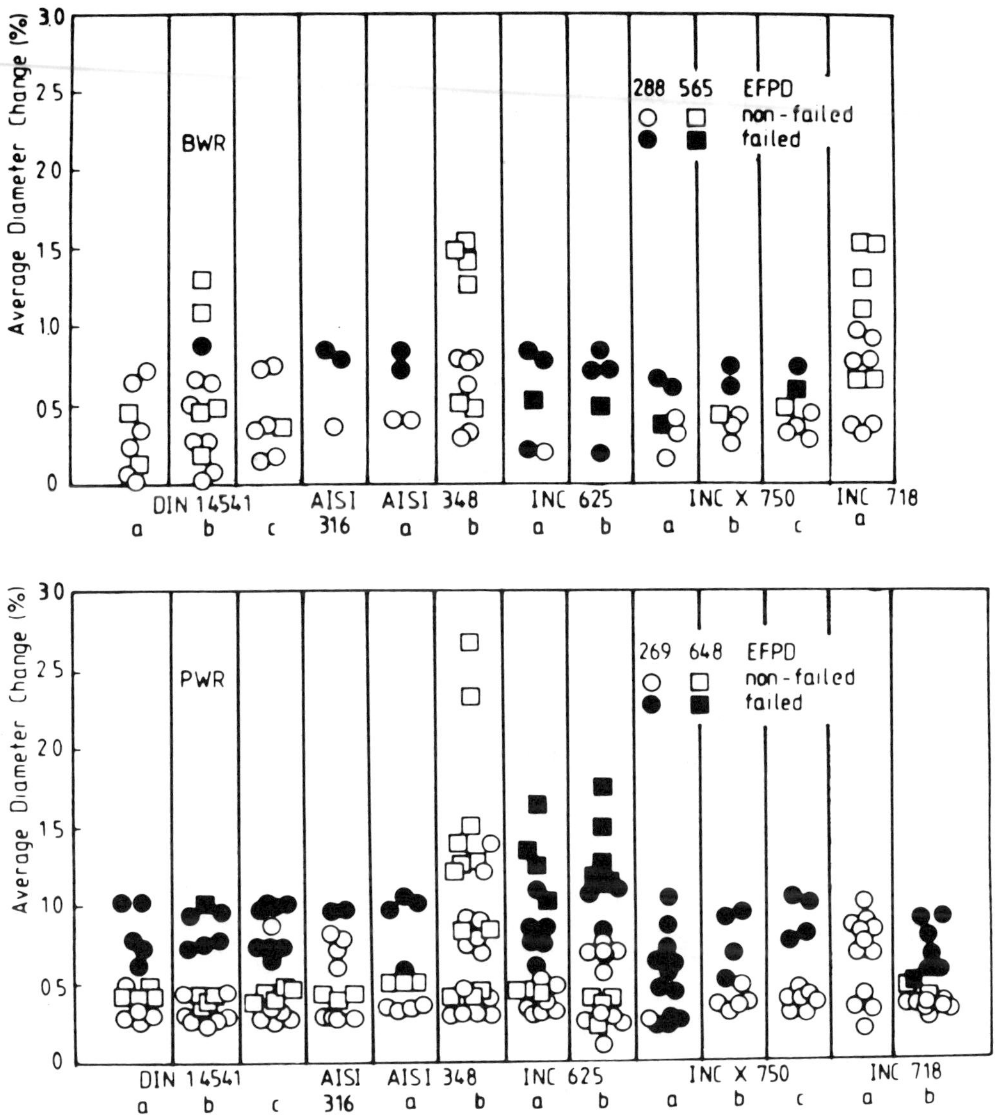

Fig. 3 IASCC of various alloys exposed to BWR and PWR core environments as a function of total strain in an in situ continuous-swelling tube test. Source: Ref 22, 23, 59

$\times 10^{22}$ n/cm^2 ($E > 1$ MeV). Stresses from the thick oxide observed in the crevice were implicated, since other (applied and residual) stresses were negligible. The primary variable from plant to plant is the average coolant conductivity, which correlates strongly with cracking incidence (Fig. 4a-c). Each point in Fig. 4 represents inspection results for one BWR plant, and data are normalized using reactor operating time (that is, percentage of components with intergranular cracks divided by months of on-line exposure). The scatter in Fig. 4(a) was attributed to variations in fluence and specific ion chemistry, as well as to limitations in the resolution of underwater visual inspection. Scatter can also result from short-term excursions in conductivity, which are not adequately reflected in the average, as identified in Fig. 4(c).

Conductivity correlations were also reported for cracking in shroud-head bolts (Fig. 4c) and

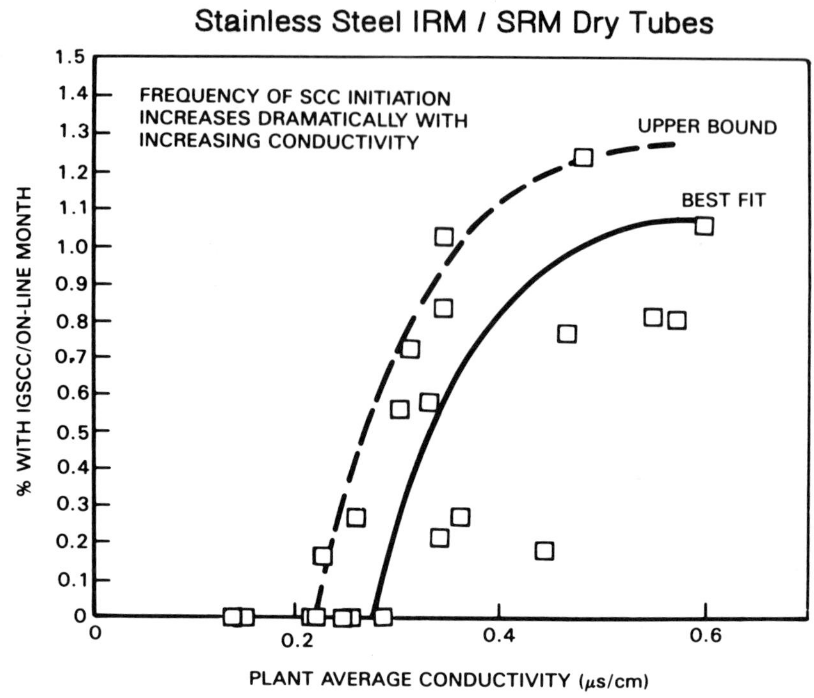

(a)

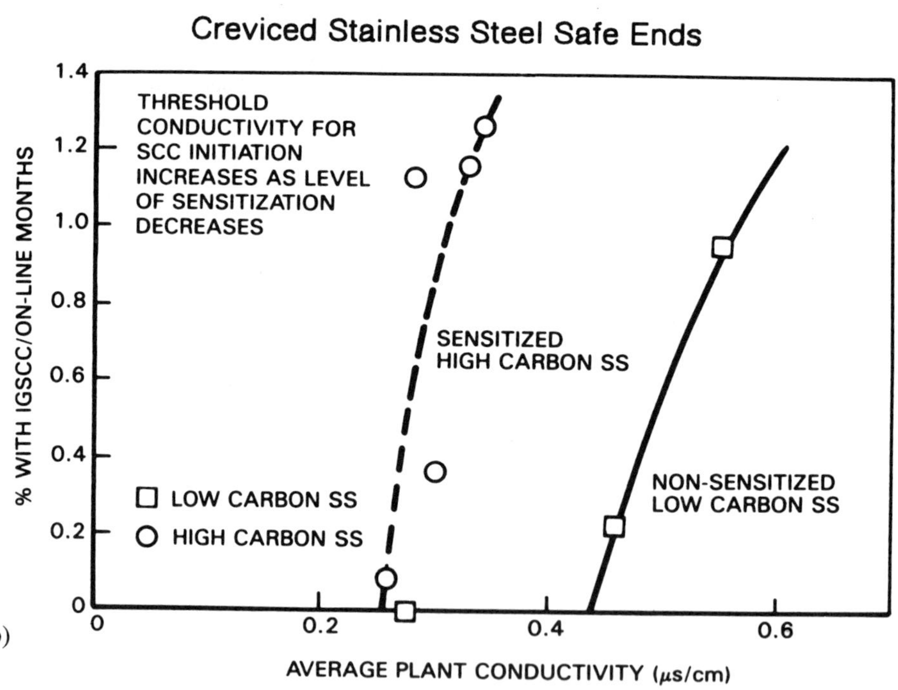

(b)

Fig. 4 Effects of average plant water purity shown in field correlations of the core-component cracking behavior for (a) stainless steel IRM/SRM instrumentation dry tubes, (b) creviced stainless steel safe ends. (continued)

creviced safe ends (Fig. 4b). The strong influence of conductivity on the cracking of stainless steel has also been shown in laboratory tests (Fig. 5) and plant recirculation piping, where predictive modeling (Ref 32-35) has been compared with field data on the operational time required to

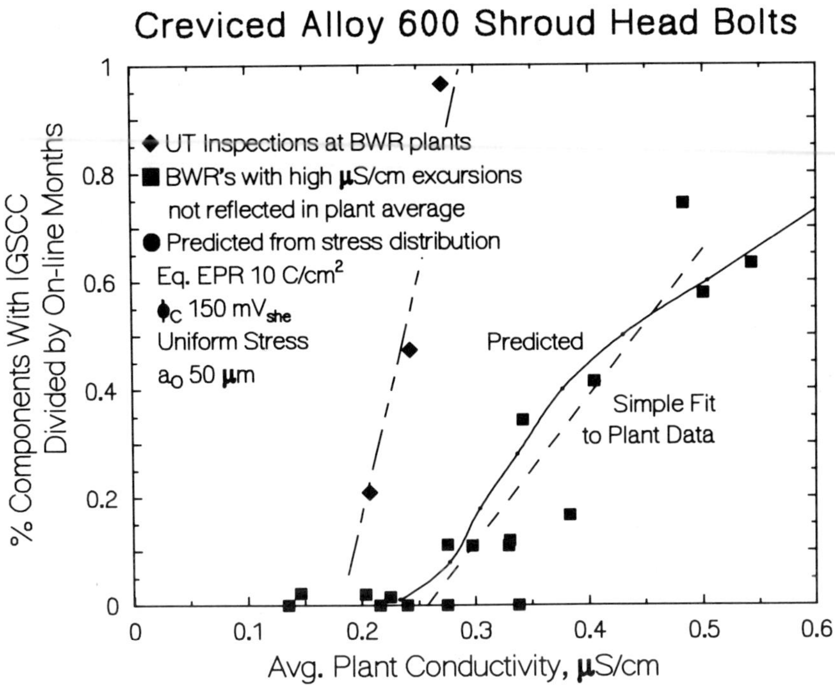

(c)

Fig. 4 Effects of average plant water purity shown in field correlations of the core-component cracking behavior for (c) creviced Inconel 600 shroud-head bolts, which also shows the predicted response versus conductivity

achieve a detectable crack, which is typically 10% of the wall thickness. Preliminary prediction of the shroud-head bolt cracking agreed reasonably well with observations (Fig. 4c).

Cracking has also occurred extensively in high-strength nickel-base alloy components (Ref 1, 5) (Table 1). Many incidents take place in lower-radiation flux regions (*e.g.*, where the end-of-life fluence is approximately below 5×10^{19} n/cm^2); an example is cracking of Inconel X-750 jet-pump beams in BWRs. Inconel X-750 cracking has also occurred frequently in PWR fuel holddown springs, which attain an end-of-life fluence of 1 to 10×10^{21} n/cm^2; this has been attributed in part to vibrational stresses. The effects of irradiation on IASCC in high-strength precipitation-hardened nickel-base alloy components are not characterized as well as in stainless steels, although in the lower flux/fluence regions the predominant effect of radiation is probably associated with the elevated corrosion potential from the radiolytic water chemistry. Overall trends and correlations from the IASCC field data are:

- Cracking is intergranular and can occur axially or circumferentially.
- No intergranular cracks related to radiation effects on solutionized stainless steel have been observed in the field at fluences approximately below 0.3×10^{21} n/cm^2.
- High stresses or dynamic strains were involved in most early incidents; however, cracking has been observed at quite low stresses at high fluences and longer operating exposures.
- An effect of corrosion potential can be inferred from the cracking of stainless steel fuel cladding and other components in BWRs and PWRs.
- Solution conductivity strongly affects cracking propensity in BWRs (Fig. 4 and 5). As noted by Brown and Gordon (Ref 7), this correlation applies equally to low- and high-flux regions and to stainless steels (Fig. 4a and b) and nickel-base alloys. Indeed, the correlation closely parallels that out-of-core (Fig. 5) (Ref 7, 32-35).
- Crevice geometries and cold work exacerbate cracking (Ref 7, 9).
- No grain-boundary carbides are required for susceptibility, although furnace-sensitized

stainless steels are clearly highly susceptible to in-core cracking.

- The fluence at which IASCC is observed is dependent on applied stress and stain, corrosion potential, solution conductivity, crevice geometry, and so on. At sufficiently high conductivities, cracking has been observed in solution-annealed stainless steel in the field (Fig. 4a and b) (Ref 7) and in the laboratory (Fig. 5) (Ref 32-35). Thus, while convenient in a practical engineering sense, the concept of a "threshold" fluence (or stress, corrosion potential, etc.) is scientifically misleading; cracking susceptibility and morphology are properly considered an interdependent continuum over many relevant parameters.

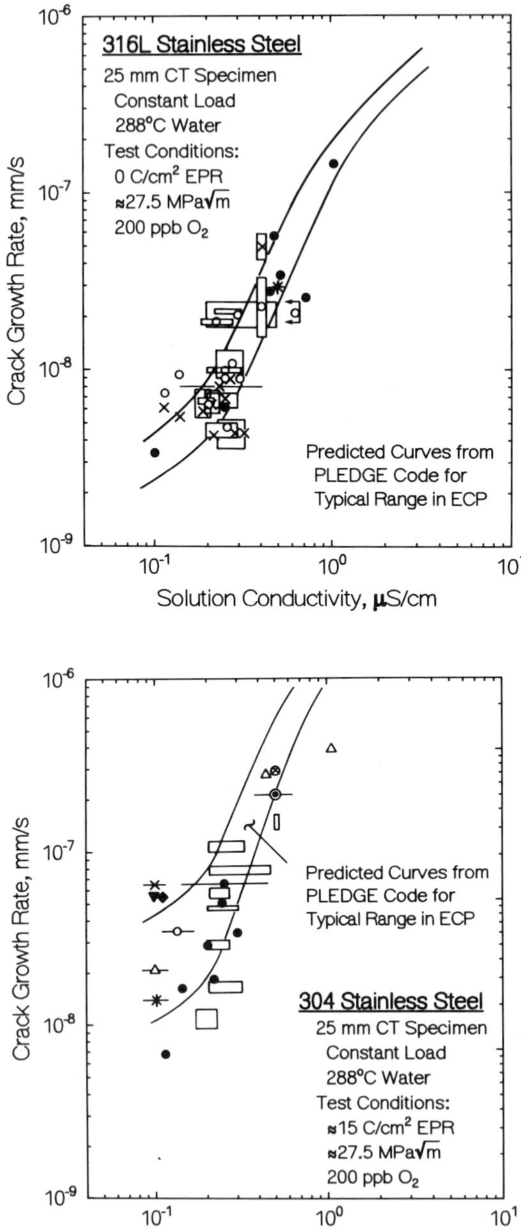

Fig. 5 Comparison between observed and predicted crack-growth rate versus solution conductivity for statically loaded type 316L and sensitized type 304 stainless steels in 288 °C (550 °F) water containing 200 ppb O_2. Source: Ref 32-35

Effects of Radiation Damage in Materials

The effects of radiation on material properties have been widely recognized and studied for decades. Early studies emphasized radiation hardening, swelling, and creep (Ref 1, 36, 37) (as well as radiation water chemistry, to be discussed later). Radiation damage is induced by a variety of high-energy particles which cause lattice atoms to be displaced to interstitial sites, leaving behind vacancies. These radiation-induced interstitials (primarily iron, chromium, and nickel) should be distinguished from atoms such as carbon and nitrogen, which normally occupy interstitial sites. Each particle-energy combination exhibits different damage characteristics, *e.g.*, depth uniformity, cascade damage localization, and so on. For example, high-energy protons and lower-energy (thermal) neutrons tend to undergo glancing angle collisions with lattice atoms, producing a series of isolated displacement events along their paths. High-energy neutrons (E > 0.1 MeV) penetrate much deeper on average into metals, but tend to undergo a single collision, producing a primary knockoff atom and subsequent cascading of atoms into adjacent atoms on a localized scale until the energy of the secondary recoil atoms falls below a threshold value of approximately 40 eV. While the damage is severe within a 20 to 50 nm region, the percentage of surviving vacancy-interstitial pairs is lower (10 to 40% survive) because of recombination during and just after the cascade event (intercascade annealing).

Neutron flux (damage rate) is reported in $n/cm^2 \cdot s$ or dpa/s and fluence (integrated damage) in n/cm^2 or dpa. Displacement damage (dpa) is a

(a)

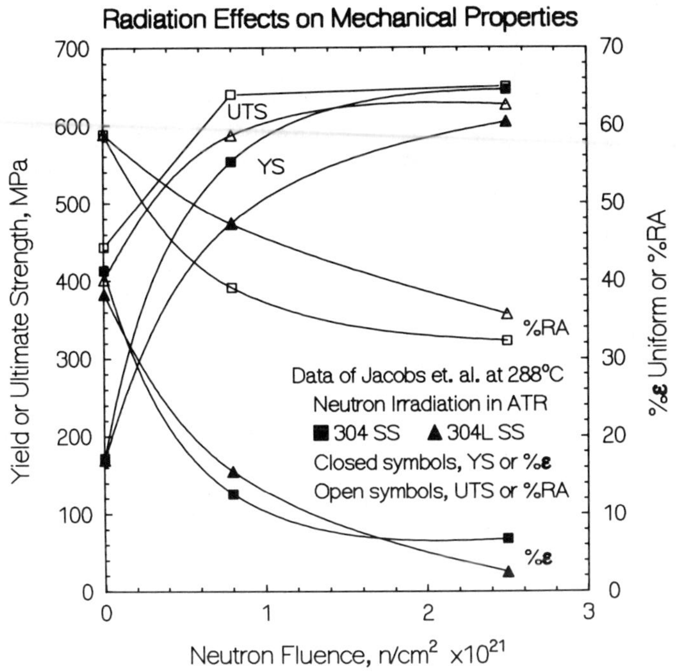

(b)

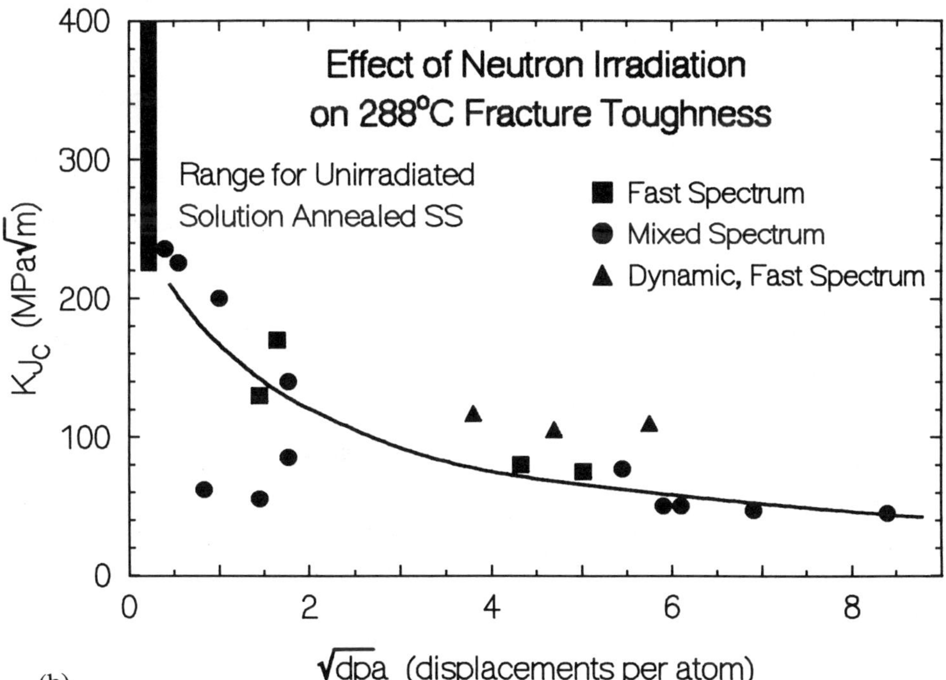

Fig. 6 Effect of fast neutron fluence under LWR conditions on mechanical properties at 288 °C (550 °F) of types 304 and 304L stainless steel. (a) Fluence dependence of yield strength, ultimate tensile strength, uniform elongation, and reduction in area (Ref 40). (b) Fluence dependence of fracture toughness (Ref 38, 39)

more fundamental and physical measure of the radiation damage to metals than is fluence; conversion between n/cm^2 and dpa is complex and depends on the neutron energy-intensity spectrum and the target material, with typical values for stainless steel in BWRs being approximately 1.5×10^{21} n/cm^2 per dpa for neutron energies above 0.1 MeV and approximately 0.7×10^{21} n/cm^2 per dpa for neutron energies above 1 MeV. Typical average displacement cross sections for stainless steel in LWRs are approximately 750 b (barns) ($E > 0.1$ MeV) and approximately 1500 b ($E > 1$ MeV). Neutron damage is relatively uniform through the thickness of most components within the core itself, although the neutron flux varies with axial and radial location in the core. Components/structures peripheral to the core (for example, the top-guide core barrel/shroud that surrounds the core) experience lower neutron fluxes, which can be significantly attenuated over their several-inch thicknesses.

Many radiation-induced vacancies and interstitials disappear by recombination, but a small fraction migrate over relatively long distances and aggregate into small clusters or disappear by absorption at sinks such as dislocation lines or grain boundaries. The microstructural evolution during irradiation is complex and not fully understood. A review by Maziasz (Ref 38, 39) identifies dislocation structures, cavity character, and precipitate types in different temperature/fluence regimes. Among the more common features that develop during irradiation are interstitial loops and vacancy clusters, formed during the migration and interaction of these species. Based on observations in older, lower-resolution electron microscopes, small clusters are often referred to as "black dot" damage.

The formation of various point and clustered defects causes an increase in hardness, yield strength, and ultimate tensile strength, a loss of ductility and fracture toughness (Ref 1, 36-40), and a shift in dislocation motion from macroscopically homogeneous to very heterogeneous, with deformation localized to submicron regions of intense shear where point and clustered defects were swept away by early dislocation motion. This causes a "channel fracture" appearance characterized by dimpled, crystallographic facets at 45° to the tensile axis. Thus, at fluences above approximately 2 dpa, the uniform elongation in tensile tests is low, but localized necking can result in relatively large reductions in area. For stainless steel, the fluence dependence on yield strength, ultimate tensile strength, and ductility is shown in Fig. 6 (a) (Ref 40), and its dependence on fracture toughness is shown in Fig. 6(b) (Ref 38, 39).

Additionally, diffusion of the vacancies and interstitial atoms to sinks causes nonequilibrium segregation and accelerated creep. The effects of accelerated creep can be either beneficial or deleterious. For example, under constant-displacement conditions load relaxation occurs (Fig. 7), whereas under constant-load conditions elevated creep rates result, and cracking is enhanced.

The possible interactions between radiation and environmental cracking are numerous, and many are poorly quantified and/or completely unknown. In addition to changes in mechanical properties, microstructural evolution, creep, and radiation-induced segregation, many associated phenomena may be influential in IASCC, including hydrogen trapping, localized shear, inhomogeneous deformation near interfaces due to hardening and radiation-induced segregation, phase transformations, transmutations (to form hydrogen, helium, etc.), and damage to the protective oxide. While many of these radiation phenomena are hypothetically important in IASCC, the paucity of and difficulty in obtaining data dictate that key radiation phenomena be identified and studied; to date, the international consensus is that key elements include radiation-induced segregation and radiation water chemistry.

Radiation-Induced Segregation

The motion of vacancies and interstitials formed during displacement damage causes nonequilibrium segregation at grain boundaries and other vacancy/interstitial sinks. The existence of this nonequilibrium segregation, known as radiation-induced segregation (RIS), was first proposed from theoretical considerations by Anthony (Ref 41) in 1968 and was later observed in the 1970s (Ref 42, 43). Vacancies and interstitial atoms can be produced at concentrations much higher than their thermal equilibrium values, and their subsequent diffusion to sinks produces concentration profiles. At higher irradiation temperatures, the thermal equilibrium of vacancies rises, and relative concentration (and associated effects) of radiation-induced vacancies and interstitials decreases (Fig. 8). At lower temperatures, the recombination rate of vacancies and interstitials increases, reducing the number of surviving species.

Because vacancies and interstitials exchange at different rates with different atomic species, a redistribution of solutes occurs in regions where persistent fluxes of interstitials or vacancies are present—near grain boundaries, for example. Solutes that migrate more rapidly via an intersti-

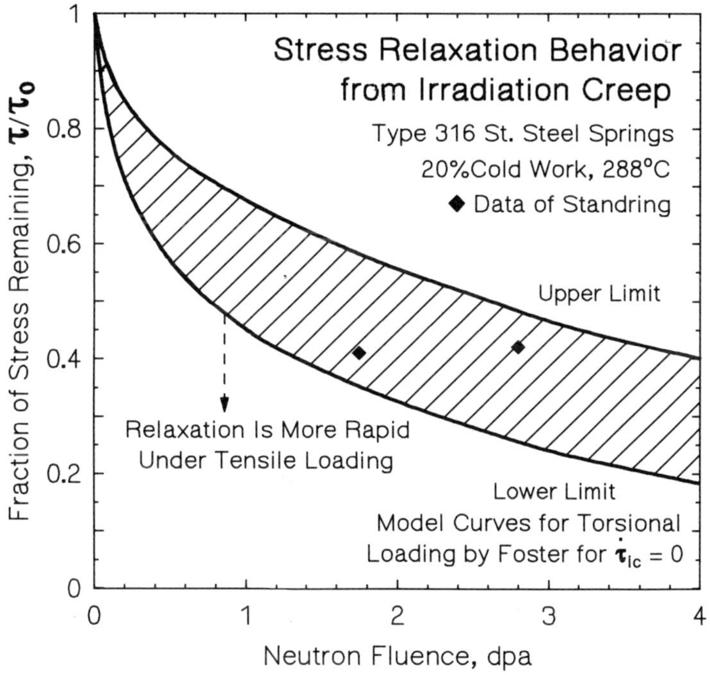

Fig. 7 Effects of radiation-induced creep on load relaxation of stainless steel at 288 °C (550 °F). Source: Ref 82

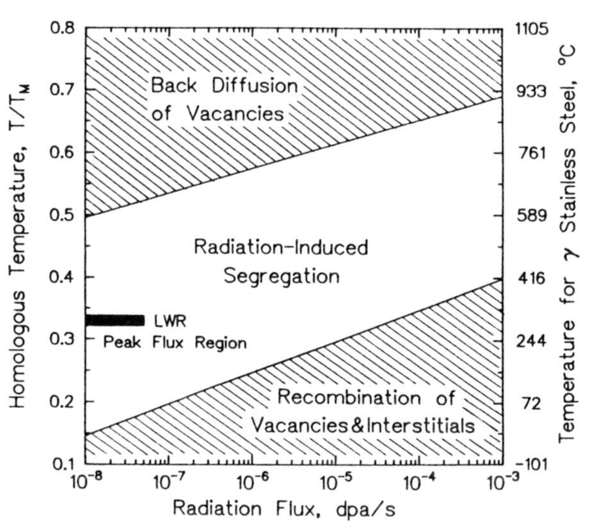

Fig. 8 Diagram of homologous temperature versus flux showing the regimes in which radiation-induced segregation is most pronounced

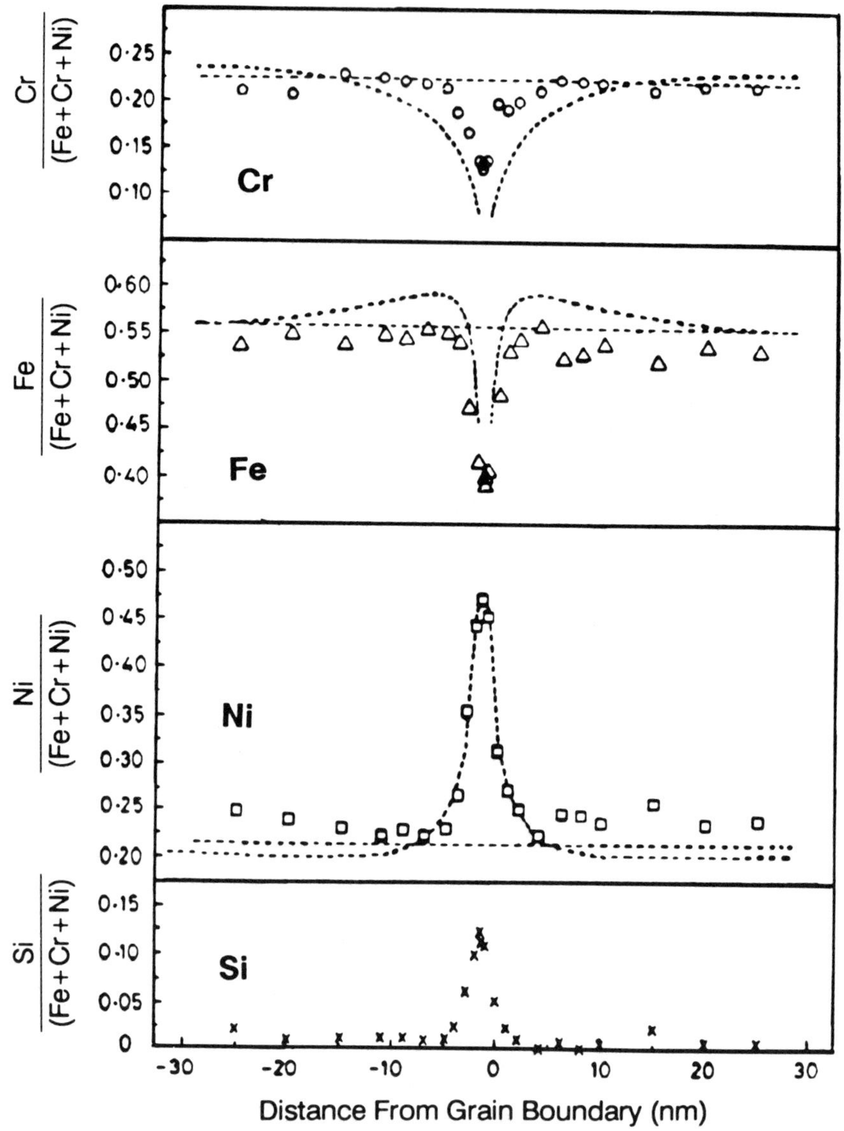

Fig. 9 Composition profiles across grain boundaries obtained by D-STEM in a 20Cr-25Ni-Nb stainless steel irradiated to 2 to 5 × 10^{21} n/cm^2 SGHWR at 288 °C (550 °F). Data are compared with predictive modeling of radiation-induced segregation. Source: Ref 24, 30

tial mechanism (*e.g.*, silicon in austenitic steels) tend to be dragged along with the interstitials and become enriched near sinks. Similarly, solutes that migrate rapidly (compared with other constituents) via a vacancy mechanism (*e.g.*, chromium in stainless steel) become depleted near grain boundaries, since the vacancy-solute exchange process causes the solute atom with the highest jump rate (or diffusivity) to move up the vacancy concentration profile. The relative contributions of interstitial and vacancy mechanisms control the degree of enrichment or depletion of a particular solute.

In broad terms, the RIS profiles near grain boundaries in stainless steel parallel those that often develop from thermal exposure (segregation/sensitization), with radiation-inducing enrichment of silicon, phosphorus, sulfur, and

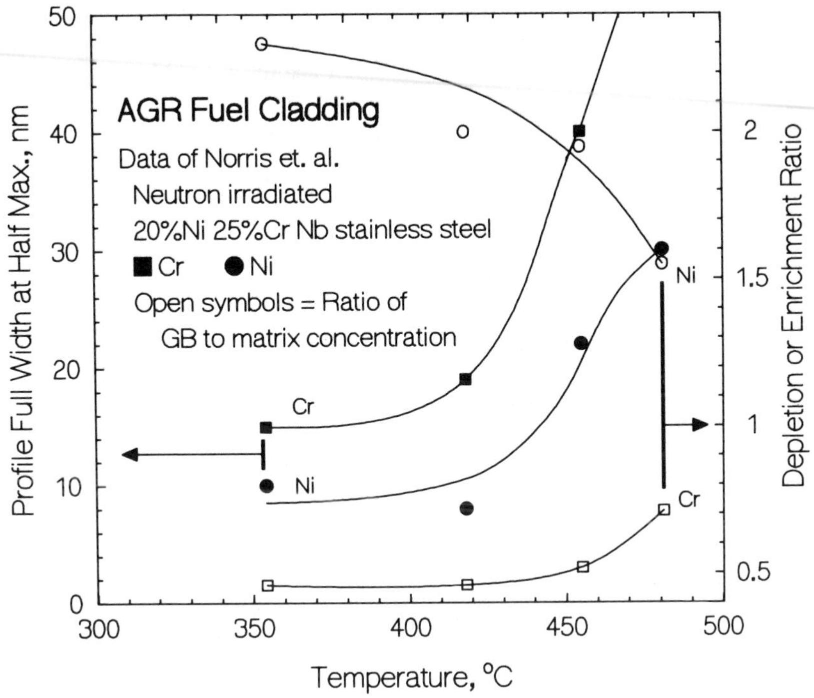

Fig. 10 Temperature dependence of the chromium profile width and observed minimum values (the actual minimum at the grain boundary is lower) for chromium and nickel. Data were obtained by D-STEM in a 20Cr-25Ni-Nb stainless steel fuel cladding irradiated to 1.03 dpa at 354 °C (670 °F), 2.16 dpa at 418 °C (785 °F), 2.77 dpa at 255 °C (490 °F), and 3.02 dpa at 481 °C (900 °F) in an AGR (Ref 30). Cracking of the AGR fuel cladding occurred during storage in a spent fuel pond.

nickel and depletion of chromium, molybdenum, and titanium. Since diffusion rates are low under LWR conditions (250 to 350 °C, or 480 to 660 °F), the profiles for major elements (*e.g.*, Chromium) are much narrower (*e.g.*, one-tenth) than for common thermal exposures (500 to 700 °C, or 930 to 1290 °F). The mechanisms of RIS obviously differ in many fundamental ways from thermal segregation and sensitization so that, for example, grain-boundary depletion of chromium by RIS occurs independently of carbide precipitation, and the thermodynamics of thermal sensitization are not applicable to RIS.

Early direct measurements of RIS were performed by Auger electron spectroscopy (AES) coupled with depth profiling by sputtering (Ref 42, 43). In general, the irradiations used heavy-metal ions, were greatly accelerated (*e.g.*, by approximately 10,000×) compared with LWR conditions, and were performed at temperatures between 440 and 770 °C (825 and 1420 °F). This resulted in comparatively broad profiles (*e.g.*, ± 40 nm) in types 304 and 316 stainless steels, with depletion from approximately 18% Cr to as low as 5%, and enrichment from approximately 9% Ni to as high as 16%. Questions were raised about the applicability of these high-flux, high-temperature conditions to LWRs, and about the possibility of carbide nucleation during irradiation and/or during the slow straining procedure used to obtain grain-boundary fracture.

Over the past 5 to 10 years, efforts to characterize compositional profiles on materials and components irradiated under LWR conditions have increased. Detailed quantitative data were obtained by Norris (Ref 30, 44) and Shepherd *et al.* (Ref 45, 46) using a high-resolution dedicated scanning transmission electron microscope (D-STEM). Figure 9 shows the narrow profiles obtained from a 20Cr-25Ni-Nb stainless steel

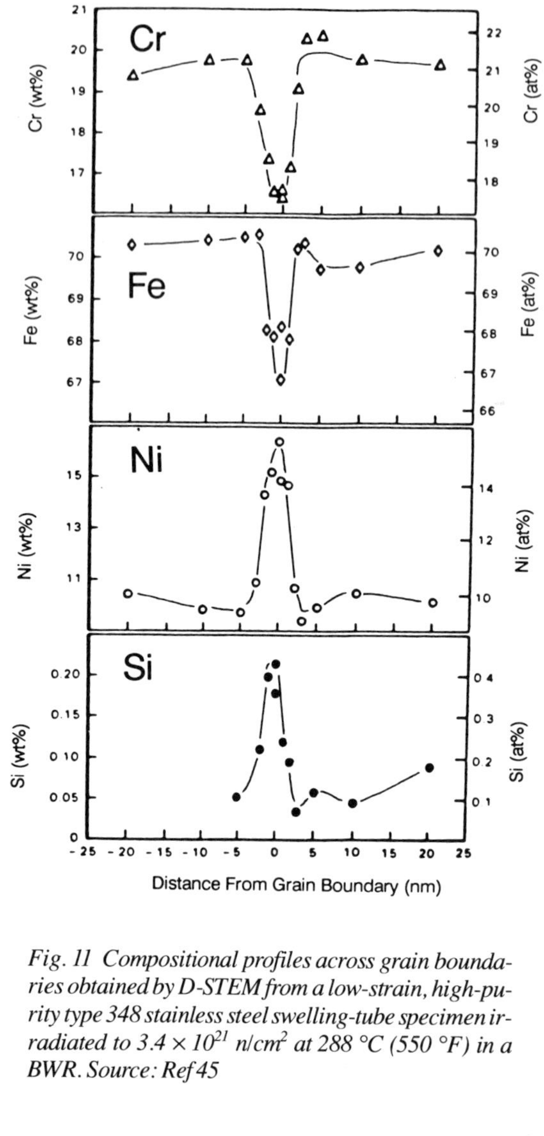

Fig. 11 Compositional profiles across grain boundaries obtained by D-STEM from a low-strain, high-purity type 348 stainless steel swelling-tube specimen irradiated to 3.4×10^{21} n/cm^2 at 288 °C (550 °F) in a BWR. Source: Ref 45

irradiated in an SGHWR at 288 °C (550 °F) (Ref 30). The curves represent model predictions, which do not agree quite as well with data at 288 °C as at temperatures greater than 350 °C (660 °F) (Ref 44). The temperature dependence of the profile width and observed minimum/maximum values (the actual values are somewhat more extreme) for chromium and nickel are shown in Fig. 10 for the same 20Cr-25Ni-Nb steel irradiated at 354 to 481 °C in an advanced gas-cooled reactor (AGR) (Ref 44). Profiles similar to the SGHWR data were also observed for types 304, 348 and high-purity 348 stainless steel irradiated in BWRs and test reactors (*e.g.*, Fig. 11 and 12) at 288 °C (550 °F). In these latter cases, less chromium depletion was observed, which appears to result from the fairly large differences in the nickel content of the standard stainless steels versus the 20Cr-25Ni-Nb alloy. An additional important factor is that the compositional homogeneity of the preirradiated material can vary, with initial grain-boundary enrichment of chromium observed in some isolated cases (Ref 46).

Theoretical modeling of RIS has developed in two branches, one dealing with the "concentrated" alloying constituents (*e.g.*, Fe-Ni-Cr) (Ref 47-52) and the other with the "dilute" (*i.e.*, < 1 at.% solute) alloying constituents (*e.g.*, silicon in nickel) (Ref 52-55). The concentrated models are comparatively simple, since the basic assumption is that the probability of an atom jumping to another site via either vacancy or interstitial motion depends only on the atom that migrates, not on the atoms surrounding it. Treatment on the vacancy motion alone may be sufficient to explain RIS in Fe-Ni-Cr alloys (Ref 44, 50, 51). Thermal-diffusion data show that nickel diffuses slowly and that chromium diffuses rapidly in these materials, which implies that nickel should become enriched near grain boundaries and chromium should become depleted. This model provides good agreement with the observed segregation behavior in these materials (see Fig. 9), although the agreement is weaker at lower temperatures and appears to be seriously in error as the nickel content is increased from 10 to 15% Ni to 35% Ni (Ref 51).

The dilute models are much more complex and involve parameters (such as binding energies between solutes and vacancies or interstitials) that cannot be independently measured. Johnson and Lam (Ref 53) first developed a phenomenological model for radiation-induced segregation in dilute alloys. Recently, a more accurate model has been developed (Ref 54, 55) based on the kinetic theory of diffusion. The theoretical models include the formation, dissociation, and motion of solute-interstitial and solute-vacancy pairs. If atoms readily form solute-vacancy or solute-interstitial pairs that are able to migrate through the lattice, then strong segregation will occur. For example, silicon binds strongly to both

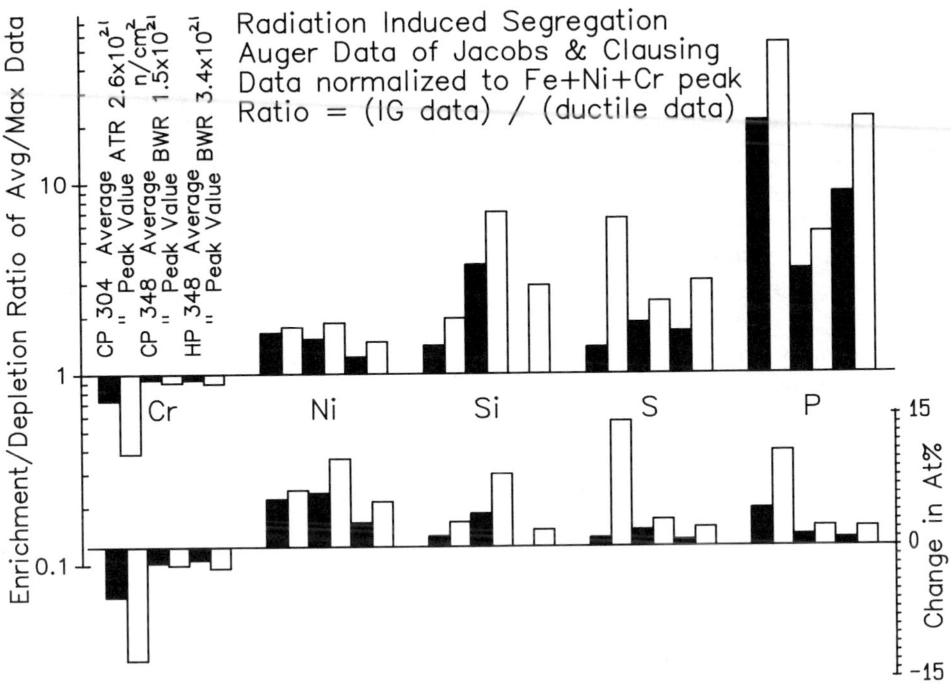

Fig. 12 Comparison using AES (Ref 45, 61) of commercial-purity type 304, commercial-purity type 348, and high-purity type 348 stainless steels irradiated to various fluences in an ATR or BWR. The average of all grain-boundary data, and the extreme reading, are plotted for each heat in terms of the enrichment or depletion ratio and the absolute change in atomic percent. Source: Ref 1

vacancies and interstitials and segregates strongly to free surfaces or grain boundaries (Ref 24, 45, 46, 52). Quantitative predictions (Ref 54, 55) agree with observations in some instances, *e.g.*, ion irradiations of phosphorus in nickel, although agreement in other systems such as silicon in stainless steel (Ref 52) is not nearly as good.

Uncertainties remain regarding the characterization, understanding, and modeling of RIS of both major and minor alloying elements. Unfortunately, the effect of RIS on IASCC is understood no better. More sophisticated theoretical models and improvements in experimental techniques and resolution in analytical characterization techniques may clarify these discrepancies and uncertainties.

Effects of Radiation-Induced Segregation on IASCC

Although radiation-induced changes in composition occur in narrow regions close to grain boundaries, they significantly increase the susceptibility of a material to intergranular cracking, as shown in Fig. 2 to 4. Direct comparison between RIS and IASCC is rare, although this is an object of several research programs (Ref 56-60). An investigation by Scott *et al.* (Ref 24) examined failed SGHWR fuel-cladding ferrules and identified the locations and morphology of cracking. No controlled experiments to evaluate cracking susceptibility were performed, and the nature of the service failures limits the characterization of the cracking kinetics versus fluence, stress, etc. However, the effect of neutron fluence is evident from the observation of cracking primarily in high-fluence regions after an accumulated fluence of greater than 2 to 5×10^{20} n/cm^2 at 288 °C (550 °F), in good agreement with the approximately 5×10^{20} n/cm^2 threshold identified by Jacobs *et al.* (Ref 8) for highly stressed components (Fig. 2). The D-STEM data (Fig. 9) showed significant depletion of chromium (from approximately 22% to less than 12%Cr) and iron (from 56% to less than 43%), and enrichment of nickel (from approximately 22% to greater than 47%Ni) and silicon (from approximately 0.5% to greater than 7%Si) at the grain boundary of the fuel-cladding ferrules that cracked. Similar observations (Ref 30) were made on the same material exposed in an AGR over a range of temperatures (Fig. 10); cracking of the AGR material occurred in spent fuel-storage ponds. Chromium depletion was implicated as the primary culprit, although the authors also expressed concern about the effects of silicon enrichment; no role in IASCC was attributed to phosphorus, sulfur or boron segregation.

Laboratory studies of RIS effects on IASCC were also performed by Jacobs *et al.* (Ref 61) on two heats of commercial-purity type 304 stainless steel of similar composition. The steels were irradiated in an advanced test reactor (ATR) to 2.2 to 3.1×10^{21} n/cm^2, then evaluated in slow-strain-rate tests, in corrosion tests, and by AES. The slow-strain-rate tests were performed at 3.3×10^{-7}/s in 288 °C (550 °F) oxygen-saturated (42 ppm) water and showed significant and similar susceptibility for both heats (*e.g.*, 92% versus 81% IASCC). The corrosion weight loss was measured in a 24-h exposure to a boiling solution of 4 *N* HNO_3 + 5 g/L of Cr^{6+} as $K_2Cr_2O_7$ and found to be approximately five times higher in the irradiated (versus unirradiated) material for exposures up to 20 h, when the weight loss in the irradiated material increased significantly, apparently from grain dropout. Auger electron spectroscopy was used to examine 12 grain boundaries from one of the irradiated heats of commercial-purity type 304; Fig. 12 summarizes the average and extreme (maximums for silicon, nickel, sulfur, phosphorus, etc. and minimums for chromium) values. In most cases (especially for sulfur and chromium) the average (and maximum) values are significantly skewed by one or a few points (see Table 5 in Ref 61).

Another study by Jacobs *et al.* (Ref 45) on high-purity and commercial-purity type 348 stainless steel characterized the swelling tubes tested in a BWR by Garzarolli *et al.* (Ref 22, 23). These tubes were filled with Al_2O_3 or Al_2O_3 + 3% B_4C to induce swelling during irradiation at a Kraftwerk Union (KWU) BWR for one or two fuel cycles, corresponding to fluences of 1.5×10^{21} n/cm^2 (for commercial-purity 348) or approximately 3.2×10^{21} n/cm^2 (for high-purity type 348). The in-reactor exposure resulted in intergranular cracking of the commercial-purity alloy (type 348, labeled "AISI 348a" in Fig. 3) above approximately 0.8% strain in the BWR plant, and above approximately 0.6% strain in the PWR plant. By contrast, the high-purity alloy (type 348, labeled "AISI 348b" in Fig. 3) exhibited no cracking at up to approximately1.6% strain in the BWR and no cracking at up to approximately 2.6% strain in the PWR. The high-purity alloy was clearly shown to be more resistant to IASCC, based on 52 different data points (Fig. 3) (Ref 22, 23, 59).

Laboratory slow-strain-rate tests, HNO3/ Cr^{6+} corrosion tests, AES (Fig. 12), and D-STEM (Fig. 11 and 13) (Ref 45) were used to characterize the material following irradiation. The slow-strain-rate tests indicate that commercial-purity type 348 is more susceptible, consistent with the KWU results (Fig. 3), although the results are complicated by cracking in only two of three commercial-purity type 348 specimens, with the two failures associated with a preexisting crack (from the KWU exposure) and creviced initiation at the loading pin. The HNO_3 /Cr^{6+} corrosion tests show greater weight loss for the commercial-purity type 348 than for high-purity type 348 in the irradiated condition, although anomalies were observed as a function of fluence and for the unirradiated material, as has been noted elsewhere (Ref 40).

The high-resolution D-STEM data (Fig. 11 and 13) show increases in silicon of 2 to 3 wt% (commercial-purity type 348) versus 0.2 to 0.3 wt% (high-purity type 348), in general agreement with AES measurements (Ref 45, 61). The nickel data differed somewhat, with STEM showing similar levels of nickel enrichment for commercial-purity type 348 versus high-purity 348 (3 to 6%), while AES showed less nickel enrichment for high-purity type 348. The STEM data show chromium depletion by 3 to 5%, somewhat more than observed by AES (Fig. 12).

In the absence of grain-boundary compositional profiles, comparisons of RIS and IASCC can be based on bulk composition if sufficient distinctions exist (*e.g.*, between high-purity ver-

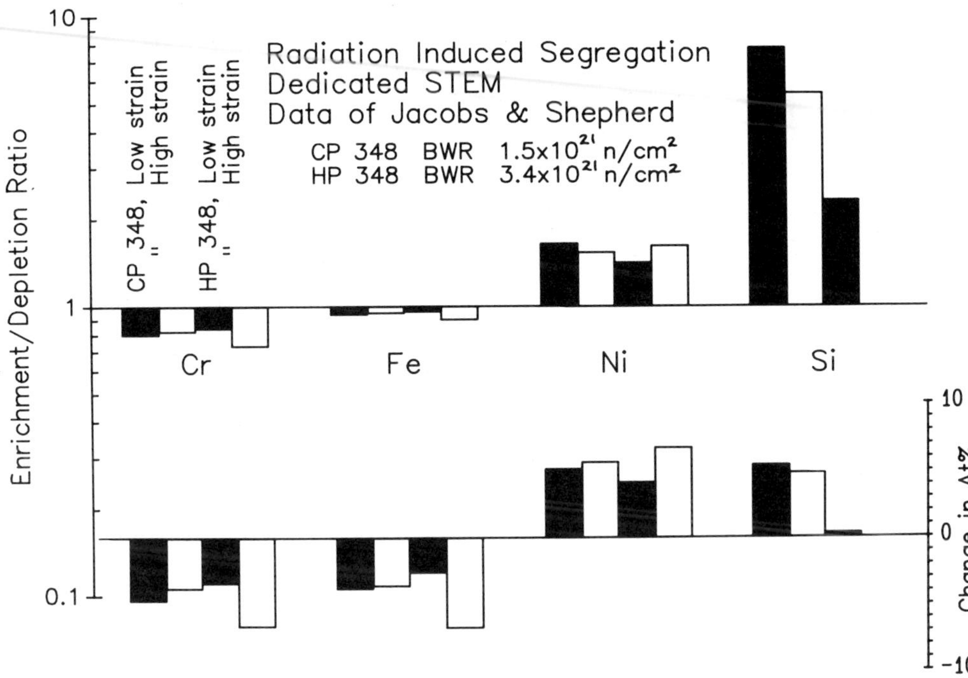

Fig. 13 Comparison of D-STEM analysis data (Ref 45) from commercial-purity type 348 and high-purity type 348 stainless steels irradiated to various fluences in a BWR. These data generally agree with the AES data in Fig. 12 on the identical materials. Source: Ref 1.

sus normal-purity materials). With the exception of the type 348 stainless steels (described above, see Ref 45), the materials used in the KWU swelling-tube experiments cannot provide such a comparison since the four titanium-bearing heats (similar to type 321 stainless steel) and one type 316 stainless steel are quite similar in their impurity concentrations. The effect of high-purity versus normal-purity type 304 stainless steels can be seen in the data of Ljungberg (Ref 58, 62) (Fig. 14). Wide variations exist in the bulk impurity levels among the four heats evaluated, with the worst-purity versus best-purity heat representing a variation of 8× in carbon, 10× in silicon, 27× in phosphorus, 6× in sulfur, and 3× in nitrogen. If a strong relationship between bulk alloy purity and irradiated grain-boundary chemistry is assumed (as is suggested in Fig. 11 to 13), Ljungberg's data indicate that the effects of impurities are minor.

There are many other studies in which the effects of radiation segregation can be inferred, although direct measurements of grain-boundary chemistry coupled with IASCC data generally do not exist. Of particular note are the constant-load data of Jacobs and Wozaldo (Ref 63) obtained in laboratory tests on pre-neutron-irradiated tensile specimens in 288 °C (550 °F) water containing 42 ppm oxygen, and *in situ* crack-growth-rate data (Ref 25-27) obtained on fracture-mechanics specimens of types 304 and 304L stainless steel exposed in the core of Nine Mile Point Unit 1 BWR. These data will be discussed later.

Additionally, tests measuring the effects of fluence on corrosion properties have been performed. Jacobs *et al.* (Ref 45, 61) (as described above) used a HNO_3/Cr^{6+} test to characterize weight loss as a function of material purity and fluence. Inazumi *et al.* (Ref 64) observed the effects of radiation-induced segregation in a neutron-irradiated titanium-modified stainless steel using a room-temperature electrochemical potentiokinetic reactivation (EPR) test. Katsura *et al.* (Ref 65) used a "double loop" EPR test and a modified Strauss test to compare neutron irradiation with thermal sensitization in type 304 stainless steel; while smaller effects of neutron

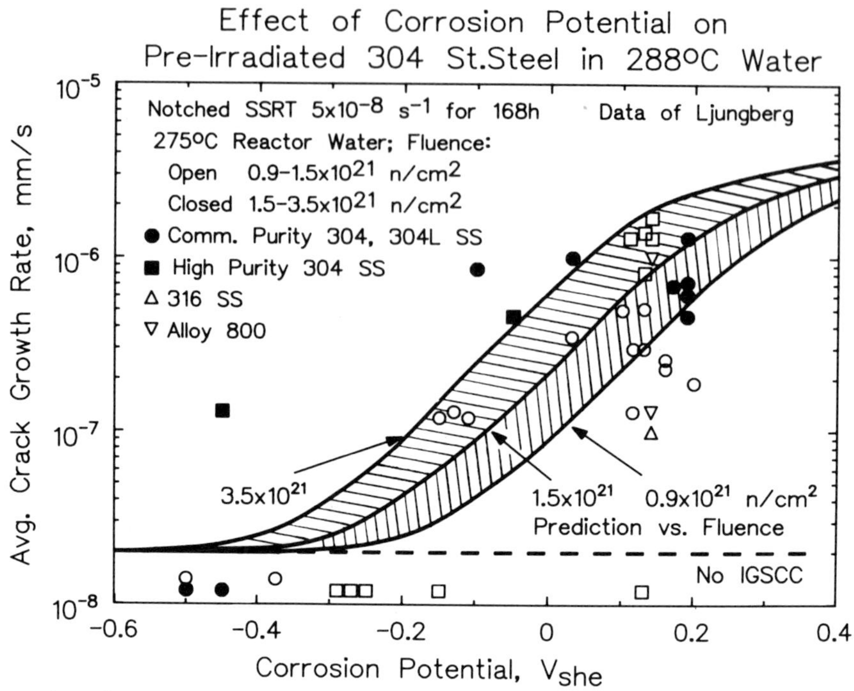

Fig. 14 Comparison of predicted and observed crack-growth rates for stainless steels irradiated in a BWR at 288 °C (550 °F) to various fluences. Notched tensile specimens were slow-strain-rate tested by Ljungberg (Ref 58, 62) in 288 °C (550 °F) pure water. Source: Ref 1

irradiation were observed, the EPR test results correlated closely with slow-strain-rate tests in 288 °C (550 °F) water. Jacobs (Ref 66) used a HNO_3/Cr^{6+} test to demonstrate the effectiveness of heat treatment in alleviating radiation effects.

Correlations between RIS and IASCC have also been performed using commercial-purity, high-purity, and phosphorus-doped high purity heats of type 304 stainless steel and commercial-purity type 316 stainless steel (Ref 83, 84). Grain boundary Cr, Fe, Ni, Mo, Si, and P profiles were obtained by high resolution STEM, and showed depletion of Cr, Fe, and Mo, and enrichment of Ni, Si, and P. The changes were monotonic with BWR neutron fluence and were generally more pronounced in the high purity materials. Enrichment of Cr was observed at low fluence, consistent with observations of Cr enrichment in unirradiated materials (Ref 46). Susceptibility in slow strain rate tests in 288 °C water containing 42 ppm oxygen increased vs. fluence (from ≈ 0.1 to 9.2×10^{21} n/cm^2), although the data were offset somewhat toward higher fluences compared to those in Figure 1, as expected from the initial Cr enrichment.

Commercial- and high-purity type 304 stainless steels were also irradiated in a BWR to fluences between 1.4 and 2.0×1021 n/cm^2 (Ref 85). Grain boundary microchemistry was determined using Auger electron spectroscopy, and IASCC susceptibility was measured in slow strain rate tests in 288 °C water containing 280 ppb O_2. Ni and Si enrichment were more pronounced in the commercial-purity steel, although Cr depletion was greatest in the high purity steel. While Cr depletion was unusually high, both the Cr measurements and IASCC response showed that the high-purity material was more susceptible.

Two heats of 304 stainless steel (high-purity and high purity doped with P) were proton irradiation and the nevaluated for grain boundary chemistry by Auger electron spectroscopy, for microstructure using TEM, and for IASCC susceptibility using slow srtain rate testing in 288 °C water of controlled purity (Ref 86). IASCC susceptibility correlated with Cr depletion, and the presence of P segregation appeared to be beneficial. Commercial- and high-purity type 304 stainless steels were evaluated following proton irradiation (Ref 87) using slow strain rate testing at $\approx 2 \times 10^{-7}\ s^{-1}$ in 287 °C, oxygen-saturated water. No IASCC was observed at 0, 0.2, or 0.4 dpa in the commercial purity steel, while the high-purity alloy exhibited IASCC at 0.4 dpa but not at 0 or 0.2 dpa. A high purity alloy with phosphorus slowed somewhat less susceptibility, again indicating a beneficial effect of P.

Recent unreported data from several ongoing programs provide no clear basis for implicating the presence or absence of any single impurity in IASCC. Many elements, including silicon, nitrogen, hydrogen, sulfur, phosphorus, carbon, and boron have been implicated (Ref 1), and in isolated cases, high-purity stainless steels have proved moderately resistant, although attempts to reproduce the behavior using nominally identical compositions and thermomechanical processing have not been successful. The author maintains the view that depletion of chromium plays an important role in IASCC susceptibility, since (1) even small depletion levels and narrow widths, obtained by thermal sensitization, can enhance intergranular cracking in hot water (Ref 67, 68); (2) significant grain-boundary chromium depletion does occur, although the extent may depend on many factors, including temperature, nickel content, impurity levels, and so on; and (3) an additional, and perhaps major, complicating factor in RIS characterization studies is that the initial compositional state of the material cannot be assumed to be uniform. Recent data (Ref 46) clearly show that, in some cases, significant grain-boundary enrichment of chromium (several percent) can exist following solution annealing. Since the initial condition of the material is presumed to be homogeneous and thus is rarely characterized, at least some of the ambiguity and variability in RIS and IASCC data can be attributed to this phenomenon.

Many questions regarding the nature, characterization, and quantitative modeling of RIS and its effect on IASCC remain unanswered, including:

- Why are the segregation profiles so much shallower and narrower in SGHWR (Fig. 9) and BWR (Fig. 11) materials than in AGR materials (Fig. 10)? The difference is not consistent with the current model predictions of the effects of temperature, nickel content, or other factors.
- Do complex, competitive interactions dominate the segregation of impurities? Similarly, can minute concentrations of impurities (for example, 5 ppm B, which is greater than the vacancy concentration induced by neutron irradiation in power reactors) affect RIS of major alloying elements, for example, by interacting with and inhibiting vacancy motion?
- Why are the correlations between cracking susceptibility and either bulk or grain-boundary compositions not more definitive? Isolated correlations have been made, for example, with chromium depletion in the SGHWR material, and with silicon (only) in type 348 stainless steel (not in type 304). However, consistent individual or multiple correlations with grain-boundary chemistry cannot yet be made.
- The absence of a clear correlation may suggest that other factors dominate, such as complex interactions among alloying elements and impurities, or among RIS, radiation hardening, and creep. Additionally, variability in the difficult and complex tests on irradiated materials, caused by poorly controlled outlet solution conductivity, surface preparation, and so on, may create experimental "noise" that complicates attempts to isolate the effects of RIS on IASCC.

Radiation Water Chemistry and Corrosion Potential

Water is decomposed by ionizing radiation into various primary species (Ref 69-72), including both radicals (*e.g.*, e^-_{aq}, H, OH, HO_2) and molecules (*e.g.*, H_2O_2, H_2, O_2), which can be oxidizing (*e.g.*, O_2, H_2O_2, HO_2) or reducing (*e.g.*, e^-_{aq}, H, H_2). Their concentrations are roughly proportional to the square root of the radiation flux in pure water; the effects of dissolved hydrogen, oxygen, or ionic species can be large. The radiation energy-intensity spectrum influences the concentration of each radiolytic species, which is described in terms of a yield, or *G* value (molecules produced per 100 eV absorbed by water). In LWRs, the *G* values for most species are within a factor of approximately 3 for fast neutron versus

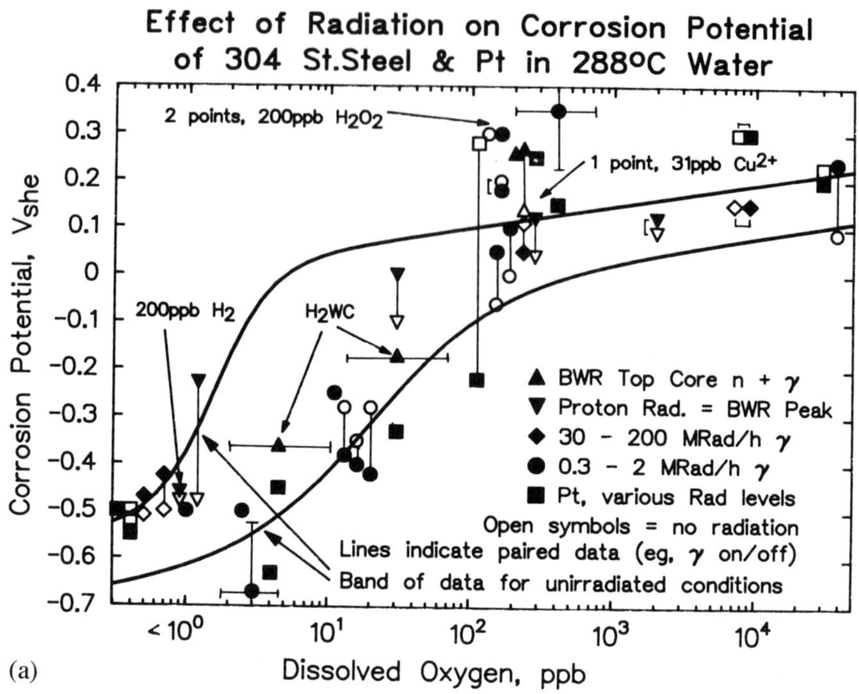

(a)

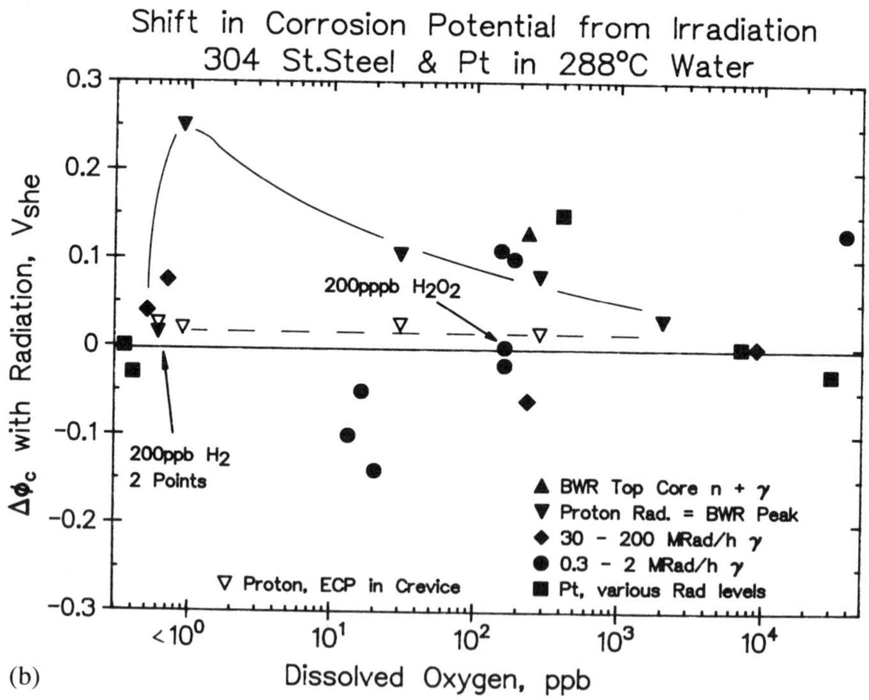

(b)

Fig. 15 (a) Effect of radiation on the corrosion potential of type 304 stainless steel in 288 °C (550 °F) water. The curves denote the range of typical values in the unirradiated corrosion-potential data (Ref 1). (b) Effect of radiation on the shift in corrosion potential from the value under unirradiated conditions for type 304 stainless steel in 288 °C (550 °F) water. With the exception of the BWR measurements, all data were obtained in controlled-radiation on/off experiments (Ref 1). Curves in (b) show the trends in the proton-irradiated data, where the effects of radiation (on/off and over a range of fluxes) were evaluated for a variety of dissolved O_2 and H_2 concentrations under otherwise identical conditions.

gamma radiation. Despite this similarity, the influence of fast neutron radiation is much stronger than gamma radiation, primarily because the energy deposition rate, or mean linear energy transfer (LET), is greater: 40 eV/nm for fast neutrons, versus 0.01 eV/mn for gamma radiation (Ref 71). Also the neutron flux in LWRs (*e.g.*, approximately 1.03×10^9 rad/h core average and approximately 1.68×10^9 rad/h peak in a BWR of 51 W/cm^3 power density) is also higher than the gamma flux (approximately 0.34×10^9 rad/h). Indeed, the moderate gamma levels present in the downcomer in the outside annulus of a BWR core actually promote recombination of hydrogen and various oxidants (Ref 69, 72). The contribution of thermal neutrons and beta particles to radiolysis is small in LWRs.

As in many electrochemical processes, the integrated effects of various oxidants and reductants on environmental cracking are best described via changes in corrosion potential, which controls the thermodynamics and influences the kinetics of most reactions. Since electrochemical potentials are logarithmically dependent on local oxidant, reductant, and ionic concentrations (via the Nernst relationship, $\phi = \phi_0 + RT/nF\ \ln[\text{products/reactants}]$), radiation–induced increases in concentration of various species by many orders of magnitude may have comparatively small effects on the corrosion potential in hot water. Further, corrosion potentials are mixed potentials involving a balance of anodic and cathodic reactions on the metal surface, which depend on the concentrations of both oxidizing and reducing species. At low oxidant concentrations, the rapid drop in corrosion potential to approximately $-0.5V_{she}$ (Fig. 15a) results from mass-transport-limited kinetics (*e.g.*, oxygen transport to the metal surface). In this regime, more pronounced shifts in corrosion potential with radiation can occur, presumably from the radiolytic formation of oxidizing species within the mass-transport-limited stagnant layer.

The relationship between dissolved oxygen and corrosion potential in hot water as a function of radiation type and flux is shown in Fig. 15(a), in which the connected points represent data obtained in controlled, radiation on/off experiments. The data from these latter experiments are shown in Fig. 15(b) in terms of a radiation-induced shift in potential. The curves in Fig. 15(a) represent the scatter band for the data obtained under unirradiated conditions. Similar scatter also exists in the irradiated corrosion potential data in Figure 15(a) and is comprised of contributions from both real effects and experimental error. These radiation measurements are quite complex and should involve careful qualification of the radiation resistance and thermodynamic response of reference electrodes. High-radiation flux experiments were performed by Taylor (Ref 73) and Miller and Andresen (Ref 29) using multiple, fundamentally different types of radiation-hardened reference electrodes. In-reactor measurements have also been performed by Indig (Ref 25, 74) using multiple, radiation-qualified silver chloride reference electrodes.

Figures 15(a) and (b) show that little, if any, elevation in corrosion potential results from irradiation sources that do not include neutrons or simulate their contribution [*e.g.*, using high-energy protons (Ref 29)]. Some studies using gamma radiation (Ref 1, 28, 73) showed a significant decrease in corrosion potential, especially in the intermediate (*e.g.*, 10 to 200 ppb) range of dissolved oxygen; this is consistent with enhanced recombination of oxidizing and reducing species such as occurs in the downcomer region of BWRs (Ref 72). In instances where neutrons or protons have been used, a consistent, significant elevation in corrosion potential is observed. This is more pronounced in hot water containing low dissolved oxygen concentrations and no dissolved hydrogen (Fig. 15b), where increases of more than +0.25 V occur. At higher inlet oxygen concentrations (*e.g.*, approximately 200 ppb), the data still show a significant shift (typically, +0.1 to 0.15 V) in corrosion potential for radiation conditions representative of peak LWR core fluxes (Fig. 15b); less increase is observed for inlet oxygen concentrations associated with, for example, air saturation (approximately 8.8 ppm O_2) or oxygen saturation (approximately 42 ppm O_2 at STP). A similar elevation in corrosion potential is observed for additions of hydrogen peroxide (200 ppb H_2O_2, Fig. 15a), which suggests that H_2O_2 may be a major factor in increasing the corrosion potential under irradiated conditions.

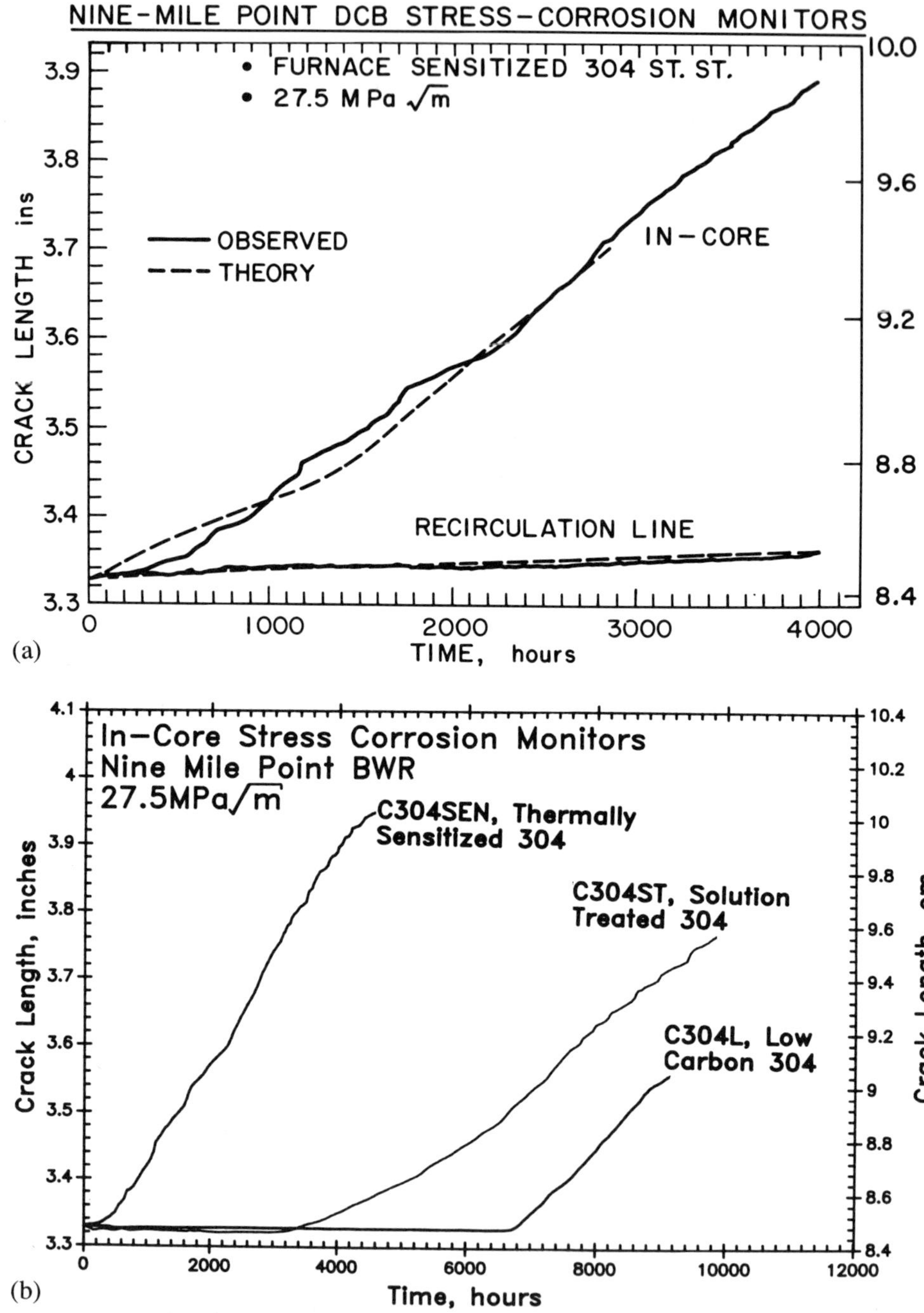

Fig. 16 Data for fracture-mechanics specimens of type 304 stainless steel exposed in the high-flux region of the core and in the recirculation line of Nine Mile Point Unit 1 BWR. All specimens were precracked and wedge loaded to an initial stress-intensity factor of approximately 27.5 MPa$\sqrt{m}$ (25 ksi$\sqrt{in.}$). (a) Comparison of predicted and observed crack length versus time for furnace-sensitized type 304 stainless steel specimens in the core and recirculation line. (b) Crack length versus time for a furnace-sensitized and two annealed specimens of type 304 stainless steel in the core (Ref 25-27)

In-core, *in situ* measurements in BWRs show that the corrosion potential, which is approximately +0.2 to +0.25 V_{she} in normal water chemistry, can be decreased by greater than 0.5 V by

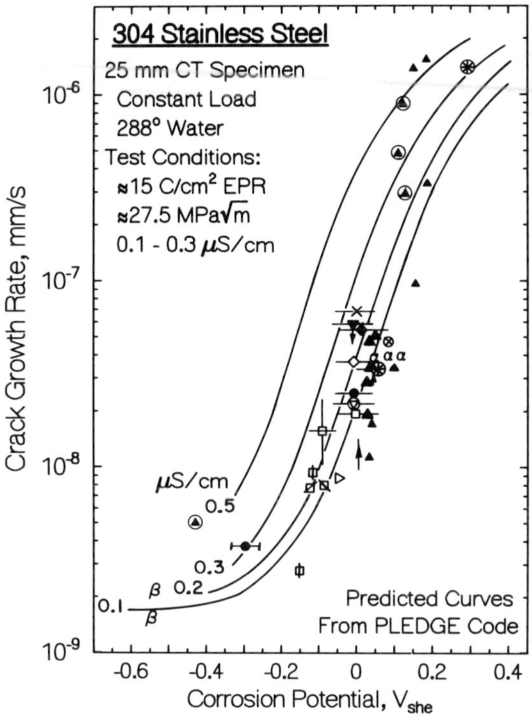

Fig. 17 Observed and predicted relationships of crack-growth rate versus corrosion potential for furnace-sensitized type 304 stainless steel at a constant K of approximately 27.5 MPa$\sqrt{m}$ (25 ksi$\sqrt{in.}$). The observed data were obtained in water of conductivity between 0.1 and 0.3 μS/cm. The predicted relationships show the sensitivity of the crack-growth rate to changes in combinations of corrosion potential and water purity (0.1 to 0.5 μS/cm) (Ref 32-35). The circled points above 0.1 V (SHE) (and at – 0.42 V, SHE were obtained in-core in a BWR or under proton irradiation simulating BWR core exposure.

sufficient additions of dissolved hydrogen in a BWR (Ref 74). This is corroborated by other measurements (Ref 29) (Fig. 15b) that show essentially no radiation-induced elevation in corrosion potential when the fully deaerated inlet water contains excess dissolved hydrogen (200 ppb H_2, 0 ppb O_2).

The effect of radiation on the corrosion potential within a crack or crevice has also been of interest, with the possibility that a net oxidizing environment in the crack could be created which could elevate the corrosion potential above the potential at the crack mouth. In the absence of radiation, measurements in high-temperature water in artificial crevices (*e.g.*, tubing) (Ref 75, 76), at the tip of growing cracks (Ref 77), and of short-crack-growth behavior (Ref 78) show that the corrosion potential remains low (*i.e.*, – 0.5 ± 0.1 V_{she} in 288 °C, or 550 °F, pure water) for all bulk oxygen concentrations, indicating that complete oxygen consumption occurs within the crack. Recent measurements of radiation effects in crevices (Ref 29) show that the elevation in corrosion potential is limited to less than 0.05 V ($\phi_c < -0.45$ V_{she}) in-core; this is consistent with the interpretation of available corrosion-potential data on free surfaces (Ref 1). These small changes will not significantly affect the ~ 0.75 V (+ 0.25 V^{mouth}_{she} – 0.5 V^{crack}_{she}) potential difference in the crack under irradiated normal BWR water chemistry conditions. The potential difference, along with other factors, controls the enhancement mechanism that can lead to an increased anion activity and altered pH at the crack tip (Ref 32-35).

Effects of Radiation Water Chemistry Corrosion Potential on IASCC

Laboratory tests have been conducted (Ref 8, 58, 62) on preirradiated material using slow-strain-rate testing in hot water with additions of oxygen and/or hydrogen peroxide to elevate the corrosion potential to simulate the effect of radiation. Tests by Jacobs *et al.* (Ref 8) on stainless steel irradiated to approximately 3×10^{21} n/cm^2 showed a strong effect of dissolved oxygen (and, by inference, corrosion potential) on IASCC (Fig. 2b). Similarly, Ljungberg (Ref 58, 62) tested a variety of preirradiated materials by slow strain rate (Fig. 14) and observed decreasing average crack-growth rates with decreasing corrosion potential.

In-situ data on fracture mechanics specimens of furnace-sensitized type 304 stainless steel exposed in Nine Mile Point Unit 1 BWR showed that the higher corrosion potentials measured in-core versus in the recirculation piping induced significantly higher measured crack-growth rates (Fig. 16a). These data are compared with other irradiated and unirradiated data in Fig. 17 based on simultaneous measurements of corrosion potential and crack-growth rate in fracture-mechanics specimens; the accompanying curves represent model predictions (Ref 1, 25-28, 32-35), which will be discussed later. Clearly, the *in situ* data compare favorably with the spectrum of unirradiated data and data obtained on a fracture-mechanics specimen of furnace-sensitized type 304 stainless steel using high-energy proton irradiation to simulate the mix of neutron and gamma radiation present in power reactors.

Predictive Modeling of IASCC

IASCC is increasingly viewed in terms of a continuum in environmental cracking behavior in high-temperature water (Ref 1, 25-28, 32-35), which provides a foundation for understanding and modeling IASCC. Its similarity to unirradiated cracking includes common dependencies on exposure variables—for example, water purity (conductivity), corrosion potential, stress, strain rate, and so on—as well as a preferential intergranular crack path defined by a known compositional gradient at grain boundaries. On this basis, modeling of unirradiated environmental cracking has been extended to IASCC, with encouraging agreement between observation and quantitative predictions of cracking rates and component lifetimes.

Fundamental modeling of crack-growth rates in unirradiated stainless steels, Inconels, and low-alloy steels exposed to hot water was developed by Ford and Andresen (Ref 1, 32-35) and was shown to yield good predictions of both crack-growth rates as a function of individual variables such as corrosion potential (Fig. 17) and conduc-

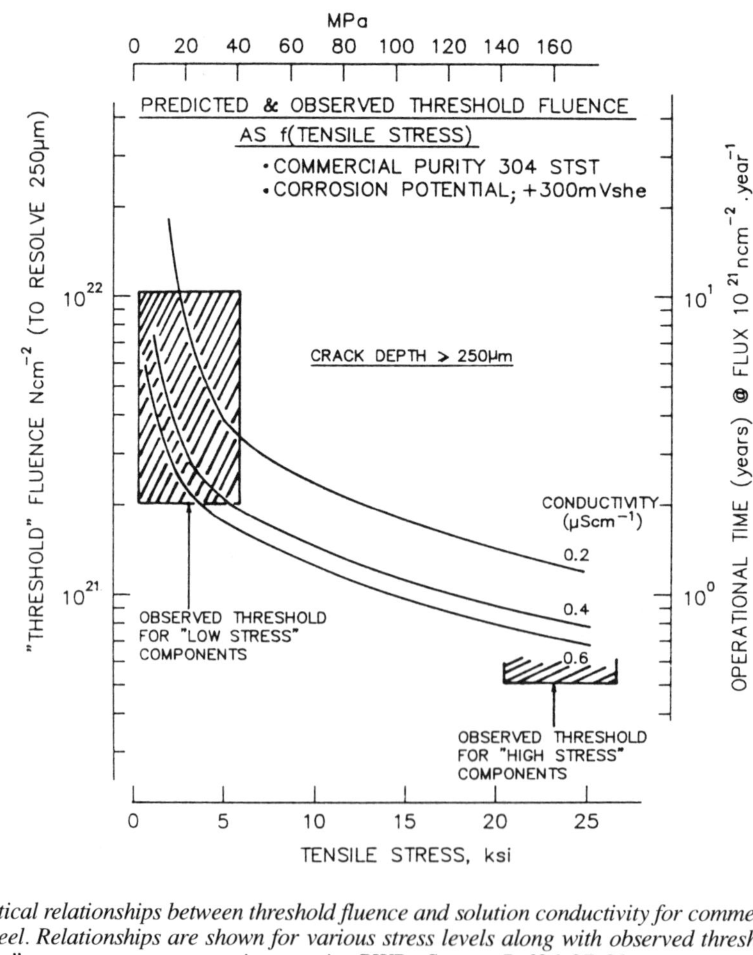

Fig. 18 Theoretical relationships between threshold fluence and solution conductivity for commercial-purity type 304 stainless steel. Relationships are shown for various stress levels along with observed threshold fluences for "high" and "low" stress core components in operating BWRs. Source: Ref 26, 27, 81

tivity (Fig. 5) and of component lifetimes (Fig. 4c). Effects on IASCC of many radiation phenomena can be theorized; of those, several critical factors were identified (Fig. 1), and their effects on cracking quantified (Ref 25-28):

- The effect of radiation on water chemistry and corrosion potential was initially derived from simultaneous measurements of fracture-mechanics crack growth rates (Ref 79) and corrosion potentials (Ref 25) obtained at in-core and recirculation piping locations at Nine Mile Point Unit 1 BWR (Fig. 17). The effect of radiation on corrosion potential was isolated by comparing the response of two identically furnace-sensitized type 304 stainless steel specimens, a condition that is not consequentially affected by irradiation, particularly at low fluence. The observed corrosion potential and crack-growth rate, and similar data obtained under proton irradiation conditions, were found to be consistent with prior modeling predictions on sensitized stainless steels (Fig. 16a and 17).
- The effects of RIS were quantified as a function of fluence by analyzing the changes in crack-growth rates versus time of the two in-core, initially solution-annealed specimens of types 304 and 304L stainless steel (Fig. 16b). Apart from differences in the time to obtain initial intergranular crack growth from the transgranular fatigue precrack, the observed growth rates (*e.g.*, at 7000 h) were identical, consistent with their similar compositions and flux histories (Ref 25, 26, 79). Thus, the integrated effects of RIS for all elements were modeled using algorithms that defined a relationship between fluence and changes in IASCC growth rates induced by radiation.

Thus, predictions of IASCC accurately accounted for, without fitting, the effects of radiation on corrosion potential in furnace-sensitized type 304 stainless steel (Fig. 16 and 17). Using the algorithms for evolution of RIS versus fluence as well as other algorithms to account for radiation-induced creep relaxation and hardening (Ref 80), predictions were also made of crack length versus operating time for reactor internal components exposed to various neutron fluxes,

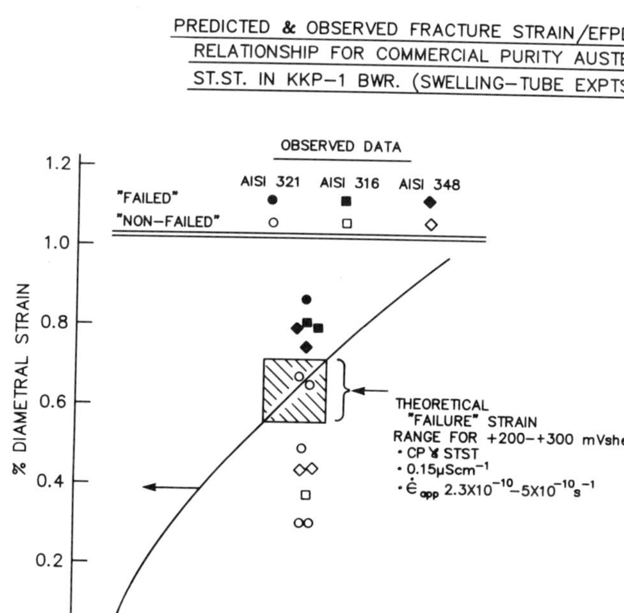

Fig. 19 Theoretical strain/effective full power day relationship for stainless steel tubes together with the theoretical boundary between "failure" and "nonfailure." Also shown are observations for commercial-purity types 321, 316, and 348 stainless steels. Source: Ref 26, 27, 81

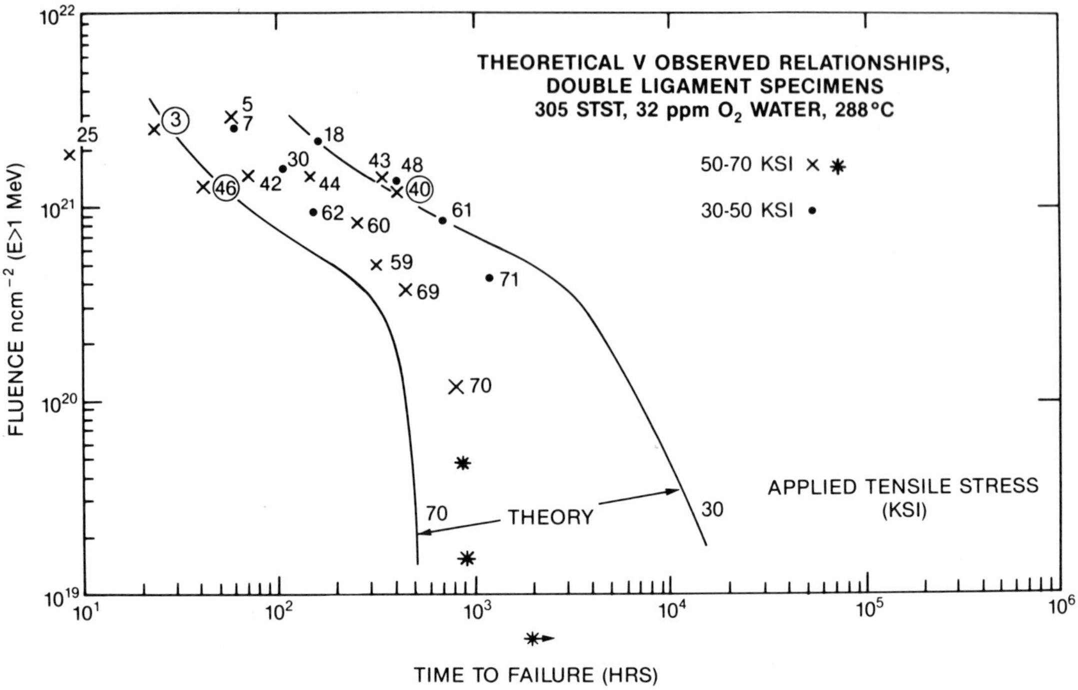

Fig. 20 Comparison between observed (Ref 63) and predicted (Ref 81) time to failure for the effect of fast neutron fluence on preirradiated type 304 stainless steel tested at constant load in the laboratory in oxygen-saturated 288 °C (550 °F) water

stressing conditions, and conductivities in 288 °C (550 °F) water (Ref 1, 25-28, 81). These predictions were in good agreement with the observed "threshold" fluence at which cracking occurs in low- and high-stress components as a function of solution conductivity (Fig. 18). Using the assumptions described in Fig. 19, the modeling accurately predicts the strain at which cracking should be observed in the KWU swelling-tube tests (Fig. 3).

Constant-load (Fig. 20) (Ref 63) and slow-strain-rate data (Ref 8, 61) of Jacobs on irradiated type 304 stainless steel are also predicted as a function of fluence and stress or strain rate. Similarly, data of Ljungberg (Ref 58, 62) (Fig. 14) are predicted reasonably well. Scatter in many instances can be related to (1) the grouping of ranges in fluence, (2) test-to-test differences, *e.g.*, in conductivity, (3) differences between the corrosion potential measured on an electrode in a separate autoclave and the potential on the actual slow-strain-rate specimen, and so on. Other predictions of cracking in shroud-head bolts (Fig. 4c) and absorber tubing (Ref 80, 81) have also been made.

Further mechanistic and quantitative refinements will be necessary to accurately understand and model the individual and collective effects of irradiation. However, the ability to model IASCC as a consistent part of the environmental cracking continuum provides importance guidance to the scientific and engineering tasks required to understand, predict, and control IASCC, both for LWR plant management and life-extension decisions, as well as in other applications.

Conclusions

IASCC is a complex phenomenon that nonetheless falls within the continuum of environmental cracking in LWRs. Neutron (and other high-energy-particle) irradiation causes a variety of changes in material properties. IASCC in LWR environments increases above a threshold fluence of approximately 3 to 5 × 10^{20} n/cm^2, which corresponds to the fluence at which significant changes occur in grain-boundary composition from RIS, mechanical properties, and microstructure.

Radiation-induced segregation of many species to and from grain boundaries has been confirmed for fast neutron irradiation under LWR conditions (as well as for high-energy proton, electron, and heavy-metal ion irradiation at temperatures from 288 to 770 °C, or 550 to 1420 °F). The fundamental basis for RIS is well understood, although current models accurately predict RIS under limited conditions of temperature and flux and for specific elements and bulk compositions. RIS affects corrosion and stress-corrosion susceptibility, although it is not completely clear which elements are damaging or whether synergisms among segregated species exist.

Effects of radiation creep depend strongly on the mode of loading, since it leads to beneficial load relaxation under constant-displacement conditions but to elevated creep rates and enhanced cracking under constant-load conditions. Effects of other radiation-induced changes in microstructure and mechanical properties (hardness, yield and ultimate tensile strength, ductility, fracture toughess, dislocation channeling, etc.) are less well understood, but are difficult to interpret as beneficial to IASCC.

Ionizing radiation causes radiolysis of water, which generates a variety of oxidizing and reducing molecules, radicals, and ions. High-energy particles such as fast neutrons are particularly efficient at causing radiolysis because their linear energy transfer rate to water is high (40 eV/nm versus 0.01 eV/nm for gamma radiation). The effect of radiolysis is best described in terms of its effect on corrosion potential, which can be significant on free metal surfaces, but appears to be quite limited in cracks and crevices.

Extensions to the unirradiated prediction algorithms were developed to account for the primary effects of radiation, including radiolysis, radiation-induced segregation, radiation creep relaxation, and radiation-induced increases in yield strength. The resulting algorithms successfully predict a variety of laboratory and field data on irradiated materials as a function of wide ranges in fluence, test technique, loading condition, water chemistry, etc.

References

1. P.L. Andresen, F.P. Ford, S.M. Murphy, and J.M. Perks, State of Knowledge of Radiation Effects on Environmental Cracking in Light Water Reactor Core Materials, *Proc. 4th Int. Conf. on Environmental Degradation of Materials in Nuclear Power Systems—Water Reactors,* National Association of Corrosion Engineers, 1990
2. International Cooperative Group on Irradiation Assisted Stress Corrosion Cracking (IGC-IASCC), L.J. Nelson, Vice-Chairman, Electric Power Research Institute
3. A.J. Jacobs, and G.P. Wazadlo, Irradiation Assisted Stress Corrosion Cracking as a Factor in Nuclear Power Plant Aging, in *Proc. Int. Conf. Nuclear Power Plant Aging, Availability Factor and Reliability Analysis,* American Society for Metals, 1985, p 173; see also A.J. Jacobs, "Letter Report and Literature Search," GE Nuclear Energy, May 1979
4. F. Garzarolli, H. Rubel, and E. Steinberg, Behavior of Water Reactor Core Materials with Respect to Corrosion Attack, in *Proc. Environmental Degradation of Materials in Nuclear Power Systems—Water Reactors,* National Association of Corrosion Engineers, 1984, p 1-24
5. H. Hanninen and I. Aho-Mantilla, Environment-Sensitive Cracking of Reactor Internals, in *Proc. Environmental Degradation of Materials in Nuclear Power Systems—Water Reactors,* American Institute of Mining, Metallurgical, and Petroleum Engineers, 1987, p 77-92
6. G.M. Gordon and K.S. Brown, Dependence of Creviced BWR Component IGSCC Behavior on Coolant Chemistry, in *Proc. 4th Int. Conf. Environmental Degradation of Materials in Nuclear Power Systems—Water Reactors,* National Association of Corrosion Engineers, 1990, p 14-46 to 14-62
7. K.S. Brown and G.M. Gordon, Effects of BWR Coolant Chemistry on the Propensity for IGSCC Initiation and Growth in Creviced Reactor Internals Components, in *Proc. Environmental Degradation of Materials in Nuclear Power Systems—Water Reactors,* American Institute of Mining, Metallurgical, and Petroleum Engineers, 1987, p 243-248
8. A.J. Jacobs, D.A. Hale, and M. Siegler, unpublished data, GE Nuclear Energy, San Jose, CA, Jan 1986
9. R.L. Cowan and G.M. Gordon, Intergranular Stress Corrosion Cracking and Grain Boundary Composition of Fe-Ni-Cr Alloys, in *Stress Corrosion Cracking and Hydrogen Embrittlement of Iron-Base Alloys,* R.W. Staehle, J. Hochmann, R.D.

McCright, and J.E. Slatern, Ed., National Association of Corrosion Engineers, 1977, p 1063-1065
10. C.F. Cheng, *Corrosion,* Vol 20, 1964, p 462
11. R.N. Duncan, W.H. Arlt, H.E. Williamson, G.S. Baroch, J.P. Hoffman, and T.J. Pashos, *Nucl. Appl.,* Vol 1, 1965, p 413
12. J.S. Armijo, J.R. Low, and U.E. Wolff, *Nucl. Appl.,* Vol 1, 1965, p 462
13. T.J. Pashos *et al.,* Failure Experience with Stainless Steel Clad Fuel Rods in VBWR, *Trans. Am. Nucl. Soc.,* Vol 7, (No. 2), 1964, p 416
14. C.F. Cheng, *Corrosion,* Vol 20, 1964, p 341
15. C.F. Cheng, *J. Nucl. Mater.,* Vol 56, 1975, p 11
16. C.F. Cheng, *Reactor Technol.,* Vol 13, 1970, p 310
17. L.D. Schaffer, Army Reactor Program Progress Report ORNL 3231, Jan 1962
18. J.B. Brown, Jr., B.W. Storhok, and J.E. Gates, Post-Irradiation Examination of the PM3A Type I Serial 2 Core, *Trans. ANS,* Vol 10, 1967, p 668-669
19. V. Pasupathi and R. W. Klingensmith, "Investigation of Stress Corrosion in Clad Fuel Elements and Fuel Performance in the Connecticut Yankee Reactor," NP-2119, Electric Power Research Institute, Nov 1981
20. I. Multer, European Operating Experience, in *Proc. Joint Topical Meeting on Commercial Nuclear Fuel Technology Today,* Toronto, 75-CNA/ANS100, April 1975
21. J.T. Storre and D.H. Locke, High Burnup Irradiation Experience in Vulcain, *Nucl. Eng. Int.,* Feb 1970, p 93-99
22. F. Garzarolli, D. Alter, P. Dewes, and, J.L. Nelson, Deformability of Austenitic Stainless Steels and Nickel-Base Alloys in the Core of a Boiling and a Pressurized Water Reactor, in *Proc. 3rd Int. Symp. Enviromental Degradation of Materials in Nuclear Power Systems—Water Reactors,* TMS-AIME, 1988, p 657-664
23. F. Garzarolli, D. Alter, and P. Dewes, Deformability of Austenitic Stainless Steels and Nickel-Base Alloys in the Core of a Boiling and a Pressurized Water Reactor, in *Proc. Fourth Int. Symp. Environmental Degradation of Materials in Nuclear Power Systems — Water Reactors,* American Nuclear Society, 1986, p 131-138
24. P.M. Scott, W. Foot, L.A. Goldsmith, S. Dumbill, T.M. Williams, J.M. Perks, C.A. English, and W.G. Burns, Irradiation Assisted Stress Corrosion of Fuel Element Components of a Steam Generating Heavy Water Reactor, *NEA/UNIPEDE Specialists Meeting on Life Limiting and Regulatory Aspects of Core Internals and Pressure Vessels,* Stockholm, Oct 14-16, 1987
25. R.A. Head, M.E. Indig, and P.L. Andresen, "Measurement of In-Core and Recirculation System Responses to Hydrogen Water Chemistry at Nine Mile Point Unit 1 BWR," Contract RP2680-5, Final Report, 1989
26. P.L. Andresen and F.P. Ford, "Modeling of Irradiation Effects on Stress Corrosion Crack Growth Rates, Paper No. 497," presented at Corrosion/89, National Association of Corrosion Engineers, 1989
27. P.L. Andresen and F.P. Ford, "Modeling of Irradiation Effects on Stress Corrosion Crack Growth in Laboratory and Field Data," Minutes, ICG-IASCC, Boston, April 1989
28. P.L. Andresen, "Effects of Irradiation on Corrosion Potential and Effects of Potential on Cracking Kinetics, Minutes," ICG-IASCC, Boston, April 1989
29. W.D. Miller and P.L. Andresen, unpublished work, GE Nuclear Energy, 1989
30. D.I.R. Norris, C. Baker, and J.M. Titchmarsh, Compositional Profiles at Grain Boundaries in 20%Cr/25%Ni/Nb Stainless Steel, in *Proc. Radiation Induced Sensitization of Stainless Steels,* Berkeley Nuclear Labs, 1986, p 86-98
31. M.P. Manahan, R. Kohli, J. Santucci, and P. Sipush, A Phenomenological Investigation of In-Reactor Cracking of Type 304 Stainless Steel Control Rod Cladding, *Nucl. Eng. Des.,* 1989, p 113
32. F.P. Ford, P.L. Andresen, H.D. Solomon, G.M. Gordon, S. Ranganath, D. Weinstein, and R. Pathania, Application of Water Chemistry Control, On-Line Monitoring, and Crack Growth Models for Improved BWR Materials Performance, in *Proc. 4th Int. Symp. Environmental Degradation of Materials in Nuclear Power Systems—Water Reactors,* National Association of Corrosion Engineers, 1990, p 4-26 to 4-51
33. F.P. Ford, D.F. Taylor, P.L. Andresen, and R.G. Ballinger, "Environmentally Controlled Cracking of Stainless and Low Alloy Steels in LWR Environments," NP-5064M (RP2006-6), Electric Power Research Institute, 1987
34 F.P. Ford and P.L. Andresen, Development and Use of a Predictive Model of Crack Propagation in 304/316L, A533B/A508 and Inconel 600/182 Alloys in 288 °C Water, in *Proc. Environmental Degradation of Materials in Nuclear Power Systems—Water Reactors,* TMS-AIME, 1988, p 789-800
35. P.L. Andresen and F.P. Ford, Life Prediction by Mechanistic Modeling and System Monitoring of Environmental Cracking of Iron and Nickel Alloys in Aqueous Systems, *Mater. Sci. Eng.,* Vol A103, 1988, p 167-184
36. *Effects of Radiation on Materials, 14th Int. Symp.,* Andover, MA, June 1988, American Society for Testing and Materials
37. *Radiation-Induced Changes in Microstructure, 13th Int. Symp., Parts I and II,* F.A. Garner, N.H. Packan, and A.S. Kumar, Ed., STP 955 and 956, American Society for Testing and Materials, 1987; see also STP 872, 870, 725, 683, 611, and 570)

38. P. Maziasz, "Effects of Helium on Microstructural Development in Type 316 Stainless Steel Under Neutron Irradiation," ORNL-6121, 1985
39. G. Odette, P. Maziasz, and J. Spitznagel, *J. Nucl. Mater.*, Vol 85/86, 1979, p 1289
40. A.J. Jacobs, G.P. Wozadlo, K. Nakata, Y. Yoshida, and I. Masaoka, Radiation Effects on the Stress Corrosion and Other Selected Properties of Types 304 and 316 Stainless Steel, in *Proc. Environmental Degradation of Materials in Nuclear Power Systems—Water Reactors,* American Institute of Mining, Metallurgical, and Petroleum Engineers, 1987, p 673-681
41. T.R. Anthony, Solute Segregation and Stresses around Growing Voids in Metals, in *Proc. Radiation-Induced Voids in Metals and Alloys,* J.W. Corbett and L.C. Ianiello, Ed., U.S. Atomic Energy Commission Symp. Series 26, CONF 710601, 1971, p 630
42. P.R. Okamoto, S.D. Harkness, and J.J. Laidler, Solute Segregation to Voids During Electron Irradiation, *ANS Trans.*, Vol 16, 1973, p 70
43. P.R. Okamoto and L.E. Rehn, Radiation-Induced Segregation in Binary and Ternary Alloys, *J. Nucl. Mater.*, Vol 83, 1979, p 2-23
44. D.I.R. Norris, C. Baker, and J.M. Titchmarsh, A Study of Radiation-Induced Sensitisation in 20Cr/25Ni/Nb Steel by Compositional Profile Measurements at Grain Boundaries, in *Proc. Materials for Nuclear Reactor Core Applications,* Vol 1, British Nuclear Engineering Society, 1987, p 277
45. A.J. Jacobs, R.E. Clausing, M.K. Miller, and C. Shepherd, Influence of Grain Boundary Composition on the IASCC Susceptibility of Type 348 Stainless Steel, in *Proc. 4th Int. Conf. Environmental Degradation of Materials in Nuclear Power Systems—Water Reactors,* National Association of Corrosion Engineers, 1990, p 14-21 to 14-45
46. C.M. Shepherd and T.M. Williams, Simulation of Microstructural Aspects of IASCC in Water Reactor Core Components, in *Proc. 4th Int. Conf. on Environmental Degradation of Materials in Nuclear Power Systems—Water Reactors,* National Association of Corrosion Engineers, 1990, p 14-11 to 14-20
47. J.R. Manning, *Phys. Rev. B,* Vol 4, (No. 4), 1971, p 1111
48. A.D. Marwick, *J. Phys. F: Met. Phys.,* Vol 8, 1978, p 1849
49. H. Wiedersich, P.R. Okamoto, and N.Q. Lam, *J. Nucl. Mater.*, Vol 83, 1979, p 98
50. J.M. Perks and S.M. Murphy, Modelling the Major Element Radiation-Induced Segregation in Concentrated Fe-Cr-Ni Alloys, in *Proc. Materials for Nuclear Reactor Core Applications,* Vol 1, British Nuclear Engineering Society, 1987, p 165
51. S. Dumbill and T.M. Williams, Irradiation-Induced Grain Boundary Segregation in Neutron-Irradiated Fe-Ni-Cr Alloys, in *Proc. Materials for Nuclear Reactor Core Applications,* Vol 1, British Nuclear Engineering Society, 1987, p 119-123
52. E.P. Simonen, L.A. Charlot, and S.M. Bruemmer, Modeling Irradiation-Induced Grain Boundary Segregation in Stainless Steels, Paper No. 39, presented at Corrosion/91, National Association of Corrosion Engineers, 1991
53. R.A. Johnson and N.Q. Lam, *Phys. Rev. B,* Vol 13, 1976, p 4364
54. S.M. Murphy, "A Model for Segregation in Dilute Alloys During Irradiation," Harwell Research Report, AERE-R.13405, Feb 1989
55. S.M. Murphy and J.M. Perks, "Analysis of Phosphorus Segregation in Ion-Irradiated Nickel," Harwell Research Report, AERE-R.13495, July 1989
56. A.J. Jacobs, personal communication, GE Nuclear Energy, March 1991
57. D. Weinstein, P.L. Andresen, R.G. Ballinger, M. Hayashi, A. Sudo, and H.D. Solomon, "Irradiation-Assisted Stress Corrosion Crack Growth Rate and ECP Measurements Under Simulated-BWR, Neutron-Irradiated Conditions," Contract RPC101-3, Electric Power Research Institute
58. L.G. Ljungberg, personal communication, ABB Atom, Sweden, March 1991
59. F. Garzarolli, D. Alter, and P. Dewes, "Influence of Irradiation and Stress on the Behavior of Structural Materials," Contract 2181-3, Electric Power Research Institute
60. T. Karlsen and C. Vitanza, "Research Program on In-Situ Irradiation-Assisted Stress Corrosion Crack Growth Rate and ECP Measurements," OECD Halden Reactor Project, Halden, Norway
61. A.J. Jacobs, R.E. Clausing, L. Heatherly, and R.M. Kruger, Irradiation Assisted Stress Corrosion Cracking and Grain Boundary Segregation in Heat Treated Type 304 Stainless Steel, in *Proc. 14th Int. Symp. Effects of Radiation on Materials,* American Society for Testing and Materials, Andover, MA, June 1988
62. L.G. Ljungberg, unpublished data, ABB Atom, Sweden, 1989; data presented at the ICG-IASCC Meeting, Boston, April 1989
63. A.J. Jacobs and G.P. Wozadlo, "Stress Corrosion Testing of Irradiated Type 304 Stainless Steel Under Constant Load," Paper No. 41, presented at Corrosion/91, National Association of Corrosion Engineers, 1991.
64. T. Inazumi, G.E.C. Bell, and A. Hishinuma, "Evaluation of Radiation-Induced Sensitization Using EPR for Austenitic Stainless Steels," Paper No. 507, presented at Corrosion/90, National Association of Corrosion Engineers, 1990

65. R. Katsura, M. Kodama, and S. Nishimura, "A DL-EPR Study on the Influence of Neutron Irradiation in Type 304 Stainless Steel," Paper No. 43, presented at Corrosion/91, National Association of Corrosion Engineers, 1991
66. A.J. Jacobs, "Heat Treatment as a Potential Preventive Measure Against IASCC," Paper No. 505, presented at Corrosion/90, National Association of Corrosion Engineers, 1990
67. P.L. Andresen, "Specific Ion Chemistry and Chromium Effects on Crack Growth Rates in 288 °C Water," Paper No. 490, presented at Corrosion/90, National Association of Corrosion Engineers, 1990
68. S.M. Bruemmer, B.W. Arey, and L.A. Charlot, "Influence of Chromium Depletion on IGSCC of Type 304 Stainless Steel," Paper No. 38, presented at Corrosion/91, National Association of Corrosion Engineers, 1991
69. C.C. Lin, An Overview of Radiation Chemistry in Reactor Coolants, in *Proc. Environmental Degradation of Materials in Nuclear Power Systems-Water Reactors,* American Nuclear Society, 1986, p 160-172
70. W.G. Burns and P.B. Moore, *Radiat. Eff.,* Vol 30, 1976, p 233
71. P. Cohen, *Water Coolant Technology of Power Reactors,* Gordon and Breach, 1969
72. *Proc. Water Chemistry of Nuclear Reactor Systems, #5,* British Nuclear Energy Society, 1989
73. D.F. Taylor, "Response of Electrochemical Sensors to Ionizing Radiation in High Temperature Aqueous Environments," Paper No. 90501, presented at Corrosion/90, National Association of Corrosion Engineers, 1990
74. B.M. Gordon, "Hydrogen Water Chemistry for BWR," Task 27, "Materials and Environmental Monitoring Within the Duane Arnold BWR," Contract RP1930-1, Electrical Power Research Institute
75. D.F. Taylor and C.A. Caramihas, High Temperature Aqueous Crevice Corrosion in Alloy 600 and 304L Stainless Steel, in *Localized Crack Chemistry and Mechanics of Environmentally Assisted Cracking,* American Institute of Mining, Metallurgical, and Petroleum Engineers, 1983
76. P. Combrade, M. Foucault, and G. Slama, About the Crack Tip Environment Chemistry in Pressure Vessel Steel Exposed to Primary PWR Coolant, in *Proc. Atomic Energy Agency Specialists Meeting on Subcritical Crack Growth,* NUREG/CP0067, Vol 2, 1985, p 201-218
77. G. Gabetta and G. Buzzanaca, Measurement of Corrosion Potential Inside and Outside Growing Crack During Environmental Fatigue Tests, in *Proc. Atomic Energy Agency Specialists Meeting on Subcritical Crack Growth,* NUREG/CP0067, Vol 2, 1985, p 219-230
78. P.L. Andresen, I.P. Vasatis, and F.P. Ford, "Behavior of Short Cracks in Stainless Steels and Inconels in 288 °C Water," Paper No. 495, presented at Corrosion/90, National Association of Corrosion Engineers, 1990: submitted for publication in *Corrosion*
79. "In-Reactor Stress Corrosion Monitor Prototype Development, Installation and Operation," Research Report EP85-37, Final Report by GE Nuclear Energy to Empire State Electric Energy Research Corp., Feb 1988
80. P.L. Andresen and F.P. Ford, unpublished work, GE Corporate Research and Development Center, 1990
81. P.L. Andresen and F.P. Ford, Application of Predictive Models for IASCC in Austenitic Alloys to Laboratory and Core Component Field Data, to be published in Proc. NACE Conf. on "Life Prediction of Corrodible Structures," Cambridge, U.K., Sept 23-26, 1991
82. J.P. Foster, Analysis of In-Reactor Stress Relaxation Using Irradation Creep Models, in *Irradiation Effects on the Microstructure and Properties of Metals,* STP 611, American Society for Testing and Materials, 1976, p 32-50
83. K.Asano, K. Fukuya, K. Nakata and M. Kodama, "Changes in Grain Boundary Composition Induced By Neutron Irradiation on Austenitic Stainless Steels", *Proc. 5th Int. Symp. on Environmental Degradation of Materials in Nuclear Power Systems-Water Reactors,* ANS, 1992, p 838-846
84. M. Kodama, S. Nishimura, J. Morisawa, S. Suzuki, S. Shima and M.Yamamoto, "Effects of Fluence and Dissolved Oxygen on IASCC in Austenitic Stainless Steels," *Proc. 5th Int. Symp. on Environmental Degradation of Materials in Nuclear Power Systems-Water Reactors,* ANS, 1992, p 948-954
85. H.M. Chung, W.E. Ruther, J.E. Sanecki and T.F. Kassner, "Irradiation Induced Sensitization and Stress Corrosion Cracking of Type 304 Stainless Steel Core Internal Components," *Proc. 5th Int. Symp. on Environmental Degradation of Materials in Nuclear Power Systems-Water Reactors,* ANS, 1992, p 795-805
86. J.R. Cookson, R.D. Carter, D.L. Damcott, M. Atzmon, G.S. Was and P.L. Andresen, "Grain Boundary Segregation and Stress Corrosion Cracking of High Energy Proton Irradiation 304L Stainless Steel," *Proc. 5th Int. Symp. on Environmental Degradation of Materials in Nuclear Power Systems-Water Reactor,* ANS, 1992, p 806-813
87. K. Fukuya, K. Nakata and A. Horie, "IASCC Study Using High Energy Ion Irradiation," *Proc. 5th Int. Symp. on Environmental Degradation of Materials in Nuclear Power Systems-Water Reactors,* ANS, 1992, p 814-820

CHAPTER 7

Stress-Corrosion Cracking of Copper Alloys

John A. Beavers, Cortest Columbus Technologies, Inc.

The phenomenon of stress-corrosion cracking (SCC) was first observed in copper-zinc alloys more than 80 years ago. The SCC failures were reported in India, where they occurred in brass cartridge cases during the rainy season (and were thus later characterized by the descriptive term "season cracking"). The root cause of the failures was traced to the practice of storing the cartridge cases in horse barns, where they came in contact with ammonia vapors. A further contributing factor was the cold-forming operation used to make the cases, which introduced high residual tensile stresses. Sperry (Ref 1) provides further discussion of these early SCC failures.

This example of season cracking conveniently identifies the three primary factors associated with SCC failure of copper-base alloys: a specific corrosive medium, a susceptible material, and tensile stresses. These factors are often presented graphically (see Fig. 1) to demonstrate that all three must be present simultaneously for SCC to occur.

This chapter is organized to feature these three primary factors. First, those environments that induce SCC in copper-base alloys are discussed, along with cracking-related environmental factors, such as potential, pH, and temperature. The next section identifies SCC-susceptible copper-base alloys and metallurgical factors that affect SCC, such as microstructure, heat treatment, and cold working. Attention is then focused on relevant mechanical factors (for example, stress and stress intensity). Testing methods are discussed, as are techniques for the mitigation of SCC (which generally involve elimination of one or more of the three primary factors). Finally, the mechanisms of SCC of copper-based alloys are briefly discussed.

Environmental Effects

Effect of Environment Composition

Aqueous ammonia is the principal environmental associated with SCC of copper-base alloys in service applications. Failures of brasses occur most commonly in moist conditions where ammonia, water, and oxygen are present. At one time it was thought that ammonia or its derivatives were the only species responsible for SCC

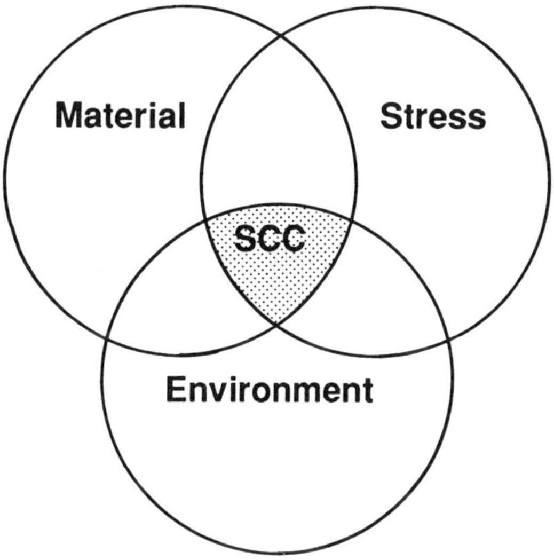

Fig. 1 Schematic representation of the three factors that must be present simultaneously for SCC to occur

Table 1 Environments Reported to Cause SCC of Copper-Base Alloys

Alloy	Environment	Ref
β-brass	Deionized water	2
70-30 brass	$NaClO_3$	3
Aluminum brass	Acidic chloride	4
Aluminum bronze	NaF	5
α-brass	Aliphatic amine	6
	$(NH_4)_2SO_4$	7
Brass	$HgNO_3$	8
	$(NH_4)_2SO_4 + CuSO_4$	9
	Outdoor environment	10
	$NaNO_2$	11
	Citrates and tartrates	12
	Moist SO_2	13
	SO_4^{-2}	14
	queous ammonia	Various
Yellow brass	Na_2SO_4 + borate	15
	$Na_2SO_4 + Na_2S$	15

of copper-base alloys. More recently, other such environments have been identified; Table 1 shows that cracking of various copper-base alloys can be caused by a variety of species, including sulfates, nitrates, nitrites, chloride-containing species, and even pure water.

Most research on SCC of copper-base alloys has been carried out in ammonia or variations of this environment. In the 1960s, Pugh and Westwood (Ref 16) demonstrated that SCC of brass would not occur under freely corroding conditions in aqueous ammonia in the absence of dissolved copper, which forms a complex ion in solution. As the copper content of the ammonia increased, cracking became more severe, and at some critical copper concentration an oxide film, or tarnish, formed (Fig. 2). Later work by Pugh and Green (Ref 17) demonstrated that cracking could be produced in copper-free ammonia solutions by anodically polarizing the brass specimen. Thus, the primary role of the cupric complex ion appears to be as an oxidant in solution that provides a cathodic reaction.

The chemistry/electrochemistry of the copper-ammonia system has been studied extensively and is reasonably well understood. It has been reviewed in detail elsewhere (Ref 18, 19) and is summarized below. On a copper surface, the primary reduction reaction in aerated, copper-free aqueous ammonia is oxygen reduction, and the primary oxidation reaction is oxidation of copper to form cuprous complex ions:

$$Cu + 2NH_3 \rightarrow Cu\,(NH_3)_2^+ + e^-$$

These ions react with oxygen to form cupric complex ions:

$$2\,Cu\,(NH_3)_2^+ + \tfrac{1}{2}\,O_2 + H_2O + 4\,NH_3 \rightarrow 2\,Cu(NH_3)_4^{+2} + 2OH^-$$

In oxygenated solutions, copper in solution exists primarily as the cupric complex ion. The exchange current for cupric ion reduction is much greater than that of oxygen reduction, and thus cupric ion reduction is the dominant reaction in copper-containing solutions. The overall process is autocatalytic in that the corrosion reaction generates more cuprous ions in solution, which subsequently form cupric ions, which in turn accelerate the rate of corrosion. This process does not continue indefinitely, because the onset of tarnishing, which occurs at some critical copper ion content, leads to a decrease in the corrosion rate, as shown in Fig. 2. On the other hand, Fig. 2 also shows that the severity of cracking continues to increase with increasing copper content in the tarnishing range.

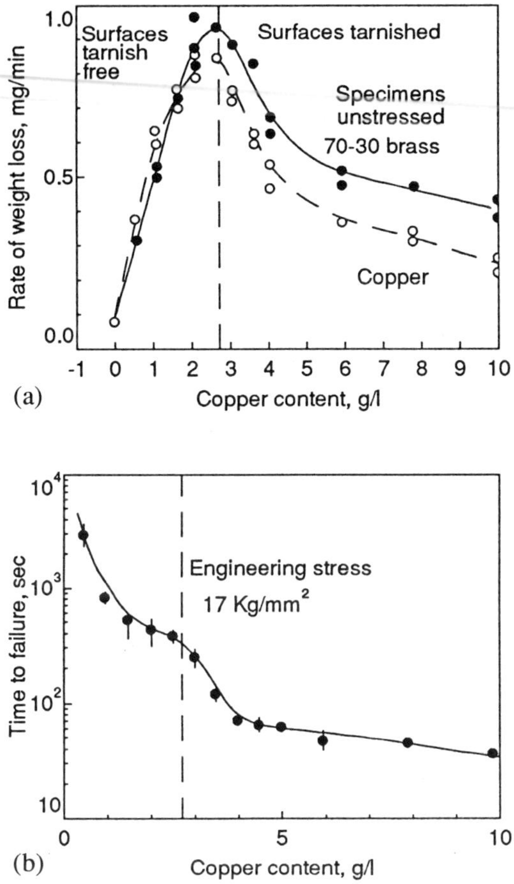

Fig. 2 Relationship between the concentration of dissolved copper in ~15 N aqueous ammonia and: (a) rate of weight loss of unstressed specimens of copper and 70-30 brass; (b) time to failure (t_F) for 70-30 brass. After Ref 16

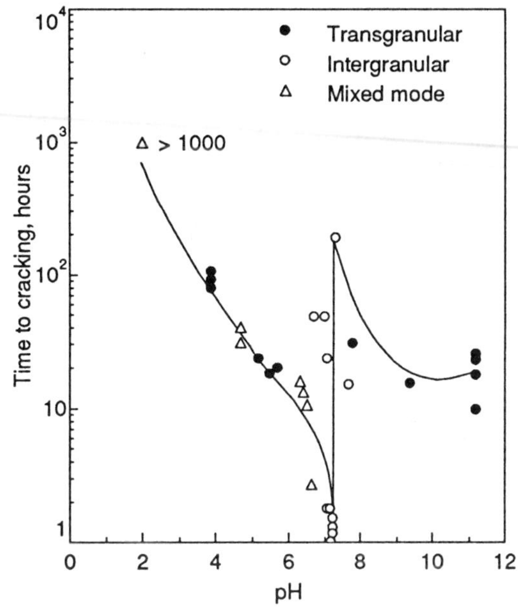

Fig. 3 Time to cracking as a function of the pH for brass in ammoniacal copper sulfate solutions (Mattsson's solutions). Note that the specimen tested at pH of 2 did not fail in 1000 h of testing. After Ref 9

Based on research on the chemistry of ammonia solutions, it can be concluded that failures in the field occur under very aggressive conditions in that the condensed layer of moisture can contain very high copper ion concentrations due to the large ratio of metal surface area to solution volume. Oxygen access to the solution is also facilitated by the thin condensed layer, leading to rapid development of high concentrations of the cupric complex ion.

Effect of Solution pH

The majority of research on pH effects on SCC of copper-base alloys has been performed in Mattsson's solutions, concentrated ammonia environments have very high pH values. Relatively little research has been performed in nonammoniacal environments. In the original milestone work performed by Mattsson (Ref 9), the effect of pH on the SCC and film-forming characteristics of Cu-37Zn was studied in ammoniacal copper sulfate solutions, which became known as Mattsson's solutions. The pH of the solutions was varied by adding the ammonia partly as ammonium hydroxide (NH_4OH) and partly as ammonium sulfate [$(NH_4)_2SO_4$]; sulfuric acid was added to obtain the lowest pH solutions. All solutions had 0.05 *M* Cu and 1 *M* NH_3.

The SCC tests were performed on U-bend specimens machined from cold-rolled (half-hard) brass sheet stock. Specimens were exposed in stoppered flasks under freely corroding conditions. Within the flasks was a small air space above the test solutions. Time to failure as a function of pH is shown in Fig. 3, where it can be seen that a minimum existed at a pH of about 7.2. A majority of the cracks were transgranular in the pH ranges of 3.9 to 5.7 and 7.8 to 11.2, whereas cracking was primarily intergranular in the pH

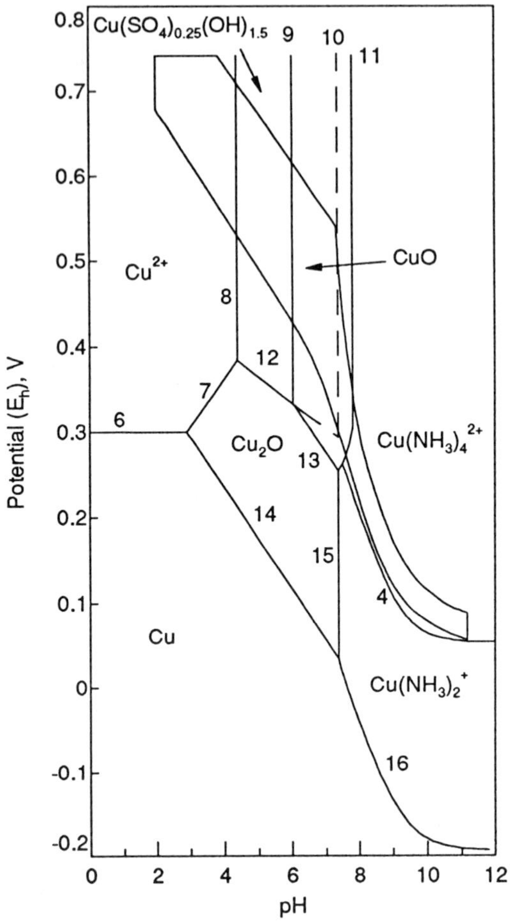

Fig. 4 Potential/pH diagram for a system of copper and a water solution with 1.0 g·mol/L of ammonia partly as ammonium sulfate and 0.05 g·atom/L of dissolved copper added as sulfate at 25 °C (77 °F). Numbers refer to equations from Ref 9. The shaded zone marks solution properties at the beginning of the tests. Source: Ref 9

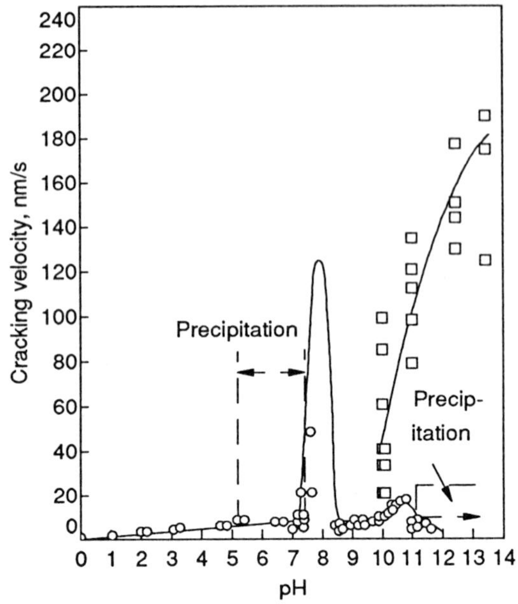

Fig. 5 Effect of pH on average SCC velocity of brass specimens tested in two ammoniacal solutions. ○, 0.88 mol/L of NH_3 + 0.05 mol/L Cu. □, 7.8 mol/L of NH_3 + 0.05 mol/L of Cu. Source: Ref 22

range of 6.3 to 7.7. In the latter pH range, a black surface film was found on the specimens. Mattsson measured the redox potential of the solutions with a platinum electrode prior to testing and related the potential and pH of the solutions on a Pourbaix diagram for the copper-ammonia system (Fig. 4). The shaded zone in Fig. 4 indicates the properties of the test solutions. The tarnish films on the specimens in the intermediate-pH range were found to be composed primarily of Cu_2O by Mattsson (Ref 9) and subsequently by Hoar and Booker (Ref 20) and by Johnson and Leja (Ref 12).

Later studies by Lahiri (Ref 21), Bhakata and Lahiri (Ref 7), and Sparks and Scully (Ref 22) confirmed the pH effects on cracking susceptibility of brasses reported by Mattsson (Ref 9). Takemoto and Hori (Ref 23) also reported a similar dependence of cracking on pH for a Cu-4Ti alloy. In the study by Sparks and Scully (Ref 22), the research was extended to consider annealed and cold worked 70-30 brass and higher-pH concentrated ammonia solutions. Testing was performed on single-edge notched cantilever-beam specimens initially loaded to a K value of 35 MPa $\sqrt{m}$ (32 ksi $\sqrt{in.}$). Results, summarized in Fig. 5, show that the high-pH concentrated solutions also produced extensive cracking, with cracking velocity increasing with increasing pH. One difference in the results of the studies of Mattsson (Ref 9) and Sparks and Scully (Ref 22) was in the fracture mode. In the latter study, cracking was primarily intergranular (with some transgranular) at all but the highest pH values tested, whereas Mattsson reported intergranular cracking only at intermediate pH values.

A limited amount of research on the effects of pH on SCC of copper-base alloys has been performed in nonammoniacal environments. Shih *et al.* (Ref 5) investigated the SCC of aluminum bronze in sodium fluoride solutions with pHs of 8, 10, and 12 and found the most severe cracking at pH 8. Torchio (Ref 24) and Torchio and Mazza (Ref 25) investigated the effect of pH on SCC of aluminum brass in acidic chloride and sulfate solutions, respectively. In the former study, cold worked U-bend specimens were exposed in 1 *M* NaCl + 0.25 *M* sodium citrate solutions with 0.5 to 1.5 g/L CuCl and pHs of 1 to 6. Maximum susceptibility to SCC was observed between a pH of 3 to 4 and for the highest copper content.

In the study by Torchio and Mazza (Ref 25), U-bend specimens were exposed to 0.1 *M* Na_2SO_4 + 0.005 *N* Cl^- with varying $CuSO_4$ concentrations and pHs. Results, summarized in Fig. 6, show that maximum cracking occurred at pH 3 at the highest $CuSO_4$ concentration tested. An interesting aspect of this study is that intergranular SCC, which occurred at pH values below 3, was associated with nontarnishing conditions, while transgranular cracking occurred in tarnishing solutions above pH 3. This behavior is opposite to that observed in Mattsson's and other ammoniacal solutions, where intergranular SCC is normally associated with tarnishing conditions.

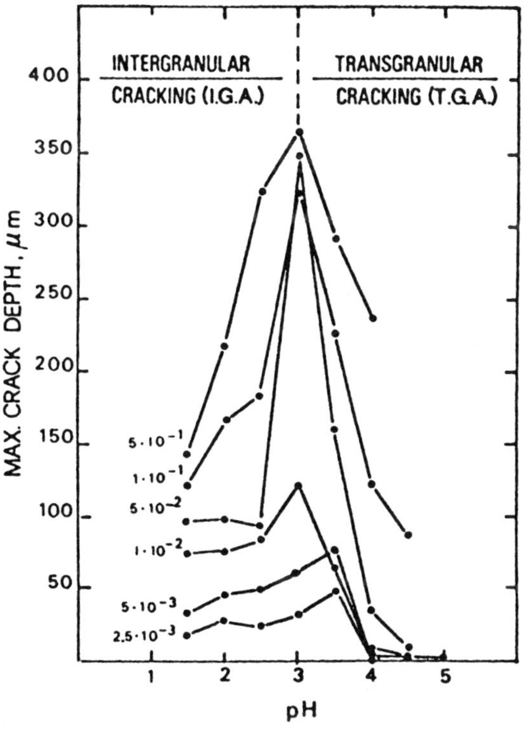

Fig. 6 Maximum crack depth as a function of pH and $CuSO_4$ content measured on U-bend specimens exposed to 0.1 M Na_2SO_4 + 0.005 N Cl^-. Data to the left of the curves refer to the $CuSO_4$ molar concentration. Test time = 240 h. Source: Ref 25

Effect of Potential

Researchers have investigated the effect of potential on the SCC of copper-base alloys in ammoniacal environments as well as in other environments that cause cracking. In the former, studies have been performed both in concentrated aqueous ammonia and in Mattsson's solution. Pugh and Green (Ref 17) studied the effect of applied potential on the time to failure of Cu-30Zn in copper-free 15 *M* aqueous ammonia. The data, summarized in Fig. 7, show that cracking susceptibility increased with increasing anodic polarization, as did the dissolution rate of the specimens. No cracking was observed at the free corrosion potential. In these tests, the specimens were tarnish free and the failure mode was transgranular. Based on this work, it was concluded that the role of the cupric complex ion in solution, which promotes SCC, is to anodically polarize the specimen. Tong (Ref 26) observed similar behavior for Cu-30Zn specimens tested under constant-

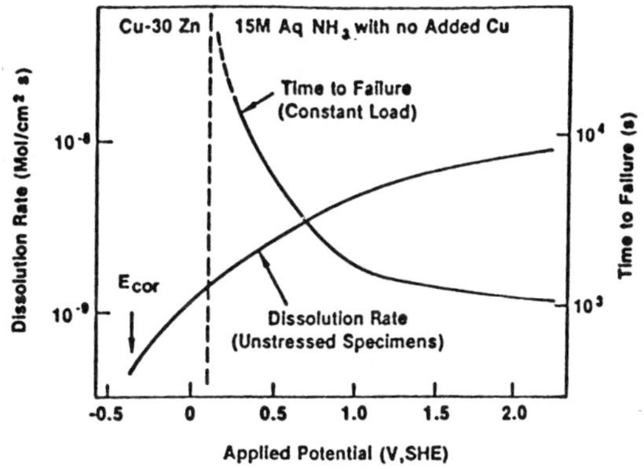

Fig. 7 Effect of applied anodic potentials on the dissolution rate and stress-corrosion life of Cu-30Zn in 15 M aqueous ammonia. Specimens were tarnish-free in all cases and SCC failure was entirely transgranular. Source: Ref 17

load conditions in a nontarnishing 1 *M* ammonia solution containing added copper. In these tests, the data exhibited maximum cracking susceptibility at an anodic overpotential of about 200 mV, with failure times increasing with greater levels of anodic polarization.

Kermani and Scully (Ref 27) studied the effect of potential on the crack velocity of annealed Cu-30Zn in a 15 *N* ammonia solution containing 6 g/L of copper, which is a tarnishing solution. The tests were performed on notched cantilever-beam specimens at three different stress-intensity levels. Results (Fig. 8) are very similar to those reported by Tong (Ref 26) for constant-load specimens, with maximum susceptibility at an anodic overpotential. The rapid decrease in the cracking velocity at more noble potentials was attributed to the onset of nontarnishing conditions. Cracking at all potentials was transgranular.

Several researchers have investigated the effect of applied potential on SCC of copper-base alloys in Mattsson's solution. As previously described, Mattsson's solution is an aqueous 1 *M* $(NH_4)_2SO_4$ solution containing dissolved copper added as $CuSO_4$. The pH of the solution is varied by adding the ammonia partly as NH_3 and partly as $(NH_4)_2SO_4$. Hoar and Booker (Ref 20) investigated the effect of potential on time to failure of α-brass wires in Mattsson's solution of pH 7.2, which is a tarnishing solution at the free corrosion potential [+0.15 to +0.25 V (SHE)]. They found that crack propagation was accelerated by some anodic polarization (up to 50 mV) and retarded by greater anodic polarization or by cathodic polarization.

Sircar *et al.* (Ref 28) performed constant-load tests on annealed Cu-30Zn specimens in Mattsson's solutions of varying copper concentration as a function of applied potential. Results (Fig. 9) show behavior similar to that observed by Hoar and Booker (Ref 20) and that in the ammonia solutions, with maximum susceptibility at an anodic applied potential. A fracture-mode transition from intergranular to transgranular was ob-

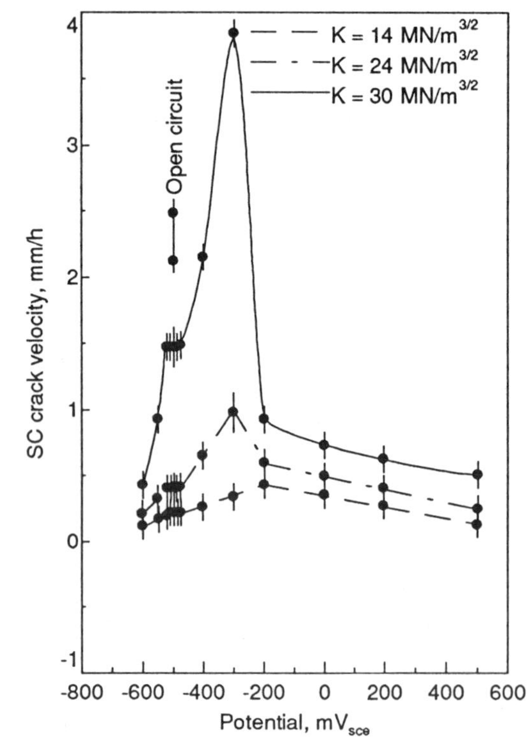

Fig. 8 Effect of potential and stress intensity on crack velocity of α-brass in 15 N aqueous ammonia containing 6 g/L Cu. Source: Ref 27

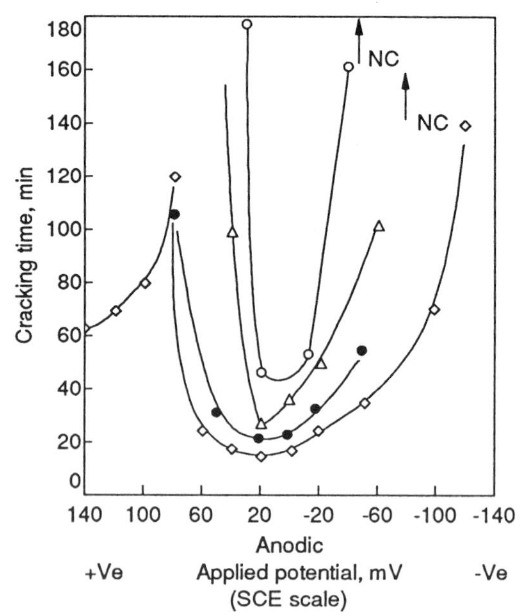

Fig. 9 Effect of applied potential on the time to fracture of α-brass in Mattsson's solutions of pH 7 containing 1 g·mol/L NH_4^+ and different compositions of dissolved copper. Stress = 13.0 kg/mm^2. Source: Ref 28

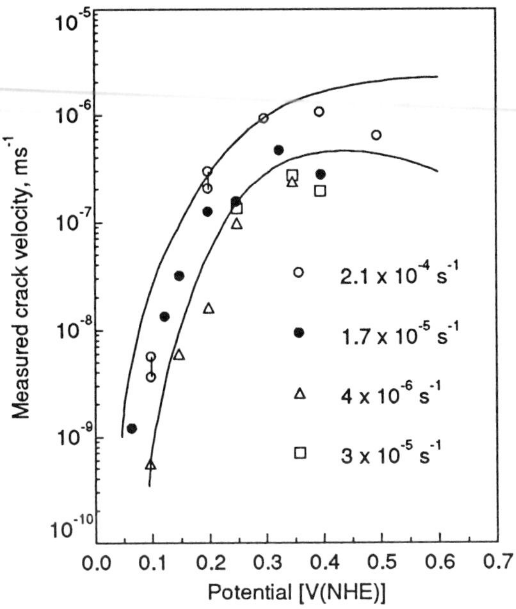

Fig. 10 Effects of potential and strain rate on crack-velocity values measured for yellow brass in deaerated 1 M $NaNO_2$ solution. Strain rates: ○, 2.1×10^{-4}/s; ●, 1.7×10^{-5}/s; △, 4×10^{-6}/s; □ 3×10^{-5}/s. Continuous lines show the crack-velocity values predicted by fast-strain-rate experiments. Source: Ref 29

served with increasing applied anodic potential. The data in Fig. 9 also suggest that the potential range over which cracking occurred narrowed with decreasing copper content of the solution, although the potential of maximum cracking was unaffected.

Alvarez *et al.* (Ref 29), Gouda *et al.* (Ref 30), and Yu and Parkins (Ref 31) investigated the effect of potential on SCC of copper-base alloys in sodium nitrite solutions. All three studies found that anodic polarization from the free corrosion potential accelerated cracking, while cathodic polarization retarded cracking. Both Alvarez *et al.* (Ref 29) and Yu and Parkins (Ref 31) observed some diminution of cracking at very noble potentials, which was attributed to excessive general corrosion of the specimen. Data by Alvarez *et al.* for annealed yellow brass tested by a slow-strain-rate technique in 1 *M* $NaNO_2$ are presented in Fig. 10. In the tests, the free corrosion potential was about 0 V (NHE) and the fracture mode was transgranular.

The effect of potential on SCC of α–brass also has been investigated in solutions of sodium chlorate (Ref 3), sodium sulfate (Ref 15), and aliphatic amine (Ref 6). Shih *et al.* (Ref 5) studied the effect of potential on the SCC of aluminum bronze in sodium fluoride solutions. In all of these environments, anodic polarization of the material promoted cracking, while cathodic polarization retarded cracking. A significant difference among the different alloy/environment systems is in the behavior at the free corrosion potential. For example, α-brass readily cracked at the free corrosion potential in the aliphatic amine solution, while aluminum bronze cracked in the sodium fluoride solution only under anodic polarization. Slight cracking was observed for a brass in aqueous sodium chlorate at the free corrosion potential; cracking was greatly accelerated by anodic polarization.

Effect of Temperature

In general, the temperature dependence of SCC of copper-base alloys appears to follow Arrhenius-type behavior. Data by Suzuki and Hisamatsu (Ref 32) for pure copper and several copper-zinc alloys in a dilute ammonium hydroxide solution over a temperature range of 40 to 70 °C (104 to 158 °F) are given in Fig. 11. The tests were performed under constant load at the free corrosion potential. Yu and Parkins (Ref 31) also observed Arrhenius behavior in slow-strain-rate tests performed on Cu-30Zn in a sodium nitrite solution. The tests were performed at – 0.1 V (SCE) over a temperature range of 20 to 100 °C (68 to 212 °F) at a strain rate of about 7×10^{-4}/ s. However, at a slower strain rate, non-Arrhenius-type behavior was observed, with a maximum cracking velocity at an intermediate temperature (see Fig. 12). The behavior observed at the slower strain rate was attributed to a change in the rate-controlling process over the temperature range studied.

Takano (Ref 33) investigated the effects of crosshead speed and temperature on the SCC of Cu-30Zn in Mattsson's solution of pH 7.2. Slow-strain-rate testing was performed over a temperature range of 25 to 50 °C (77 to 122 °F). It was found that at high crosshead speeds (above 10^{-3} mm/min), cracking susceptibility increased with

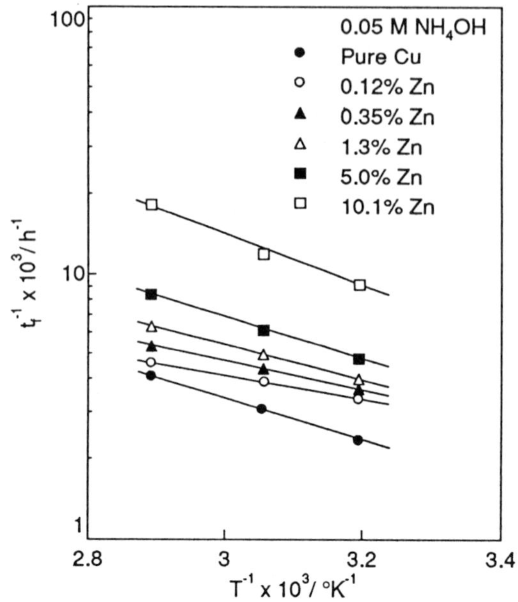

Fig. 11 Arrhenius plot of reciprocal of time-to-failure (t_F^{-1}) for specimens tested under constant load in 0.05 M NH_4OH solution. Source: Ref 32

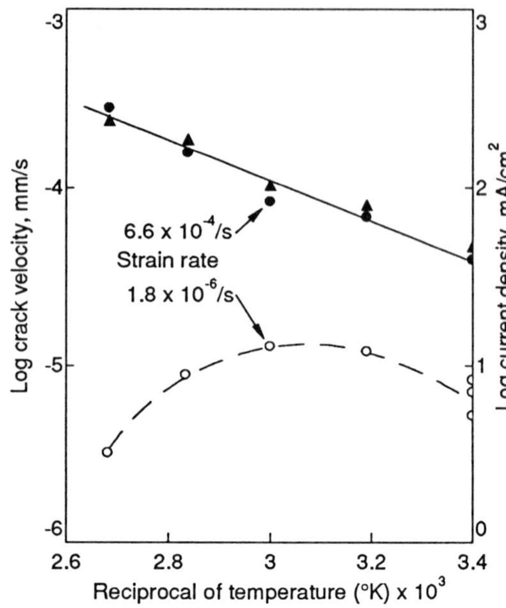

Fig. 12 Effects of temperature and strain rate on cracking velocity of α-brass in 1 M $NaNO_2$. Also shown is the maximum current density from rapid straining tests at different temperatures. Source: Ref 31

increasing temperature; however, susceptibility decreased with increasing temperature at slower crosshead speeds. This behavior was attributed to the interplay between the corrosion rate and the creep rate at the crack tip. In any case, the findings are qualitatively similar to those of Yu and Parkins (Ref 31).

Metallurgical Effects

Effects of Alloy Composition

Most of the literature on SCC of copper-base alloys has been concerned with copper-zinc alloys in ammoniacal solutions. However, a number of other copper-base alloys are susceptible to SCC (see Table 2). In general, the susceptible alloys are solid-solution α-phase copper alloys containing other elements such as manganese, aluminum, and even gold. The rate and path of failure (intergranular or transgranular) depend on the nature of the solute and the environment.

Several researchers have investigated the effect of zinc content of binary Cu-Zn alloys on SCC susceptibility in ammoniacal solutions (*e.g.*, Ref 43, 44). Data from Ref 44, presented in Fig. 13, show that susceptibility to SCC in constant-load tests increased with increasing zinc content for both tarnishing and nontarnishing solutions. These data also show that the fracture-mode was intergranular for tarnishing solutions, while a fracture-mode transition from intergranular to transgranular occurred at approximately 15% Zn for the nontarnishing solutions. This transition has been attributed to a change in the dislocation structure as solute addition increases.

While SCC susceptibility of Cu-Zn alloys increases monotonically with increasing zinc content, susceptibility appears to reach a maximum with increasing alloy content for other α-phase copper alloys. Data showing the effect of aluminum and nickel on time to failure of binary alloys in moist ammonia are given in Fig. 14. The Cu-Zn alloys also are more susceptible to SCC than the other binary alloys, as comparison of the data in Fig. 13 and 14 demonstrates.

The effects on SCC of minor alloy additions to Cu-Zn alloys also have been investigated. Syrett and Parkins (Ref 45) studied the effects of

Table 2 Copper-Base Alloys That Have Experienced SCC

Alloy	Environment	Ref
Pure copper	Ammonia	32
	Sodium nitrite	34
DHP copper	Sodium nitrite	35
OFHC copper	Sodium nitrite	35
Cu-Zn alloys	Ammonia	Many authors
Cu-10Mn	Mattsson's solution	Many authors
Cu-5Zn-5Mn	Mattsson's solution	36
Cu-5Zn-10Mn	Mattsson's solution	36
Cu-20Zn-5Ni	Mattsson's solution	36
Aluminum brass	Acidic chloride-citrate	37
Cu-2Be	Ammonia	38
Cu-4Ti	Ammonia	23
Cu-6Mn	Ammonia	39
Cu-12Mn	Ammonia	39
Cu-18Mn	Ammonia	39
Cu-4Al	Ammonia	39
Cu-8Al	Ammonia	39
Cu-Au	Aqua regia	40
	$FeCl_3$	40
	Ammonia	40
Cu-Al	Ammonia	41
Cu-Ga	Ammonia	41
Cu-Ge	Ammonia	41
Cu-Zn	Ammonia	41
	Water	2
Cu-Zn-Sn	Water	2
β-brass	Water	2
Cusil	Mattsson's solution	42
Cu-72Ag	Mattsson's solution	42

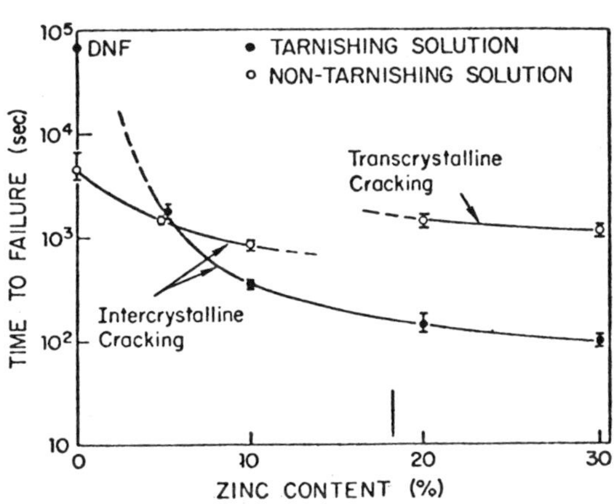

Fig. 13 Effect of zinc content on time to failure for specimens tested at 10% proof stress in tarnishing and nontarnishing ~15 N ammoniacal solutions containing 8 g/L Cu. Source: Ref 44

tin and arsenic on the SCC of Cu-20Zn alloys in Mattsson's solutions of pH 7.3 and 11.3 and found that additions of both elements increased cracking resistance. These tests, like those described above, were performed under freely corroding conditions. It also was observed that the alloy additions promoted a fracture-mode transition from intergranular to transgranular. The results were attributed to the effects of the solute on the passivation characteristics of the alloys, as opposed to a change in their stacking-fault energy. On the other hand, Torchio (Ref 24) found that additions of arsenic and phosphorus to aluminum brass decreased the SCC resistance of the alloy in an acidified citrate chloride solution. The crack path was transgranular in the presence of arsenic, in agreement with the studies of Syrett and Parkins (Ref 45), but was intergranular when only phosphorus was present. Thus, alloying effects in copper-base alloys appear to be environment specific. This conclusion greatly complicates alloy selection for SCC resistance.

A significant finding of SCC research on copper-base alloys over the past 15 years is that pure copper undergoes SCC (Ref 32, 34) (pure copper was long thought to be immune to cracking). Cracking was transgranular and was observed in sodium nitrite solutions. More recently, Cassagne *et al.* (Ref 46) have reported transgranular SCC of pure copper in sodium acetate solutions as well.

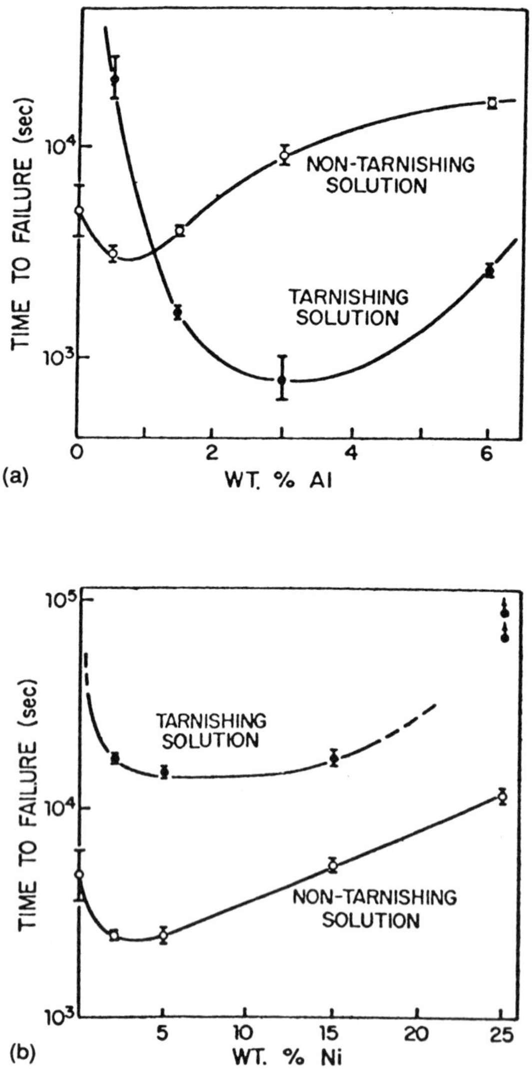

Fig. 14 Effect of aluminum content (a) or nickel content (b) on time to failure of copper-base alloys tested in tarnishing and nontarnishing ~15 N aqueous ammonia containing 8 g/L Cu. Source: Ref 44

Effects of Microstructure and Heat Treatment

Relatively little research has been devoted to the effects of microstructure and heat treatment on SCC of copper-base alloys. This may reflect the fact that extensive thermomechanical processing is not normally performed on α-phase copper alloys for most engineering applications (in contrast to low-alloy steels and aluminum alloys). Several researchers have studied the effect of grain size on cracking susceptibility. Edmunds (Ref 47) evaluated the effect of grain size on time to failure of 70-30 brass specimens exposed to moist ammonia vapor under constant load. Results, shown in Fig. 15, indicate a monotonic decrease in time to failure with increasing grain size. In a later study by Birley and Tromans (Ref 48), the effect of grain size on the fracture stress of Cu-30Zn was evaluated by slow-strain-rate testing in Mattsson's solutions of pH 5.9 to 8. It was found that the fracture stress followed a Petch relationship:

$$\sigma_f = \sigma_o + KD^{-1/2}$$

where σ_f is fracture stress, σ_o and K are constants, and D is grain size, as shown in Fig. 16. The

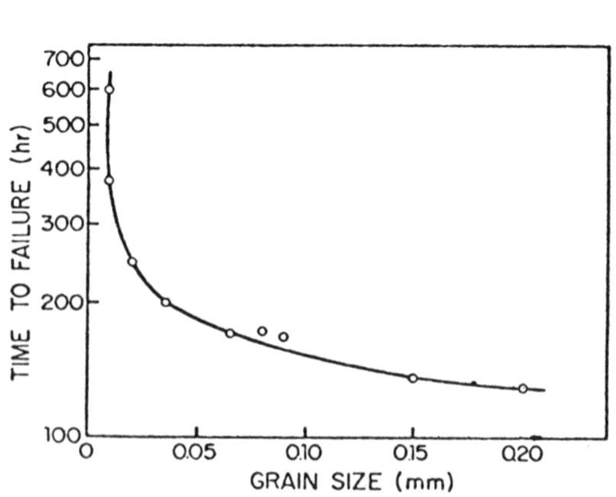

Fig. 15 Effect of grain size on time to failure of 70-30 brass tested in moist ammonia vapor. After Ref 47

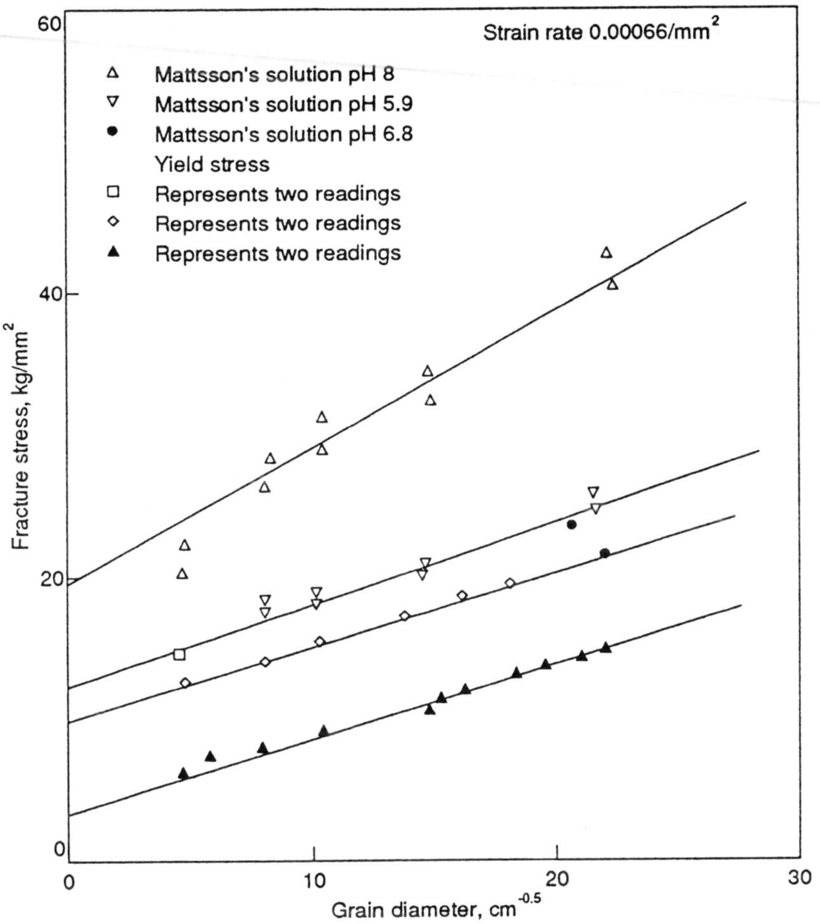

Fig. 16 Effects of grain diameter and solution pH on the stress required to initiate cracking of α-brass in Mattsson's solution in slow-strain-rate tests. Source: Ref 48

authors thus concluded that grain-size effects during SCC can be accounted for in terms of their effects on flow stress. In the pH 5.9 and 6.8 solutions, the onset of cracking occurred at plastic strains of about 7 and 4.5%, respectively, which were independent of grain size. In the pH 8 solution, the plastic strain necessary for crack initiation increased with decreasing grain size. This behavior is reflected by the largest slope in Fig. 16, compared to the slopes observed for the lower-pH solutions or for the Petch relationship.

More recently, Yu *et al.* (Ref 49) evaluated the effect of grain size of 70-30 brass on the initiation of stress-corrosion cracks in a sodium nitrate solution by means of the slow-strain-rate technique. The findings were similar to those reported by Birley and Tromans (Ref 48), with the stress for crack initiation following a Petch-type relationship. The plastic strain at crack initiation was about 1% and increased slightly with decreasing grain size.

Popplewell *et al.* (Ref 50) investigated the effects of postforming heat treatments on the SCC susceptibility of cold-worked specimens of four copper-base alloys (CDA alloys 230, 619, 638, and 688) in moist ammonia and in Mattsson's solution (pH 7.2). Short-term (30-min) heat treatments were performed over a temperature range of 120 to 315 °C (250 to 600 °F) on preformed U-bend specimens, which were subsequently stressed elastically (under constant deflection) and exposed to the cracking environments. This sequence of forming and heat treating was designed to simulate service conditions where a component is formed and stress-relieved. For alloy 230 (85Cu-15Zn), the heat treatments reduced cracking susceptibility by reducing the residual stress. For the remaining three alloys, SCC susceptibility of heat-treated specimens was influenced by competing processes: stress relief, which reduced SCC susceptibility, and ordering, which increased susceptibility. For alloys 638 (95Cu-2.8Al-1.5Si-0.25Co) and 688 (73.5Cu-22.7Zn-3.4Al-0.4Co), optimum temperature ranges for heat treating were identified; for alloy 619 (86.5Cu-4Fe-9.5Al), cracking susceptibility increased with increasing annealing temperature.

Scully (Ref 51) evaluated the effects of annealing and crosshead speed on the SCC of 70Cu-30Zn in an ammoniacal cupric sulfate solution of pH 6.8. The as-received material had a Vickers hardness of 115 and a 0.1% proof stress of 275 MPa (39.9 ksi), while the annealed material had a Vickers hardness of 86 and a proof stress of 205 MPa (29.7 ksi). The cracking velocity of the as-received material was found to be nearly an order of magnitude higher than that of the annealed material over a broad range of crosshead speeds. This effect may not be attributable to residual stresses, because the material is plastically deformed in slow-strain-rate testing.

Mazza and Torchio (Ref 37) evaluated the effects of a sensitization heat treatment and the phosphorus and arsenic content of an aluminum brass on SCC susceptibility in an acidic chloride-citrate solution. The sensitization heat treatment consisted of annealing 50% cold-rolled sheet at 600°C (1110 °F) for 15 min, then sensitizing the specimens at 350 °C (660 °F) for 30 min. The heat treatment was optimized for sensitization by means of exposure tests performed on unstressed specimens in acidic synthetic seawater, followed by metallographic examination for intergranular attack. The SCC tests were performed under constant load, and time to failure was recorded. It was found that phosphorus and arsenic had a dominant effect on SCC susceptibility but that the sensitization heat treatment increased susceptibility to cracking at a given alloy composition. The authors speculated that the increased susceptibility may have resulted from ordering, as was concluded by Popplewell *et al.* (Ref 50).

Effect of Cold Work

It is well established that residual stresses introduced into components by cold forming can increase susceptibility to SCC. The SCC failure of brass electrical connectors in an outdoor environment, reported in Ref 10, is a good practical example of this type of failure. The heads on the connectors were formed by cold forging, which led to rapid failure in the field application. Previously, Landgren and Mattsson (Ref 52) had shown that cold-drawn Cu-37Zn cups failed by SCC in a rural environment, whereas annealed cups did not fail in 10 years under similar exposure conditions.

The effect of cold work on residual stresses described above should be distinguished from the effect of cold work on the inherent SCC susceptibility of copper-base alloys. Unfortunately, it is very difficult to separate these two factors experimentally, which may be one reason for the contradictory literature on this subject. This problem of contradiction is probably best exhibited by the work of Erzurum and Yeh (Ref 53), who studied SCC of C36000 (Cu-37Zn-3Pb). This alloy contains some β phase in addition to lead precipitates. SCC testing was performed in concentrated ammonia, preconcentrated with copper, using slow-strain-rate and constant-load techniques. In the slow-strain-rate tests, SCC susceptibility increased in the liquid phase, but decreased in the vapor phase with increasing cold work. In the constant-load tests, susceptibility to SCC decreased with increasing cold work for both vapor and liquid exposures (see Fig. 17 and 18). With both test techniques, cracking was transgranular and susceptibility was greater in the vapor phase. The authors attributed the cold-work effect in the constant-load tests and in the slow-strain-rate tests in the vapor phase to the preferred orientation imparted during cold working. The contradictory results found for the slow-strain-rate tests performed in the liquid phase were attributed to a change in the rate-limiting step for cracking as a result of the less aggressive nature of the test solution compared to the vapor.

Scully (Ref 51) reported results for slow-strain-rate tests on Cu-30Zn in an ammoniacal solution of pH 6.8 that were similar to those reported by Erzurum and Yeh (Ref 53). Over a broad range of strain rates, the as-received material, which was cold worked, exhibited higher cracking velocities than the annealed material. The cracking mode was transgranular.

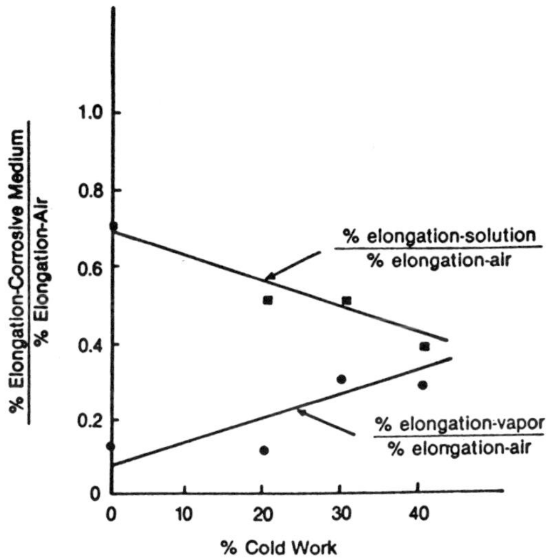

Fig. 17 Effect of percent of cold work and phase (solution and vapor) on elongation of alloy C36000 in 15 N aqueous ammonia containing 6 g/L dissolved copper in slow-strain-rate tests (1.6 × 10^{-5}/s). Source: Ref 53

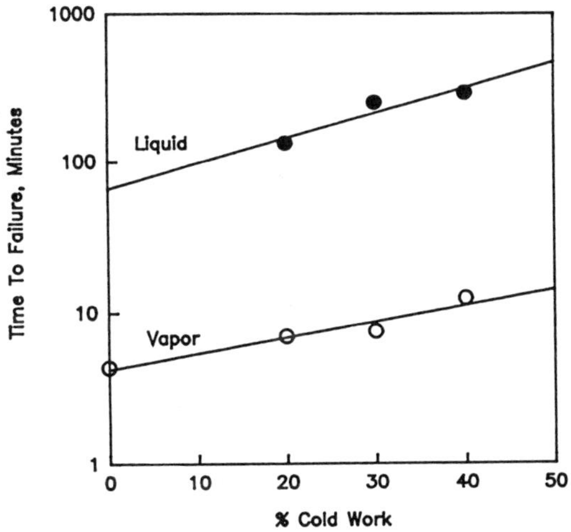

Fig. 18 Effect of percent of cold work and phase (solution and vapor) on time to failure of α-brass in 15 N aqueous ammonia containing 6 g/L dissolved copper in constant load tests at 50% of the yield strength. Source: Ref 53

Sircar *et al.* (Ref 6) studied the SCC of Cu-32Zn in methyl amine solution containing dissolved copper over a range of potentials, using a constant-load test technique. Results of the study, summarized in Fig. 19, show that the annealed material was more susceptible to cracking than the cold-worked material, which is qualitatively similar to results obtained by Erzurum and Yeh (Ref 53) for concentrated ammoniacal solutions. Figure 19 also reveals that the potential at maximum susceptibility is somewhat more positive for the cold-worked material.

Mechanical and Geometrical Effects

Effects of Stress and Stress Intensity

Several researchers have investigated the relationship between engineering stress and time to failure of copper-zinc alloys in ammoniacal environments. Typical data by Pugh *et al.* (Ref 44) are presented in Fig. 20. These data suggest that a threshold, below which cracking does not occur, is not present. However, the data do indicate two distinct slopes. Hoar and Booker (Ref 20) have

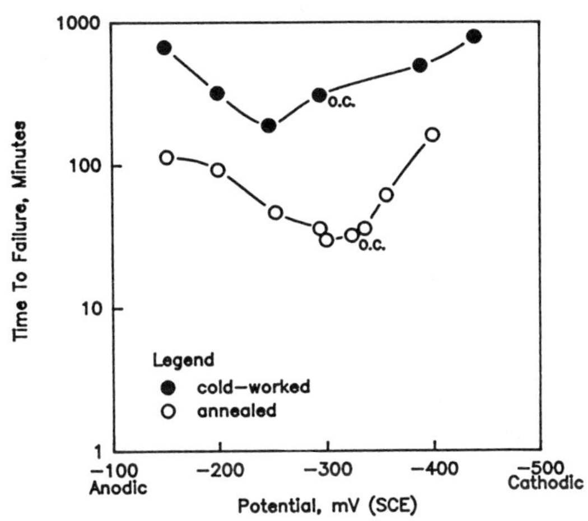

Fig. 19 Effect of potential and cold work on time to failure of α-brass in 2.2 g·mol/L methyl amine at pH 12.5 tested under constant load. Source: Ref 6

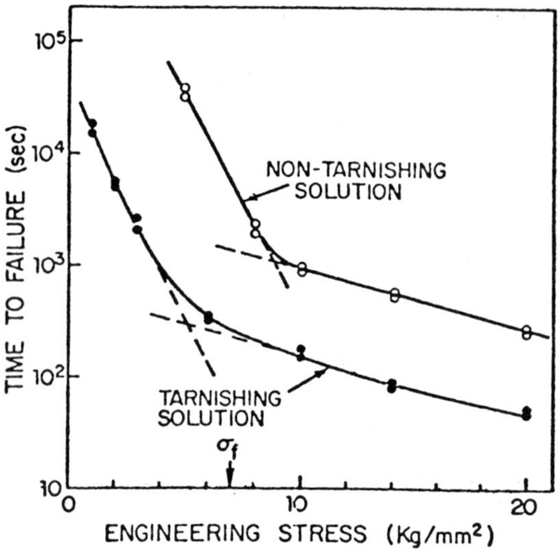

Fig. 20 Influence of engineering stress on time to failure of 70-30 brass tested in tarnishing and nontarnishing ~15 N aqueous ammonia containing dissolved copper. Source: Ref 44

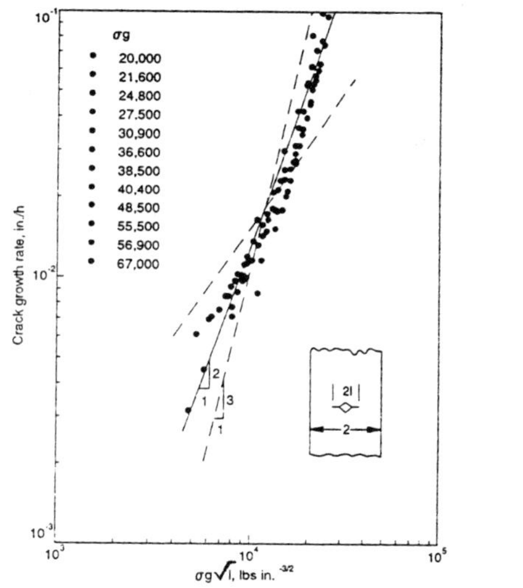

Fig. 21 Rate of stress-corrosion crack propagation in cold-rolled brass exposed to 0.05 M $CuSO_4$ + 0.48 M $(NH_4)_2SO_4$ (pH 7.25). Source: Ref 54

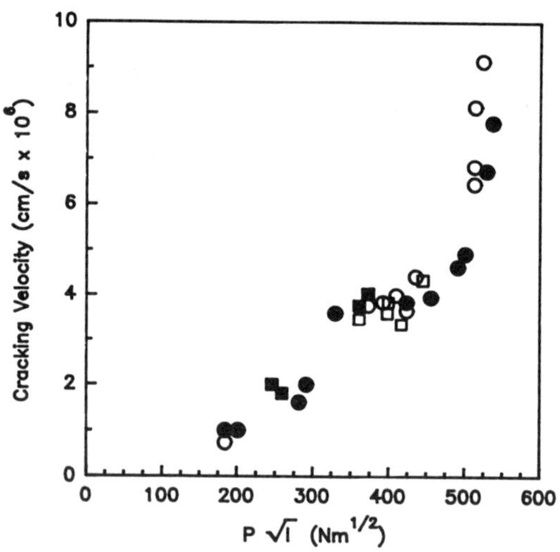

Fig. 22 Cracking velocity as a function of $P\sqrt{l}$ for notched tensile specimens tested in tarnishing 15 N aqueous ammonia. P, applied load: l, crack length. Different symbols indicate different specimens. Source: Ref 55

suggested that the change in slope occurs at the flow stress of the material.

The total time to failure of a specimen, represented by the data in Fig. 20, contains both initiation and propagation stages of cracking. McEvily and Bond (Ref 54) studied the crack-propagation stage of cold-rolled Cu-30Zn in Mattsson's tarnishing solution of pH 7.2. Sheet specimens containing center through-notches were tested under various constant loads, and crack growth was measured optically. Results (Fig. 21) indicate a relationship of the form

$$\frac{dl}{dt} = A\sigma_g^2 l = B\,K^2$$

over a wide range of crack velocities, where l is one-half the crack length, σ_g is the maximum applied gross cross-section stress, t is test time, and A and B are constants. This equation is valid for slope 2 in Fig. 21. As indicated by the above equation, the parameter $\sigma_g\sqrt{l}$ is proportional to the stress intensity (K) at the crack tip.

Beavers (Ref 55) studied the effect of stress intensity on crack growth in annealed admiralty brass in a 15 N tarnishing ammoniacal solution. Testing was performed on single-edge notched tensile specimens under constant load. Results, summarized in Fig. 22, show the classical three stages of cracking: stage 1, at low values of stress intensity (low values of $P\sqrt{l}$), where crack velocity is dependent upon stress intensity; stage 2, where crack velocity is independent of stress intensity; and stage 3, where crack velocity is again stress-intensity dependent. Note that $P\sqrt{l}$ is pro-

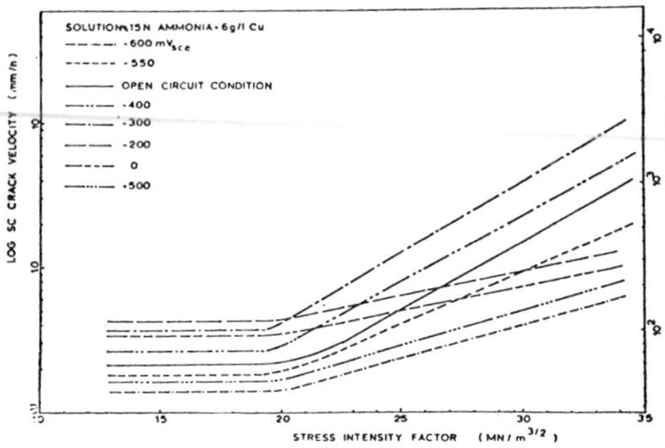

Fig. 23 Effect of stress intensity (K) and potential on crack velocity of α-brass in aqueous 15 N ammonia containing 6 g/L Cu. Source: Ref 27.

portional to K. In such a curve, K_{ISCC} is defined as the highest value of K at which measurable cracking was not observed. In the study by Beavers (Ref 55), crack branching occurred in stage 2. This is typical and is attributed to the fact that the rate-limiting process is independent of stress or stress intensity. The onset of stage 3 cracking is associated with general yielding of the specimen, which was the case in this study. Comparison of these results with those previously reported by McEvily and Bond (Ref 54) suggests that the data in Fig. 21 represent stage 3 cracking.

Kermani and Scully (Ref 27) investigated crack growth in edge-notched cantilever-beam specimens of Cu-30Zn in 15 N ammoniacal solutions and observed behavior similar to that reported by Beavers (Ref 55), although their testing did not provide stage 1 data. It was found that the crack-growth rates were a function of solution composition and applied potential. Typical data are presented in Fig. 23.

SCC Test Methods for Copper-Base Alloys

Standard stress-corrosion test methods are commonly used to study SCC of copper-base alloys. These include constant-strain techniques (employing C-ring, U-bend, and bent-beam specimens), constant-load techniques, slow-strain-rate techniques, and fracture-mechanics techniques. Applicable standards include:

- ASTM G 30: "Standard Practice for Making and Using U-Bend Stress-Corrosion Test Specimens"
- ASTM G 38: "Standard Practice for Making and Using C-Ring Stress-Corrosion Test Specimens"
- ASTM G 39: "Standard Practice for Preparing and Using Bent-Beam Stress-Corrosion Test Specimens"
- ASTM G 49: "Standard Practice for Preparation and Use of Direct Tension Stress-Corrosion Test Specimens"
- ASTM G 58: "Standard Practice for Preparation of Stress-Corrosion Test Specimens for Weldments"

Specific standards for fracture-mechanics techniques for SCC testing are not available, but the test techniques used frequently follow the following two standards:

- ASTM E 399: "Standard Test Method for Plane Strain Fracture Toughness of Metallic Materials"
- ASTM E 616: "Standard Terminology Related To Fracture Testing"

Similarly, a specific standard is not currently available for slow-strain-rate testing, but guidance can be found in two ASTM publications:

- ASTM STP 665: *Stress Corrosion Cracking — The Slow Strain Rate Technique*, G.M. Ugiansky and J.H. Payer, Ed., 1979
- ASTM STP 821: *Environment-Sensitive Fracture: Evaluation and Comparison of Test Methods*, S.W. Dean, E.N. Pugh, and G.M. Ugiansky, Ed., 1984

A unique aspect of SCC testing of copper-base alloys is the test environment. Several ammoniacal solutions are commonly used: 1 N or 15 N NH_3, preconcentrated with cupric ions, and Mattsson's solution. The preconcentrated solutions generally are prepared by dissolving electrolytic copper dust in reagent-grade ammonia under oxygen pressure, typically about 7×10^4 Pa (10 psig). Procedures for the preparation of Mattsson's solution are described in ASTM G 37: "Standard Test Method for Use of Mattsson's Solution of pH 7.2 To Evaluate the Stress-Corrosion Cracking Susceptibility of Copper-Zinc Alloys."

Table 3 Various Techniques That Have Been Used for the Study of SCC of Copper-Base Alloys

Technique	Alloy	Ref
Three point bend	α-brass	63
	Aluminum bronze	58
C-rings	α-brass	60
	Aluminum brass	4
Constant load	C36000 brass	53
	α-brass	6, 20, 28, 59
	Cu-Mn	39
	Cu-Al alloys	39
	70-30 brass	16, 61
	Cu, Cu-Zn alloys	32
Constant strain	Cu-Zn-Sn alloys	45
Fracture mechanics	α-brass	54, 62
	70-30 brass	22
Slow strain rate	70-30 brass and copper	31
	brass	8
	C36000 brass	53
	70-30 brass	51, 65
	Cu-30Zn	56
	Admiralty brass	57
	α-brass	14
	72-28 brass	3
	Yellow brass	15
	Cu-Ag eutectic	42
	Cu-Zn-Sn alloys	45
	Aluminum bronze	5
U-bend	α-brass	12
	Cu-Zn	36
	Cu-Zn-Mn	36
	Cu-Zn-Ni alloys	36
	Aluminum brass	4
	70-30 brass	64
	63-37 brass	9

SCC test methods are discussed more extensively in Chapter 17 of this book and in Volume 13 of the 9th Edition of *Metals Handbook* ("Evaluation of Stress-Corrosion Cracking," by D.O. Sprowls). Table 3 lists references in which various SCC techniques have been used.

Mitigation of SCC of Copper-Base Alloys

Successful mitigation of SCC of copper-base alloys generally involves modifying one or more of the three primary factors associated with SCC: material susceptibility, the specific corrosive environment, and tensile stresses. These factors are shown graphically in Fig. 1. Elimination of any one of these factors will mitigate SCC. In the example of the season cracking of brass cartridge cases described in the introductory section of this chapter, it was found that the source of the cracking environment (ammonia) was the horses in the barns where the cartridge cases were stored. Storing them elsewhere mitigated the problem. Another example where elimination of a cracking environment helped to mitigate SCC involves copper refrigeration tubing. Occasional SCC failures occurred in insulated copper tubing used in commercial refrigeration systems. The root cause of the failure was found to be ammonia, which was a by-product of the manufacturing process for the insulation. Limits were established for the acceptable level of ammonia in the insulation, and this appears to have lessened the problem.

In the example of season cracking, it was also found that the cold-forming operation used to make the cartridge cases introduced high residual tensile stresses and that annealing the cases after forming also eliminated the problem. Another example of reducing residual stresses to mitigate SCC is described by Han *et al.* (Ref 10). In this study, it was found that cold-worked brass electrical connectors failed in an outdoor environment unless properly stress-relieved. A mercurous nitrate solution is often used in industry as a quality-control test of copper alloy articles for residual stresses. This environment promotes liquid-metal embrittlement (from the mercury) if residual tensile stresses are present.

The third primary option for mitigating SCC is to change the alloy to one that is less susceptible to SCC in a given environment. An example of the application of this option is the SCC of brass condenser tubes in power plants. As reported by Beavers *et al.* (Ref 66), a measurable percentage of the failures of copper alloy condenser tubes in the 1960s and 1970s occurred as a result of SCC. The air-removal section of the condenser is particularly prone to SCC as a result of the concentration of oxygen and ammonia on the steam side of the condenser. The most common method for lessening the SCC is to use a more cracking-resistant alloy, such as CDA 706 (Cu-10Ni), for this section of the condenser.

It should be cautioned that mitigation of SCC of copper-base alloys by means of modifying one of the three primary factors may not always be as straightforward as the above discussion implies. For example, Fig. 20 suggests that there is no threshold stress below which SCC will not occur for 70-30 brass in aqueous ammonia. Thus, reducing the stress on a component may prolong its life, but may not prevent SCC from ultimately causing failure. Changing to an alternate copper-base alloy also may not be fully effective in mitigating SCC in some very aggressive cracking environments, since even pure copper may be susceptible to cracking. For these situations, a "belt and suspenders" approach may be prudent, modifying two or more of the primary factors simultaneously.

Cracking Mechanisms

Much of the research into SCC of copper-base alloys has focused on the mechanism of cracking, and a detailed discussion of this subject is beyond the scope of this chapter. Several balanced reviews on the topic have been written by Pugh *et al.* (Ref 19, 44, 67) and are summarized here. The reader is referred to these articles for further details.

The ammonia-brass system has dominated mechanistic studies of SCC of copper-base alloys (*e.g.*, Ref 68). Such studies generally have been divided according to fracture mode: intergranular and transgranular. There is no consensus as to whether these modes correspond to two different mechanisms or whether there is a common mechanism; Pugh and his coworkers favor the former theory.

Intergranular SCC

The characteristics of intergranular SCC are generally considered to be consistent with the film-rupture mechanism proposed by Champion (Ref 69) and Logan (Ref 70). According to this mechanism, the crack initiates by film rupture as a result of plastic deformation at the crack tip. The crack propagates by localized anodic dissolution as repassivation is inhibited by further plastic deformation at the crack tip. The mechanism, as originally proposed, is essentially continuous, with the key elements being the repassivation behavior and the deformation behavior of the alloy/environment system. Characteristics of intergranular SCC of copper-zinc alloys that are consistent with the film-rupture mechanism include the association of cracking with a thin passive layer at the crack tip, the smooth nature of the intergranular fracture surfaces, and the absence of acoustic emission during crack propagation. In addition, based on

Faraday's law, the observed cracking velocities are consistent with a dissolution process.

On the other hand, the originally proposed film-rupture mechanism was a transgranular mechanism, and the issue of crack mode has not been fully explained. It has been postulated that the grain-boundary repassivation is slow because of segregation, but the nature of the segregation has not been established. Moreover, intergranular SCC of brass has been observed in nontarnishing solutions, where the presence of a passive film has not been confirmed.

Transgranular SCC

Over the past 15 years, the mechanism of transgranular SCC in the ammonia-brass system has received considerable attention. Characteristics of this cracking mode that must be accounted for in the mechanism include its discontinuous nature, the cleavagelike appearance of the fracture surfaces, and the observation that cracking can occur in the absence of anodic dissolution. Beavers and Pugh (Ref 71) demonstrated that cracking in admiralty brass occurs on parallel 110 planes and that discrete pulses of acoustic emission are emitted during the cracking process. Several researchers have confirmed the presence of crack-arrest markings on the fracture surfaces, and Newman and Sieradzki (Ref 72) have observed discontinuous current transients during cracking. Bertocci *et al.* (Ref 65) first reported that cracking could occur in the absence of anodic dissolution, and this was later confirmed by others.

The dissolution models such as the film-rupture mechanism described previously are not consistent with the above characteristics and are not considered applicable to transgranular SCC. Similarly, models based on adsorption of a specific chemical species at the crack tip are generally thought to be continuous models and thus are not applicable. Other mechanisms that have been proposed include hydrogen embrittlement, cleavage of a thick tarnish (tarnish rupture) or a dealloyed layer, and film-induced cleavage (FIC). The tarnish-rupture model was discredited in the 1970s when it was shown that the thick tarnish does not extend to the crack tip. A model based on rupture of a dealloyed layer does not adequately explain transgranular SCC of pure copper. More recently, the hydrogen-embrittlement mechanism has been shown to be improbable based on thermodynamic considerations.

The FIC mechanism was first proposed by Sieradzki and Newman (Ref 73). According to this mechanism, the crack initiates and propagates in a brittle film on the metal surface. Traveling at a high velocity and on an orientation favorable for cleavage, the crack propagates some distance into the ductile metal prior to blunting. The major difference between this model and the tarnish-rupture model (or rupture of a dealloyed layer) is the propagation of the crack into the ductile matrix. For the brass-ammonia system, the brittle film is thought to be a copper-rich dealloyed layer. Newman *et al.* (Ref 74) demonstrated a key aspect of the model when they showed that the presence of a dealloyed layer on a Cu-30Zn sample could induce significant crack propagation (about 12 μm) in the bulk alloy. On the other hand, dealloying of copper-base alloys occurs in systems where cracking has not been observed, such as chlorides. Thus, the film apparently must possess other characteristics for SCC to occur. Moreover, SCC of pure copper suggests, at the least, that other types of films also may promote crack propagation in the ductile matrix, assuming the model is correct. Clearly, further research is needed to resolve these issues.

References

1. E.S. Sperry, *Brass World*, Vol 2, 1906, p 39
2. S.S. Birley and D. Tromans, Stress Corrosion Cracking of Beta Brass in Water, *Metall. Trans. A*, Vol 12, 1981, p 1215
3. V.K. Gouda *et al.*, Stress Corrosion Cracking of 72/28 Brass in Sodium Chlorate Solution, *Corrosion*, Vol 38 (No. 12), 1982, p 609
4. S. Torchio, Stress Corrosion Cracking of Aluminum Brass in Acidic Chloride Solutions, *Corros. Sci.*, Vol 21, 1981, p 59
5. H.C. Shih, R.J. Tzou, and H.L. Wang, "Susceptibility of Aluminum Bronze to SCC in Fluoride Environments," Paper No. 183, presented at Corrosion/87, National Association of Corrosion Engineers, 1987
6. S.C. Sircar *et al.*, Stress Corrosion Cracking of Alpha Brass in Aliphatic Amine Solutions, *Br. Corros. J. (Q.)*, No. 1, 1974, p 47

7. U.C. Bhakta and A.K. Lahiri, Stress Corrosion Cracking of Alpha Brass in Ammonium Sulphate Solution, *Br. Corros. J.*, Vol 8, 1973, p 182
8. S.A. Bradford and T. Lee, Effect of Strain Rate on Stress Corrosion Cracking of Brass, *Corrosion*, Vol 34 (No. 3), 1978, p 96
9. E. Mattsson, Stress Corrosion in Brass Considered Against the Background of Potential/pH Diagrams, *Electrochim. Acta*, Vol 3, 1961, p 279
10. M.K. Han, J.A. Beavers, and W. Goins, Stress Corrosion Cracking Failure of Brass Electrical Connectors in an Outdoor Atmospheric Environment, *Mater. Perform.*, Vol 26 (No. 7), 1987, p 20
11. A. Kawashima, A.K. Agrawal, and R.W. Staehle, Effect of Oxyanions and Chloride Ions on the Stress Corrosion Cracking Susceptibility of Admiralty Brass in Nonammoniacal Aqueous Solutions, in *Stress Corrosion Cracking—The Slow Strain Rate Technique*, STP 665, G.M. Ugiansky and J.H. Payer, Ed., American Society for Testing and Materials, 1979, p 266
12. H.E. Johnson and J. Leja, Surface Chemical Factors in the Stress-Corrosion Cracking of Alpha Brass, *Corrosion*, Vol 22 (No. 6), 1966, p 178
13. R.G. Johnston, *Sheet Metal Inst.*, Vol 14, 1940, p 1197
14. H.W. Pickering and P.J. Byrne, Stress Corrosion of Alpha Brass in an Acid Sulfate Solution, *Corrosion*, Vol 29 (No. 8), 1973, p 325
15. M.G. Alvarez *et al.*, Evaluation of Stress Corrosion Cracking of Alpha-Brass in Non-Ammoniac Environments, *Corrosion*, Vol 46 (No. 9), 1990, p 717
16. E.N. Pugh and A.R.C. Westwood, Complex Ions and Stress-Corrosion Cracking in Alpha Brass, *Philos. Mag.*, Vol 13 (No. 121), 1966, p 167
17. E.N. Pugh and J.A.S. Green, The Role of Cupric Complex Ions in the Dissolution and Stress-Corrosion Cracking of α-Brass in Nontarnishing Aqueous Ammonia, *Metall. Trans.*, Vol 2, 1971, p 3129
18. U. Bertocci and E.N. Pugh, Chemical and Electrochemical Aspects of SCC of Alpha-Brass in Aqueous Ammonia, in *4th Inter. Congr. Metallic Corrosion*, National Research Council of Canada, 1984, p 144
19. U. Bertocci, E.N. Pugh, and R.E. Ricker, Environment-Induced Cracking of Copper Alloys, in *Environmentally-Induced Cracking of Metals*, R. Gangloff and M. Ives, Ed., NACE Reference Book No. 10, National Association of Corrosion Engineers, 1989, p 273
20. T.P. Hoar and C.J.L. Booker, The Electrochemistry of the Stress-Corrosion Cracking of Alpha Brass, *Corros. Sci.*, Vol 5, 1965, p 821
21. A.K. Lahiri, Stress-Corrosion Cracking of Alpha Brass in $CuSO_4/(NH_4)_2SO_4$ Solution, *Br. Corros. J.*, Vol 3, 1968, p 289
22. G.M. Sparks and J.C. Scully, Stress Corrosion Crack Propagation in α-1 Brass in Ammoniacal Solutions, *Corros. Sci.*, Vol 14, 1974, p 619
23. T. Takemoto and S. Hori, The Effect of pH, Copper Content and Ammonia Concentration on the Stress Corrosion Cracking Susceptibility of an Aged Cu-4%Ti Alloy, *Corros. Sci.*, Vol 18 (No. 4), 1978, p 323
24. S. Torchio, The Influence of Arsenic and Phosphorus on the Stress Corrosion Cracking of Aluminum Brass in Chloride-Citrate Solution, *Corros. Sci.*, Vol 21, 1981, p 425
25. S. Torchio and F. Mazza, The Influence of Chloride Ions on the Stress Corrosion Cracking of Al-Brass in Acidic Sulphate Solutions, *Corros. Sci.*, Vol 26, 1986, p 813
26. H.S. Tong, Ph.D. thesis, University of Illinois at Urbana-Champaign, 1977
27. M. Kermani and J.C. Scully, Stress Corrosion Crack Propagation in α-Brass in 15N Ammoniacal Solutions, *Corros. Sci.*, Vol 18, 1978, p 883
28. S.C. Sircar *et al.*, Mechanism of SCC of α-Brass in Mattsson's Solution Under Potentiostatic Conditions, *Corros. Sci.*, Vol 12, 1972, p 217
29. M.G. Alvarez *et al.*, Anodic Rate-Controlling Steps in Transgranular Stress Corrosion Cracking of α-Brass in $NaNO_2$ Solutions, *Corros. Sci.*, Vol 24 (No. 9), 1984, p 769
30. V.K. Gouda, S.M. Sayed, and H.A. El-Sayed, Stress Corrosion Cracking Behavior of Brass in Sodium Nitrite Solutions, in *4th Int. Congr. Metallic Corrosion*, National Research Council of Canada, 1984, p 153
31. J. Yu and R.N. Parkins, Stress Corrosion Crack Propagation in α-Brass and Copper Exposed to Sodium Nitrite Solutions, *Corros. Sci.*, Vol 27 (No. 2), 1987, p 159
32. Y. Suzuki and Y. Hisamatsu, Stress Corrosion Cracking of Pure Copper in Dilute Ammoniacal Solutions, *Corros. Sci.*, Vol 21, 1981, p 353
33. M. Takano, K. Teramoto, and T. Nakayama, The Effects of Crosshead Speed and Temperature on the Stress Corrosion Cracking of Cu-30%Zn Alloy in Ammoniacal Solution, *Corros. Sci.*, Vol 21 (No. 6), 1981, p 459
34. S. P. Pednekar *et al.*, Transgranular Cracking of Copper in 1M $NaNO_2$ Solution, *J. Electrochem. Soc.*, Vol 126, 1979, p 701
35. L.A. Benjamin, D. Hardie, and R.N. Parkins, Investigation of the Stress Corrosion Cracking of Pure Copper, *KBS Technical Report 83-06*, Swedish Nuclear Fuel and Waste Management Co., 1983
36. A. Parthasarathi and N.W. Polan, Stress Corrosion Cracking of Copper Alloys: The Effects of Corrosion Potential and Tarnishing Characteristics, *Corrosion*, Vol 43 (No. 12), 1987, p 747

37. F. Mazza and S. Torchio, Factors Influencing the Susceptibility to Intergranular Attack, Stress Corrosion Cracking and De-Alloying Attack of Aluminum Brass, *Corros. Sci.*, Vol 23 (No. 10), 1983, p 1053
38. R.C. Newman and G.T. Burstein, Surface Films Formed on a Cu-2 wt% Be Alloy in Ammoniacal Cu(II) Solutions, *Corros. Sci.*, Vol 20 (No. 3), 1980, p 375
39. M. Islam, An Investigation into the Mechanistic Aspects of the Stress Corrosion Cracking of Binary Copper-Manganese and Copper-Aluminum Alloys, in *4th Int. Congr. Metallic Corrosion*, National Research Council of Canada, 1984, p 161
40. L. Graff, Stress Corrosion Cracking in Homogeneous Non-Supersaturated Alloys Containing Noble or No Noble Metals, in *Proc. Conf. Fundamental Aspects of Stress Corrosion Cracking*, R.W. Staehle, A.J. Forty, and D. Van Rooyen, Ed., National Association of Corrosion Engineers, 1969, p 187
41. N. Ohtani and R.A. Dodd, Mechanisms of Transgranular Stress Corrosion Cracking of Solid Solution Alloys, *Corrosion*, Vol 21, 1965, p 161
42. N.R. Sorensen, The Environmentally Assisted Failure of Cusil in Mattsson's Solution, *Corrosion*, Vol 42 (No. 5), 1986, p 299
43. D.H. Thompson and A.W. Tracy, Influence of Composition on the Stress-Corrosion Cracking of Some Copper-base Alloys, *Trans. AIME*, Vol 185, 1949, p 100
44. E.N. Pugh, J.V. Craig, and A.J. Sedriks, The Stress-Corrosion Cracking of Copper, Silver, and Gold Alloys, in *Proc. Conf. Fundamental Aspects of Stress Corrosion Cracking*, R.W. Staehle, A.J. Forty, and D. Van Rooyen, Ed., National Association of Corrosion Engineers, 1969, p 118
45. B.C. Syrett and R.N. Parkins, The Effect of Sn and As on the Stress Corrosion of Cu-Zn Alloys, *Corros. Sci.*, Vol 10, 1970, p 197
46. T.B. Cassagne, J. Kruger, and E.N. Pugh, Role of the Oxide Film in the Transgranular Stress Corrosion Cracking of Copper, in *Environmentally Assisted Cracking: Science and Engineering*, STP 1049, W.B. Lisagor, T.W. Crooker, and B.N. Leis, Ed., American Society for Testing and Materials, 1990, p 59
47. G. Edmunds, Season Cracking of Brass, in *Symposium on Stress-Corrosion Cracking of Metals*, ASTM-AIME, 1945, p 67
48. S.S. Birley and D. Tromans, Ammoniacal Stress Corrosion Cracking of Alpha-Brass, *Corrosion*, Vol 27 (No. 7), 1971, p 297
49. J. Yu, N.J.H. Holroyd, and R.N. Parkins, Application of Slow-Strain-Rate Tests to Defining the Stress for Stress Corrosion Crack Initiation in 70/30 Brass, in *Environment Sensitive Fracture: Evaluation and Comparison of Test Methods*, STP 821, S.W. Dean, E.N. Pugh, and G.M. Ugiansky, Ed., American Society for Testing and Materials, 1984, p 288
50. J.M. Popplewell, R.P.M. Procter, and J. A. Ford, On the Relation Between Stress-Relief Annealing, Ordering, and Stress-Corrosion Cracking Susceptibility in Three Copper Alloys, *Corros. Sci.*, Vol 12, 1972, p 193
51. J.C. Scully, Propagation of Stress Corrosion Cracks Under Constant Strain-Rate Conditions, in *Stress Corrosion Cracking—The Slow Strain Rate Technique*, STP 665, G.M. Ugiansky and J.H. Payer, Ed., American Society for Testing and Materials, 1979, p 237
52. W. Landgren and E. Mattsson, Stress Corrosion of Brass: Field Tests in Different Types of Atmosphere, *Br. Corros. J.*, Vol 11 (No. 21), 1976, p 80
53. S. Erzurum and H.C. Yeh, The Effect of Environment, Cold Work, and Crystallography on the Stress Corrosion Cracking of C36000 Alloy, *Corrosion*, Vol 39 (No. 5), 1983, p 161
54. A.J. McEvily and A.P. Bond, On the Initiation and Growth of Stress Corrosion Cracks in Tarnished Brass, *J. Electrochem. Soc.*, Vol 112 (No. 2), 1965, p 131
55. J.A. Beavers, "Stress Corrosion Cracking of Admiralty Metal in Aqueous Ammonia," Ph.D. thesis, University of Illinois at Urbana-Champaign, 1977
56. M. Takano, Relationship Between Stress-Corrosion Cracking and Strain Rate in Alpha Brass, *Corros. Sci.*, Vol 11, 1971, p 813
57. W.K. Blanchard, D.A. Koss, and L.A. Heldt, Slow Strain Rate Stress Corrosion Cracking Under Multiaxial Deformation Conditions: Technique and Application to Admiralty Brass, *Corrosion*, Vol 40 (No. 3), 1984, p 101
58. J.F. Klement, R.E. Maersch, and P.A. Tully, Stress Corrosion Crack Paths in Alpha Aluminum Bronze in Ammonia and Steam Atmospheres, *Corrosion*, Vol 15, 1959, p 295t
59. E.N. Pugh, W.G. Montague, and A.R.C. Westwood, On the Role of Complex Ions in the Season-Cracking of Alpha-Brass, *Trans. ASM*, Vol 58 (No. 4), 1965, p 665
60. C.A. Loto and R.A. Cottis, Electrochemical Noise Generation During Stress Corrosion Cracking of Alpha-Brass, *Corrosion*, Vol 43 (No. 8), 1987, p 499
61. S.C. Sircar, U.K. Chatterjee, and G.M. Sherbini, The Inhibitive Effect of Extraneous Ion Addition on Stress Corrosion Cracking of Alpha-Brass, *Corros. Sci.*, Vol 23, 1983, p 777
62. M. Kermani and J.C. Scully, Fractographic Aspects of the Stress Corrosion Cracking of α-Brass in 15N Ammoniacal Solutions, *Corros. Sci.*, Vol 19 (No. 7), 1979, p 489

63. C. Edeleanu and A.J. Forty, Some Observations on the Stress-Corrosion Cracking of α–Brass and Similar Alloys, *Philos. Mag.*, Vol 5, 1960, p 1029

64. W. Lynes, Corrosion of Copper and Brass in Ammonia Containing Chromate, *Corrosion*, Vol 22, 1966, p 113

65. U. Bertocci, F.I. Thomas, and E.N. Pugh, Stress Corrosion Cracking of Brass in Aqueous Ammonia in the Absence of Detectable Anodic Dissolution, *Corrosion*, Vol 40 (No. 8), 1984, p 439

66. J.A. Beavers, A.K. Agrawal, and W.E. Berry, "Corrosion Related Failures in Power Plant Condensers," EPRI NP-1468, Electric Power Research Institute, 1980

67. E.N. Pugh, The Stress-Corrosion Cracking of Copper Alloys, in *Stress Corrosion Research*, H. Arup and R.N. Parkins, Ed., Sijthoff and Noordhoff, 1979, p 177

68. E.N. Pugh, The Mechanism of Stress Corrosion Cracking of Alpha-Brass in Aqueous Ammonia, in *Theory of Stress Corrosion Cracking in Alloys*, J.C. Scully, Ed., NATO Science Affairs Division, 1971, p 418

69. F.A. Champion, Effects Associated With Internal Stresses: Discussion, in *Symp. Internal Stresses in Metals and Alloys*, Institute of Metals, 1948, p 468

70. H. L. Logan, Film-Rupture Mechanism of Stress Corrosion, *J. Res. Natl. Bur. Stand.*, Vol 48 (No. 2), 1952, p 99

71. J.A. Beavers and E.N. Pugh, The Propagation of Transgranular Stress-Corrosion Cracks in Admiralty Metal, *Metall. Trans. A*, Vol 11, 1980, p 809

72. R.C. Newman and K. Sieradzki, Correlation of Acoustic and Electrochemical Noise in the Stress-Corrosion Cracking of α-Brass, *Scr. Metall.*, Vol 17, 1983, p 621

73. K. Sieradzki and R.C. Newman, Brittle Behavior of Ductile Metals During Stress-Corrosion Cracking, *Philos. Mag.*, Vol 51A, 1985, p 95

74. R.C. Newman, T. Shahrabi, and K. Sieradzki, Film-Induced Cleavage of α-Brass, *Scr. Metall.*, Vol 23, 1989, p 71

CHAPTER 8

*Stress-Corrosion Cracking of Aluminum Alloys**

Aluminum, as indicated by its position in the electromotive force (emf) series, is a thermodynamically reactive metal; among structural metals, only beryllium and magnesium are more reactive. Aluminum owes its excellent corrosion resistance and its usage as one of the primary metals of commerce to the barrier oxide film that is bonded strongly to its surface and that, if damaged, reforms immediately in most environments. On a surface freshly abraded and then exposed to the air, the barrier oxide film is only 1 nm (10 Å) thick but is highly effective in protecting the aluminum from corrosion.

The oxide film that develops in normal atmospheres grows to thicknesses much greater than 1 nm (10 Å) and is composed of two layers (Ref 1). The inner oxide next to the metal is a compact amorphous barrier layer whose thickness is determined solely by the temperature of the environment. At any given temperature, the limiting barrier thickness is the same in oxygen, dry air, or moist air. Covering the barrier layer is a thicker, more permeable outer layer of hydrated oxide. Most of the interpretation of aluminum corrosion processes has been developed in terms of the chemical properties of these oxide layers.

The natural film can be visualized as the result of a dynamic equilibrium between opposing forces—those tending to form the compact barrier layer and those tending to break it down. If the destructive forces are absent, as in dry air, the natural film will consist only of the barrier layer and will form rapidly to the limiting thickness. If the destructive forces are too strong, the oxide will be hydrated faster than it is formed, and little barrier will remain. Between these extremes, where the opposing forces reach a reasonable balance, relatively thick (20 to 200 nm, or 200 to 2000 Å) natural films are formed (Ref 2).

The conditions for thermodynamic stability of the oxide film are expressed by the Pourbaix (potential versus pH) diagram shown in Fig. 1. As shown by this diagram, aluminum is passive (is protected by its oxide film) in the pH range of about 4 to 8.5. The limits of this range, however, vary somewhat with temperature, with the specific form of oxide film present, and with the presence of substances that can form soluble complexes or insoluble salts with aluminum. The relative inertness in the passive range is further illustrated in Fig. 2, which gives results of weight loss measurements for alloy 3004-H14 specimens exposed in water and in salt solutions at various pH values.

Beyond the limits of its passive range, aluminum corrodes in aqueous solutions because its oxides are soluble in many acids and bases, yielding Al^{3+} ions in the former and AlO_2^- (aluminate) ions in the latter. There are, however, instances when corrosion does not occur outside the passive range, for example, when the oxide film is not soluble when the film is maintained by the oxidizing nature of the solution (Ref 4).

This chapter will focus on stress-corrosion cracking (SCC) of aluminum, and related corrosion phenomena. Simplified ratings of resistance to general corrosion and to SCC for wrought alu-

*Adapted from "Corrosion of Aluminum and Aluminum Alloys," in *Corrosion*, Volume 13 of the 9th Edition of *Metals Handbook* (ASM International, 1987)

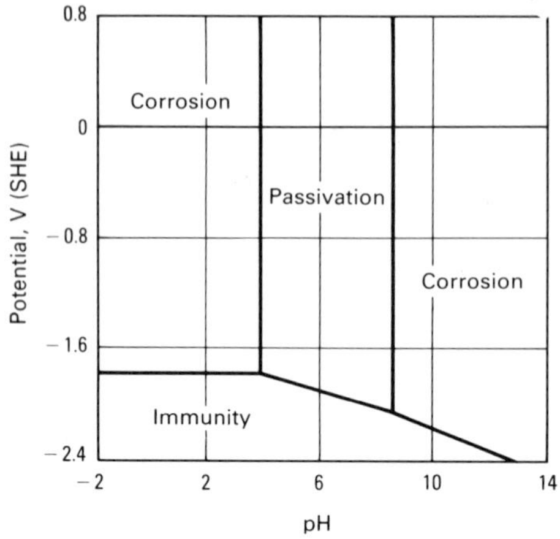

Fig. 1 Pourbaix diagram for aluminum with an $Al_2O_3 \cdot 3H_2O$ film at 25 °C (75 °F). Potential values are for the standard hydrogen electrode (SHE) scale. Source: Ref 3

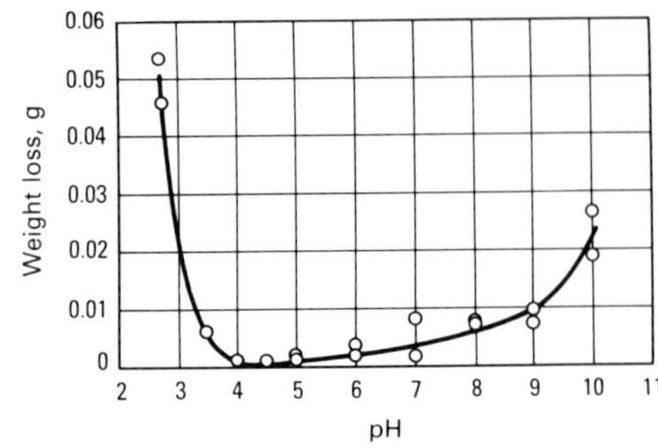

Fig. 2 Weight loss of alloy 3004-H14 exposed 1 week in distilled water and in solutions of various pH values. Specimens were 1.6 × 13 × 75 mm (0.06 × 0.5 × 3 in.). The pH values of solutions were adjusted with HCl and NaOH. Test temperature was 60 °C (140 °F).

minum alloys are presented in Table 1. Ratings for cast aluminum alloys are given in Tables 2 and 3. These ratings may be useful in evaluating and comparing alloy/temper combinations for corrosion service (more detailed ratings of resistance to SCC for high-strength wrought aluminum alloys are given later in this chapter in Table 9 and in Ref 5).

Solution Potentials

Because of the electrochemical nature of most corrosion processes, relationships among solution potentials of different aluminum alloys, as well as between potentials of aluminum alloys and those of other metals, are of considerable importance. Furthermore, the solution-potential relationships among the microstructural constituents of a particular alloy significantly affect its corrosion behavior. Compositions of solid solutions and additional phases, as well as amounts and spatial distributions of the additional phases, may affect both the type and extent of corrosion.

The solution potential of an aluminum alloy is primarily determined by the composition of the aluminum-rich solid solution, which constitutes the predominant volume fraction and area fraction of the alloy microstructure (Ref 6). Solution potential is not affected significantly by second-phase particles of microscopic size, but because these particles frequently have solution potentials differing from that of the solid-solution matrix in which they occur, localized galvanic cells may be formed between them and the matrix.

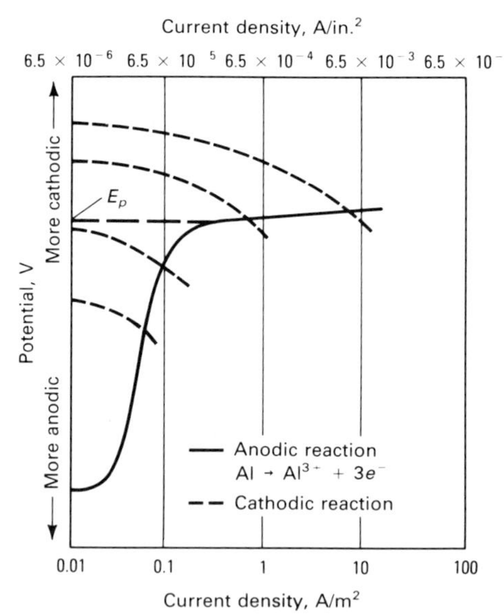

Fig. 3 Effects of principal alloying elements on the electrolytic-solution potential of aluminum. Potentials are for solution-treated and quenched high-purity binary alloys in a solution of 53 g/L NaCl plus 3 g/L H_2O_2 at 25 °C (75 °F).

Table 1 Relative ratings of resistance to general corrosion and to SCC of wrought aluminum alloys

Alloy	Temper	Resistance to corrosion General(a)	SCC(b)
1060	All	A	A
1100	All	A	A
1350	All	A	A
2011	T3, T4, T451	D(c)	D
	T8	D	B
2014	O	...	...
	T3, T4, T451	D(c)	C
	T6, T651, T6510, T6511	D	C
2017	T4, T451	D(c)	C
2018	T61	...	...
2024	O	...	...
	T4, T3, T351, T3510, T3511, T361	D(c)	C
	T6, T861, T81, T851 T8510, T8511	D	B
	T72	...	...
2025	T6	D	C
2036	T4	C	...
2117	T4	C	A
2218	T61, T72	D	C
2219	O	...	...
	T31, T351, T3510, T3511, T37	D(c)	C
	T81, T851, T8510, T8511, T87	D	B
2618	T61	D	C
3003	All	A	A
3004	All	A	A
3105	All	A	A
4032	T6	C	B
5005	All	A	A
5050	All	A	A
5052	All	A	A
5056	O, H11, H12, H32, H14, H34	A(d)	B(d)
	H18, H38	A(d)	C(d)
5056 (cont.)	H192, H392	B(d)	D(d)
5083	All	A(d)	B(d)
5086	O, H32, H116	A(d)	A(d)
	H34, H36, H38, H111	A(d)	A(d)
5154	All	A(d)	A(d)
5252	All	A	A
5254	All	A(d)	A(d)
5454	All	A	A
5456	All	A(d)	B(d)
5457	O	A	A
5652	All	A	A
5257	All	A	A
6053	O	...	...
	T6, T61	A	A
6061	O	B	A
	T4, T451, T4510, T4511	B	B
	T6, T651, T652, T6510, T6511	B	A
6063	All	A	A
6066	O	C	A
	T4, T4510, T4511 T6, T6510, T6511	C	B
6070	T4, T4511, T6	B	B
6101	T6, T63, T61, T64	A	A
6151	T6, T652	...	...
6201	T81	A	A
6262	T6, T651, T6510 T6511, T9	B	A
6463	All	A	A
7001	O	C(c)	C
7075	T6, T651, T652 T6510, T6511	C(c)	C
	T72, T7351	C	B
7178	T6, T651, T6510, T6511	C(c)	C

(a) Ratings are relative and in decreasing order of merit, based on exposure to NaCl solution by intermittent spraying or immersion. Alloys with A and B ratings can be used in industrial and seacoast atmospheres without protection. Alloys with C, D, and E ratings generally should be protected, at least on faying surfaces. (b) SCC ratings are based on service experience and on laboratory tests of specimens exposed to alternate immersion in 3.5% NaCl solution. A, no known instance of failure in service or in laboratory tests; B, no known instance of failure in service; limited failures in laboratory tests of short transverse specimens; C, service failures when sustained tension stress acts in short-transverse direction relative to grain structure; limited failures in laboratory tests of long transverse specimens; D, limited service failures when sustained stress acts in longitudinal or long-transverse direction relative to grain structure. (c) In relatively thick sections, the rating would be E. (d) This rating may be different for material held at elevated temperatures for long periods.

The effects of principal alloying elements on solution potential of high-purity aluminum are shown in Fig. 3. For each element, the significant changes that occur do so within the range in which the element is completely in solid solution. Further addition of the same element, which forms a second phase, causes little additional change in solution potential.

Most commercial aluminum alloys contain additions of more than one of these elements; effects of multiple elements in solid solution on solution potential are approximately additive. The amounts retained in solid solution, particularly for more highly alloyed compositions, depend highly on fabrication and thermal processing so that heat treatment and other processing variables

Table 2 Relative ratings of resistance to general corrosion and to SCC of aluminum casting alloys

Alloy	Temper	Resistance to corrosion General(a)	SCC(b)
208.0	F	B	B
224.0	T7	C	B
240.0	F	D	C
242.0	All	D	C
A242.0	T75	D	C
249.0	T7	C	B
295.0	All	C	C
319.0	F, T5	C	B
	T6	C	C
355.0	All	C	A
C355.0	T6	C	A
356.0	T6, T7, T71, T51	B	A
A356.0	T6	B	A
443.0	F	B	A
512.0	F	A	A
513.0	F	A	A
514.0	F	A	A
520.0	T4	A	C
535.0	F	A	A
B535.0	F	A	A
705.0	T5	B	B
707.0	T5	B	C
710.0	T5	B	B
712.0	T5	B	C
713.0	T5	B	B
771.0	T6	C	C
850.0	T5	C	B
851.0	T5	C	B
852.0	T5	C	B

(a) Relative ratings of general corrosion resistance are in decreasing order of merit, based on exposures to NaCl solution by intermittent spray or immersion. **(b)** Relative ratings of resistance to SCC are based on service experience and on laboratory tests of specimens exposed to alternate immersion in 3.5% NaCl solution. A, no known instance of failure in service when properly manufactured; B, failure not anticipated in service from residual stresses or from design and assembly streses below about 45% of the minimum guaranteed yield strength given in applicable specifications; C, failures have occurred in service with either this specific alloy/temper combination or with alloy/temper combinations of this type; designers should be aware of the potential SCC problem that exists when these alloys and tempers are used under adverse conditions.

influence the final electrode potential of the product. Tables 4 to 7 present representative solution potentials of commercial aluminum alloys and of several other metals and alloys.

The amounts of second phases present in aluminum and aluminum alloy products vary from nearly zero in those of aluminum 1199 and some others that also are nearly pure solid solutions to over 20% in hypereutectic aluminum-silicon casting alloys, such as 392.0 and 393.0. These phases are generally intermetallic compounds of binary, ternary, or higher-order compositions, although some elements in excess of their solid solubility are present as elemental phases. Electrode potentials of some of the simpler second-phase constituents have been measured and are presented in Table 8. Measurement of the solution potentials of aluminum alloys has been standardized in ASTM G 69 (Ref 7).

Solution-potential measurements are useful for the investigation of heat-treating, quenching, and aging practices, and they are applied principally to alloys containing copper, magnesium, or zinc. In aluminum-copper and aluminum-copper-magnesium (2*xxx*) alloys, potential measurements can determine the effectiveness of solution heat treatment by measuring the amount of copper in solid solution. Also, by measuring the potentials of grain boundaries and grain bodies separately, the difference in potential responsible for intergranular corrosion, exfoliation, and SCC can be quantified. Solution-potential measure-

Table 3 Relative ratings of resistance to general corrosion and to stress-corrosion cracking of aluminum permanent mold, die casting, and rotor metal alloys

Alloy	Temper	Resistance to corrosion General(a)	SCC(b)
Permanent mold castings			
242.0	T571, T61	D	C
308.0	F	C	B
319.0	F	C	B
	T6	C	C
332.0	T5	C	B
336.0	T551, T65	C	B
354.0	T61, T62	C	A
355.0	All	C	A
C355.0	T61	C	A
356.0	All	B	A
A356.0	T61	B	A
F356.0	All	B	A
A357.0	T61	B	A
358.0	T6	B	A
359.0	All	B	A
B443.0	F	B	A
A444.0	T4	B	A
513.0	F	A	A
705.0	T5	B	B
707.0	T5	B	C
711.0	T5	B	A
713.0	T5	B	B
850.0	T5	C	B
851.0	T5	C	B
852.0	T5	C	B
Die castings			
360.0	F	C	A
A360.0	F	C	A
364.0	F	C	A
380.0	F	E	A
A380.0	F	E	A
383.0	F	E	A
384.0	F	E	A
390.0	F	E	A
392.0	F	E	A
413.0	F	C	A
A413.0	F	C	A
C443.0	F	B	A
518.0	F	A	A
Rotor metal(c)			
100.1		A	A
150.0		A	A
170.1		A	A

(a) Relative ratings of general corrosion resistance are in decreasing order of merit, based on exposures to NaCl solution by intermittent spray or immersion. **(b)** Relative ratings of resistance to SCC are based on service experience and on laboratory tests of specimens exposed to alternate immersion in 3.5% NaCl solution. A, no known instance of failure in service when properly manufactured; B, failure not anticipated in service from residual stresses or from design and assembly stresses below about 45% of the minimum guaranteed yield strength given in applicable specifications; C, failures have occurred in service with either this specific alloy/temper combination or with alloy/temper combinations of this type; designers should be aware of the potential SCC problem that exists when these alloys and tempers are used under adverse conditions. **(c)** For electric motor rotors

ments of alloys containing copper also show the progress of artificial aging as increased amounts of precipitates are formed and the matrix is depleted of copper.

Potential measurements are valuable with zinc-containing (7*xxx*) alloys for evaluating the effectiveness of the solution heat treatment, for following the aging process, and for differentiat-

Table 4 Solution potentials of nonheat-treatable commercial wrought aluminum alloys
Values are the same for all tempers of each alloy

Alloy	Potential(a), V
1060	−0.84
1100	−0.83
3003	−0.83
3004	−0.84
5050	−0.84
5052	−0.85
5154	−0.86
5454	−0.86
5056	−0.87
5456	−0.87
5182	−0.87
5083	−0.87
5086	−0.85
7072	−0.96

(**a**) Potential versus standard calomel electrode

Table 5 Solution potentials of heat-treatable commercial wrought aluminum alloys

Alloy	Temper	Potential(a), V
2014	T4	−0.69(b)
	T6	−0.78
2219	T3	−0.64(b)
	T4	−0.64(b)
	T6	−0.80
	T8	−0.82
2024	T3	−0.69(b)
	T4	−0.69(b)
	T6	−0.81
	T8	−0.82
2036	T4	−0.72
2090	T8E41	−0.83
6009	T4	−0.80
6010	T4	−0.79
6151	T6	−0.83
6351	T5	−0.83
6061	T4	−0.80
	T6	−0.83
6063	T5	−0.83
	T6	−0.83
7005	T6	−0.94
X7016	T6	−0.86
X7021	T6	−0.99
X7029	T6	−0.85
X7146	T6	−1.02
7049	T73	−0.84(c)
	T76	−0.84(c)
7050	T73	−0.84(c)
	T76	−0.84(c)
7075	T6	−0.83(c)
	T73	−0.84(c)
	T76	−0.84(c)
7475	T6	−0.83(c)
	T73	−0.84(c)
	T76	−0.84(c)
7178	T6	−0.83(c)

(**a**) Potential versus standard calomel electrode. (**b**) Varies ±0.01 V with quenching rate. (**c**) Varies ±0.02 V with quenching rate

Table 6 Solution potentials of aluminum cast alloys

Alloy	Temper	Type of mold(a)	Potential(b), V
208.0	F	S	−0.77
238.0	F	P	−0.74
295.0	T4	S or P	−0.70
	T6	S or P	−0.71
	T62	S or P	−0.73
296.0	T4	S or P	−0.71
308.0	F	P	−0.75
319.0	F	S	−0.81
	F	P	−0.76
355.0	T4	S or P	−0.78
	T6	S or P	−0.79
356.0	T6	S or P	−0.82
443.0	F	S	−0.83
	F	P	−0.82
514.0	F	S	−0.87
520.0	T4	S or P	−0.89
710.0	F	S	−0.99

(**a**) S, sand; P, permanent. (**b**) Potential versus standard calomel electrode

ing among the various artificially aged tempers. These factors can affect corrosion behavior. In the magnesium-containing (5*xxx*) alloys, potential measurements can detect low-temperature precipitation and are useful in qualitatively evaluating stress-corrosion behavior. Potential measurements can also be used to follow the diffusion of zinc or copper in alclad products, thus determining whether the sacrificial cladding can continue to protect the core alloy (Ref 8).

Intergranular Corrosion

Intergranular (intercrystalline) corrosion is selective attack of grain boundaries or closely adjacent regions without appreciable attack of the grains themselves. Intergranular corrosion is a generic term that includes several variations associated with different metallic structures and thermomechanical treatments (Fig. 4). It is an

Table 7 Solution potentials of some nonaluminum base metals

Metal	Potential(a), V
Magnesium	−1.73
Zinc	−1.10
Cadmium	−0.82
Mild carbon steel	−0.58
Lead	−0.55
Tin	−0.49
Copper	−0.20
Bismuth	−0.18
Stainless steel(b)	−0.09
Silver	−0.08
Nickel	−0.07
Chromium	−0.40 to +0.18

(a) Potential versus standard calomel electrode. (b) Series 300, type 430

Table 8 Solution potentials of some second-phase constituents in aluminum alloys

Phase	Potential(a), V
Si	−0.26
Al_3Ni	−0.52
Al_3Fe	−0.56
Al_2Cu	−0.73
Al_6Mn	−0.85
Al_8Mg_5	−1.24

(a) Potential versus standard calomel electrode

important issue in regard to both general corrosion and SCC.

Intergranular corrosion is caused by potential differences between the grain-boundary region and the adjacent grain bodies (Ref 9). The location of the anodic path varies with the different alloy systems. In 2*xxx* series alloys, it is a narrow band on either side of the boundary that is depleted in copper; in 5*xxx* series alloys, it is the anodic constituent Mg_2Al_3 when that constituent forms a continuous path along a grain boundary; in copper-free 7*xxx* series alloys, it is generally considered to be the anodic zinc- and magnesium-bearing constituents on the grain boundary; and in the copper-bearing 7*xxx* series alloys, it appears to be the copper-depleted bands along the grain boundaries (Ref 10, 11). The 6*xxx* series alloys generally resist this type of corrosion, although slight intergranular attack has been observed in aggressive environments. The electrochemical mechanism for intergranular corrosion proposed by E.H. Dix has been verified (Ref 12) and related to the pitting potentials of aluminum (Ref 13).

Because intergranular corrosion is involved in SCC of aluminum alloys, it is often presumed

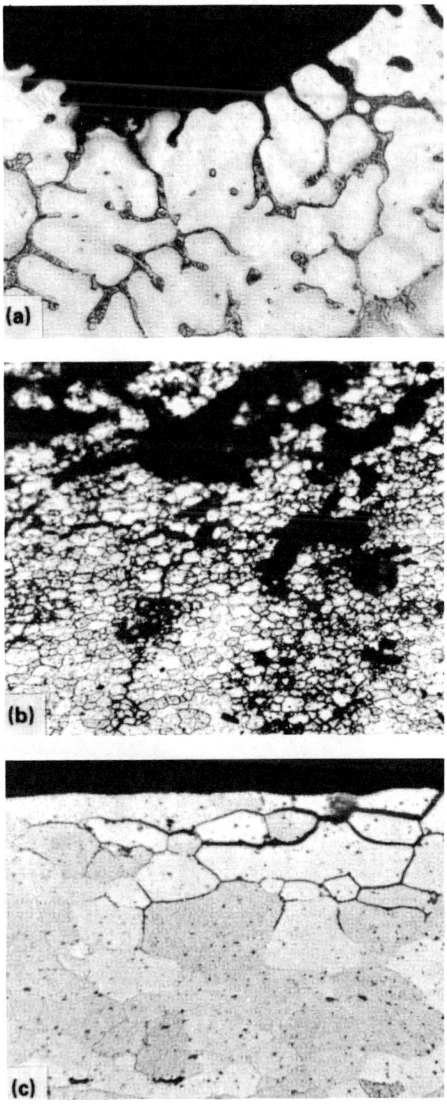

Fig. 4 Various types of intergranular corrosion. (a) Interdendritic corrosion in a cast structure. (b) Interfragmentary corrosion in a wrought, unrecrystallized structure. (c) Intergranular corrosion in a recrystallized wrought structure. All etched with Keller's reagent. 500×

to be more deleterious than pitting or general corrosion. However, in alloys that are not susceptible to SCC—for example, the 6*xxx* series alloys—intergranular corrosion is usually no more severe than pitting corrosion, tends to decrease with time and, for equal depth of corrosion, its effect on strength is no greater than that of pitting corrosion, although fatigue cracks may be more likely to initiate at areas of intergranular corrosion than at random pits.

SCC Susceptibility

Only aluminum alloys that contain appreciable amounts of soluble alloying elements, primarily copper, magnesium, silicon, and zinc, are susceptible to SCC. For most commercial alloys, tempers have been developed that provide a high degree of immunity to SCC in most environments.

The electrochemical theory of stress corrosion, which was developed in about 1940, describes certain conditions required for SCC of aluminum alloys (Ref 9, 14, 15). Further research showed inadequacies in this theory, and the complex interactions among factors that lead to SCC of aluminum alloys are not yet fully understood (Ref 16). However, there is general agreement that for aluminum the electrochemical factor predominates the electrochemical theory continues

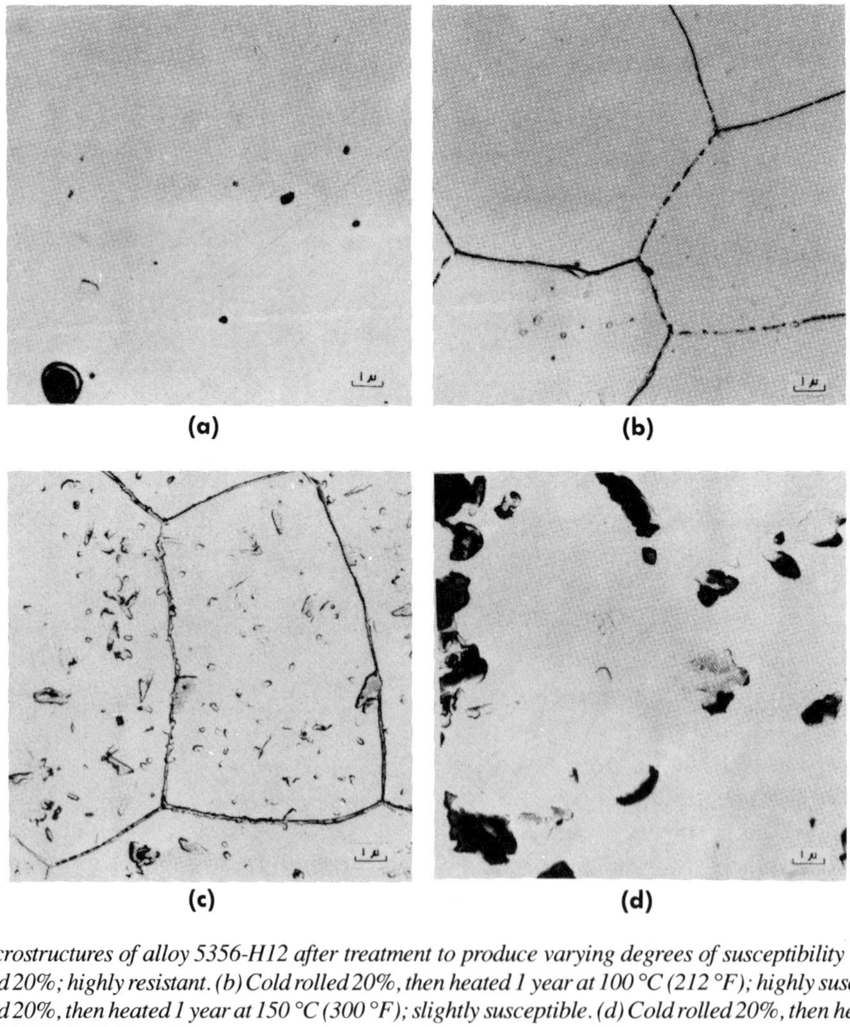

Fig. 5 Microstructures of alloy 5356-H12 after treatment to produce varying degrees of susceptibility to SCC. (a) Cold rolled 20%; highly resistant. (b) Cold rolled 20%, then heated 1 year at 100 °C (212 °F); highly susceptible. (c) Cold rolled 20%, then heated 1 year at 150 °C (300 °F); slightly susceptible. (d) Cold rolled 20%, then heated 1 year at 205 °C (400 °F); highly resistant

to be the basis for developing aluminum alloys and tempers resistant to SCC (Ref 17).

Stress-corrosion cracking in aluminum alloys is characteristically intergranular. According to the electrochemical theory, this requires a condition along grain boundaries that makes them anodic to the rest of the microstructure so that corrosion propagates selectively along them. Such a condition is produced by localized decomposition of solid solution, with a high degree of continuity of decomposition products, along the grain boundaries. The most anodic regions may be either the boundaries themselves (most commonly, the precipitate formed in them) or regions adjoining the boundaries that have been depleted of solute.

In 2*xxx* alloys, the solute-depleted regions are the most anodic; in 5*xxx* alloys, it is the Mg_2Al_3 precipitate along the boundaries. The most anodic grain-boundary regions in other alloys have not been identified with certainty. Strong evidence for the presence of anodic regions, and of the electrochemical nature of their corrosion in aqueous solutions, is provided by the fact that SCC can be greatly retarded, if not eliminated, by cathodic protection (Ref 15).

Figure 5 shows four different microstructures in an alloy containing 5% Mg. These microstructures represent degrees of susceptibility to SCC ranging from high susceptibility to high resistance, depending on heat treatment. The treatments that provide high resistance to cracking are those that produce microstructures either free of precipitate along grain boundaries (Fig. 5a) or with precipitate distributed as uniformly as possible within grains (Fig. 5d). In the latter case, corrosion along boundaries is minimized because the presence of precipitate or depleted regions throughout the microstructure increases the ratio of the total area of anodic regions to that of cathodic ones, thereby reducing the corrosion current on each anodic region. For alloys requiring microstructural control to avoid susceptibility, resistance is obtained by using treatments that produce precipitate throughout the microstructure, because precipitate always forms first along boundaries, and its formation there usually cannot be prevented.

According to electrochemical theory, susceptibility to intergranular corrosion is a prerequisite for susceptibility to SCC, and treatment of aluminum alloys to improve resistance to SCC also improves their resistance to intergranular corrosion. For most alloys, however, optimum levels of resistance to these two types of failure require different treatments, and resistance to intergranular corrosion is not a reliable indication of resistance to SCC.

In many cases, susceptibility of SCC of an aluminum alloy cannot be predicted reliably by examining its microstructure. Many observations have been made of the progressive changes in dislocation network, precipitation pattern, and other microstructural features that occur as an alloy is treated to improve its resistance to SCC, but these changes have not been correlated quantitatively with susceptibility.

Effect of Stress

Whether or not SCC develops in a susceptible aluminum alloy product depends on both magnitude and duration of tensile strength acting at the surface. The effects of the factors have been established most commonly by means of accelerated laboratory tests; results of one set of such tests are reflected in the shaded bands in Fig. 6. Despite introduction of fracture mechanics techniques capable of determining crack growth rates, such tests continue to be the basic tools used

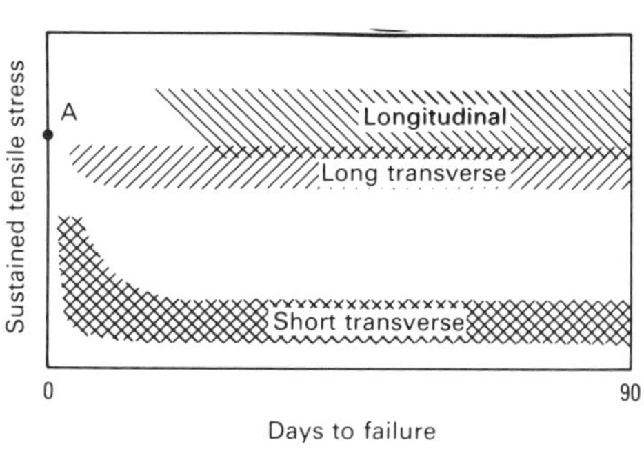

Fig. 6 SCC of alloy 7075-T651 plate. Shaded bands indicate combinations of stress and time known to produce SCC in specimens intermittently immersed in 3.5% NaCl solution. Point A is minimum yield strength in the long-transverse direction for a 75 mm (3 in.) thick plate.

in evaluating resistance of aluminum alloys to SCC. These tests suggest a minimum (threshold) stress that is required for cracking to develop.

Although empirical in nature, the threshold value provides a valid measure of the relative susceptibilities of aluminum alloys to SCC under the specific conditions of a particular test or environment. Also, for some alloy/temper combinations, results of accelerated laboratory tests reliably predict stress-corrosion performance in service; for example, results of an 84-day alternate immersion test of alloy 7075 and alloy 7178 products correlated well with performance of these products in a seacoast environment.

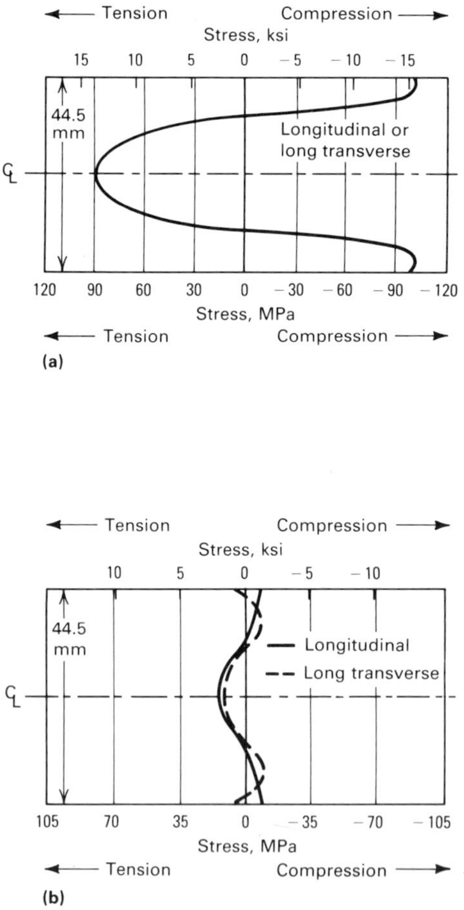

Fig. 7 Comparison of residual stresses in a thick, constant cross-section 7075-T6 aluminum alloy plate before and after stress relief. (a) High residual stresses in the solution-treated and quenched alloy. (b) Reduction in stresses after stretching 2%. Source: Ref 18

Stress Relieving

Residual stresses are induced in aluminum alloy products when they are solution heat treated and quenched. Figure 7(a) shows the typical distribution and magnitude of residual stresses in thick high-strength material of constant cross section. Quenching places the surfaces in compression and the center in tension. If the compressive surface stresses are not disturbed by subsequent fabrication practices, the surface has an enhanced resistance to SCC because a sustained tensile stress is necessary to initiate and propagate this type of corrosion.

On the other hand, one of the most common practices associated with SCC problems is machining into the residual high tensile stress areas of material that has not been stress relieved. If the exposed tensile stresses are in a transverse direction or have a transverse component and if a susceptible alloy or temper is involved, the probability of SCC is present (Ref 18).

Aluminum products of constant cross section are stress relieved effectively and economically by mechanical stretching. The stretching operation must be done after quenching and, for most alloys, before artificial aging. Note the low magnitude of residual stresses after stretching (Fig. 7b) as compared to the as-quenched material in Fig. 7(a). Federal specifications for rolled and extruded products provide for stress relieving by stretching on the order of 1 to 3%.

Thus, the use of the stress-relieved temper for heat-treated mill products will minimize SCC problems related to quenching stresses. The stress-relieved temper for most alloys is identified by the designation T*x*5*x* or T*x*5*xx* after the alloy number, for example, 2024-T351 or 7075-T6511 (Ref 19).

Effects of Grain Structure and Stress Direction

Many wrought aluminum alloy products have highly directional grain structures (Fig. 8). Such products are highly anisotropic with respect to resistance to SCC (Fig. 6). Resistance, which is measured by magnitude of tensile stress required to cause cracking, is highest when the stress is applied in the longitudinal direction, lowest in the

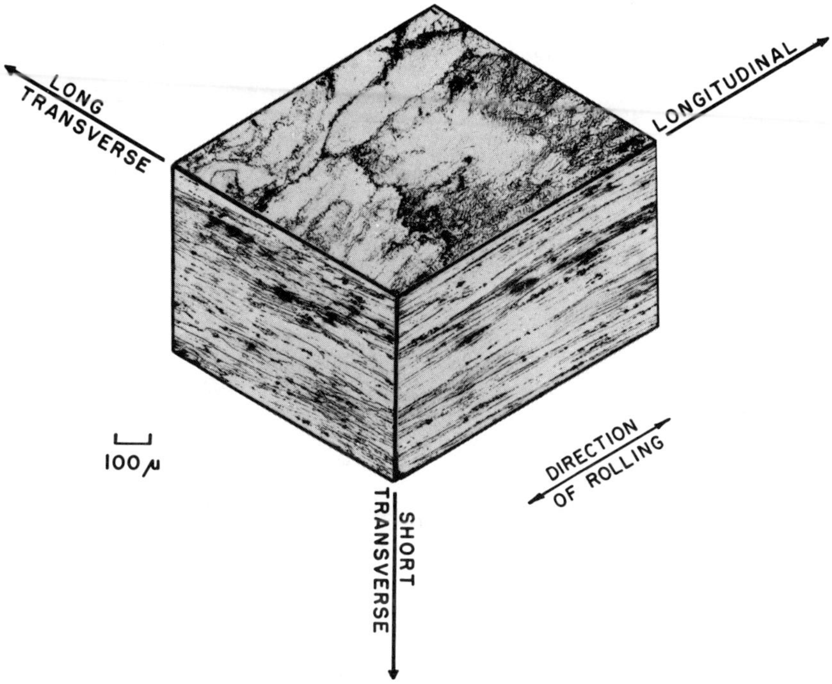

Fig. 8 Composite micrograph showing grain structure of a 38 mm (1.5 in.) alloy 7075-T6 plate

short-transverse direction, and intermediate in the other directions. These differences are most noticeable in the more susceptible tempers, but are usually much lower in tempers produced by extended precipitation treatments, such as T6 and T8 tempers for 2*xxx* alloys and T73, T736, and T76 tempers for 7*xxx* alloys.

Thus, direction and magnitude of stresses anticipated under conditions of assembly and service may govern alloy and temper selection. For products of thin section, applied in ways that induce little or no tensile stress in the short-transverse direction, resistance of 2*xxx* alloys in T3 or T4 tempers or of 7*xxx* alloys in T6 tempers may suffice. Resistance in the short-transverse direction usually controls application of products that are of thick section or are machined or applied in ways that result in sustained tensile stresses in the short-transverse direction. More resistant tempers are preferred in these cases.

Effects of Environment

Research indicates that water or water vapor is the key environmental factor required to produce SCC in aluminum alloys. Halide ions have the greatest effects in accelerating attack. Chloride is the most important halide ion because it is a natural constituent of marine environments and is present in other environments as a contaminant. Because it accelerates SCC, Cl^- is the principal component of environments used in laboratory tests to determine susceptibility of aluminum alloys to this type of attack. In general, susceptibility is greater in neutral solutions than in alkaline solutions and is greater still in acidic solutions.

Stress-Corrosion Ratings

A system of ratings of resistance to SCC for high-strength aluminum alloy products has been developed by a joint task group of ASTM and the Aluminum Association to assist alloy and temper selection, and has been incorporated into Ref 5. Definitions of these ratings, which range from A (highest resistance) to D (lowest resistance), are as follows:

A: Very high. No record of service problems; SCC is not anticipated in general applications.

Table 9 Relative SCC ratings for wrought products of high-strength aluminum alloys

Resistance ratings are as follows: A, very high; B, high; C, intermediate; D, low. See text for more detailed explanation of these ratings.

Alloy and temper(a)	Test direction(b)	Rolled plate	Rod and bar(c)	Extruded shapes	Forgings
2011-T3, -T4	L	(d)	B	(d)	(d)
	LT	(d)	D	(d)	(d)
	ST	(d)	D	(d)	(d)
2011-T8	L	(d)	A	(d)	(d)
	LT	(d)	A	(d)	(d)
	ST	(d)	A	(d)	(d)
2014-T6	L	A	A	A	B
	LT	B(e)	D	B(e)	B(e)
	ST	D	D	D	D
2024-T3, -T4	L	A	A	A	(d)
	LT	B(e)	D	B(e)	(d)
	ST	D	D	D	(d)
2024-T6	L	(d)	A	(d)	A
	LT	(d)	B	(d)	A(e)
	ST	(d)	B	(d)	D
2024-T8	L	A	A	A	A
	LT	A	A	A	A
	ST	B	A	B	C
2024-T851	L	A	(d)	(d)	(d)
	LT	A	(d)	(d)	(d)
	ST	B	(d)	(d)	(d)
2124-T851	L	A	(d)	(d)	(d)
	LT	A	(d)	(d)	(d)
	ST	B	(d)	(d)	(d)
2219-T3, -T37	L	A	(d)	A	(d)
	LT	B	(d)	B	(d)
	ST	D	(d)	D	(d)
2219-T6, -T8	L	A	A	A	A
	LT	A	A	A	A
	ST	A	A	A	A
6061-T6	L	A	A	A	A
	LT	A	A	A	A
	ST	A	A	A	A
7005-T53, -T63	L	(d)	(d)	A	A
	LT	(d)	(d)	A(e)	A(e)
	ST	(d)	(d)	D	D
7039-T63, -T64	L	A	(d)	A	(d)
	LT	A(e)	(d)	A(e)	(d)
	ST	D	(d)	D	(d)
7049-T73	L	A	(d)	A	A
	LT	A	(d)	A	A
	ST	A	(d)	B	A
7049-T76	L	(d)	(d)	A	(d)
	LT	(d)	(d)	A	(d)
	ST	(d)	(d)	C	(d)
7149-T73	L	(d)	(d)	A	A
	LT	(d)	(d)	A	A
	ST	(d)	(d)	B	A
7050-T736	L	A	(d)	A	A
	LT	A	(d)	A	A

(continued)

(a) Ratings apply to standard mill products in the types of tempers indicated and also in T*x*5*x* and T*x*5*xx* (stress-relieved) tempers and may be invalidated in some cases by use of nonstandard thermal treatments, or mechanical deformation at room temperature, by the user. **(b)** Test direction refers to orientation of direction in which stress is applied relative to the directional grain structure typical of wrought alloys, which for extrusions and forgings may not be predictable on the basis of the cross-sectional shape of the product: L, longitudinal; LT, long transverse; ST, short transverse. **(c)** Sections with width-to-thickness ratios equal to or less than two, for which there is no distinction between LT and ST properties. **(d)** Rating not established because product not offered commercially. **(e)** Rating is one class lower for thicker sections; extrusions, 25 mm (1 in.) and thicker; plate and forgings, 38 mm (1.5 in.) and thicker

Table 9 Relative SCC ratings for wrought products of high-strength aluminum alloys (continued)
Resistance ratings are as follows: A, very high; B, high; C, intermediate; D, low. See text for more detailed explanation of these ratings.

Alloy and temper(a)	Test direction(b)	Rolled plate	Rod and bar(c)	Extruded shapes	Forgings
	ST	B	(d)	B	B
7050-T76	L	A	A	A	(d)
	LT	A	B	A	(d)
	ST	C	B	C	(d)
7075-T6	L	A	A	A	A
	LT	B(e)	D	B(e)	B(e)
	ST	D	D	D	D
7075-T73	L	A	A	A	A
	LT	A	A	A	A
	ST	A	A	A	A
7075-T736	L	(d)	(d)	(d)	A
	LT	(d)	(d)	(d)	A
	ST	(d)	(d)	(d)	B
7075-T76	L	A	(d)	A	(d)
	LT	A	(d)	A	(d)
	ST	C	(d)	C	(d)
7175-T736	L	(d)	(d)	(d)	A
	LT	(d)	(d)	(d)	A
	ST	(d)	(d)	(d)	B
7475-T6	L	A	(d)	(d)	(d)
	LT	B(e)	(d)	(d)	(d)
	ST	D	(d)	(d)	(d)
7475-T73	L	A	(d)	(d)	(d)
	LT	A	(d)	(d)	(d)
	ST	A	(d)	(d)	(d)
7475-T76	L	A	(d)	(d)	(d)
	LT	A	(d)	(d)	(d)
	ST	C	(d)	(d)	(d)
7178-T6	L	A	(d)	A	(d)
	LT	B(e)	(d)	B(e)	(d)
	ST	D	(d)	D	(d)
7178-T76	L	A	(d)	A	(d)
	LT	A	(d)	A	(d)
	ST	C	(d)	C	(d)
7079-T6	L	A	(d)	A	A
	LT	B(e)	(d)	B(e)	B(e)
	ST	D	(d)	D	D

(a) Ratings apply to standard mill products in the types of tempers indicated and also in T*x*5*x* and T*x*5*xx* (stress-relieved) tempers and may be invalidated in some cases by use of nonstandard thermal treatments, or mechanical deformation at room temperature, by the user. **(b)** Test direction refers to orientation of direction in which stress is applied relative to the directional grain structure typical of wrought alloys, which for extrusions and forgings may not be predictable on the basis of the cross-sectional shape of the product: L, longitudinal; LT, long transverse; ST, short transverse. **(c)** Sections with width-to-thickness ratios equal to or less than two, for which there is no distinction between LT and ST properties. **(d)** Rating not established because product not offered commercially. **(e)** Rating is one class lower for thicker sections; extrusions, 25 mm (1 in.) and thicker; plate and forgings, 38 mm (1.5 in.) and thicker

B: High. No record of service problems; SCC not anticipated at stresses of the magnitude caused by solution heat treatment. Precautions must be taken to avoid high sustained tensile stresses (exceeding 50% of the minimum specified yield strength) produced by any combination of sources including heat treatment, straightening, forming, fit-up, and sustained service loading.

C: Intermediate. Stress-corrosion cracking not anticipated if total sustained tensile stress is maintained below 25% of minimum specified yield strength. This rating is designated for the short-transverse direction in products used primarily for high resistance to exfoliation corrosion in relatively thin structures, where appreciable stresses in the short-transverse direction are unlikely.

D: Low. Failure due to SCC is anticipated in any application involving sustained tensile stress in the designated test direction. This rating is currently designated only for the short-transverse direction in certain products.

These stress levels are not to be interpreted as threshold stresses and are not recommended for design. Documents such as MIL-HANDBOOK-5, MIL-STD-1568, NASC SD-24, and MSFC-SPEC-552A should be consulted for design recommendations.

The relative ratings of resistance to SCC for high-strength wrought aluminum alloys are presented in Table 9. These ratings, assigned primarily by alloy and temper, also distinguish among test directions and product types.

***2xxx* Alloys.** Thick-section products of *2xxx* alloys in the naturally aged T3 and T4 tempers have low ratings of resistance to SCC in the short-transverse direction. Ratings of such products in other directions are higher, as are ratings of thin-section products in all directions. These differences are related to the effects of quenching rate (largely determined by section thickness) on the amount of precipitation that occurs during quenching. If *2xxx* alloys in T3 and T4 tempers are heated for short periods in the temperature range used for artificial aging, selective precipitation along grain or subgrain boundaries may further impair their resistance.

Longer heating, as specified for T6 and T8 tempers, produces more general precipitation and significant improvements in resistance to SCC. Precipitates are formed within grains at a greater number of nucleation sites during treatment to T8 tempers. These tempers require stretching, or cold working by other means, after quenching from the solution heat treatment temperature and before artificial aging. These tempers provide the highest resistance for SCC and the highest strength in the *2xxx* alloy.

Some studies on aluminum-copper-lithium alloys indicate that these alloys have their highest resistance to SCC at or near peak-aged tempers (Ref 20-22). Underaging of these alloys (For example, 2090) is detrimental; overaging decreases resistance only slightly. The susceptibility of the underaged microstructure has been attributed to the precipitation of an intermetallic constituent, AL_2CuLi, on grain boundaries during the early stages of artificial aging. This constituent is believed to be anodic to the copper-rich matrix of an underaged alloy, causing preferential dissolution and SCC. As aging time increases, copper-bearing precipitates form in the interior of the grains, thus increasing the anode-cathode area ratio in the microstructure to a more favorable value that avoids selective grain-boundary attack. Similar studies of stress-corrosion behavior are being conducted on aluminum-lithium-copper-magnesium alloys (for example, 8090) (Ref 23).

***5xxx* alloys** are not considered heat treatable and do not develop their strength through heat treatment. However, these alloys are processed to H3 tempers, which require a final thermal stabilizing treatment to eliminate age softening, or to H2 tempers, which require a final partial annealing. The H116 or H117 tempers are also used for high-magnesium *5xxx* alloys and involve special temperature control during fabrication to achieve a microstructural pattern of precipitate that increases the resistance of the alloy to intergranular corrosion and SCC. The alloys of the *5xxx* series span a wide range of magnesium contents, and the tempers that are standard for each alloy are primarily established by the magnesium content and the desirability of microstructures highly resistant to SCC and other forms of corrosion.

Although *5xxx* alloys are not heat treatable, they develop good strength through solution hardening by the magnesium retained in solid solution, dispersion hardening by precipitates, and strain-hardening effects. Because the solid solutions in the higher-magnesium alloys are more highly supersaturated, the excess magnesium tends to precipitate out as Mg_2Al_3, which is anodic to the matrix. Precipitation of the phase with high selectivity along grain boundaries, accompanied by little or no precipitation within grains, may result in susceptibility to SCC.

The probability that a susceptible microstructure will develop in a *5xxx* alloy depends on magnesium content, grain structure, amount of strain hardening, and subsequent time/temperature history. Alloys with relatively low magnesium contents, such as 5052 and 5454 (2.5 and 2.75% Mg, respectively), are only mildly supersaturated; consequently, their resistance to SCC is not affected by exposure to elevated temperatures. In

Fig. 9 Microstructures of alloy 5083-O plate stretched 1%. (a) As-stretched. (b) After heating 40 days at 120 °C (250 °F)

contrast, alloys with magnesium contents exceeding about 3%, when in strain-hardened tempers, may develop susceptible structures as a result of heating or even after very long times at room temperature. For example, the microstructure of alloy 5083-O (4.5% Mg) plate stretched 1% (Fig. 9a) is relatively free of precipitate (no continuous second-phase paths), and the material is not susceptible to SCC. Prolonged heating below the solvus, however, produces continuous precipitate, which results in susceptibility (Fig. 9b).

6*xxx* Alloys. The service record of 6*xxx* alloys shows no reported cases of SCC. In laboratory tests, however, at high stresses and in aggressive solutions, cracking has been demonstrated in 6*xxx* alloys of particularly high alloy content, containing silicon in excess of the Mg_2Si ratio and/or high percentages of copper.

7*xxx* Alloys Containing Copper. The 7*xxx* series alloy that has been used most extensively and for the longest period of time is 7075, an aluminum-zinc-magnesium-copper-chromium alloy. Introduced in 1943, this aircraft construction alloy was initially used for products with thin sections, principally sheet and extrusions. In these products, quenching rate is normally very high, and tensile stresses are not encountered in the short-transverse direction; thus, SCC is not a problem for material in the highest-strength (T6) tempers. When 7075 was used in products of greater size and thickness, however, it became apparent that such products heat treated to T6 tempers were often unsatisfactory. Parts that were extensively machined from large forgings, extrusions, or plate were frequently subjected to continuous stresses, arising from interference misfit during assembly or from service loading, that were tensile at exposed surfaces and aligned in unfavorable orientations. Under such conditions, SCC was encountered in service with significant frequency.

The problem resulted in the introduction (in about 1960) of the T73 tempers for thick-section 7075 products. The precipitation treatment used to develop these tempers requires two-stage artificial aging, the second stage of which is done at a higher temperature than that used to produce T6 tempers. During the preliminary stage, a fine high-density precipitation dispersion is nucleated, producing high strength. The second stage is then used to develop resistance to SCC and exfoliation. Extensive accelerated and environmental testing has demonstrated that 7075-T73 resists SCC even when stresses are oriented in the least favorable direction, at stress levels of at least 300 MPa (44 ksi). Under similar conditions, the maximum stress at which 7075-T6 resists cracking is about 50 MPa (7 ksi). The excellent test results for 7075-T73 have been confirmed by extensive service experience in various applications.

The additional aging treatment required to produce 7075 in T73 tempers, which have high resistance to SCC, reduces strength to levels below those of 7075 in T6 tempers. Alloy 7175, a variant of 7075, was developed for forgings. In the T736 temper, 7175 has strength nearly comparable to that of 7075-T6 and has better resistance to SCC. Other newer alloys—such as 7049 and 7475, which are used in the T73 temper, and 7050, which is used in the T736 temper—couple high strength with very high resistance and improved fracture toughness.

The T76 tempers, which also require two-stage artificial aging and which are intermediate to the T6 and T73 tempers in both strength and resistance to SCC, are developed in copper-containing 7*xxx* alloys for certain products. Comparative ratings of resistance for various products of all these alloys, as well as for products of 7178, are given in Table 9.

The microstructural differences among the T6, T73, and T76 tempers of these alloys are differences in size and type of precipitate, which changes from predominantly Guinier-Preston (GP) zones to T6 tempers to η', the metastable transition form of $\eta(MgZn_2)$, in T73 and T76 tempers. None of these differences can be detected by optical metallography. In fact, even the resolutions possible in transmission electron microscopy are insufficient for determining whether the precipitation reaction has been adequate to ensure the expected level of resistance to SCC. For quality assurance, copper-containing 7*xxx* alloys in T73 and T76 tempers are required to have specified minimum values of electrical conductivity and, in some cases, tensile yield strengths that fall within specified ranges. The validity of these properties as measures of resistance to SCC is based on many correlation studies involving these measurements, laboratory and field stress-corrosion tests, and service experience.

Copper-free 7*xxx* Alloys. Wrought alloys of the 7*xxx* series that do not contain copper are of considerable interest because of their good resistance to general corrosion, moderate-to-high strength, and good fracture toughness and formability. Alloys 7004 and 7005 have been used in extruded form and, to a lesser extent, in sheet form for structural applications. More recently introduced compositions, including 7016, 7021, 7029, and 7146, have been used in automobile bumpers formed from extrusions or sheet.

As a group, copper-free 7*xxx* alloys are less resistant to SCC than other types of aluminum alloys when tensile stresses are developed in the short-transverse direction at exposed surfaces. Resistance in other directions may be good, particularly if the product has an unrecrystallized microstructure and has been properly heat treated. Products with recrystallized grain structures are generally more susceptible to cracking as a result of stress induced by forming or mechanical damage after heat treatment. When cold forming is required, subsequent solution heat treatment or precipitation heat treatment is recommended. Applications of these alloys must be carefully engineered, and consultation among designers, application engineers and product producers, or suppliers is advised in all cases.

Casting Alloys. The resistance of most aluminum casting alloys to SCC is sufficiently high that cracking rarely occurs in service. The microstructures of these alloys are usually nearly isotropic: consequently, resistance to SCC is unaffected by orientation of tensile stresses.

Relative ratings of cast alloys, based primarily on accelerated laboratory tests, are listed in Tables 2 and 3. It has been indicated by accelerated and natural-environment testing and verified by service experience that alloys of the aluminum-silicon 4*xx.x* series, 3*xx.x* alloys containing only silicon and magnesium as alloying additions, and 5*xx.x* alloys with magnesium contents of 8% or lower have virtually no susceptibility to SCC. Alloys of the 3*xx.x* group that contain copper are rated as less resistant, although the numbers of castings of these alloys that have failed by SCC have not been significant.

Significant SCC of aluminum alloy castings in service has occurred only in the highest-strength aluminum-zinc-magnesium 7*xx.x* alloys and in the aluminum-magnesium alloy 520.0 in the T4 temper. For such alloys, factors that require careful consideration include casting design, assembly and service stresses, and anticipated environmental exposure.

Specifications and Tests

Several aluminum alloy product specifications require defined levels of performance with respect to resistance to SCC. Standard tests used to measure such performance are described in methods standards and are referenced in materials specifications. Among these are tests for evaluating resistance to SCC of 2*xxx* alloys and of 7*xxx* alloys that contain copper by alternate immersion in 3.5% NaCl solution (Ref 24,25). Lot acceptance criteria for products of 7*xxx* copper-containing alloys in T76, T73, and T736 tempers are based on combined requirements for tensile strength and electrical conductivity.

References

1. M.S. Hunger and P. Fowle, Naturally and Thermally Formed Oxide Films on Aluminum, *J. Electrochem. Soc.*, Vol 103, 1956, p 482
2. H.P. Godard, W.B. Jepson, M.R. Bothwell, and R.L. Kane, *The Corrosion of Light Metals*, John Wiley and Sons, 1967
3. M. Pourbaix, *Atlas of Electrochemical Equilibria in Aqueous Solutions*, Pergamon Press, 1966, p 171
4. J.E. Hatch, Ed., Chapter 7, in *Aluminum: Properties and Physical Metallurgy*, American Society for Metals, 1984
5. "Classification of Resistance to Stress-Corrosion Cracking of High-Strength Aluminum Alloys," G64, *Annual Book of ASTM Standards*, American Society for Testing and Materials
6. R.H. Brown, W.L. Fink, and M.S. Hunter, Measurement of Irreversible Potentials As a Metallurgical Research Tool, *Trans. AIME*, Vol 143, 1941, p 115
7. "Standard Practice for Measurement of Corrosion Potentials of Aluminum Alloys," G 69, *Annual Book of ASTM Standards*, American Society for Testing and Materials
8. M.S. Hunter, A.M. Montgomery, and G.W. Wilcox, Microstructure of Alloys and Products, in *Aluminum*, Vol 1, K.R. Van Horn, Ed., American Society for Metals, 1967, p 77
9. E.H. Dix, Acceleration of the Rate of Corrosion By High Constant Stresses, *Trans. AIME*, Vol 137, 1940, p 11
10. W.L. Fink and L.A. Willey, Quenching of 75S Aluminum Alloy, *Met. Technol.*, Vol 14 (No. 8), 1947, p 5
11. M.S. Hunter, G.R. Frank, and D.L. Robinson, in *Proceedings of Conference: Fundamental Aspects of Stress-Corrosion Cracking*, R.W. Staehle, Ed., National Association of Corrosion Engineers, 1969, p 497
12. H. Kaesche, Pitting Corrosion of Aluminum and Intergranular Corrosion of Aluminum Alloys, in *Localized Corrosion*, B.F. Brown, J. Kruger, and R.W. Staehle, Ed., National Association of Corrosion Engineers, 1974, p 516
13. J.R. Galvele and S.M. de Micheli, Mechanism of Intergranular Corrosion of Al-Cu Alloys, *Corros. Sci.*, Vol 10, 1970, p 795
14. R.B. Mears, R.H. Brown, and E.H. Dix, Jr., A Generalized Theory of the Stress-Corrosion Cracking of Alloys, in *Symposium on Stress-Corrosion Cracking of Metals*, American Society for Testing and Materials and American Institute of Mining and Metallurgical Engineers, 1945, p 323
15. D.O. Sprowls and R.H. Brown, Stress-Corrosion Mechanism for Aluminum Alloys, in *Fundamental Aspects of Stress-Corrosion Cracking*, R.W. Staehle, A.J. Forty, and D. Van Rooyen, Ed., National Association of Corrosion Engineers, 1969, p 466
16. M.O. Speidel, Hydrogen Embrittlement of Aluminum Alloys, in *Hydrogen in Metals*, L.M. Bernstein and A.W. Thompson, Ed., American Society for Metals, 1974, p 249
17. V.A. Marichev, The Mechanism of Crack Growth in Stress Corrosion Cracking of Aluminum Alloys, *Werkst. Korros.*, Vol 34, 1983, p 300
18. E.H. Spuhler and C.L. Burton, "Avoiding Stress-Corrosion Cracking in High Strength Aluminum Alloy Structures," Green Letter, Alcoa, 1970
19. *Aluminum Standards and Data*, The Aluminum Association, 1984, p 12
20. J.G. Rinker, M. Marek, and T.H. Sanders, Jr., Microstructure, Toughness and SCC Behavior of 2020, in *Aluminum-Lithium Alloys*, T.H. Sanders, Jr. and E.A. Starke, Jr., Ed., American Institute of Mining, Metallurgical, and Petroleum Engineers, 1983, p 597
21. A.K. Vasudevan, P.R. Ziman, S.C. Jha, and T.H. Sanders, Jr., Stress-Corrosion Resistance of Al-Cu-Li-Zr Alloys, in *Aluminum-Lithium Alloys*, Vol III, C. Baker, P.J. Gregson, S.J. Harris, and C.J. Peel, Ed., The Institute of Metals, 1986, p 303
22. E.L. Colvin, S.J. Murtha, and R.K. Wyss, The Effect of Aging Time on the Stress-Corrosion Cracking Resistance of 2090-T8E41, published in the proceedings of the International Conference on Aluminum Alloys, Charlottesville, VA, 15-20 June 1986
23. N.J.H. Holroyd, A. Gray, G.M. Scamans, and R. Herman, Environment-Sensitive Fracture of Al-Li-Cu-Mg Alloys, in *Aluminum-Lithium Alloys*, Vol III, C. Baker, P.J. Gregson, S.J. Harris, and C.J. Peel, Ed., The Institute of Metals, 1986, p 310
24. "Standard Recommendation Practice of Alternate Immersion Stress Corrosion Testing in 3.5% Sodium Chloride Solution," G 44, *Annual Book of ASTM Standards*, American Society for Testing and Materials
25. "Standard Recommended Practice for Determining Susceptibility to Stress-Corrosion Cracking of High-Strength Aluminum Alloy Products," G 47, *Annual Book of ASTM Standards*, American Society for Testing and Materials

CHAPTER 9

Stress-Corrosion Cracking of Magnesium Alloys

William K. Miller, Metallurgy Department,
General Motors Research and Environmental Staff

Magnesium possesses the lowest density of all structural metals, having about one-fourth the density of iron and one-third that of aluminum. Because of this low density, both cast and wrought magnesium alloys have been developed and used for a wide variety of structural applications in which low weight is important, if not a requirement.

The magnesium industry grew rapidly during the period of the two world wars in response to the need for lightweight military aircraft engines, airframes, wheels, and other parts. After World War II, the nonmilitary market began using significant amounts of magnesium alloys in power tools, luggage, lawn mower decks, sporting equipment, camera cases, conveyors, and a variety of automotive components such as brackets, covers, housings, and wheels.

Stress-corrosion cracking (SCC) in magnesium alloys was first identified and studied during its early use in military hardware, and investigations have continued to the present day. This chapter reviews all known work that has been published on this subject since the late 1930s and summarizes the material, mechanical, and environmental factors that influence SCC susceptibility in magnesium alloys, as well as SCC crack morphologies, proposed SCC mechanisms, accounts of service failures, and recommended preventive measures.

Early History

The earliest reported work on the SCC of magnesium alloys was conducted in Germany during the 1930s (Ref 1-3). However, its importance was not widely recognized until the early 1940s, when new applications of magnesium alloys in aircraft structures focused attention on the subject (Ref 4). Interest in SCC was heightened by the cracking of airplane booms that were being stored outside after they had buckled during static load tests; leading to a major study of the SCC resistance of various commercial wrought magnesium alloys (Ref 5).

Wrought Versus Cast Alloys

The early German studies claimed that, while wrought magnesium alloys could be very susceptible to SCC, cast alloys showed practically no SCC sensitivity when loaded up to their yield strengths (Ref 1, 2). Other reports that followed also indicated that the cast alloys were much less susceptible, if not immune, to SCC (Ref 6-11). However, no cast magnesium data were reported in the later references, so these statements are presumed to have been based on the conclusions of the earlier German work. It has since been pointed out that the German investigations were very limited, with most tests conducted above the yield strengths of the alloys (Ref 12). Therefore, the basis for these claims is questionable. Also, as discussed in the section "Service Failures" in this chapter, a number of SCC failures of cast magnesium components have since been documented.

Recent laboratory studies have also demonstrated that cast magnesium alloys can be susceptible to SCC when statically loaded *below* their yield strengths during outdoor atmospheric expo-

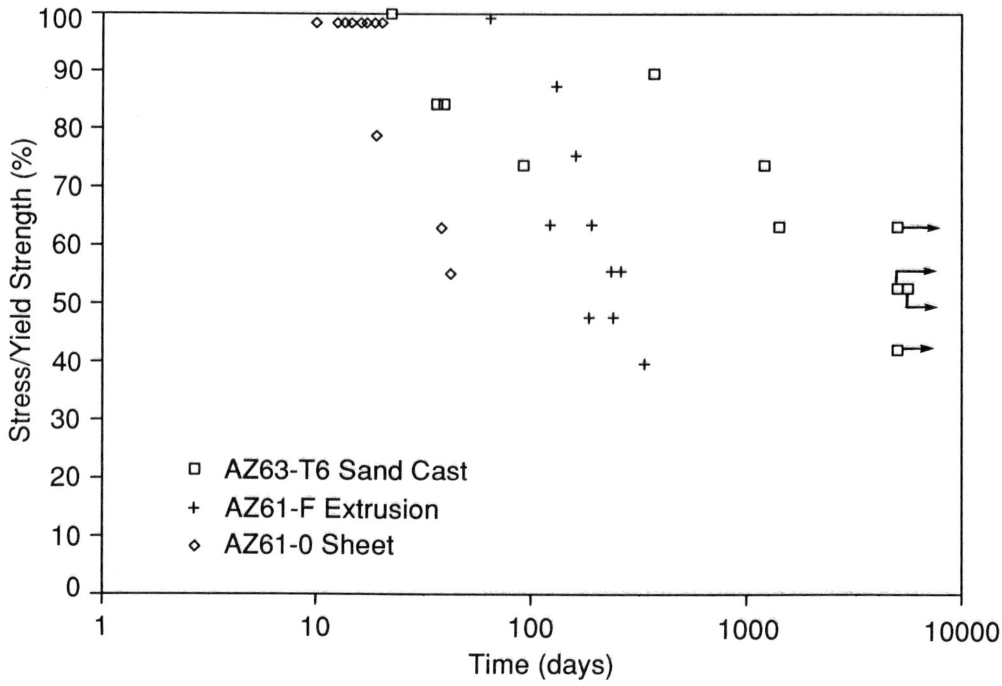

Fig. 1 Data from Ref 13 comparing similar cast and wrought magnesium alloys during long-term SCC tests in a rural environment

sure or immersion in distilled water (Ref 11-14). Long-term rural-atmosphere SCC data taken from one of these investigations are shown in Fig. 1, comparing similar-composition AZ61 sheet, extruded AZ61, and sand-cast AZ63 (Ref 13).* Although there is a large amount of scatter in these data, all three materials exhibited similar behavior at the higher stress levels. At lower stresses, the cast alloy became more resistant to SCC.

Effects of Alloy Composition

Pure magnesium is not susceptible to SCC when loaded up to its yield strength in atmospheric and most aqueous environments (Ref 1, 2, 7, 15, 16). The only reports of SCC of pure magnesium have emanated from laboratory tests in which specimens were immersed in very severe SCC solutions (Ref 17-19).

Aluminum-containing magnesium alloys have the highest SCC susceptibility (Ref 8, 10, 20), with the sensitivity increasing with increasing aluminum content (Ref 5, 8, 11, 20, 21), as illustrated in Fig. 2. An aluminum content above a threshold level of 0.15 to 2.5% is reportedly required to induce SCC behavior (Ref 9, 11, 22), with the effect peaking at approximately 6% Al (Ref 8).

Zinc also induces SCC susceptibility in magnesium alloys (Ref 2, 21), so it is not surprising that the aluminum- and zinc-bearing AZ alloys, which are the most commonly used magnesium alloys, have the greatest susceptibility to SCC. Alloys with higher aluminum content, such as AZ61, AZ80, and AZ91, can be very susceptible to SCC in atmospheric and more severe environments, while lower-aluminum AZ31 is generally

*The letter designations of these magnesium alloys identify aluminum (A) and zinc (Z) as the principal alloying elements. (Other letter designations include: E, rare earth; H, thorium; K, zirconium; L, lithium; M, manganese; and Q, silver.) The numerical part of the designation gives the rounded-off nominal percentages of these principal elements; for example, AZ61 contains 6% Al and 1% Zn.

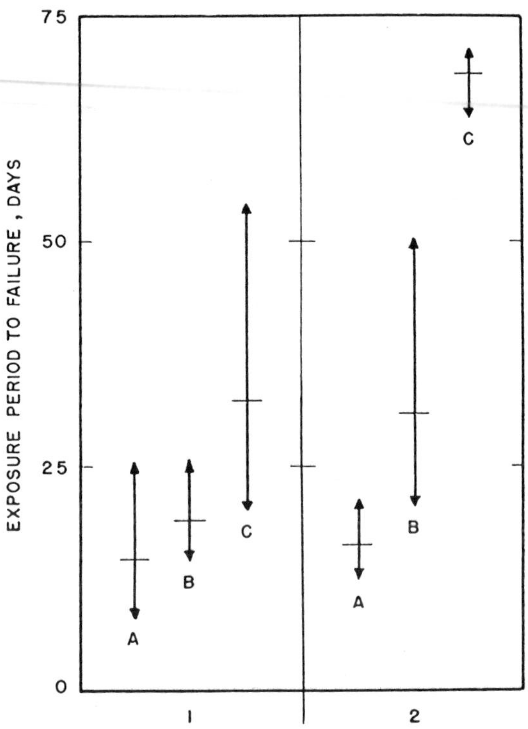

Fig. 2 SCC behavior of three magnesium sheet alloys with decreasing aluminum contents: A, AZ61; B, AZ51; C, AZ31. Data in two environments are shown: 1, intermittent immersion in 0.01% NaCl solution; 2, outdoor exposure. In both environments, SCC susceptibility decreases as aluminum content is reduced. Source: Ref 5

more resistant. However, it too can suffer SCC under certain conditions (Ref 4-7, 13, 20, 23).

Magnesium-zinc alloys that are alloyed with either zirconium or rare earths, but not with aluminum, such as ZK60 and ZE10, have intermediate SCC resistance (Ref 13, 20, 24), and in some cases SCC has not been a serious problem (Ref 8, 10). However, SCC can still occur in atmospheric environments at stresses as low as 50% of the yield strength, although life is significantly longer than for Mg-Al-Zn alloys (Ref 5, 13).

Magnesium alloys that contain neither aluminum nor zinc are the most SCC resistant. Magnesium-manganese alloys, such as M1, are among the alloys with the highest resistance to SCC, and they are generally considered to be immune when loaded up to the yield strength in normal environments (Ref 1, 2, 4-7, 13, 20, 23). In fact, SCC of Mg-Mn alloys has been reported only in tests involving stresses higher than the yield strength (Ref 5, 13) and/or exposure to very severe laboratory environments (Ref 24). Alloys QE22, HK31, and HM21 are also resistant to SCC, exhibiting SCC thresholds at approximately 70 to 80% of the yield strength in rural-atmosphere tests (Ref 13).

Magnesium-lithium alloys are of commercial interest because of their higher stiffness and lower density compared with other magnesium alloys. Tests in humid air have resulted in SCC failures of Mg-Li-Al alloys, but SCC did not occur during testing of Mg-Li alloys strengthened with zinc, silicon, and/or silver instead of aluminum (Ref 25).

Iron is found in commercial magnesium alloys in small, residual amounts and, although iron is known to reduce general corrosion resistance, its effect on SCC remains unclear. Early investigations indicated that iron was associated with decreased SCC resistance in Mg-Al-Zn alloys (Ref 22, 24, 26). Later studies, however, found that iron had minimal or no effect (Ref 8, 12, 20, 27, 28).

The effects of other alloying elements have had limited study. Shvartz and Kristal (Ref 21) reported that additions of cadmium increase SCC susceptibility, while cerium, tin, lead, copper, nickel, and silicon have no effect. Timonova (Ref 20), however, has indicated that the addition of cerium or tin to Mg-Al-Mn alloys somewhat increases SCC susceptibility.

Effects of Microstructure

Relatively few studies have been conducted to determine the effects of microstructural variables on the SCC susceptibility of magnesium alloys, and these studies have been restricted to Mg-Al alloys tested primarily in salt-chromate solutions.

It has been proposed that transgranular SCC is related to preferential corrosion of the Mg-Al matrix, a result of the galvanic couple created with cathodic FeAl precipitates that are distributed within the more active, anodic matrix (Ref 22, 24, 26, 29). However, there is no direct evidence relating the crack path with this precipitate, and detailed thin-film electron microscopy examinations by Fairman and Bray (Ref 27) have suggested that transgranular attack is not related to precipitates of any kind.

Intergranular SCC has been related to localized galvanic attack of the matrix when coupled with cathodic $Mg_{17}Al_{12}$ grain-boundary precipitates (Ref 26, 28-30). Fairman and Bray (Ref 28) have suggested that this preferential corrosion also produces a stress concentration that ruptures the protective surface film and promotes further rapid corrosion of the matrix.

Unlike SCC in wrought aluminum and titanium, preferred crystallographic orientation does not appear to affect the SCC of wrought magnesium alloys (Ref 3, 10, 16).

Effects of Cold Work

The effects of cold work on the SCC susceptibility of magnesium alloys are ambiguous. A few limited studies have shown deformation processes, such as the cold rolling of sheet, to reduce the SCC resistance of magnesium alloys (Ref 3, 15, 23, 31), while other tests have indicated that deformation increases SCC resistance (Ref 5, 13). However, this subject has not been investigated in any detail, and there have been no attempts to separate residual stress and microstructural effects.

Effects of Heat Treatment

As with cold work, the reported effects of heat treatment on SCC susceptibility are varied and sometimes contradictory. Heat treatment has been found to have an insignificant effect on magnesium alloy SCC threshold stress intensities and crack-growth rates (Ref 32), and the SCC resistance of Mg-Al-Zn alloys has been shown to be unaffected by solution treating, variations in aging temperature (Ref 2), or differences in the solution treating temperature (Ref 24). In other studies, however, heat treating of cold-worked magnesium alloys reduced SCC susceptibility because of assumed reductions in residual stress and the formation of a more resistant microstructure (Ref 4, 31). Data reported by Logan and Hessing (Ref 5) show this to be true for extruded AZ80, and Timonova (Ref 20) reported developing SCC immunity in a very SCC-susceptible Mg-8Al alloy by using a heat treatment that promoted the formation of a microstructure with very fine, uniformly distributed intermetallic compounds. In contrast to these results, others have found that annealing of cold-rolled AZ31 (Ref 5) and AZ61 (Ref 7) and solution treating and room-temperature aging of a Mg-Li-Al alloy (Ref 25) increased SCC susceptibility, although artificial aging of the Mg-Li-Al alloy then im-

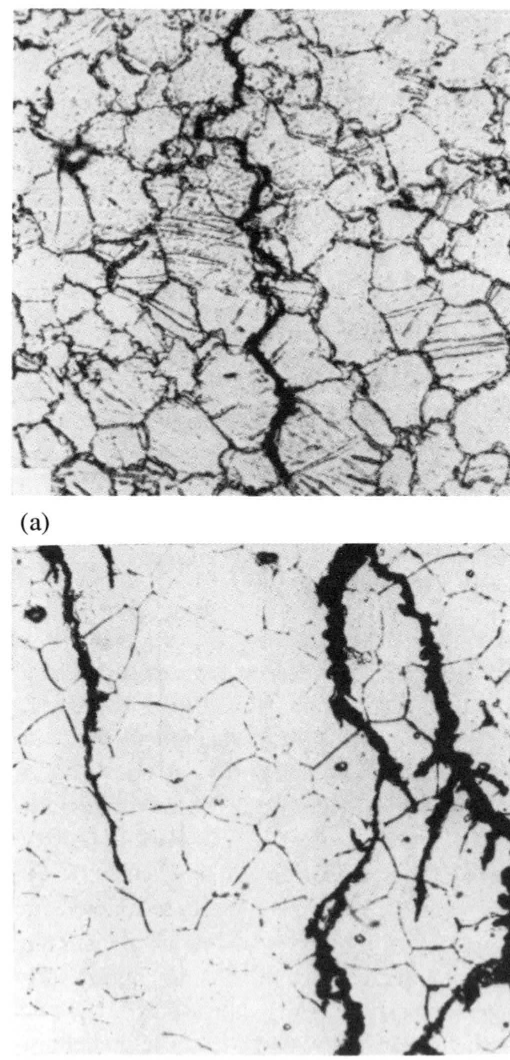

(a)

(b)

Fig. 3 SCC in an extruded Mg-6Al-1Zn alloy tested in a salt-chromate solution, showing intergranular crack propagation (top) in the furnace-cooled alloy and transgranular propagation (bottom) in the water-quenched material. Source: Ref 26 proved its resistance (Ref 25). Slow cooling from the solution-treating temperature, along with certain thermomechanical treatments, can produce $Mg_{17}Al_{12}$ grain-boundary precipitates, which, as shown in Fig. 3, increase the susceptibility of an alloy to intergranular SCC (Ref 26, 28-30).

The effect of heat treatment on SCC may also depend on the SCC environment. Artificial aging of cold-rolled AZ61 sheet improved SCC resistance when tested in salt-chromate solution or atmospheric exposure. However, the same treatment reduced SCC resistance in a 3% NaCl solution (Ref 7).

Effects of Environment

Stress-corrosion cracking of magnesium alloys can occur in many environments. In general, the only solutions that do not induce SCC are either those that are nonactive to magnesium, such as dilute alkalies, concentrated hydrofluoric acid, and chromic acid, or those that are highly active, in which general corrosion predominates (Ref 7, 15).

Air Environments. Exposure to normal atmospheric environments has been shown to produce SCC in susceptible magnesium alloys (Ref 1, 2, 14, 33-35); such environments include rural (Ref 6, 13), urban (Ref 5), industrial (Ref 6), and coastal (Ref 5, 6, 21) conditions. Tests of identical materials at different sites — for example, seacoast versus rural (Ref 7) and rural versus industrial (Ref 6) — have produced nearly identical results, indicating that SCC severity is not necessarily associated with the corrosion severity of the environment.

Rainfall, dew, and high humidity accelerate SCC of magnesium alloys during atmospheric exposure (Ref 7, 36), with failures often occurring during the drying period after a rain (Ref 5). Laboratory tests have shown that SCC occurs in indoor air only when the relative humidity exceeds 85 to 98% (Ref 20, 25, 36), with additions of oxygen or carbon dioxide slightly reducing this threshold (Ref 36).

Water Environments. Distilled water produces SCC in magnesium alloys during spraying or full, partial, or intermittent immersion (Ref 3, 7, 12, 21, 24, 33). Dissolved oxygen accelerates SCC (Ref 12, 24, 37), and deaeration considerably retards or even prevents it (Ref 24). Rapid SCC has also been observed during seawater immersion (Ref 36).

Aqueous Solutions. The desire for an accelerated test to study SCC of magnesium alloys led to the development by Sager *et al.* (Ref 23) of an aqueous NaCl + K_2CrO_4 solution that produced very rapid cracking and a ranking of alloy susceptibility that appeared to correlate with atmospheric tests. The aggressiveness of this solution is thought to result from the partial passivation of the magnesium by the chromate, which retards general corrosion but allows, or even promotes, local attack by the chloride. This electrolyte has remained very popular in laboratory studies of magnesium SCC, even though it has since been found to correlate poorly with service experience (Ref 5, 6). In fact, no accelerated laboratory test has been developed that adequately predicts service life or the relative susceptibility of different alloys (Ref 9, 11, 36, 38).

The addition of nitrate ions (Ref 27, 39) or carbonate ions (Ref 27) to salt-chromate solutions inhibits SCC. This is believed to be associated with the formation of a stronger, more stable or more readily repaired passive film.

Magnesium SCC is also produced in many other *dilute* aqueous solutions, including (in approximate decreasing severity): NaBr, Na_2SO_4, NaCl, $NaNO_3$, Na_2CO_3, $NaC_2H_3O_2$, NaF, and Na_2HPO_4 (Ref 32, 36, 40). Accelerated SCC has also been reported in dilute solutions of KF, KHF_2, HF (Ref 20, 24, 41), KCl, CsCl, NaI (Ref 36), KI (Ref 32), $MgCO_3$ (Ref 24), NaOH, and H_2SO_4, HNO_3, and HCl acids (Ref 20).

Although pH can affect the general corrosion of magnesium alloys, a pH of between 1.05 and 11.5 has been found to have no effect on SCC susceptibility in a salt solution (Ref 42). However, when pH is greater than 12, magnesium alloys become very resistant to SCC (Ref 7, 23). The pH of salt-chromate solutions has been reported by Mears *et al.* (Ref 43) to have an effect on whether the SCC crack path is intergranular or transgranular, although other studies have shown the path to be independent of pH (Ref 26, 29, 30).

Cathodic polarization has been shown in many studies to reduce, if not prevent, SCC of magnesium alloys in salt-chromate and other aqueous solutions (Ref 18, 26, 43-47). In a few cases, however, cathodic polarization was found to accelerate SCC (Ref 12, 48).

Anodic polarization generally has been found to increase the SCC susceptibility of magnesium and magnesium alloys (Ref 18, 46, 47), although mixed behavior has also been reported (Ref 27, 45). In single-phase Mg-Al alloys, Fairman and Bray (Ref 27) showed that high anodic potentials can produce a passive film that prevents SCC in a salt-chromate solution. In a multiphase alloy, however, the same conditions do not produce a protective film, and the electrochemical potential difference between the $Mg_{17}Al_{12}$ phase and matrix solid solution can promote pitting (Ref 28).

Effects of Temperature

Susceptibility to SCC of magnesium alloys has been found to increase with increasing temperature during tests in atmospheric environments (Ref 36), water (Ref 49), and a H_2SO_4 + NaCl solution (Ref 45). In a passivating solution, however, SCC can be significantly reduced as temperature increases within a restricted range, due to improved passivation kinetics (Ref 45, 50).

Temperature also affects the amount of creep deformation that occurs in a material. As will be discussed in the next section, creep can lead to a reduction in stress. Therefore, under some conditions, SCC susceptibility may be reduced by the additional creep that occurs at higher temperatures.

Effects of Strain Rate and Stress

As with other metals, the SCC susceptibility of magnesium alloys increases with increasing tensile stress. It has been recommended that constant working stresses applied for long periods of time be limited to 30 to 50% of the yield strength to prevent SCC in normal atmospheric environments (Ref 14, 51-53). However, this design guideline may not always be conservative enough. For example, SCC of AZ61-H24 sheet occurred in less than 9 months in an outdoor rural environment when spring loaded in tension to 23% of the yield strength (Ref 13).

Graf (Ref 3) suggested that the SCC threshold stress is associated with the onset of plastic deformation, that is, the elastic limit of the alloy. The elastic limit is difficult to measure unambiguously, however, and the 0.2% yield strength (the stress at which 0.2% plastic deformation occurs) is generally used instead. It is interesting to note, therefore, that the 30% yield-strength limit recommended for die-cast alloy AZ91 (Ref 52) correlates well with the elastic limit of this material being equal to approximately one-third of the yield strength (Ref 54).

Constant-load SCC tests have been shown to be more severe than constant-deflection tests (Ref 29). Under a constant load, stress increases as the cross section is reduced by cracking or corrosion. However, this condition produces decreasing stress when deflection is fixed. Creep, which can occur at low stresses at room temperature (Ref 55), can also produce stress reduction through relief of both residual stresses (Ref 13) and stresses induced by an imposed strain (Ref 6, 11, 12, 55).

A number of investigations have found strain rate to be an important factor in the SCC of magnesium alloys (Ref 36, 44, 47, 56). In laboratory constant-strain-rate tests in a salt-chromate solution, SCC took place only within a restricted range of strain rates, with ductile failure occurring at both higher and lower rates (Ref 47, 56). These results have been used to support a film-rupture SCC mechanism, discussed later in this chapter.

Crack Morphology

Although both transgranular and intergranular crack propagation have been observed, SCC in magnesium alloys is usually transgranular, with significant secondary cracking (branching). Initiation of these cracks invariably occurs at corrosion pits (Ref 39, 47, 57, 58).

Transgranular SCC fractures in wrought magnesium alloys are typically crystallographic in appearance (Fig. 4), with coarse, parallel markings corresponding to a stepped, faceted topography within each grain. The crystallographic orientation of the crack has been heavily investigated. Siebel (Ref 1) and Beck (Ref 2) proposed

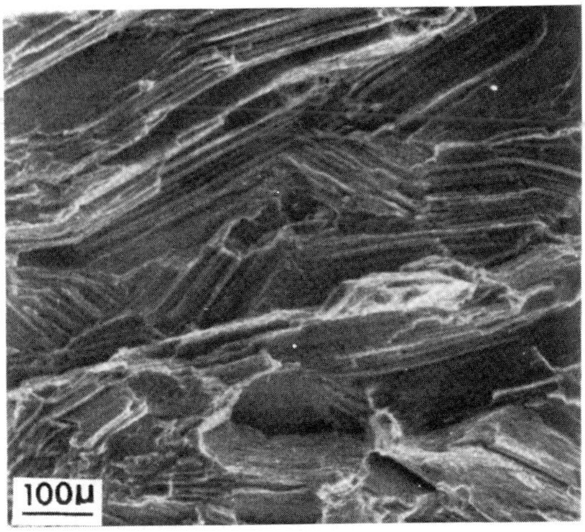

Fig. 4 Scanning electron micrograph of a transgranular SCC fracture surface in Mg-7.6Al sheet tested in salt-chromate solution. Source: Ref 61

Fig. 5 Electron micrograph of a surface replica showing fine, parallel striations (region A) on a transgranular SCC fracture. Source: Ref 57

that the crack follows twinning planes. More recently, Oryall and Tromans (Ref 59) reported that deformation twinning ahead of the crack tip led to the subsequent passage of the crack along the twin-boundary interface. Others, however, have argued that SCC crack propagation is not related to twinning (Ref 3, 27). Propagation along the basal slip plane has been proposed (Ref 26) and then refuted (Ref 46, 59, 60). Metzger (Ref 41) found the SCC crack to follow, for the most part, a $(10\bar{1}0)$ prism plane, with occasional branching along the $(\bar{1}101)$ pyramidal slip plane, where chemical attack had produced V-shape grooves.

Pugh and coworkers (Ref 57, 58, 61) reported that transgranular SCC occurs by discontinuous cleavage on $\{31\bar{4}0\}$ planes, with the coarse stepped or pleated topography consisting of alternate faces produced by $\{31\bar{4}0\}$ cleavage and ductile rupture parallel to (0001). Meletis and Hochman (Ref 17) concurred that the crack propagates along cleavage planes, but found the fracture surfaces to consist of $\{2\bar{2}03\}$ facets separated by $\{2\bar{2}03\}$ steps that were formed by the fracture of thin regions separating parallel propagating cracks. They suggested that the difference between this cleavage plane and that reported by Pugh, as well as the crystallographic systems observed by others, may result from the use of analytic techniques that provide an overall macroscopic crack orientation, rather than the orientation of the individual SCC facets. Lynch and Trevena (Ref 19) found SCC crack growth to follow cleavage planes similar to those identified by Meletis and Hochman, but with cleavagelike facets or secondary cracks on {0001} planes. The fracture surfaces also contained very fine flutes or dimples, which led Lynch and Trevena to conclude that SCC occurred by a very localized microvoid-coalescence process, rather than by cleavage.

Very fine, parallel striations, spaced approximately 1 μm or less apart, have been observed during transmission electron microscopy (TEM) examination of transgranular SCC fracture surfaces (Fig. 5). The striations are oriented perpendicular to the direction of cracking, occur within the coarser markings described above, and have been associated with successive positions of the crack during its discontinuous propagation (Ref 27, 46, 57, 58). However, Oryall and Tromans (Ref 59) have suggested that these striations do not correspond to crack-arrest fronts, but rather to the propagation of the crack along very small twin-boundary steps.

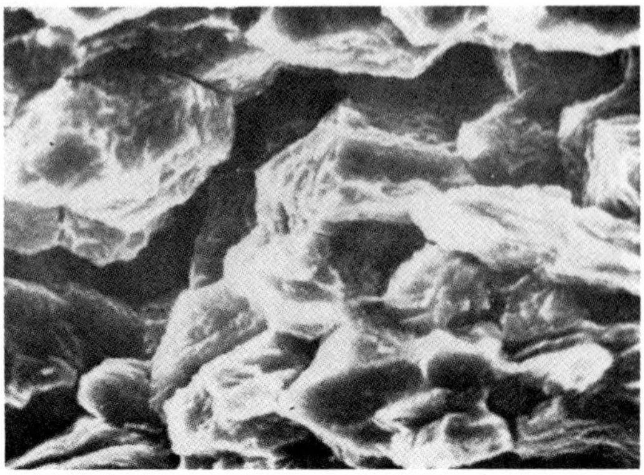

Fig. 6 Scanning electron micrograph of an intergranular SCC fracture surface in AZ61 wire tested in salt-chromate solution. Source: Ref 28

Mixed transgranular and intergranular crack propagation (and, occasionally, totally intergranular cracking) have also been observed during magnesium SCC (Fig. 3 and 6). The role of grain-boundary FeAl precipitates in intergranular crack propagation has already been discussed. Intergranular cracking is also reportedly promoted by a small grain size (Ref 26, 29) and/or a low residual iron content (Ref 29). Intergranular crack propagation may also be characteristic of specific alloys or alloy/environment combinations. For example, SCC testing of a Mg-Zn-Zr alloy in water (Ref 24), a Mg-Li-Al alloy in humid air (Ref 25), and alloy AZ61 in a 75 °C (167 °F) 5% NaCl solution (Ref 49) produced predominantly intergranular fractures.

SCC Mechanisms

Transgranular SCC

Over the years, many mechanisms have been proposed to explain transgranular SCC in magnesium. These mechanisms can be classified into one of two general groups, depending on whether crack propagation occurs primarily through a process of dissolution or of brittle fracture.

Dissolution Models. An electrochemical dissolution model was first proposed by Mears *et al.* (Ref 43) and received early support because of the observation that cathodic protection could prevent or reduce SCC in magnesium alloys. In work that followed, a number of specific dissolution processes were proposed, based on the mechanisms discussed below.

Preferential Attack. Accelerated dissolution is produced by the electrochemical cell that is created around FeAl, $Mg_{17}Al_{12}$, or possibly other precipitates that are cathodic to the magnesium alloy matrix (Ref 20, 22, 24, 26). Stress contributes to the process by producing a stress concentration that increases the local corrosion rate (Ref 20), or by rupturing the protective surface film and controlling its reformation so that the electrochemical process can proceed (Ref 20, 26).

Film Rupture. The protective surface film is ruptured by localized plastic deformation, producing an electrochemical cell between the anodic film-free and the cathodic filmed areas (Ref 37, 44, 62, 63). This mechanism could produce either continuous or discontinuous crack growth, depending on the rate at which reformation of the surface film occurs (Ref 62). Constant-strain-rate SCC tests have demonstrated that crack propagation can depend on the competing rates at which bare metal is created and a new surface film is formed (Ref 47, 56).

Tunneling. Based on their observation of fine corrosion tunnels in certain SCC systems, Pickering and Swann (Ref 64) proposed that a mechanically weak, tubular pitted structure is produced along active slip planes. Eventually, the thin ligaments that remain between these tunnels rupture.

The validity of any dissolution-based magnesium SCC model has been challenged on two points. First, calculations have shown that prohibitively high corrosion current densities would be required to account for observed SCC cracking rates (Ref 46). In addition, corrosion and local plastic deformation that are implicit in these processes are inconsistent with the flat, interlocking facets and shallow surface steps that are commonly observed on transgranular SCC fracture surfaces (Ref 58).

Brittle Fracture Models. Mechanisms in which SCC crack propagation occurs through a brittle fracture process are consistent with the observed crack morphology and thus have been the focus of the most recent studies in this area. The fracture processes that have been proposed fall into the two general categories of cleavage and hydrogen embrittlement.

Cleavage Processes. Pardue *et al.* (Ref 29) were the first to suggest that stress-corrosion cracks in magnesium alloys propagate by a mechanical fracturing process. They proposed that the role of corrosion was to produce pits or other stress concentrators that trigger a crack and then to remove obstacles that stop the crack. Fairman and coworkers (Ref 27, 30) proposed that the passage of dislocations on slip planes ruptures the surface film, allowing development of a corrosion pit which then initiates cleavage. The process is then repeated after the crack is stopped by barriers, a reduction in stress, or blunting of the crack. They also suggested that the effect of aluminum on magnesium SCC resulted from the formation of a defective, more easily cleaved Al,$Mg(OH)_2$ surface film (Ref 27). Pugh *et al.* (Ref 46) proposed that fracture occurs via an embrittled surface layer, either an oxide or a porous layer formed by selective dissolution of magnesium or aluminum. The crack then becomes blunted and stops when it enters the ductile substrate. The process repeats after the film reforms.

Hydrogen Embrittlement. Liu (Ref 65) first suggested that cathodically generated hydrogen was related to magnesium SCC, and other investigators have since contributed to this model (Ref 17-19, 47, 57, 58, 61-68). Experimental evidence supporting a hydrogen-embrittlement mechanism includes:

- SCC crack initiation and propagation are accompanied by hydrogen evolution (Ref 57, 58)
- Immersion in a cracking solution before stress is applied produces a fracture similar to a SCC fracture (Ref 66, 68)
- The effect of preimmersion in a cracking solution is reversed by vacuum annealing (Ref 66) or exposure to room-temperature air (Ref 18)
- Testing in gaseous hydrogen results in the same crack characteristics produced in aqueous solution tests (Ref 47)
- SCC occurs at crack velocities at which only adsorbed hydrogen should be present at the crack tip (Ref 19)

A number of specific hydrogen-embrittlement mechanisms have been proposed for magnesium. A weak, stress-induced magnesium hydride may form (Ref 58, 65), a theory supported by observations of an approximately 1 μm thick brittle MgH_2 film on a magnesium SCC fracture (Ref 58). The SCC cleavage plane may be the hydride habit or cleavage plane (Ref 17, 58, 66). High hydrogen concentrations could produce stress concentrations on the cleavage plane (Ref 17). Hydrogen in solid solution may lower the cleavage strength of magnesium (Ref 18). Finally, surface-adsorbed hydrogen could weaken the interatomic bonds at the crack tip (Ref 19).

At one time, a hydrogen-embrittlement SCC model was thought to be inapplicable to magnesium because of the effects that anodic and cathodic polarization have on magnesium SCC. That is, cathodic polarization, which generates hydrogen, generally inhibits SCC crack growth, while anodic polarization, in which hydrogen is not generated, usually accelerates cracking (Ref 46). However, it has been demonstrated that anodic polarization can, within a restricted potential range, discharge more hydrogen than cathodic polarization — a result of the so-called "negative difference effect" (Ref 18, 47). One explanation proposed for this phenomenon is that atomic hydrogen can enter and embrittle magnesium only when it is generated by local electrochemical reactions at active, film-free pit surfaces. Because cathodic polarization prevents pitting, SCC cannot occur. During anodic polarization, however, pitting is promoted, hydrogen is adsorbed/absorbed, and embrittlement proceeds.

It has been suggested that hydrogen embrittlement may not always be the primary mechanism for SCC crack growth (Ref 67). While hydrogen embrittlement can occur in passivating solutions because dissolution is effectively suppressed, high dissolution rates produced in more aggressive solutions could retard the hydrogen-embrittlement reaction.

Intergranular SCC

Because intergranular SCC is not the major fracture mode in magnesium, it has received much less attention than transgranular SCC. As already discussed, intergranular cracking has been attributed to the occurrence of a grain-boundary $Mg_{17}Al_{12}$ phase that produces galvanic attack of the adjacent matrix.

Service Failures

Despite the SCC sensitivity shown by magnesium alloys in laboratory tests, it has often been reported that service failures are rare and normally result from excessive residual stress produced during fabrication (Ref 6, 7, 9, 52, 69). More recently, however, failures of a number of forged AZ80 French aircraft components were reported, apparently resulting from excessive assembly and residual stresses (Ref 51). Service failures of cast + forged South African magnesium aircraft wheels have also been described (Ref 35). Speidel (Ref 34), in a comprehensive review of more than 3000 unclassified failure reports from aerospace companies, government agencies, and research laboratories in the United States and five Western European countries, estimated that approximately 10 to 60 magnesium aerospace component SCC service failures occurred each year from 1960 to 1970. Of this total, more than 70% involved either cast alloy AZ91-T6 or wrought alloy AZ80-F.

Prevention of SCC

As with other metals, the methods used to control SCC in magnesium alloys involve changing one of the three simultaneous conditions that SCC requires: the loading of a *susceptible alloy* above a *critical stress* during exposure to a SCC-producing *environment*.

Selection of a Resistant Alloy. As has been discussed, the magnesium alloys with the greatest SCC susceptibility are the Mg-Al alloys, their susceptibility increasing with increasing aluminum content. Magnesium-zinc alloys have intermediate susceptibility, and alloys that contain neither aluminum nor zinc are the most SCC resistant. Unlike aluminum and some other metal alloy systems, no special heat treatments have been found that will reliably reduce or eliminate magnesium SCC. However, cladding susceptible magnesium alloys with alloy M1 has produced SCC-resistant sheet materials.

Control of Stress. The constant working stress supported by the magnesium must be kept below a critical level, which, as discussed previously, has been reported to be 30 to 50% of the tensile yield strength of the alloy. Bolted or riveted joints can also produce locally high stresses, so proper joint design and construction techniques are important (Ref 52) — for example, using preformed parts, avoiding the overtorquing of bolts, and providing adequate spacing and edge margins for rivets (Ref 53). Because cast-in inserts may cause SCC due to local residual stresses created in the surrounding magnesium, it has been recommended that inserts with a wall thickness greater than 1.25 mm (0.05 in.) be preheated before casting (Ref 53).

Many early magnesium SCC studies found welding residual stresses to be particularly dangerous; as a result, a low-temperature thermal stress-relief treatment has become standard recommended practice for welded assemblies. Shot peening and other mechanical processes that create favorable compressive surface residual stresses may also effectively increase SCC resistance (Ref 7, 20, 69).

Control of the Environment. Properly controlled cathodic polarization may reduce, or even prevent, SCC of magnesium alloys in aqueous solutions. In most cases, however, polarization of the magnesium is not feasible, and little else can be done to alter the environment. The only remaining option is to separate the environment and the magnesium by using an organic or inorganic coating. Studies of the effectiveness of coatings have shown them to extend life but not to totally prevent SCC, with breaks in the coating producing an expected reduction in this protection (Ref 13, 20, 51-53). One laboratory study found an inorganic coating even accelerated SCC under certain conditions (Ref 70).

References

1. G. Siebel, The Influence of Stress on the Corrosion of Electron Metals, *Jahrbuch der deutschen Luftfahrtforschung*, Part 1, 1937, p 528-531
2. A. Beck, *The Technology of Magnesium and Its Alloys*, 2nd ed., F.A. Hughes and Co., 1940, p 294-297
3. L. Graf, Concerning the Problem of Stress-Corrosion, *Luftwissen*, Vol 7, 1940, p 160-169
4. M.A. Hunter, The Stress Corrosion of Magnesium Alloys, in *Metals Handbook*, American Society for Metals, 1948, p 234-236
5. H.L. Logan and H. Hessing, Stress Corrosion of Wrought Magnesium Base Alloys, *J. Res. Natl. Bur. Stand.*, Vol 44, 1950, p 233-243

6. W.S. Loose and H.A. Barbian, Stress-Corrosion Testing of Magnesium Alloys, in *Symp. Stress-Corrosion Cracking of Metals*, American Society for Testing and Materials, 1945, p 273-292
7. W.S. Loose, Magnesium and Magnesium Alloys, in *The Corrosion Handbook*, H.H. Uhlig, Ed., John Wiley and Sons, 1948, p 232-250
8. E.F. Emley, *Principles of Magnesium Technology*, Pergamon Press, 1966, p 720-721
9. H.L. Logan, Magnesium Alloys, in *The Stress Corrosion of Metals*, John Wiley and Sons, 1966, p 217-237
10. I.J. Polmear, *Light Alloys: Metallurgy of the Light Metals*, American Society for Metals, 1981, p 154-159
11. *Metals Handbook*, 9th ed., Vol 13, *Corrosion*, ASM International, 1987, p 745
12. W.K. Miller, Stress Corrosion Cracking of Magnesium Die Casting Alloys, in *Materials Stability and Environmental Degradation*, Materials Research Society Symp. Proc., Vol 125, A. Barkatt, E.D. Verink, Jr., and L.R. Smith, Ed., 1988, p 253-259
13. "Exterior Stress Corrosion Resistance of Commercial Magnesium Alloys," Report Mt 19622, Dow Chemical USA, March 8, 1966
14. Y. Kawase and T. Tomoda, "Evaluation of Stress Corrosion Cracking of MDCIA Magnesium Die Casting Alloy for the Development of Magnesium Steering Upper Bracket," Toyota Motor, 1985
15. N.D. Tomashov, *Theory of Corrosion and Protection of Metals* (transl.), B.H. Tytell, I. Geld, and H.S. Preiser, Ed., MacMillan, 1966, p 626
16. M.J. Blackburn and M.O. Speidel, The Influence of Microstructure on the Stress Corrosion Cracking of Light Alloys, in *Electron Microscopy and Structure of Materials*, G. Thomas, Ed., University of California Press, 1972, p 905-919
17. E.I. Meletis and R.F. Hochman, Crystallography of Stress Corrosion Cracking in Pure Magnesium, *Corrosion*, Vol 40 (No. 1), 1984, p 39-45
18. R.S. Stampella, R.P.M. Procter, and V. Ashworth, Environmentally-Induced Cracking of Magnesium, *Corros. Sci.*, Vol 24 (No. 4), 1984, p 325-341
19. S.P. Lynch and P. Trevena, Stress Corrosion Cracking and Liquid Metal Embrittlement in Pure Magnesium, *Corrosion*, Vol 44 (No. 2), 1988, p 133-124
20. M.A. Timonova, Corrosion Cracking of Magnesium Alloys and Methods of Protection Against It, in *Intercrystalline Corrosion and Corrosion of Metals Under Stress* (transl.), I.A. Levin, Ed., Consultants Bureau, 1962, p 263-282
21. G.L. Shvartz and M.M. Kristal, *Corrosion of Chemical Apparatus* (transl.), Consultants Bureau, 1959, p 175-176, 197
22. R.D. Heidenreich, C.H. Gerould, and R.E. McNulty, Electron Metallographic Methods and Some Results for Magnesium Alloys, *Trans. AIME*, Vol 166, 1946, p 15-36
23. G.F. Sager, R.H. Brown, and R.B. Mears, Tests for Determining Susceptibility to Stress-Corrosion Cracking, in *Symp. Stress-Corrosion Cracking of Metals*, American Society for Testing and Materials, 1945, p 255-272
24. E.C.W. Perryman, Stress-Corrosion of Magnesium Alloys, *J. Inst. Met.*, Vol 78, 1951, p 621-642
25. J.C. Kiszka, Stress Corrosion Tests of Some Wrought Magnesium-Lithium Base Alloys, *Mater. Protect.*, Vol 4 (No. 2), 1965, p 28-29
26. D.K. Priest, F.H. Beck, and M.G. Fontana, Stress-Corrosion Mechanisms in a Magnesium-Base Alloy, *ASM Trans.*, Vol 47, 1955, p 473-487
27. L. Fairman and H.J. Bray, Transgranular SCC in Mg-Al Alloys, *Corros. Sci.*, Vol 11, 1971, p 533-541
28. L. Fairman and H.J. Bray, Intergranular Stress-Corrosion Crack Propagation in Magnesium Aluminum Alloys, *Br. Corros. J.*, Vol 6, 1971, p 170-174
29. W.M. Pardue, F.H. Beck, and M.G. Fontana, Propagation of Stress-Corrosion Cracking in a Magnesium-Base Alloy as Determined by Several Techniques, *Trans. ASM*, Vol 54, 1961, p 539-548
30. L. Fairman and J.M. West, Stress Corrosion Cracking of a Magnesium Aluminum Alloy, *Corros. Sci.*, Vol 5, 1965, p 711-716
31. M.A. Timonova, L.I. D'yachenko, Yu.M. Dolzhanskii, M.B. Altman, N.V. Sakharova, and A.A. Blyablin, Influence of Annealing on Stress Corrosion Cracking Resistance of VMD10-1 Magnesium Alloy, *Protect. Met.*, Vol 19 (No. 1), 1983, p 99-102
32. M.O. Speidel, M.J. Blackburn, T.R. Beck, and J.A. Feeney, Corrosion Fatigue and Stress Corrosion Crack Growth in High Strength Aluminum Alloys, Magnesium Alloys and Titanium Alloys Exposed to Aqueous Solutions, in *Corrosion Fatigue: Chemistry, Mechanics and Microstructure*, National Association of Corrosion Engineers, 1972, p 324-342
33. B.C. Madden, Jr., Test Methods and Progress in the Stress-Corrosion Investigation at Wright Field, in *Symp. Stress-Corrosion Cracking of Metals*, American Society for Testing and Materials, 1945, p 233-254
34. M.O. Speidel, Stress Corrosion Cracking of Aluminum Alloys, *Metall. Trans.A*, Vol 6, 1975, p 631-651
35. J.J. Lourens, Failure Analysis as a Basis for Design Modification of Military Aircraft, in *Fracture and Fracture Mechanics Case Studies*, R.B. Tait and G.G. Garrett, Ed., Pergamon Press, 1985, p 47-56
36. M.A. Pelensky and A. Gallaccio, Stress Corrosion of Magnesium Alloys—Environmental Factors, in *Stress Corrosion Testing*, STP 425, American Society for Testing and Materials, 1967, p 107-115

37. N.D. Tomashov and V.N. Modestova, Effect of Stress on the Corrosion and Potentials of Alloys of the Magnesium-Manganese System, in *Intercrystalline Corrosion and Corrosion of Metals Under Stress* (transl.), I.A. Levin, Ed., Consultants Bureau, 1962, p 251-262
38. L. Whitby, Magnesium and Its Alloys, in *Corrosion Resistance of Metals and Alloys, Part I*, 2nd ed., F. LaQue and H.R. Copson, Ed., Reinhold, 1963, p 173
39. R.P. Frankenthal, The Inhibition of Pitting and Stress-Corrosion Cracking of Mg-Al Alloys by NO_3^-, *Corros. Sci.*, Vol 7, 1967, p 61-62
40. V.E. Belyakov and V.V. Romanov, Influence of Anions on the Long-Term Strength of MA2-1 Magnesium Alloy, *Sov. Mater. Sci.*, Vol 7 (No. 4), 1971, p 400-402
41. M. Metzger, discussion of Crack Propagation During Stress Corrosion, in *Physical Metallurgy of Stress Corrosion Fracture*, T.N. Rhodin, Ed., Interscience, New York, 1959, p 94-95
42. V.E. Belyakov, S.V. Pushkina, and V.V. Romanov, Effect of pH of Working Media on Long-Time Strength of a Magnesium Alloy MA2-1, *Sov. Mater. Sci.*, Vol 6 (No. 5), 1970, p 539-541
43. R.B. Mears, R.H. Brown, and E.H. Dix, Jr., A Generalized Theory of Stress Corrosion of Alloys, in *Symp. Stress-Corrosion Cracking of Metals*, American Society for Testing and Materials, 1945, p 233-254
44. H.L. Logan, Mechanism of Stress-Corrosion Cracking in the AZ31B Magnesium Alloy, *J. Res. Natl. Bur. Stand.*, Vol 61 (No. 6), 1958, p 503-508
45. V.V. Romanov, Effect of Some Factors on the Susceptibility of Magnesium Alloys to Corrosion Cracking, in *Intercrystalline Corrosion and Corrosion of Metals Under Stress* (transl.), I.A. Levin, Ed., Consultants Bureau, 1962, p 283-287
46 E.N. Pugh, J.A.S. Green, and P.W. Slattery, On the Propagation of Stress-Corrosion Cracks in a Magnesium-Aluminum Alloy, in *Fracture 1969, Proc. 2nd Int. Conf. Fracture*, Chapman and Hall, 1969, p 387-395
47. K. Ebtehaj, D. Hardie, and R.N. Parkins, The Influence of Chloride-Chromate Solution Composition on the Stress Corrosion Cracking of a Mg-Al Alloy, *Corros. Sci.*, Vol 28 (No. 8), 1988, p 811-829
48. V.A. Marichev, Hydrogen Embrittlement of Structural Materials in Solutions of Chrome Anhydride, *Sov. Mater. Sci.*, Vol 24 (No. 1), 1988, p 92-95
49. S. Orman, A Rapid Test for Stress-Corrosion Cracking, *Corros. Sci.*, Vol 9, 1969, p 849-851
50. D.A. Vermilyea, A Theory for the Propagation of Stress Corrosion Cracks in Metals, *J. Electrochem. Soc.*, Vol 119, 1972, p 405-407
51. M. Vialatte, Study of the SCC Behavior of the Alloy Mg-8% Al, in *Symp. Engineering Practice to Avoid Stress Corrosion Cracking*, NATO, 1970, p 5-1 to 5-10
52. *Magnesium: Designing Around Corrosion*, Dow Chemical Co., 1982, p 16
53. E. Groshart, Magnesium, Part I—The Metal, *Met. Finish.*, Vol 83 (No. 10), 1985, p 17-20
54. W.K. Miller and E.F. Ryntz, Magnesium for Automotive Application: A State-of-the Art Assessment, *SAE Trans.*, Vol 92, 1983, p 2.524-2.552
55. W.K. Miller, Creep of Die Cast AZ91 Magnesium at Room Temperature and Low Stress, *Metall. Trans. A*, Vol 22, 1991, p 873-877
56. W.R. Wearmouth, G.P. Dean, and R.N. Parkins, Role of Stress in the Stress Corrosion Cracking of a Mg-Al Alloy, *Corrosion*, Vol 29 (No. 6), 1973, p 251-258
57. D.G. Chakrapani and E.N. Pugh, The Transgranular SCC of a Mg-Al Alloy: Crystallographic, Fractographic and Acoustic-Emission Studies, *Metall. Trans. A*, Vol 6, 1975, p 1155-1163
58. A.J. Bursle and E.N. Pugh, On the Mechanism of Transgranular Stress-Corrosion Cracking, in *Mechanisms of Environment Sensitive Cracking of Materials*, P.R. Swann, F.P. Ford, and A.R.C. Westwood, Materials Society (London), 1977, p 471-481
59. G. Oryall and D. Tromans, Transgranular Stress Corrosion Cracking of Solution Treated and Quenched Mg-8.6Al Alloy, *Corrosion*, Vol 27 (No. 8), 1971, p 334-341
60. H.L. Logan, Stress-Corrosion Cracking of the AZ31B Magnesium Alloy, *J. Res. Natl. Bur. Stand., Eng. Instrum.*, Vol 65C (No. 3), 1961, p 165-169
61. D.G. Chakrapani and E.N. Pugh, On the Fractography of Transgranular Stress Corrosion Failures in a Mg-Al Alloy, *Corrosion*, Vol 31 (No. 7), 1975, p 247-252
62. H.L. Logan, Film-Rupture Mechanism of Stress Corrosion, *J. Res. Natl. Bur. Stand.*, Vol 48 (No. 2), 1952, p 99-105
63. D.K. Priest, Electrochemical Aspects of Stress Corrosion, *J. Electrochem. Soc.*, Vol 106 (No. 4), 1959, p 358-360
64 H.W. Pickering and P.R. Swann, Electron Metallography of Chemical Attack Upon Some Alloys Susceptible to Stress Corrosion Cracking, *Corrosion*, Vol 19 (No. 11), 1963, p 373t-389t
65. H.W. Liu, Stress-Corrosion Cracking and the Interaction Between Crack-Tip Stress Field and Solute Atoms, *Trans. ASME, J. Basic Eng.*, Vol 92, 1970, p 633-638
66. D.G. Chakrapani and E.N. Pugh, Hydrogen Embrittlement of a Mg-Al Alloy, *Metall. Trans. A*, Vol 7, 1976, p 173-178
67. V.A. Marichev and S.A. Shipilov, Influence of Electrochemical Polarization on Crack Growth in Corrosion Cracking and Corrosion Fatigue of

Magnesium Alloys, *Sov. Mater. Sci.*, Vol 33 (No. 3), 1986, p 240-244

68. G.L. Makar, J. Kruger, and K. Sieradzki, Environment-Assisted Cracking of Rapidly Solidified Mg-Al Alloys, in *47th Annual World Magnesium Conf. Proc.*, International Magnesium Association, 1990, p 37- 42
69. J.D. Hanawalt, Joint Discussion on Aluminum and Magnesium, in *Symp. Stress-Corrosion Cracking of Metals*, American Society for Testing and Materials, 1945, p 310
70. Y.B. Yakovlev, L.P. Trutneva, N.I. Isaev, and G. Nemetch, Influence of Protective Films on Kinetics of Stress Corrosion Cracking of Magnesium Alloy MA2-1, *Protect. Met.*, Vol 20 (No. 3), 1984, p 300-306

CHAPTER 10

Stress-Corrosion Cracking of Titanium Alloys

R.W. Schutz, RMI Titanium Company

Titanium alloys have long been recognized for their superior corrosion resistance in many aggressive environments compared to common engineering metals. This resistance stems from the ability of titanium, as a reactive metal, to readily passivate in oxygen-containing media by forming a tenacious, coherent, self-healing oxide surface film (primarily TiO_2) (Ref 1, 2). As a result, these alloys resist a variety of conditions, from mildly reducing to highly oxidizing, over a wide range of pH and temperature, even in the presence of halides. This includes enhanced resistance to all common modes of metal corrosion, such as general, crevice, pitting, erosion, and stress corrosion, as outlined in other reference sources (Ref 1, 2).

As titanium alloys began to be considered for structurally critical applications with stringent "fail-safe" criteria, it became necessary to define cracking resistance of the alloys when exposed to a variety of aggressive service environments while highly stressed in tension. Although most commercial titanium alloys exhibit excellent resistance to stress-corrosion cracking (SCC) in most aerospace and industrial service environments, several unique service conditions that especially stimulate SCC have been discovered over the years. In the mid-1950s, numerous titanium alloys were found to be susceptible to SCC in red-fuming nitric acid (Ref 3,4). In the same period, laboratory creep testing of titanium alloy turbine blades revealed SCC under chloride salt deposits at temperatures above 250 °C (480 °F). In the late 1950s, contact of cadmium-plated fasteners with hot-stressed titanium alloy jet engine components (Ref 3, 4) produced cracking failures.

Catastrophic failures of titanium alloy missile tanks pressurized with either anhydrous methanol or N_2O_4 were also traced back to SCC in the early 1960s (Ref 3-5). Interest in titanium alloy utilization in supersonic aircraft and deep-sea submersibles in the mid-1960s led to the discovery that numerous medium- to high-strength titanium alloys in certain metallurgical conditions experience a significant degradation in fracture properties in saltwater/seawater (Ref 3, 4, 6). Since the early 1980s, downhole tubular applications for the more resistant high-strength titanium alloys in geothermal brine and deep sour gas well service (Ref 7, 8) have necessitated identification of temperature thresholds for alloy SCC in high-temperature (150 to 275 °C, or 300 to 525 °F) sweet and sour concentrated brines.

This chapter reviews and defines the particular environmental conditions that are now known to stimulate SCC in commercial titanium alloys. Typical alloy SCC data are presented and referenced. Attention is given to the mechanical and metallurgical factors that have been identified as influencing SCC behavior in each case, and possible mechanisms are discussed. Finally, practical guidelines and strategies for inhibiting and/or preventing titanium alloy SCC are presented.

Commercial Titanium Alloys

The designation, nominal composition, and alloy type of commercial and semicommercial ti-

Table 1 Designations and Nominal Compositions of Selected Commercial Titanium Alloys

Common alloy designation	UNS designation	Nominal composition, %	ASTM grade	Alloy type
Grade 1	R50250	Unalloyed titanium	1	α
Grade 2	R50400	Unalloyed titanium	2	α
Grade 3	R50550	Unalloyed titanium	3	α
Grade 4	R50700	Unalloyed titanium	4	α
Ti-Pd	R52400/R52250	Ti-0.15Pd	7/11	α
Grade 12	R53400	Ti-0.3Mo-0.8Ni	12	Near-α
Ti-3-2.5	R56320	Ti-3Al-2.5V	9	Near-α
Ti-6-4	R56400	Ti-6Al-4V	5	α-β
Ti-6-2-1-8	R56210	Ti-6Al-2Nb-1Ta-0.8Mo	...	Near-α
Ti-5-2.5	R54520	Ti-5Al-2.5Sn	6	α
Ti-8-1-1	R54810	Ti-8Al-1V-1Mo	...	Near-a
Ti-6-2-4-2	R54620	Ti-6Al-2Sn-4Zr-2Mo	...	Near-a
Ti-4-3-1	R56430	Ti-4Al-3Mo-1V	...	a-b
Ti-550	...	Ti-4Al-2Sn-4Mo-0.5Si	...	a-b
Ti-6-6-2	R56620	Ti-6Al-6V-2Sn-0.6Fe-0.6Cu	...	a-b
Ti-7-4	R56740	Ti-7Al-4Mo	...	a-b
Ti-6-2-4-6	R56260	Ti-6Al-2Sn-4Zr-6Mo	...	a-b
Ti-8Mn	R56080	Ti-8Mn	...	b
Beta III	R58030	Ti-11.5Mo-6Zr-4.5Sn	...	b
Ti-3-8-6-4-4	R58640	Ti-3Al-8V-6Cr-4Zr-4Mo	...	b
Ti-13-11-3	R58010	Ti-3Al-13V-11Cr	...	b
Ti-8-8-2-3	R58820	Ti-8V-8Mo-3Al-2Fe	...	b
Ti-15-5-3	...	Ti-15Mo-5Zr-3Al	...	β

tanium alloys addressed in this chapter are listed in Table 1. These alloys are typically categorized into four basic types: alpha (α), near-α, alpha-beta (α/β), and beta (β).

The α alloys, exhibiting a hexagonal close-packed (hcp) crystal structure, include the unalloyed grades and those alloys consisting primarily of α-phase-stabilizing elements, such as aluminum, tin, and the interstitials: oxygen, nitrogen, and carbon. With strength derived from solid-solution hardening, α alloys are generally not heat treatable to higher strengths and are usually used in the mill-annealed (MA) condition. As more β-stabilizing elements are added to α titanium, increasing amounts of the β phase (body-centered cubic, or bcc structure) are stabilized at ambient temperature. Beta-stabilizing elements commonly include vanadium, molybdenum, chromium, manganese, iron, niobium and tantalum. Stabilization of a small proportion of β phase produces near-α alloys, exhibiting little or no heat treatability. As β alloying additions and β phase volume fraction increase, heat treatability and strength level increase as a result of two-phase and solid-solution strengthening mechanisms. The α/β alloys may be α/β or β mill annealed, α/β duplex annealed (DA) for a favorable balance in strength, ductility, and toughness, or α/β or β solution treated and aged (STA) to boost strength levels. Beta alloys, produced by adequate β-stabilizing element additions, are characterized by their ability to retain 100 % β phase after rapid cooling to room temperature from temperatures above the β transus. Most commercial β titanium alloys are "metastable" β and not "fully stable" β, such that solution treating and aging (to precipitate fine α phase within the β matrix) will produce very high strengths and deep section hardenability.

Environments That Promote SCC in Titanium Alloys

Overall, titanium alloys can be considered to be highly stress-corrosion resistant compared to other common engineering metals and alloys in the vast majority of environments encountered in nature and in industrial and aerospace service. This superior resistance is the result of the ability of titanium, as a reactive metal, to instantaneously form and maintain a tenacious, protective surface oxide in oxygen- or water-containing media (Ref 1, 2). In such passivating environments, titanium alloys may be either immune to SCC altogether or, as is often the case with some of the more highly alloyed grades, may exhibit stage II (but not stage I) subcritical cracking behavior. This is the situation in most common aqueous environments, including seawater and brines, where alloy composition and metallurgical condition dictate the susceptibility of a titanium alloy to SCC.

Titanium alloys may also be exposed to a few unique environments in which mechanical or chemical breakdown of the protective oxide film may occur and repassivation is not possible, such as environments devoid of oxygen or water species required for passivation and/or those where severe embrittlement ensues. In these situations, stage I along with stage II subcritical cracking behavior in titanium alloys is generally observed, such that no exact, well-defined crack threshold stress level can be identified. These environments include certain liquid and solid metals, dry methanol, molten halide salts, and hot halide salt residues.

Table 2 Environments Known To Promote Stress-Corrosion Cracking of Commercial Titanium Alloys

Environment: Medium	Temperature °C	Temperature °F	Titanium alloys with known susceptibility
Oxidizers			
Nitric acid (red-fuming)	Room temperature		Ti, Ti-8Mn, Ti-6Al-4V, Ti-5Al-2.5Sn
Nitrogen tetroxide (no excess NO)	30-75	86-167	Ti-6Al-4V
Organic compounds			
Methyl alcohol (anhydrous)	Room temperature		Ti-6Al-4V, grade 2, grade 4, Ti-4Al-3Mo-1V, Ti-3Al-8V-6Cr-4Zr-4Mo, Ti-8Al-1Mo-1V, Ti-13V-11Cr-3Al, Ti-5Al-2.5Sn
Methyl chloroform	370	700	Ti-8Al-1Mo-1V, Ti-6Al-4V, Ti-5Al-2.5Sn, Ti-13V-11Cr-3Al
Ethyl alcohol (anhydrous)	Room temperature		Ti-8Al-1Mo-1V, Ti-5Al-2.5Sn
Ethylene glycol	Room temperature		Ti-8Al-1Mo-1V
Trichloroethylene	370, 620, 815	700, 1150, 1500	Ti-8Al-1Mo-1V, Ti-5Al-2.5Sn
Trichlorofluorethane	788	1450	Ti-8Al-1Mo-1V, Ti-5Al-2.5Sn, Ti-6Al-4V, Ti-13V-11Cr-3Al
Chlorinated diphenyl	315-370	600-700	Ti-5Al-2.5Sn
Hot salt			
Chloride and other halide salts/residues	230-430	450-805	Most commercial alloys, except grades 1, 2, 7, 11, 12, and 9
Metal embrittlement			
Cadmium (solid + liquid)	25-600	77-1110	Ti-8Mn, Ti-13V-11Cr-3Al, grade 2, Ti-6Al-4V
Mercury (liquid)	Room temperature		Grade 4, Ti-6Al-4V
	370	700	Ti-13V-11Cr-3Al, Ti-8Al-1Mo-1V
Silver (solid) and AgC	232-480	450-900	Ti-7Al-4Mo, Ti-5Al-2.5Sn
Ag-5Al-2.5Mn (braze alloy)	340	645	Ti-6Al-4V, Ti-8Al-1Mo-1V
Miscellaneous			
Seawater/NaCl solution	Room temperature		Unalloyed Ti (with >0.3%O), Ti-2.5Al-1Mo-11Sn-5Zr-0.2Si (IMI-679), Ti-3Al-11Cr-13V, Ti-5Al-2.5Sn, Ti-8Mn, Ti-6Al-4V, Ti-6Al-6V-2Sn, Ti-7Al-2Nb-1Ta, Ti-4Al-3Mo-1V, Ti-8Al-1Mo-1V, Ti-6Al-Sn-4Zr-6Mo
Distilled water	Room temperature		Ti-8Al-1Mo-1V, Ti-5Al-2.5Sn, Ti-11.5Mo-6Zr-4.5Sn
Chlorine gas	288	550	Ti-8Al-1Mo-1V
10% HCl	35, 340	95, 645	Ti-5Al-2.5Sn, Ti-8Al-1Mo-1V
LiCl, KBr, and Na_2SO_4 solution (0.6 *M*)	Room temperature		Ti-6Al-4V, Ti-6Al-6V-2Sn
Molten chloride/bromide salts	300-500	570-930	Ti-8Al-1Mo-1V

Table 2 provides an overview of the types of environments that are known to stimulate SCC in certain titanium alloys, along with the specific alloys that have exhibited finite susceptibility in laboratory and service exposures. It should be notes that the alloys listed in Table 2 represent only those reported in the literature; it should not be assumed that other titanium alloys are necessarily resistant to SCC under similar conditions. It also should be noted that the alloys listed were susceptible to SCC in at least one metallurgical condition (not indicated in the table). This means that changes in alloy composition (within specifications) and/or metallurgical conditions (*i.e.*, processing and/or final heat treatment) may or may not eliminate susceptibility to SCC. These influencing factors are discussed in detail for each type of environment in the following sections.

Aqueous Environments

By far, the vast majority of current SCC data on titanium alloys has been generated in 3.5% NaCl and seawater media at ambient temperature. This reflects the significant interest and effort over the past 25 years toward designing these alloys into critical aircraft and submersible components operating in marine environments. The primary factors that have been found to influence SCC behavior in titanium alloys can be conveniently categorized as mechanical, metallurgical (composition and structure), or environmental in nature.

Mechanical Effects. A full range of susceptibility of titanium alloys to SCC in aqueous solutions and halides is observed and can be readily categorized into four behavioral types:

- Type 0 behavior reflects alloys that are immune to aqueous SCC, such as the low-alloy industrial alloys, including commercial-purity grades 1, 2, and 3, and alloy grades 7, 9, 11, and 12.
- Type 1 behavior reflects alloys that are fairly resistant to SCC and that require a loaded fatigue crack to indicate susceptibility. Loaded smooth or notched configurations will not exhibit SCC. An example is mill-annealed or duplex-annealed Ti-6Al-4V in neutral brines.
- Type 2 behavior reflects mildly susceptible alloys that may have either a loaded fatigue crack or a machined notch to indicate susceptibility. Mill-annealed Ti-8Al-1V-1Mo and Ti-5Al-2.5 Sn in neutral salt solutions are examples.
- Type 3 behavior reflects alloys that are highly susceptible to SCC, for which loaded smooth sample configurations will produce susceptibility in the absence of a stress riser. Step-cooled (SC) Ti-8Al-1Mo-1V, and Ti-8Mn and Ti-13V-11Cr-3Al in neutral salt solutions are good examples.

It is important to recognize that most commercial titanium alloys in standard metallurgical conditions display type 0 and 1 behavior in aqueous halide media. This reflects an elevated resistance to SCC in loaded smooth or notched component configurations and often may only be revealed at very high stress intensities, such as those associated with highly stressed, precracked components. As a result, the SCC susceptibility identified for most titanium alloys in the laboratory is seldom observed in actual field applications. The few alloys that do exhibit type 2 or 3 behavior are generally in sensitized metallurgical conditions, in which susceptibility may be alleviated by heat treatment and/or minor adjustments in alloy composition.

In addition to these stress-concentration effects, material thickness, orientation, and loading mode and rate can influence alloy SCC behavior (Ref 4). It has generally been observed that SCC susceptibility in α and α/β alloys decreases with decreasing material thickness, as shown for Ti-6Al-4V and Ti-8Al-1Mo-1V in Fig. 1 (Ref 9). There appears to be a critical thickness (Ref 9, 10) for each alloy, below which plane stress prevails and SCC does not occur. This critical thickness value is dependent on the alloy and its heat treatment, orientation, and loading rate. Because of the significant anisotropy (crystallographic texture) developed during α or α/β forging and/or rolling of α phase, α and α/β titanium alloys generally exhibit a pronounced preferred orientation for the stress-corrosion fracture path. A cleavage plane at approximately 15° to the basal (0001) plane predominates and is commonly observed. With basal planes aligned parallel to the rolling direction and normal to sheet or plate surfaces,

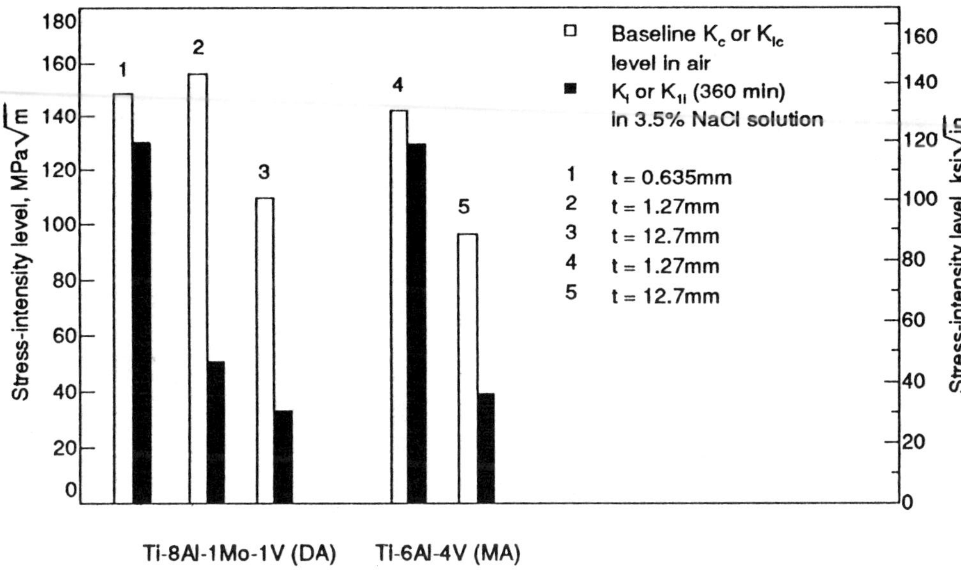

Fig. 1 Effect of specimen thickness on the fracture toughness of Ti-8Al-1Mo-1V (DA) and Ti-6Al-4V (MA) tested in air and in 3.5% NaCl. Source: Ref 9

highly textured (basal transverse) Ti-8Al-1Mo-1V, for example, will exhibit the least resistance to SCC in the transverse direction (*i.e.*, resistance: TL < SL < LT). This anisotropy is much less pronounced in highly susceptible (type 3) α or α/β alloys, or in β alloys where the β phase cleavage plane is typically (100). Note that anisotropy and preferred crack orientation are also minimal in near-α and α/β alloys with fully transformed β microstructures.

In many alloys, the K_{ISCC} value will increase as loading rate decreases (Ref 4). This effect is highly dependent on alloy composition and is less pronounced in either nonsusceptible or highly susceptible alloys (*i.e.*, types 0 and 3). Type 1 and 2 alloys also exhibit higher K_{ISCC} values if loaded in air prior to solution exposure. This has been demonstrated with Ti-6Al-4V and grade 4 alloys tested in ambient 3.5% NaCl solution. Preloading above the K_{ISCC} value may similarly elevate resultant K_{ISCC} values in these alloys of intermediate susceptibility (Ref 4).

Metallurgical Factors. In α titanium alloys, SCC behavior is primarily composition dependent, with grain morphology, size, and texture playing a secondary role. SCC behavior is especially sensitive to aluminum and oxygen content in the alloy, and also to nitrogen, carbon, and/or tin levels. Alloy SCC susceptibility increases at aluminum levels of ≥ 5 wt%, which stems from the increase tendency to form the highly ordered Ti_3Al (α_2) phase. This low-ductility ordered phase forms in the 400 to 700 °C (750 to 1290 °F) temperature range and increases crack velocity and decreases K_{ISCC} as volume fraction increases. Oxygen levels above 0.20 to 0.25 wt% also promote SCC, again because of a transition from wavy to planar slip. This means that although grades 1 and 2 titanium fully resist aqueous SCC, certain higher interstitial chemistries of the grade 3 (or grade 4) alloy may not, as indicated by the following ambient saltwater data:

Grade 2 (0.12% O_2) →

$$K_{Ic} = 66\ \text{MPa}\sqrt{\text{m}}\ (60\ \text{ksi}\sqrt{\text{in.}})$$
$$K_{ISCC} = 66\ \text{MPa}\sqrt{\text{m}}\ (60\ \text{ksi}\sqrt{\text{in.}})$$

Grade 4 (0.38% O_2) →

$$K_{Ic} = 110\ \text{MPa}\sqrt{\text{m}}\ (100\ \text{ksi}\sqrt{\text{in.}})$$
$$K_{ISCC} = 38\ \text{MPa}\sqrt{\text{m}}\ (35\ \text{ksi}\sqrt{\text{in.}})$$

The influence of aluminum and oxygen on titanium alloy SCC is synergistic in a negative fashion, as indicated in Fig. 2 (Ref 4). Tin additions also promote ordered structures and thus are also negative in this regard, as evidenced by the

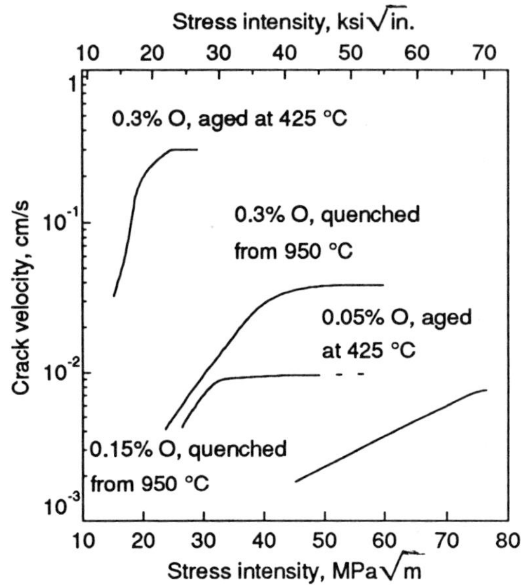

Fig. 2 Influence of oxygen and heat treatment on crack velocity (V) versus stress intensity (K) relationships for Ti-6Al binary alloys tested in 0.6 M KCl at 24 °C (75 °F). Source: Ref 4

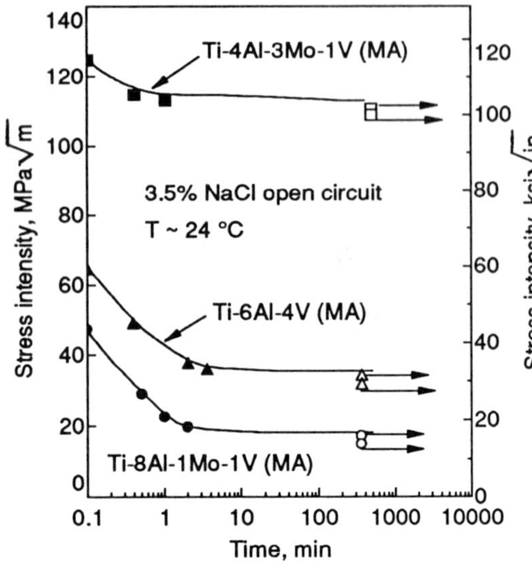

Fig. 3 Variation of time to failure with stress intensity, K, for three mill-annealed commercial α/β alloys tested in 3.5% NaCl. Source: Ref 12, 13

susceptible Ti-5Al-2.5Sn alloy. Zirconium additions, in contrast, are somewhat beneficial.

Compositionally susceptible α alloys generally remain so, whether the structure is equiaxed α or martensitic (quenched from the β-phase field). In general, SCC susceptibility of equiaxed α alloys diminished with decreasing grain size (Ref 4, 6, 11). Depending on the alloy, transformation to martensitic phase structure may or may not improve SCC resistance (Ref 6, 11). The SCC fracture mode in α alloys is transgranular, with the fracture path highly oriented in unidirectionally processed (textured) material (Ref 11) and more random in martensitic structures.

In α/β alloys, it is also the α phase that can exhibit SCC susceptibility in aqueous media. Thus, dependence of alloy SCC susceptibility on aluminum, tin, and O_2 content is still relevant. The strong negative effect of increasing aluminum content is evident in the α/β alloy K_{ISCC} data plotted in Fig. 3 (Ref 12, 13). Similarly, oxygen levels above 0.10 to 0.12 wt.% generally diminish α/β alloy SCC resistance significantly, as shown in Fig. 4 (Ref 4, 14, 15). It is for optimized K_{Ic} and K_{ISCC} values that extra-low-interstitial (ELI) grades of certain α/β alloys, such as ELI Ti-6Al-4V, are often specified for critical saltwater applications. On the other hand, most isomorphous β phase stabilizer additions, such as molybdenum, vanadium, niobium, and tantalum, tend to reduce or eliminate SCC susceptibility in aqueous halide solutions. The β eutectoid additions, such as iron, manganese, chromium, silicon, and copper, however, tend to degrade SCC resistance. Increasing hydrogen content may also reduce SCC resistance in α/β alloys (Ref 4, 11).

Phase structure is also vital in dictating SCC behavior in α/β alloys. The degree of susceptibility is directly related to the grain size, volume fraction, and mean free path of the susceptible α phase (Ref 6, 11). Increasing α grain size increases susceptibility. Because the β phase acts as a ductile, SCC-resistant crack arrester (Ref 6, 11), structures that have an increased β-phase volume fraction and that break up or disperse α phase tend to be beneficial. Thus, fine, equiaxed β phase resulting from processing in the α/β field is much less effective than the lamellar, randomly oriented β phase produced by β processing. Beta processing and/or final β heat treatment improves both air and aqueous toughness by improving

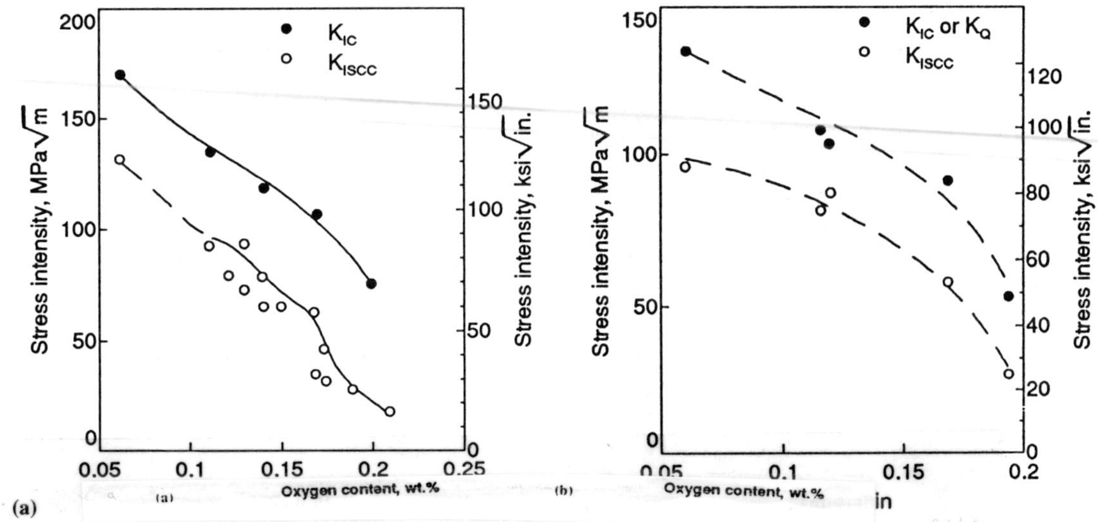

Fig. 4 Variations of K_{Ic} (or K_Q) and K_{ISCC} with oxygen content in Ti-6Al-4V (RW orientation, 3.5% NaCl). (a) Optimum properties for β-ST + MA treatment. (b) Optimum properties for β-STA material (1065 °C AC + 940 °C WQ + 650 °C WQ + 650 °C AC for 4 h). Source: Ref 14, 15

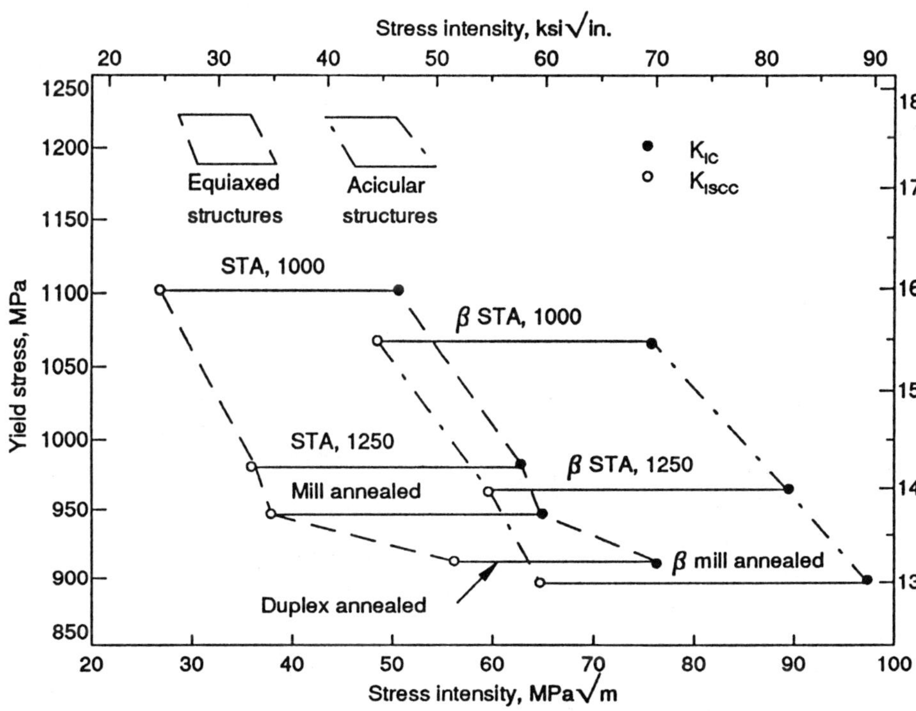

Fig. 5 Range of K_{Ic} and K_{ISCC} values for Ti-6Al-4V in various metallurgical conditions in 3.5% NaCl solution at 24 °C (75 °F). Source: Ref 4

chemical homogeneity, randomizing crystallographic texture, and providing a structure with greatest crack-path tortuosity.

In other words, transformed β microstructures, either martensitic or Widmanstätten (basket weave), improve both air (K_{Ic}) and saltwater

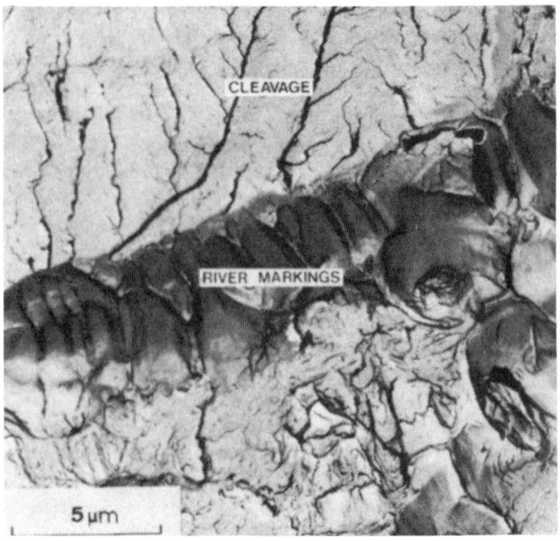

Fig. 6 Fractograph revealing typical transgranular cleavage and ductile river markings and flutes associated with aqueous chloride SCC in α/β titanium alloys. Source: Ref 6

toughness (K_{ISCC}) values relative to equiaxed α/β structures. The magnitude of this microstructural effect is depicted in Fig. 5 for Ti-6Al-4V in air and saltwater (Ref 4). Furthermore, fine α platelets surrounded by a continuous film of β phase and smaller prior-β grains exhibit superior SCC resistance compared to coarse, long α platelets in transformed β structures. The martensitic phases (α′ or α″) formed upon quenching tend to be highly resistance to SCC (Ref 6, 11) and become less so when they are tempered to form α phase.

As discussed previously for α alloys, α/β alloys containing higher aluminum levels (≥5 wt%) may form Ti_3Al (α_2) phase (Ref 11) when aged in the 400 to 700 °C (750 to 1290 °F) range. Increasing oxygen and/or tin levels further aggravates the phenomenon, which can severely sensitize the alloy to SCC (see Fig. 2). Slow or step-cooled Ti-8Al-1Mo-1V is a classic example of an α/β alloy sensitized to SCC by this effect (see Table 3).

The SCC fracture mode in α/β alloys consists of transgranular cleavage through the α phase and ductile tearing (dimples and tear ridges) through the β phase. A typical fractograph is presented in Fig. 6. Crack velocities are high, and in the range of 1 to 50 mm/min (0.04 to 2 in./min) (Ref 6, 11).

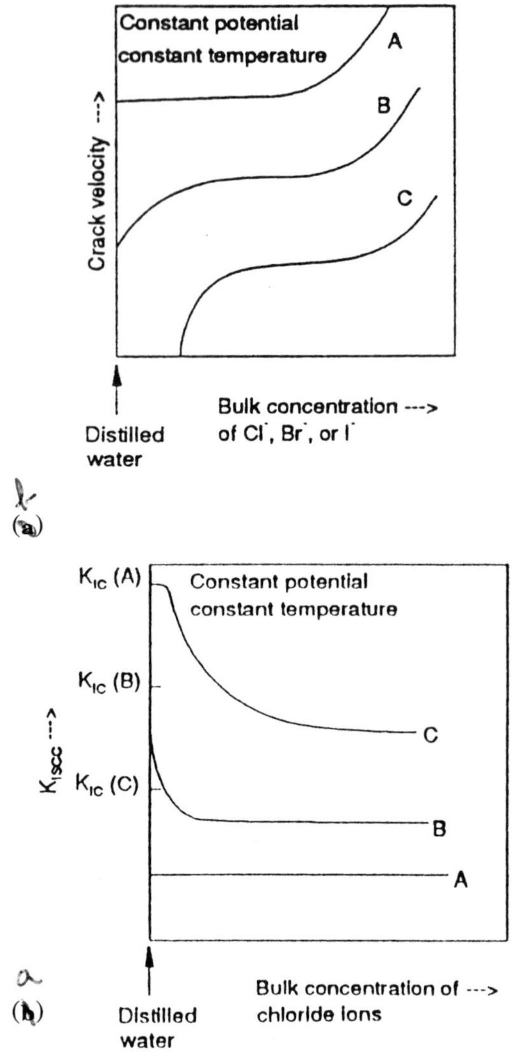

Fig. 7 Influence of halide ion concentration on (a) K_{ISCC} and (b) crack velocityof titanium alloys in aqueous solutions

In β titanium alloys, the β phase may be susceptible to either transgranular or intergranular SCC, depending on alloy composition and microstructure. Intergranular cracking has been observed only in a few aged β alloys (Ref 6, 11), particularly with fine α precipitates formed at lower aging temperatures. Beta phase stabilized by either molybdenum, vanadium, niobium, or tantalum is immune to SCC. Exceptions include β phase stabilized by the eutectoid elements, manganese and chromium. Transgranular cleavage of β phase is known to occur in the Ti-8Mn al-

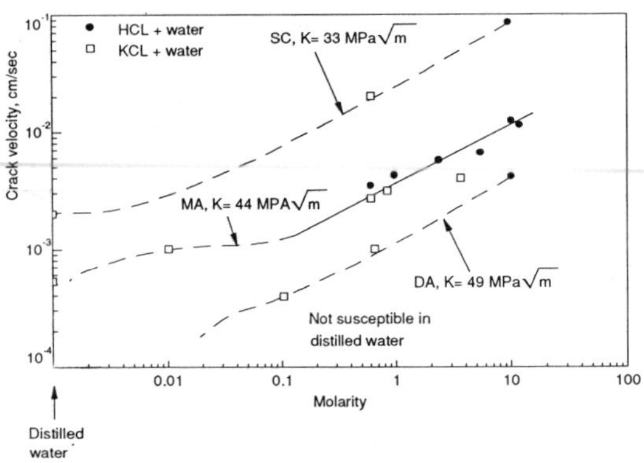

Fig. 8 Effect of increasing chloride concentration on region II cracking velocity in Ti-8Al-1Mo-1V in 24 °C (75 °F) aqueous solution. Source: Ref 16

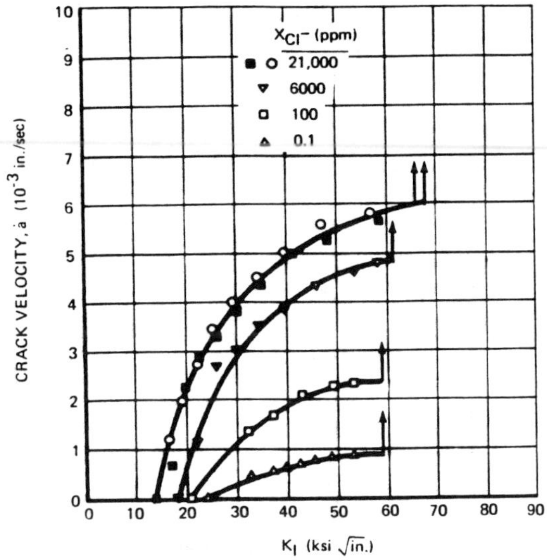

Fig. 9 Influence of halide concentration on the SCC behavior of Ti-8Al-1Mo-1V in aqueous solution at 25 °C (77 °F). Source: Ref 17

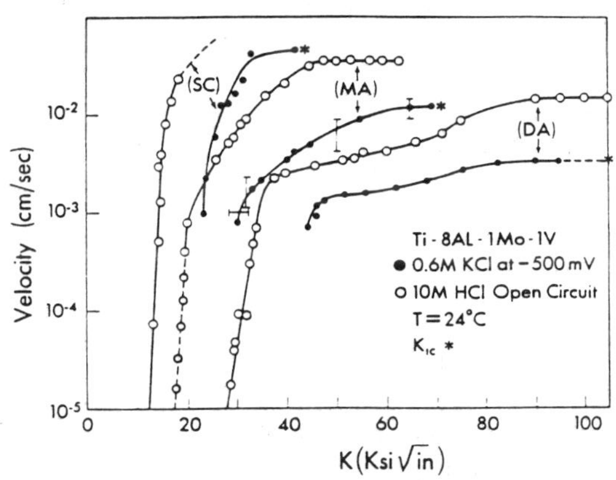

Fig. 10 Influence of acidity on the cracking behavior and velocity of Ti-8Al-1Mo-1V in chloride solution at 24 °C (75 °F). Source: Ref 16

loy with a $\beta + \alpha$ structure, and in solution-treated Ti-13V-11Cr-3Al (Ref 6). SCC susceptibility in this case is mitigated by decreasing the grain size and mean free path of the β phase. Although $\beta + \omega$ phase structures appear to be highly resistant, the precipitation of compounds, such as $TiCr_2$, with the β phase can be expected to degrade cracking resistance.

Environmental Factors. The susceptibility of titanium alloys to SCC in aqueous media is influenced by the type and concentration of species in solution; the pH, temperature, and viscosity of the solution; and the metal potential in the solution. In certain metallurgical conditions, a few titanium alloys have even exhibited SCC susceptibility in distilled water (Ref 4). These alloys include Ti-8Al-1Mo-1V (step cooled), Ti-5Al-2.5Sn, and Ti-11.5Mo-6Zr-4.5Sn (STA). Addition of halide ions, such as Cl^-, Br^-, and I^-, increases susceptibility in these alloys and can induce susceptibility in other alloys that are immune to SCC in distilled water. Figures 7(a) and (b) show that increasing halide concentration tends to increase cracking velocity and decrease K_{ISCC} in alloys with low to medium susceptibilities (types 1 and 2), but has little effect on highly susceptible alloys (type 3) (Ref 4). This effect of chloride concentration on SCC behavior is revealed in Fig. 8 (Ref 16) and Fig. 9 (Ref 17) for Ti-8Al-1Mo-1V. Cracking velocity appears to vary directly with halide concentration to the 1/4 to 1/2 power in concentrated solutions (Ref 18, 19).

Other ionic species in solution can have a neutral or even an inhibitive effect on alloy SCC if the alloy is not highly susceptible. These species include NO_3^-, SO_4^{-2}, F^-, OH^-, CrO_4^{-2}, and PO_4^{-3} (Ref 20, 21). The inhibitive influence diminishes as halide levels increase. Little influence of nonoxidizing cations, such as Na^+, K^+ and Li^+, is noted.

However, oxidizing cations, such as Cu^{+2}, may raise K_{ISCC} values (Ref 4, 22), depending on alloy heat treatment.

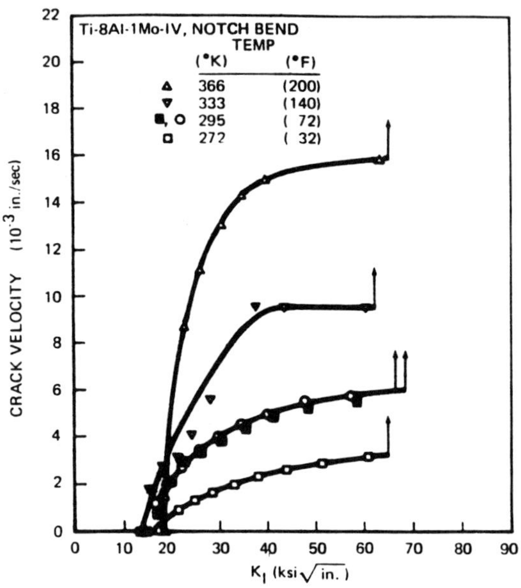

Fig. 11 Effect of temperature on SCC velocity of Ti-8Al-1Mo-1V (notch bend specimens) in 3.5% NaCl solution. Source: Ref 17

Because titanium alloys exhibit no stage I type crack growth in neutral solutions (*i.e.*, slowest crack velocities measured are approximately 10^{-3} cm/s, or 4×10^{-4} in./s), it may be concluded that true K_{ISCC} thresholds exist, below which cracks will not propagate. This is not the case in highly acidic solutions, where both stage I and II cracking behavior is observed (Ref 4,22). Increasing acidity generally reduces K_{ISCC} and increases stage II cracking velocity, as indicated by data for Ti-8Al-1Mo-1V in Fig. 10 (Ref 16). Increasing alkalinity appears to have no obvious or significant effect on SCC behavior relative to neutral conditions. As hydroxide concentrations exceed 1 *M* or so, increasing inhibition may be expected (Ref 20, 23).

Only limited published data on the influence of temperature on titanium alloy SCC behavior in aqueous media are available. Data for the Ti-8Al-1Mo-1V alloy suggest that K_{ISCC} values may be relatively unaffected by temperatures as high as 93 °C (200 °F) in neutral salt solutions (Ref 17, 22). However, stress cracking velocity does increase sharply with increasing temperature, as shown in Fig. 11 (Ref 17). A few titanium alloys that exhibit negligible SCC tendencies in near-ambient saltwater may exhibit well-defined temperature thresholds for SCC susceptibility in

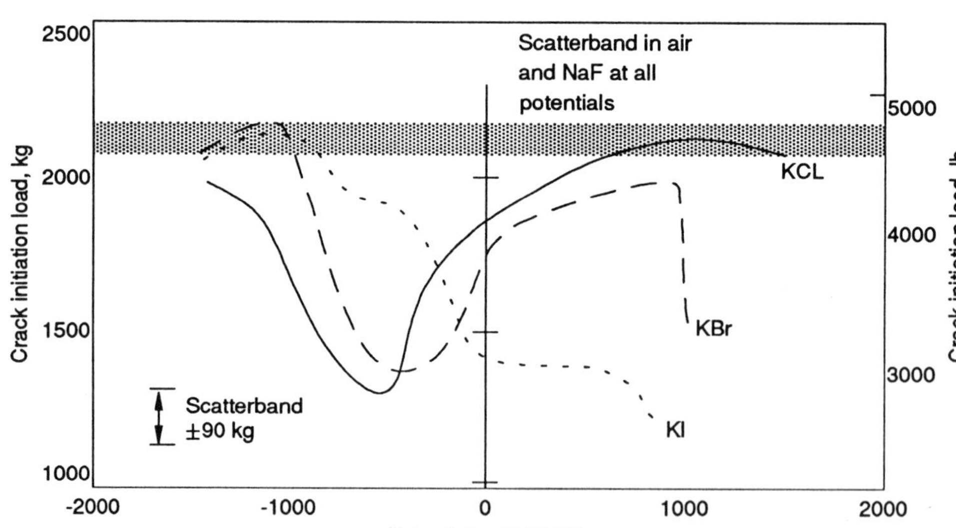

Fig. 12 Effect of potential on crack initiation stress for α/β titanium alloys in various halide solutions at 25 °C (77 °F). Source: Ref 27

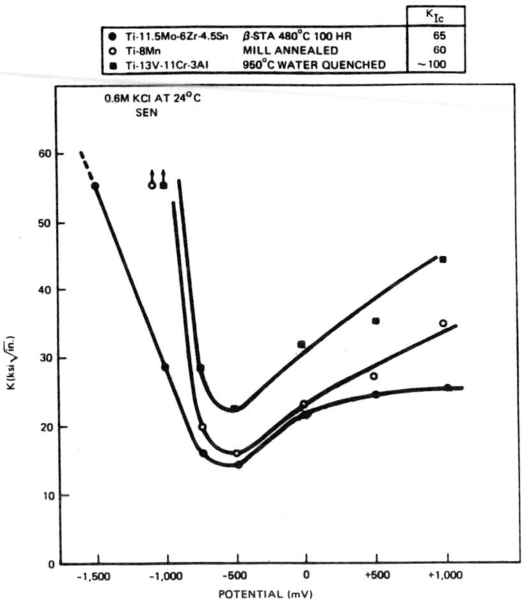

Fig. 13 Influence of potential on the K_{ISCC} of several β and near-β titanium alloys in 0.6M KCl solutions at 24 °C (75 °F). Source: Ref 22

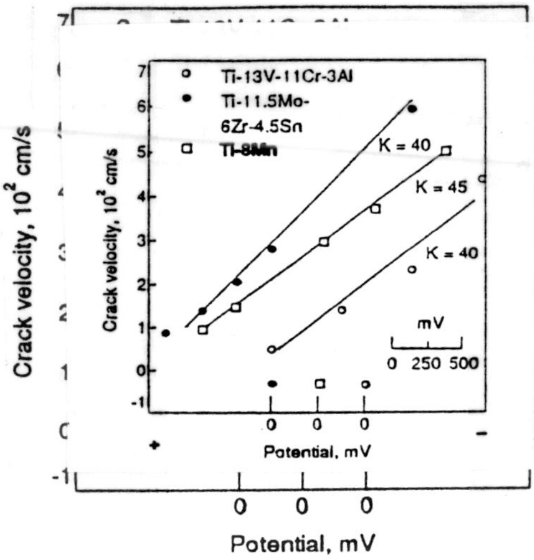

Fig. 14 Influence of potential on cracking velocity in β titanium alloys in 0.6M KCl solution at 24 °C (75 °F). Source: Ref 22

high-temperature brines. For example, the Ti-3Al-8V-6Cr-4Zr-4Mo alloy (STA condition) resists SCC in near-neutral NaCl brines up to approximately 180 °C (355 °F), but can exhibit stress cracking during C-ring and slow-strain-rate testing above this temperature (Ref 7). The presence of hydrogen sulfide and/or sulfur appears to impart little effect. In this case, SCC threshold temperatures can be raised significantly (by 15 to 80 °C, or 27 to 145 °F) by microstructural refinement and/or minor (≤0.1%) alloy additions of platinum group metals (for example, palladium) (Ref 7). Resistance to SCC in sour brines to temperatures as high as 260 °C (500 °F) has similarly been achieved for the Ti-6Al-4V and Ti-6Al-2Sn-4Zr-6Mo alloys by minor palladium additions (Ref 24). However, most titanium alloys immune to ambient saltwater SCC, such as grades 1, 2, 7, 9, and 12, remain so in halide solutions to temperatures in excess of 200 °C (390 °F). For example, titanium grades 2 and 12 exhibit resistance to SCC in slow-strain-rate tests in natural groundwaters and saturated NaCl brines to temperatures as high as 250 °C (480 °F) (Ref 25, 26).

In most SCC-susceptible titanium alloys, anodic or cathodic polarization tends to inhibit SCC and increase K_{ISCC} values. This is illustrated in Fig. 12 and 13 for several alloys in ambient halide solutions (Ref 22, 27). Within a narrow potential region (typically centered around –500 mV SCE), K_{ISCC} reaches a minimum value that depends on the alloy and its metallurgical condition. Somewhat similar behavior has been reported for Ti-6Al-4V-0.05Pd in a sour brine at 232 °C (450 °F) (Ref 24). The results shown in Fig. 12 suggest that cathodic protection can be achieved at potentials negative to –1.0 V (SCE) or anodic to 0 V (SCE) in neutral chloride and bromide solutions. The anodic and cathodic polarization inhibition phenomenon is not as apparent in highly susceptible alloys, such as step-cooled Ti-8Al-1Mo-1V (Ref 22).

Increasing potential also serves to increase stress cracking velocity in neutral halide solutions. This correlation is generally linear, as shown in Fig. 14 (Ref 19, 22). In highly acidic solutions, however, stage II crack velocity becomes independent of applied potential, as shown in Fig. 15 (Ref 4, 22). As a result, inhibition via cathodic polarization is not achievable in high acidic solutions. Decreasing solution viscosity has also been shown to linearly increase cracking velocity and

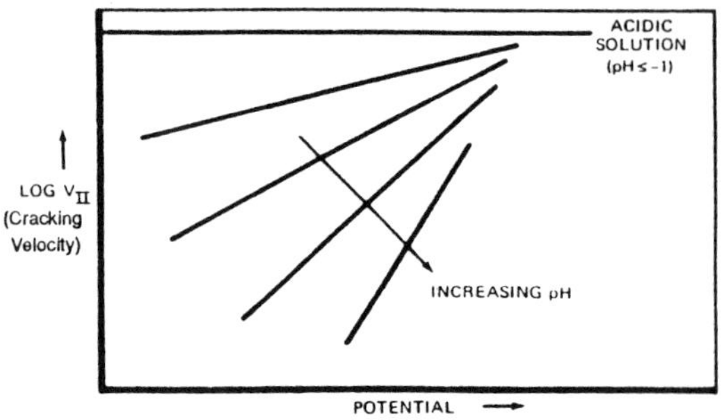

Fig. 15 Influence of potential and acidity on region II cracking velocity for titanium alloys in aqueous halide solutions. Source: Ref 4, 22

decrease time to failure in certain aqueous solutions (Ref 22, 28).

K_{ISCC} Data. K_{ISCC} values for various commercial titanium alloys from referenced sources are presented in Table 3. These data were generated in ambient neutral 3.5% NaCl solution. It should be cautioned that these K_{ISCC} values are highly dependent on alloy composition, metallurgical condition (*i.e.*, processing and heat treatment), and product form and thickness; therefore, they may or may not be representative of commercially available alloy product materials.

Aqueous SCC Mechanism. It has been postulated by Wanhill and others (Ref 6, 11) that initiation of aqueous SCC in susceptible titanium alloys results from planar slip and formation of wide slip steps that rupture the protective surface oxide film. If rapid film repassivation does not occur, a minute active "trench" may form from localized anodic dissolution at the metal surface. This active trench, if sufficiently localized, may then initiate the stress crack growth. This initiation step is highly dependent on (1) the degree of alloy slip planarity and slip step width, which concentrates deformation into relatively few slip bands, (2) oxide film repassivation kinetics, and (3) metal deformation (*i.e.*, strain) rate. Planar slip and slip step width are enhanced by precipitation of ordered $Ti_3(Al, Sn)$ phases, increased alloy interstitial content, microstructures providing large mean free paths of susceptible α phase, and mechanical strcss concentrators (notches, precracks). The relationship between repassivation rate and deformation rate is especially critical for stress crack growth, because a balance must be met where localized active corrosion conditions can be sustained within the newly formed trench.

Propagation of SCC from active surface sites ("trenches") on titanium alloys has been postulated to occur via two basic, but highly debated, mechanisms (Ref 4, 6, 11, 22). One mechanism, anodic-assisted cracking, basically purports anodic metal dissolution at highly localized sites of stress concentration (such as at the crack tip), which when aided by applied tensile stresses, propagates cracks into the metal. This assumes a delicate balance between critical crack-tip stress intensity, crack-tip environment, and crack-tip corrosion rate. For example, excessive localized active corrosion could blunt the advancing crack tip and halt SCC. Highly reducing acidic conditions involving pH levels as low as 1 to 2 can develop within advancing cracks. Similar to the crevice corrosion mechanism proposed for titanium alloys (Ref 2, 32), these highly reducing and halide conditions can depassivate the metal and concentrate halide (Cl^-, Br^-, I^-) and hydronium (H_3O^+) ion species well within the crack. Note that the presence of certain anodic (less resistant) phases in certain titanium alloys, such as Ti_3Al (α_2), silicides, and $TiCr_2$, tend to enhance cracking via anodic dissolution.

The other prominent mechanism is based on a hydrogen-assisted cracking phenomenon. Hydrogen absorption may result from localized anodic dissolution and generation of film-free surfaces within the advancing crack. Absorbed hydrogen can subsequently migrate and concentrate in the vicinity of the crack tip due to existing tensile stress gradients. Localized hydrogen embrittlement of the metal ahead of the crack tip then facilitates continued stress crack growth. This effect is prominent in the α phase of α and α/β alloys, where local precipitation of brittle titanium hydride precipitates ahead of the advancing crack promotes low-energy (cleavage) fracture. Because hydrogen embrittlement is highly sensitive to alloy composition and microstructure, correlation with titanium alloy SCC behavior can be expected and has been observed in most cases (Ref 6, 11, 22).

Given the diversity and inconsistencies in stress cracking observations, it appears that no single mechanism alone can explain aqueous

Table 3 Fracture Toughness Data for Various Titanium Alloys Tested in Air and 3.5% NaCl Solution at 25 °C (77 °F)

Alloy	Thickness mm	Thickness in.	Heat treatment (a)	Yield strength MPa	Yield strength ksi	K_{Ic} or K_C MPa$\sqrt{m}$	K_{Ic} or K_C ksi$\sqrt{in.}$	K_{ISCC} or K_{SCC} MPa$\sqrt{m}$	K_{ISCC} or K_{SCC} ksi$\sqrt{in.}$
Grade 2	...	...	αA-AC	...	...	66	60	66	60
Grade 2 (0.06 O_2)	19.0	0.75	αA-AC	...	...	58	53	57	52
Grade 3 (0.20 O_2)	19.0	0.75	αA-AC	...	...	79	72	75	68
Grade 4 (0.40 O_2)	19.0	0.75	αA-AC	...	...	99	90	58	53
Grade 4	12.7	0.50	αA-AC	572	83	135	123	36	33
	12.7	0.50	αA-WQ	586	85	141	128	37	34
	12.7	0.50	STA	565	82	120	113	43	39
	12.7	0.50	βA-WQ	524	76	115	105	77	70
	12.7	0.50	β-STA	531	77	106	96	53	48
Ti-5Al-2.5Sn	12.7	0.50	αA-AC	869	126	97	88	33	30
	12.7	0.50	αA-WQ	855	124	113	103	30	27
	12.7	0.50	STA	889	129	91	83	25	23
	12.7	0.50	βA-WQ	869	126	131	119	41	37
Ti-5Al-2.5Sn (ELI)	9.7	0.38	αA-AC	827	120	108	98	45	41
Ti-8Al-1Mo-1V	1.27	0.05	MA	1000	145	82	75	33	30
	1.27	0.05	DA	930	135	176	160	55	50
	12.7	0.50	MA	1000	145	53	48	22	20
	12.7	0.50	DA	930	135	110	100	35	32
	12.7	0.50	MA-WQ	841	122	>110	>100	46	42
	12.7	0.50	β-ST-WQ	869	126	>110	>100	>110	>100
Ti-6Al-2Sn-4Zr-2Mo	12.7	0.50	STA	1048	152	58	53	30	27
Ti-6Al-4V (standard grade)	1.5	0.06	MA	945	137	165	150	120	110
	1.5	0.06	DA	917	133	165	150	120	110
	1.5	0.06	STA	1100	160	104	95	72	65
	1.5	0.06	β-STA	1070	155	104	95	72	65
	12.7	0.50	MA	945	137	66	60	38	35
	12.7	0.50	DA	917	133	77	70	57	52
	12.7	0.50	STA	1100	160	52	47	27	25
	12.7	0.50	β-STA	1070	155	77	70	49	45
Ti-6Al-4V (0.05 O_2)	19.0	0.75	α/β rolled	820	119	...	...	71-119	65-108
	19.0	0.75	β-rolled	738	107	...	...	85-110	77-100
Ti-6Al-4V (0.07 O_2)	19.0	0.75	a/β rolled	862	125	...	...	75-84	68-76
	19.0	0.75	β-rolled	786	114	...	...	82-98	75-89
Ti-6Al-4V (0.11 O_2)	19.0	0.75	α/β rolled	910	132	...	...	53-64	48-58
	19.0	0.75	β-rolled	807	117	...	...	76-86	69-78
Ti-6Al-4V (0.15 O_2)	19.0	0.75	α/β rolled	945	137	...	...	45-51	41-46
	19.0	0.75	β-rolled	869	126	...	...	59-62	54-56
Ti-4Al-3Mo-1V	12.7	0.05	MA	862	125	187	170	143	130
	12.7	0.05	DA	862	125	209	190	160	145
	12.7	0.05	β-STA	965	140	132	120	110	100
	12.7	0.50	MA	862	125	126	115	115	105
	12.7	0.50	DA	862	125	137	125	132	120
	12.7	0.50	β-STA	965	140	104	95	77	70
Ti-7Al-4Mo	12.7	0.50	MA	993	144	80	73	34	31
	12.7	0.50	STA	1151	167	40	36	29	26
IMI-680	12.7	0.50	βF+DA	1144	166	48	44	31	28
Ti-6Al-6V-2Sn	12.7	0.50	MA	1083	157	66	60	22	20
	12.7	0.50	DA	1007	146	88	80	27	25
	12.7	0.50	STA	1170	170	49	45	33	30
	12.7	0.50	β-STA	1048	152	77	70	49	45
Ti-6Al-2Sn-4Zr-6Mo	12.7	0.50	MA	1100	160	60	55	22	20
	12.7	0.50	DA	1035	150	88	80	49	45
	7.6	0.30	MA	965	140	57	52	29	26

(continued)

(a) αA, alpha annealed; βA, beta annealed; AC, air cooled; DA, duplex annealed; MA, mill annealed; ST, solution treated; STA, solution treated and aged; WQ, water quenched. Source: Ref 4, 29-31

Table 3 Fracture Toughness Data for Various Titanium Alloys Tested in Air and 3.5% NaCl Solution at 25 °C (77 °F) (continued)

Alloy	Thickness mm	in.	Heat treatment (a)	Yield strength MPa	ksi	K_{Ic} or K_C MPa$\sqrt{m}$	ksi$\sqrt{in.}$	K_{ISCC} or K_{SCC} MPa$\sqrt{m}$	ksi$\sqrt{in.}$
Ti-8Mo-8V-3Al-2Fe (continued)	7.6	0.30	β-STA	1170	170	89	81	49	45
Ti-11.5Mo-6Zr-4.5Sn	12.7	0.50	β-STA	1035	150	71	65	27	25
Ti-8Mo-8V-3Al-2Fe	12.7	0.50	β-STA	1248	181	55	50	34	31
	12.7	0.50	β-STA	1035	150	67	61	42	38
	12.7	0.50	β-STA	1227	178	49	45	31	28
	12.7	0.50	β-STA	1415	205	22	20	22	20
Ti-3Al-8V-6Cr-4Zr-4Mo	12.7	0.50	β-STA	1331	193	55	50	41	37
	14.2	0.56	β-STA	1262	183	38	35	38	35
	>6.4	>0.25	β-STA	1140	165	53	48	53	48
	>6.4	>0.25	β-STA	986	143	56	51	56	51
Ti-13V-11Cr-3Al	3.3	0.13	β-ST	883	128	98	89	29	26
	12.7	0.50	β-ST	827	120	>110	>100	38	35
	12.7	0.50	β-STA	1100	160	77	70	33	30
	5.1	0.20	β-STA	1055	153	71	65	35	32

(a) αA, alpha annealed; βA, beta annealed; AC, air cooled; DA, duplex annealed; MA, mill annealed; ST, solution treated; STA, solution treated and aged; WQ, water quenched. Source: Ref 4, 29-31

SCC of titanium alloys. Both anodic and cathodic mechanisms may be acting simultaneously and/or interdependently in some cases. More detailed discussions of these mechanisms can be found elsewhere (Ref 6, 11, 22).

Methanol and Other Alcohols

Unlike other engineering metals, titanium and zirconium alloys are highly susceptible to SCC in methanol liquid and vapor. SCC susceptibility is primarily limited to methanol for most commercial titanium alloys, although a few highly susceptible alloys may also be affected by other low-molecular-weight alcohols. The most practical categorization of titanium alloy SCC behavior in methanol solutions is based on alloy stress fracture mode: intergranular versus transgranular. The fracture mode exhibited is highly dependent on alloy composition and applied stress level (stress intensity).

Intergranular failure in methanol is primarily observed in titanium alloys not normally susceptible to aqueous SCC, such as grade 1 and 2 alloys, and in certain β alloys, such as Ti-13V-11Cr-3Al. This fracture mode is also commonly observed in almost all titanium alloys during stage I cracking at low stress levels. Intergranular cracking in methanol generally involves anodic dissolution and requires little or no stress to propagate. Application of stress accelerates cracking, but this mode of propagation remains independent of stress level. Typical fractographic features are pictured in Fig. 16.

This mode of cracking requires the present of at least trace levels of halides or halogens, such as ≥0.3 ppm Cl^- or > 10^{-6} *N* HCl for unalloyed titanium (Ref 33). Water represents a potent cracking inhibitor when added to methanol above a critical level. This inhibiting effect is indicted in Fig. 17 and 18 for unalloyed titanium in ambient methanol-water solutions (Ref 33, 34). Increased acidity and halide concentration increases the minimum water level required for complete inhibition, as these figures show. Similar effects are noted for Ti-6Al-4V, as shown in Fig. 19 (Ref 35). A safe practical minimum water content for precluding methanol SCC in commercial-purity titanium and Ti-6Al-4V is approximately 1.5 wt%.

The effect of other various species added to water-lean methanol solutions has also been studied. It has been found that, although sulfuric and formic acids accelerate cracking to a lesser extent than halides, phosphoric and acetic acid additions will not cause cracking (Ref 36). Oxidizing metal

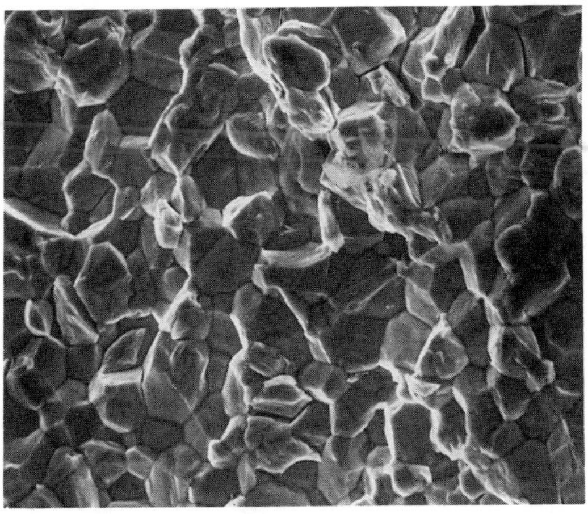

Fig. 16 Fractograph revealing intergranular SCC of unalloyed titanium in anhydrous methanol containing bromides. 500 ×

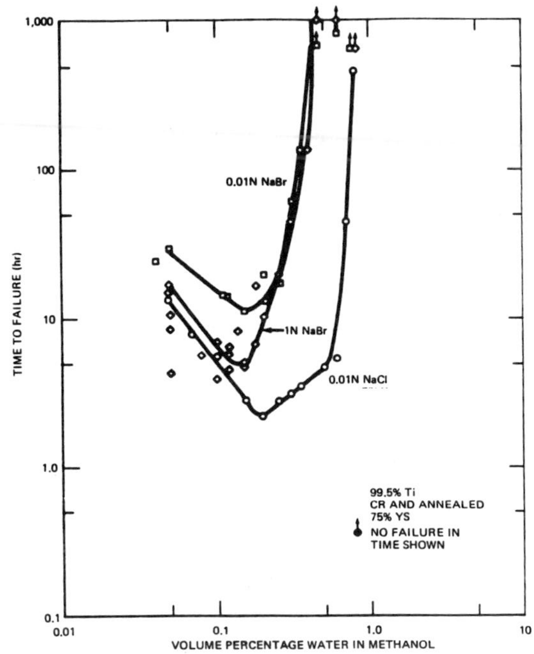

Fig. 17 Influence of water and halide contents in methanol on the SCC susceptibility of unalloyed titanium at 25 °C (77 °F). Source: Ref 33, 34

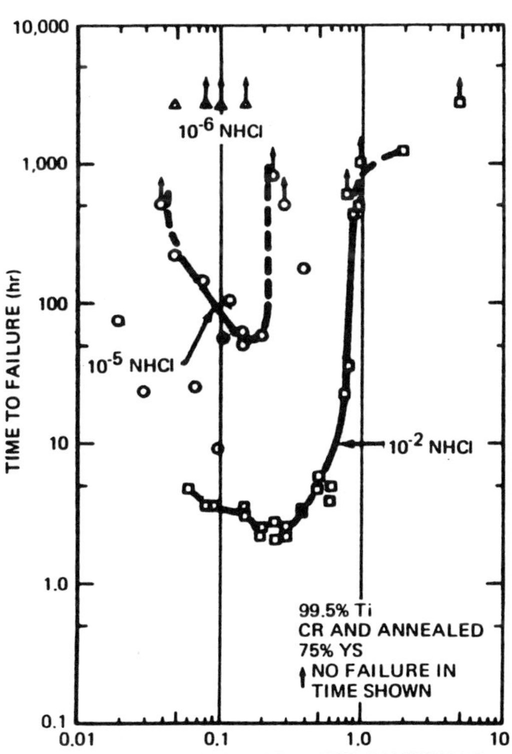

Fig. 18 Effect of water content and acidity on the SCC susceptibility of unalloyed titanium at 25 °C (77 °F). Source: Ref 33, 34

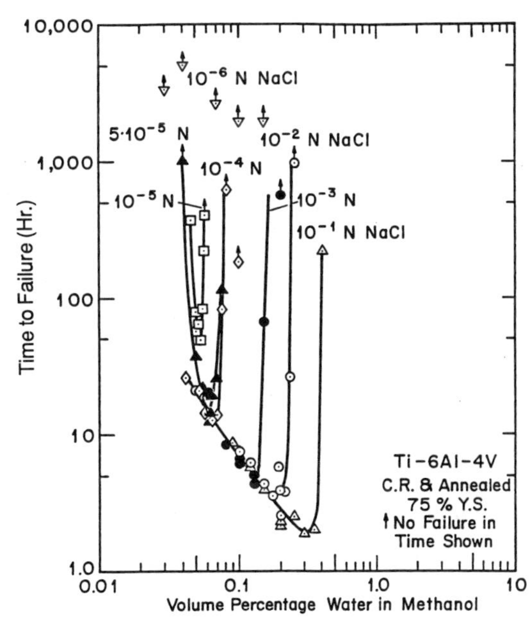

Fig. 19 Effect of chloride concentration and water content on the SCC susceptibility of Ti-6Al-4V in methanol. Source: Ref 35

ion additions, including those of gold, iron, copper, mercury, and especially palladium, accelerated both the anodic dissolution and cracking rates (Ref 37, 38). Also, additions of methyl ethyl ketone (MEK) and acetone in excess of several weight percent inhibit methanolic SCC of unalloyed titanium if halide/halogen levels are not significant (Ref 39).

The influence of impressed potentials on intergranular SCC in these alloys is pronounced. Anodic polarization tends to accelerate cracking rate regardless of methanol solution pH. On the other hand, cathodic polarization inhibits cracking tendencies, such that potentials below –250 mV (SCE) completely suppress cracking of unalloyed titanium in neutral methanol solutions (Ref 40).

Increasing temperature has been shown to accelerate cracking rate (Ref 41). Metallurgically speaking, an ultrafine grain size and reduced interstitial level appear to enhance SCC resistance in unalloyed titanium.

Transgranular Failure. Transgranular SCC is generally observed in titanium alloys that are susceptible to SCC in aqueous solutions. Typical examples include the more highly alloyed titanium alloys containing greater than 5% Al (*i.e.*, Ti-6Al-4V and Ti-8Al-1Mo-1V) and/or high interstitial levels, which tends to exhibit coplanar slip. Although several exceptions exist, these alloys tend to exhibit intergranular stage I cracking and transgranular stage II cracking, indicating that cracking mode is highly dependent on stress level. The general appearance of transgranular cleavage fracture and α–phase cleavage plane indices are essentially the same as those observed in aqueous solutions (see Fig. 6). The exceptions to this behavior include the beta- III and fine-grained Ti-3Al-8V-6Cr-4Zr-4Mo β alloys, both of which exhibit intergranular stage II cracking (Ref 22, 42). On the other hand, coarser-grained structures of the aged Ti-3Al-8V-6Cr-4Zr-4Mo alloy predominately exhibit transgranular stage I and II cracking, as revealed in Fig. 20. Grain size and phase morphology are highly influential factors in β alloy SCC behavior.

As with intergranular methanol SCC, alloys exhibiting transgranular SCC in methanol can experience accelerated cracking with halogen/halide additions, and inhibitive effects with increasing water content. SCC is generally difficult to observe in absolutely pure methanol, but becomes favored at HCl levels above $10^{-6}M$. The minimum level of water required for full SCC inhibition depends on alloy composition, metallurgical condition, temperature, halide level, acidity, and other species present. Alloys susceptible to SCC in distilled water, such as step-cooled Ti-8Al-1Mo-1V, cannot be inhibited regardless of water content. SCC inhibition in neutral halide-containing methanol solutions can be achieved with 1.0 to 1.5 wt% and 2.5 to 3.0 wt% water additions for the Ti-6Al-4V and Ti-3Al-8V-6Cr-4Zr-4Mo alloys, respectively. Reductions in minimum water level required may be realized in α/β alloys by reducing aluminum content and/or interstitial levels (*i.e.*, ELI grades). The negative influences of halide addition and sensitivity to metallurgical condition on cracking rate are graphically depicted for Ti-8Al-1Mo-1V in Fig. 21 (Ref 22).

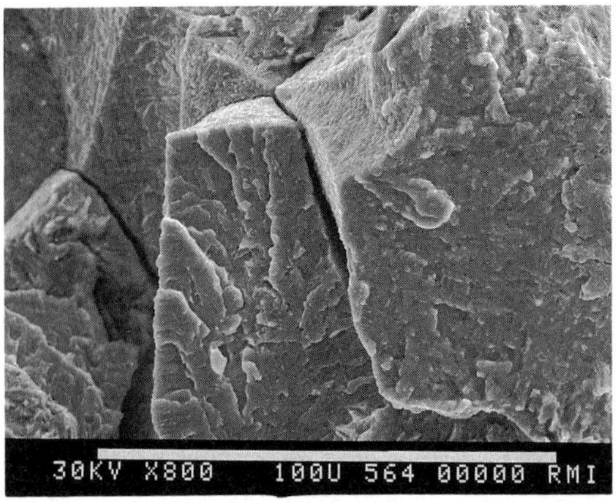

Fig. 20 Fractograph revealing transgranular SCC of Ti-3Al-8V-6Cr-4Zr-4Mo in anhydrous methanol. A similar fracture appearance is observed in NaCl brine above 180 °C (355 °F). 800×

Numerous species have been found that inhibit transgranular (stage II) stress cracking in methanol, depending on halide level. These include nitrate and sulfate ions (Ref 27), NaF, and 0.1*M* concentrations of Al^{+3}, Zr^{+4}, Cd^{+}, and Sn^{+2} metallic ions (Ref 43). The effectiveness of a minor nitrate addition on Ti-3Al-8V-6Cr-4Zr-4Mo

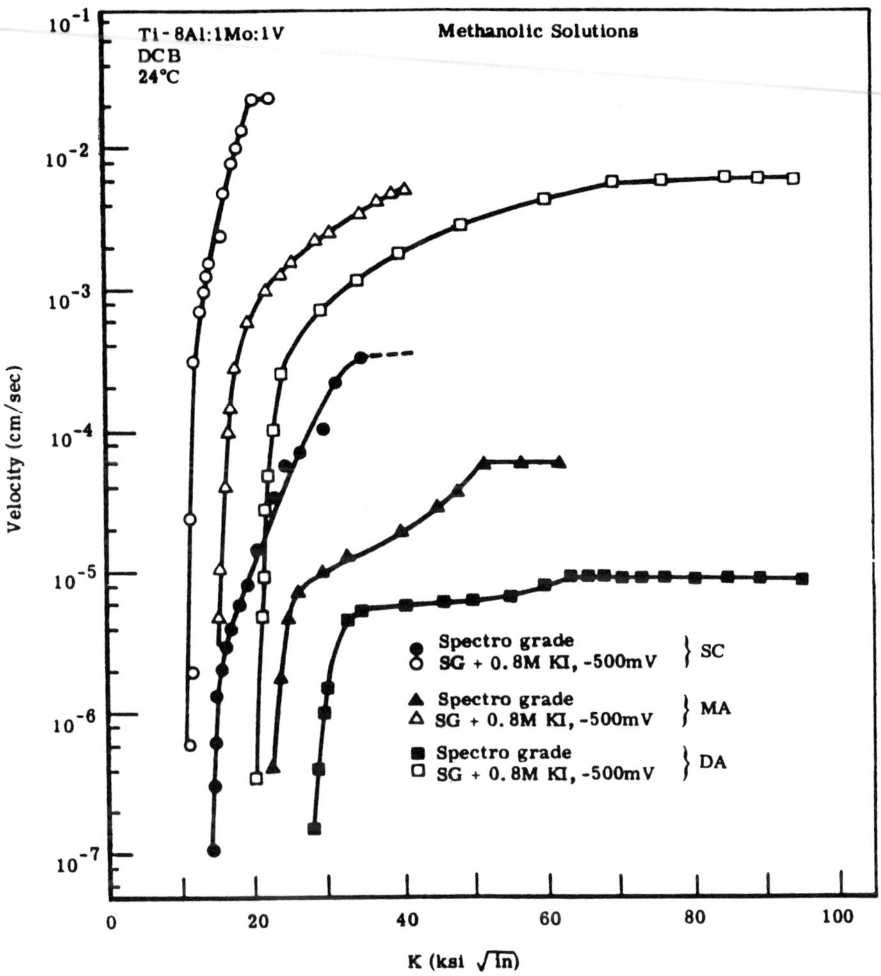

Fig. 21 Effects of halide addition and alloy metallurgical condition on SCC behavior of Ti-8Al-1Mo-1V (double cantilever beam specimens) in methanol at 24 °C (75 °F). Source: Ref 22

alloy cracking resistance in a methanol/HCl solution is illustrated in Fig. 22 (Ref 44).

Increasing temperature increases stage II cracking velocity (Ref 16) and appears to reduce K_{ISCC} values in pure methanol. Studies on Ti-3Al-8V-6Cr-4Zr-4Mo in methanol-water solutions (Ref 8, 42), however, suggest that the minimum water level required for SCC inhibition may actually diminish with increasing temperature. Thus, inhibitor effectiveness may increase with temperature and offset the normal deleterious influence of temperature on anodic dissolution and cracking.

Parallel to the behavior observed with intergranular SCC, transgranular stress cracking in methanol is linearly accelerated by anodic polarization, but inhibited by cathodic polarization. Investigators (Ref 22, 40) have demonstrated that titanium alloy SCC in neutral methanol/halide solutions can be fully arrested by applied cathodic potentials of −1.0 to −1.5 V (SCE).

Other Alcohols. Titanium alloy susceptibility rapidly diminishes with increasing alcohol carbon atom chain size (molecular weight) (Ref 4, 22). Most common industrial titanium alloys, including unalloyed titanium and Ti-6Al-4V, are

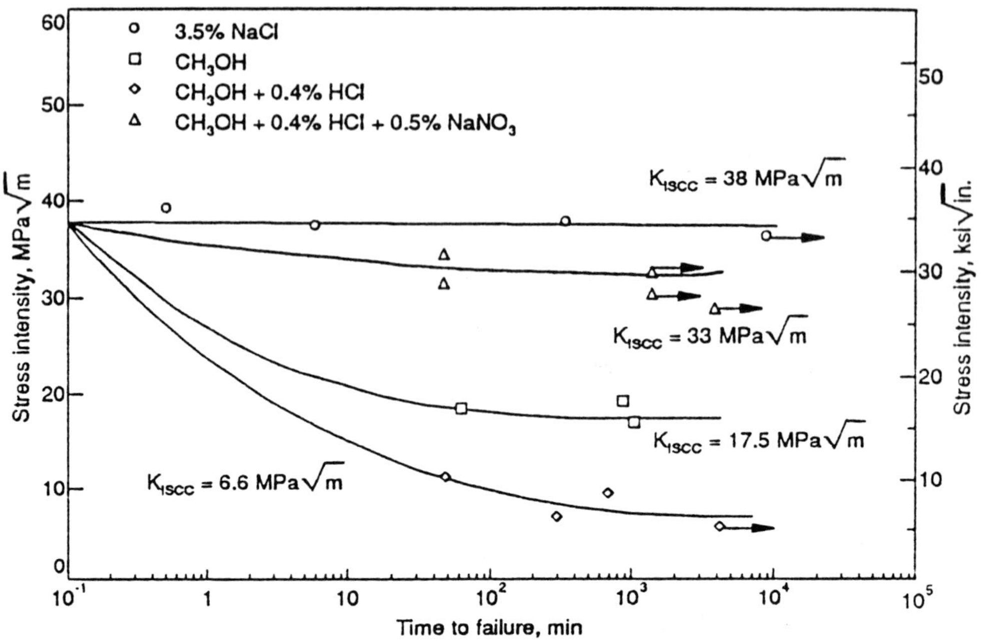

Fig. 22 Effect of acid and nitrate additions on the K_{ISCC} values of STA Ti-3Al-8V-6Cr-4Zr-4Mo in methanol at 25 °C (77 °F). Source: Ref 44

generally considered to be resistant to SCC in all alcohols other than methanol. Note, however, that the addition of halogens, such as Cl_2, Br_2, or I_2, or other (non-oxygen-containing) strong oxidizers (*e.g.*, $FeCl_3$), to various anhydrous alcohols can induce SCC in all titanium alloys, even the unalloyed grades (Ref 45-47). Depending on the alloy and the oxidizer concentration, much higher water levels are required for SCC inhibition. The highly susceptible Ti-8Al-1Mo-1V alloy uniquely experiences cracking in non-halogen/halide-containing anhydrous ethanol and ethylene glycol (Ref 48). This phenomenon seems to disappear in this alloy when the alcohol carbon atom chain size exceeds three.

SCC Mechanism. It is generally agreed that initiation of SCC is associated with local disturbance and/or rupture of the protective oxide film on titanium, typically where slip bands surface during metal deformation. This locally active site, if surface-film repassivation is not readily achieved, may suffer local anodic dissolution and become a stress riser. Because TiO_2 film formation is not possible in anhydrous or water-lean methanol, metal dissolution proceeds to solubilize titanium and form rather nonprotective titanium methylate (or methoxide) surface films according to reactions 1 and 2:

$$Ti + 4CH_3OH \xrightarrow{\text{no } H_2O} Ti(OCH_3)_4 + 4H\cdot \quad (1)$$

$$Ti + 3CH_3OH + H_2O \xrightarrow[\text{H2O}]{\text{lean}} TiOH(OCH_3)_3 + 4H\cdot \quad (2)$$

$$Ti + 2H_2O \rightarrow TiO_2 + 4H\cdot \quad (3)$$

Once local dissolution achieves a critical flaw size under applied tensile stress, crack growth occurs and these interactive overload/dissolution cycles continue. Note that atomic hydrogen is a by-product of dissolution reactions 1 and 2, and can be absorbed and concentrate (in a stress field) in the vicinity of the crack tip. It has been proposed that stress crack growth is facilitated by local hydrogen embrittlement (*i.e.*, α–phase hydride precipitation) ahead of the crack tip. When sufficient water- or oxygen-containing oxidizing species are present, full repassivation via TiO_2 film formation is possible (reaction 3) and no SCC occurs.

The interplay between the anodic dissolution reaction and by-product hydrogen absorption has clouded identification of the controlling mechanism for titanium alloy SCC in methanol. As with aqueous SCC, contradictory observations have made this a highly debated issue over the years (Ref 4, 22). It would seem likely that the anodic mechanism predominates at higher temperatures and in intergranular failures, whereas the hydrogen effects are amplified in titanium alloys that exhibit transgranular failure (and coplanar slip) at lower temperatures. Stimulation of stress cracking of unalloyed and alloyed titanium in various near-anhydrous alcohols in the presence of strong non-oxygen-containing oxidizing agents (*i.e.*, Cl_2, Br_2, I_2, or $FeCl_3$) appears to be a good example of anodic failure. Studies (Ref 45-47) reveal that the onset of SCC corresponds with metal potentials exceeding the alloy anodic pitting potential in methanol. These oxidizing agents, if added in sufficient concentration, shift the corrosion potential of the alloy above its meager pitting potential in the methanol solution, thereby stimulating anodic pitting and intergranular attack.

Other Organic Solvents

Although very little SCC testing of titanium alloys has been conducted, it is generally assumed that most commercial titanium alloys are resistant to SCC in most nonhalogenated organic solvents. This includes ketones, aldehydes, ethers, hydrocarbons (alkanes, alkenes), and almost all alcohols (except for methanol). Worst-case testing involving anhydrous benzene and ethyl ether solvents containing 0.5% Br_2 additions revealed no SCC with unalloyed titanium, but indicated accelerated cracking with the highly sensitive Ti-8Al-1Mo-1V alloy (Ref 46). Although Ti-6Al-4V also resisted the benzene, brominated benzene and ethyl ether, stress cracking was noted in the 0.5% Br_2/ethyl-ether solution. Thus, even the more SCC-sensitive titanium alloys (those containing more than 5 wt% Al) exhibit SCC only in these common organic solvents under highly modified, extreme conditions.

Chlorinated Solvents. Most commercial titanium alloys are considered to be highly resistant to SCC in chlorinated organic solvents, such that loaded smooth or notched specimens will not experience failure. Fracture toughness testing in several of these solvents, however, has revealed that, although unalloyed titanium is immune, certain high-aluminum titanium alloys may be susceptible. More specifically, Ti-8Al-1Mo-1V and Ti-5Al-2.5Sn have been reported to exhibit SCC in carbon tetrachloride(CCl_4), methylene chloride (CH_2Cl_2), methylene iodide (CH_2I_2), trichloroethane (CH_3CCl_3), and trichloroethylene (C_2HCl_3) (Ref 4, 22).

Investigation of SCC behavior of Ti-8Al-1Mo-1V in CCl_4 reveals characteristics almost identical to those obtained in 3.5% saltwater media. In particular, cleavage fracture morphology, cracking direction (approximately 15° from basal plane), cracking velocity, and K_{ISCC} were essentially the same (Ref 16, 22, 27). As in saltwater media, a loaded precrack is necessary to produce SCC. Iodine additions accelerate cracking and reduce K_{ISCC} values as in methanol. Although water additions imparted little effect in CCl_4, it did significantly decrease stage II cracking velocities for Ti-8Al-1Mo-1V in CH_2Cl_2. These overall cracking traits suggest a mechanism similar to that operative in aqueous media, but still leaves unanswered the long-debated question as to whether SCC is anodically (chloride) or cathodically (hydrogen) controlled.

Freons include a large family of fluorinated hydrocarbons used as refrigerants and solvents. Even though most titanium alloys are highly resistant to these highly stable compounds at normal service temperatures, susceptibility to SCC has been exhibited by certain alloys. Table 4 outlines K_{ISCC} data developed for susceptible alloys in three grades of Freon at ambient temperatures (Ref 49-51). The TF grade ($C_2Cl_3F_3$) had the least negative influence on K_I values compared to the MF (CCl_3F) and C318 (C_4F_8) grades tested. Cracking velocities are high and similar to those in methanolic solutions, and stage I type cracking may preclude identification of true threshold K_{ISCC} values.

Miscellaneous Solvents. K_{ISCC} values have been determined for Ti-8Al-1Mo-1V and Ti-6Al-4V in monomethylhydrazine (Ref 51). The Ti-6Al-4V alloy (STA condition) exhibits a drop from a K_{Ic} value of 44 MPa$\sqrt{m}$ (40 ksi$\sqrt{in.}$) to a K_{ISCC} value of approximately 33 MPa$\sqrt{m}$ (30

Table 4 K_{ISCC} Values for Titanium Alloys in Various Grades of Freon

Freon grade	Alloy	Heat Treatment (a)	K_{Ic} MPa$\sqrt{m}$	K_{Ic} ksi$\sqrt{in.}$	K_{ISCC} MPa$\sqrt{m}$	K_{ISCC} ksi$\sqrt{in.}$
TF	Ti-6Al-4V	STA	55	50	55	50
TF	Ti-6Al-4V	STA	44	40	36	33
TF	Ti-6Al-4V	MA	65	59	57	52
TF	Ti-5Al-2.5Sn	MA	74	67	33	30
TF	Ti-8Al-1Mo-1V	STA	88	80	48	44
TF	Ti-8Al-1Mo-1V	MA	33	30	< 25	< 23
MF	Ti-6Al-4V	STA	44	40	25	23
MF	Ti-6Al-4V	MA	65	59	< 27	< 25
MF	Ti-5Al-2.5Sn	MA	74	67	25	23
MF	Ti-8Al-1Mo-1V	MA	33	30	25	23
C318	Ti-5Al-2.5Sn	MA	74	67	< 54	< 49

(a) MA, mill annealed; STA, solution treated and aged. Source: Ref 49-51

ksi$\sqrt{in.}$) in this solvent at 27 to 40 °C (80 to 105 °F).

Hot Salt SCC

Hot salt SCC is a specialized case of halide-induced stress cracking of certain titanium alloys that occurs at temperatures above 220 °C (430 °F) or so in the presence of surface halide salt residues. The phenomenon was first discovered in 1955 during hot laboratory creep testing of a titanium alloy that cracked prematurely because of NaCl residues associated with specimen surface fingerprints. Although no actual cases of hot-service component failures have been assigned to hot salt SCC to date despite extensive hot service of titanium alloys in jet engines, this phenomenon still remains a well-studied laboratory curiosity.

Hot salt SCC may occur on highly stressed titanium alloy components with halide-containing surface residues or deposits after exposure in the 220 to 510 °C (430 to 950 °F) temperature range. Stress cracking is primarily influenced by temperature, stress level, exposure time, salt residue type and composition (Ref 52, 53), and the specific alloy. As indicated in Fig. 23 (Ref 52), hot salt SCC susceptibility increases with increasing stress, temperature, and exposure time. As this Larson-Miller-type relationship suggests, SCC threshold stress level of an alloy is reduced with increasing exposure time and/or temperature. Some measured SCC threshold stress levels for several common titanium alloys are presented in Table 5 (Ref 10).

The nature of surface salt residue/deposits and environmental conditions are also critical factors. Hot salt SCC has only been reported for chloride, bromide, and iodide salts and does not occur with fluoride or hydroxide salts. With chloride salts, attack severity increase with cation type as follows: $MgCl_2 < SrCl_2 < CsCl < CaCl_2 < KCl < BaCl_2 < NaCl < LiCl$ (Ref 54). Thinner surface oxide films and salt coatings both tend to enhance susceptibility to this form of cracking. The presence of both water and oxygen is necessary for attack as well (Ref 10, 55, 56). Humid air, moist salt deposits, or even water of hydration (*i.e.*, $MgCl_2{\cdot}6H_2O$) in various salts are viable sources of water for SCC. An excess amount of water can retard cracking tendencies, indicating a critical window for moisture level. In the absence of at least traces of oxygen, water by itself does not appear to be sufficient to promote cracking. Note that fully dried salts and/or total environmental pressures below 10 μ (Ref 10, 57) both effectively preclude SCC in susceptible alloys.

Metallurgical Aspects. Hot salt SCC many not only manifest itself as diminished ductility or premature fracture under load during hot exposure, but may also produce post-hot-exposure embrittlement at room temperature. Thus, testing for hot salt SCC includes observation for stress cracking during hot exposure, as well as tensile or

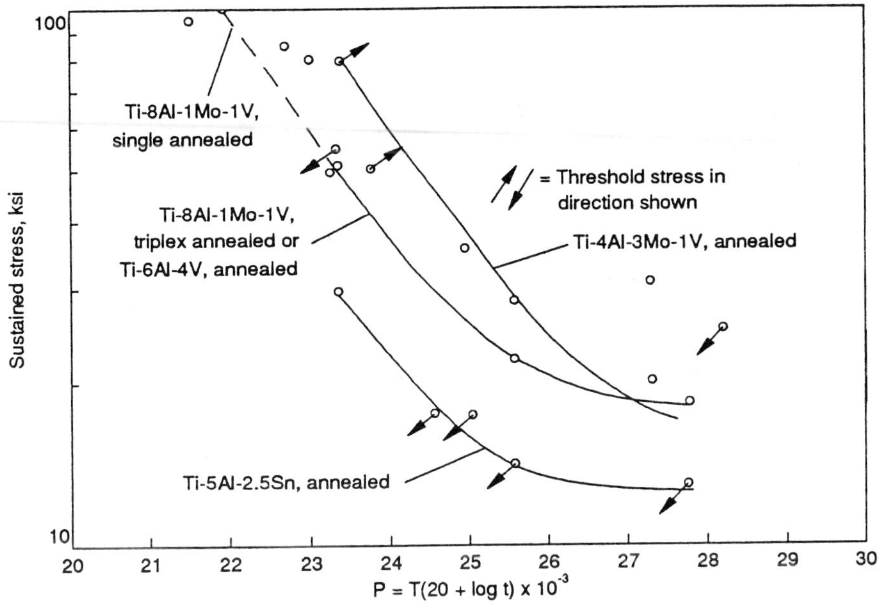

Fig. 23 Larson-Miller-type relationship for hot salt SCC of various titanium alloys. Source: Ref 52

Table 5 Hot Salt SCC Threshold Stress Levels for Several Titanium Alloys After 100 h Exposure (NaCl-Coated)

Alloy	Condition	Threshold stress, MPa (ksi) 288 °C	(550 °F)	426 °C	(800 °F)
Unalloyed Ti	Annealed	None		None	
Ti-4Al-3Mo-1V	Aged	579	(84)	338	(49)
Ti-6Al-4V	Annealed	345	(50)	165	(24)
Ti-5Al-2.5Sn	Annealed	193	(28)	138	(20)
Ti-8Al-1Mo-1V	Annealed	172	(25)	124	(18)

Source: Ref 10

bend properties at room temperature after exposure (Ref 22), as described in ASTM Test Specification G-41. For certain α and α/β titanium alloys, more dramatic reductions in ductility have been noted at low tensile strain rates tested after hot exposure (Ref 58). Because cracking does not initiate at precracked sites, smooth test specimens are effective and most widely used.

Hot salt SCC produces a high multiplicity of cracks characterized by extensive branching. Cracking is stepwise and highly discontinuous, with typical propagation rates on the order of 10^{-4} to 10^{-5} cm/s (4×10^{-5} to 4×10^{-6} in./s) (Ref 22). Most α and α/β alloys exhibit a predominantly intergranular mode of fracture, with some cases of mixed transgranular/intergranular cleavage reported in some alloys (*e.g.*, Ti-5Al-2.5Sn).

Susceptibility to hot salt SCC is highly alloy dependent. The unalloyed titanium grades containing less than 0.25 wt% O and grades 7, 9, 11, and 12 are considered immune to attack. Susceptibility increases with increasing aluminum content above 3 wt%, suggesting that crystallographic ordering and Ti_3Al (α_2) phase formation may be detrimental factors. Zirconium and most β-isomorphous alloying elements, such as molybdenum, tantalum, and vanadium, improve resistance. Minor silicon additions may also be beneficial.

Table 6 Relative Susceptibility of Titanium Alloys to Hot Salt SCC

Alloy	Susceptibility
Ti-8Al-1Mo-1V	High
Ti-5Al-2.5Sn	High
Ti-13V-11Cr-3Al	Intermediate
Ti-6Al-6V-2Sn	Intermediate
Ti-6Al-4V	Intermediate
Grade 4	Intermediate
Ti-8Mo-8V-2Fe-3Al	More resistant
Ti-11.5Mo-6Zr-4.5Sn	More resistant
Ti-4Al-3Mo-1V	More resistant
Ti-2.25Al-1Mo-11Sn-5Zr-0.25Si	More resistant
Grades 1 and 2	Immune
Grades 7 and 11	Immune
Grade 9	Immune
Grade 12	Immune

Source: Ref 59

These relative compositional effects are apparent in the comparative alloy SCC susceptibilities listed in Table 6 (Ref 59).

Some limited studies on near-α and α/β alloys (Ref 60) suggest that β processing to produce Widmanstätten structures may be beneficial. For example, β processing and/or final β annealing of Ti-8Al-1Mo-1V and Ti-4Al-3Mo-1V alloys produced significant improvements in hot salt cracking resistance over standard mill annealing.

Mechanism. A number of mechanisms for hot salt SCC of titanium alloys have been proposed over the years (Ref 4, 10, 22, 58-61). Though much debated, it is generally agreed that SCC stems from pyrohydrolytic formation of a hydrogen halide from its corresponding halide salt (*i.e.*, HCl, HBr, HI) that subsequently attacks the metal. This attack produces titanium chloride corrosion products and atomic hydrogen. This hydrogen is partially absorbed and diffuses into the metal under stress, causing local embrittlement and stepwise crack growth. Titanium chloride by-products may also undergo pyrohydrolytic decomposition in the presence of moisture to regenerate hydrogen halides (and titanium oxide products), thus sustaining the metal attack.

The following equations outline the proposed reaction steps for hot salt SCC with NaCl-coated specimens (Ref 10, 60, 61):

$$2\,NaCl + Ti\,(metal) + 1/2\,O_2 + H_2O \xrightarrow{\Delta} TiCl_2 + 2NaOH \qquad (1)$$

$$TiCl_2 + H_2O \xrightarrow{\Delta} TiO + 2HCl \qquad (2)$$

$$TiCl_n + {}_nH_2O \xrightarrow{\Delta} Ti\,(OH)_n + nHCl \qquad (2)$$

$$Ti\,(metal) + 2HCl \xrightarrow{\Delta} TiCl_2 + 2H\cdot \qquad (3)$$

Reaction 1 reveals the necessity of both moisture and oxygen (as cathodic depolarizer) in the formation of a titanium chloride corrosion product at the hot salt deposit/metal interface. Moisture is again critical in reaction 2 in order to form the corrosive hydrogen halide species, which attacks the metal in reaction 3. Here the attack autocatalytically regenerates titanium chlorides necessary for continued generation of HCl and the embrittling atomic hydrogen species.

This model is generally consistent with experimental observations. Crack propagation is based on the premise that absorbed atomic hydrogen from reaction 3 diffuses and concentrates within stress gradients in the metal, thereby locally embrittling the metal at a stress concentration site (*i.e.*, notch or crack tip). The dependence of hydrogen as the embrittling species is supported by the observation that post-hot-salt exposure embrittlement may be removed by hot vacuum annealing or by post-exposure tensile testing above 90 °C (195 °F) (Ref 58). Finally, this local hydrogen concentration phenomenon may also explain why no cases of hot salt SCC in hot titanium alloy components within aircraft gas turbine engines have been observed to date. Typical aircraft flights are of relatively short duration and may not exceed crack initiation times. Critical local hydrogen concentration may not be achieved during these periods and may be noncumulative since it can dissipate from these local stress sites between flights (Ref 60).

Oxidizing Nitrogen-Oxide Compounds

Although titanium alloys are highly corrosion resistant to most nitrogen-oxide compounds and gases because of their generally oxidizing nature, two specific compounds have been observed to induce titanium alloy SCC under special conditions: red nitrogen tetroxide (N_2O_4) and red-fuming nitric acid (RFNA). Both compounds represent powerful oxidizing agents used in rocket and missile propulsion systems.

Nitrogen Tetroxide. In 1964, a Ti-6Al-4V pressure vessel containing a high-purity grade of liquid N_2O_4 failed during proof testing, prompting a significant research effort to define this cracking phenomenon. Early work (Ref 62) revealed that stress cracking was possible when stressed smooth titanium alloy samples contacted liquid (but not vapor) red (or oxygenated) N_2O_4, but did not occur with green N_2O_4. The higher-purity red N_2O_4 contains a higher free oxygen content (typically 0.006% versus 0.002%) and no nitric oxide (NO) compared to the green liquid form (Ref 22, 63).

Cracking of Ti-6Al-4V in red N_2O_4 has been reported over the 275 to 620 MPa (40 to 90 ksi) stress level (Ref 64) and in Ti-8Al-1Mo-1V at stress intensities as low as 12 $\text{MPa}\sqrt{\text{m}}$ (11 $\text{ksi}\sqrt{\text{in.}}$) (Ref 65). Increasing temperature over the tested range of 21 to 71 °C (70 to 160 °F) resulted in decreased time to failure of smooth specimens (Ref 66) and a decrease in K_{ISCC} values (typically 70 to 80% of K_{Ic}) Ref 67) for Ti-6Al-4V. For example, Ti-6Al-4V, stressed to 620 MPa (90 ksi), cracked in 200 h at 30 °C (86 °F) and after only 12 h at 71 °C (160 °F). Increasing stress level, on the other hand, does not significantly influence time to failure, although it does dramatically increase crack frequency and density. It has been postulated that increased stress level produces an increase in slip step density at the surface, which correspondingly provides increased sites for crack initiation. Typical cases of titanium alloy SCC in red N_2O_4 exhibit magnitudes of crack density greater than that produced by SCC in other media.

Stress cracking in red N_2O_2 is primarily intergranular in α titanium alloys and becomes increasing transgranular (mixed mode) in α/β alloys as β-phase volume fraction increases (Ref 68, 69). Cracking velocity is considerably slower in N_2O_4 than in salt water, but has not been precisely measured.

Of particular significance is the highly potent inhibiting influence of the NO and water levels in liquid N_2O_4 and the related absence of SCC susceptibility in green N_2O_4. The key to inhibition is the presence of a small excess of NO species in the N_2O_4 liquid. This means reduction of free oxygen levels so that the NO species are not destroyed via reaction 1:

$$NO + \tfrac{1}{2}O_2 \leftrightarrow NO_2 \quad (1)$$

The addition of water further inhibits SCC by reducing the oxygen level according to reaction 2 and by producing NO species per reactions 3 and 4:

$$N_2O_4 + \tfrac{1}{2}O_2 + H_2O \leftrightarrow 2HNO_3 \quad (2)$$

$$N_2O_4 + H_2O \leftrightarrow HNO_3 + HNO_2 \quad (3)$$

$$3HNO_2 \leftrightarrow HNO_3 + 2NO + H_2O \quad (4)$$

A typical level of NO in commercial grades of green N_2O_4 is 0.7 to 0.8% (Ref 22).

An electrochemical mechanism for SCC of titanium alloys in red N_2O_4 is not probable because the liquid is highly nonconductive. Variations in chloride level in N_2O do not significantly influence SCC behavior, dispelling thoughts of strong anodic control. Hydrogen generation/embrittlement explanations are also not likely because of the severely oxidizing nature of this compound. Fundamentally, surface oxide film breakdown and/or formation of an (unknown) nonprotective film by chemical means appears to be the cause of stress cracking (Ref 63, 64).

Red-Fuming Nitric Acid. The first reported incident of titanium alloy SCC occurred in dry RFNA (Ref 64, 65). Titanium alloys can exhibit susceptibility both to severe intergranular attack (severe weight loss), forming highly pyrophoric surface deposits, and to stress cracking in anhydrous or water-lean RFNA. The critical window for the occurrence of intergranular pyrophoric reactions on titanium relative to NO_2 and H_2O content in RFNA has been well defined (Ref 69). The two modes of corrosion attack are not interdependent, but often occur simultaneously in stressed metal exposures. Both phenomena are aggravated by increasing NO_2 levels and are inhibited by increasing water content (Ref 65, 69-71). A water addition of at least 1.5 to 2.0% fully inhibits both corrosion attack and cracking in all titanium alloys.

Stress cracking has been observed on stressed smooth specimens of unalloyed titanium in anhydrous (<0.7% H_2O) RFNA containing 20% NO_2 (Ref 65, 70, 71). Reducing the NO_2 content to 0 to 10% eliminates susceptibility in unalloyed grades, but not in the Ti-5Al-2.5Sn, Ti-6Al-4V, and Ti-8Mn alloys. The cracking mode in all alloys tested has been highly intergranular. Cracking tendencies may be diminished by exposure of RFNA to air/oxygen and especially by minor $CuSO_4$ additions (Ref 71). Because current military specifications for RFNA generally require more than 2% H_2O, titanium alloys are considered "safe" from an SCC standpoint.

The mechanism of SCC of titanium alloys in RFNA is probably similar to that proposed for red N_2O_4 liquid. Chemical formation of a nonprotective film in the absence of sufficient water activity (to form TiO_2) is a probably scenario. A possible corrosion reaction is (Ref 70):

$$Ti + 8HNO_3 \xrightarrow[H2O]{lean} 4NO_2 + 4H_2O + Ti(NO_3)_4$$

Molten Salts

Molten salts may be encountered in heat treatments baths, fast-breeder reactors, descaling baths, and battery research. To date, Ti-8Al-1Mo-1V has been the only titanium alloy evaluated with respect to stress cracking in molten salts. Studies have focused on pure alkali chloride or bromide eutectics (*i.e*, KCl + LiCl, KBr + LiBr) at 350 to 500 °C (660 to 930 °F), alkali nitrates with melting points below 125 °C (260 °F), and fused hydroxides. Smooth specimens of Ti-8Al-1Mo-1V have failed in a LiCl-KCl melt at 375 °C (710 °F) when loaded to 69 MPa (10 ksi) (K_{ISCC} <10 MPa$\sqrt{m}$, or <9 ksi$\sqrt{in.}$) (Ref 54.). These chloride/bromide eutectics produce the stage I and II subcritical crack growth behavior typified in Fig. 24 (Ref 16). Similar to methanol solutions and liquid mercury, a mixture of intergranular separation and cleavage is associated with stage I cracking, with stage II exhibiting transgranular cleavage (Ref 22, 54, 72). Although increasing temperature did increase stage II crack velocity, varying water content (10 to 100 ppm) and microstructure played little role in overall cracking behavior (Ref 16). Cathodic protection was effective for inhibiting SCC when impressed potentials were negative to approximately –1.2V

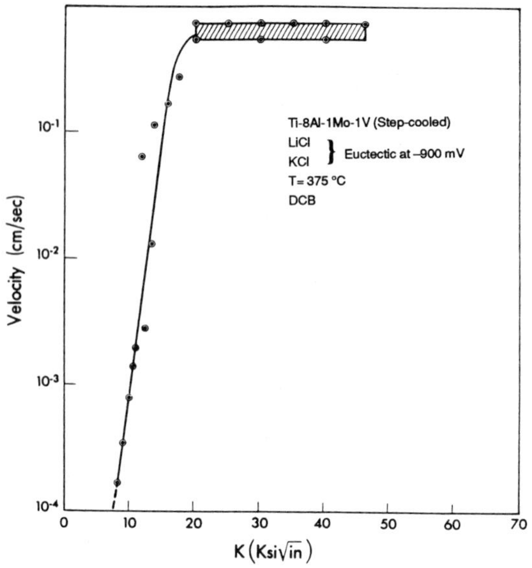

Fig. 24 SCC behavior of step-cooled Ti-8Al-1Mo-1V (double cantilever beam specimens) in fused LiCl/KCl eutectic (-900 mV) at 375 °C (705 °F). Source: Ref 16

(Ag/AgCl) (Ref 16). Note that rapid anodic dissolution of titanium occurs at potentials noble to – 0.9 V.

Pure fused nitrate salts did not produce cracking in Ti-8Al-1Mo-1V up to at least 450 °C (840 °F), but did stimulate cracking at temperatures as low as 300 °C (570 °F) when Cl^-, Br^-, or I^- (but not F^-) ions were added (Ref 22). Increasing halide concentration increases stage II crack velocities as expected. Although the influence of potential on crack behavior is rather complex (Ref 22), cracking inhibition is indicated at very cathodic potentials. Significantly reduced cracking velocities were noted in fused sodium hydroxide containing halides. The studies to date strongly suggest an anodic halide dissolution mechanism for SCC. A cathodic explanation involving hydrogen does not fit the molten salt SCC behavior observed.

Liquid and Solid Metal Embrittlement

Titanium alloys may experience premature failure and brittle fracture when stressed while in intimate contact with certain metals in liquid and/or solid forms (Ref 4, 22, 62, 73-75). These metals include cadmium, mercury, silver, cesium, gallium, gold, and possibly zinc.

Liquid metal embrittlement (LME) in liquid cadmium has been reported for the Ti-8Mn, Ti-13V-11Cr-3Al, and grade 2 titanium alloys. Dramatic reductions in both elongation and strength were observed in the 325 to 400 °C (620 to 750 °F) range, which were very dependent on strain rate (Ref 22, 76). Penetration of the TiO_2 oxide film by the liquid cadmium appears to be the critical factor, with cracking predominantly intergranular.

Liquid mercury has produced embrittlement in the grade 4, Ti-13V-11Cr-3Al, Ti-6Al-4V, Ti-8Al-1Mo-1V and Ti-8Mn alloys (Ref 22, 77, 78). As shown in Fig. 25 for Ti-8Al-1Mo-1V (Ref 16), both stage I and II cracking behavior is exhibited, with microstructural influences paralleling those observed in methanol and aqueous media. A mixed mode of fracture is noted, with intergranular and transgranular cracking associated with stage I and II cracking regions, respectively. The transgranular cleavage plane is on or near the (0001) plane, also similar to α-phase SCC in aqueous media.

Exposure of stressed U-bend and C-ring samples of grades 2 and 12 and Ti-3Al-8V-6Cr-4Zr-4Mo (Ref 8) in liquid mercury reveal no cracking tendencies to temperatures as high as 230 °C (445 °F). These static tests have shown that the surface oxide film of titanium is not wetted by liquid mercury, thereby precluding amalgamation and embrittlement. However, cracking failure may be induced when oxide films of stressed samples are severely mechanically damaged (heavily scraped or scratched) while immersed in the liquid metal. Enhanced oxide film barriers, achieved through anodizing or thermal oxidation, may offer improved resistance to LME in mercury, unless the film is physically compromised. The protective influence of a thermally nitrided, nonwetting titanium surface has similarly been demonstrated in mercury to temperatures as high as 370 °C (698 °F) (Ref 79).

LME has also been reported for titanium exposed to silver brazing alloys (Ref 80, 81) and molten zinc (Ref 22). The Ti-13V-11Cr-3Al alloy has exhibited transgranular cleavage in molten zinc. Similar to high-temperature mercury, degradation of titanium alloy properties in molten zinc, cesium, and gallium is generally the result of titanium dissolution and diffusion of the metal along grain boundaries to form brittle phases.

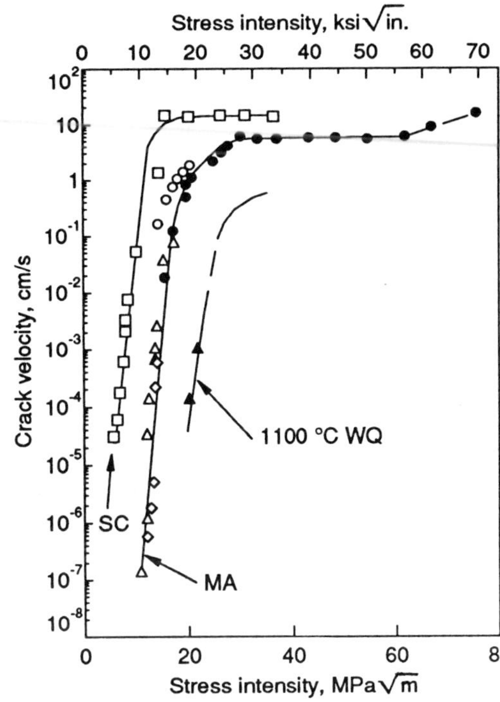

Fig. 25 SCC behavior of Ti-8Al-1Mo-1V (double cantilever beam specimens) in liquid mercury at 24 °C (75 °F). Source: Ref 16

Solid Metal Embrittlement (SME). Early reported incidences of SME stemmed from investigations of cracked titanium alloy jet engine compressor disks involving cadmium- and silver-plated fasteners (Ref 80). Studies that followed confirmed that embrittlement could be induced at temperatures well below the melting point of the metal (Ref 22, 74, 75, 82), as long as very intimate metal contact with the stressed titanium alloy was achieved. In the case of cadmium, SME has been shown to occur in Ti-6Al-4V, Ti-13V-11Cr-3Al, and Ti-8Al-1Mo-1V from room temperature up to the melting point of cadmium (321 °C, or 610 °F) (Ref 74). The cadmium must be smeared or pressed (beyond the yield point) into the highly

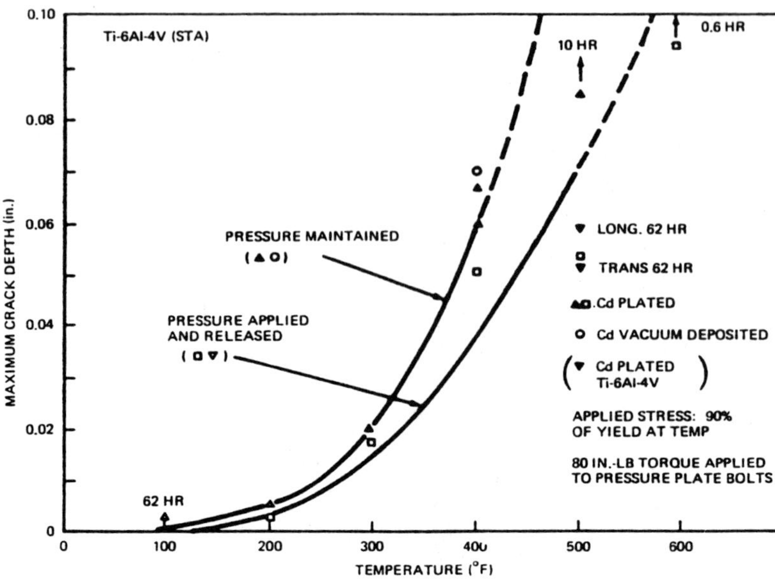

Fig. 26 Stress cracking depth on a (STA) Ti-6Al-4V sample stressed in tension and in bearing contact with solid cadmium for various temperatures and times. Source: Ref 74

stressed titanium alloy surface at loads exceeding the yield stress for brittle failure to occur. Simply stressing cadmium-coated titanium is not sufficient to initiate cracking. Increasing temperature and stress level decrease time to cracking and increase crack depth, as shown in Fig. 26 (Ref 74).

In contrast to interstitials, cadmium does not diffuse into the surface of titanium and produce brittle layers, but rather diffuses along fresh metal or oxide-covered surfaces of the crack. The dependence of crack rate on time and temperature is consistent with a cadmium surface diffusion-controlled mechanism. The cracks produced are generally shallow and fine, high in density, and visually difficult to detect. Fracture mode is mixed, with a predominance of intergranular cracking. Similar to SCC in mercury, the cleavage plane was near the basal (0001) direction.

Any barrier film or coating that inhibits direct bearing contact of cadmium onto titanium alloy surfaces can avert cadmium embrittlement. For example, sprayed or vapor-deposited aluminum coatings, organic coatings, dry film lubricants, and/or oxidized surfaces on titanium may provide effective protection, depending on bearing loads.

Intergranular embrittlement of Ti-6Al-6V-2Sn has also been reported in contact with cadmium, gold, and silver (Ref 73). Thresholds for silver and gold were 204 and 232 °C (400 and 450 °F), respectively. No embrittlement was stimulated by nickel or copper metal to temperatures as high as 287 °C (550 °F). Again, metal surface diffusion over the titanium oxide film to the crack tip was indicated as the mechanism. Pure silver and silver brazing alloys have also induced SME of Ti-8Al-1Mo-1V, Ti-7Al-4Mo, and Ti-5Al-2.5Sn (Ref 80) at temperatures of 468 and 343 °C (875 and 650 °F), respectively. The presence of silver chloride stimulates SME more aggressively in titanium alloys than in pure silver. The perceived SME threshold for Ti-6Al-4V in contact with silver is approximately 340 °C (645 °F).

Compatibility of titanium alloys in contact with solid zinc metal has not been well defined or documented. There are vague indications that some susceptibility to SME may be possible (Ref 22).

It should be pointed out that, practically speaking, the likelihood of incurring SME of titanium alloys in most applications is very low. This is because the problem metal must be smeared at extreme bearing loads and plastically embedded into the titanium surface while the titanium is loaded in tension to well above 50% of the yield strength.

Gaseous Environments

Very limited published data exists on subcritical crack growth of titanium in gaseous environments. The few cases where susceptibility has been reported for certain titanium alloys in gases include moist chlorine, dry HCl, and dry hydrogen.

The Ti-8Al-1Mo-1V alloy exhibits cracking in moist chlorine gas at 288 °C (550 °F) (Ref 83). Various binary titanium alloys are also known to experience cracking in this gas at 427 °C (800 °F) (Ref 84). U-bend specimen testing of Ti-8Al-1Mo-1V in HCl gas at 1 atm revealed SCC over the temperature range of 240 to 400 °C (465 to 750 °F) (Ref 85). Time to crack initiation decreased with increasing temperature and with a water addition of 5 to 6 mol%. Cracking characteristics were similar to those produced by hot salt SCC.

Hydrogen gas exposures may also result in titanium alloy embrittlement, depending on gas temperature, pressure, and purity. In relatively pure, dry hydrogen atmospheres, the oxide film can break down under stress and not reform, allowing direct ingress of hydrogen via diffusion into the metal. Thus, as temperature and hydrogen pressures increase, hydrogen ingress increases. At low temperatures, such as in liquid hydrogen, slow uniform hydriding of α and α/β titanium alloys occurs, such that spalling and associated metal loss under stress is possible (Ref 86, 87). The maximum degrading influence of absorbed hydrogen occurs at 0 °C (32 °F) for Ti-6Al-4V (Ref 22). In this temperature range, sustained load cracking (subcritical crack growth) is observed for most α and α/β alloys (Ref 88) and is generally attributable to hydride precipitation within the α phase under stress. Fracture characteristics resemble those obtained in aqueous SCC failures, with cracking velocities less than or equal to 10^{-3} cm/s (4×10^{-4} in./s) (Ref 89). At higher temperatures and pressures, absorption and diffusion of hydrogen may become so great that rapid embrittlement may occur.

The presence of oxygen or water in the hydrogen gas may sustain the protective metal oxide film, thereby inhibiting hydrogen absorption and embrittlement. The inhibitive influence of at least 2.0 wt% water, for example, was noted in commercial-purity titanium exposed to 2750 to 5500 kPa (400 to 800 psi) hydrogen gas at 150 to 300 °C (300 to 570 °F) (Ref 2). The application of enhanced surface-oxide barrier films, produced by thermal oxidation or anodizing, also helps to thwart hydrogen absorption (Ref 2).

Applicable SCC Test Methods

Test methods for assessing the SCC resistance of titanium alloys can be grouped into three basic categories:

- *Category 1 tests* use smooth, statically loaded specimens, such as U-bend, C-ring, bent beam, and dead-loaded tensile specimens (described in ASTM Test Specifications G30, G35, G36, G38, and G47, respectively).
- *Category 2 tests* are fracture mechanics tests that use notched and precracked specimens that are statically or dynamically loaded, such as cantilever beam, compact tension, and double cantilever beam specimens, and are conducted per ASTM E 399 guidelines.
- *Category 3 tests* use smooth or notch tensile specimens that are dynamically loaded at relatively low strain rates; the slow-strain-rate test (SSRT) is the primary example.

Test Selection. Selection of a suitable SCC test method for titanium alloys is primarily determined by the nature of the stress cracking data or design information desired and by the chemical nature of the test environment relative to titanium alloy passivation. For simple go/no-go assessment of SCC with no quantitative cracking parameters, the least costly category 1 tests may be adequate if SCC initiation (incubation) periods are not too long. The data are typically considered in terms of sample stress versus time to failure, as shown in Fig. 27(a). Slow-strain-rate test results are generally assessed as comparisons or ratios of test sample ductility and/or time-to-failure values to those from a reference sample (tested in air or inert gas). However, it is vital to include thorough examination of sample fracture surfaces for indications of SCC (*i.e.*, cleavage or intergranular cracking).

If fracture mechanics design information (*i.e.*, K_{ISCC} or stage I or II crack growth parameters) is required, then the more sophisticated, costly category 2 tests are the obvious choice. K_{ISCC} results for titanium alloys are stress and time dependent, such that K_{ISCC} values may be plotted as a function of time to failure to reveal true, conservative K_{ISCC} values (see Fig. 27b). Sufficient test periods (typically > 6 to 10 h) at each stress-intensity level tested should be allowed to ensure that the "plateau" shown in Fig. 27(b) has been reached. Slow-strain-rate testing of titanium alloys can be considered accelerated, worst case, very conservative, and sometimes unrealistic from a design/use standpoint, because the surface-oxide film is continuously being broken during straining to failure. Discriminating strain rates for titanium alloys in aqueous and methanolic solutions lie within the 10^{-5} to 10^{-7}/s window.

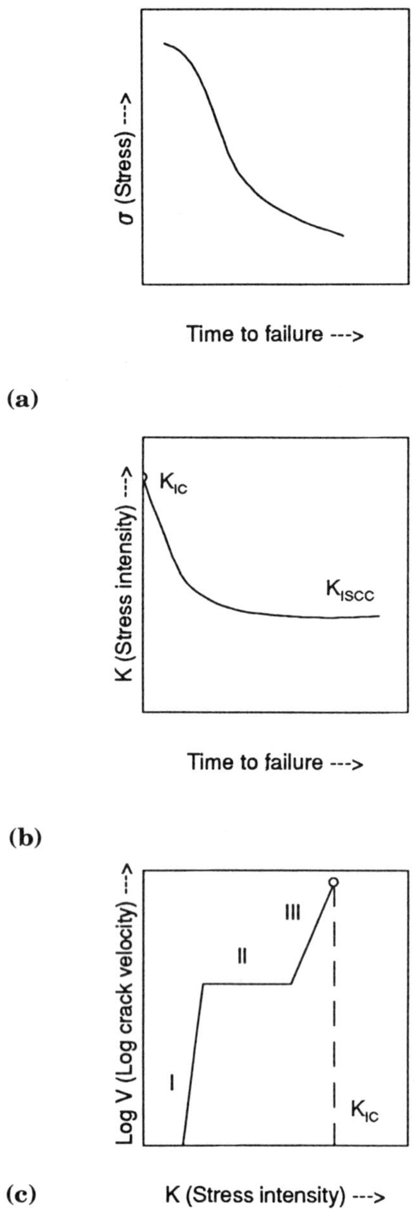

Fig. 27 Various types of SCC data presentation. (a) Typical for category 1 testing. (b) and (c) Typical for category 2 testing

The SCC test method for titanium alloys may also be based on the particular type of environment to be tested. In environments where alloy repassivation (oxide film formation) is favored and where stage I subcritical cracking is generally not observed, category 2 and 3 tests are more discriminating than category 1 tests. This is the case for aqueous media, including salt water and brines, and chlorinated organic solvents in which cracking failure of category 1 specimens is generally not observed at near-ambient temperatures. One must provide initial or continued local oxide film breakdown, such as in category 2 and 3 tests, to expeditiously initiate and manifest potential SCC tendencies in these alloys. Exceptions to this rule are those very few titanium alloys that may exist in highly sensitized metallurgical conditions. For example, alloys susceptible to SCC in ambient neutral salt water in smooth or notched samples include step-cooled Ti-8Al-1Mo-1V and the Ti-8Mn and Ti-13V-11Cr-3Al alloys. Highly stressed category 1 tests specimens may also be applicable for testing in hot aqueous halides or chlorinated solvents, where higher temperatures may overcome SCC activation barriers in certain alloys after sufficient exposure periods.

For environments where stage I subcritical cracking behavior occurs and repassivation is not favored, category 1 tests are often preferred over category 2 and 3 tests for practical reasons. These environments include methanol, red-fuming nitric acid, N_2O_4, liquid/solid metals, molten salts, and hot salt. Note that category 2 testing is not physically applicable to the evaluation of SME and hot salt SCC in these alloys.

Strategies for SCC Prevention in Titanium Alloys

Prevention of SCC may be achieved by control or alteration of mechanical, metallurgical, and/or environmental factors. Control of mechanical factors is fundamental and common to all metals and includes reducing imposed tensile stress levels, avoiding stress concentrators in component design, relieving residual stresses from fabrication and other sources, and introducing surface compressive stresses (by shot peening, for example). The metallurgical and environmental strategies for prevention of titanium alloy SCC, however, are highly dependent on the particular alloy and environment under consideration. In an effort to offer some practical direction toward SCC mitigation and inhibition, the following sections provide strategies for key environments known to promote SCC in titanium alloys. Note that one or more of the specific strategies listed may be required for SCC prevention, and the prior text with references should be consulted for details.

Aqueous Media. For α and α/β alloys, strategies include:

- If possible, reducing section thicknesses of components below the critical thickness for plane strain
- Promoting basal-normal over basal-transverse crystallographic texture via processing
- Reducing grain size
- Beta processing and/or heat treating to produce β-transformed microstructures

If SCC limited, consider alloys with one or more of the following compositional traits:

- An aluminum content of less than 5 wt%
- Higher levels of β-isomorphous alloying elements (molybdenum, niobium, tantalum, vanadium)
- Reduced levels of β-eutectoid alloying elements (silicon, copper, manganese, iron and chromium)
- Reduced levels of interstitials (oxygen, nitrogen, and carbon); use ELI grades
- Reduced tin content

Other strategies include:

- Considering SCC-resistant low-alloy grades, such as grades 1, 2, 7, 9, and 12, as alternatives to higher-strength alloys
- Reducing grain size
- Applying cathodic potentials in neutral solutions to a limit where excessive hydrogen absorption can be avoided

Methanol Solutions. Strategies include:

- Adding sufficient water to inhibit (alloy dependent)
- Adding small (≥0.5%) additions of nitrates (*e.g.*, $NaNO_3$, KNO_3, $AgNO_3$)
- Applying cathodic potentials
- Eliminating halogens or halides from methanol
- Reducing alloy grain size and interstitial content
- If possible, using higher-molecular-weight alcohols or other solvents in place of anhydrous methanol

Halogenated Organic Solvents. Strategies include:

- Considering similar alloy compositional, microstructural, and textural guidelines offered for aqueous media
- Considering low-alloy grades, such as grades 1, 2, 7, 9, and 12, as alternatives to higher-strength alloys

Hot Salt Deposits/Residues. Strategies for susceptible titanium alloys include:

- Avoiding formation of halide salt deposits or residues on stressed metal surfaces if metal temperatures exceed approximately 230 °C (445 °F); rinsing metal surfaces free of halides with a halide-free solvent
- If halide deposits/residues exist, heating only in dry and/or oxygen-free environments if temperatures exceed approximately 230 °C (445 °F)
- Considering similar alloy compositional and microstructural guidelines offered for aqueous media
- Plating samples with nickel metal

Also, SCC-resistant low-alloy grades, such as grades 1, 2, 7, 9, and 12, should be considered as alternatives to higher-strength alloys.

Nitrogen Tetroxide. Strategies include:

- Converting from the red form to the "SCC-safe" green form
- Increasing water and nitric oxide content and limiting oxygen levels in N_2O_4

Red-Fuming Nitric Acid. Strategies include:

- Converting from the red form to the "safe" white form
- Reducing NO_2 content to RFNA
- Maintaining water content at a minimum of 1.5 to 2.0 wt%

- Considering minor $CuSO_4$ additions as an inhibitor

Liquid/Solid Metal Embrittlement. Strategies include:

- Avoiding mechanical damage to titanium surfaces (oxide film damage) and plastic deformation of titanium when contacting solid or liquid cadmium and liquid mercury
- Avoiding the plating of cadmium onto titanium surfaces in highly stressed component situations

Where intimate bearing contact of stressed titanium is anticipated with cadmium, mercury, or silver (above 200 °C, or 390 °F) or with gold (above 230 °C, or 450 °F), consider the following strategies with regard to barrier coatings:

- Using solid film lubricants (*e.g.*, MoS_2 or graphite) for solid metal contact
- Thermally oxidizing titanium surfaces
- Thermally nitriding titanium surfaces
- Anodizing titanium surfaces
- Spraying or vapor depositing aluminum metal onto titanium surfaces
- Plating titanium surfaces with nickel or copper
- Applying thin plastic (*e.g.*, Teflon) films/coatings to titanium surfaces

References

1. R.W. Schutz, Titanium, *Process Industries Corrosion—The Theory and Practice*, National Association of Corrosion Engineers, 1986, p 503
2. *Metals Handbook*, 9th ed., Vol 13, *Corrosion*, ASM International, 1987, p 669
3. *Proceedings of an International Symposium on Stress Corrosion Mechanisms in Titanium Alloys*, 27-29 Jan 1971, cosponsored by Georgia Institute of Technology and National Association of Corrosion Engineers
4. B.F. Brown, "Stress Corrosion Cracking in High Strength Steels and in Titanium and Aluminum Alloys," Naval Research Laboratory, 1972
5. "Accelerated Crack Propagation of Titanium in Methanol, Halogenated Hydrocarbons, and Other Solutions," DMIC Memorandum 228, Defense Metals Information Center, Battelle Memorial Institute, 6 March 1967
6. R.J.H. Wanhill, Aqueous Stress Corrosion in Titanium Alloys, *Br. Corros. J.*, Vol 10 (No. 2), 1975, p 69-78
7. R.W. Schutz, M. Xiao, and T.A. Bednarowicz, "Stress Corrosion Behavior of the Ti-3Al-8V-6Cr-4Zr-4Mo Titanium Under Deep Sour Gas Well Conditions," Paper 51, presented at NACE Corrosion '92 (Nashville), 27 April - 1 May 1992
8. D.R. Klink and R.W. Schutz, "Engineering Incentives for Utilizing Ti-3Al-8V-6Cr-4Zr-4Mo Titanium Tubulars in Highly Aggressive Deep Sour Gas Wells," Paper 63, presented at NACE Corrosion '92 (Nashville), 27 April-1 May 1992
9. D.E. Piper, S.H. Smith, and R.V. Carter, Corrosion Fatigue and Stress-Corrosion Cracking in Aqueous Environments, *ASM Met. Eng. Q.*, Vol 8, 1968, p 50
10. A.J. Hatch, H.W. Rosenberg, and E.F. Erbin, Effects of Environment on Cracking in Titanium Alloys, *Stress-Corrosion Cracking of Titanium*, STP 397, ASTM, 1966, p 122
11. J. Brettle, Stress Corrosion of Titanium and Its Alloys in Aqueous Chloride Environments, *Met. Mater.*, Oct 1972, p 442-451
12. T.L. MacKay, C.B. Gilpin, and N.A. Tiner, "Stress-Corrosion Cracking of Titanium Alloys at Ambient Temperatures in Aqueous Solutions," Contract NAS 7-488, Report SM-49105-F1, McDonnell Douglas Corp., July 1967
13. P. Finden, "Comparative Data—Titanium Alloy Screening Tests," D6-24541-TN, Boeing Co.
14. R.H. Olsen, R.R. Boyer, H.A. Moreen, A.L. Wingert, W.E. Quist, W.F. Spurr, F.L. Parkinson, and P. Finden, "Metallurgical and Fractographic Analysis of Ti-6Al-4V," MLER G-8842-70-24 (Part I), Boeing Co., Dec 1970
15. R.R. Boyer, H.A. Moreen, R.H. Olsen, and A.L. Wingert, "Metallurgical and Fractographic Analysis of Ti-6Al-4V," MLER 6-8842-71-3 (Part II), Boeing Co., Feb 1971
16. T.R. Beck, M.J. Blackburn, W.H. Smyrl, and M.O. Speidel, "Stress-Corrosion Cracking of Titanium Alloys: Electrochemical Kinetics, SCC Studies with Ti:8-1-1, SCC and Polarization Curves in Molten Salts, Liquid Metal Embrittlement, and SCC Studies with Other Titanium Alloys," Contract NAS 7-489, Quarterly Progress Report 14, Dec 1969
17. J.D. Boyd, P.J. Moreland, W.K. Boyd, R.A. Wood, D.N. Williams, and R.I. Jaffee, "The Effect of Composition on the Mechanism of Stress-Corrosion Cracking of Titanium Alloys in N_2O_4 and Aqueous and Hot-Salt Environments," Contract NASr-100 (09), 26 Aug 1969
18. D.T. Powell and J.C. Scully, Stress-Corrosion Cracking of Alpha Titanium Alloys at Room Temperature, *Corrosion*, Vol 24 (No. 6), 1968, p 151
19. T.R. Beck, Stress-Corrosion Cracking of Titanium Alloys: II. An Eletrochemical Mechanism, *J. Electrochem. Soc.*, Vol 115, 1968, p 890
20. T.R. Beck, "Stress-Corrosion Cracking of Titanium Alloys. Preliminary Report on Ti-8Al-1Mo-1V Alloy and Proposed Electrochemical Mechanism," D1-82-0554, Boeing Co., July 1965

21. N.G. Feige and T. Murphy, Fracture Behavior of Titanium Alloys in Aqueous Environments, *Met. Eng. Q.*, Vol 7 (No. 1), 1967, p 53
22. M.J. Blackburn, J.A. Feeney, and T.R. Beck, "State-of-the-Art of Stress-Corrosion Cracking of Titanium Alloys," Part 4 of Monograph Review, Boeing Co., sponsored by the Advanced Research Projects Agency, ARPA Order No. 878 and NAS7-489, June 1970
23. D.A. Litvin and B. Hill, Effect of pH on Sea-Water Stress-Corrosion Cracking of Ti-7Al-2Cb-1Ta, *Corrosion*, Vol 26 (No. 3), 1979, p 89
24. M. Ueda *et al.*, "Corrosion Behavior of Titanium Alloys in a Sulfur-Containing H_2S-CO_2-Cl^- Environment," Paper 271, presented at NACE Corrosion '90 (Las Vegas), 23-27 April 1990
25. J.A. Ruppen, R.S. Glass, and M.A. Molecke, Titanium Utilization in Long-Term Nuclear Waste Storage, *Titanium for Energy and Industrial Applications,* American Institute of Mining, Metallurgical, and Petroleum Engineers, 1981, p 355
26. S.G. Pitman, "Enviromechanical Testing of Ti Grade 2 and Ti Grade 12 in Basalt Groundwater," *Industrial Applications of Ti and Zr: 3rd Conference*, STP 830, ASTM, 1984, p 5-18
27. T.R. Beck and M.J. Blackburn, Stress-Corrosion Cracking of Titanium Alloys, *AIAA J.*, Vol 6 (No. 2), 1968, p 326
28. T.R. Beck, M.J. Blackburn, and M.O. Speidel, "Stress-Corrosion Cracking of Titanium Alloys: SCC of Aluminum Alloys, Polarization of Titanium Alloys in HCl and Correlation of Titanium and Aluminum SCC Behavior," Contract NAS 7-489, Quarterly Progress Report 11, March 1969
29. S.R. Seagle, R.R. Seeley, and G.S. Hall, The Influence of Composition and Heat Treatment on the Aqueous Stress Corrosion of Titanium, *Applications Related to Phenomena in Titanium Alloys,* STP 432, ASTM, 1967, p 170
30. R.W. Judy, Jr., and R.J. Goode, "Investigation of Processing Factors and Interstitial Oxygen Content on the Stress-Corrosion-Cracking Resistance of Ti-6Al-4V Plate," NRL Memorandum Report 2145, Naval Research Laboratory, July 1970
31. R.A. Wood, "Beta Titanium Alloys," MCIC Report 72-11, Metals and Ceramics Information Center, Battelle Columbus Laboratories, Sept 1972
32. R.W. Schutz, "Understanding and Preventing Crevice Corrosion of Titanium Alloys," Paper 162, presented at NACE Corrosion '91 (Cincinnati), 11-15 March 1991
33. E.G. Haney and W.R. Wearmouth, Effect of "Pure" Methanol on the Cracking of Titanium, *Corrosion*, Vol 25 (No. 2), 1969, p 87
34. E.G. Haney and W.R. Wearmouth, "Investigation of Stress-Corrosion Cracking of Titanium Alloys," Report No. 6, Research Grant NGR-39-008-014, National Aeronautics and Space Administration, May 1969
35. E.G. Haney and P. Fugassi, Titanium Behavior in Methanol-Water-Chloride Solutions—Effect of Cations, *Proceedings of an International Symposium on Stress Corrosion Mechanisms in Titanium Alloys,* 27-29 Jan 1971, cosponsored by Georgia Institute of Technology and National Association of Corrosion Engineers
36. K. Mori, A. Takamura, and T. Shimose, Stress-Corrosion Cracking of Ti and Zr in HCl-Methanol Solutions, *Corrosion*, Vol 22 (No. 2), Feb 1966, p 29-31
37. A.J. Sedriks and J.A.S. Green, Stress-Corrosion Cracking and Corrosion Behavior of Titanium in Methanol Solutions: Effect of Metal Ions in Solution, *Corrosion*, Vol 25, (No. 8), 1969, p 324
38. E.G. Haney, "Investigation of Stress-Corrosion Cracking of Titanium Alloys," Report No. 7, Research Grant NGR-39-008-014, National Aeronautics and Space Administration, Feb 1970
39. E.G. Haney and W.R. Wearmouth, "Investigation of Stress Corrosion Cracking of Titanium Alloys," Report No. 4, Research Grant NGR-39-008-014, National Aeronautics and Space Administration June 1968
40. B.S. Hickman, J.C. Williams, and H.L. Marcus, Transgranular and Intergranular Stress-Corrosion Cracking of Titanium Alloys, *Austral. Inst. Met.*, Vol 14 (No. 3), 1969, p 138
41. E.G. Haney and W.R. Wearmouth, "Investigation of the Stress-Corrosion Cracking of Titanium Alloys," Report No. 5, Research Grant NGR-39-008-014, National Aeronautics and Space Administration, Nov 1968
42. R.W. Schutz and M. Xiao, "Stress Corrosion Cracking Resistance of Beta-C^{TM} Titanium Springs in Methanol Solutions," paper presented at 11th Offshore Mechanics and Arctic Engineers Conference (Calgary, Canada), 7-11 June 1992
43. E.G. Haney and P. Fugassi, Inhibition of Methanol Cracking of Ti-6Al-4V Foil: Effect of Pretreatments with Aqueous Solutions of Electrolytes, *Corrosion*, Vol 27, Jan 1971, p 46-48
44. W.F. Czyrklis and M. Levy, Stress Corrosion Susceptibility of β Titanium Alloy 38-6-44, *Corrosion*, Vol 32 (No. 3), March 1976, p 99-102
45. A. Cerquetti and F. Mazza, Electrochemical Behavior and SCC of Titanium in Alcoholic Solutions, *Proceedings of an International Symposium on Stress-Corrosion Mechanisms in Titanium Alloys*, 27-29 Jan 1971, cosponsored by Georgia Institute of Technology and National Association of Corrosion Engineers
46. C.M. Chen, H.B. Kirkpatrick, and H.L. Gegel, Cracking of Titanium Alloys in Methanolic and Other Media, *Procedings of an International Sym-*

posium on Stress Corrosion Mechanisms in Titanium Alloys, 27-29 Jan 1971, cosponsored by Georgia Institute of Technology and National Association of Corrosion Engineers

47. A.J. Sedricks and J.A.S. Green, Stress Corrosion of Titanium in Organic Liquids, *J. Met.*, April 1971, p 48-54
48. G. Sandoz, Subcritical Crack Propagation in Ti-8Al-1Mo-1V Alloy in Organic Environments, Salt Water, and Inert Environments, *Proceedings of Conference—Fundamental Aspects of Stress-Corrosion Cracking,* National Association of Corrosion Engineers, 1969, p 684
49. C.C. Seastrom and R.A. Gorski, The Influence of Fluorocarbon Solvents on Titanium Alloys, *Accelerated Crack Propagation of Titanium by Methanol, Halogenated Hydrocarbons, and Other Solutions,* DMIC Memorandum 228, Defense Metals Information Center, Battelle Memorial Institute, 6 March 1967, p 20
50. S.V. Glorioso, "Lunar Module Pressure Vessel Operating Criteria," Specification SE-V-0024, NASA/MSC, Oct 1968
51. C.F. Tiffany and J.N. Masters, "Investigation of the Flaw Growth Characteristics of Ti-6Al-4V Titanium Used in Apollo Spacecraft Pressure Vessels," CR-65586, National Aeronautics and Space Administration, March 1967
52. R.V. Turley and C.H. Avery, Elevated-Temperature Static and Dynamic Sea-Salt Stress Cracking of Titanium Alloys, *Stress-Corrosion Cracking of Titanium*, STP 397, ASTM, 1966, p 1
53. S.P. Rideout, R.S. Ondrejcin, and M.R. Louthan, Hot-Salt Stress-Corrosion Cracking of Titanium Alloys, *The Science, Technology and Application of Titanium*, Pergamon Press, 1970
54. H.L. Logan, Studies of Hot-Salt Cracking of the Titanium-8% Al-1% Mo-1% V Alloy, *Proceedings of Conference—Fundamental Aspects of Stress-Corrosion Cracking*, National Association of Corrosion Engineers, 1969, p 662
55. H.L. Logan, M.J. McBee, C.J. Bechtoldt, B.T. Sanderson, and G.M. Ugiansky, Chemical and Physical Mechanisms of Salt Stress-Corrosion Cracking in the Titanium 8-1-1 Alloy, *Stress Corrosion Cracking of Titanium*, STP 397, ASTM, 1966, p 215
56. S.P. Rideout, R.S. Ondrejcin, M.R. Louthan, and D.E. Rawl, The Role of Moisture and Hydrogen in Hot-Salt Cracking of Titanium Alloys, *Proceedings of Conference—Fundamental Aspects of Stress-Corrosion Cracking*, National Association of Corrosion Engineers, 1969, p 650
57. G.J. Heimerl, D.N. Braski, D.M. Royster, and H.B. Dexter, Salt Stress Corrosion of Ti-8Al-1Mo-1V Alloy Sheet at Elevated Temperatures, *Stress-Corrosion Cracking of Titanium*, STP 397, ASTM, 1965, p 194
58. H.R. Gray, Hot-Salt Stress Corrosion of a Titanium Alloy: generation of Hydrogen and Its Embrittling Effect, *Corrosion*, Vol 25, (No. 8), 1969, p 337
59. W.K. Boyd, Stress-Corrosion Cracking of Titanium and Its Alloys, *Proceedings of Conference—Fundamental Aspects of Stress-Corrosion Cracking,* National Association of Corrosion Engineers, 1969, p 593
60. J.R. Myers and J.A. Hall, Hot Salt Stress Corrosion Cracking of Titanium Alloys: An Improved Model for the Mechanism, *Corrosion*, Vol 33 (No. 7), 1977, p 252-257
61. R.S. Ondrejcin *et al.*, The Hot-Salt Cracking of Titanium Alloys, *Proceedings of an International Symposium on Stress Corrosion Mechanisms in Titanium Alloys,* 27-29 Jan 1971, cosponsored by Georgia Institute of Technology and National Association of Corrosion Engineers
62. J.D. Jackson and W.K. Body, "Corrosion of Titanium," DMIC Memorandum 218, Defense Metals Information Center, Battelle Memorial Institute, 1 Sept 1966
63. A.Z. Conner, Stress Corrosion Cracking of Ti-6Al-4V in Liquid Dinitrogen Tetroxide, *Proceedings of an International Symposium on Stress Corrosion on Mechanisms in Titanium Alloys*, 27-29 Jan 1971, cosponsored by Georgia Institute of Technology and National Association of Corrosion Engineers
64. W.K. Boyd, Stress Corrosion of Titanium and Its Alloys, *Proceedings of an International Symposium on Stress Corrosion Mechanisms in Titanium Alloys*, 27-29 Jan 1971, cosponsored by Georgia Institute of Technology and National Association of Corrosion Engineers
65. D.W. Stough, F.W. Fink, and R.S. Peoples, "The Stress Corrosion and Pyrophoric Behavior of Titanium and Titanium Alloys," TML Report 84, Battelle Memorial Institute, 15 Sept 1957
66. J.D. Jackson, W.K. Boyd, and R.W. Staehle, "Stress Corrosion of Ti-6Al-4V in Liquid Nitrogen Tetroxide," DMIC Technical Note, Defense Metals Information Center, Battelle Memorial Institute, 11 April 1966
67. C.F. Tiffany, J.N. Masters, and R.C. Shah, "Fracture Mechanics Assessment of Apollo Launch Vehicle and Spacecraft Pressure Vessels," D2-114248-1, Boeing Co., Jan 1969
68. P.J. Moreland and W.K. Boyd, The Corrosion and Stress Corrosion Cracking Behavior of Ti and Ti-Al Base Alloys in Dinitrogen Tetroxide, *Proceedings of an International Symposium on Stress Corrosion Mechanisms in Titanium Alloys*, 27-29 Jan 1971, cosponsored by Georgia Institute of Technol-

ogy and National Association of Corrosion Engineers

69. L.L. Gilbert and C.W. Funk, Explosions of Titanium and Fuming Nitric Acid Mixtures, *Met. Prog.*, Nov 1956, p 93-96
70. J.B. Rittenhouse, The Corrosion, Pyrophoricity, and Stress Corrosion Cracking of Titanium Alloys in Fuming Nitric Acid, *Trans. ASM*, Vol 51, 1957
71. H.B. Bomberger, Titanium Corrosion and Inhibition in Fuming Nitric Acid, *Corrosion*, Vol 13 (No. 5), May 1957, p 17-21
72. W.H. Smyrl and M.J. Blackburn, Stress Corrosion Cracking in Molten Salts, *Proceedings of an International Symposium on Stress Corrosion Mechanisms in Titanium Alloys*, 27-29 Jan 1971, cosponsored by Georgia Institute of Technology and National Association of Corrosion Engineers
73. R.E. Stoltz and R.H. Stulen, Solid Metal Embrittlement of Ti-6Al-6V-2Sn by Cadmium, Silver, and Gold, *Corrosion*, Vol 35 (No. 4), April 1979, p 165
74. D.N. Fager and W.F. Spurr, Solid Cadmium Embrittlement: Titanium Alloys, *Corrosion*, Vol 26 (No. 10), Oct 1970, p 409
75. D.A. Meyn, Solid Cadmium Cracking of Titanium Alloys, *Corrosion*, Vol 29 (No. 5), May 1973, p 192
76. H.A. Johnson, "Stress Cracking of Titanium," Technical Memorandum WCRT TM56-07, Wright Air Development Center, Wright Patterson Air Force Base
77. *Nuclear Fuels and Materials Development*, 2nd ed., Report TID-11295, U.S. Army Engineering Corps, Sept 1962
78. J.B. Hollowell, J.G. Dunleavy, and W.K. Boyd, "Liquid-Metal Embrittlement," DMIC Technical Note, Defense Metals Information Center, Battelle Memorial Institute, 9 April 1965
79. J.Y.N. Wang, Titanium and Titanium Alloys in Mercury—Some Observations on Corrosion and Inhibition, *Nucl. Sci. Eng.*, Vol 18, 1964, p 18-30
80. R.E. Duttweiler, R.R. Wagner, and K.C. Antony, An Investigation of Stress-Corrosion Failures in Titanium Compressor Components, *Stress-Corrosion Cracking of Titanium*, STP 397, ASTM, 1966, p 152
81. G. Martin, Investigation of Long-Term Exposure Effects Under Stress of Two Titanium Structural Alloys, *Stress-Corrosion Cracking of Titanium*, STP 397, ASTM, 1965, p 95
82. D.N. Fager and W.F. Spurr, "Solid Cadmium Embrittlement: Part I, Titanium Alloys," D6-22691, Boeing Co., Seattle WA
83. D.N. Braski, "Preliminary Investigation of Effect of Environmental Factors on Salt Stress-Corrosion Cracking of Ti-8Al-1Mo-1V at Elevated Temperatures," Technical Memorandum TMX-1048, National Aeronautics and Space Administration, Dec 1964
84. R.E. Adams and E. Von Tiesenhausen, Study of Stress Cracking of Commercial Titanium Alloys, *Proceedings of Conference—Fundamental Aspects of Stress Corrosion Cracking*, National Association of Corrosion Engineers, 1969, p 691
85. R.S. Ondrejcin and M.R. Louthan, "Role of Hydrogen Chloride in Hot-Salt Stress-Corrosion Cracking of Titanium-Aluminum Alloys," Report DP (NASA)-1130, E.I. du Pont de Nemours, Dec 1967
86. D.N. William and D.J. Maykuth, "Reactions of Titanium With Gaseous Hydrogen at Ambient Temperatures," DMIC Technical Note, Defense Metals Information Center, Battelle Memorial Institute, 4 Feb 1966
87. B.G. Koehl, D.N. Williams, and E.S. Bartlett, *J. Less-Common Met.*, 1969
88. R.J. Walter and W.T. Chandler, "Effects of High-Pressure Hydrogen on Metals at Ambient Temperatures," Final Report NAS 8-19, National Aeronautics and Space Administration, Oct 1969
89. H.G. Nelson and D.P. Williams, "Environmental Hydrogen Embrittlement of Ti-6Al-4V," paper presented at AIME Spring Meeting (Las Vegas), American Institute of Mining, Metallurgical, and Petroleum Engineers, May 1970

CHAPTER 11

Stress-Corrosion Cracking of Zirconium Alloys

Te-Lin Yau, Teledyne Wah Chang Albany

Zirconium is a highly reactive metal, as evidenced by its redox potential of −1.53 V (NHE) at 25 °C (77 °F). It has a great affinity for oxygen. When zirconium is exposed to an oxygen-containing environment, an adherent, protective oxide film forms on its surface. This film is formed spontaneously in air or water at ambient temperature and below. Moreover, this film is self-healing and protects the base metal from chemical and mechanical attack at temperatures up to 300 °C (570 °F). As a result, zirconium exhibits excellent general corrosion resistance in many environments, including most mineral and organic acids, strong alkalies, saline solutions, and some molten salts. Zirconium also resists attack by certain oxidizing media, unless halides are present.

Zirconium is characterized by a high melting point, adequate strength, good thermal conductivity, low thermal expansion, great transparency to thermal neutrons, and excellent corrosion resistance. For more than 40 years, zirconium and its alloys have been successfully used in a variety of applications, including fuel cladding and structural components in nuclear reactors, chemical equipment, getters, and alloying.

Zirconium resists stress-corrosion cracking (SCC) in certain strong agents, such as H_2S, NaOH, and $MgCl_2$, which would induce SCC in other common metals and alloys. Zirconium service failures due to SCC are very limited in chemical applications. The high SCC resistance of zirconium can probably be attributed to its high repassivation rate. Any break in the surface film quickly heals, provided that there is sufficient oxygen present. Despite these excellent corrosion properties, zirconium and its alloys are susceptible to SCC in such environments as Fe^{3+}- or Cu^{2+}-containing solutions, CH_3OH + halides, concentrated HNO_3, halogen vapors, and liquid mercury or cesium.

This chapter deals with the environmental, metallurgical, and mechanical aspects of SCC of zirconium and its alloys. Possible mechanisms are discussed, and methods for prevention or control of SCC are presented. Commercially available grades of zirconium alloys for nuclear and chemical applications are listed in Table 1.

Environmental Aspects

Environments known to induce SCC in zirconium and its alloys are listed in Table 2. It should be noted that SCC does not normally occur in certain of these environments, but it will occur if additional specific conditions are met. For example, SCC of zirconium and its alloys in iodide vapor at temperatures up to 200 °C (390 °F) is observed only in precracked specimens. Moreover, SCC in neutral chloride solutions requires anodic polarization.

Aqueous Solutions

Oxidizing Chloride Solutions. It has long been known that zirconium is one of the few metals that resist attack by highly reducing corrosives such as HCl (Ref 21,22). Zirconium is susceptible to localized corrosion in oxidizing chloride solutions such as $FeCl_3$, $CuCl_2$, wet chlorine, and aqua regia (Ref 21, 23, 24).

It is not surprising that the earliest documented case of SCC of zirconium occurred in $FeCl_3$ solutions (Ref 1). Figure 1 illustrates the effect of $FeCl_3$ concentration on the time to failure of Zircaloy-4 at two stress levels, 207 and 276 MPa (30 and 40 ksi) (Ref 2). The SCC susceptibility of Zircaloy-4 increases rapidly in ≤5% $FeCl_3$ with increasing $FeCl_3$ concentration. Further increase in $FeCl_3$ concentration has little effect on SCC susceptibility.

The presence of small amounts of oxidizing ions such as Fe^{3+} in acidic chloride solutions such as HCl also significantly increases the SCC susceptibility of zirconium (Ref 4). It has been found that zirconium is susceptible to transgranular SCC in < 24% HCl and to intergranular SCC in ≥ 24% HCl. It is clear that altering the environment can change the cracking mode, even when metallurgical and mechanical factors remain constant.

Neutral Halides. Zirconium has a high resistance to SCC in many halide solutions including NaCl, $MgCl_2$, and seawater (Ref 25, 26), the latter two being particularly strong SCC-causing agents for common metals and stainless steels. Moreover, the presence of sulfides and/or iron salts does not increase the SCC susceptibility of

Table 1 Chemical Requirements for Zirconium Alloys

	Composition, %								
Alloy	Zr + Hf (min)	Hf (max)	Sn	Nb	Fe	Cr	Ni	Fe + Cr	Fe + Cr + Ni
Nuclear grades									
Zircaloy-2	...	0.010	1.20-1.70	...	0.07-0.20	0.05-0.15	0.03-0.08	...	0.18-0.38
Zircaloy-4	...	0.010	1.20-1.70	...	0.18-0.24	0.07-0.13	...	0.28-0.37	...
Zr-2.5Nb	...	0.010	...	2.40-2.80	...	...	...	...	...
Chemical grades									
Zr702	99.2	4.5	...	...	...	...	...	0.2 max	...
Zr704	97.5	4.5	1.0-2.0	...	...	...	...	0.2-0.4	...
Zr705	95.5	4.5	...	2.0-3.0	...	...	...	0.2 max	...

Table 2 Environments Known To Induce SCC in Zirconium and Its Alloys

Environment	Temperature, °C (°F)(a)	Test methods(b)	Cracking mode(c)	Ref
Aqueous solutions				
Oxidizing chloride solutions	RT-BP	U, T, SR, SSRT	IG, TG	1-4
Nitric acid	RT-204 (400)	U, C, SSRT	TG	5-7
Neutral halides (NaCl, KCl, NaBr, etc.)	RT-154 (310) (Needs polarization)	SR, DCB, T	TG	3, 8, 9
Organic liquids				
Methanol with I_2, Br_2, Cl^-	RT-65(150)	U, SR, DCB, T	IG, TG (for DCB)	3, 10-15
Higher alcohols with similar additives	RT	U	IG, TG	10
Chlorinated hydrocarbons (e.g. CCl_4, $CHCl_3$)	RT		IG	3
Freons	RT	DCB	TG	16
Halogen vapors				
I_2, Br_2, Cl_2	RT	DCB	TG	3, 11, 12, 17, 18
I_2	250-500 (480-930)	DCB, SR	TG, IG	17, 19, 20
Hot and fused salts				
Chloride salts and nitrate salts + chloride salts	300-450 (570-840)	DCB, SR	IG, TG	11, 12, 18
Metals				
Hg, Cd, Cs	RT-400 (750)	U, SR, DCB	TG	11, 12

(a) RT, room temperature; BP, boiling point. (b)U, U-bend,; T, tensile; SR, split ring; SSRT, slow-strain-rate tensile; C, C ring; DCB, double cantilever beam. (c)IG, intergranular; TG, transgranular

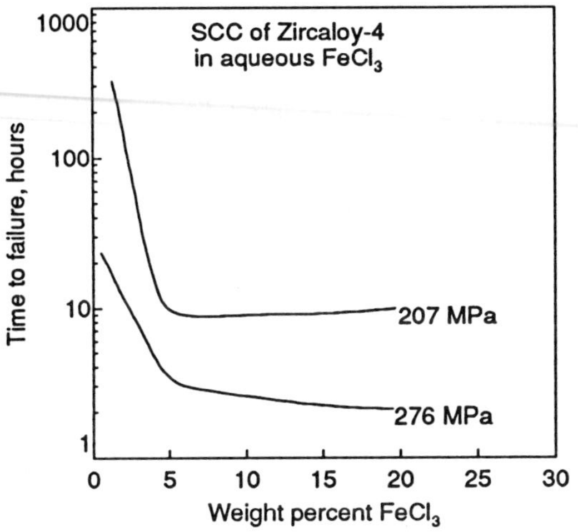

Fig. 1 Time to failure of Zircaloy-4 at two stress levels in $FeCl_3$ solutions. Source: Ref 2

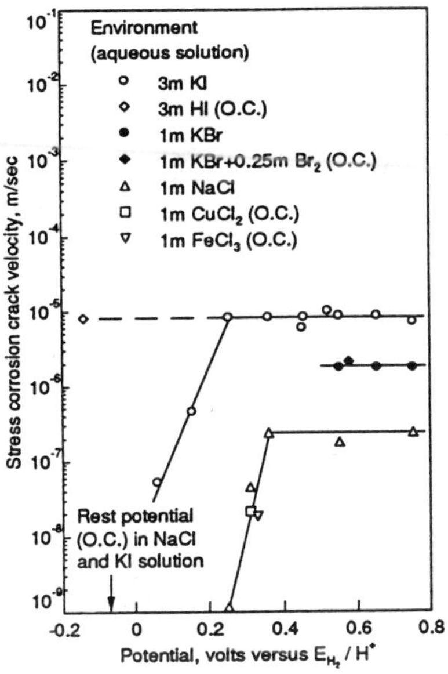

Fig. 2 Crack-growth rate of Zircaloy-4 in aqueous halide solutions as a function of applied potential. OC, open circuit. Source: Ref 31

zirconium (Ref 27-29). Unlike most metals, zirconium has a very low affinity for sulfide ions, and their presence in solutions ranging from acidic to caustic is not known to induce sulfide stress cracking in zirconium (Ref 27-30). The presence of heavy-metal ions in neutral halide solutions has little effect on SCC susceptibility (Ref 27-29).

However, zirconium may be susceptible to SCC in neutral halide solutions if an anodic potential is applied, either by an external current or by coupling with a noble metal. The effect of anodic potential on the SCC of Zircaloy-4 in halide solutions is shown in Fig. 2 (Ref 31), which indicates that the material is susceptible to SCC in NaCl solutions if the electrode potential exceeds 0.25 V (SHE). The rest potential of Zircaloy-4 in NaCl solutions is well below this level; therefore, SCC does not occur under normal conditions. In contrast, the rest potential of Zircaloy-4 in $FeCl_3$ or $CuCl_2$ solutions is noble enough to induce SCC.

Nitric Acid. Zirconium exhibits excellent resistance to general corrosion in nitric acid (Ref 7, 21). It has been successfully used in nitric acid production for cooler condensers and reboilers and in spent-fuel reprocessing for dissolvers.

Zirconium has a high affinity for oxygen, and nitric acid has a high passivating power. The use of conventional sustained stress or strain methods can, with some difficulty, detect zirconium susceptibility to SCC in nitric acid. For example, it takes hundreds of hours for U-bend specimens of Zr702 or Zircaloy-4 to fail in 90% HNO_3 (Ref 5, 6). The required time is much longer in $\leq 70\%$ HNO_3. This susceptibility is easily revealed by slow-strain-rate testing (Ref 5, 7, 32). However, the crack-propagation rate is very low (less than 2×10^{-9} m/s) in $\leq 70\%$ HNO_3. Therefore SCC of zirconium is of much greater concern in $>70\%$ HNO_3.

Titanium is also susceptible to SCC in nitric acid, particularly in the presence of NO_2 (Ref 33). However the presence of NO_2 in nitric acid has an insignificant (Ref 5) to beneficial (Ref 6) effect on the SCC of zirconium and its alloys. As discussed in Ref 34, NO_2 is not as oxidizing as HNO_3. Nitrogen dioxide moves potential to the active direction, which is beneficial to zirconium but harmful to titanium.

Organic Liquids

Zirconium is susceptible to SCC in certain commercially pure organic liquids such as metha-

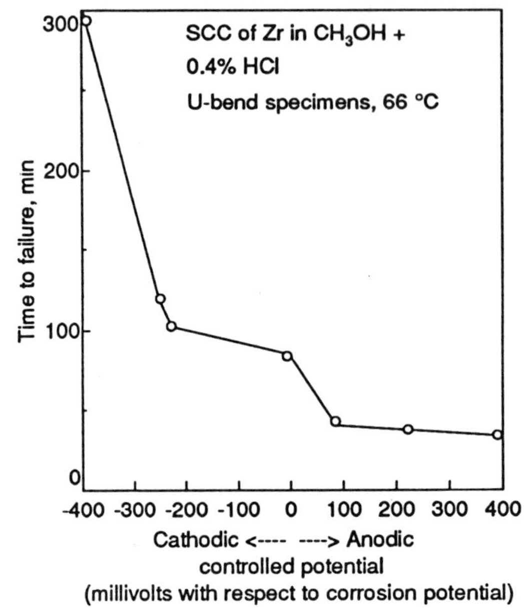

Fig. 3 Effect of HCl concentration in methanol on time to failure of zirconium. A, U-bend specimens, 66 °C (150 °F); B, split-ring specimens, 25 °C (77 °F); C, U-bend specimens, 25 °C (77 °F). Source: Ref 3, 10, 14.

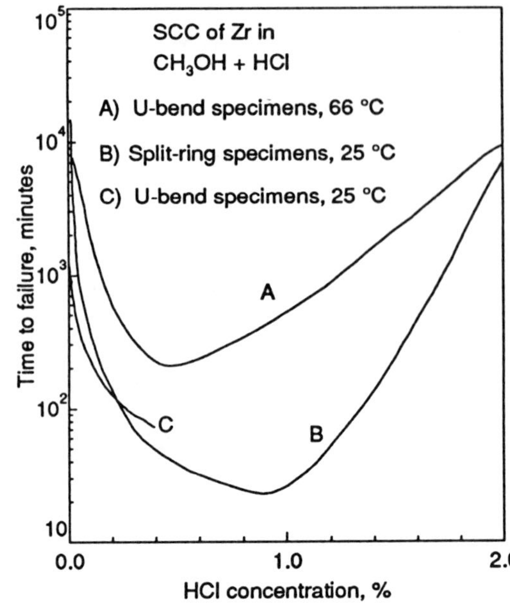

Fig. 4 Effect of applied potential on the time to failure of zirconium in methanol + 0.4% HCl. U-bend specimens, 66 °C (150 °F). Source: Ref 14

nol (CH_3OH), CCl_4, $CHCl_3$, and Freon. It has been speculated that traces of halide are responsible (Ref 12).

The most investigated case is SCC of zirconium in methanol-HCl solutions. The effects of HCl concentration and applied potentials are shown in Fig. 3 and 4 (Ref 3, 10, 14). Time to failure reaches a minimum between 0.4 and 1% HCl (Fig. 3), indicating the importance of chloride concentration and water content. Increasing HCl concentration simultaneously increases water content. Consequently, the detrimental effect of halide ions can be offset by sufficient water content. Alternatively, as observed in aqueous solutions, cathodic potentials lengthen the time to failure (Fig. 4).

As indicated in Table 2, in the presence of iodine, zirconium becomes susceptible to SCC in a large number of organic liquids. Time to failure is strongly dependent on iodine concentration (Fig. 5) (Ref 3, 15).

Halogen Vapors

It is important to study the SCC of zirconium alloys in halogen vapors, particularly iodine vapor, because fuel-cladding failures in nuclear reactors are most likely caused by iodine as a fission product (Ref 17, 19, 20).

The effect of temperature on SCC of Zircaloy-2 in iodine is illustrated in Fig. 6 (Ref 17). There is an apparent lower temperature limit for the occurrence of SCC. Below this limit, smooth specimens do not fail in halogen vapors. However, SCC of precracked specimens has been observed in halogen vapors at room temperature (Ref 3, 31). This observation agrees with findings in iodine-containing organic liquids: DCB specimens may crack but not smooth specimens. Organic solvents do not form complexes with iodine and contain essentially free iodine molecules. Nevertheless, it should be noted that zirconium alloys may pit in wet vapors of chlorine or bromine.

Hot and Fused Salts

A condition of "dry-out" is a possible occurrence in nuclear reactors. Dissolved salts in water could precipitate out under this condition. Certain SCC tests have been performed on zirconium alloys in possibly aggressive salts (Ref 11, 12). Re-

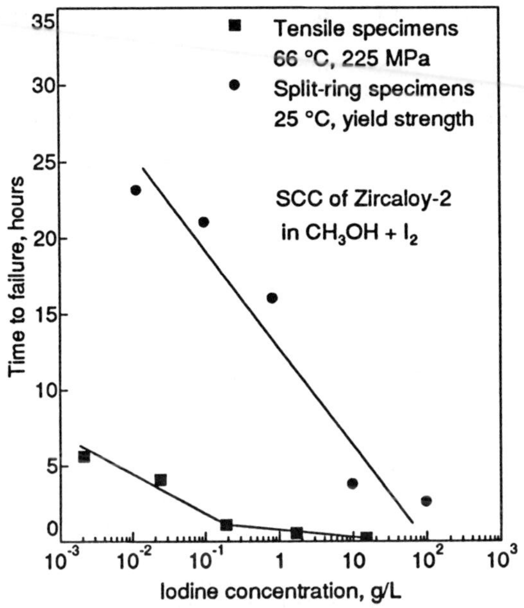

Fig. 5 Effect of iodine concentration on time to failure of Zircaloy-2 in methanol. Source: Ref 3, 15

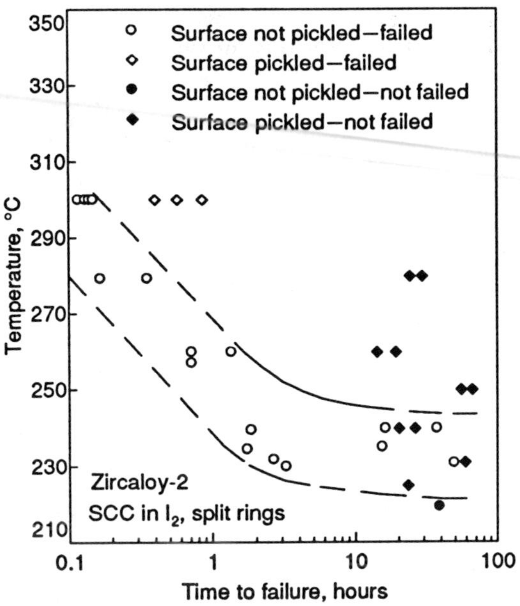

Fig. 6 Effect of temperature on time to failure of stress split rings (stress = 310 MPa, or 45 ksi) of Zircaloy-2 with an iodine content of 3×10^{-3} g /cm^2. Source: Ref 17

sults indicated that split-ring specimens of Zircaloys could be cracked in 350 °C (660 °F) air when covered with a eutectic mixture of LiCl and KCl (melting point of 349 °C, or 660 °F). It became more difficult to crack specimens when they were coated with NaCl (melting point of 801 °C, or 1474 °C). Also, cracking was possible with a coating of low-purity CsI (melting point of 626 °C, or 1159 °F) (Ref 11, 12).

Metals

Zirconium alloys showed transgranular cracking in both mercury and cesium at room temperature under dynamic deformation conditions (Ref 11, 12). An important factor in inducing cracking is the attainment of good wetting of the zirconium surface with the liquid metal. The presence of surface oxide films makes wetting difficult. It is more difficult to wet zirconium in cesium than in mercury because of the rapid formation of cesium oxide. Results of DCB tests indicated that crack-propagation rates were high (appproximately 10 cm/s) for Zircaloy-2 in mercury (Ref 3).

Metallurgical Aspects

Metallurgical factors are important in the SCC of metals and alloys. This section discusses the effects of alloy composition, texture, residual stress, and irradiation.

Alloy Composition. All zirconium alloys have small amounts of various alloying elements (Table 1). They are more or less susceptible to SCC in nearly all the environments listed in Table 2. Often, factors such as strength level and texture may be more important than alloy composition in observed differences among the SCC resistances of zirconium alloys. In fact, alloy composition plays a minor role in the SCC of zirconium alloys compared with other alloy systems.

Nevertheless, the effect of alloy composition on zirconium SCC is not straightforward. As reported in Ref 6, the presence of niobium in Zr705 has significantly lowers its SCC resistance in nitric acid. On the other hand, alloying with niobium may be beneficial to the SCC resistance of zirconium in CH_3OH + HCl environments, as shown in Table 3 (Ref 35). The beneficial effect of

Table 3 Results of Slow-Strain-Rate-Testing (Strain Rate = 2.5×10^{-6}/s) of Zirconium Alloys in CH_3OH + HCl Solutions at Room Temperature

	Maximum stress, MPa/Maximum strain, %		
Alloy(a)	CH_3OH	CH_3OH + 0.4% HCl	CH_3OH + 0.4% HCl + 3% H_2O
Zr702 L	432/26.4	297/3.5	446/28.8
Zr702 T	398/10.8	246/4.1	456/29.8
Zr704 L	471/10.7	368/6.4	413/8.0
Zr704 T	432/15.0	329/5.9	382/5.8
Zr705 L	559/12.0	536/8.3	595/25.0
Zr705 T	540/20.0	522/12.6	546/21.2

(a) L: longitudinal- cut specimen, T: transverse-cut specimen
Source: Ref 35

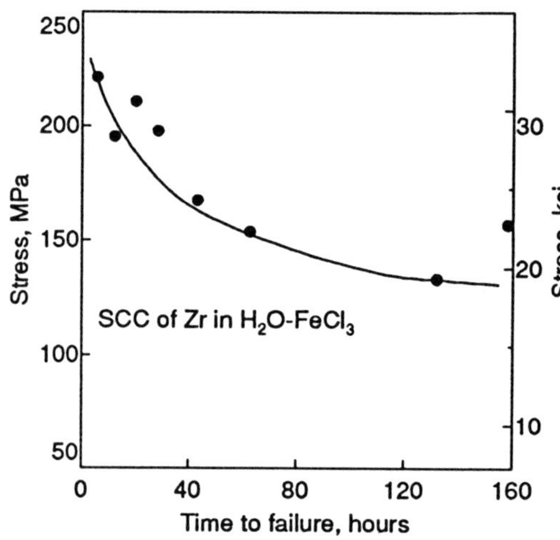

Fig. 7 Effect of stress on time to failure of zirconium in 25% $FeCl_3$ solution at ambient temperature. Source: Ref 1

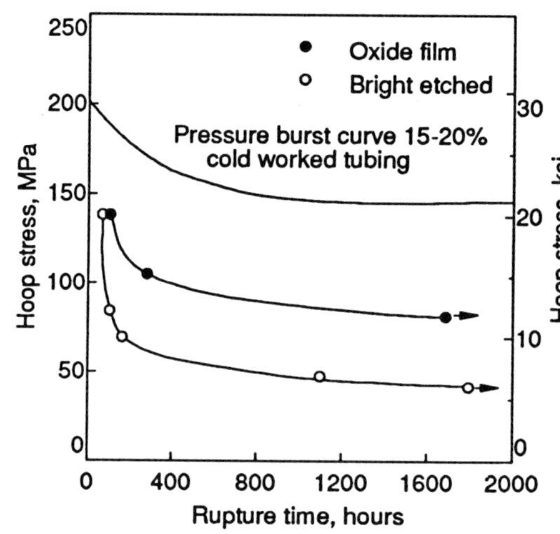

Fig. 8 Effect of hoop stress on the rupture time of cold-worked Zircaloy-4 tubing with the addition of 8×10^{-3} g/cm² of iodine at 427 °C (800 °F). Source: Ref 42

niobium is apparent in the most sensitive condition, *i.e.*, CH_3OH + 0.4% HCl.

Moreover, the addition of 2% Cr caused zirconium to become immune to SCC in CH_3OH + 0.4% HCl (Ref 36). However, immunity of systems such as this cannot be assumed in other SCC-inducing environments.

Texture. Zirconium is a hexagonal close-packed (hcp) metal at normal temperatures. Because of the shortage of slip systems and the strict orientation relationships for slip and twinning in hcp metals, a strong texture can result from deformation and subsequent heat treatment. When a metal is textured, the operable deformation systems cause the anisotropy of physical and mechanical properties and, consequently, SCC resistance.

Edstrom and Lagerberg (Ref 37) have reported that Young's modulus in a zirconium single crystal is highest in the direction of the c-axis (about 27% higher than that in the direction of the a-axis) and is lowest in the direction about 40° from the a-axis (about 12% lower than that in the direction of the a-axis). Therefore, it would be easier for a crack to proceed on or near the basal plane of the hexagonal crystal lattice. Highly textured polycrystalline materials should behave much like a single crystal. Thus, texture is a very important factor in transgranular SCC, as indicated in Ref 5, 6, and 38 to 41 and in Table 3.

Residual stress, which results from welding and fabrication operations, is important in the SCC of metals and alloys. Treatments that lower the tensile residual stress, such as heat treatment and shot peening, reduce the likelihood of SCC.

Neutron irradiation improves the strength of zirconium alloys but degrades fracture toughness. Irradiation also increases the SCC susceptibility of zirconium alloys in iodine (Ref 17). Annealing for several hours at 600 to 700 °C (1110 to 1290 °F) after irradiation eliminates its harmful effect (Ref 17).

Mechanical Aspects

As with other metals and alloys, SCC of zirconium depends strongly on stress, stress intensity, and strain rate. These factors have been investigated by conventional as well as specially developed methods.

Stress. The magnitude of tensile stress has a definite effect on time to failure, as illustrated in Fig. 7 (Ref 1). The higher the stress, the shorter the time to failure. When the applied stress is low enough, a threshold may exist. In 25% $FeCl_3$, the threshold stress appears to be near 130 MPa (19 ksi), which is about 50% of its yield strength (Fig. 7).

Weinberg (Ref 42) performed tests simulating conditions in fuel cladding. Tubes of Zircaloy-4 were pressurized with argon and iodine at 427 °C (800 °F). Test results are plotted in Fig. 8, which shows that iodine additions reduce the failure strength in a manner similar to the trend shown in Fig. 7. Time to failure is markedly reduced when the hoop stress in the tubes is increased. Oxide film formed in water at 360 °C (680 °F) for 3 days resulted in minor improvement of the SCC resistance of Zircaloy-4 at iodine levels of 8 mg/cm^2.

Determination of threshold stress may help to prevent SCC. Threshold stress can be affected by many factors, such as alloy composition, texture, and environment. For example, the threshold stress appears to be near 90% of the yield strength of Zr702 in 90% HNO_3 at room temperature (Ref 32), but it is higher than the yield strength in boiling 70% HNO_3 (Ref 6).

Strain Rate. Because of the time-dependent nature of SCC, strain rate is an important factor in determining the occurrence of SCC and the mode of cracking. Figure 9 demonstrates the effect of strain rate on SCC of zirconium in CH_3OH +

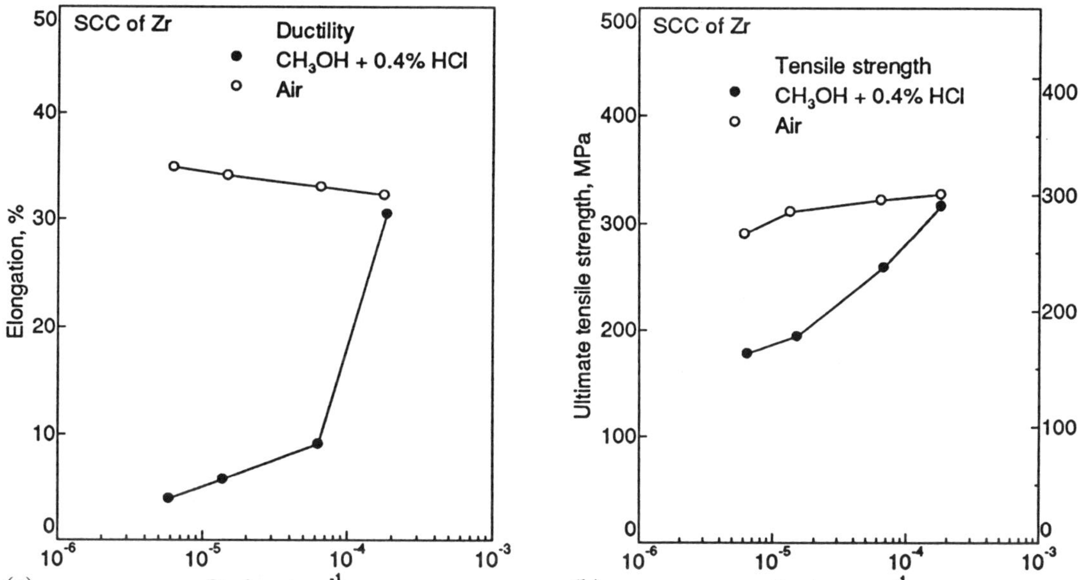

*Fig. 9 Effect of strain rate on the mechanical properties of zirconium in CH_3OH + 0.4% HCl at room temperature. **(a)** Ductility. **(b)** Tensile strength. Source: Ref 14*

0.4% HCl (Ref 14). Stress-corrosion cracking becomes pronounced when the strain rate is between 10^{-6} and 10^{-5}/s and disappears when the strain rate is above 10^{-4}/s.

Similarly, the effect of strain rate on SCC of zirconium in 90% HNO_3 is shown in Fig. 10 (Ref 32). It can be seen that zirconium is not susceptible to SCC at a high strain rate of 2.5×10^{-4}/s. SCC susceptibility increases with decreasing strain rate down to 7.5×10^{-7}/s. At a further lower strain rate of 2.5×10^{-7}/s, susceptibility decreases again.

These cases illustrate the competition among strain rate, dissolution rate, and repassivation rate. A high strain rate favors mechanical failure, and a low strain rate favors passivity. Stress-corrosion cracking cannot occur when the corrosion rate is too high; it will occur only when there is a critical balance among strain rate, dissolution rate, and repassivation rate.

Cracking modes of Zircaloy in iodine at 225 to 550 °C (435 to 1020 °F) depend on strain rate as follows (Ref 43):

- **Transgranular cracking** occurs at strain rates from 1×10^{-7} to 2×10^{-4}/s.
- **Intergranular cracking** occurs at strain rates from 2×10^{-5} to 5×10^{-4}/s.
- Ductile failure occurs at strain rates from 8×10^{-5} to 5×10^{-2}/s and beyond.

Stress intensity affects the rate of crack propagation, as illustrated in Fig. 11 for Zircaloy-4 in bromine water (Ref 31). The cracking mode is intergranular. For a fully annealed material, the crack-propagation rate is independent of specimen orientation. Figure 11 shows that the crack-propagation rate strongly depends on the applied stress intensity up to about 20 MPa$\sqrt{m}$ (18 ksi$\sqrt{in.}$). At higher values, the crack-propagation rate remains constant and independent of stress intensity. In the stress-intensity-dependent region, cracks grow singly and straight; in the stress-intensity-independent region, they branch.

SCC Mechanisms

Identification of mechanisms is the most difficult aspect of SCC study. The summary of the editors on consensus issues for the 1988 Kohler

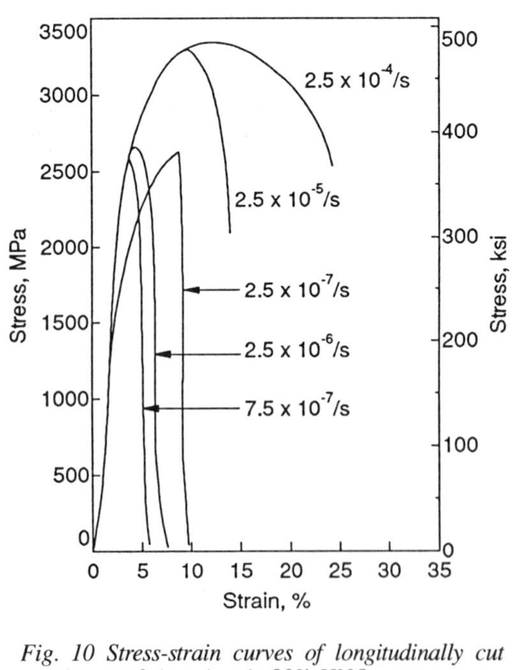

Fig. 10 Stress-strain curves of longitudinally cut specimens of zirconium in 90% HNO_3 at room temperature. Source: Ref 32

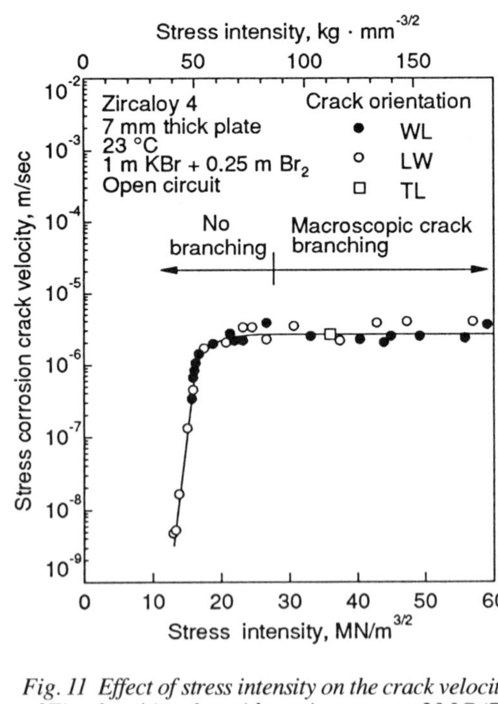

Fig. 11 Effect of stress intensity on the crack velocity of Zircaloy-4 in a bromide environment at 23 °C (73 °F). Source: Ref 31

Conference on Environment-Induced Cracking of Metals (Ref 44) stated that:

> Identification of consensus issues is undoubtedly controversial. Many of the written comments from participants suggested that no true consensus was reached on any particular aspect of environmental cracking! This camp believes that while we may have progressed somewhat in our appreciation of the various environmental cracking phenomena since the Firminy meeting, we have not reached any true understanding of the phenomena as a whole. This view proposes that continued, often unsubstantiated and qualitative, debate on the details of a variety of mechanisms hinders our progress toward material and component performance predictions. This situation, more than anything else, makes for frustration within the engineering design community.
>
> A second camp of participants concludes that substantial progress has been made since Firminy. Much of the speculation of the 1970s has been replaced by quantitative experimental evidence. Quantitative and testable models are emerging, with focus on specific embrittlement mechanisms that are relevant to classes of alloys and environments. Meaningful life predictions may be developed to include environmental cracking. The critical importance of crack chemistry and crack-tip process-zone deformation and fracture has spawned new models and experimental probes. Those with this view do not believe that all problems are solved, but rather that the field is generally on the right track. Useful information is being made available for the user, but improvements and new ideas are, of course, still required.

Controversy surrounds the topic of mechanisms for SCC of zirconium alloys. This can be attributed to two distinct characteristics of zirconium: it is a highly passive metal because of its strong affinity for oxygen, and it is a hydride-forming metal. Both characteristics result from the high reactivity of zirconium with many elements. Consequently, three major mechanisms have been proposed for SCC of zirconium alloys: (1) anodic dissolution (Ref 15), (2) hydrogen embrittlement (Ref 2, 45), and (3) stress sorption (Ref 11).

Controversy also exists as to whether different mechanisms are needed to account for intergranular and transgranular cracking. Cox (Ref 46) indicates that the mechanism for intergranular cracking is established relatively easily. In most cases, it appears to be a dissolution process, and weight losses are generally observed. One exception is the phenomenon observed in fused salts. Here the specimen undergoes a weight gain, although the fracture surface is apparently fully intergranular. Transgranular cracking appears to be much more complicated.

However, SCC of zirconium alloys can be intergranular, transgranular, or both. The cracking mode can be altered by changing acid concentration (Ref 4), heat treatment (Ref 4), stress intensity (Ref 16), and so on. Cracking mode is apparently the consequence of the combined effect of environmental, metallurgical, and mechanical factors, and it is not necessary to establish different mechanisms to account for it.

Anodic Dissolution. In most cases of SCC, hydrogen embrittlement is not implicated in the mechanism. Evidence includes:

- Zirconium experiences a passive to transpassive transition with increasing potential in most chemical environments (Ref 47). The commonly observed "active nose" for many metals and alloys is not observed for zirconium. Zirconium prefers a reducing condition that produces stable passive films. Cathodic polarization often improves the SCC resistance of zirconium (Ref 4, 14, 32), and anodic polarization often increases its susceptibility (Ref 3-5,8,9).
- Transgranular cracking proceeds along planes that are inconsistent with the habit planes of zirconium hydrides (Ref 5).
- Zirconium hydrides generally cannot be found (Ref 4-6). The potential range in which cracking has been observed is more noble than the hydrogen reversible potential (Ref 32). In certain cases, zirconium hydrides have been found (*e.g.*, in NaCl solutions) (Ref 48, 49). However they do not seem to be necessary for cracking to occur.
- The formation of zirconium hydrides resulting from hydrogen charging in HCl did not cause brittle failure (Ref 4).
- The addition of hydrogen poisoners, such as sulfide and gallium ions, does not increase the susceptibility of zirconium to brittle failure (Ref 25-30, 50).

According to the anodic dissolution mechanism, the occurrence of SCC requires some critical balance between active and passive behaviors, since activity promotes general corrosion, while passivity significantly retards any corrosion process. Hence, SCC is likely to occur in the active-passive transitional areas of an anodic polarization curve (Ref 51). It is not necessary that SCC be observed in all transitional zones for all metals in all media, since the critical balance between active and passive behaviors cannot be achieved in all cases. In other words, SCC of metals should be observed only within one or more potential ranges for a specific metal/environment system. When the potential of metals is controlled outside these ranges, SCC will not occur.

Hydrogen Embrittlement. Hydrogen has very low solubility in zirconium at low temperatures. For example, at room temperature, the terminal solubility of hydride in zirconium is less than 0.05 ppm (Ref 52). The presence of hydrides can cause brittle failure. There is circumstantial evidence to support the hydrogen-embrittlement mechanism:

- Alloys susceptible to SCC are also susceptible to a static fatigue type of hydrogen embrittlement.
- Hydrogen absorption can occur in certain aqueous solutions, such as H_2SO_4, HCl, and HF.
- Addition of arsenite has increased hydrogen uptake and SCC susceptibility in some instances (Ref 53).

However, the hydrogen-embrittlement mechanism cannot explain the effects of potential, the inconsistent presence of hydrides, and the fact that hydrides result in ductile failures. In the author's opinion, the observation of hydrides in certain cases can be considered the consequence of cracking rather than the reason for cracking. During the cracking process, hydrogen may be generated by corrosion reactions. Because zirconium is not protected by passive film upon cracking, the reaction between zirconium and hydrogen to form hydrides becomes more likely.

Zirconium hydrides are certainly responsible for one type of failure: delayed hydride cracking (DHC). Delayed hydride cracking in high-strength zirconium alloys, such as Zr705, is almost entirely transgranular and is possible only in the presence of zirconium hydrides (Ref 54-56). This mechanism will not be discussed further, since the requirements for DHC are small amounts of zirconium hydrides and high stress gradients. Zirconium hydrides are not necessarily the result of hydrogen absorption from the environment.

The stress sorption mechanism has been proposed to cover liquid-metal embrittlement and those SCC environments where neither an electrochemical nor a hydrogen-embrittlement process appears to be possible (Ref 11). This mechanism may be broadly applicable because of its simplicity. For example, anodic polarization is necessary for SCC of zirconium in NaCl solutions. The anodic condition may produce free chlorine at the crack tip in a form suitable for such a stress-sorption mechanism.

SCC Prevention and Control

Many conventional measures for the prevention and control of SCC are applicable to zirconium and its alloys. There are also measures that have been developed specifically for these materials.

Surface Modifications. Surface condition plays an important role in the initiation of cracks. Good surface finishing is beneficial in preventing SCC. As indicated in Fig. 12, certain polishing treatments increase the rest potential of zirconium in 10% HCl + 500 ppm Fe^{3+} (Ref 57)—an unfavorable development, since zirconium favors conditions that are more reducing. Pickling appears to be the best surface finishing treatment for the prevention of localized corrosion, including SCC.

Shot peening improves the time to failure considerably in hot and fused salts (Ref 3). It can control propagation after cracks have initiated. However, peening that uses common shots such as glass and cast iron can easily damage and contaminate the surface of zirconium.

Recently, it has been found that the undesirable effects of surface damage and contamination can be avoided by peening with ZrO_2 shot (Ref 35). For example, U-bend specimens of Zr702 failed in 80% or 90% HNO_3 at room temperature within 14 h (Ref 6). ZrO_2-shot-peened U-bend specimens did not fail in the same conditions after a 126-day exposure.

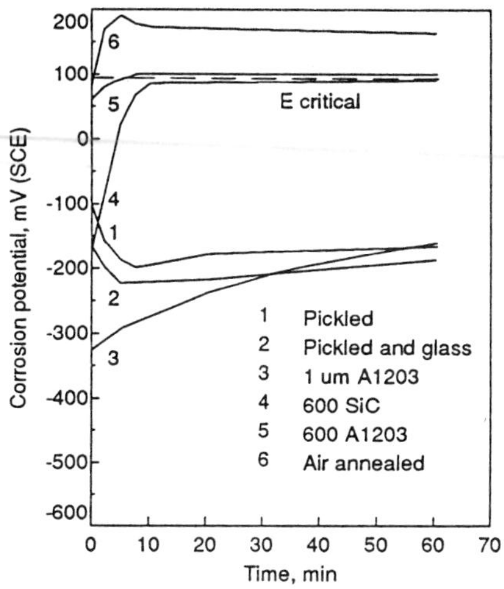

Fig. 12 Rest potential versus time for zirconium with different surface conditions in 10% HCl + 500 ppm Fe^{3+} at 30 °C (86 °F). Source: Ref 57

In nuclear applications, lining Zircaloy tubing with pure zirconium (Ref 58, 59) or coating it with graphite (Ref 59) can prevent iodine SCC.

Stress. Tensile stresses are necessary in a given SCC system and can take the form of residual, applied, thermal, welding, or corrosion product stress. High stresses promote corrosion, initiating as well as propagating cracks. There is a threshold stress below which cracking does not occur in an extended period of time. Lowering stresses below the threshold value can prevent SCC; methods include annealing to reduce residual stresses, thickening the section to lower applied stresses, and reducing the load.

Neutron irradiation increases the strength of zirconium and its susceptibility to iodine SCC (Ref 17). Annealing for several hours at 600 to 700 °C (1110 to 1290 °F) after irradiation removes its detrimental effect (Ref 17).

Normally, zirconium alloys can be stress relieved at 565 °C (1050 °F) for 1 h per 25 mm (1 in.) of thickness. Heat treating at higher temperatures (760, 880, and 1000 °C, or 1400, 1615, and 1830 °F) does not improve SCC resistance to 90% HNO_3, as revealed by slow-strain-rate testing (Ref 32). It should be noted that high-temperature treatments may have a pronounced effect on crack initiation, because a more homogeneous structure can be produced.

Texture plays an important role in the transgranular SCC of zirconium, and information regarding anisotropy and texture can be used to control SCC. The ideal texture is one in which the basal planes are perpendicular to the potential plane for crack propagation. However, texture is a parameter that is sometimes very difficult to control (*e.g.*, weld metals).

Galvanic Coupling. It was stated earlier that zirconium becomes susceptible to SCC in neutral halide solutions if coupled with a noble metal. On the other hand, SCC can be prevented by coupling with a less noble metal. For example, coupling with type 304 stainless steel prevented SCC of a zirconium alloy in 8 *N* HNO_3 + Ce^{4+} (Ref 60). This coupling resulted in a drop of more than 300 mV in corrosion potential.

Electrochemical Protection. Zirconium has a well-defined passive region in most corrosives. With increasing acid concentration or the presence of an oxidant, the corrosion potential of zirconium moves toward the transpassive region, thus developing a condition favorable to localized corrosion, including SCC. Use of a potentiostat to impress a small cathodic potential makes zirconium become more passive again. An arbitrary value 50 to 100 mV below the corrosion potential has been effective in controlling SCC of zirconium in oxidizing HCl solutions (Ref 4). A value of up to 500 mV below the corrosion potential can be used for zirconium in HNO_3 solutions (Ref 32). Cathodic polarization is beneficial for zirconium in CH_3OH + HCl environments (Ref 14).

Inhibitors. Reducing agents capable of lowering the corrosion potential of zirconium, such as hydrogen and stannous ions, should improve its SCC resistance. In this regard, the unusual beneficial effect of hydrogen fluoride on the SCC of zirconium in 90% HNO_3 has been studied (see Table 4) (Ref 32). Zirconium corrodes in HF solutions at high rates; these rates can be lowered by adding zirconium sponge, which forms stable complex ions.

The most famous inhibitor for zirconium in organic liquids is water (Ref 10). Zirconium, like all passive metals, needs the presence of an oxygen-containing compound such as water for repassivation. Consequently, zirconium can

Table 4 Results of Slow-Strain-Rate Testing (Strain Rate = 2.5×10^{-6}/s) in 90% HNO_3 at Room Temperature

Additional ingredient	Maximum stress, MPa/Maximum strain, % E_{BS}, mV(SCE)/E_{BF}, mV(SCE) (a) Longitudinal	Transverse	Weld
None	390/6.7	441/19.7	357/6.4
	980/640	780/700	830/670
220 ppm HF	462/26.9	455/25.8	378/14.6
	350/-80	705/280	300/-160
880 ppm Zr sponge + 200 ppm HF	475/27.9	461/25.0	415/24.6
	370/160	500/200	450/85
800 ppm Zr sponge	398/9.0	459/14.0	356/9.9
	740/600	760/640	685/580

(a) BS, before straining; BF, before fracturing
Source: Ref 32

experience corrosion problems, including SCC, in methanol and certain other organic liquids. The easiest solution is to add sufficient amounts of water.

References

1. J.T. Dunham and H. Kato, "Stress Corrosion Cracking Susceptibility of Zirconium in Ferric Chloride Solution," U.S. Bureau of Mines Report, BM-R1-5784, 1961
2. K.C. Thomas and R.J. Allio, *J. Nucl. Appl.*, Vol 1, 1965, p 252
3. B. Cox, Chemistry and Materials Division Progress Reports AECL-4065, 4146, 4204, 4259, 4351, and 4509, Atomic Energy of Canada Ltd.
4. T.L. Yau and M. Maguire, *Corrosion*, Vol 40, 1984, p 289; Vol 41, 1985, p 397
5. J.A. Beavers, J.C. Griess, and W.K. Boyd, *Corrosion*, Vol 36, 1981, p 292
6. T.L. Yau, *Corrosion*, Vol 39, 1983, p 167
7. T.L. Yau, *STP-917*, American Society for Testing and Materials, 1986, p 57
8. G.A. Cragnolino, "Stress Corrosion of Zirconium and Its Alloys in the Presence of Chlorides," doctoral thesis, University of Buenos Aires, 1975
9. B. Cox, *Corrosion*, Vol 29, 1973, p 157
10. K. Mori, A. Takamura, and T. Shimose, *Corrosion*, Vol 22, 1966, p 29
11. B. Cox, *Corrosion*, Vol 28, 1972, p 207
12. B. Cox, Reports AECL-3551, 3612, and 3799, Atomic Energy of Canada Ltd.
13. P.K. De, K. Elayaperumal, and J. Balachandra, *SAEST Trans.*, Vol 5, 1970, p 15
14. K. Elayaperumal, P.K. De, and J. Balachandra, *Corros. Sci.*, Vol 11, 1971, p 579
15. K. Elayaperumal, P.K. De, and J. Balachandra, *J. Nucl. Mater*, Vol 45, 1972/73, p 323
16. B. Cox, *STP 551*, American Society for Testing and Materials, 1974, p 419
17. J.C. Wood, *J. Nucl. Mater.*, Vol 45, 1972/73, p 105
18. B. Cox, *Oxid. Met.*, Vol 3, 1971, p 399
19. H.S. Rosenbaum, J.H. Davies, and J.Q. Pon, *Electrochem. Technol.*, Vol 4, 1966, p 153
20. G. Ostberg, AB Atomenergi Report, S 406, 1969
21. L.B. Golden, I.R. Lane, and W.L. Acherman, *Ind. Eng. Chem.*, Vol 44, 1952, p 1930; Vol 45, 1953, p 782
22. W.E. Kuhn, *Corrosion*, Vol 15, 1959, p 103t; Vol 16, 1960, p 136t
23. W.E. Lusby, Jr., *Corrosion*, Vol 13, 1957, p 654t
24. P.J. Gegner and W.L. Wilson, *Corrosion*, Vol 15, 1959, p 341t
25. T.L. Yau, "Corrosion Properties of Zirconium in Chloride Salt Solutions," Paper No. 26, presented at Corrosion/83, National Association of Corrosion Engineers, 1983
26. T.L. Yau, "Zirconium for Seawater and Chloride Solutions," presented at 4th Asian-Pacific Corrosion Control Conf., Tokyo, 26-31 May 1985
27. T.L. Yau, *Corrosion*, Vol 38, 1982, p 615
28. T.L. Yau, "Zirconium Versus Corrosive Species in Geothermal Fluids," Paper No. 140, presented at Corrosion/84, National Association of Corrosion Engineers, 1984
29. T.L. Yau, "Corrosion Behavior of Zirconium in H_2S - Containing Environments," presented at 5th Asian-Pacific Corrosion Control Conf., Melbourne, 23-27 Nov 1987
30. T.L. Yau, *TAPPI J.*, Vol 74 (No. 3), 1991, p 149
31. M.O. Speidel, "Subcritical Crack Growth in Zirconium Alloys," presented at ASTM Committee E-24-Sub IV, Los Angeles, 1971
32. T.L. Yau, *Corrosion*, Vol 44, 1988, p 745
33. L.L. Gilbert and C.W. Funk, *Met. Prog.*, Vol 70 (No. 11), 1956, p 93
34. T.L. Yau, Zirconium and Titanium: Corrosion Resistance of Contrasts, in *Proc. 1990 Int. Conf. Titanium Products and Applications*, Titanium Development Association, 1990, p 325

35. Tests conducted at Teledyne Wah Chang Albany
36. P.K. De, K. Elayaperumal, and J. Balachandra, *Indian J. Technol.*, Vol 11, 1973, p 605
37. J.O. Edstrom and G. Lagerberg, *Nucl. Eng.*, Vol 13 (No. 143), 1968, p 348
38. B. Cox, *Corrosion*, Vol 30, 1974, p 191
39. B.L. Adams, D.L. Baty, and K.L. Murthy, *Scr. Metall.*, Vol 12, 1978, p 1151
40. E. Smith, *J. Nucl. Mater.*, Vol 89, 1980, p 87
41. D.B. Knorr and R.M. Pelloux, *Metall. Trans. A*, Vol 13, 1982, p 73
42. J.G. Weinberg, "Iodine SCC of Zirca loy-4 Tubing," Report WAPD-TM 1048, Feb 1974
43. J.C. Wood, B.A. Surette, I.M. London, and J. Baird, *J. Nucl. Mater.*, Vol 57 (No. 2), 1975, p 155
44. *Environment-Induced Cracking of Metals*, R.P. Gangloff and M.B. Ives, Ed., NACE-10, National Association of Corrosion Engineers, 1990
45. J.C. Scully and D.T. Powell, *Corros. Sci.*, Vol 10, 1970, p 719
46. B. Cox, *Langmuir*, Vol 3 (No. 6), 1987, p 867
47. T.L. Yau and R.T. Webster, Corrosion of Zirconium and Hafnium, in *Metals Handbook*, 9th ed., Vol 13, *Corrosion*, ASM International, 1987, p 707
48. B. Cox, Report AECL-4589, Atomic Energy of Canada Ltd., 1973
49. R.A. Ploc, Report AECL-4509, Atomic Energy of Canada Ltd, 1973
50. L.L. Migai, *et al.*, *Nauch. Tr. N-i.i Proekt. In-t Redkomet, Prom-sti*, Vol 106, 1981, p 123
51. R.W. Staehle, *Stress Corrosion Cracking and Hydrogen Embrittlement of Iron-Base Alloys*, NACE-5, National Association of Corrosion Engineers, 1977, p 180
52. J.J. Kearns, *J. Nucl. Mater.*, Vol 22, 1967, p 292
53. J.A.S. Green and A.J. Sedriks, *Corrosion*, Vol 28, 1972, p 226
54. C.E. Ells, *Can. Inst. Metall. Annu. Vol.*, Vol 17, 1978, p 32
55. C.E. Coleman and J.F.R. Ambler, *STP 633*, American Society for Testing and Materials, 1976, p 589
56. B. Cox, *STP 681*, American Society for Testing and Materials, 1978, p 306
57. T.L. Yau and M. Maguire, Control of Localized Corrosion of Zirconium in Oxidizing Chloride Media, in *Advances in Localized Corrosion*, NACE-9, National Association of Corrosion Engineers, 1990, p 311
58. H.S. Rosenbaum *et al.*, "Large Scale Demonstration of Barrier Fuel, "in *Proc. ANS Meeting on LWR Fuel Performance*, Orlando, Vol 2, April 1985, p 7-63
59. K. Inone *et al.*, An Overview of the Joint Development Work on PCI Remedy Fuel, in *Proc. ANS Meeting on LWR Fuel Performance*, Orlando, Vol 2, April 1985, p 6-1
60. H. Kajimura and K. Yamanaka, "Prevention of Stress Corrosion Cracking of Zirconium or Its Alloys, "Jpn. Kokai Tokkyo Koho Jp 01 92, 348, 11 Apr 1989

CHAPTER 12

Environmentally Assisted Cracking of Uranium Alloys

Kenneth H. Eckelmeyer, Sandia National Laboratories

Uranium and uranium alloys are used in a variety of nonnuclear applications requiring metals of very high density. One of the problems encountered in the use of uranium is its very low resistance to oxidation and corrosion. Alloying allows substantial increases in corrosion resistance as well as improvements in mechanical properties. Uranium alloys, however, are frequently susceptible to environmentally assisted cracking, particularly when processed to high strength levels. Cracking has occurred in atmospheres as innocuous as moist or dry air, so in engineering applications its avoidance must be given serious attention.

Previous reviews of environmentally assisted cracking of uranium alloys have concentrated on presenting detailed data for stress-corrosion cracking (SCC) (Ref 1-4) or hydrogen embrittlement (Ref 2, 5-7) in specific alloys and alloy systems. In addition to incorporating more recent data, this chapter will describe and discuss the relative advantages and disadvantages of various testing and characterization methods, integrate the effects of external environments and internal hydrogen, describe trends in cracking phenomenology and relate them to the physical metallurgy of uranium alloys, and provide a guide to the selection of compositions and heat treatments for applications requiring materials of various strengths.

The corrosion and environmentally assisted cracking behaviors of uranium alloys are closely related to alloy content, microstructure, and strength level. This introduction will briefly summarize these topics. Additional details on the physical metallurgy and mechanical properties of uranium alloys can be found in several reviews (Ref 8-11).

Uranium is most commonly alloyed with transition-metal elements from groups 4A through 6A, particularly titanium, niobium, molybdenum, and zirconium. These alloying elements must be dissolved in solid solution for maximum benefit to corrosion resistance and/or strength. All of these elements are extremely soluble in the high-temperature γ phase ($T > 771$ °C, or 1420 °F), but exhibit very limited solubilities in the intermediate-temperature β phase (771 °C, 1420 °F $< T <$ 665 °C) or the low-temperature α phase ($T <$665 °C, or 1230 °F). As a result, annealed alloys typically exhibit two-phase microstructures consisting of essentially unalloyed α-uranium and alloy-enriched second phases (intermetallic compounds in all but the U-Nb system). These annealed alloys are somewhat stronger than unalloyed uranium (due to the presence of the second phases), but exhibit little or no improvement in corrosion resistance (because their primary uranium phases contain little or no alloy content in solid solution).

More commonly, uranium alloys are solution heat treated at 800 to 850 °C (1470 to 1560 °F) and cooled to room temperature at rates sufficient to suppress equilibrium transformations to two-phase microstructures, thereby producing metastable variants of either the α phase (in dilute alloys) or the γ phase (in concentrated alloys) in which the alloying elements are retained in super-

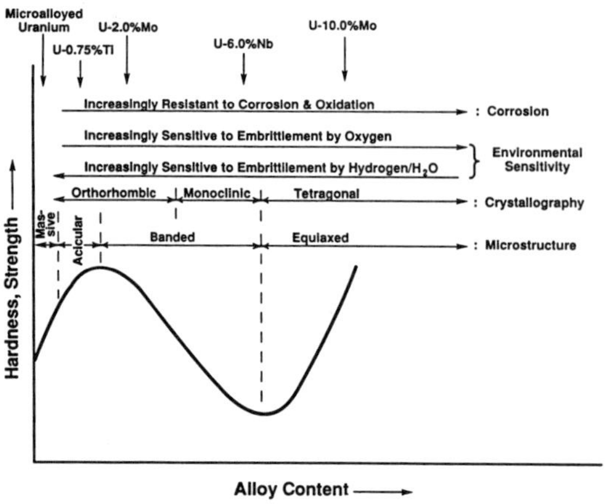

Fig. 1 Effect of composition on physical metallurgy, general corrosion resistance, and environmentally assisted cracking trends in quenched uranium alloys. Source: Ref 11

saturated solution. These metastable solid solutions exhibit substantial improvements in both corrosion resistance and mechanical properties. The severe quenches frequently required to obtain these phases (particularly in dilute alloys), however, introduce substantial residual stresses, which can provide the driving force for subsequent environmentally assisted cracking.

The microstructures and properties of as-quenched uranium alloys are summarized in Fig. 1. General corrosion and oxidation resistance increase with increasing alloy content (provided the quench rate is sufficient to retain all of the alloying elements in solid solution). The effects of alloy content on strength and microstructure are more complex. Initial alloying additions result in a regime of substantial strengthening. In the most dilute portion of this regime, the microstructures consist of irregular grains similar to those in unalloyed uranium, but slightly more concentrated alloys exhibit acicular martensitic microstructures. Further increases in alloy content result in a second regime, in which strength decreases with increasing alloy additions. The microstructures in this regime are typical of highly twinned thermoelastic or banded martensites. The decrease in strength with increasing alloy content appears to be associated with increasing mobility of the twin boundaries. Still further increases in alloy content result in a third regime of alloy strengthening, in which the microstructures are variants of the elevated-temperature γ phase. Quenched alloys in all three regimes are susceptible to environmentally assisted cracking, and the degree of susceptibility increases with increasing strength level. Therefore, unalloyed uranium and moderately concentrated alloys, such as U-6Nb, are the most resistant to premature failure, whereas fairly dilute alloys, such as U-0.75Ti and U-2Mo, and highly concentrated alloys, such as U-10Mo and U-7.5Nb-2.5Zr, are more susceptible. In addition, dilute and concentrated alloys are sensitive to somewhat different environments. Unalloyed uranium and dilute alloys are degraded by internal hydrogen or atmospheric moisture (which produces hydrogen by reaction with the surface of the uranium). As alloy content increases, sensitivity to hydrogen decreases, but sensitivity to atmospheric oxygen increases. Alloys of nearly all compositions are susceptible to aqueous environments that contain chloride ions.

Because the metastable structures in all three composition regimes are supersaturated with alloying elements, they are amenable to subsequent age hardening. Age hardening occurs at temperatures below approximately 450 °C (840 °F) due to very small microstructural changes such as the formation of coherent precipitates. These increases in strength make the materials inherently more susceptible to environmentally assisted cracking; however, the quench-induced residual stresses that frequently drive the cracking processes are also progressively relieved during aging. Overaging occurs above approximately 450 °C (840 °F) by cellular decomposition of the metastable microconstituents to the equilibrium phases. Overaged alloys are much less susceptible because of their lower strengths, their two-phase microstructures, and the complete relieving of quench-induced residual stresses that occurs during the overaging process.

Manifestations and Methods of Characterization

Environmentally assisted cracking in uranium and uranium alloys manifests itself as reductions in ductility during short-term tensile testing and as time-dependent formation and/or growth of cracks under longer-term conditions of sustained loading. These manifestations provide the basis for testing and characterizing stress-corrosion and hydrogen-embrittlement phenomena in these materials.

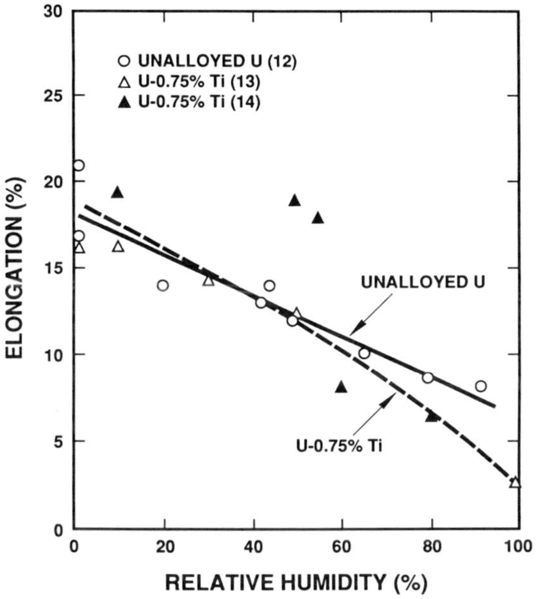

Fig. 2 Effect of humidity on tensile ductilities of unalloyed uranium (Ref 12) and age-hardened U-0.75%Ti (Ref 13, 14)

Reduction of tensile ductility is a useful indicator of environmentally assisted fracture because it is quick, simple, and can be used for all alloys, even those whose low yield strengths preclude crack formation and growth in the linear elastic regime. For example, the decrease in the ductility of unalloyed uranium with increasing relative humidity, shown in Fig. 2, documents the deleterious effect of atmospheric moisture on the fracture behavior of this material (Ref 12). A similar, but more pronounced, effect is seen in higher-strength age-hardened U-0.75Ti (Ref 13, 14). Changes in reduction in area generally provide a more sensitive indicator than do changes in elongation, but frequently exhibit substantial variations due to differences in specimen geometry. Despite the fact that this technique is a fairly sensitive detector of susceptibility, its usefulness is limited by the fact that results are difficult to quantify; *i.e.*, while decrease in ductility is a useful screening tool for detecting potential problems with particular material/environment combinations it does not provide threshold data that can be incorporated into engineering designs to ensure that failures will not occur.

Measurement of the times to failure of smooth samples loaded either in tension or in four-point bending is another commonly used method for

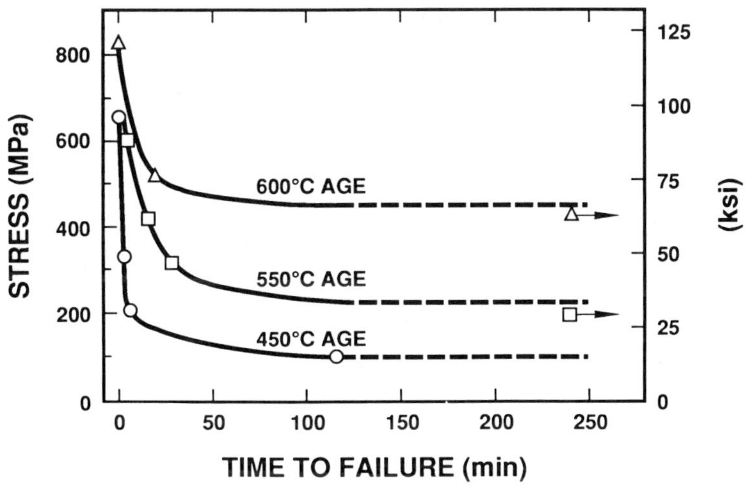

Fig. 3 Effects of overaging and initial outer fiber stress on times to failure of U-7.5Nb-2.5Zr four-point-bend specimens. Source: Ref 17

characterizing environmentally assisted cracking in uranium alloys (Ref 15-17). The data are typically presented as plots of failure time (sometimes on a log scale) versus initial stress, as shown in Fig. 3. A plateau is usually observed in the data, and the stress level below which failures are not observed is sometimes designated as the threshold stress for cracking (for the pertinent combination of alloy, heat treatment, and environment). One characteristic of this approach is that it measures the combined times for crack initiation and propagation. In cases where these effects have been separated, the time required for crack initiation typically exceeds that required for propagation; thus, the results are frequently dominated by initiation phenomena (Ref 17). Although this test method is useful for comparing materials and/or heat treatments, the use of threshold stresses in engineering design is nonconservative. It discounts the possible presence of flaws or other stress risers in engineering components that can preclude the necessity of crack initiation and can also result in the growth of environmentally assisted cracks at lower macroscopic stress levels.

An alternative static loading technique employs precracked specimens and measures times to failure as functions of initial stress intensity (Ref 18-21). The plots are similar to those discussed above, except that the more fundamental threshold stress intensity is determined, rather than a potentially nonconservative threshold stress. In addition, because the specimens are precracked, the times to failure generally approximate the more conservative times required for crack propagation, rather than the typically much longer times required for crack initiation in smooth samples. In some instances, alloys that had been designated as "immune" to SCC based on smooth bar tests were later shown to be susceptible when tested using this fracture-mechanics-based approach. Figure 4 shows an example of the use of this method to characterize the effect of chloride-ion concentration on SCC susceptibility in U-7.5Nb-2.5Zr (Ref 18). The stress-intensity thresholds provided by these tests are more useful in engineering design and can be used to calculate the flaw sizes that must be inspected to ensure that environmentally assisted crack growth will not occur (for given combinations of stress and environment). Typical threshold stress intensities for various alloys, heat treatments, and environments are given in Table 1. Although threshold stress intensities provide the most conservative design criteria, they are frequently rather insensitive measures of the effects of alloy composition, heat treatment, or environment on susceptibility to environmentally as-

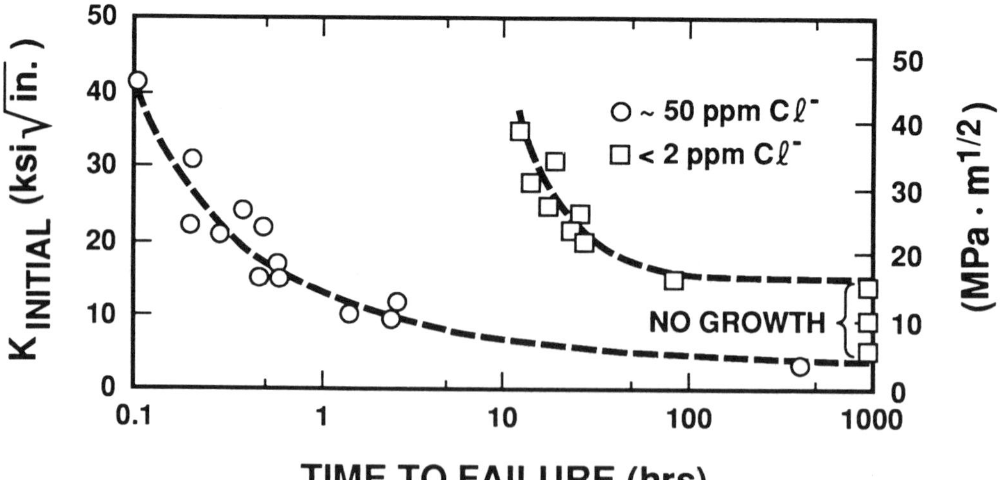

Fig. 4 Effects of chloride-ion concentration and initial stress intensity on times-to-failure of U-7.5Nb-2.5Zr. Source: Ref 18

Table 1 Mechanical Properties of Common Uranium Alloys

Alloy	Condition	Yield strength MPa	Yield strength ksi	Ultimate tensile strength MPa	Ultimate tensile strength ksi	Elongation, %	Plain-strain fracture toughness (K_Ic) MPa√m	Plain-strain fracture toughness Dry ksi√in.	Threshold stress intensity (K_ISCC) MPa√m (ksi√in.) Moist air	Threshold stress intensity 50 ppm air	Threshold stress intensity High Cl⁻	Cl⁻(c)	Ref
Unalloyed	As-cast	205	30	450	65	6	...	...	...	...	...	...	11
	Cast and outgassed	295	43	700	102	22	...	...	...	...	...	...	11
	Rolled and outgassed	270	39	720	104	31	...	...	...	...	...	...	11
U-0.75Ti	Aged 380 °C/6 h	965	140	1565	227	19	45	41	38 (35)	25 (23)	16 (15)	16 (15)	20
U-4.5Nb	As-quenched	310	45	1035	150	23	~62	~56	22 (20)	23 (21)	...	...	18
	Aged 260 °C/80 h	1070	155	1270	184	8	~49	~45	21 (19)	22 (20)	20 (18)	0(a)	18
U-6Nb	As-quenched	170	25	790	115	31	~80	~73	44 (40)	22 (20)	18 (16)	...	31
	Aged 200 °C/2 h	360	52	800	116	27	~75	~68	...	17 (15)	12 (11)	...	31
U-7.5Nb -2.5Zr	Aged 150 °C/1 h	630	91	905	131	20	~57	~52	20 (18)	20 (18)	(4.5)	0(a)	2
U-8Mo	Cast and homogenized	770	112	835	121	23	37	34	5 (4.5)(b)		...	...	38
U-10Mo	Cast and homogenized	875	127	920	133	23	35	32	8 (7)(b)		...	...	38
U-10Mo-1Ti	Cast and homogenized	880	128	950	138	43	66	60	8 (7)(b)		...	...	38

(a) Crack growth observed at zero applied load due to corrosion product wedging. **(b)** Relative humidity not controlled or reported. **(c)** Cl^- concentrations greater than 500 ppm

sisted cracking at higher stress intensities. For example, increases in environmental Cl^- concentration from less than 2 ppm to 50 ppm had little effect on K_{ISCC} of age-hardened U-4.5Nb; however, at applied stress intensities 50% greater than K_{ISCC}, this increase in Cl^- concentration decreased failure time from 270 min to 3 min (Ref 19). Accordingly, comparison of failure times at a particular stress intensity provides a more sensitive basis for order-ranking susceptibilities than do threshold stress intensities.

The most sophisticated characterization method measures crack-growth rate as a function of stress-intensity factor (for given combinations of alloy, heat treatment, and environment) (Ref 22-24). This enables determination of not only the threshold stress intensity but also the kinetics of crack growth at higher stress intensities, which in turn enables calculation (at least in principle) of the lifetimes of engineering components containing growing subcritical cracks. In practice, more conservative design philosophies are usually employed to preclude the presence of active stress-corrosion cracks in structural parts. Crack-growth kinetic measurements, however, are commonly used to characterize the effects of temperature and the partial pressures of various gaseous species on cracking rates in various uranium alloys. This method is well suited for fun-damental studies attempting to establish quantitative understanding of the mechanisms and kinetics of environmentally assisted cracking (Ref 22-24). Figure 5 shows an application of this

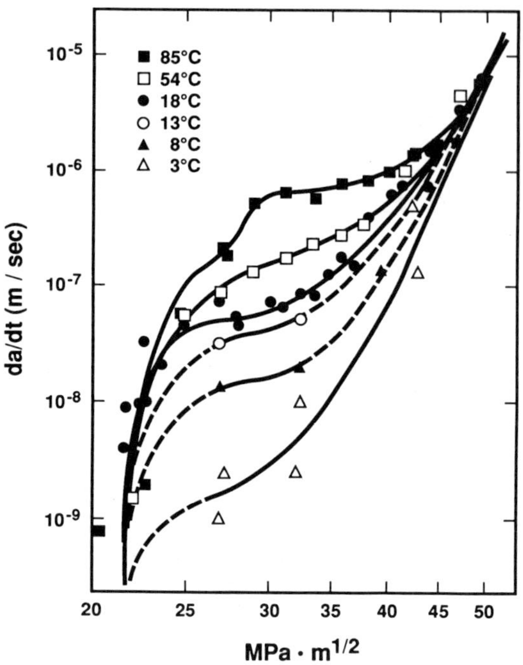

Fig. 5 Effects of temperature and stress intensity on crack-growth rates of U-4.5Nb tested in moist air. Source: Ref 22

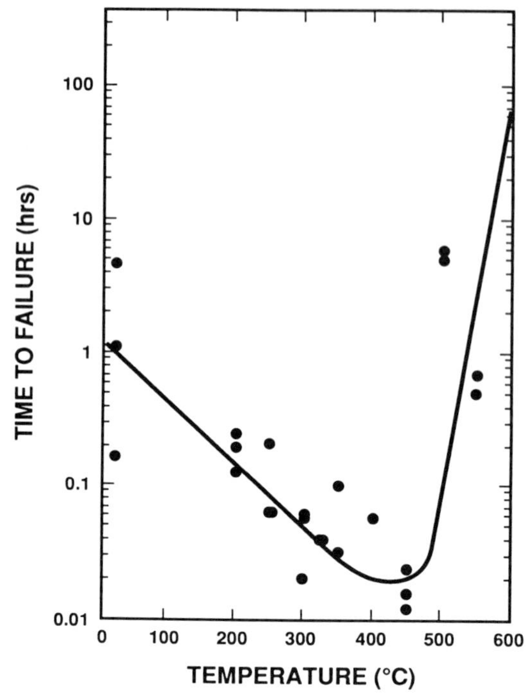

Fig. 6 Effect of aging temperature on times to failure of U-5Nb stressed to 275 MPa (40 ksi) in an aqueous solution containing 10^{-4} N KCl. Source: Ref 15

method to determine the effect of temperature on crack-propagation kinetics in U-4.5Nb (Ref 22).

Material Variables

A number of interrelated material variables, such as composition, heat treatment, microstructure, and strength level, strongly influence the propensity of uranium alloys toward environmentally assisted cracking.

Strength level is the most important single variable. Comparisons of different alloys with a variety of as-quenched strengths and particular alloys age hardened to various strength levels indicate that the propensity for premature failure increases with increasing strength. For example, among as-quenched alloys, U-6Nb, with a yield strength of 170 MPa (25 ksi), exhibits far lower susceptibility to environmentally assisted cracking than do U-0.75Ti, U-2Mo, U-7.5Nb-2.5Zr, or U-10Mo, whose yield strengths range from 620 to 875 MPa (90 to 127 ksi). Similarly, environmentally assisted cracking is promoted when any particular alloy is age hardened to higher strength levels (Ref 15, 17, 20) and is retarded by the reduction in strength that occurs during higher-temperature overaging (Ref 15, 17), as shown for U-5Nb in Fig. 6.

Strength levels being equal, the compositions of quenched or quenched and aged alloys influence the susceptibility of the materials to specific environments more than they do the overall propensity of the materials toward environmentally assisted cracking. Total alloy concentration appears to be the dominant factor in this regard, rather than particular alloying elements or combinations of alloying elements. Although alloys of nearly all compositions are highly susceptible to chloride ions, significant variations in susceptibility to hydrogen and oxygen occur with alloy composition, as outlined in Fig. 1. These variations will be discussed more fully in the next section.

Again at equal strength levels, the coarser two-phase microstructures produced by anneal-

Table 2 Comparisons of Underaged and Overaged U-2.3Nb Tested In 10^{-4} *N* KCl Solutions

Aging treatment	Yield strength MPa	Yield strength ksi	Time to failure at 280 MPa (41 ksi), h(a)
Underaged (230 °C/24 h)	930	135	255
Overaged (430 °C/24 h)	930	135	2021

(a) Average of three samples

Table 3 Effect of Retained γ Phase on Sensitivity to Atmospheric Moisture of U-2Mo

Heat treatment	Phases	Yield strength(a) MPa	Yield strength(a) ksi	Elongation(a), % 25% RH	Elongation(a), % 100% RH
Quenched from 800 °C	α′	640	93	14	4
+ Aged 500 °C/5 h	α + U_2Mo	723	105	16	2
Quenched from 660 °C	α + γ	400	58	12	20
+ Aged 275 °C/5 h	α + γ	793	115	10	17
+ Aged 325 °C/5 h	α + γ	906	131	6	10
+ Aged 550 °C/5 h	α + U_2Mo	572	83	24	7

(a) Average of two samples

ing or overaging are less susceptible to environmentally assisted cracking than the single-phase or very fine two-phase microstructures typical of materials in the as-quenched or quenched-and-underaged heat treatment condition, respectively. For example, U-2.3Nb samples underaged at 230 °C (445 °F) to yield strengths of 930 MPa (135 ksi) and loaded to 30% of their yield strength failed in one-eighth the time of comparable samples overaged at 430 °C (805 °F) to the same strength level, as shown in Table 2 (Ref 25). In addition, much lower residual stresses are typically present in annealed or overaged materials, thus frequently providing the additional benefit of a lower driving force for environmentally assisted cracking than is typical of materials in the quenched or underaged conditions (Ref 26). Caution must be exercised in prescribing annealed or overaged heat treatments, however, because general corrosion resistance is greatly reduced, and many alloys, including U-0.75Ti and U-6Nb, have inherently low ductility and fracture toughness in these conditions.

In some cases, the microstructures of uranium alloys can be purposely modified in order to decrease sensitivity to a particular environment. An example of this is shown in Table 3 (Ref 27). U-2Mo, a dilute alloy whose ductility is normally very sensitive to atmospheric moisture after quenching from the γ-phase field, was heat treated in a two-phase field and quenched to produce a lamellar microstructure of essentially pure α-uranium and alloy-enriched γ phase. The presence of the retained γ phase with its high solubility for hydrogen apparently provided a sink for hydrogen, thus preventing it from entering the much more sensitive α phase, where the hydrogen would have caused embrittlement when this material was tested in high-humidity air. Consistent with this, the material lost its immunity to degradation by atmospheric moisture after the γ phase was decomposed by a subsequent heat treatment.

Environmental and Testing Variables

There are three distinctly different environmental causes of premature fracture in uranium alloys: (1) aqueous solutions containing chloride ions and dissolved oxygen, (2) hydrogen (either internal hydrogen or hydrogen produced by reaction of atmospheric moisture with the surface of the uranium), and (3) atmospheric oxygen.

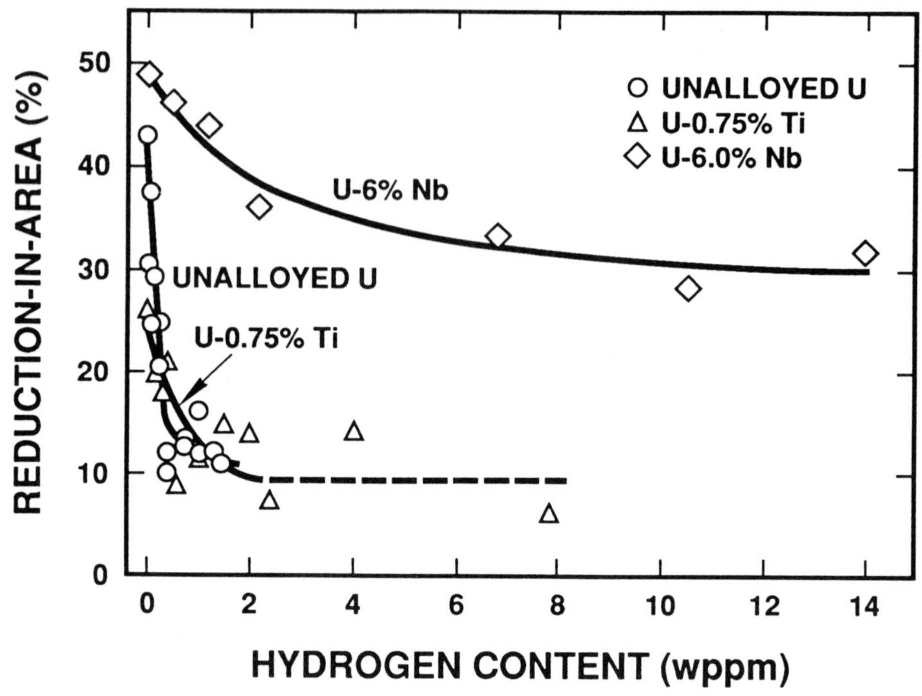

Fig. 7 Effect of internal hydrogen content on tensile ductilities of unalloyed uranium, age-hardened U-0.75Ti, and as-quenched U-6Nb. Source: Ref 7

Aqueous solutions containing chloride ions and dissolved oxygen are the most devastating environments for SCC over a wide range of uranium alloy compositions. Chloride-ion-related cracking occurs by an anodic dissolution mechanism and has been observed in both smooth and precracked specimens, frequently at very low stresses or stress intensities. Cracking initiates when passive films on the metal surface are destroyed by chloride ions, and proceeds via the establishment of a differential aeration cell in which the anodic region corresponds to the oxygen-depleted zone at the center of a water drop (during initial crack formation) or at the tip of the crack (after the crack has formed) (Ref 28, 29). The cracks grow along prior-γ grain boundaries, resulting in characteristically intergranular fracture surfaces (Ref 18, 30).

Times to failure, threshold stresses, and stress intensities for cracking generally decrease with increasing chloride-ion concentration (Ref 18). In several cases, particularly highly alloyed materials such as U-7.5Nb-2.5Zr, Cl^- concentrations as low as a few parts per million have been shown to cause intergranular cracking. All alloys characterized to date exhibit stress-corrosion thresholds below 20 MPa$\sqrt{m}$ (18 ksi$\sqrt{in.}$) in solutions containing 50 ppm Cl^- (see Table 1). The fracture surfaces of dilute alloys that failed in such low Cl^- concentrations, however, frequently resembled the transgranular fractures typical of oxygen-induced subcritical cracks, suggesting that oxygen, rather than Cl^-, was the chemical species responsible for cracking (Ref 31). At higher Cl^- concentrations some alloys crack even in the absence of stress. This self-propagating crack growth is apparently driven by wedging forces produced as the lower-density (higher-volume) corrosion products form in the confined space between the opposite surfaces of the crack (Ref 19).

Hydrogen causes substantial problems in unalloyed uranium and quenched dilute alloys. These materials exhibit very low solubilities for hydrogen and are seriously degraded either by internal hydrogen (typically introduced into the material during prefabrication heating in molten salt baths or heat treatment in atmospheres containing moisture) or by moisture in the testing environment (Ref 6, 7, 12-14, 32-35). The general belief is that these two effects occur by the same mechanism (the view espoused in this chapter); however, some investigators have proposed a more fundamental difference between the internal hydrogen and external moisture effects (Ref 33). According to the predominant view, the environmental moisture effect involves the reaction of water with the uranium surface ($U + 2H_2O \rightarrow UO_2 + 4H$) to produce atomic hydrogen, which enters the material and causes degradation identical to that caused by *in situ* internal hydrogen. Both effects have been clearly documented in unalloyed uranium, where either internal hydrogen or external humidity reduces tensile ductility, as shown in Fig. 2 and 7. Dilute alloys, such as U-0.75Ti, exhibit similar effects (also shown in Fig. 2 and 7). Because the dilute alloys have higher strengths than unalloyed uranium, however, the consequences can be more dramatic than simple decreases in tensile ductility. Residual stresses introduced by quenching or welding frequently cause delayed cracking in dilute alloys that contain excessive concentrations of internal hydrogen, or even in materials of low hydrogen content when they are exposed to high-humidity environments (Ref 27, 36).

Hydrogen-assisted failure in unalloyed uranium generally occurs by intergranular cracking along the α grain boundaries, in some cases mixed with transgranular cleavage (Ref 33). In U-0.75Ti, fracture occurs primarily along the martensite plate boundaries (Ref 35), which are the counterparts of the α grain boundaries in unalloyed uranium. Similar instances of low-ductility intergranular tensile fracture and delayed cracking have been observed in quenched or quenched and aged U-2Mo and U-2.4Nb that had not been vacuum solution heat treated to remove internal hydrogen (Ref 27, 37). These alloys, however, exhibited intergranular fracture along the prior-γ grain boundaries, apparently because the boundaries between adjacent plates in the thermoelastic or banded martensite are low-energy twin boundaries, making the prior-γ boundaries the primary surfaces of high-energy crystallographic misorientation in the material. In effect, then, all of these hydrogen-related fractures are fundamentally intergranular, but the particular boundaries along which cracking occurs, and the associated fracture surface appearances, vary with alloy content and microstructure.

Hydrogen-related cracking in uranium and dilute uranium alloys is typically a time-dependent process. For example, the substantial ductility losses associated with low-strain-rate tensile testing of materials containing small amounts of internal hrydrogen, or with testing in humid environments can be overcome by testing at high rates, as shown in Fig. 8 (Ref 6, 35). In addition, sustained-load cracking due to either atmospheric moisture or internal hydrogen occurs in a delayed fashion (Ref 20, 36). This indicates that, in cases of low initial internal hydrogen content, embrittlement requires hydrogen migration to, and concentration at, preferred fracture sites (microstructural discontinuities, such as grain boundaries, in regions of high triaxial stress). During low-strain-rate and/or sustained-loading tests, sufficient time is available for such hydrogen movement, and premature failure occurs. In high-strain-rate tests, however, insufficient time is available for hydrogen movement, so premature fracture is prevented. In cases where high internal hydrogen contents provide for prestress hydrogen saturation of (and possible fine hydride precipitation on) preferred fracture sites, how-

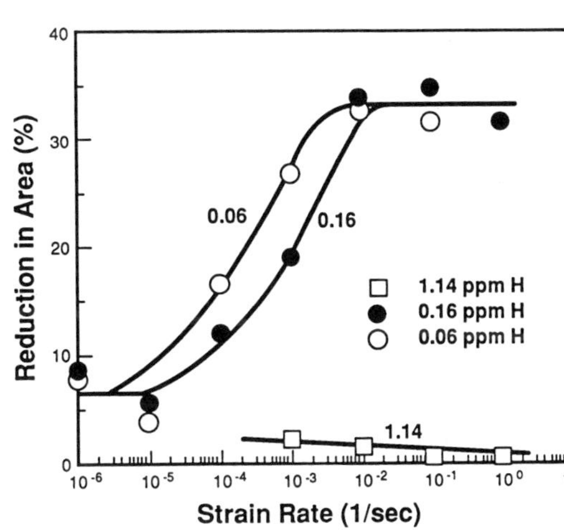

Fig. 8 Effects of internal hydrogen content and strain rate on tensile ductility of U-0.75Ti tested in dry air. Source: Ref 35

ever, brittle behavior is observed at all strain rates.

Solubility for hydrogen increases with increasing alloy content. Correspondingly, sensitivity to both internal hydrogen and atmospheric moisture decreases as alloy content increases. As a result, intermediate-concentration alloys, such as U-6Nb, are far more tolerant of internal hydrogen (as shown in Fig. 7) and do not exhibit dramatic ductility losses when tested in moist air.

A third type of environmentally assisted fracture is caused by oxygen present in environments such as dry or humid air, aerated water, or very dilute aqueous chloride solutions (Ref 2, 19, 38). Oxygen-related subcritical crack growth is distinct in that it occurs transgranularly, resulting in quasi-cleavage-appearing fracture surfaces (Ref 19). This transgranular failure has been observed in alloys as dilute as U-2.25Nb (Ref 25), but is generally difficult to initiate in alloys of dilute and intermediate compositions; therefore, it occurs primarily in precracked samples of these alloys and not in smooth specimens (Ref 31). As a result, oxygen-related cracking is an infrequent problem in engineering components made of dilute or intermediate-concentration alloys such as U-0.75Ti or U-6Nb. Sensitivity to oxygen cracking increases with increasing alloy content, however, so concentrated alloys, such as U-7.5Nb-2.5Zr and U-10Mo, are more susceptible (Ref 19, 39). Uranium-molybdenum alloys seem particularly prone to oxygen cracking at intermediate and high alloy concentrations and frequently experience this type of failure even in smooth (nonprecracked) specimens. The presence of atmospheric moisture has been reported to improve the tensile ductilities of some U-Mo alloys (Ref 39), perhaps because the moisture condenses to produce oxygen-depleted water at crack tips and prevents the freshly exposed metal from coming into immediate contact with the embrittling oxygen. The low threshold stress intensities associated with this mode of cracking in concentrated alloys, particularly those containing molybdenum, make them highly susceptible to stress corrosion in very innocuous normal atmospheric environments and are strong impediments to the structural application of these materials.

Avoiding and Overcoming Environmentally Assisted Cracking

The three options for avoiding environmentally assisted cracking are (1) protection of the material from damaging environments, (2) reduction of the stresses that drive crack growth, and (3) selection of a material with sufficient resistance to cracking.

Metallic coatings, such as electroplated nickel and ion-plated aluminum, are sometimes applied to unalloyed uranium and dilute uranium alloys to increase their resistance to general corrosion (Ref 4). Nickel plating is effective, provided the coating is free of imperfections and truly impermeable to the environment. The nickel is not cathodically protective of the underlying uranium, however, and when it fails corrosion frequently proceeds even more rapidly. Little evaluation has been done of the degree of protection provided by nickel plating against SCC. The ductility of U-0.75Ti is degraded to some extent by the plating process, either because of surface roughening done by preetching to enhance adhesion of the nickel or because of the introduction of hydrogen into the material during plating (Ref 40). In addition, nickel plating is ineffective in preventing ductility losses in this material during tensile testing; it ruptures as the metal deforms, thus allowing moisture to come into contact with the underlying metal (Ref 41). Electroplated nickel would probably not provide high-reliability protection against galvanic SCC in engineering components of complex geometry, because it is not cathodically protective and tends to be thinnest and least perfect at inside corners and grooves, where cracks are most likely to start. In addition, if small flaws are present, electroplated coatings can sometimes obscure them and trap potentially damaging chemical species in them, thus increasing the danger of subsequent SCC.

Conversely, aluminum coatings deposited by a variety of techniques have been shown to significantly increase stress-corrosion failure times in age-hardened uranium alloys containing 4 to 5% Nb that are subjected to aqueous chloride environments (Ref 42, 43). These beneficial effects were retained even when deliberate flaws were introduced into the coatings, apparently because the aluminum provided cathodic protection for the underlying metal (Ref 43). Unfortunately, aluminum coatings are not efficient cathodic protectors when applied to more dilute alloys, such as U-0.75Ti. In addition, they require excellent surface cleanliness prior to deposition and, if not properly applied, can delaminate when subjected to moderate temperature excursion.

Evidence obtained to date indicates that the best protective coating for use on uranium alloys is electrolytically deposited nickel with a zinc overcoat (Ref 44). The zinc cathodically protects the underlying uranium against galvanic corrosion even when the coating is flawed. Therefore, this coating might be expected to provide good protection against SCC in chloride-containing environments, but would protect against the moisture and oxygen effects only as long as it remained perfectly impervious to these environments. Galvanic effects occurring in the vicinity of flawed coatings might also result in the production of hydrogen, which could cause hydrogen embrittlement in dilute alloys. Specific tests to evaluate the degree of protection provided by nickel-zinc coatings against the various forms of environmentally assisted cracking have not been conducted.

Control of stresses is vital to minimizing the probability of environmentally assisted cracking in uranium alloy parts. Stresses are sometimes caused by external loads, but more commonly the major stresses present in uranium alloy components are residual stresses introduced by forming processes (particularly those performed at temperatures below about 500 °C, or 930 °F), heat treatment (especially when quenching is involved), or welding (Ref 26). The magnitudes of these residual stresses are often in the vicinity of the yield strength of the material, more than sufficient in many cases to cause environmentally assisted cracking. In some instances, such as quenched parts of simple geometry, the residual stresses at the surface are compressive and are beneficial in preventing cracking. When extensive machining is done on these parts or when more complex geometries are quenched, however, highly undesirable tensile stresses can exist at portions of the surface. Welded parts commonly exhibit surface tensile stresses whose magnitudes are limited only by the yield strength of the material. These frequently result in delayed cracking in welds of high-strength alloys (Ref 36).

Residual stresses can be reduced either mechanically or thermally (Ref 26). Mechanical stress relieving is most applicable to simple-geometry components, such as quenched cylinders, which frequently exhibit substantial compressive surface stresses and tensile interior stresses. This "stress leveling" is accomplished by compressive upsetting to 1 to 2% plastic strain. As illustrated in Fig. 9, the magnitudes of both the tensile and compressive residual stresses are substantially reduced. Thermal stress relieving can be done in the 250 to 600 °C (480 to 1110 °F) temperature range and is applicable to parts of all geometries. Because age hardening also occurs in this temperature range, however, changes in the tensile properties and stress-corrosion thresholds of previously quenched materials frequently also occur during stress-relieving heat treatments. Low-temperature thermal stress relieving typically increases strength and reduces ductility and threshold stress intensity. An analysis for U-

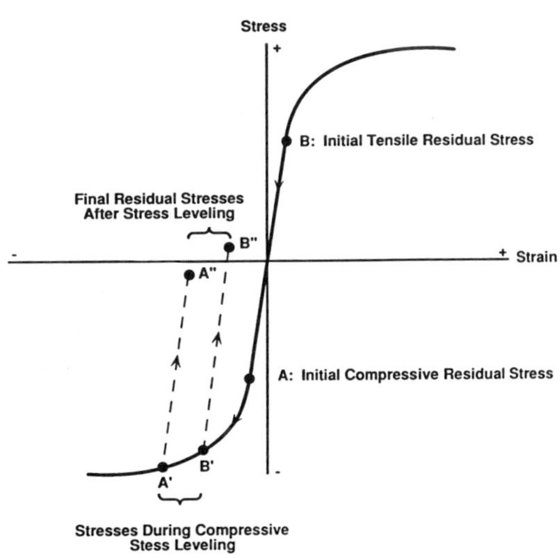

Fig. 9 Stress-strain curve of quenched uranium alloy illustrating initial residual stresses at the surface (A) and interior (B) and showing how compressive mechanical stress relief reduces residual stress magnitudes. Source: Ref 11

0.75Ti indicates that the reduction in residual stresses overrides the reduction in threshold stress intensity, such that the net effect of combined stress relieving and age hardening is decreased susceptibility to SCC (Ref 45). Overaging results in complete relief of residual stresses (Ref 45); therefore, alloys such as U-2Mo and U-2.4Nb, which exhibit attractive combinations of strength and ductility in extensively overaged conditions, can be used to advantage in situations where SCC resistance is of paramount importance. These alloys must not be rapidly cooled following overaging, or residual stresses can be reintroduced. Unfortunately, the more common U-0.75Ti and U-6Nb alloys exhibit poor ductility in the overaged condition, making complete thermal stress relieving an unattractive option. Additional details regarding sources of residual stresses and methods of relieving these stresses in uranium alloys can be found in a recent review (Ref 26).

Selection of materials resistant to environmentally assisted cracking is best enabled by a discussion of the cracking characteristics of various uranium alloys, presented in the following section.

Environmentally Assisted Cracking Characteristics of Common Uranium Alloys

Unalloyed uranium (Ref 2, 4-7, 12, 32-34) is a material of relatively low strength that exhibits little tendency for in-service cracking. Components that have been deformation processed and/or welded have withstood long periods in normal atmospheric conditions without delayed cracking. The primary shortcoming of unalloyed uranium in terms of environmentally assisted cracking is that its tensile ductility is substantially reduced when it is strained at relatively low rates in moist environments or when it contains high concentrations of internal hydrogen (which can be removed by vacuum heat treatment [Ref 7, 11]).

U-0.75Ti (Ref 1, 2, 4, 6-9, 11, 13, 14, 20, 35-37, 40, 41, 44-46) is an age-hardenable alloy that is well suited for use in applications requiring high strength (yield strengths greater than 800 MPa, or 116 ksi) and good ductility and toughness. It must be severely quenched to obtain age-hardenable martensitic microstructures, and the as-quenched martensite has a high yield strength (~650 MPa, or 94 ksi). As a result, high residual stresses are commonly encountered. This alloy is moderately susceptible to cracking due to internal hydrogen, atmospheric moisture, or chloride ions, particularly if age hardened to high strength levels. The combination of high residual stresses and moderate susceptibility to both internal hydrogen embrittlement and environmental stress-corrosion makes cracking a dangerous possibility if the residual stresses at portions of the surface are tensile.

Subcritical quenches produce even higher-strength microconstituents with lower stress-corrosion thresholds and potentially higher residual stresses. Welds in U-0.75Ti are particularly prone to delayed cracking because of high tensile stresses produced by the welding process and subcritical cooling rates that result in highly susceptible microconstituents in the weld zone. The likelihood of cracking due to quench-induced residual stresses can be reduced by minimizing hydrogen content (by vacuum-solution heat treatment prior to quenching), by mechanical stress relieving (for parts of simple geometry), or by thermal reduction of residual stresses (by age hardening as soon as possible after the quench). Overaging to completely eliminate residual stresses is not an attractive option, because the material exhibits low ductility and toughness in the overaged condition. No stress-corrosion data are available for other U-Ti compositions, but cursory information suggests that susceptibility increases with increasing titanium content (perhaps in part because of increasing quench-rate sensitivity and the related likelihood of producing highly susceptible microconstituents due to subcritical quenching).

U-6Nb (Ref 1-4, 7, 9-11, 19, 21, 30, 31, 47-50) is well suited for applications requiring good general corrosion resistance and high ductility, but limited yield strength (less than 350 MPa, or 51 ksi). It is markedly less sensitive to quench rate than U-0.75Ti and quenches to a much lower yield strength (~170 MPa, or 25 ksi). Thus, residual stresses are typically low. In the most common as-quenched or slightly aged conditions (where general corrosion resistance and ductility are both high), this alloy is very resistant to degradation of tensile ductility and to sustained-load cracking caused by internal hydrogen or atmospheric moisture. In principle, oxygen can cause crack growth, but this is an unlikely engineering problem, because oxygen-related transgranular cracks have never been observed to initiate in smooth specimens and the low yield strengths limit residual stress magnitudes, resulting in large critical flaw sizes (of the order of 1 cm for as-quenched material and 1 mm for 200 °C, or 390 °F, aged material). The presence of chloride ions, however, can cause both crack initiation and growth. Overall, low residual stresses combined with good resistance to environmentally assisted cracking create a very low danger of cracking in this material, provided chloride-containing environments are avoided. As evidence of this, U-6Nb can be welded with relative ease, and the welds are typically free from delayed cracking. Resistance to cracking in environments that contain moisture and especially chloride ions decreases with age hardening and then increases as the material is overaged, but this alloy is not commonly used in these aged or overaged conditions because of suboptimum combinations of strength and ductility.

Leaner U-Nb alloys, such as U-4.5Nb (Ref 1-4, 8, 15, 16, 18, 19, 21, 22, 31, 42, 43) and U-2.4Nb (Ref 2, 7, 21, 25, 26, 31, 45, 46), exhibit better combinations of age-hardened strength and ductility than U-6Nb and are sometimes considered as alternatives to U-0.75Ti for applications requiring moderate- or high-strength age-hardenable materials. Susceptibility generally increases with decreasing niobium content and increasing amounts of age hardening, and decreases substantially with overaging. Stress-corrosion cracking in uranium alloys containing 4 to 5% Nb has received substantial attention, particularly in various age-hardened conditions. These compositions behave similarly to U-6Nb, being sensitive primarily to chloride ions and secondarily to oxygen. Overaging increases stress-corrosion thresholds and eliminates residual stresses, but also results in low ductility. U-2.4Nb in the as-quenched or underaged condition exhibits increased sensitivity to hydrogen. When either quenched-and-overaged or simply annealed, however, it exhibits two-phase microstructures with attractive combinations of moderate yield strength (550 to 750 MPa, or 80 to 109 ksi), high ductility, very low residual stresses, and high resistance to both internal hydrogen embrittlement and environmental SCC.

U-Mo alloys (Ref 1, 2, 8, 11, 23, 37-39, 51-53) can be more easily melted and cast than their U-Nb counterparts, and compositions ranging from 2 to 10% Mo have been considered for various applications, particularly ones requiring thick-wall castings. U-2Mo is typically considered for use in the annealed or overaged condition, where its microstructure and properties are similar to those of U-2.4Nb — except for its ductility, which is substantially lower. U-2Mo can also be quenched and age hardened, but generally exhibits lower ductility, general corrosion resistance, and stress-corrosion resistance than comparably aged U-0.75Ti. Quenched alloys containing 4 to 5% Mo are somewhat similar to U-6Nb in physical metallurgy and tensile properties, but suffer much greater susceptibility to oxygen-induced cracking. Concentrated alloys, such as U-10Mo, are sometimes considered for applications requiring high general corrosion resistance and high strength (yield strength in the vicinity of 875 MPa, or 127 ksi). U-10Mo exhibits very low quench-rate sensitivity and does not need to be rapidly cooled in order to retain a variant of the elevated-temperature γ structure. However, concentrated U-Mo alloys are very prone to SCC in oxygen-containing environments and are infrequently used for this reason. The oxygen sensitivity of these alloys has been reported to decrease with decreasing carbon content. This finding has led to the development of U-Mo-Ti alloys containing 8 to 10% Mo and 0.5 to 1% Ti. During melting of these alloys, the titanium combines with carbon to form titanium carbides, which float to the surface and can be slagged off. These reduced-carbon ternary alloys are somewhat less prone to cracking in oxygen than their binary counterparts.

U-7.5Nb-2.5Zr (sometimes termed "Mulberry") (Ref 1-4, 6, 7, 11, 17-19, 28-30, 54, 55) is similar to U-10Mo in its quench-rate insensitivity, phase structure, and general corrosion resistance, but exhibits somewhat lower unaged yield strength (550 to 630 MPa, or 80 to 91 ksi). In this condition it exhibits substantially better stress-corrosion resistance to oxygen than does U-10Mo, but is very sensitive to chloride ions. Cracking resistance can be improved by overaging, but the material is not generally used in this condition because of low ductility.

Higher-order alloys involving combined contents of molybdenum, niobium, titanium, zirconium, and in some cases, vanadium, with total alloy concentrations ranging from approximately 3 to 7%, have been characterized to some extent (Ref 1, 2, 56-59). Unfortunately, virtually all of the stress-corrosion characterization of these alloys has been done on as-fabricated (non-heat-treated) material with suboptimum general corrosion resistance and mechanical properties. While some investigations have been conducted on the relationships among heat treatment, microstructure, and tensile properties, little or no data are available on the stress-corrosion behavior of these materials in the quenched and age-hardened conditions.

Acknowledgments

The author appreciates the careful manuscript reviews and helpful comments of Drs. R.H. Jones, A.D. Romig, Jr., J.W. Koger, and J.J. Dillon. This work was performed at Sandia National Laboratories, which is operated for the U.S. Department of Energy under Contract DE-AC04-76DP00789.

References

1. N.J. Magnani, Stress Corrosion Cracking of Uranium Alloys, in *Physical Metallurgy of Uranium Alloys*, J.J. Burke *et al.*, Ed., Brook Hill, 1976, p 935-947
2. N.J. Magnani, Hydrogen Embrittlement and Stress Corrosion Cracking of Uranium and Uranium Alloys, in *Advances in Corrosion Science and Technology*, Vol 6, M.G. Fontana and R.W. Staehle, Ed., Plenum, 1976, p 89-161
3. J.W. Koger, An Overview of Corrosion, Corrosion Protection, and Stress Corrosion Cracking of Uranium and Uranium Alloys, in *Metallurgical Technology of Uranium and Uranium Alloys*, American Society for Metals, 1981, p 751-793
4. L.J. Weirick, Corrosion of Uranium and Uranium Alloys, in *Metals Handbook*, 9th ed., Vol 13, *Corrosion*, ASM International, 1987, p 813-822
5. H. Inouye and A.C. Schaffhauser, "Low Temperature Ductility and Hydrogen Embrittlement of Uranium — A Literature Review," ORNL-TM-2563, Oak Ridge National Laboratory, July 1969
6. G.L. Powell and J.B. Condon, Hydrogen in Uranium Alloys, in *Physical Metallurgy of Uranium Alloys*, J.J. Burke *et al.*, Ed., Brook Hill, 1976, p 427-461
7. G.L. Powell, Overview of Hydrogen Embrittlement of Uranium and Uranium Alloys, in *Metallurgical Technology of Uranium and Uranium Alloys*, American Society for Metals, 1981, p 877-899
8. K.H. Eckelmeyer, Aging Phenomena in Dilute Uranium Alloys, in *Physical Metallurgy of Uranium Alloys*, J.J. Burke *et al.*, Ed., Brook Hill, 1976, p 463-509
9. K.H. Eckelmeyer, Diffusional Transformations, Strengthening Mechanisms, and Mechanical Properties of Uranium Alloys, in *Metallurgical Technology of Uranium and Uranium Alloys*, American Society for Metals, 1981, p 129-201
10. R.A. Vandermeer, Martensitic Transformations, Shape Memory Effects and Other Curious Mechanical Effects in Uranium Alloys, in *Metallurgical Technology of Uranium and Uranium Alloys*, American Society for Metals, 1981, p 251-293
11. K.H. Eckelmeyer, Uranium and Uranium Alloys, in *Metals Handbook*, 10th ed., Vol 2, *Properties and Selection: Nonferrous Alloys and Special-Purpose Materials*, ASM International, 1990, p 670-682
12. G.A. Whitlow and R. Willows, "The Effect of Humidity on the Ductility of Uranium," Report 0-46/66, Atomic Weapons Research Establishment (Aldermaston, UK), 1966
13. R.J. Jackson, "Effect of Humidity on Tensile Ductility of Uranium-Titanium Alloys," RFP-2048, Dow Chemical Rocky Flats Division, 1972
14. V.C. Hemperly, Martin Marietta Y-12 Plant, unpublished research, 1973
15. B.D. McLaughlin, L.L. Stephenson, and C.J. Miglionico, Influence of Aging Time and Temperature on the Susceptibility of Gamma-Quenched U-5% Nb Alloy to SCC, *Corrosion*, Vol 28 (No. 1), 1972, p 35-38
16. J.M. Macki and R.L. Kochen, Stress Corrosion Cracking of Aged U-4.5%Nb, *Corrosion*, Vol 29 (No. 4), 1973, p 153-156
17. L.J. Weirick, Effect of Heat Treatment Upon the Stress Corrosion Cracking of U-7.5%Nb-2.5%Zr, *Corrosion*, Vol 31 (No. 1), 1975, p 5-14
18. N.J. Magnani, The Effect of Chloride Ions on the Cracking Behavior of U-7.5%Nb-2.5%Zr and U-4.5%Nb, *J. Nucl. Mater.*, Vol 42, 1972, p 271-277

19. N.J. Magnani, "The Effects of the Environment on the Cracking Behavior of Selected Uranium Alloys," Paper No. 58, presented at Corrosion/72, National Association of Corrosion Engineers, 1972
20. N.J. Magnani, The Effects of Environment, Orientation, and Strength Level on the Stress Corrosion Behavior of U-0.75% Ti, *J. Nucl. Mater.*, Vol 54, 1974, p 108-116
21. A.R. Miller, The Stress Corrosion Cracking Behavior of U-Nb Alloys in Air, *Corrosion*, Vol 33 (No. 5), 1977, p 186-187
22. N.J. Magnani, "The Correlation Between Oxidation and Stress Corrosion Cracking of U-4.5%Nb," Paper No. 19, presented at Corrosion/77, National Association of Corrosion Engineers, 1977
23. J. Corcos, A.M. Nomine, and D. Miannay, Gaseous Oxygen and Hydrogen Embrittlements of a U-10%Mo Alloy, *Fracture 1977*, Vol 2A, D.M.R. Taplin, Ed., 1977, p 261-267
24. D. Lepoutre, A.M. Nomine, and D. Miannay, Comportement en Corrosion Sous Contrainte Dans L'Oxygene et L'Hydrogene Gazeux D'un Alliage U-7.5%Nb-2.5%Zr, *J. Less Common Met.*, Vol 121, 1986, p 521-530
25. J.M. Macki and R.L. Kochen, "The Stress Corrosion Cracking of Underaged and Overaged U-2.3%Nb in an Aqueous Chloride Environment," Report RFP-1904, Dow Chemical Rocky Flats Division, 1972
26. K.H. Eckelmeyer, "Residual Stresses in Uranium and Uranium Alloys," SAND85-1427, Sandia National Laboratories, 1985
27. K.H. Eckelmeyer, Sandia National Laboratories, unpublished research, 1989
28. S.W. Zehr, A Study of the Intergranular Cracking of U-7.5%Nb-2.5%Zr Alloy in Aqueous Chloride Solutions, *Corrosion*, Vol 28 (No. 5), 1972, p 196-205
29. L.J. Weirick and C.W. Schoenfelder, Effect of Oxygen, Chloride Ion, and Water Vapor on Crack Initiation in U-7.5%Nb-2.5%Zr, *Corrosion*, Vol 30 (No. 5), 1974, p 169-174
30. J.W. Koger, "Stress Corrosion Cracking of Uranium Alloys," Y-DA-5624, Union Carbide Y-12 Plant, 1973
31. N.J. Magnani, "Stress Corrosion Cracking of Uranium-Niobium Alloys," SAND78-0439, Sandia National Laboratories, 1978
32. A.N. Hughes, S. Orman, and G. Picton, Environmental Factors Affecting the Mechanical Properties of Uranium, Part I: The Effect of Water Vapor, *J. Nucl. Mater.*, Vol 33, 1969, p 159-164
33. A.N. Hughes, S. Orman, G. Picton, and M.A. Thorne, Environmental Factors Affecting the Mechanical Properties of Uranium, Part 2: The Mechanism of Water Embrittlement of Uranium, *J. Nucl. Mater.*, Vol 33, 1969, p 165-172
34. P. Adamson, S. Orman, and G. Picton, The Effects of Hydrogen on the Tensile Properties of Uranium When Tested in Different Environments, *J. Nucl. Mater.*, Vol 33, 1969, p 215-224
35. B.C. Odegard, K.H. Eckelmeyer, and J.J. Dillon, The Embrittlement of U-0.8%Ti by Absorbed Hydrogen, in *Hydrogen Effects on Material Behavior*, N.R. Moody and A.W. Thompson, Ed. TMS, 1990, p 775-784
36. J.W. Whittaker and M.W. Richey, The Detection of Hydrogen Assisted Crack Formation in U-0.8%Ti Alloy Electron Beam Weldments, *J. Acoust. Emiss.*, Vol 6, 1987, p 257-260
37. K.H. Eckelmeyer and F.J. Zanner, The Effect of Aging on the Mechanical Behaviors of U-0.75%Ti and U-2.0%Mo, *J. Nucl. Mater.*, Vol 62, 1976, p 37-49
38. A.M. Nomine, D. Bedere, and D. Miannay, The Influences of Physio-Chemical Parameters on the Mechanical Properties of Some Isotropic Uranium Alloys, in *Physical Metallurgy of Uranium Alloys*, J.J. Burke *et al.*, Ed., Brook Hill, 1976, p 657-700
39. C.A.W. Peterson and R.R. Vandervoort, "Stress Cracking in the U-10%Mo Alloy," UCRL-7769, Lawrence Livermore National Laboratory, 1964
40. A.S. Nagelberg, "Effects of Storage Environment and Nickel Plating on the Mechanical Properties of U-0.75%Ti," SAND80-8227, Sandia National Laboratories, 1980
41. S.W. Zehr, H.R. Johnson, J.E. Smugeresky, K.A. Pashman, and A.S. Nagelberg, "Effects of Testing and Storage Environments on Mechanical Properties of Ni-Plated and Bare U-0.75%Ti," SAND79-8504, Sandia National Laboratories, 1979
42. B.D. McLaughlin and N.T. Panousis, Protective Metallic Coatings for Gamma-Quenched U-5%Nb Alloy, *J. Nucl. Mater.*, Vol 43, 1972, p 343-346
43. J.M. Macki, G. Mah, R.L. Kochen, and C.W. Nordin, The Stress Corrosion Cracking of Aluminum-Coated U-4.5%Nb, *J. Nucl. Mater.*, Vol 47, 1973, p 173-176
44. L.J. Weirick, "Evaluation of Metallic Coatings for the Corrosion Protection of a U-0.75%Ti Alloy," SLL-73-5024, Sandia National Laboratories, 1973
45. K.H. Eckelmeyer, Residual Stresses and Stress Relieving in Uranium Alloys, in *Residual Stresses and Stress Relaxation*, E. Kula and V. Weiss, Ed., Plenum, 1982, p 389-407
46. G.L. Powell and K.A. Thompson, Hydrogen Embrittlement in Lean Uranium Alloys, in *Hydrogen Effects on Material Behavior*, N.R. Moody and A.W. Thompson, Ed., TMS, 1990, p 765-773
47. J.W. Koger, A.M. Ammons, and J.E. Ferguson, "Effect of Aging on the General Corrosion and Stress Corrosion Cracking of U-6%Nb Alloy," Y-1999, Union Carbide Y-12 Plant, 1975

48. R.L. Kochen, R. Mah, and R.J. Jackson, "Stress Corrosion Cracking of a U-6%Nb Alloy," RFP-2429, Rockwell International Rocky Flats Plant, 1976

49. P.E. Armstrong, J.B. Straw, and G.L. Powell, "The Effect of Hydrogen on the Mechanical Properties of a U-5.7%Nb Alloy," LA-9870, Los Alamos National Laboratory, 1983

50. G.L. Powell and W.G. Northcutt, Jr., Internal Hydrogen Embrittlement of U-5.7%Nb Alloy, *J. Nucl. Mater.*, Vol 132, 1985, p 47-51

51. J.W. Pridgeon, "Stress Corrosion Cracking in U-Mo Alloys," Y-1417, Union Carbide Y-12 Plant, 1963

52. M.A. Beaubron, Influence de la Teneur en Carbon Sur la Corrosion Sous Tension d'un Alliage U-10%Mo Dans L'Air Humide a 20C, *J. Nucl. Mater.*, Vol 43, 1972, p 351-353

53. D. Bedere and D. Miannay, Analyse de L'Emission Acoustique Produite Au Cours de la Fissuration de Corrosion Sous Contrainte de L'Alliage U-10%Mo, *J. Nucl. Mater.*, Vol 48, 1973, p 183-192

54. J.W. Koger, "Variables Which Influence the Stress Corrosion Cracking of U-7.5%Nb-2.5%Zr Alloy," Y-1965, Union Carbide Y-12 Plant, 1975

55. D. Lepoutre, A.M. Nomine, and D. Miannay, Stress Corrosion Cracking Behavior in Gaseous Oxygen of a U-7.5%Nb-2.5%Zr Alloy, in *Application of Fracture Mechanics to Materials Structures*, Martinus Nijhoff, 1983, p 907-917

56. W.F. Czyrklis and M. Levy, Stress Corrosion Cracking Behavior of Uranium Alloys, *Corrosion*, Vol 30 (No. 5), 1974, p 181-187

57. J. Greenspan, D.A. Colling, and F.J. Rizzitano, Polynary Uranium Alloys, in *Physical Metallurgy of Uranium Alloys*, J.J. Burke, *et al.* Ed., Brook Hill, 1976, p 701-724

58. J.P. Drolet, W.H. Erickson, and H.P. Tardif, A Review of DREV Uranium Research, in *Physical Metallurgy of Uranium Alloys*, J.J. Burke *et al.*, Ed., Brook Hill, 1976, p 725-746

59. K.H. Eckelmeyer, The Effects of Heat Treatment on the Microstructure and Mechanical Behavior of U-0.75%TMo-0.75%Nb-0.75%Zr-0.5%Ti, *J. Nucl. Mater.*, Vol 68, 1977, p 92-99

CHAPTER 13

Stress-Corrosion Cracking of Amorphous Alloys

Koji Hashimoto and Asahi Kawashima,
Institute for Materials Research, Tohoku University

Amorphous alloys consist of a chemically homogeneous single phase; hence, amorphous iron-base alloys that contain certain amounts of strongly passivating elements such as chromium possess an extremely high corrosion resistance (Ref 1). The mechanical behavior of amorphous alloys prepared by rapid quenching from the liquid state (by melt spinning, for example) is also remarkable. They are very strong and stiff, but tough; for example, amorphous iron-base alloys have a fracture strength of 2.45 to 3.92 GPa (355 to 569 ksi) and a Young's modulus of 127 to 166 GPa (18.4 to 24.1 $\times$ 10^3 ksi) (Ref 2). These alloys are capable of plastic deformation on shear bands, although in tension the plastic elongation to fracture is generally less than 2%.

Amorphous alloys are characterized by a high toughness and have no specific crystallographic slip planes. Consequently, they are assumed to be insensitive to the selective corrosion attack of slip planes (as occurs in crystalline alloys) that leads to stress-corrosion cracking (SCC). The plastic deformation of amorphous alloys occurs immediately before tensile fracture, forming localized shear bands, and fracture takes place along the bands (Ref 3). Accordingly, active-path corrosion of amorphous alloys requires instantaneous dissolution of the shear bands within the very short period of time between the onset of plastic deformation and tensile failure. Such a situation cannot occur unless the alloys have very low corrosion resistance.

General expression of the corrosion behavior of amorphous alloys is impossible, because corrosion resistance varies greatly among them. In particular, their corrosion resistance differs markedly depending on whether or not strongly passivating elements such as chromium are present. As mentioned before, active-path corrosion can occur only when the shear bands are instantaneously corroded immediately before mechanical failure. The effect of stress on the corrosion behavior of amorphous alloys thus depends on alloy composition.

The amorphous alloys can be prepared by various methods, including melt spinning, sputter deposition, mechanical alloying, electrodeposition, electroless deposition, and ion implantation. Mechanical alloying forms powders, and the other methods form amorphous alloys covering solid substrates. Only melt spinning can form specimens with measurable mechanical properties. Accordingly, investigation of the interaction of corrosive attack and mechanical stress has been restricted to melt-spun amorphous alloys.

Hydrogen Embrittlement

A detailed investigation of the interaction of applied stress and corrosive action in the corrosion-resistant amorphous alloys has been performed by Kawashima *et al.* (Ref 4). It is generally difficult to prepare amorphous alloys with thicknesses of several tens of micrometers and smooth side, top, and bottom surfaces by

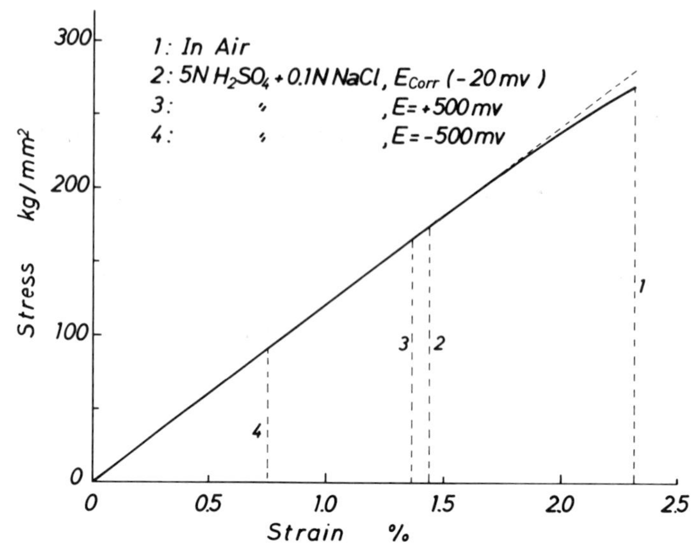

Fig. 1 Stress-strain curve of amorphous Fe-23Ni-7.5Cr-13P-7C alloy in air and in 5 N H_2SO_4 + 0.1 N NaCl at various potentials and a strain rate of 4.2×10^{-6}/s. Source: Ref 4

rapid quenching from the liquid state. Instead, they prepared ribbon-shaped amorphous Fe-Ni-Cr-P-C alloys* with shiny surfaces by twin-roll quenching of melts. In general, the surface composition of a melt-spun ribbon is not the same as its bulk composition, although the compositions of both surfaces of ribbons prepared by twin-roll quenching are identical. In order to obtain chemically homogeneous specimens of uniform thickness and width, both surfaces and both sides of the specimen were polished with SiC paper. Constant-extension-rate testing (CERT) was performed in 5 *N* H_2SO_4-NaCl and HCl solutions at various strain rates and applied potentials after the specimens had been left for 2 h in the solutions without stress.

As shown in Fig. 1, fracture stresses at anodic, cathodic, and corrosion potentials are lower than that obtained in air. The lowering of the fracture stress during cathodic polarization is particularly prominent. Failure in air occurred on a plane at an angle of about 45° to the tensile axis because of a type of plane stress-shear fracture (Ref 5). The fracture surface was composed of a smooth region produced by large local plastic shearing and a veinlike region produced by plastic instability (Ref 6). In contrast, the failure under cathodic polarization grew normal to the applied tensile stress, and the fracture surface showed a brittle mode characterized by granular structure (Ref 7). As Fig. 2 illustrates, in the potential region less noble than the equilibrium potential of hydrogen evolution, cracking susceptibility increases with potential decrease in all solutions. This indicates that the amorphous alloys suffer hydrogen embrittlement.

Hydrogen embrittlement of amorphous iron-base alloy ribbons was also studied via bend and tensile testing of hydrogen-charged specimens (Ref 8, 9). It was found (Ref 8) that the diffusion constants of hydrogen in these alloys were much smaller than those in iron or steel and were rather close to those in covalent crystals.

As shown in Fig. 2, the amorphous Fe-Ni-Cr-P-C alloy at open-circuit or higher potentials in 2 *N* and 5 *N* H_2SO_4 was insensitive to corrosion: the fracture stresses were 90 to 100% of that in air, and the fracture surfaces were the same as that in air. In contrast, fracture stress in chloride-containing sulfuric acid tends to decrease with increase in potential from the corrosion potential. Fracture surfaces exhibited a mixture of granular

*Amorphous alloy composition is generally given in atomic percent.

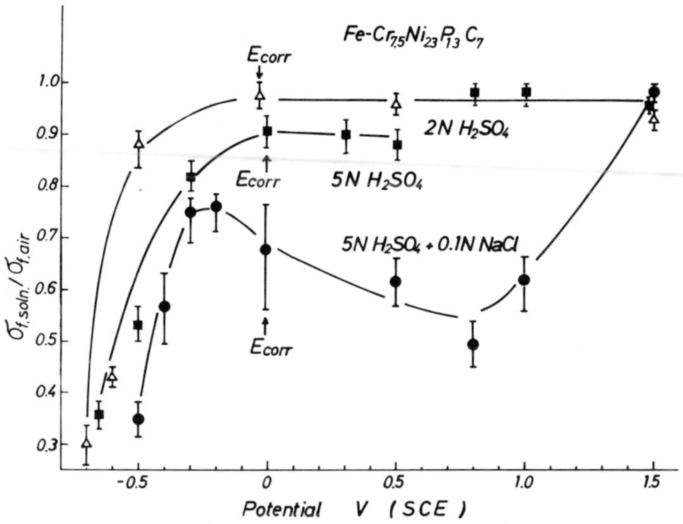

Fig. 2 Variation of SCC susceptibility of amorphous Fe-23Ni-7.5Cr-13P-7C alloy as a function of applied potential in 2 N and 5 N H_2SO_4 and 5 N H_2SO_4 + 0.1 N NaCl at a strain rate of 5.6×10^{-6}/s. Source: Ref 4

Table 1 Amount of Hydrogen Absorbed in Amorphous Fe-20Ni-7Cr-14P-6C Alloy (a)

Environment	Hydrogen absorption, ppm: Unstressed specimens, E_{corr}	Unstressed specimens, −0.5 V	Stressed specimens, E_{corr}	Stressed specimens, +0.5 V
Air (18.0 ppm)				
5 *N* H_2SO_4	...	...	37.2	37.8
5 *N* H_2SO_4 + 0.5 *N* NaCl	25.4	138.3	43.3	52.4
5 *N* H_2SO_4 + 3 *N* NaCl	28.4	...	...	...

(a) Unstressed specimens were exposed for 2 h under open-circuit conditions and for 1 h at − 0.5 V (SCE). Traction testing of specimens was performed up to a stress of about 1.08 GPa (156 ksi) for approximately 1.5 h in solutions.
Source: Ref 4

and veinlike modes. The crack-nucleation sites, from which cracks propagated into the specimen, had granular patterns, as did those in cathodically polarized specimens. Failure at corrosion and anodic potentials was also regarded as hydrogen embrittlement.

As Fig. 3 illustrates, the fracture stress in air is not affected by previous exposure of the specimen to the solution. When traction testing was performed first in the solution and then in air, fracture stress decreased with increasing time of traction testing in the solution. Decrease in fracture stress thus occurred only when traction testing was carried out in the corrosive solution. Nevertheless, when the specimen, which was previously tensile tested in the solution before failure, was left in warm oil or in air, the fracture stress recovered to that in air. Results similar to those shown in Fig. 3 were obtained when traction testing was performed at 500 mV (SCE).

Because cracks formed by active-path corrosion cannot be recovered in warm oil or in air, cracking during traction testing in the corrosive solution in the passive region can be attributed to hydrogen embrittlement. Measurements were taken of the hydrogen absorbed in the specimens (Table 1). The amounts of hydrogen absorbed during tensile testing in the corrosive solution, where cracking occurred, are high enough to induce local embrittlement in comparison with those in crack-free specimens traction tested in 5 *N* H_2SO_4 without chloride.

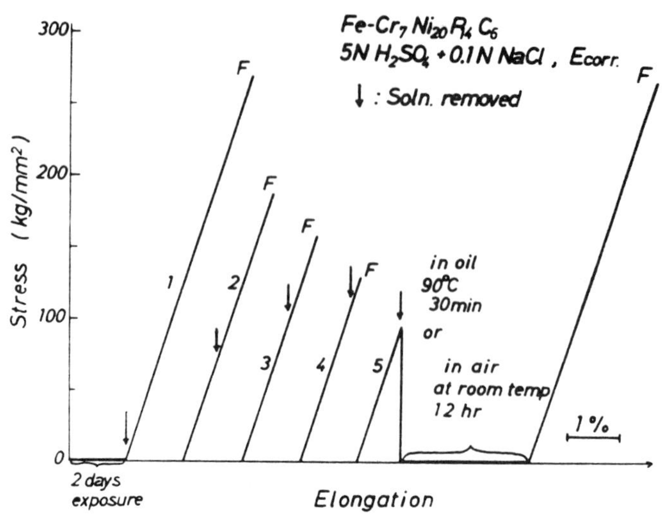

Fig. 3 Tensile stress-elongation of amorphous Fe-20Ni-7Cr-14P-6C alloy in 5 N H_2SO_4 + 0.1 N NaCl under open-circuit conditions at a strain rate of 5.6×10^{-6}/s. F denotes fracture of specimen. Arrows indicate the time when the solution was removed from the cell and the specimen rinsed with distilled water. Subsequently, traction testing was continued in air. In experiments No. 1, the unstressed specimen, installed in a CERT machine, was exposed for 2 days in the solution. After the specimen was rinsed with distilled water, traction testing in air began. In experiment No. 5, traction testing in the solution was stopped at the time indicated by the arrow, and the crosshead of the CERT machine was returned to its original position. After specimens were rinsed and heated in oil at 90 °C (195 °F) for 30 min or stored in air at room temperature (RT) for 12 h, traction testing was resumed in air. Source: Ref 4

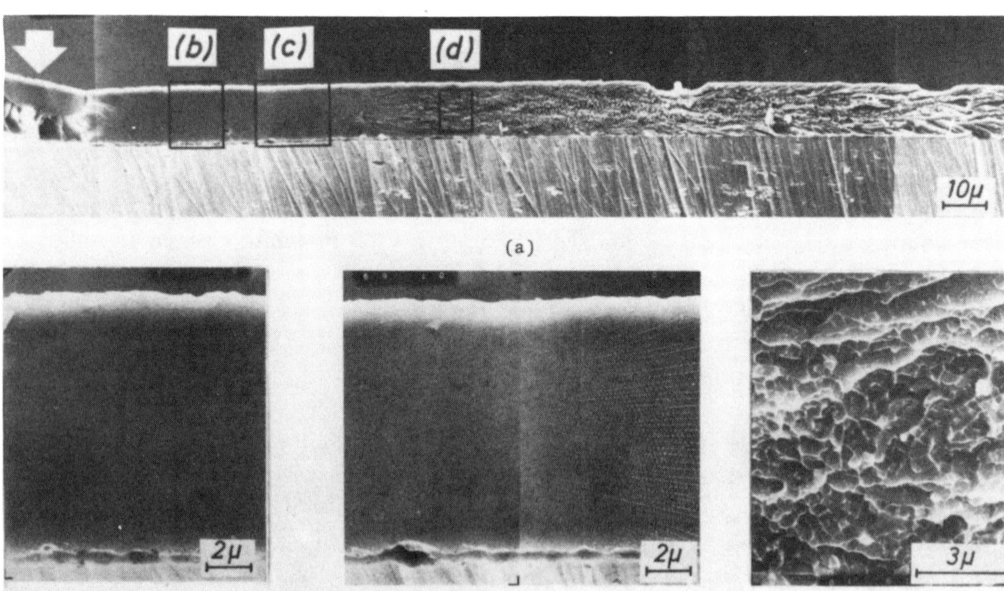

Fig. 4 Fracture surface of amorphous Fe-10Cr-12Mo-18C alloy that failed by tensile testing at a strain rate of 5.9×10^{-6}/s during galvanostatic cathodic polarization at 1 mA/cm^2 in 1 N H_2SO_4 solution. (b) to (d) High magnification of the areas boxed in (a). Source: Ref 10

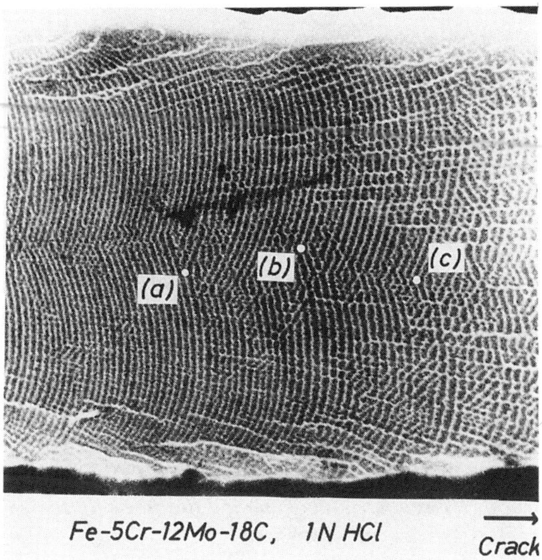

Fig. 5 Fracture surface of amorphous Fe-15Cr-12Mo-18C alloy that failed by tensile testing in 1 N HCl at a strain rate of 5.9×10^{-6}/s. Source: Ref 2

Hydrogen embrittlement under anodic polarization has been interpreted as follows. If a tensile stress is applied in relatively strong acidic solutions with chloride ions, a sharp notch on the specimen will be produced by stress-assisted corrosion or by concentration of stress on preexisting surface defects. Because film-free sites on an amorphous alloy are highly reactive, the solution condition will be suitable for reduction of hydrogen ions, resulting in penetration of hydrogen into the alloy and instantaneous catastrophic failure. The amount of hydrogen absorbed by cathodic polarization is also remarkably high. The disordered structure of amorphous alloys, including the free volume (that is, the space unoccupied by atoms), is responsible for large hydrogen absorption.

Crack propagation based on hydrogen embrittlement has also been studied. As shown in Fig. 4, detailed observation of fracture surface reveals an interesting feature. Near the origin of fracture, a smooth and almost featureless region can be seen (Fig. 4b). As the crack propagates, a shell-like pattern appears (Fig. 4c), and then ridges and tear marks are formed (Fig. 4d), after which the fracture surface becomes more roughened. This is very similar to the fracture surface of inorganic glasses. The shell-like pattern has also been observed on the fracture surfaces of bent specimens immersed in HCl. This pattern appears to be Wallner lines, which are characteristic of the fracture surfaces of completely brittle solids (Ref 11). Because the crack cannot propagate at sonic velocity, stress waves reflecting from the specimen surface perturb the propagating crack to generate the characteristic Wallner lines.

Wallner lines depend on the relative velocity of the crack front and the elastic stress wave in the material, and the crack velocity can be derived from them. Figure 5 shows the fracture surface of an iron-base amorphous alloy specimen that failed under open-circuit conditions (Ref 2). The crack velocity calculated from Fig. 5 was in the range of 0.4 to 0.5 C (1120 to 1400 m/s), where C is the velocity of transverse elastic wave (Ref 2, 10). Hydrogen formed by corrosion of the tensile-stressed alloy is absorbed and accumulates locally in a distorted region. This accumulation leads to local embrittlement of the alloy and to very fast failure upon tensile loading, whereupon surface microcracks or microcrevices caused by localized corrosion may act as notches. Time to failure, therefore, is governed primarily by the time of local enrichment of hydrogen in the alloy as a result of corrosion. Similar studies have been carried out for amorphous Ni-36Fe-14Cr-12P-6B (Ref 12) and Ni-20Co-12Cr-6Mo-6Fe-16B (Ref 13) alloys.

Cracking based on hydrogen embrittlement under open-circuit conditions results from hydrogen formation in a corrosion process. Consequently, hydrogen embrittlement of amorphous alloys under open-circuit conditions can be avoided by improving corrosion resistance. As shown in Fig. 6 (Ref 2), increasing the chromium content of amorphous Fe-Cr-Mo-metalloid alloys enhances corrosion resistance and prevents hydrogen embrittlement.

Alloys Without Strongly Passivating Elements

The effect of deformation of amorphous alloys on corrosion behavior was first examined by Pampillo (Ref 14). He found that localized shear bands formed by compressive deformation of 2 mm (0.08 in.) diam rods of amorphous Pd-

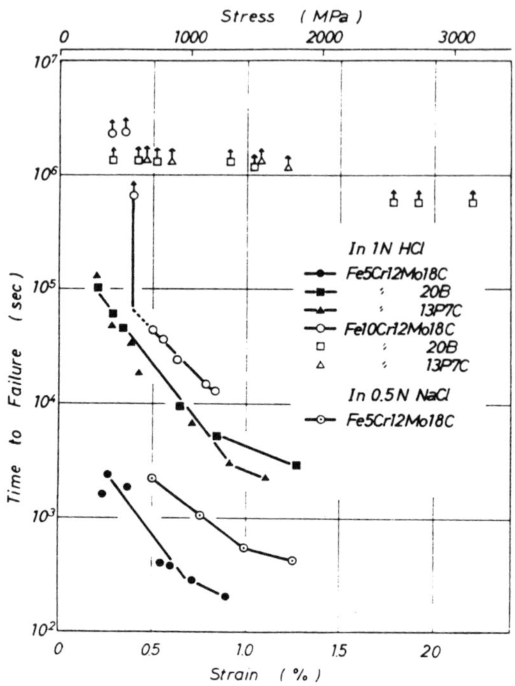

Fig. 6 Variation of time to failure with applied strain or stress for amorphous Fe-Cr-Mo-metalloid alloys measured by bend testing in 1 N HCl and 0.5 N NaCl at 30 °C (86 °F). Source: Ref 2

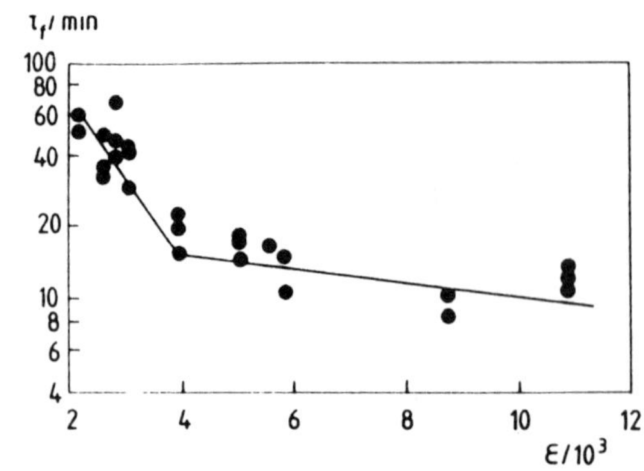

Fig. 7 Time to failure of amorphous Fe-40Ni-20B alloy measured by bend testing in 0.39 M $FeCl_3$ at 30 °C (86 °F) as a function of applied strain (ε). Reprinted from Ref 16 with permission

16.5Si-6Cu alloy were susceptible to preferential etching. After a specimen was deformed about 10%, a flat face was cut and polished mechanically, then electropolished with a mixture of acetic acid and perchloric acid, and finally electroetched with a solution of 25% nitric acid in methanol at below 0 °C (32 °F). Shear bands were preferentially etched by this procedure. Raising the temperature of the electroetching solution to about 10 °C (50 °F) produced perfectly hemispherical pits more or less randomly but also along the path of the slip band. The etching effect disappeared when the specimen was annealed for 1 h at 325 °C (615 °F), which was 50 °C (90 °F) below the glass transition temperature. Pampillo interpreted the preferential etching of the shear bands in terms of a compositional disordering, such as the breaking of a short-range ordered structure.

The first stress-corrosion experiment on the amorphous alloy was also performed by Pampillo (Ref 15). Samples of ribbon-shaped amorphous Ni-29Fe-14P-6B-2Al alloys loaded in tension were immersed in a 3.5 *N* NaCl solution at 26 °C (79 °F). A sample failed after 117 h under an applied stress, 75% of the fracture stress in air. The cross section showed numerous stress-corrosion cracks growing normal to the applied tensile stress on one surface of the ribbon and no cracks on the other surface. The deepest crack, through about 25% of the specimen thickness, led to fast final mechanical failure. The fast failure occurred by the pseudocleavage fracture mode on a plane at an angle of about 60° to the tensile axis. This departure from the 45° angle was assumed to be caused by the notch effect of the stress-corrosion crack. Pampillo observed that nucleation of the stress-corrosion cracks seemed to occur at an embrittled surface layer that showed a fine crack network similar to "dry mud cracks" and that the stress-corrosion cracks grew on the portion of these fine cracks that ran normal to the tensile axis.

There are a number of interesting facts about this study: (1) cracks were formed on only one surface of the ribbon specimen; (2) formation of the embrittled surface layer was found in the cracked surface; (3) a fine crack network was found in the embrittled surface layer; and (4) cracks called "stress-corrosion cracks" seemed to be formed by corrosion of the fine cracks. Despite the fact that the term "stress-corrosion cracks"

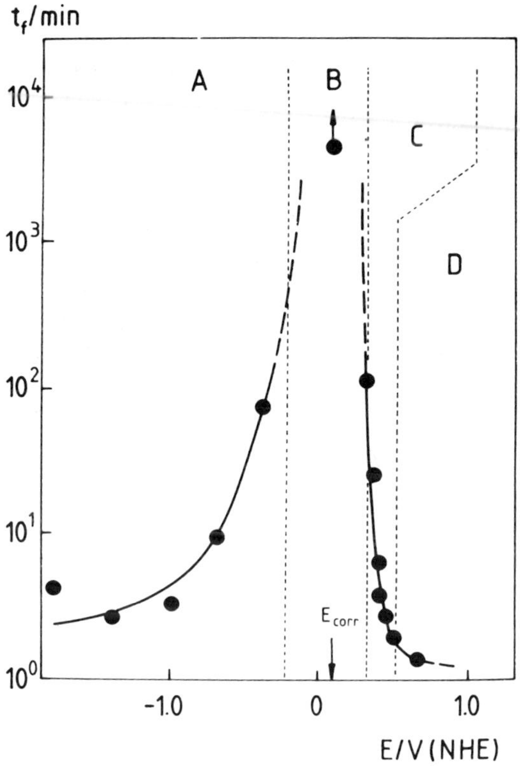

Fig. 8 Time to failure of amorphous Fe-40Ni-20B alloy measured by bend testing in 1 M HCl at 30 °C (86 °F) as a function of potential. Initial strain: ε = 0.0056. Reprinted from Ref 16 with permission

was used, these facts suggest no synergistic effect of stress and corrosion, and final fracture occurred by the mechanical failure of the unembrittled layer after cracking of the embrittled layer formed by corrosion.

Archer and McKim (Ref 16) investigated the stress-corrosion behavior of amorphous Fe-40Ni-20B alloy in ferric chloride, polythionic acid, and hydrochloric acid by bend testing. In 0.39 *M* $FeCl_3$ at 30 °C (86 °F), the specimen surface became duller within 1 min, and after about 10 min, a number of randomly positioned pits were formed. These pits, with a diameter of about 5 µm, developed into cracks that propagated slowly in a direction normal to the applied strain. As shown in Fig. 7, above a certain threshold strain, time to failure in the ferric chloride solution is nearly independent of applied strain. There was a marked depletion of nickel in the crack wall because of preferential dissolution of nickel from the freshly formed crack wall.

In polythionic acid, the alloy was corroded, and stressed specimens eventually failed by cracking, although the cracks were fewer in number and much longer than those formed in ferric chloride. At the open-circuit potential in 1 *M* HCl, no SCC occurred. However, both cathodic and anodic polarization in the hydrochloric acid led to fracture, as shown in Fig. 8 (Ref 16). Hydrogen embrittlement was reported as responsible for the failure in zone A and stress corrosion for the failure in zone C.

Concluding Remarks

Failure seemingly induced by hydrogen embrittlement has so far been found only for amorphous iron-base alloys and not for alloys such as palladium-base and nickel-base alloys. Amorphous alloys have not yet been used as corrosion-resistant metals. Their resistance against hydrogen embrittlement is easily improved by adding highly passivating elements. However, amorphous alloys are currently used for various applications, although their corrosion resistance is not always taken into account. Because amorphous iron-base alloys without high corrosion resistance suffer hydrogen embrittlement, the SCC susceptibility of any amorphous alloy should be examined before it is put into service.

References

1. M. Naka, K. Hashimoto, and T. Masumoto, Corrosion Resistance of Amorphous Iron Alloys Containing Chromium, *J. Jpn. Inst. Met.*, Vol 38, 1974, p 835-841
2. A. Kawashima, K. Hashimoto, and T. Masumoto, Hydrogen Embrittlement of Amorphous Fe-Cr-Mo Alloys, *Corrosion*, Vol 36, 1980, p 577-583
3. S. Takayama and R. Maddin, Tensile Deformation of Amorphous Nickel-Palladium-Phosphorus Alloys, *Metall. Trans.*, Vol 7A, 1976, p 1065-1072
4. A. Kawashima, K. Hashimoto, and T. Masumoto, Stress Corrosion Cracking of Amorphous Iron Base Alloys, *Corros. Sci.*, Vol 16, 1976, p 935-943
5. T. Masumoto and R. Maddin, Structural Stability and Mechanical Properties of Amorphous Metals, *Mater. Sci. Eng.*, Vol 19, 1975, p 1-24
6. H.J. Leamy, H.S. Chen, and T.T. Wang, Plastic Flow and Fracture of Metallic Glass, *Metall. Trans.*, Vol 3, 1972, p 699-703

7. C.A. Pampillo and D.E. Polk, The Strength and Fracture Characteristics of Fe, Ni-Fe and Ni-Base Glasses at Various Temperatures, *Acta Metall.*, Vol 22, 1974, p 741-749

8. M. Nagumo and T. Takahashi, Hydrogen Embrittlement of Some Fe-Base Amorphous Alloys, *Mater. Sci. Eng.*, Vol 23, 1976, p 257-259

9. R.K. Viswanadham, J.A.S. Green, and W.G. Montague, Hydrogen Embrittlement of an Fe-Based Amorphous Metal, *Scr. Metall.*, Vol 10, 1976, p 229-230

10. A. Kawashima, K. Hashimoto, and T. Masumoto, Fractographic Study of Amorphous Iron-Base Alloys Embrittled by Hydrogen and Heat-Treatment, *Scr. Metall.*, Vol 14, 1980, p 41-45

11. D.G. Holloway, *The Physical Properties of Glass*, Wykeham Publisher Ltd., 1973

12. H.S. Tong and J.E. Macur, Hydrogen Embrittlement Behavior of the Amorphous Alloy $Fe_{32}Ni_{36}Cr_{14}P_{12}B_6$, *Corrosion*, Vol 38, 1982, p 464-467

13. T.K.A. Namboodhiri, Corrosion of an Amorphous $Ni_{40}Co_{20}Cr_{12}Mo_6Fe_6B_{16}$ Alloy, in *Chemistry and Physics of Rapidly Solidified Materials*, B.J. Berkowizt and R.O. Scattergood, Ed., TMS-AIME, 1983, p 197-209

14. C.A. Pampillo, Localized Shear Deformation in a Glassy Metal, *Scr. Metall.*, Vol 6, 1972, p 915-917

15. C.A. Pampillo, Review — Flow and Fracture in Amorphous Alloys, *J. Mater. Sci.*, Vol 10, 1975, p 1194-1227

16. M.D. Archer and R.J. McKim, The Stress Corrosion Behavior of Glassy Fe-40Ni-20B Alloy in Aqueous Acidic Media, *Corrosion*, Vol 39, 1983, p 91-98

CHAPTER 14

Stress-Corrosion Cracking of Glasses and Ceramics

Stephen W. Freiman, Ceramics Division,
National Institute of Standards and Technology

Stress-corrosion cracking (SCC) of glasses and ceramics is in many ways similar to, and in others quite different from, ostensibly the same phenomenon in metals. The similarity lies in the shape of crack-velocity/stress-intensity curves for the two classes of materials. However, the mechanisms underlying these curves for glasses and ceramics can differ markedly from those for metals. A primary difference lies in the fact that almost all ceramic materials are brittle at ambient temperatures. Unlike most metals, there is no active dislocation motion that can act to reduce stress concentrations at flaw tips, nor are dislocations involved in the stress-corrosion process. Because of the absence of plastic deformation, resistance to flaw growth (that is, critical fracture toughness, K_{Ic}) in glasses and most ceramics is low (of the order of 1 to 4 MPa$\sqrt{m}$, or 0.9 to 3.6 ksi$\sqrt{in.}$), which means that relatively small flaws (20 to 50 μm in radius) can act as sources of failure. These flaws can be processing defects, such as pores or inclusions, or surface cracks produced during machining or handling. Also, because their critical fracture toughness is low, flaw sizes in polycrystalline ceramics are of the order of grain sizes, leading to further complexities.

The focus of this chapter is on the propensity of flaws in almost every glass and ceramic to grow slowly in the presence of certain environments, especially water, at stresses well below those that would produce immediate failure. This stress-corrosion or subcritical crack-growth behavior leads to what is known as "static fatigue," in which parts fail after periods of time under a constant load. Although cycling the load can introduce other phenomena, the processes with which this chapter will be concerned should not be affected.

Environmentally enhanced crack-growth behavior in a number of glass and ceramic materials will be reviewed and the mechanisms by which this process occurs briefly discussed. Techniques for using laboratory data in the design of parts that can safely withstand service stresses and environments will also be discussed. Finally, experimental crack-growth data will be presented for a number of materials.

General Mechanisms

Stress corrosion in glasses and ceramics can be characterized by a plot of the logarithm of crack velocity versus the stress intensity, K_I, at the crack tip, as shown in Fig. 1. As can be seen, the shape of such a curve is quite complex; each straight-line region represents a different rate-controlling mechanism, as summarized below*:

- *Region I*: The regime of primary importance in terms of engineering design. Flaws spend most of their lifetimes in this portion of the curve. The crack-growth rate is governed by the rate of reaction of an environmental molecule with a crack-tip bond (Ref 1, 2)
- *Region IA*: A regime seen primarily in soda-lime-silica glass. It is attributed to an approach to a stress corrosion limit (Ref 3), al-

*In some ways, this curve is similar to that observed for aluminum alloys, for example.

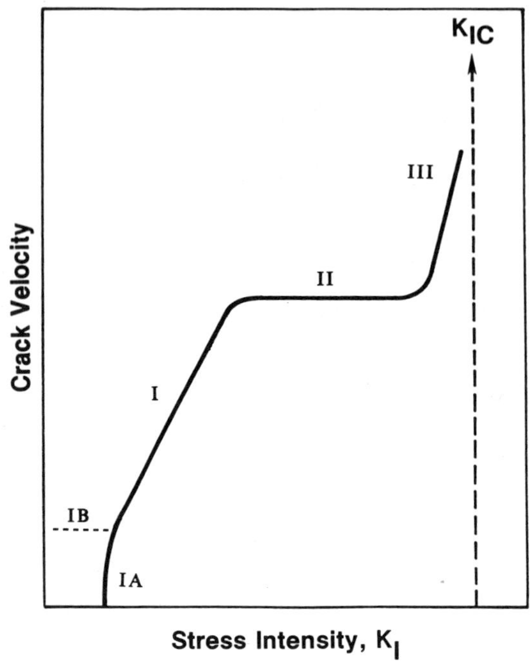

Fig. 1 Schematic of a crack velocity versus K_I plot showing the various regimes of crack-growth behavior

though the exact mechanism by which this limit is reached is unknown

- *Region IB*: A regime observed in binary alkali-silicate glasses, for example, Na_2O-$2SiO_2$ (Ref 4), in which wedging stresses produced by a reaction product at the crack tip add to the driving force
- *Region II*: A regime in which the rate-governing step is the diffusion of the active species to the crack tip (Ref 5)
- *Region III*: A regime in which crack growth is governed by electrostatic interactions (Ref 5). Even at these relatively high velocities, crack growth is sensitive to changes in the dielectric constant of different gaseous or liquid environments (Ref 5, 6).

As noted above, region I of the crack-growth diagram is of primary importance from an engineering point of view as well as of greatest scientific interest. A number of years ago, Wiederhorn (Ref 1) established that rates of crack growth in region I are controlled by the rate of reaction of the environment with the highly strained bonds of the solid at the crack tip. This reaction may or may not involve the formation of corrosion products (thus, the term "stress corrosion" may be a misnomer). Wiederhorn (Ref 1) also used chemical reaction-rate theory to develop a rate equation for crack growth in solids. By assuming that crack velocity is directly proportional to reaction rate, the following expression is obtained (Ref 2):

$$V = V_0 a(H_2O)\exp[(-E* + bK_I)/RT] \qquad \text{(Eq 1)}$$

where $a(H_2O)$ is the activity of water, all of the non-stress-related terms are included in the activation energy ($E*$), and b is proportional to the activation volume for the chemical reaction.

In deriving such an expression, the chemical activity of the water is found to be an important variable in determining the reaction rate. Because the key variable is the chemical activity of the water (that is, its partial pressure relative to saturation in a given environment) rather than its absolute concentration, completely inert environments are almost impossible to achieve. Notice that crack velocities are directly proportional to activity and, therefore, to the relative humidity in a given environment. The fact that crack-growth rates are dependent on relative humidity rather than concentration means that *no liquid can be considered completely inert*, even though the absolute quantity of water dissolved in it is quite small.

Chemical rate theory provides no details of the specific chemical reaction between water and the solid. In fact, fairly detailed models have been developed (Ref 7, 8) to explain why water almost always enhances crack growth in glasses and ceramics as well as why environmental molecules similar in chemistry to water (for example, NH_3 and N_2H_4) also can participate in the crack-tip reaction. The particular crack growth model that is applicable seems to depend on the type of atomic bonding in the solid. In oxides such as silica (SiO_2) and alumina (Al_2O_3), there is a direct reaction between the water molecule and the solid that produces a transfer of the proton on the water molecule to the solid (Ref 7). The ability of the water to donate both electrons and a proton is the key to its effectiveness in promoting fracture of the crack-tip bonds; other molecules such as NH_3 also react in this way. In materials such as magnesium fluoride and gallium arsenide, the dielectric properties of the liquid seem to be more important (Ref 8, 9). As a consequence, the presence of an actual liquid rather than simply an environmental molecule seems to be needed at the crack tip for cracks to grow subcritically in these materials.

As seen in Fig. 2, the pH of aqueous solutions also affects crack growth in silicate glasses (Ref 10). Note that acidic solutions behave like water in their enhancement of crack growth, while slopes of crack-growth curves are significantly reduced in high-pH, basic solutions, making crack growth more extensive. This behavior suggests that in acidic solutions a water molecule is the key participant, while in basic environments the attack of an OH^- ion controls crack-growth rate. Along this same line, work reported by White *et al.* (Ref 11) indicates that blocking OH^- from the crack tip (for instance, by forcing it to complex with a counter-ion such as lithium) can eliminate its deleterious effect on crack growth. As shown in Fig. 3, crack-growth data obtained in 1 *M* LiOH solutions are identical to those obtained in water, while data taken in 1 *M* NaOH, KOH, and CsOH lie at a lower slope and a higher crack velocity.

Experimental Procedures

A number of procedures can be used to measure crack-growth rates in a material (Ref 12). These procedures fall into three general categories: (1) classical fracture-mechanics techniques that employ a macroscopic crack, (2) measurements of strength as a function of stressing rate, known in the ceramics community as "dynamic fatigue,"* and (3) procedures in which microscopic cracks introduced into the specimen surface are measured directly. In general, the fracture-mechanics techniques give the most quantitative and detailed information regarding crack velocity as a function of stress intensity. Drawbacks of fracture-mechanics tests are the

* Dynamic fatigue refers to subcritical crack growth under the influence of nonstatic stresses.

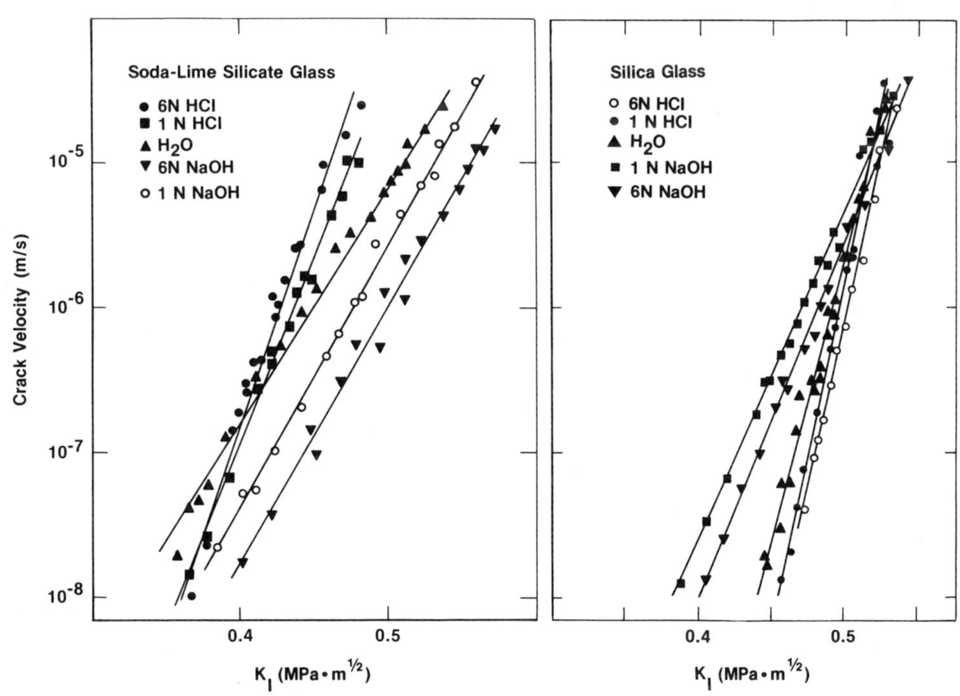

Fig. 2 Crack velocity as a function of pH for soda-lime-silica glass (left) and vitreous silica (right). Note that slopes in high pH are similar to that in water. After Ref 10

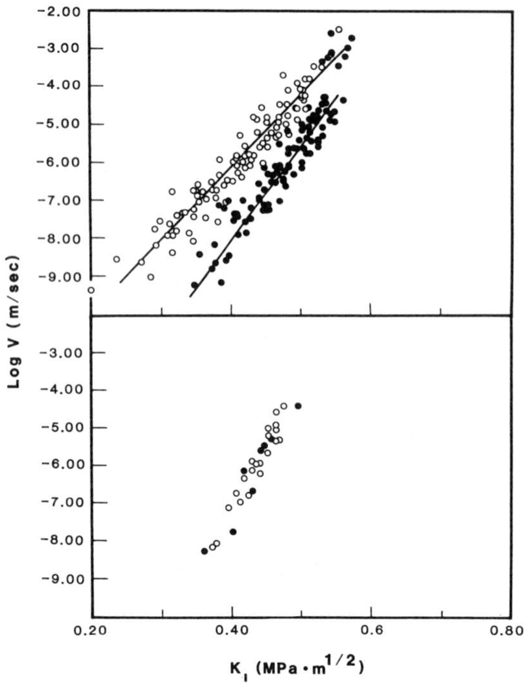

Fig. 3 Crack velocity as a function of K_I for vitreous silica in 1 N KOH (○) and 1 N LiOH (•) (top) and 1 N LiOH (○) and water (•) (bottom). After Ref 11

quantity of material needed to prepare specimens of the right size and the data scatter obtained for polycrystalline ceramics, particularly at low crack velocities.

Strength-measurement techniques, particularly those that employ an artificially induced crack in the specimen surface, are favored for construction of design diagrams for a material, because the severity of the flaw is known quantitatively and can be controlled through the indentation load. As discussed in the next section, flexural specimens are loaded at stressing rates that vary over three orders of magnitude or greater, and their failure strength is determined. Flexural specimens can also be used to measure times to failure as a function of applied load. A significant advantage of the dynamic-fatigue procedure is that the flaws used to obtain the laboratory data are of the same size and type as those present in actual components.

Finally, when only small amounts of material are available (for example, single crystals of a new compound), some idea of its stress-corrosion behavior can be obtained by comparing indentation-induced crack lengths in water to those in a more moisture-free environment, such as dry silicone oil. Under some circumstances, quantitative measures of environmentally enhanced crack growth can be obtained in this way (Ref 13).

Design Diagrams

As noted earlier, the extent to which moisture enhances the rate of crack growth can be expressed in terms of the crack velocity, V, as a function of the stress-intensity factor, K_I, at the crack tip. For engineering purposes, ease of manipulation of equations suggests that crack velocity be plotted as a power function of K_I:

$$V = V_0 (K_I/K_{Ic})^N \qquad \text{(Eq 2)}$$

where V_0 and N are constants that depend on the material and the environment. The smaller the value of N, the greater the susceptibility of the material to subcritical crack growth and the greater the care that must be taken in the design of components. Based on Eq 2, it is easy to derive Eq 3, which shows that the strength of a specimen, σ_f, is related to the rate at which it is loaded, $\dot{\sigma}$:

$$\sigma_f = (\lambda_d \dot{\sigma})^{1/(N+1)} \qquad \text{(Eq 3)}$$

where λ_d is a constant that depends on the critical fracture toughness, initial flaw size, and other material parameters. This so-called "dynamic-fatigue expression" allows determination of the parameters necessary for the prediction of component lifetime. A logarithmic plot of σ_f versus $\dot{\sigma}$ will yield a straight line of slope N. An example of a dynamic-fatigue plot is given in Fig. 4. It is important in such a test to measure strength over as many orders of magnitude of stressing rates as possible (three being a minimum) to obtain accurate determinations of N.

When model flaws are introduced into a specimen surface using a Vickers indenter, residual stresses are associated with the induced crack. These stresses mean that the slope of the dynamic-fatigue curve is N′ rather than N (Ref 14), where:

$$N = (4/3)N' - 2/3 \qquad \text{(Eq 4)}$$

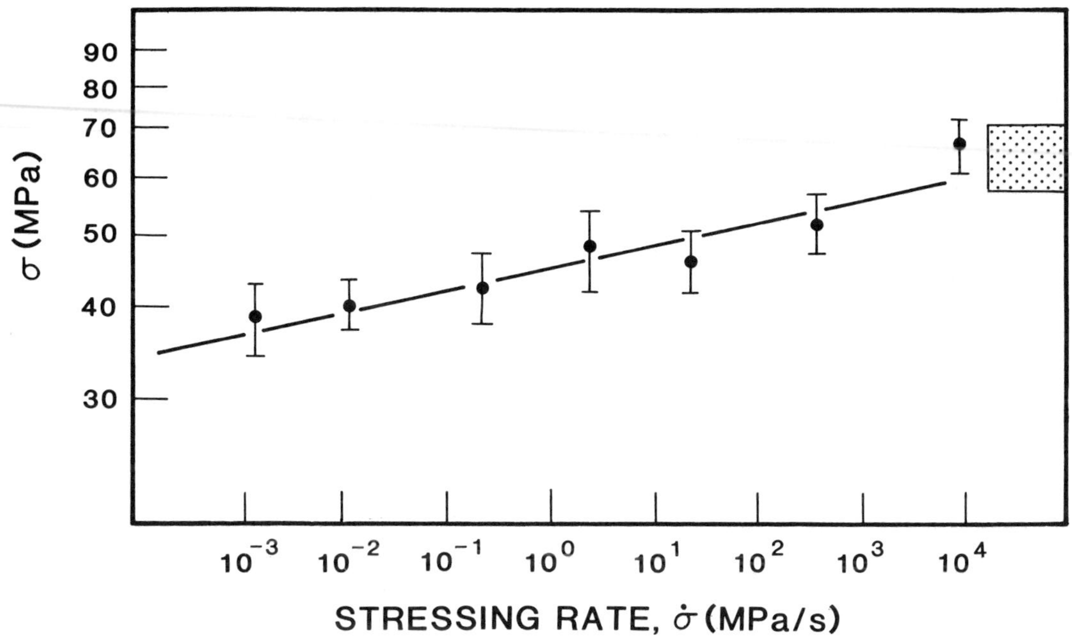

Fig. 4 Dynamic-fatigue plot for a Z5U capacitor material. Flaws were introduced using a 5 N Vickers indenter. An N value of 67 was calculated from the slope of this plot.

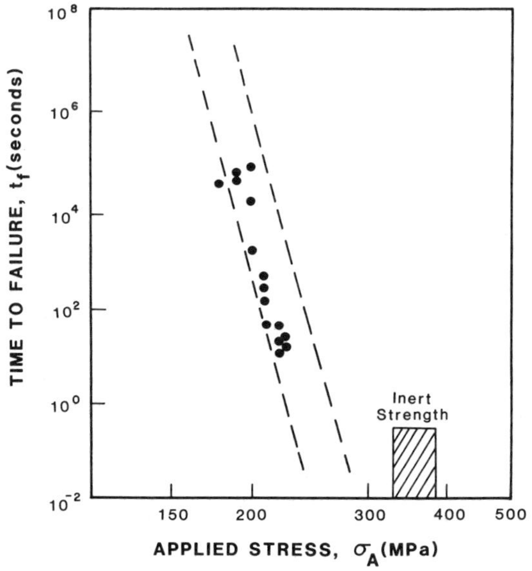

Fig. 5 Plot of predicted time to failure as a function of applied stress for a 96% Al_2O_3. The slope and position of the dashed lines were determined from dynamic-fatigue data. The points represent median times to failure of groups of 10 specimens under static stresses. After Ref 16

N′ and $\lambda_{d'}$ can then be used to predict the lifetime of a component subjected to a service stress, σ:

$$\text{Time to failure } (t) = (\lambda_{d'}/(N' + 1)\,\sigma^{-N'} \qquad \text{(Eq 5)}$$

where $\lambda_{d'}$ is now a measure of the flaw severity in the component. N′ is independent of flaw severity as long as the general flaw shape remains unchanged (Ref 14). It is obviously important that the design diagram be constructed using flaw sizes in the specimen that are similar to those in the component. The component designer can determine changes in the position of the strength-stressing rate curve for the flaw distribution in the actual part by using the following expression (Ref 15):

$$\lambda_{d'2} = \lambda_{d'1}\,(\sigma_{i2}/\sigma_{i1})^{N'-2} \qquad \text{(Eq 6)}$$

where σ_i is the initial strength of the specimens before any environmental degradation has occurred.

Given the above information, a design diagram can then be constructed such as that shown in Fig. 5 for aluminum oxide (Ref 16). The width of the bar representing the initial strength distri-

Table 1 Crack-Growth Exponents, N, for Glasses and Glass-Ceramics
See Ref 19 for a more complete listing.

Material(a)	Environment	Technique (b)	N	Ref
SiO_2	Air	DF	38	20
	Water	DCB	30	7
SiO_2-TiO_2, ULE (CGW 7971)	Air; water	DCB	35	21
	Water	DF	27	21
	Water	DF	11–16	22
Na_2O-CaO-SiO_2	Air; water	DCB	16–21	23
Al_2O_3-B_2O_3-SiO_2 (CGW 7740)	Water	DCB	25	24
	Water	DF	33	24
Na_2O-2SiO_2	Water	DCB	21	4
	Air		93	
Li_2O-2SiO_2	Water	DCB	11	25
Fluoride (e.g., ZBLAN)	Air	DF	75	26
As_2S_3	Air	DCB	17	27
Zerodur	Water	DF	31	28
	Air		59	
Li_2O-Al_2O_3-SiO_2 glass-ceramics	Water	DF	24–28	15

(**a**) ULE, ultralow expansion; CGW, Corning Glass Works. (**b**) DF, dynamic fatigue; DCB, double cantilever beam

Table 2 Crack Growth Exponents, N, for Crystalline Ceramics

Material	Form(a)	Environment	Technique(b)	N	Ref
GaAs	SC	Water	DCB	36	9
	SC	Water	ICG	41	13
Al_2O_3	SC	Water	DCB	67	8
	PC	Water	DF	28–47(c)	32
CaF_2	SC	Water; air	DCB	>50	30
MgF_2	SC	Water	DCB	11	8
	PC	Air	DCB	9	34
Si	SC/PC	Water	DCB/ICG	∞	31
ZnS	PC	Air	DF	76	26
ZnSe	PC	Water; air	DCB	40	33
$BaTiO_3$	PC	Water	DF	70	35
PZT	PC	Water	DCB/DF	60–70	36

(**a**) SC, single crystal; PC, polycrystalline. (**b**) DCB, double cantilever beam; ICG, indentation crack growth; DF, dynamic fatigue. (**c**) Many sources of Al_2O_3

bution was used as the uncertainty in the lifetime at a given stress level. Notice the relatively good agreement between predictions based on dynamic-fatigue data and the results of static-fatigue experiments. To achieve greater accuracy in predictability, the uncertainty in the value of N must be taken into account. This is accomplished by covariance analysis (Ref 17). Also, any change in initial flaw size conditions will lead to a shift in the lifetime curve.

Proof testing (Ref 18), in which parts are overstressed in order to eliminate those with very low strength, can also be used to ensure that no in-service failures will occur. However, there are a number of complexities associated with proof testing. For example, the part must be loaded in the proof test exactly as it will be in service. In addition, the proof test must be conducted in an environment with humidity as low as possible in order to avoid subcritical crack growth, and the *unloading* rate in the proof test must be as rapid as possible to ensure that truncation of the strength distribution curve takes place.

Crack-Growth Data

Water-enhanced crack-growth behavior has been determined for a wide variety of glasses and ceramics, ranging from oxide and heavy-metal fluoride glasses to sapphire and polycrystalline Al_2O_3, ZnS, ZnSe, and MgF_2. All of the data come from crack growth in region I.

Glasses. Because they are transparent, homogeneous, and isotropic on a microscopic scale, silicate glasses were one of the first ceramic materials in which subcritical crack growth was measured. As seen in Table 1, the chemical composition of a glass significantly affects the slope of its crack-growth curve. On the other hand the fracture toughness of glasses or the position of the crack-growth curve is relatively insensitive to composition. Unfortunately, the compositional dependence of N even for relatively minor variations in glass chemistry still cannot be predicted.

Crack growth has also been measured in nonoxide glasses. Materials such as As_2S_3 are relatively susceptible to crack growth (Table 1). Heavy-metal fluoride glasses (ZBLAN) demonstrate a complex crack-growth behavior (Ref 29). Nonetheless, crack-growth exponents can be determined for these glasses by dynamic-fatigue procedures (Ref 26).

Glass-ceramics, such as Zerodur, a trademarked lithium-aluminum-silicate, which are made as glasses and then crystallized in a thermal treatment, also exhibit moisture-induced crack growth (Ref 28). Crack-growth data for this material are also included in Table 1.

Crystalline Ceramics. Materials in single-crystal form, such as GaAs and Al_2O_3 (sapphire), are used as semiconducting or optical components. The major difference in behavior between these materials and glasses is that their bonding properties and, therefore, their elastic properties and resistance to crack growth depend on crystal orientation relative to the crack.

In polycrystalline ceramics, the easiest fracture plane varies from grain to grain (that is, crystal to crystal), and the presence of grain boundaries complicates the rupture process as well. Table 2 gives selected examples of exponents, N, obtained for crystalline ceramics. These values will change markedly with small variations in grain-boundary chemistry. For example, it has been shown (Ref 37) that variations in N determined by dynamic-fatigue tests on 96% Al_2O_3 prepared from different starting powders can be directly correlated with changes in the CaO and MgO concentrations in the glassy phase at the grain boundaries. Cracks grow through grains (transgranular fracture) or through boundaries (intergranular fracture), depending on the chemistry of each, and can change from one to fracture mode to the other with crack velocity. As an example, MgF_2 has been observed to undergo a change in fracture mode from intergranular (in the regime in which the crack is moving slowly due to moisture enhanced bond rupture) to transgranular (after the critical crack velocity has been surpassed) (Ref 38).

References

1. S.M. Wiederhorn, Influence of Water Vapor on Crack Propagation in Soda-Lime Glass, *J. Am. Ceram. Soc.*, Vol 50, 1967, p 407-415
2 S.W. Freiman, Effects of Chemical Environments on Slow Crack Growth in Glasses and Ceramics, *J. Geophysical Res.*, Vol 89, 1984, p 4072-4076
3 T.A. Michalske, The Stress Corrosion Limit: Its Measurement and Implications, in *Fracture Mechanics of Ceramics*, Vol 5, R.C. Bradt, A.G. Evans, D.P.H. Hasselman, and F.F. Lange, Ed., Plenum, 1983, p 277-284
4. C.J. Simmons and S.W. Freiman, Effect of Corrosion Processes on Subcritical Crack Growth in Glass, *J. Am. Ceram. Soc.*, Vol 64 (No. 11), 1981, p 683-686
5. S.W. Freiman, Effect of Alcohols on Crack Propagation in Glass, *J. Am. Ceram. Soc.*, Vol 57 (No. 8), 1974, p 350-353
6. S.M. Wiederhorn, S.W. Freiman, E.R. Fuller, Jr., and C.J. Simmons, Effect of Water and Other Dielectrics on Crack Growth, *J. Mater. Sci.*, Vol 17, 1982, p 3460-3478
7. T.A. Michalske and S.W. Freiman, A Molecular Mechanism for Stress Corrosion in Vitreous Silica, *J. Am. Ceram. Soc.*, Vol 66 (No. 4), 1983, p 284-288
8. T.A. Michalske, B.C. Bunker, and S W. Freiman, Stress Corrosion of Ionic and Mixed Ionic/Covalent Solids, *J. Am. Ceram. Soc.*, Vol 69 (No. 10), 1986, p 721-724
9. G.S. White, S.W. Freiman, and A.M. Wilson, Environmentally Enhanced Fracture of GaAs, *Comm. Am. Ceram. Soc.*, Vol 72, 1989, p 2193-2195
10. S.M. Wiederhorn and H. Johnson, Effect of pH on Crack Propagation in Glass, *J. Am. Ceram. Soc.*, Vol 56 (No. 4), 1973, p 192-197
11. G.S. White, S.W. Freiman, S.M. Wiederhorn, and T.D. Coyle, Effects of Counter Ions on Crack Growth in Vitreous Silica, *J. Am. Ceram. Soc.*, Vol 70, 1987, p 891-895

12. S.W. Freiman, Brittle Fracture Behavior of Ceramics, *Am. Ceram. Soc. Bull.*, Vol 67, 1988, p 392-402
13. G.S. White, S.W. Freiman, and A.M. Wilson, Indentation Determination of Crack Growth Parameters in Gallium Arsenide, *J. Am. Ceram. Soc.*, Vol 74, 1991, p 419-421
14. E.R. Fuller, Jr., B.R. Lawn, and R.F. Cook, Theory of Fatigue for Brittle Flaws Originating from Residual Stress Concentrations, *J. Am. Ceram. Soc.*, Vol 66, 1983, p 314-321
15. R.F. Cook, S.W. Freiman, and T.L. Baker, The Effect of Microstructure on Reliability Predictions for Glass Ceramics, *Mater. Sci. Eng.*, Vol 77, 1986, p 199-212
16. A.C. Gonzalez, H. Multhopp, R.F. Cook, B.R. Lawn, and S.W. Freiman, Fatigue Properties of Ceramics with Natural and Controlled Flaws: A Study on Alumina, in *STP 844*, American Society for Testing and Materials, 1984, p 43-56
17. S.M. Wiederhorn, E.R. Fuller, Jr., A.G. Evans, and J. Mandel, An Error Analysis of Failure Prediction Techniques Derived from Fracture Techniques, *J. Am. Ceram. Soc.*, Vol 59, 1976, p 403
18. A.G. Evans and S.M. Wiederhorn, Proof Testing of Ceramic Materials—An Analytical Basis for Failure Prediction, *Int. J. Fract. Mech.*, Vol 10, 1974, p 379
19. S.W. Freiman, T.L. Baker, and J.B. Wachtman, Jr., A Computerized Fracture Mechanics Data Base for Oxide Glasses, *Bull. Am. Ceram. Soc.*, Vol 64, 1985, p 1452-1455; NBS Technical Note 1212, 1985
20. J.E. Ritter, Jr. and C.L. Sherbourne, *J. Am. Ceram. Soc.*, Vol 54, 1971, p 601-605
21. S.M. Wiederhorn, A.G. Evans, E.R. Fuller, Jr., and H. Johnson, Application of Fracture Mechanics to Space-Shuttle Windows, *J. Am. Ceram. Soc.*, Vol 57, 1974, p 319-323
22. D.C. Cranmer and D.J. Speece, Dynamic Fatigue of Polished and Unpolished Ultralow Expansion Glass, *Adv. Ceram. Mater.*, Vol 2, 1987, p 219-221
23. S.W. Freiman, G.S. White, and E.R. Fuller, Jr., Environmentally Enhanced Crack Growth in Soda-lime Glass, *J. Am. Ceram. Soc.*, Vol 68 (No. 3), 1985, p 108-112
24. J.E. Ritter, Jr. and J. Manthuruthil, Static Fatigue of Silicate Glasses, *J. Glass Technol.*, Vol 14, 1973, p 60-64; S.M. Wiederhorn and C.H. Bolz, Stress Corrosion and Static Fatigue of Glass, *J. Am. Ceram. Soc.*, Vol 53 (No. 10), 1970, p 543-548
25. S.W. Freiman, unpublished data
26. D.C. Cranmer, S.W. Freiman, G.S. White, and A.S. Raynes, Moisture- and Water-Induced Crack Growth in Optical Materials, in *Optical Surfaces Resistant to Severe Environments*, SPIE 1330, Society of Photo-optical Instrumentation Engineers, 1990, p 152-163
27. S.W. Freiman, unpublished data
28. M.J. Viens, "Fracture Toughness and Crack Growth of Zerodur," NASA Tech. Mem. 4185, 1990
29. A.C. Gonzalez and C.A. Pantano, in preparation
30. P.F. Becher and S.W. Freiman, Crack Propagation in Alkaline-Earth Fluorides, *J. Appl. Phys.*, Vol 49, 1978, p 3779-3783
31. G.S. White, S.W. Freiman, E.R. Fuller, Jr., and T.L. Baker, Effects of Crystal Bonding on Brittle Fracture, *J. Mater. Res.*, Vol 3, 1988, p 491-497; G.S. White and S.W. Freiman, unpublished results
32. R.F. Cook, B.R. Lawn, and C.J. Fairbanks, Microstructure-Strength Properties in Ceramics: II, Fatigue Relations, *J. Am. Ceram. Soc.*, Vol 68 (No. 11), 1985, p 616-623
33. S.W. Freiman, J.J. Mecholsky, R.W. Rice, and J.C. Wurst, Influence of Microstructure on Crack Propagation in ZnSe, *J. Am. Ceram. Soc.*, Vol 58 (No. 9-10), 1975, p 406-409
34. S.W. Freiman, Effect of Environment on Fracture of Ceramics, *Ceramurgia*, Vol 2, 1976, p 111-118
35. R.F. Cook, S.W. Freiman, B.R. Lawn, and R.C. Pohanka, Fracture of Ferroelectric Ceramics, *Ferroelectrics*, Vol 50, 1983, p 267-272
36. S.W. Freiman, L. Chuck, J.J. Mecholsky, D.L. Shelleman, and L. Storz, Fracture Mechanisms in Lead Zirconate Titanate Ceramics, in *Fracture Mechanics of Ceramics*, Vol 8, R.C. Bradt, A. G. Evans, D.P.H. Hasselman, and F.F. Lange, Ed., Plenum, 1986, p 175-185
37. J.R. Hellmann, J. Matsko, S.W. Freiman, and T.L. Baker, Microstructure — Mechanical Property Relationships in 94% Alumina Ceramics, in *Tailoring Multiphase and Composite Ceramics*, R.E. Tressler, G.L. Messing, C.G. Pantano, and R.E. Newnham, Ed., Plenum Press, 1986, p 367-380
38. J.J. Mecholsky and S.W. Freiman, Fractographic Analysis of Delayed Failure in Ceramics, in *STP 733*, American Society for Testing and Materials, 1981, p 246-258

CHAPTER 15

Stress-Corrosion Cracking of Weldments in Boiling Water Reactor Service

Joseph C. Danko, Center for Materials Processing, University of Tennessee

For more than a decade, boiling water reactor (BWR) piping systems fabricated from austenitic stainless steel have suffered from stress-corrosion cracking (SCC). This has resulted in loss of plant availability, increased cost of electrical energy, and increased exposure to radiation of skilled personnel involved in inspection and repair operations (Ref 1). This chapter will review the phenomenon of SCC in welded BWR pipe and present methods for its mitigation.

BWR Piping Systems

Most of the BWRs worldwide are of a General Electric design, so this model will be used to describe plant piping systems and materials of construction. Of the six different designs that have evolved over the years, four are BWR models (models 3 to 6). Their piping systems are essentially the same, fabricated of either AISI type 304, 305 or 316 austenitic stainless steel (most plants use type 304).

BWR pipes are produced either by extrusion or by plate rolling and seam welding. Pipes of 304 mm (12 in.) diam or less are generally extruded, while larger pipes are plate rolled and welded. Extruded pipe is usually delivered in the mill-annealed condition, which varies from one pipe fabricator to another. As a result, some pipes are delivered in the "sensitized" condition (for a discussion of sensitization of stainless steels, see the chapter "Stress-Corrosion Cracking of Stainless Steels" in this book). For plate-rolled and seam-welded pipes, solution heat treatment following welding removes the weld sensitization, including any residual sensitization in the base material and weld residual stresses.

Incidents of SCC have been reported in BWR weldments of austenitic stainless steel pipe ranging from 76 to 710 mm (3 to 28 in.) in diameter. Most of the cracking has occurred in the recirculation piping system, but systems involving the core spray, reactor water cleanup, residual heat removal, control-rod return lines, and isolation condenser lines have also been affected. A schematic of a recirculation piping system representative of BWR models 3 to 6 is shown in Fig. 1. Piping systems in a typical BWR may have 150 to 200 welded joints, depending on the model. More than 1000 SCC incidents in BWR pipes have occurred worldwide. However, there has been no reported occurrence of a pipe rupture. Thus SCC is not considered to represent a hazard to the public safety (Ref 2).

Stress-Corrosion Cracking of Pipe Weldments

Cracking in types 304 and 316 stainless steel pipe weldments has been confined largely to the heat-affected zone (HAZ). The cracking morphology is intergranular, following the sensitized grain boundaries in the HAZ. In some of the pipe weldments, intergranular SCC penetrates the weld metal, but there have been no reported weld failures. In general, intergranular SCC propaga-

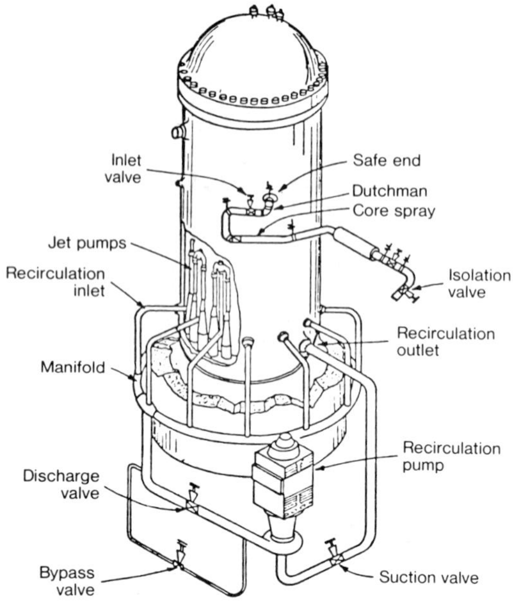

Fig. 1 Schematic of BWR recirculation piping system

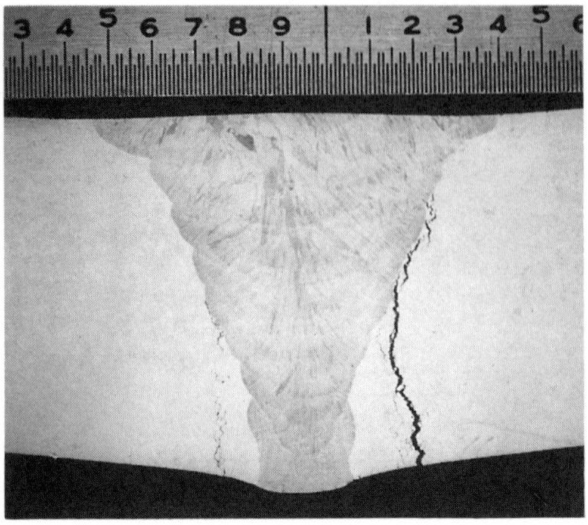

Fig. 2 Photomacrograph of intergranular SCC in a type 304 stainless steel pipe weldment

tion in the weld is very limited because of the weld microstructure and mechanical properties of the weld. The technical basis for this will be discussed in the next section of this chapter. Intergranular SCC is predominantly oriented along the circumference of the girth or butt weld in the HAZ. Axial stresses are responsible for this crack orientation. Although there have been a few observations of axially oriented intergranular SCC, these are rare and result from pipe hoop stresses. Axially oriented cracks are also confined to the HAZ because of the barriers on each side of it — namely, the weld metal and the base material, both of which resist intergranular SCC. Intergranular SCC in a type 304 pipe weldment is shown in Fig. 2. Cracking is in the HAZ and originates on the internal diameter of the pipe.

Intergranular SCC initiates on the inside of the pipe, where the three key conditions necessary for its promotion exist: tensile stresses, sensitization in the HAZ, and an aqueous environment provided by the BWR coolant water. Once intergranular SCC is initiated, the crack will continue to propagate, provided sufficient tensile stresses are present. Initiation may occur in several locations in the HAZ, depending on the local conditions that promote intergranular SCC. Both initiation and propagation are highly variable processes that depend on the level of sensitization, the chemistry of the BWR coolant, and the state of stress.

Model for Intergranular SCC

The phenomenon of intergranular SCC in welded austenitic stainless steel pipes has been the subject of much research and development (Ref 3, 4), and a model for the mechanism of intergranular SCC has been developed (Ref 5). The basis for this model is the coexistence of three major contributors in the weldment: (1) tensile stresses above the yield stress of the base material, (2) a sensitized microstructure, and (3) the BWR water coolant.

Applied tensile stresses acting on the pipe weldment at or above the yield stress of the base material are needed for the initiation and propagation of intergranular SCC. These tensile stresses include stresses generated by plant operation and residual stresses from welding, pipe installation, and pipe fabrication.

Sensitization is defined as a metallurgical condition of chromium depletion in the austenite grain boundaries that results when chromium-rich carbides ($Cr_{23}C_6$) precipitate during the thermal cycles of the welding process. When the chromium content in the austenite grain bounda-

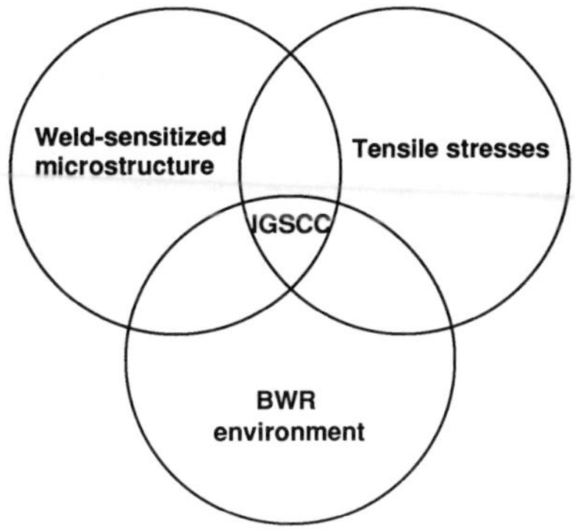

Fig. 3 Three key contributors necessary for intergranular SCC in welded types 304 and 316 stainless steel pipes

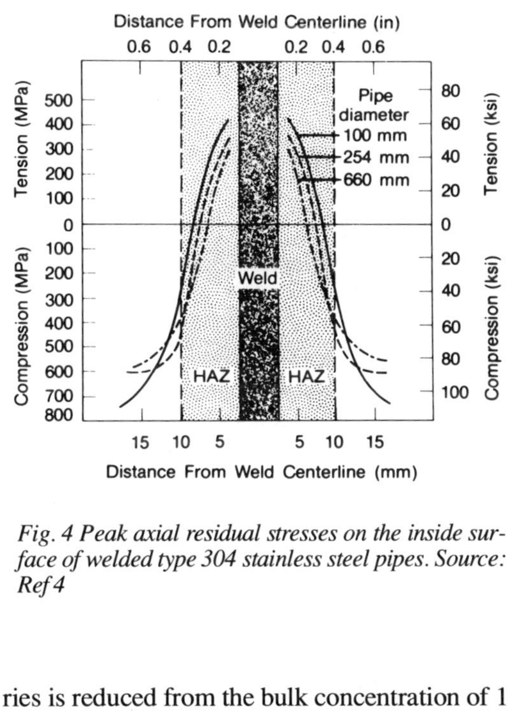

Fig. 4 Peak axial residual stresses on the inside surface of welded type 304 stainless steel pipes. Source: Ref 4

ries is reduced from the bulk concentration of 18 wt% to 12 wt% or less, passivation or resistance to corrosion is lost. This leads to initiation sites for intergranular SCC, and crack propagation follows the grain boundary.

The BWR coolant is high-purity water with a neutral pH. At the BWR operating temperature of 288 °C (550 °F), the water contains small amounts of impurities and 100 to 300 ppb of dissolved oxygen resulting from radiolytic dissociation of the water. In this environment, the electrochemical potential (ECP) of welded types 304 and 316 stainless steel is in the range for intergranular SCC (Ref 6).

A schematic representation of the three key factors contributing to intergranular SCC is shown in Fig. 3. The degree to which any given factor must be present for intergranular SCC to occur depends on the level of intensity and overlap with the other two factors. For example, with a high level of sensitization and high tensile stresses, a relatively unaggressive BWR environment is necessary for intergranular SCC initiation and propagation. On the other hand, elimination of one of the factors will suppress cracking. Thus, the model provides three direct technical approaches for the mitigation of intergranular SCC in BWR weldments. These approaches, commonly referred to as "pipe remedies" or "countermeasures," can be categorized as stress related, sensitization related, or environment related.

Stress-Related Remedies

Piping systems in BWRs are designed to meet code requirements. For earlier plants, ANSI Code for Pressure Piping B31.1 was used; newer plants were designed to Section III of the ASME Boiler Code for Nuclear Power Plant Components. Using these codes, the design stresses of the piping systems were well below the material yield strength. However, the code piping stress analysis did not include allowances for intergranular SCC and residual stresses from pipe fabrication, fit-up, and welding. Of these various stresses, those from welding are a major contributor to the total stresses acting on pipe weldments. This phenomenon will be discussed first in this section, followed by five remedies for it.

Weld Residual Stresses

Weld residual stresses, both surface and through-wall, were measured in a variety of type 304 stainless steel pipe sizes using strain-gage and X-ray methods. Pipe girth welds were prepared using field welding procedures. A plot of peak axial stresses for welded pipes with diameters of 100, 254, and 660 mm (4, 10, and 26 in.) is shown in Fig. 4. Axial stresses are responsible for the initiation of intergranular SCC of circumferential orientation. As shown in Fig. 4, very high

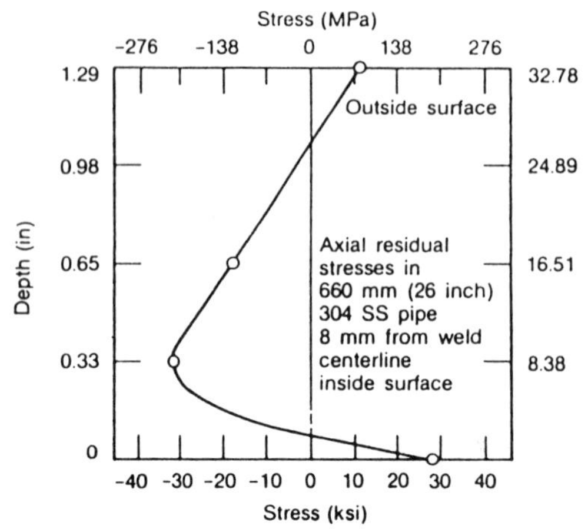

Fig. 5 Through-wall distribution of weld residual stresses in a 660 mm (26 in.) diam type 304 stainless steel pipe. Source: Ref 7

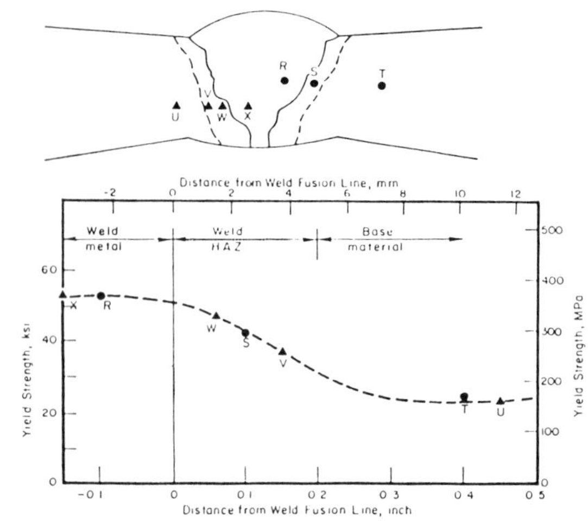

Fig. 6 Yield strength profile at 288 °C (550 °F) across a 660 mm (26 in.) diam type 304 stainless steel pipe weldment. Source: Ref 4

tensile surface residual stresses exist in the sensitized HAZ. Clearly, these residual stresses are a major contributor to the overall tensile stresses acting on the pipe weldments.

Measurements of through-wall residual stresses on the same three pipes showed tensile stresses at the inside surface and partial through-wall for the 100 and 254 mm (4 and 10 in.) diam pipes. However, the 660 mm (26 in.) pipe showed compressive residual stresses just beneath the surface that extended well into the pipe wall (Fig. 5).

These residual stress patterns help explain the intergranular SCC behavior of the BWR pipe welds. When these stresses are added to operating stresses, the yield stress of the pipe material is likely exceeded. The high peak surface axial tensile residual stresses are a major contributor to the initiation of intergranular SCC. The through-wall axial tensile residual stresses contribute to crack propagation. This view is consistent with actual plant incidents. The propensity for intergranular SCC is greater for the 100 mm (4 in.) pipes and decreases with increasing pipe diameter. In large-diameter pipes of 660 mm (26 in.), the lower surface tensile residual stresses increase initiation time while the subsurface compressive stresses decrease the rate of crack propagation. Hence, much longer plant operating times are needed for intergranular SCC in large-diameter pipes.

To complete this discussion of the role of residual stresses in intergranular SCC, three other important factors must be considered: the permanence of the weld residual stresses, the mechanical properties of the weldment, and the effect of the pipe weld preparation method.

Permanence of Weld Residual Stresses. The high residual stresses in austenitic stainless steel weldments were considered to lessen during plant operation at 288 °C (550 °F). For this reason, weld residual stresses were not considered in code piping analysis. However, this theory was dispelled when residual stress measurements were made on a 610 mm (24 in.) type 304 pipe weldment removed from a BWR after 10 years of service (Ref 8). Residual stress patterns were obtained that were quite similar to those measured in as-welded 660 mm (26 in.) type 304 pipe (Fig. 5). Subsequently, other pipes removed from operating plants showed similar permanence of the weld residual stresses. These results support the position that weld residual stresses need to be factored into the pipe system design stress analysis.

Mechanical properties of weldments depend on a number of factors. Only the more important ones will be considered here. The chemistry of the weld filler material and the cooling rate during welding determine the ratio of austenite to ferrite in the weld metal. Ferrite is

Table 1 Maximum Tensile Surface Residual Stresses Caused by Surface Treatments of Type 304 Stainless Steel

Sample	Surface preparation	Maximum tensile surface residual stress(a), MPa (ksi) Parallel to lay	Perpendicular to lay
Pipe, inside surface	Machined	550 (80) T	70 (10) T
Pipe, inside surface	Light grind (hand grinder)	550 (80) T	0 (0)
Flat coupon, inside surface	Heavy grind (hand grinder)	550 (80) T	120 (17) T
Flat coupon, inside surface	Ground (hand grinder)	760 (110) T	275 (40) T

(a) T, tension; C, compression

much stronger than austenite; as the ferrite level increases, so does the strength of the weld. Toughness, on the other hand, decreases with increasing ferrite level. The morphology and fineness of the microstructure also affect mechanical properties. A filler material of 308 composition is normally used for type 304 stainless steel. A minimum of 5% ferrite in the as-deposited condition is specified to avoid hot cracking of the weld (Ref 9). The presence of the ferrite at this level also increases resistance to intergranular SCC.

The welding process affects the final weld microstructure and thus mechanical properties. For shop welds, submerged arc welding (SAW) is generally used, depending on the pipe joining configuration. Field welding may use shielded metal arc welding (SMAW) and gas tungsten arc welding (GTAW). Typically, manual GTAW is used for the root pass, followed by either SMAW or automated GTAW. The best balance of mechanical properties is produced by GTAW followed by SMAW. The SAW method produces an acceptable but less desirable balance of strength and toughness (Ref 10).

The mechanical properties of a weldment will vary from the weld centerline to the base material. Microtensile samples removed from the weldment of a 660 mm (26 in.) type 304 stainless steel pipe were used to measure tensile properties. A plot of the yield strength across the weldment at 288 °C (550 °F) is shown in Fig. 6. The highest yield strength is in the weld metal, with a continuous drop in yield strength occurring as the base material is reached. The reason for this variation is the presence of δ ferrite in the weld and the warm working in the HAZ that results from the shrinkage of the weld metal during solidification. A greater warm-working strain occurs near the weld fusion line and decreases as the base material is approached.

Two very significant pieces of technical information emerge from these results. One is the high yield strength of the weldment at the BWR operating temperature of 288 °C (550 °F). This explains why the residual stresses are above the yield strength of the type 304 base material at room temperature and at the BWR operating temperature. The second relates to the region of intergranular SCC in the pipe weldments. From a yield-strength basis, the weakest region in the weldment is the base material. However, this region is not susceptible to intergranular SCC because it is not sensitized. The presence of δ ferrite in the weld metal provides additional resistance to intergranular SCC. Therefore, the HAZ is the area with a susceptible microstructure and a yield stress less than the weld metal. Initiation sites for intergranular SCC in the HAZ depend on the intensity and overlap of tensile stress and degree of sensitization (assuming the environmental factor is constant across the weldment).

Effect of Weld Preparation Method. In preparing a pipe for welding, the inside surface of the pipe is machined or ground to match the two pipe pieces. Depending on the pipe diameter, postweld grinding may also be used to clean up the weld for

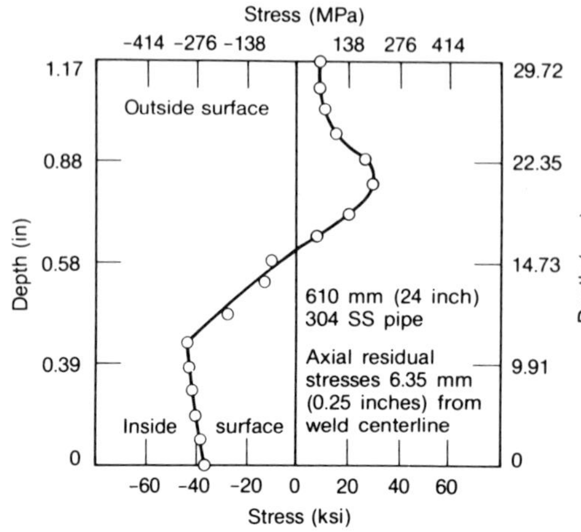

Fig. 7 Through-wall axial residual stresses for LPHSW of a 610 mm (24 in.) diam type 304 stainless steel pipe. Source: Ref 12

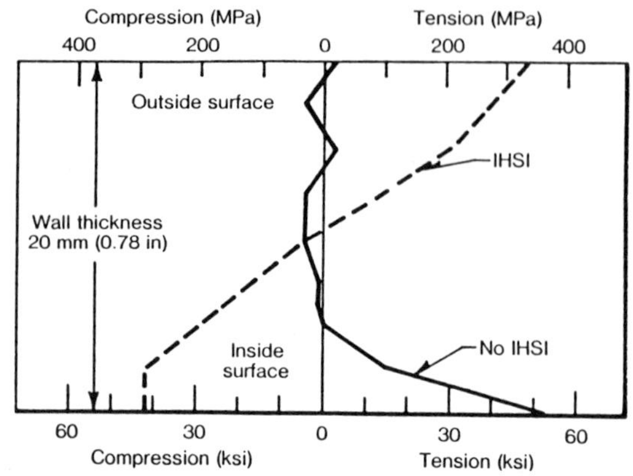

Fig. 8 Through-wall axial residual stresses in IHSI treated and nontreated 410 mm (16 in.) diam type 304 stainless steel pipe weldments. Source: Ref 13

inspection. The method of surface preparation will affect the surface residual stresses. Heavy machining and grinding will result in cold working, high residual stresses, and areas of stress intensification. These cold-worked areas will also result in surface recrystallization during welding. All of these changes enhance the initiation process of intergranular SCC.

Surface residual stresses caused by surface preparation methods are presented in Table 1. These data show significant tensile residual stresses associated with machining and grinding. These stresses, added to the weld residual stresses, contribute to the initiation of intergranular SCC. The high surface residual stresses of weld preparation also contribute significantly to recrystallization during welding. This process exacerbates intergranular SCC initiation.

Heat-Sink Welding (HSW)

In the HSW process, flowing or sprayed water is used on the inside of the pipe after the root pass and several layers have been applied to provide structural support (Ref 11). This process produces compressive residual stresses on the inside surface of the pipe and partially through-wall. This remedy has rarely been used in pipe repair/replacement operations because of the need for cooling water in the pipe, which presents a major problem during plant shutdown.

Last-Pass Heat-Sink Welding (LPHSW)

In the LPHSW process, the cooling water is introduced during the last pass or crown-weld deposit (Ref 12). A high heat input is used during the crown-weld deposit to establish a temperature gradient across the pipe wall thickness. This process results in very favorable residual stresses, as shown in Fig. 7 for a 610 mm (24 in.) type 304 stainless steel pipe. The compressive residual stress extended well into the pipe wall; with this stress distribution, initiation of intergranular SCC is avoided and, in pipes with shallow cracks, propagation is retarded or arrested. LPHSW is a relatively new process and consequently has had limited application. It can be used to treat welds in operating plants as well as for repair/replacement operations.

Induction Heating Stress Improvement (IHSI)

The IHSI process was conceived, developed, and first implemented in Japan (Ref 13, 14). Induction heating is used to raise the outside surface of the pipe weldment to 500 to 550 °C (930 to 1020 °F) while maintaining the inside surface at 100 °C (212 °F) with flowing water. The resultant temperature gradient across the pipe wall produces small plastic strains, thereby altering the as-welded residual stresses. Residual stresses imparted to a 406 mm (16 in.) type 304 pipe weld-

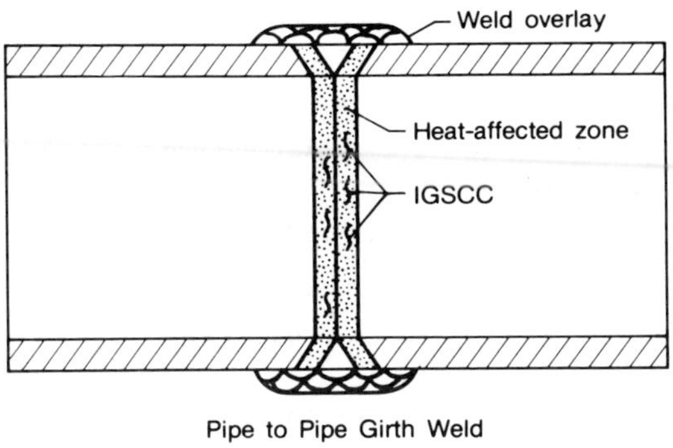

Fig. 9 Schematic of weld overlay for repair of a pipe girth weld

ment are shown in Fig. 8. These results are similar to those produced by LPHSW (Fig. 7). IHSI has been used for repair/replacement of pipelines in many operating BWR plants and in new plant construction.

Weld Overlay

Weld overlay was developed as a short-term remedy for welds in types 304 and 316 stainless steel pipe that had experienced intergranular SCC. Weld overlays were designed to provide structural reinforcement to meet code designs. Analysis of pipes treated with weld overlays showed favorable compressive residual stresses in the pipe weld, an additional benefit of this process. Weld overlay consists of a weld buildup on the outer diameter of the pipe using 308L filler material with a minimum ferrite level of 8% (FN8). At this level of δ ferrite, the propagation of intergranular SCC is usually arrested. Currently, weld overlays are applied by orbital GTAW. Where possible, weld overlay is applied with reactor coolant present in the piping. This is an important factor in minimizing the cost of the operation, because the piping system does not have to be drained. A schematic of weld overlay on a pipe weldment is shown in Fig. 9.

Mechanical Stress Improvement (MSI)

The MSI process uses a simple mechanical tool consisting of a mechanical clamp connected with circumferential studs. These are tightened by a hydraulic tensioner to apply compressive forces, producing deformation of the pipe that extends throughout the weldment. Reduction of the pipe diameter imparts residual compressive stresses at the inner surface in both axial and hoop directions. The residual compressive hoop stresses extend through the entire pipe wall. The MSI process can be accomplished with or without water in the pipe and can be applied in less than 1 h once the device is positioned over the pipe weldment. The process can be accurately controlled and the cycle repeated (Ref 15). The MSI process has had limited application in BWR piping systems because it is the most recently developed stress-related pipe remedy.

Sensitization-Related Remedies

The objectives of remedies related to sensitization are to eliminate weld sensitization or to protect the sensitized HAZ from the BWR coolant environment. Remedies in this category include solution heat treatment, corrosion-resistant cladding, and alternative pipe materials.

Solution Heat Treatment

Solution heat treatment is used for shop welds. It eliminates weld sensitization, residual stresses, and the effects of machining and grinding. At present, approximately 40% of the welds in recirculation piping systems are solution heat treated. In this process, the pipe welds are solution annealed at 1065 °C (1950 °F) for 15 min per 25 mm (1 in.) of thickness to dissolve all the grain-boundary chromium carbides. Solution heat treating also removes weld residual stresses. After the heat treatment, the pipe welds are water quenched to avoid reprecipitation of the chromium carbides. It is important to note that no intergranular SCC incidents have been reported in the welds of plate-rolled and seam-welded pipes that were solution heat treated.

Corrosion-Resistant Cladding

In this process, a corrosion-resistant cladding is applied on the inner surface of the pipe to protect the HAZ from the environment (Ref 16). Two versions of corrosion-resistant cladding have been developed and qualified. A field version consists of cladding the inside surface of the pipe with 308L weld metal with a minimum δ ferrite content of 8%. The pipe weld is then made; the only materials in contact with the BWR coolant in the region of high tensile residual stresses are the weld metal and the cladding. In the shop version, the cladding is applied to the pipe and then solution heat treated. Following heat treatment, the pipe is machined for welding and the pipe weld made. Both versions of the corrosion-resistant cladding process have been applied to the repair/replacement of BWR pipes.

Alternative Pipe Materials

To avoid weld sensitization in types 304 and 316 stainless steel, the carbon content must be reduced to levels less than the solid solubility limit in the austenite. Thus, reducing the carbon levels of conventional types 304 and 316 to a maximum of 0.02 wt% C prevents weld sensitization. Overcoming the reduction in yield stress because of the loss in carbon content was accomplished by adding 0.06 to 0.10 wt% N (maximum). Types 304 and 316 with controlled carbon and nitrogen contents are classified as nuclear grades (NG) (Ref 17). These NG stainless steels have found application in pipe repair/replacement and in new plant construction.

Environmental Remedies

At normal operating conditions, the BWR recirculation coolant, a high-purity neutral-pH water, contains 100 to 300 ppb of dissolved oxygen. The source of this oxygen is the radiolytic decomposition of the water, which also contains 20 ppb of dissolved hydrogen. At a level of 100 to 300 ppb of dissolved oxygen, intergranular SCC will occur in the sensitized HAZ of types 304 and 316 stainless steel (Ref 18). If the oxygen content is reduced to levels below 20 ppb, intergranular SCC is extremely difficult to initiate. This reduction can be accomplished by hydrogen water chemistry, which controls the oxygen level in the entire coolant system through additions of hydrogen to the feed water. Laboratory and in-reactor experiments have revealed that the conductivity of the water (the content of ionic impurities) must be closely controlled to a level of 0.3 μS/cm or less to suppress intergranular SCC in sensitized types 304 and 316 components and pipe welds (Ref 19). The electrochemical potential, a measure of the dissolved oxygen level, must be less than –230 mV (SHE). Hydrogen water chemistry is undergoing field trials, and a number of utilities plan to implement this remedy.

Qualification of Pipe Remedies

In order to qualify the various pipe remedies for field service, full-size welded pipes were tested. This was necessary because when small laboratory test specimens were machined from the pipe welds, residual stresses were relieved. The effect of these stresses, an important contributor to intergranular SCC, thus could not be evaluated (Ref 20). Special facilities were constructed to test pipes ranging from 100 to 610 mm (4 to 24 in.) in diameter. A typical 100 mm (4 in.) pipe test specimen and test stand are shown in Fig. 10. With 11 pieces in the pipe test specimen, 22 HAZs could be tested at one time. This testing capability was needed to meet the requirements of a statistical test plan, the objective of which was to compare the times to failure of "remedied" pipes with those of untreated or reference type 304 stainless steel pipe welds. A pipe remedy showing an improvement in time to failure of a factor of 20 over the reference pipe welds is equivalent to a 40-year plant lifetime. Many of the pipe remedies tested did in fact achieve this factor of improvement.

Application to Other Systems

The results of research on SCC of stainless steel pipe welds in BWRs are applicable to other industries. When austenitic stainless steel welds are exposed to the conditions required for SCC, cracks and failure must be expected. Weld residual stresses are an important factor in the performance of welded joints. These are often overlooked in the design/stress analysis of a system. Failures of austenitic stainless steel welds have occurred in systems operating at ambient temperatures, such as service water lines and wind tunnel piping. In many cases, it is the presence of chlorides in proper concentrations that leads to SCC of the components.

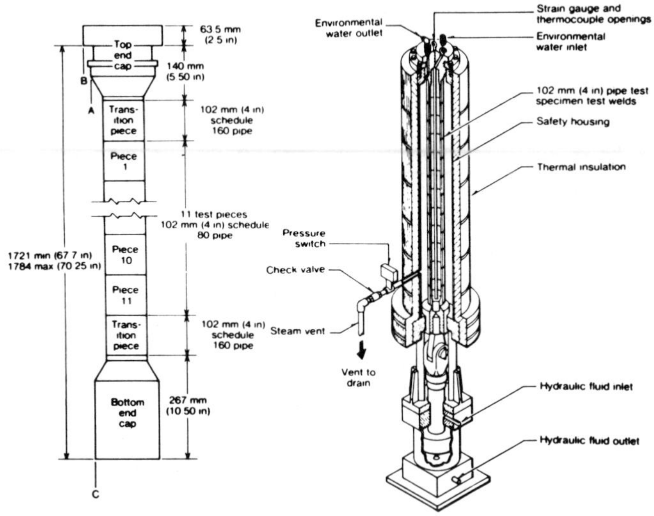

Fig. 10 Welded pipe test specimen and typical test stand

The remedies developed for SCC of BWR pipes are also applicable to other industries. Use of a specific remedy depends on the contributing factor. If it is stress, then redesign, stress-relieving treatments, alternative materials, or introduction of compressive residual stresses should be considered. Selection of a particular remedy depends on economics, ease of application, reliability, and whether a short- or long-term repair is required.

Conclusions

Based on the information presented in this chapter, the following conclusions can be drawn:

- Types 304 and 316 stainless steel pipe weldments in BWR piping systems are susceptible to intergranular SCC.
- Over the past 15 years, intergranular SCC incidents have resulted in significant loss of plant availability and economic losses of more than several billion dollars.
- Intergranular SCC occurs in the weld HAZ; cracking is generally oriented circumferentially.
- Tensile stresses, weld sensitization, and the BWR coolant environment are the major contributors to intergranular SCC.
- Weld residual stresses are not considered in the code pipe system analysis, but contribute significantly to the overall applied stresses on pipe weldments.
- Weld residual stresses may be altered to favorable compressive stress by HSW, LPHSW, IHSI, MSI, and weld overlay pipe remedies.
- Sensitization-related pipe remedies include solution heat treatment, corrosion-resistant cladding, and use of alternative materials such as types 304 and 316 NG stainless steel.
- Hydrogen water chemistry, an environmental pipe remedy, has the potential for protecting austenitic piping systems from intergranular SCC.
- Methods for mitigating intergranular SCC have been developed, qualified, and implemented in many BWR pipe repair and replacement operations.
- Intergranular SCC of weldments in austenitic stainless steel BWR piping systems is currently under control.

References

1. R.E. Smith, Progress in Reducing Stress Corrosion Cracking in BWR Piping, in *American Power Conf. Proc.*, Vol 39, 1977, p 232-240

2. "Investigation and Evaluation of Cracking in Austenitic Stainless Steel Piping of Boiling Water Reactor Plants," Tech. Report NUREG 75/067, U.S. Nuclear Regulatory Commission, Oct 1975

3. H.H. Klepfer *et al.*, Investigation of Causes of the Cracking in Austenitic Stainless Steel Piping, in *NEDO 21000*, Vol 1 and 2, General Electric Co., July 1975

4. "Studies on AISI Type 304 Stainless Steel Piping Weldments for Use in BWR Applications," EPRI

NP-944 Final Report on Project RP-449-2, Electric Power Research Institute, Dec 1979

5. R.E. Hanneman, Model for Intergranular Stress Corrosion Cracking of Welded Type 304 Stainless Steel Piping in BWRs, EPRI WS-74-174, Vol 1, Electric Power Research Institute, May 1980
6. J.C. Danko and M.E. Indig, In-Reactor and Ex-Reactor Electrochemical Methods for Corrosion and Stress Corrosion Cracking Testing, in *Predictive Methods for Assessing Corrosion Damage to BWR Piping and PWR Steam Generator Proceedings*, National Association of Corrosion Engineers, 1982
7. Measurement of Residual Stresses in Type-304 Stainless Steel Piping Butt Weldments, EPRI NP-1413, Electric Power Research Institute, 1980
8. H. Flache and R. Ettemeyer, Piping Repair Case Study at KRB Unit A, *Trans. Am. Nucl. Soc.*, Vol 35, 1980, p 487
9. F.C. Hull, Effect of Delta Ferrite on the Hot Cracking of Stainless Steel, *Weld. J.*, Vol 46 (No. 91), 1967, p 3995-4095
10. "Toughness of Austenitic Stainless Steel Pipe Welds," EPRI NP-4768 on Project 1238-2, Electric Power Research Institute, Oct 1986
11. R. Sasaki *et al.*, "Mitigation of Inside Surface Residual Stresses of Type 304 Stainless Steel Pipe Welds by Inside Water Cooling Method," EPRI WS-79-174, Electric Power Research Institute, May 1980
12. "Last Pass Heat Sink Welding," EPRI NP-3479, Electric Power Research Institute, March 1984
13. D.C. Bertossa, "Induction Heating Stress Improvement Qualification," EPRI WS-79-174, Electric Power Research Institute, May 1980
14. S. Tanka and T. Umemto, "Residual Stress Improvement by Means of Induction Heating," EPRI WS-79-174, Electric Power Research Institute, May 1980
15. J.S. Porowski *et al.*, *Nucl. Eng.*, April 1985, p 56-58
16. N.R. Hughes, "Qualification of Solution Heat Treatment, Corrosion Resistant Clad and Heat Sink Welding," EPRI WS-79-174, Electric Power Research Institute, May 1980
17. J.E. Alexander, "Alternate Piping Alloy Qualification," EPRI WS-79-174, Electric Power Research Institute, May 1980
18. B.M. Gordon and G.M. Gordon, Corrosion in Boiling Water Reactors, in *Metals Handbook*, 9th ed., Vol 13, *Corrosion*, ASM International, 1987, p 929-937
19. R.L. Jones, Prevention of Stress Corrosion Cracking in Boiling Water Reactors, *Mater. Perform.*, Feb 1991, p 70-73
20. J.C. Danko *et al.*, A Pipe Test Method for Evaluating the Stress Corrosion Cracking Behavior of Welded Type 304 Stainless Steel Pipes, in *Properties of Steel Weldments for Elevated Temperature Pressure Containment Applications*, MPC-9, American Society of Mechanical Engineers, 1978

CHAPTER 16

Detection and Sizing of Stress-Corrosion Cracks in Boiling Water Reactor Environments

Soung-Nan Liu and Stan M. Walker, Electric Power Research Institute

Over the past decade, significant advances have been made in nondestructive evaluation (NDE) technology as applied to boiling water reactor (BWR) piping systems. This success is a direct outgrowth of efforts to overcome the difficult examination problem of detection and sizing of intergranular stress-corrosion cracking (SCC) in stainless steel piping components in BWR plants.

Intergranular SCC in BWR Piping Systems

In 1982, intergranular SCC was found in large-diameter stainless steel piping, which prompted a reevaluation of the effectiveness of ultrasonic examination. The reevaluation showed conventional ultrasonic examination methods to be inadequate, but advanced techniques to be successful. This resulted in a Nuclear Regulatory Commission (NRC) requirement for training and qualification of ultrasonic inspectors and systems.

Major advances made in the last decade in intergranular SCC detection and sizing in stainless steel piping systems include the introduction of special transducers for intergranular SCC detection, the use of advanced sizing techniques based on crack-tip diffraction, and the increased level of personnel training and qualification. In addition, a number of commercially available automated examination systems have been developed that have proved successful in intergranular SCC detection and sizing.

Chronology of Events. Intergranular SCC presented a particularly difficult NDE problem because of its treelike branching pattern and its location in the heat-affected zone (HAZ) of a weld. Conventional examination techniques were based on the assumption that cracks reflected ultrasound such that the reflected amplitude was proportional to the crack size. Under this assumption, the smaller reflected response from the jagged, branching intergranular SCC gave a distorted estimate of its true size and was often confused with other geometrical reflectors in the weld region.

Intergranular SCC in stainless steel piping systems was a known problem, but its occurrence was associated only with small-diameter piping prior to 1982. The problem became a serious concern in March 1982 when, during a primary system hydrostatic test by Niagara Mohawk at Nine Mile Point Unit 1, plant personnel detected leakage at two of the ten large furnace-sensitized recirculation-system safe ends. Further examination revealed three pinholes and one 12.5 mm (½ in.) long axial indication, all of which were located in the HAZs of the welds. The leaking areas had been ultrasonically examined nine months prior to the hydrostatic test, and no reportable indications were identified. Because of this, the NRC Office of Inspection and Enforcement (I&E) issued IE Bulletin 82-03 (Ref 1) to:

- Provide a reasonable level of assurance that state-of-the-art examinations were sufficient to detect cracking in BWR thick-wall recirculation piping welds
- Determine the generic significanceof the Nine Mile Point Unit 1 piping degradation

A qualification test of ultrasonic methodology was outlined in the IE Bulletin. Plant licenses were required to demonstrate the effectiveness of their ultrasonic detection methodology on representative pipe samples cracked in service before returning their plants to service. The "generic issue" concern was to be evaluated by the NRC following the licensee written reports, as required by the bulletin, by December 1, 1982.

In March 1983, with utility-submitted data revealing cracking of the main recirculation-system piping in five of the seven plants examined, the NRC issued IE Bulletin 83-02 (Ref 2). This bulletin emphasized the "generic issue" concern and required actions similar to IE Bulletin 82-03.

The NRC reviewed available data and concluded that the intergranular SCC problem was a generic issue for all BWRs. In June 1984, the NRC issued Generic Letter 84-11 (Ref 3). This document required a reinspection program for piping susceptible to intergranular SCC and outlined the minimum requirements of such a program. The Generic Letter also required all levels II and III ultrasonic examiners to demonstrate competence in accordance with IE Bulletin 83-02 and level I examiners to demonstrate field performance capability.

Although these were not the first instances of intergranular SCC in BWR piping, they were the first cases of its appearance in large-diameter piping. In smaller piping, intergranular SCC had been found in BWR bypass and core spray lines for several years prior to 1975. The NRC required the shutdown and examination of these piping systems in all BWRs in 1975 (Ref 4). Intergranular SCC was discovered in these smaller-diameter lines in nine plants.

Subsequently, the NRC required that piping systems shown to be susceptible to intergranular SCC on the basis of field experience receive examination beyond the requirements of the American Society of Mechanical Engineers (ASME) Boiler and Pressure Vessel Code (Ref 5). By 1982, more than 200 incidents of intergranular SCC had been reported in U.S. plants, but none was associated with large-diameter piping until the Nine Mile Point Unit 1 occurrence.

Detection of Intergranular SCC

When intergranular SCC was first reported in BWR stainless steel piping, conventional ultrasonic detection techniques were found to be unreliable. Since then, detection of intergranular SCC has significantly improved as a result of extensive research and personnel training and qualification. Research continues to improve detection reliability by making full use of signal features to identify intergranular SCC and to differentiate it from benign reflectors.

A cross section of a BWR piping weld, removed from service, is shown in Fig. 1. Various geometric properties of the weldment make detection of intergranular SCC difficult. The weld root and the counterbore could produce ultrasonic signals that might be confused with cracking. Additionally, the large weld crown (resulting from a weld repair) could lead to the misconception that the root is located at the center of the weld crown, thereby causing the intergranular SCC to be at the assumed root location and possibly resulting in a

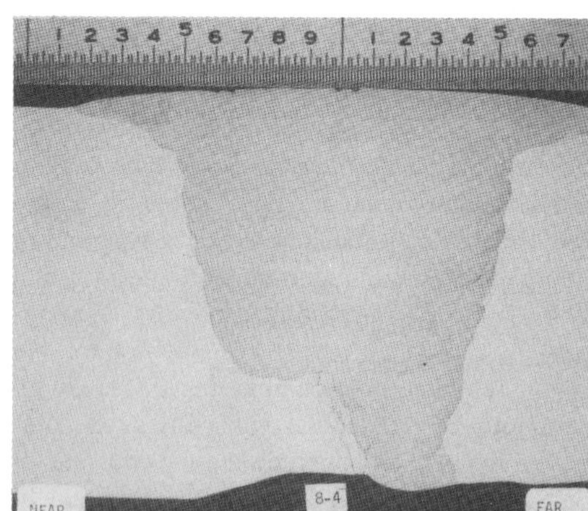

Fig. 1 Cross section of a BWR piping weld removed from service

false dismissal of the crack. The interference of the weld crown with the travel of the search unit could also prevent detection. Discrimination should be based on a combination of signal amplitude, signal shape, reflector location, signal behavior, and the number of indications.

Significant improvements in search units, personnel training, indication evaluation, and scanning techniques have each added to the level of confidence in the detection of intergranular SCC. These improvements are discussed below.

The 1.5 MHz Examination Frequency. Examination procedures underwent intense scrutiny in an effort to improve the reliability of intergranular SCC detection. Evaluation of the frequency characteristics of the reflected wave form was of prime concern. The small size and tightness of the cracks in the BWR piping suggested that they could best be detected using relatively high frequencies. However, investigation proved otherwise. Frequency spectrum analysis of the reflected ultrasonic pulse indicated that the cracks were most reflective to a frequency of 1.5 MHz.

The cracked pipe specimens had been removed from BWR systems and examined using a set of specially developed search units. Southwest Research Institute (SwRI) reported in June 1976 that the 1.5 MHz search units in the set provided 10 times the amplitude obtained with 2.25 MHz units (Ref 6).

Field application of the 1.5 MHz examination frequency was not without problems. Examinations of small-diameter pipes were attempted using 6.4 mm (¼ in.) diam search units, which, because of increased beam spread, increased the recording of geometric reflectors. The beam spread also generated surface waves from the leading edge of the angle-beam wedge, adding to the confusion. Increasing the search unit size to a diameter of 9.5 mm (E3/8 in.) solved most of this problem. Because of the enhanced amplitude obtained with a frequency of 1.5 MHz, numerous reexaminations were required to verify the measurements (Ref 7).

Single- Versus Dual-Element Angle-Beam Search Units. A dual-element angle-beam search unit was developed and found to substantially improve intergranular SCC detection in laboratories as a result of increased sensitivity and improved signal-to-noise ratio. These improvements were reported to be most significant on large-diameter heavy-wall piping and on piping with unusual outside surface geometries (Ref 8).

Subsequent evaluations found that the dual-element search units failed to provide consistently better results in the field. When 1.5 MHz 45° single- and dual-element search units were compared on cracks, the dual-element unit yielded a broader range of amplitudes than the single-element unit. The focusing effect of the dual-element search unit limited the depth of field and prevented extension of the examination beyond a reflection from the inside surface. Also, the use of dual-element search units on piping systems required highly specialized unit design. This specialization, while practical under laboratory conditions, became exceedingly difficult under field conditions. The need for specially designed search units to accommodate the wide range of pipe diameters and wall thicknesses encountered necessitated having a very large number of special search units available for each inspection. Because there was no convincing evidence that the dual-element search unit improved detection of intergranular SCC, its use was discontinued (Ref 7).

The scanning techniques used to detect intergranular SCC in thin-wall piping (less than 25 mm, or 1 in., thick) were not as effective when applied to heavy-wall piping (greater than 25 mm thick). The stresses found in larger-diameter heavy-wall piping result in cracks that are closer to the weld root and that tend to be axially oriented. A scanning method in which the search-unit beam is skewed under the weld crown from the base metal was found to be more reliable for these examinations. In conjunction with the skewed examination, higher-angle beams have been incorporated. Several examination agencies have elected to use 50° to 55° refracted shear waves in order to allow ultrasonic impingement at the weld root on joints with wide weld crowns. The higher beam angle combined with a higher-frequency, broad-banded search unit has resulted in increased resolution and thus greater detectability of intergranular SCC near the weld root (Ref 7, 9).

Analysis. Interpretation of the ultrasonic A-scan presentation for the probable source of intergranular SCC has become increasingly important. The earlier application of beam tracing alone has been found to be unreliable because of the close proximity of cracks to geometric reflectors. Identification of the A-scan presentation due to noted reflector characteristics (such as signal sharpness, number and shape of signals in close proximity, and effect of search-unit skew) has proved useful for distinguishing crack indications from nearby geometric reflectors (Ref 9).

Training. The amount of training required for the successful performance of intergranular SCC examinations varies greatly depending on the examiner's background and ultrasonic examination experience. For thin-wall piping, examiners must be alert to low-amplitude indications, which are sometimes located in areas removed slightly from normally presumed areas of interest (for example, farther away from the weld centerline than in thicker-wall pipes).

For thick-wall piping, there are additional scanning requirements for skewed examinations. Also, the nature of the cracks (in circumferential scanning) is such that their resulting ultrasonic indications are difficult to discern from weld-root geometric indications.

Two modes of training are appropriate. Classroom instruction concerning the background and expected characteristics of intergranular SCC is necessary, as are laboratory sessions in which examiners receive "hands-on" examination experience. The lab sessions should use actual field-removed samples whenever possible (Ref 9).

Crack Height Sizing

As intergranular SCC progresses in BWR piping systems, estimation of the height of progression through the pipe wall became important so that an informed choice could be made between repair and replacement. In May 1983, several utilities expressed concern about the

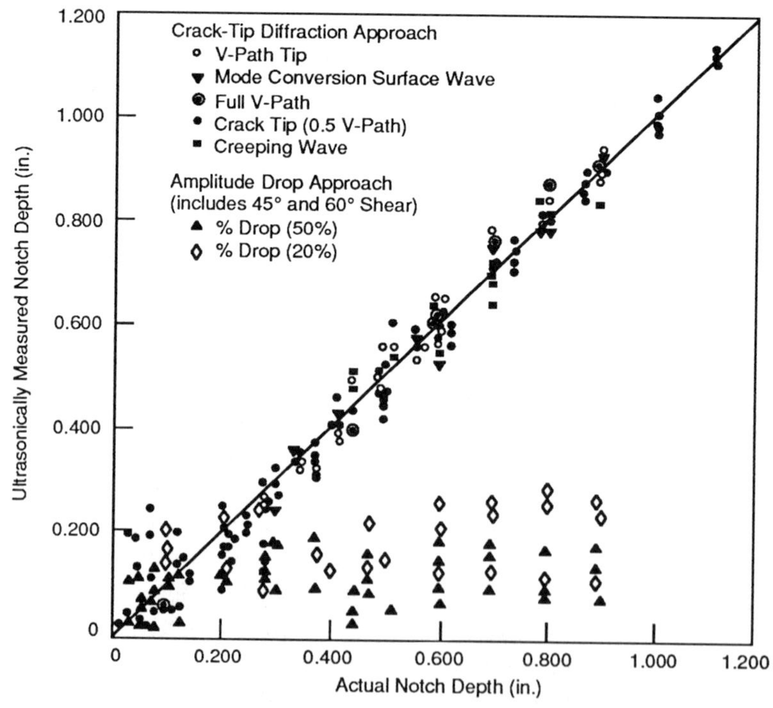

Fig. 2 Comparison of sizing results obtained with crack-tip-diffraction and amplitude-drop sizing techniques. Source: Ref 10

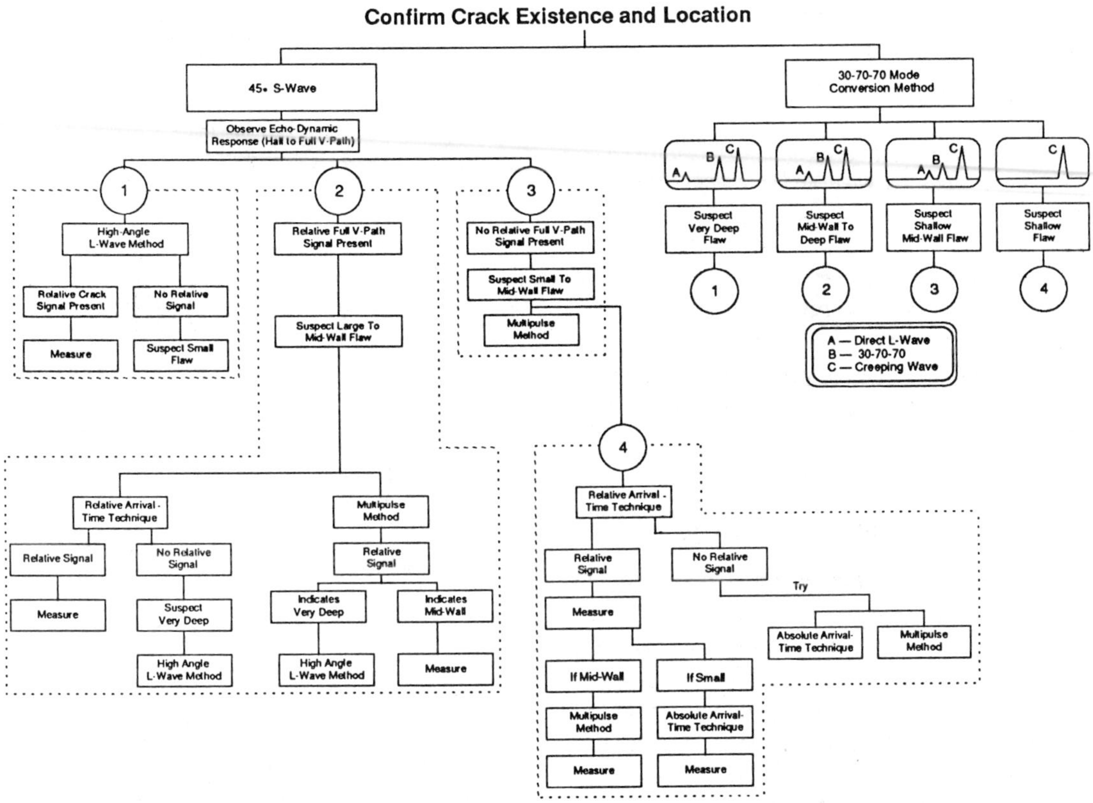

Fig. 3 Flow chart for applying complementary flaw-height sizing techniques for intergranular SCC. Source: Ref 11

adequacy of the ultrasonic techniques being used to size the cracking and requested that the Electric Power Research Institute (EPRI) evaluate their effectiveness.

An intergranular SCC height-sizing round-robin exercise was conducted, and the results clearly demonstrated that commonly used amplitude-based ultrasonic methodology was not effective. Techniques using the crack-tip diffraction method, however, provided good results. Figure 2 shows a plot of ultrasonically measured height versus actual height for different implementations of amplitude-drop and tip-diffraction sizing techniques applied to notches in stainless steel specimens (Ref 10). The crack-tip data clearly show more accuracy and less scatter. The problem definition for ultrasonic sizing methodology was established and a promising avenue for improvement identified (Ref 11).

In late 1983, EPRI and the Boiling Water Reactor Owners Group II (BWROG II) established a "coordination plan" with the NRC to continue the performance demonstration qualification activities. This plan was formally implemented in 1984, with EPRI serving as the administrating agency. The plan not only addresses detection and discrimination of intergranular SCC, but also the technology for sizing the cracks. The plan also contains provisions for periodic requalification.

The demand for a program to train examiners to size flaws was even stronger for the detection course. The sizing techniques shown to be effective were not generally familiar to the examination community. In addition, the coordination plan mandated a sizing performance demonstration, an exercise that fit naturally in a training course structure. The EPRI NDE Center developed the sizing course format and examination parameters and wrote the course materials, first made available in November 1983.

The course presents a sequential approach for sizing, using an initial crack-tip diffraction tech-

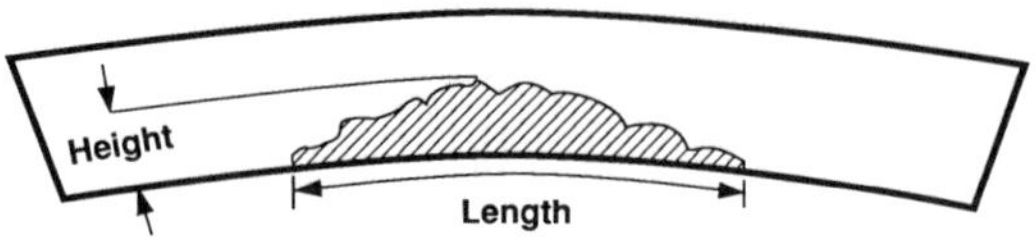

Fig. 4 Cross-sectional representation of intergranular SCC in pipe showing length and height dimensions

nique with supplementary time-of-flight techniques to assist in evaluation when tip signals are not sufficiently distinct. Participants are taught to use all techniques in appropriate combination for sizing (Ref 12). This methodology is depicted as a flow chart in Fig. 3.

Crack Length Sizing

Acceptance of a flawed pipe for continued service using the evaluation procedures of ASME Boiler and Pressure Vessel Code Section XI, IWB-3000, requires flaw size determination. In piping, crack sizing consists of determining the length of a crack (its circumferential or axial extent) and its height (the through-wall penetration). This is shown schematically in Fig. 4. Because of the uncertainty associated with some ultrasonic sizing techniques, erroneous conclusions may be reached.

Several international round-robin evaluations have measured the effectiveness of ultrasonic examination teams. In these studies, conventional and advanced NDE techniques were evaluated using full-scale mockups containing service-induced cracks as well as intentionally introduced defects. These studies focused on height sizing, however, and little data are available on the accuracy of various length-sizing techniques.

Because the amplitude of the signal reflected from intergranular SCC is the first signal characteristic to be noticed during ultrasonic examination, most crack-length measurements are based on the amplitude of the returned ultrasonic signal. Many codes and standards specify material acceptance criteria based on the reflected signal amplitude.

Figure 5 shows the basic amplitude-drop procedure. The search unit is positioned to obtain the maximum amplitude signal from the crack. From this position, the unit is moved laterally (parallel to the crack) until the signal amplitude decreases to some threshold level specified by the relevant procedure — for example, 50% DAC, 20% DAC, 50% of maximum amplitude, total loss of signal, or to the last maximum prior to decreasing to the noise level.

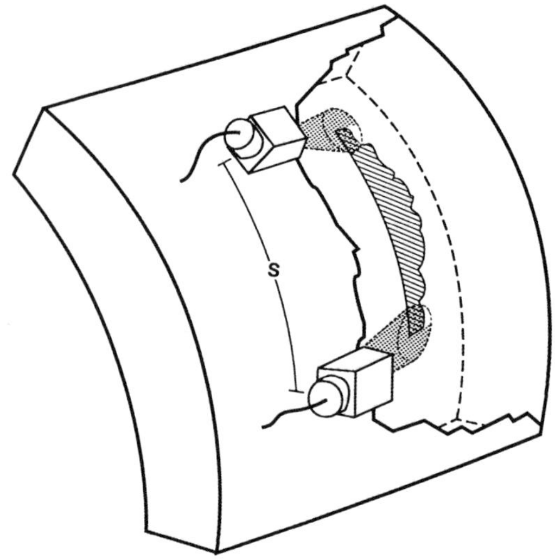

Fig. 5 Measuring the size of a crack larger than the beam diameter using the amplitude-drop technique

Cracks larger than the beam can be accurately sized using this procedure. The specific procedural requirement of measuring the length to various amplitude levels results in slight differences in measured length if the crack amplitude is greater than 100% DAC and drops smoothly to the noise level without intermediate peaks. However, it is important to measure crack length consistently so that changes indicating crack growth can be detected.

Due to the nature of intergranular SCC, the indications do not smoothly decrease to the noise level. Many intergranular SCC indications are less than 50% DAC and may have several intermediate peaks along their length. This results in extreme undersizing when the end points are measured to 50% DAC. Experience has shown that measuring the end points to loss of signal or the last maximum prior to loss of signal provides estimates that are close to the true length and that should be within a beam width of the detectable length. It may not be necessary to obtain length measurements more accurate than this, as the re-

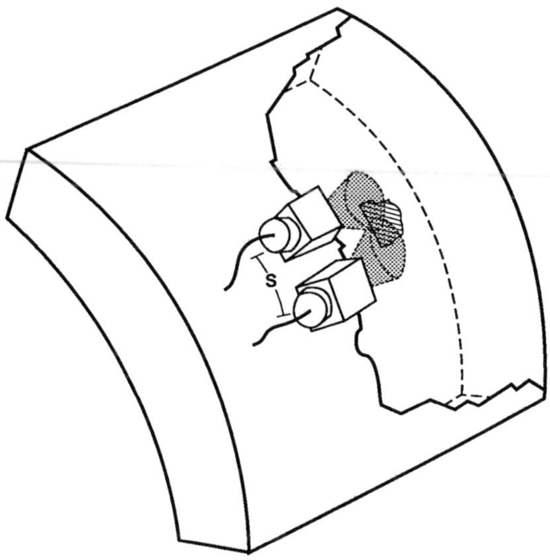

Fig. 6 Measuring the size of a crack smaller than the beam diameter using the amplitude-drop technique

pair/replace decision will not be affected by slight variations in crack length. Consistent measurements are necessary, however, for monitoring intergranular SCC for continued growth.

If smaller flaws are sized (those whose length is smaller than the beam diameter), the search-unit lateral movement (shown in Fig. 6) will include some fraction of the beam diameter in addition to the crack length, resulting in inaccurate measurements.

Conclusion

Safe operation of nuclear power plants is the primary driving force behind NDE of these facilities. The outstanding safety record of this industry directly attests to the success of these efforts. NDE technologies are constantly being upgraded to accommodate evolving materials technology. Legal and regulatory requirements for nuclear utilities impose periodic NDE. Safe, profitable operation of nuclear power plants requires NDE services to ensure early detection of potential problems.

NDE technology as applied to BWR piping examinations has made significant advances over the past decade. This success is a direct outgrowth of efforts to overcome the difficult examination problems of detecting and sizing intergranular SCC.

References

1. "Stress Corrosion Cracking in Thick-Wall, Large-Diameter, Stainless Steel Recirculation System Piping at BWR Plants," NRC IE Bull. No. 82-03, U.S. Nuclear Regulatory Commission, Office of Inspection and Enforcement, Oct 14, 1982
2. "Stress Corrosion Cracking in Large-Diameter Stainless Steel Recirculation System Piping at BWR Plants," NRC IE Bull. No. 83-02, U.S. Nuclear Regulatory Commission, Office of Inspection and Enforcement, March 4, 1983
3. "Inspection of BWR Stainless Steel Piping," NRC Generic Letter 84-11, U.S. Nuclear Regulatory Commission, Office of Nuclear Reactor Regulation, 1984
4. "Failures in 4-Inch Bypass Piping of Dresden 2," NRC IE Bull. No. 74-10B, U.S. Nuclear Regulatory Commission, Office of Inspection and Enforcement, 1975
5. "ASME Boiler and Pressure Vessel Code," Section XI, American Society of Mechanical Engineers, 1983
6. J.T. McElroy, "Detailed Analysis of the Fundamental Ultrasonic Response Data From Stainless Steel Stress Corrosion Crack Specimens," EPRI NP-676-SR, Electric Power Research Institute, Feb 1978
7. S.M. Walker, Detection and Confirmation of Intergranular Stress-Corrosion Cracking in Heavy Wall Stainless Steel Piping, in *Proc. 12th Nuclear Power Educational Seminar*, Southwest Research Institute, April 1982
8. M.E. Lapides, "Improved Ultrasonic Inspection Method for Stainless Steel Piping," EPRI NP-1153-SR, Electric Power Research Institute, Aug 1979
9. "UT Operator Training for Detection of Intergranular Stress-Corrosion Cracking (IGSCC)," EPRI NDE Center, Competency Area 910, March 1988
10. G.J. Dau and M.M Behravesh, Status of Intergranular Stress Corrosion Crack Depth Sizing, in *Non-Destructive Examination for Pressurized Components*, Elsevier Applied Science, 1984
11. "UT Operator Training for Sizing of Intergranular Stress-Corrosion Cracking (IGSCC)," EPRI NDE Center, Competency Area 911, March 1988
12. H. Stephens, Training for Inservice Inspection of Boiling Water Reactor (BWR) Piping Welds with Intergranular Stress Corrosion Cracking (IGSCC), in *Proc. 8th Int. Conf. NDE in the Nuclear Industry*, ASM International, 1987

CHAPTER 17

Evaluation of Stress-Corrosion Cracking

Donald O. Sprowls, Consultant

Stress corrosion cracking (SCC) is one of the two most subtle corrosion-related causes of premature fracture of structures. Long ago, engineers learned to expect fatigue and corrosion fatigue failures of cyclically loaded components and to plan for the replacement of fatigue-critical parts before they failed in service. In the absence of cyclic operating loads, it is not difficult to detect severely corroded parts that are liable to fail as a result of simple mechanical overload stresses. Sometimes, however, critical parts can fail long before fatigue effects would be expected and can result in major in-service catastrophes, as when an airplane wing-attachment fitting fractures. Parts can even fail before they ever see service, as when a shiny new machined hydraulic cylinder with a large crack is discovered by an aircraft assembly line worker.

Figure 1 suggests that such failures must result from either SCC or hydrogen stress cracking. These two processes differ from corrosion fatigue and tensile overload in that the metal must be in a vulnerable metallurgical condition. In certain alloy systems, hydrogen embrittlement may also contribute to both SCC and corrosion fatigue.

Susceptibility to SCC is not a property of a material in the same sense as a mechanical property, but it is nonetheless a very important factor in determining the serviceability of metallic materials. Evaluating SCC is far more complicated than simply performing a series of mechanical property tests.

This chapter is structured according to the following broad outline:

- General state of the art
- Selection of a test method
- Static loading of smooth specimens
- Static loading of precracked specimens
- Dynamic loading
- Selection of test environments
- Interpretation of test results

Most crucial to any SCC evaluation program are selection of an appropriate test method and interpretation of test results.

General State of the Art

Historically, service failures caused by SCC have been identified with sustained tensile stress; thus, SCC testing has developed around the use of static loading. In some situations, it is advantageous to use an actual structural component for testing. However, this is usually not practical; more often it is necessary to select smaller specimens that afford the required predictive capability.

Before about 1965, only constant-load or constant-strain tests of smooth and notched test specimens of various configurations were used to assess SCC. During the 1960s, two accelerated testing techniques based on different mechanical approached emerged. One technique tests and analyzes statically loaded, mechanically precracked test specimens by using linear elastic fracture mechanics concepts. The second technique consists of constant (slow) strain rate tests on smooth or precracked specimens. Laboratory

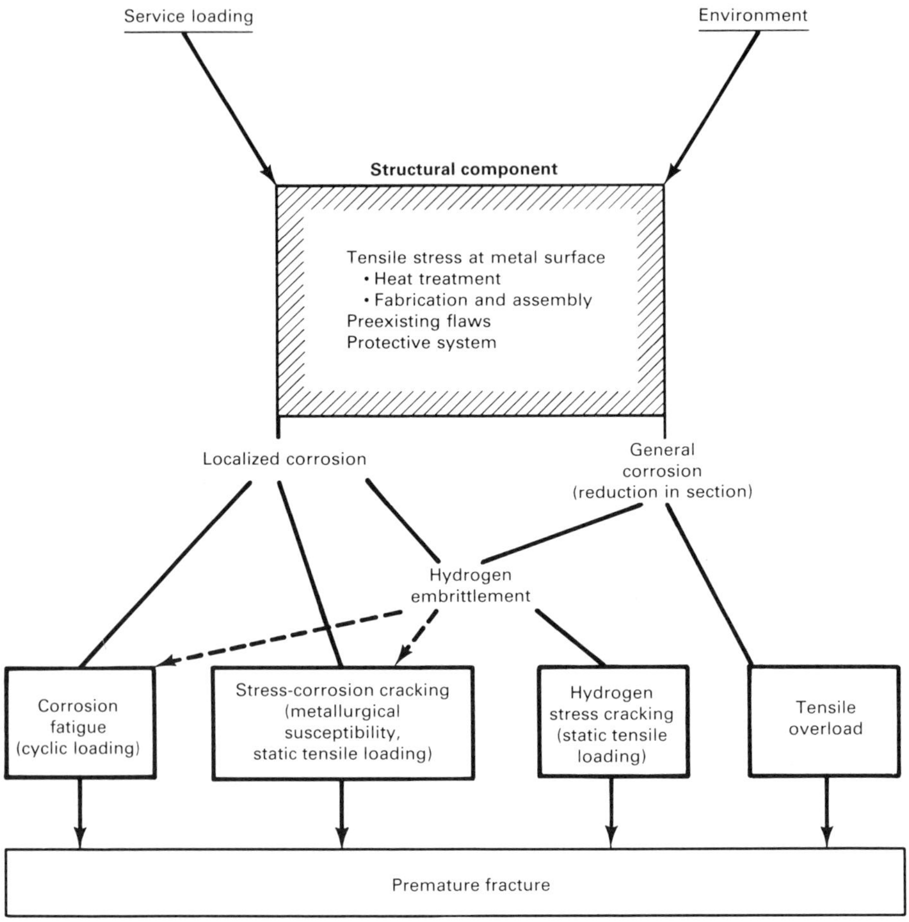

Fig. 1 Causes of premature fracture influenced by the corrosion of a structural component.

testing with these techniques frequently produces SCC where older, traditional tests may not.

Today, more different test methods are available than ever before. This chapter will attempt to dispel some of the confusion surrounding the selection of appropriate test methods.

Initiation Versus Propagation of SCC. The process of SCC is frequently discussed in terms of initiation (incubation and nucleation) and propagation, and illustrations similar to Fig. 2 are often found in the literature. However, the details of an accepted model have not been established. There may be a gradual transition from localized corrosion to crack initiation and growth with no separation of stages, or there may be a repeated succession of short steps of initiation and growth. In any event, from an engineering standpoint, it is convenient to view the process in two generic stages: initiation and propagation. Such terminology will be used throughout this chapter.

Two basic electrochemical corrosion reactions, anodic and cathodic, dominate the SCC process in conjunction with mechanical tensile stress. The chemical composition of the environment, including pH and the presence of hydrogen recombination poisons that affect the cathode reaction product, and the composition and metallurgical condition of the metal determine which

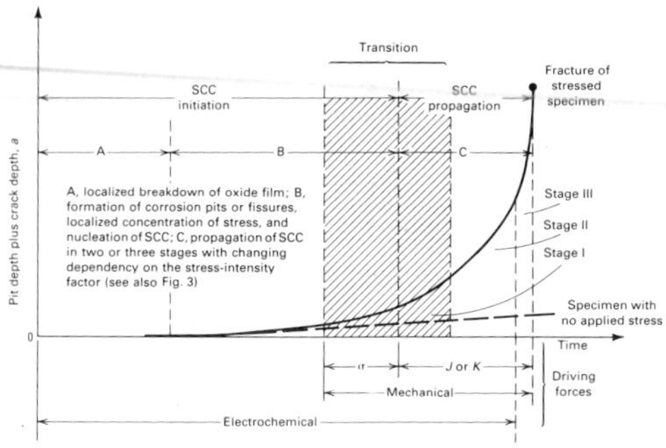

Fig. 2 The relative influences of electrochemical and mechanical factors in the corrosion and SCC damage of a susceptible material. The shaded area represents the transition of driving force from dominance by electrochemical factors to chiefly mechanical factors. Precise separation of initiation and propagation stages is experimentally difficult. Stimulation of cracking by atomic hydrogen may also become involved in this transition region. Source: Ref 1

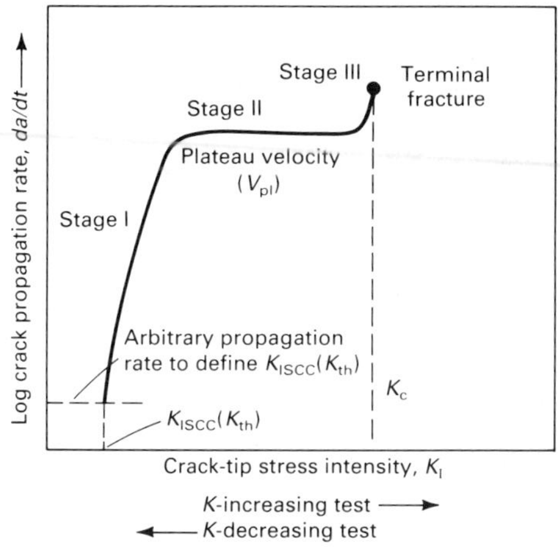

Fig. 3 Effect of stress intensity on the kinetics of SCC. Stages I and II may not always be straight lines but may be strongly curved, and one or the other may be absent in some systems. Stage III is of little interest and is absent in K-decreasing tests. Source: Ref 2

of the two partial corrosion reactions is dominant. Anodic SCC (active-path corrosion) involves the dissolution of metal during the initiation and propagation of cracks. Cathodic SCC (embrittlement by corrosion product hydrogen) involves the deposition of hydrogen at cathodic sites on the metal surface or on the walls of a fissure or crack and its subsequent absorption into the metal lattice. More information on the mechanisms of SCC is available in Chapter 1.

Figure 2 also suggests the relative influences of the electrochemical and mechanical driving forces in the SCC process. There is a change as SCC proceeds, with the role of stress being negligible at first and then becoming dominant as subcritical cracking advances. Environmental action must always be involved, although it may be dominant only at first. The pre-existence of a mechanical flaw or crack in the stressed metal may of course alter the initiation stage. Application of the fracture-mechanics-based stress-intensity factor (J for elastic-plastic fracture mechanics, K for linear elastic fracture mechanics) as a driving force for the propagation of SCC is illustrated schematically in Fig. 2 and 3.

Standardization of SCC tests in the United States was initiated in the 1960s by the American Society for Testing and Materials (ASTM), the National Association of Corrosion Engineers (NACE), and the federal government. Standard tests have also been developed in Europe (Ref 3), and uniform testing standards are currently under development on a broader basis through the International Organization for Standardization (ISO). Reference will be made throughout this chapter to available standards and to useful publications for details on test methods.

Selection of Test Method

Before a program of SCC testing can be undertaken, an appropriate test method must be selected. Selection depends largely on the information required.

Test Objectives. Typical objectives of SCC testing programs include (Ref 4):

Determination of the risk of SCC for a given application

- Comparison of alloys and mill products
- Comparison of environmental factors
- Evaluation of fabrication practices for structural components
- Evaluation of protective systems
- Prediction of service life

Alloy development and improvement

- Screening tests to compare variables in composition and metallurgical processing

- Establish claims for SCC performance of improved mill products
- Quality control

Study of SCC Mechanisms

General Considerations. The aim of SCC testing usually is to provide information more quickly than can be obtained from service experience, and at the same time attempt to predict service behavior. The most common approaches employed to achieve this are the use of high stresses, slow continuous straining, precracked specimens, higher concentration of species in the test environment than in the service environment, increased temperature, and electrochemical stimulation. It is important, however, that these procedures be controlled in such a way that the details of the failure mechanism are not changed. The primary challenge was expressed well in the following statement, written more than 40 years ago: "While it is relatively easy to determine if a product is susceptible to SCC, it is far more difficult to determine if it possesses a 'degree of susceptibility' which will restrict its general usefulness" (Ref 5).

A number of "standard tests" have been found to provide reasonable guidance as to likely service behavior for given specific applications. However, these laboratory tests are appropriate only to service conditions where experience has shown a relevant relationship, however empirical, to exist. The fact that a given alloy passes or does not pass a test previously found useful in relation to another alloy may or may not be significant, and a test that discriminates correctly between alloys used for a given application will not necessarily provide safe guidance if the exposure conditions are different. The use of a standard test beyond the point for which there is experience therefore requires validation (Ref 3).

The SCC test method selected should not be so severe that it rejects a material that would be adequate for a particular application; on the other hand, the test should not be so mild that it passes materials that will fail in service. For tests of new and unfamiliar materials or environments, it is useful to perform more than one type of test. Although standardized tests, which can be specified for screening tests in alloy or process development or quality control, are an essential link between research and engineering, there is a need on the other hand for much freer choice of test conditions for research studies of SCC mechanisms.

As metal producers constantly seek to develop stronger and tougher structural alloys, there is an increasing pressure on corrosion investigators to be able to determine their "degree of susceptibility" to SCC. With the new test procedures available today it it possible to investigate not only the probability of the initiation of SCC but also to observe the rate of SCC propagation for alloy/environment systems of interest. Therefore it is incumbent on the investigator to use test methods and to provide an interpretation of the test results that will enable the materials users to make a proper assessment of the risk of SCC in a given application.

Static Loading of Smooth Specimens

Tests for predicting the stress-corrosion performance of an alloy in a particular service application should be conducted with a stress system similar to that anticipated in service. Table 1 lists the numerous sources of sustained tension that are known to have initiated SCC in service and the applicable methods of stressing. Most SCC

Table 1 Stressing Methods Applicable to Various Sources of Sustained Tension in Service

Source of sustained tension in service	Constant strain	Constant load
Residual stress		
Quenching after heat treatment	X	
Forming	X	
Welding	X	
Misalignment (fit-up stresses)	X	
Interference fasteners	X	
Interference bushings		
Rigid	X	
Flexible		X
Flareless fittings	X	
Clamps	X	
Hydraulic pressure	X	X
Dead weight		X
Faying surface corrosion	X	X

Note: The greatest hazard arises when residual, assembly and operating stresses are additive.

service problems involve tensile stresses of unknown magnitude that are usually very high, approaching the elastic limit of the material.

Results are strongly influenced by mechanical aspects of a test, such as method of loading and specimen size. These mechanical aspects can control initiation and propagation lifetimes and can influence estimates of a threshold stress. Therefore, an apparent threshold stress for SCC is not a material property, and threshold estimates must be qualified with regard to the test conditions and the level of statistical significance.

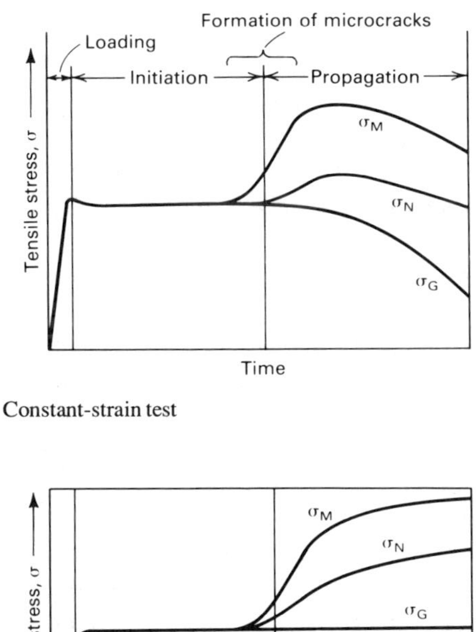

(a) Constant-strain test

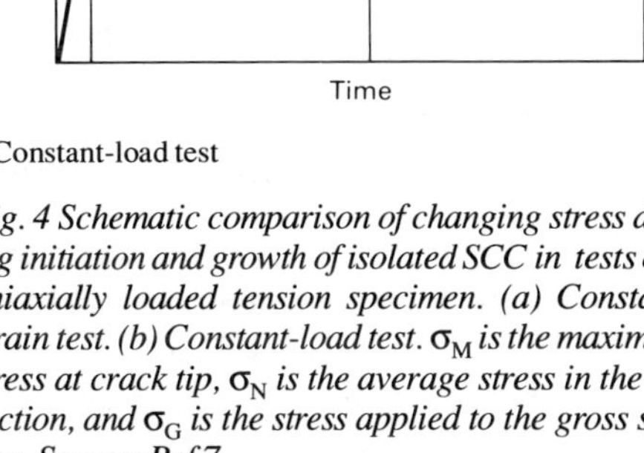

(b) Constant-load test

Fig. 4 Schematic comparison of changing stress during initiation and growth of isolated SCC in tests of a uniaxially loaded tension specimen. (a) Constant-strain test. (b) Constant-load test. σ_M is the maximum stress at crack tip, σ_N is the average stress in the net section, and σ_G is the stress applied to the gross section. Source: Ref 7

Constant-Strain Versus Constant-Load Tests

Constant-strain (fixed-displacement) tests are widely used, primarily because a variety of simple and inexpensive stressing jigs can be devised. However, reproducibility of the exposure stress is poor with some of these techniques. Therefore, sophisticated procedures have been developed to improve this facet of testing.

Constant-strain tests are sometimes called decreasing-load tests, because after the onset of SCC in small test specimens the gross section exposure stress decreases. This results from the opening of the crack (or cracks) under the high stress concentration at the crack tip (or tips) and causes some of the applied elastic strain to change to plastic strain, with an attendant reduction in the initial load. Such trends in changing stress during crack growth are shown in Fig. 4(a).

Comparison of the stress trends for a constant-strain test (Fig. 4a) with those for a constant-load test (Fig. 4b) reveals that neither method of loading provides a *constant-stress* test after growth of microcracks has occurred. True constant-load tests results in increasing stress levels as cracking progresses, and are more likely to lead to earlier failure with complete fracture and lower estimates of a threshold stress than constant-strain tests. Figures 4(a) and (b) illustrate basic trends that can be applied to all types of test specimens, including precracked specimens. Specific curves, however, will differ depending on other test conditions.

The stiffness of the combined stressing frame/test specimen system can significantly affect materials comparisons if identical specimen stressing procedures are not used (Ref 6, Part 1). Many so-called constant-strain tests, particularly if a spring is included in the stressing system, are not actually constant-strain tests, because a significant amount of elastic strain energy may be contained in the stressing system. Depending on the "softness" of the spring or the elasticity of the stressing frame, the stiffness (compliance) of the stressing system can vary greatly between zero stiffness (dead load) and infinite stiffness (true constant total strain). Figure 5 shows the typical change in net section stress with the onset of SCC in an intermediate-stiffness stressing frame.

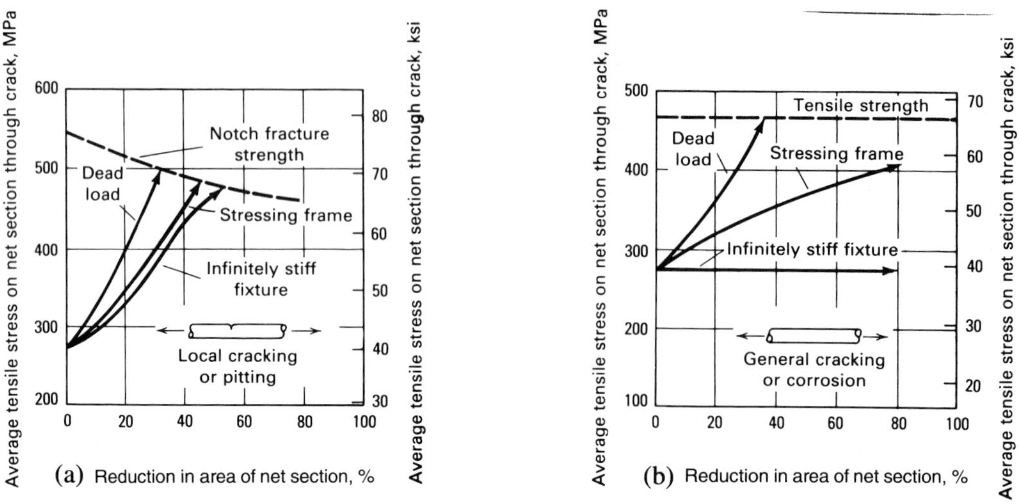

Fig. 5 Effect of loading method and extent of cracking or corrosion pattern on average net section stress in a uniaxially loaded tension specimen. Behavior is generally representative, but curves will vary with specific alloys and tempers. (a) Localized cracking. (b) General cracking. Source: Ref 8 (G 49)

The corrosion pattern in the test specimen, particularly the number and distribution of cracks, can impair the precision of results obtained by either constant-strain or constant-load tests. When localized stress-corrosion cracks propagate in a specimen stressed by either method, the average tensile stress on the net section increases rapidly until the notch fracture strength is reached and the specimen breaks (Fig. 5a). Less penetration (reduction in area of net section) is required for fracture of specimens under dead load; this indicates that specimen life is shorter with lower-stiffness stressing frames.

With general cracking, crack propagation is more strongly influenced by frame stiffness. When microcracks initiate close to one another, their individual stress concentrations interact and are relaxed. Consequently, there may not be a sufficient stress concentration in the true constant-strain test (infinitely stiff fixture) to propagate further SCC, and the specimen will not break (Fig. 5b). Under a constant load, however, the growth of many cracks continues, and the specimen ultimately breaks. Therefore, SCC comparison of specimens tested at stress levels just above their thresholds is complicated by random variations in the cracking pattern, particularly when tested with relatively stiff stressing systems.

Although constant-load stressing appears to be advantageous for testing materials with relatively high resistance to SCC, difficulties arise when small-diameter specimens are utilized to avoid the use of massive loads or lever systems. In some test environments, highly stressed specimens may fail from general or pitting corrosion and an attendant increase in the effective stress. Such non-SCC failures complicate interpretation of test results, unless failure by SCC is confirmed by metallographic examination. Such extraneous failures are less likely to occur with specimens loaded under constant strain.

Therefore, small test specimens, which are generally preferred for laboratory screening tests and research studies, must be used with caution when estimates of serviceability are required. To determine serviceability, larger specimens should be used, and a stressing system should be selected that best duplicates the anticipated service conditions.

Bending Versus Uniaxial Tension

Historically, the most extensively used stressing systems have incorporated constant-deformation specimens stressed by bending. This method is versatile because of the variety of simple techniques that can be used to test most metal products in all types of corrosive environments. The state of stress in a bend specimen, however, is much more complex than in a tension specimen. Theoretically, tensile stress is uniform throughout the cross section in the tension specimen, except at corners in rectangular sections, but the tensile stress in bend specimens varies through the specimen thickness.

Tensile stress is at a maximum on the convex surface and decreases steeply to zero at the neutral axis. It then changes to a compressive stress, which reaches a maximum on the concave surface. Thus, only about 50% of the metal surface is under tension. As SCC penetrates the metal, the stress gradient through the section thickness produces changes in stresses and strains that are different from those in a uniaxial tension specimen. This tendency yields significantly different SCC responses for the two types of stressing (Fig. 6).

Bending stress specimens experience other sources of variability in stress that are not present with direct tension stressing. Variations occur in the principal longitudinal stress along the length and across the width of the specimen as well as with the presence of biaxial stresses, both of which are influenced by the design of the specimen. Therefore, just as in the case of constant-load stressing, optimal control of stress and more severe testing conditions are provided by uniaxial tension stressing.

Statically loaded, smooth test specimens for SCC tests can be divided into three general categories: elastic strain specimens, plastic strain

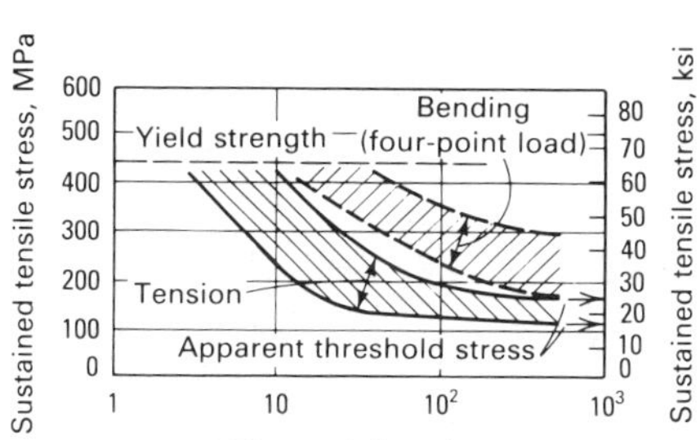

Fig. 6 Comparison of the SCC response under constant load with bending versus direct tension stressing for Al-5.3Zn-3.7Mg-0.3Mg-0.1Cr T6 temper alloy sheet. Tested to failure in 3% NaCl plus 0.1% H_2O_2. Source: Ref 9

specimens, and residual stress specimens. The commonly used specimen geometries for each of these categories are discussed below.

Elastic Strain Specimens

To control the surface tensile stress applied by deformation loading, strain must be restricted to the elastic range for the test material. The magnitude of the applied stress can then be calculated from the measured strain and modulus of elasticity. In constant-load stressing, the load typically is measured directly, and the stress is calculated by using the appropriate formula for the specimen configuration and the method of loading. Load cells or calibrated springs may be useful for applying and monitoring possible changes in load during the test. Commonly used types of specimens for tests under elastic-range stress are described below. Note that if an attempt is made to employ very high stresses by loading any of these specimens beyond the elastic limit of the test material, the actual magnitude of the applied stress will not be known precisely (see the section on C-rings).

Bent-beam specimens can be used to test a variety of product forms. The bent-beam configuration is primarily used for sheet, plate, or flat extruded sections, which conveniently provide flat specimens of rectangular cross section, but it may also be used for cast materials, rod, pipe, or machined specimens of circular cross section. This method is applicable to specimens of any metal that are stressed to levels less than the elastic limit of the material; therefore, the applied stress can be calculated or measured accurately.

Stress calculations by this method are not applicable to plastically stressed specimens. Bent-beam specimens are usually tested under constant-strain conditions, but constant-load conditions can also be used. In either case, local changes in the curvature of the specimen when cracking occurs result in change in stress and strain during crack propagation. The "test stress" is taken as the highest surface tensile stress existing at the start of the test, that is, before the initiation of SCC.

Several configurations of bent-beam specimens and stressing systems are illustrated in Fig. 7 and are described in detail in ASTM G 39 (Ref 8

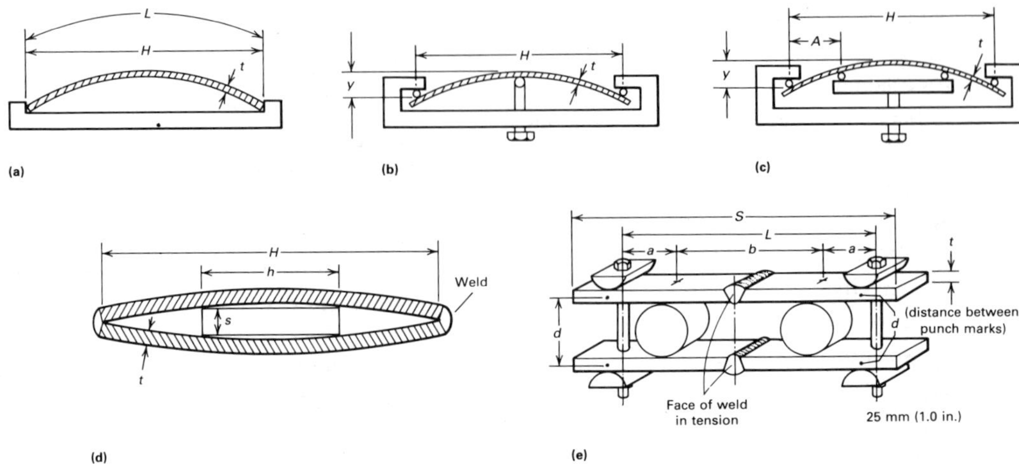

Bolt loaded double-beam specimen dimensions for various plate thicknesses

t		a		b		L		S	
mm	in.	mm	in.	mm	in.	mm	in.	mm	in.
3.2	0.125	100	4.0	50	2.0	250	10.0	305	12.0
6.4	0.25	100	4.0	50	2.0	250	10.0	305	12.0
9.5	0.375	120	4.75	90	3.5	330	13.0	380	15.0
13	0.5	120	4.75	90	3.5	330	13.0	380	15.0
19	0.75	140	5.5	150	6.0	430	17.0	480	19.0
25	1.0	150	6.0	200	8.0	510	20.0	560	22.0
38	1.5	165	6.5	305	12.0	635	25.0	685	27.0

Fig. 7 Schematic specimen and holder configurations for bent-beam stressing. (a) Two-point loaded specimen. (b) Three-point loaded specimen. (c) Four-point loaded specimen. (d) Welded double-beam specimen. (e) Bolt-loaded double-beam specimen. Formula for stressing specimen (e): $\Delta d = 2fa/3Et(3L - 4a)$, where Δd is deflection (in inches), f is nominal stress (in pounds per square inch), and E is modulus of elasticity (in pounds per square inch). Source: Ref 10

and in Part 2 of Ref 6). When specimens are tested at elevated temperatures, the possibility of stress relaxation should be investigated. More information on stress relaxation is available in the article "Creep, Stress-Rupture, and Stress-Relaxation Testing" in Volume 8 of the *ASM Handbook*.

Two-point loaded specimens can be used for materials that do not deform plastically when bent to $(L - H)/H = 0.01$. The specimens should be approximately 25 × 250 mm (1 × 10 in) flat strips cut to appropriate lengths to produce the desired stress after bending, as shown in Fig. 7(a). The maximum stress occurs at the midlength of the specimen and decreases to zero at specimen ends.

Three-point loaded specimens are flat strips that are typically 25 to 51 mm (1 to 2 in.) wide and 127 to 254 mm (5 to 10 in.) long. The thickness of a specimen is usually dictated by the mechanical properties of the material and the available product form. The specimen should be supported at the ends and bent by forcing a screw (equipped with a ball or knife-edge tip) against it at a point halfway between the end supports, as shown in Fig. 7(b). In a three-point loaded specimen, the maximum stress occurs at the midlength of the specimen and decreases linearly to zero at the outer supports.

Two- and four-point loaded specimens are often preferred over the three-point loaded specimen, because crevice corrosion often occurs at its central support. Because this corrosion site is very close to the point of highest tensile stress, it may cathodically protect the specimen and prevent possible crack formation, or it may cause hydrogen embrittlement. Furthermore, the pressure of the central support at the point of highest load introduces biaxial stresses at the area of contact and can introduce tensile stresses where compressive stresses are normally present.

Four-point loaded specimens are flat strips that are typically 25 to 51 mm (1 to 2 in.) wide and 127 to 254 mm (5 to 10 in.) long. The thickness of a specimen is usually dictated by the mechanical properties of the material and the available product form. The specimen is supported at the ends and is bent by forcing two inner supports against it, as shown in Fig. 7(c). The two inner supports are located symmetrically around the midpoint of the specimen.

In a four-point loaded specimen, the maximum stress occurs between the contact points of the inner supports; the stress is uniform in this area. From the inner supports, the stress decreases linearly toward zero at the outer supports. The four-point loaded specimen is preferred over the three-point and two-point loaded specimens, because it provides a large area of uniform stress.

Welded double-beam specimens consist of two flat strips 25 to 51 mm (1 to 2 in.) wide and 127 to 254 mm (5 to 10 in.) long. The strips are bent against each other over a centrally located spacer until both ends touch. The strips are held in position by welding the ends together, as shown in Fig. 7(d).

In a welded double-beam specimen, the maximum stress occurs between the contact points of the spacer; the stress is uniform in this area except for possible notch effects due to the weld bead. From contact with the spacer, the stress decreases linearly toward zero at the ends of the specimen.

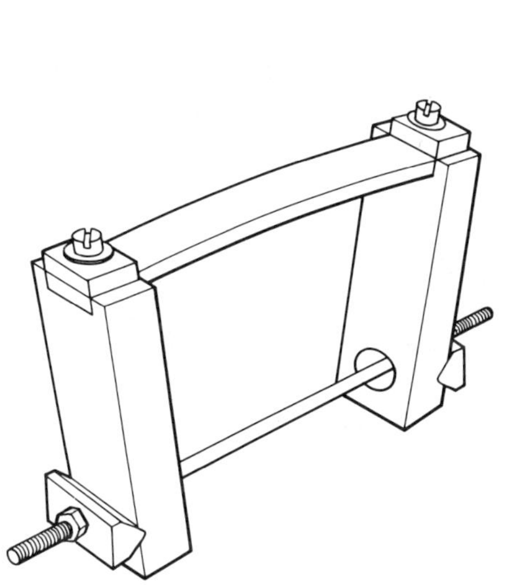

Fig. 8 Bent beam designed to produce pure bending. Source: Ref 11

A bolt-loaded double-beam specimen is shown in Fig. 7(e), along with suggested specimen dimensions for various thicknesses of plate and the formula for stressing such specimens (Ref 10). The beam deflections required to develop the intended tensile stress are calculated with the formula and are then applied by bolting the ends of the beams together. The deflections are measured with a dial gage to within ±0.0127 mm (±0.0005 in.). Thus, the error in stress application—if the beams are of homogeneous material and if the cross sections are uniform—is within 2%. The precision of the deflection measurement is within 0.5%, and the error in determining the modulus of elasticity, E, is within 1%.

Constant-moment beam specimens are designed such that a constant moment exists from one end to the other when the specimen is bent in the manner shown in Fig. 8. This bending produces equal stress along the length of the specimen. The width-to-thickness ratio is less than 4 so that biaxial stresses are eliminated.

This type of specimen offers the advantage of a relatively large area of material under a uniform stress. Such specimens can be use when the dimensions of the specimen are too small for other bent-beam specimens—for example, when specimens are taken in the short-transverse direction in plate (see Fig. 18c). The elastic stress, σ, in the convex surface is calculated by:

$$\sigma = \frac{4Ety}{h^2} \quad \text{(Eq 1)}$$

where h is the distance between the inner edges of the supports, y is the maximum deflection between the inner edges of the supports, t is the thickness of the specimen, and E is the modulus of elasticity.

C-Ring Specimens. As discussed in ASTM G 39 (Ref 8) and in Part 5 of Ref 6, the C-ring is a versatile, economical specimen for quantitatively determining the susceptibility to SCC of all types of alloys in a wide variety of product forms. It is particularly well suited for testing tubing and for making short-transverse tests on various product forms, as shown in Fig. 9. The sizes of C-rings can be varied over a wide range, but rings with outside diameters of less than about 16 mm (5/8 in.) are not recommended because of increased difficulties in machining and decreased precision in stressing.

The C-ring is typically a constant-strain specimen with tensile stress produced on the exterior of the ring by the tightening of a bolt centered on the diameter of the ring. However, an almost constant load can be developed by placing a calibrated spring on the loading bolt. C-rings can also be stressed in the reverse direction by spreading the ring and creating a tensile stress on the inside surface. These methods of stressing are illustrated in Fig. 10.

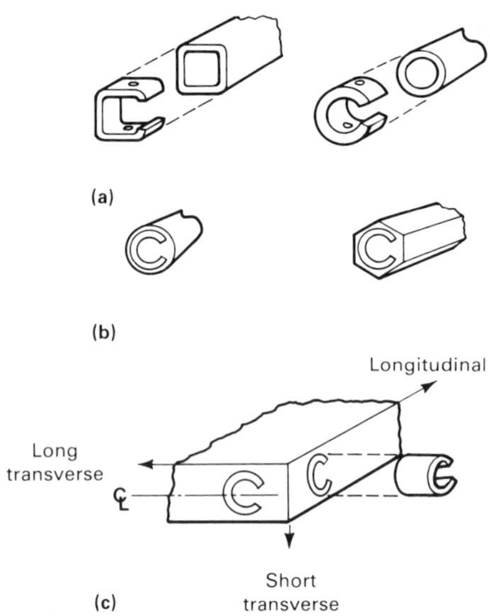

Fig. 9 Sampling procedure for testing various products with C-rings. (a) Tube. (b) Rod and bar. (c) Plate

Circumferential stress is of principal interest in the C-ring specimen. This stress is not uniform (Ref 12), as discussed previously in the section "Elastic Strain Specimens" in this chapter. The stress varies around the circumference of the C-ring from zero at each bolt hole to a maximum at the middle of the arc opposite the stressing bolt. In a notched C-ring, a triaxial stress state is present adjacent to the root of the notch (Ref 13). For all notches, the circumferential stress at the root of the notch is greater than the nominal stress and can generally be expected to be in the plastic range.

Generally, the C-ring can be stressed with high precision. The most accurate stressing procedure consists of attaching circumferential and transverse electrical strain gages to the surface stressed in tension, followed by tightening the bolt until the strain measurements indicate the desired circumferential stress.

The amount of compression required on the C-ring to produce elastic straining and the degree of elastic strain can be predicted theoretically. Therefore, C-rings can be stressed by calculating the deflection required to develop a desired elastic stress. In notched specimens, a nominal stress is estimated using a ring outside diameter measured at the root of the notch and by taking into

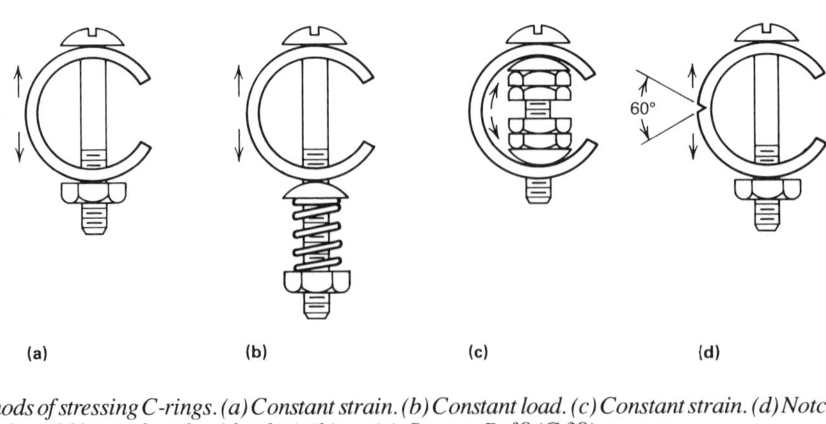

Fig. 10 Methods of stressing C-rings. (a) Constant strain. (b) Constant load. (c) Constant strain. (d) Notched C-ring; a similar notch could be used on the side of (a), (b), or (c). Source: Ref 8 (G 38)

consideration the stress-concentration factor, K_t, for the specific notch.

When C-rings are stressed by straining to high levels exceeding the elastic limit of the test material, which may be of interest for certain industrial applications, the maximum stress developed in the C-ring will not be known unless (1) a strain gage is attached to the C-ring during loading and reference is made to a stress-strain diagram for the test material or (2) an acceptable allowance factor for the plastic straining has been predetermined for the particular test material.

O-Ring specimens (Fig. 11) are used to develop a hoop stress in a particular part—for example, a cylindrical die forging in which a critical end-grain structure associated with the parting plane of the forging exists only at the surface of the forging. A relatively large surface area of metal is placed under a uniform tensile stress, and the O-ring stressing plug assembly simulates service conditions in structures containing interference-fit components. Stressed O-rings have also been used to evaluate protective treatments for the prevention of SCC (Ref 14).

An O-ring is stressed by pressing it onto an oversized plug that is machined to a predetermined diameter to develop the desired stress at the outside surface of the ring. The nominal dimensions of this specimen can be varied to suit the part being tested, but certain characteristics should be observed to achieve adequate control of the stresses. The ring width should not be more than four times the wall thickness in order to ensure maximum uniformity of the hoop stress from the centerline to the edges of the ring. The tensile stress varies through the thickness of the ring and is highest at the inside surface. Interference required for stressing an O-ring can be calculated by using:

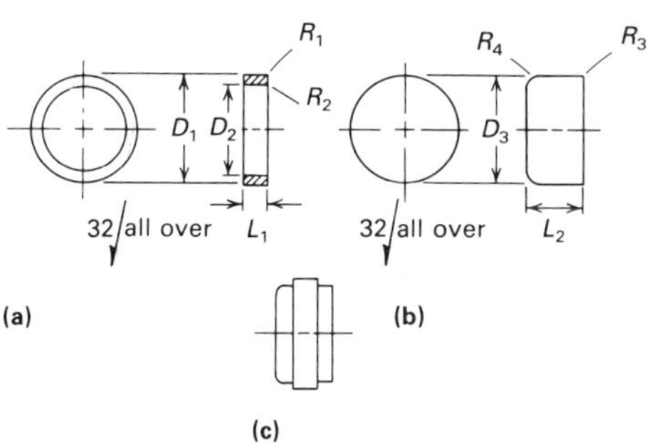

Fig. 11 O-ring SCC test specimen (a) and stressing plug (b). The O-ring is stressed by pressing it onto the plug, as shown in (c). Source Ref 14

$$I = \frac{F(\text{OD})^2}{E(\text{ID})} \quad \text{(Eq 2)}$$

where I is the interference (on the diameter) between the O-ring and the plug, E is the modulus of elasticity, ID is the inside diameter, OD is the outside diameter, and F is the circumferential stress desired on the outside surface. Additional information regarding the design and stressing of O-ring specimens is given in Ref 15.

Tension Specimens. Specimens used to determine tensile properties in air are well suited and easily adapted to SCC, as discussed in ASTM G 49 (Ref 8) and in Part 4 of Ref 6. When uniaxially loaded in tension, the stress pattern is simple and uniform, and the magnitude of the applied stress can be accurately determined. Specimens can be quantitatively stressed by using equipment for application of either a constant load, a constant strain, or an increasing load or strain.

This type of test is one of the most versatile methods of SCC testing because of the flexibility permitted in the type and size of the test specimens, the stressing procedures, and the range of stress level. It allows the simultaneous exposure of unstressed specimens (no applied load) with stressed specimens and subsequent tension testing to distinguish between the effects of true SCC and mechanical overload.

A wide range of test specimen sizes can be used, depending primarily on the dimensions of the product to be tested. Stress-corrosion test results can be significantly influenced by the cross section of the test specimen. Although large specimens may be more representative of most structures, they often cannot be prepared from the available product forms being evaluated. They also present more difficulties in stressing and handling in laboratory testing.

Smaller cross-sectional specimens are widely used. They have a greater sensitivity to SCC initiation, usually yield test results rapidly, and permit greater convenience in testing. However, the smaller specimens are most difficult to machine, and test results are more likely to be influenced by extraneous stress concentrations resulting from nonaxial loading, corrosion pits, and so on. Therefore, use of specimens less than about 10 mm (0.4 in.) in gage length and 3 mm (0.12 in.) in diameter is not recommended, except when testing wire specimens.

Tension specimens containing machined notches can be used to study SCC and hydrogen embrittlement. The presence of a notch induces a triaxial stress state at the root of the notch, in which the actual stress will be greater by a concentration factor that is dependent on the notch geometry. The advantages of such specimens include the localization of cracking to the notch region (usually) and acceleration of failure. However, unless directly related to practical service conditions, the results may not be relevant.

Tension specimens can be subjected to a wide range of stress levels associated with either elastic or plastic strain. Because the stress system is intended to be essentially uniaxial (except in the case of notched specimens), great care must be exercised in the construction of stressing frames to prevent or minimize bending or torsional stresses.

The simplest method of providing a constant load consists of a dead weight hung on one end of the specimen. This method is particularly useful for wire specimens. For specimens of large cross section, however, lever systems such as those used in creep-testing machines are more practical. The primary advantage of any dead-weight loading device is the constancy of the applied load.

A constant-load system can be modified by the use of a calibrated spring, such as that shown in Fig. 12. The proving ring, as used in the calibrated of tension testing machines, has also been adapted to SCC testing to provide a simple, compact, easily operated device for applying axial load (Fig. 13). The load is applied by tightening a nut on one of the bolts and is determined by carefully measuring the change in ring diameter.

Constant-strain SCC tests are performed in low-compliance tension-testing machines. The specimen is loaded to the required stress level, and the moving beam is then locked in position. Other laboratory stressing frames have been used, generally for testing specimens of smaller cross section. Figure 14(a) shows an exploded view of such a stressing frame, and Fig. 14(b) illustrates a special loading device developed to ensure axial loading with minimal torsion and bending of the specimen.

For stressing frames that do not contain any mechanism for the measurement of load, the stress level can be determined from measurement of the strain. However, only when the intended stress is below the elastic limit of the test material is the average linear stress (σ) proportional to the average linear strain (ε), $\sigma/\varepsilon = E$, where E is the modulus of elasticity.

When tests are conducted at elevated temperatures with constant-strain-loaded specimens,

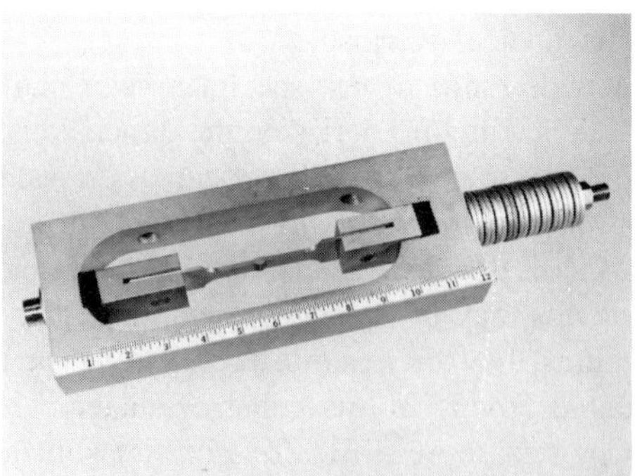

Fig. 12 Spring-loaded fixture used to stress 3.2 mm (0.125 in.) thick sheet tensile specimens in direct tension. Source: Ref 10

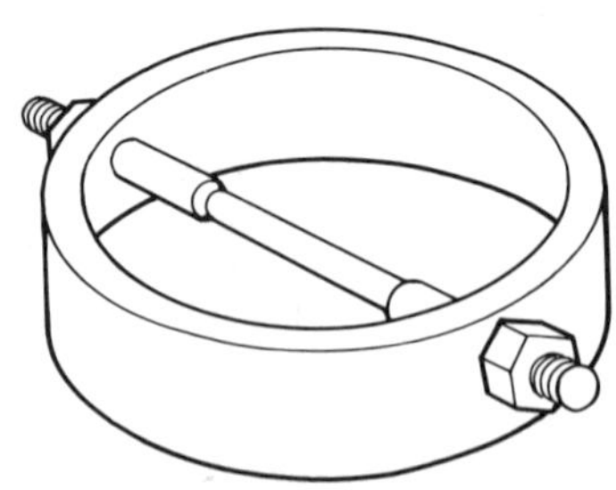

Fig. 13 Ring-stressed tension specimen for field testing. Source: Ref 3

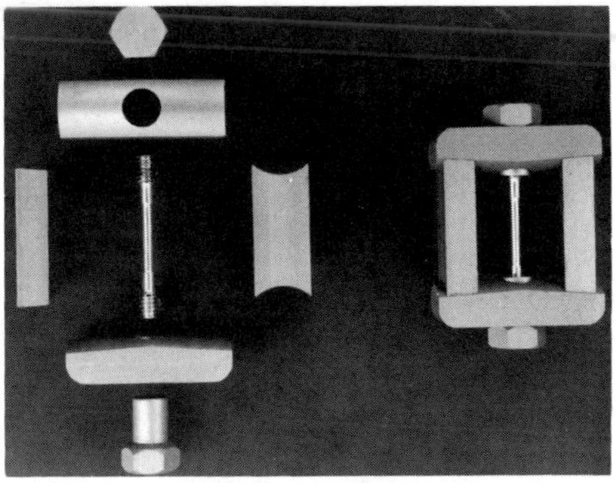

(a)

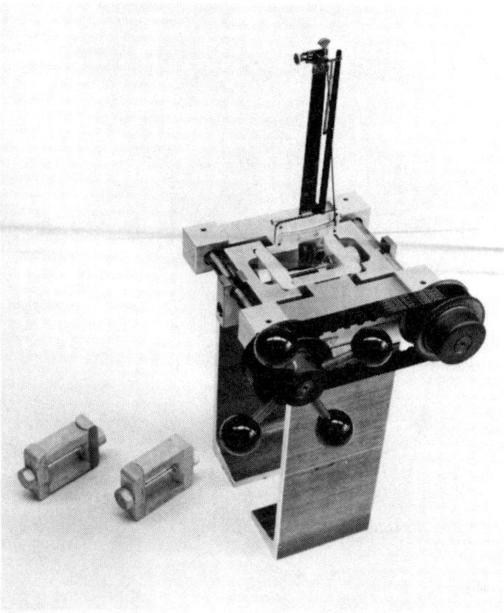

(b)

Fig. 14 (a) "Constant-strain" SCC testing frame. Exploded view (left) showing the 3.2 mm (0.125 in.) diameter tension specimen and various parts of the stressing frame. Final stressed assembly (right). Source: Ref 16. (b) Synchronous loading device used to stress specimens. The specimen is loaded to a prescribed strain value determined from a clip-on gage. The applied stress is given by the product of the strain and the material elastic modulus. A stressed assembly and one assembled finger-tight ready for stressing are shown. Source: Ref 8 (G 49)

consideration should be given to the possibility of stress relaxation. When stress relaxation or creep occurs in the test specimen, some of the elastic strain is converted to plastic strain and the nominal applied test stress is reduced. This effect is particularly important when the coefficients of thermal expansion are different for the specimen and stressing frame. Frequently, nonmetallic (plastic) insulators are used between the specimen and stressing frame to avoid galvanic action. If such plastic insulators are part of the stress-bearing system, creep (even at room temperature) can significantly alter the applied load on the specimen.

Even though eccentricity in loading can be minimized to levels acceptable for tension-testing machines, tensile stress around the circumference of round specimens and at the corners of sheet-type specimens varies to some extent. Several factors may introduce bending moments on specimens, such as longitudinal curvature and misalignment of threads on threaded-end round specimens. These factors have a greater effect on specimens with smaller cross sections. Tests should be made on specimens with strain gages affixed to the specimen surface around the circumference in 90° or 120° intervals to verify strain and stress uniformity and to determine whether machining practices and stressing jigs are of adequate tolerance and quality.

When SCC occurs, it generally results in complete fracture of the specimen, which is easy to detect. However, when testing relatively ductile materials at stress levels close to the threshold of susceptibility, fracture may not occur during the period of exposure. The presence of SCC in such cases must be determined by mechanical tests or by metallographic examination, as discussed previously.

To study trends in SCC susceptibility, such as in alloy development research, it is often necessary to detect small differences in susceptibility. For this purpose, it is advantageous to use replicate sets of specimens stressed at several levels, including zero applied stress. The sets are then removed for metallographic examination or tension tests after appropriate periods of exposure.

Figure 15 illustrates the use of this procedure with samples of 7075-T651 aluminum alloy that have been given different thermal treatments to decrease susceptibility to SCC. Analysis of these breaking stress data by statistical procedures enables calculation of survival probabilities and the estimation of a threshold stress, without depend-

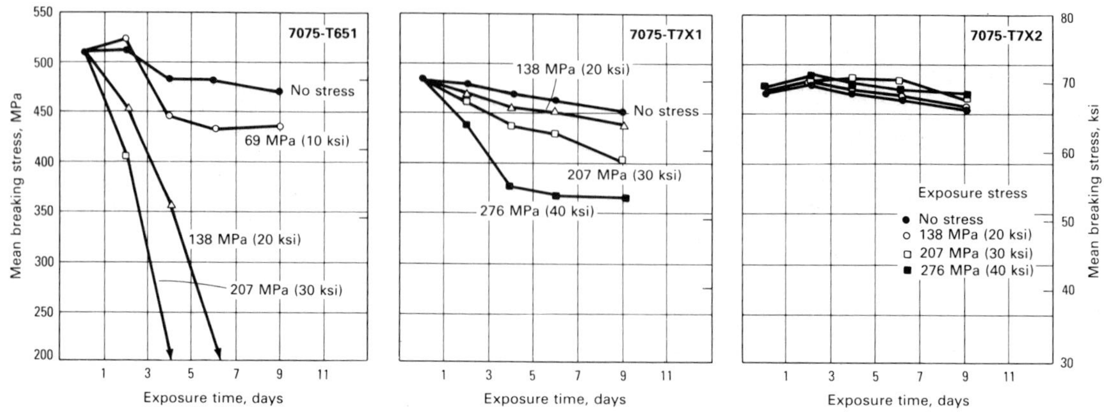

Fig. 15 Mean breaking stress versus exposure time for short-transverse 3.2 mm (0.125 in.) diameter aluminum alloy 7075 tension specimens tested according to ASTM G 44 at various exposure stress levels. Each point represents an average of five specimens. Source: Ref 1

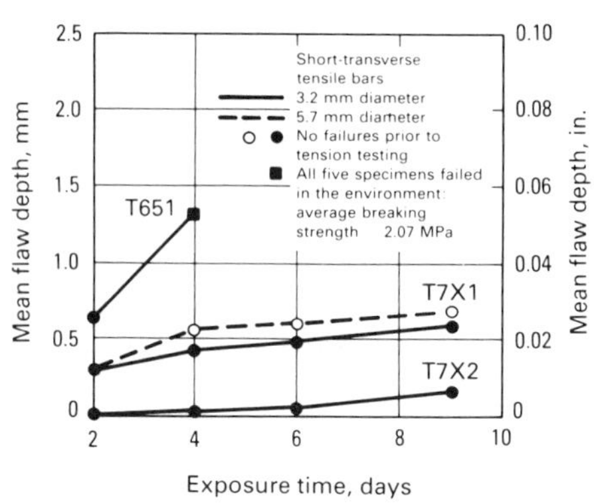

Fig. 16 Effect of temper on SCC performance of aluminum alloy 7075 subjected to alternate immersion in 3.5% NaCl solution at a stress of 207 MPa (30 ksi). Mean flaw depth was calculated from the average breaking strength of five specimens subjected to identical conditions. Source: Ref 17

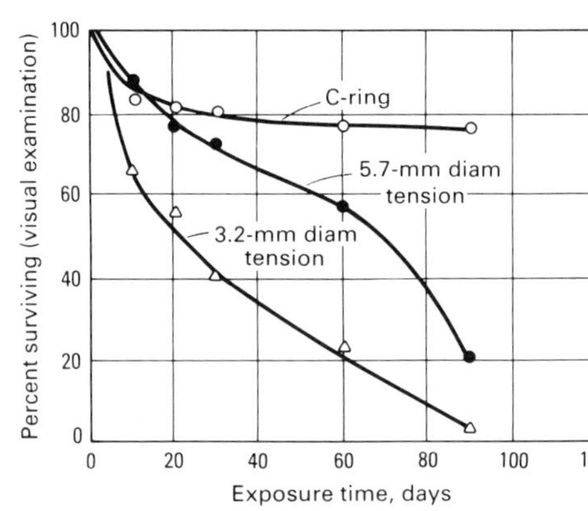

Fig. 17 Influence of specimen configuration on SCC test performance (alternate immersion in 3.5% NaCl per ASTM G 44). Aluminum alloy 7075-T7X51 specimens stressed 310 MPa (45 ksi); each point represents 60 to 90 specimens. Source: Ref 18

ing on failures during exposure. By using an elastic-plastic fracture mechanics model, an effective flaw size is calculated from the mean breaking stress, the strength, and the fracture toughness of the test material. The effective flaw size corresponds to the weakest link in the specimen at the time of the tension test, and it therefore represents the maximum penetration of the SCC. An advantage to using flaw depth to examine SCC performance is that the effects of specimen size and alloy strength and toughness can be normalized. In contrast, the specimen lifetime and breaking strength are biased by those mechanical (non-SCC) factors.

Mean trends in the 207 MPa (30 ksi) exposure data for the three temper variants of aluminum alloy 7075 examined in Fig. 15 are shown in Fig. 16. These results clearly illustrate that the thermal treatments used to reduce the SCC susceptibility of the 7075-T651 decreased the SCC penetration (Ref 17). The equivalent performance of the 7075 -T7X1 3.2 and 5.7 mm (0.125 and 0.225 in.) diameter specimens is evident. In contrast, Fig. 17 shows the specimen biases in SCC ratings obtained by traditional pass-fail methods (Ref 18). At the time of this writing, an ASTM standard is being developed by Subcommittee G.01.06 on Stress-Corrosion Cracking and Corrosion Fatigue for "Determining the SCC Resistance of High Strength Aluminum Alloy Products Using the Breaking Load Method."

Although this test method was developed for use with high strength aluminum alloys, the concept which uses residual strength as a measure of damage evolution can in principal, be applied to any alloy system in any environment. More infor-

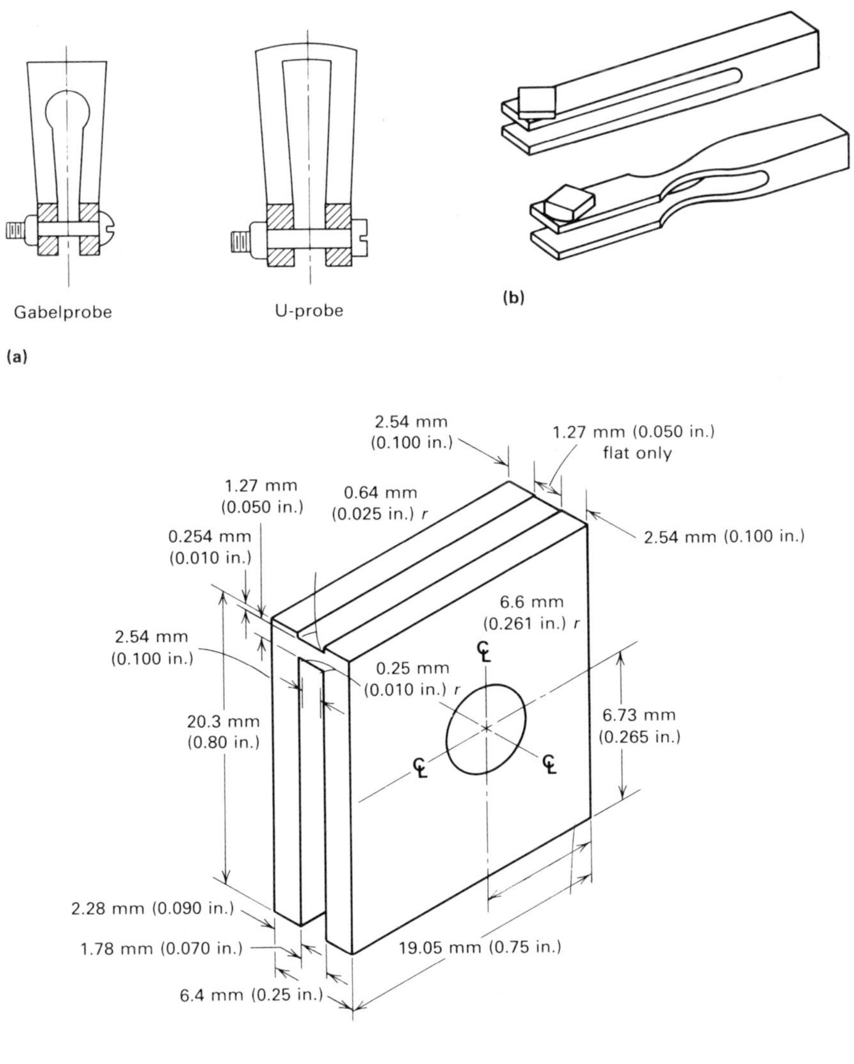

Fig. 18 Typical tuning fork SCC test specimens. (a) Source: Ref 19. (b) Source: Ref 4. (c) Source: Ref 21

mation on this promising new method including the use of a Box-Cox transformation used in the statistical analysis of the breaking load data is available in a paper by E.L. Colvin and M.R. Emptage, "The Breaking Load Method: Results and Statistical Modifications from the ASTM Interlaboratory Test Program", *New Methods for Corrosion Testing of Aluminum Alloys*, ASTM STP 1134, ASTM, 1992, p 82.

Tuning fork specimens are special-purpose specimens with numerous modifications (Fig. 18). In Europe, the metal is strained into the plastic range, and stresses and strains are usually not measured (Ref 19, 20). In the United States, however, these specimens have been used with measured strains in the elastic and plastic ranges. Specimens of the type shown in Fig. 18(b) are convenient when a small, self-contained specimen is required that will afford some insight into the applied stresses. Such a specimen is particularly well suited for testing thin plate material in the longitudinal or long-transverse direction while keeping the original mill-finished surface intact.

Tuning fork specimens are stressed by closing the specimen tines and restraining them in the closed position with a bolt placed at the tine ends. The amount of closure is determined from Eq 3, which was derived from data obtained with strain gages placed at the base of the tines on calibration specimens (Ref 4):

$$S = A\Delta t \qquad \text{(Eq 3)}$$

where S is the maximum tension stress in the outer fiber of either tine, A is the calibration constant, Δ is the total amount of closure at the tine ends, and t is the thickness of the tines.

The stress on tuning forks with straight tines is greatest in a small area at the base of the tines. In tuning forks with tapered tines, the maximum stress extends uniformly along the tapered section. Tuning forks must be given the same consideration with regard to biaxial stresses as other flexurally loaded specimens.

The miniature tuning fork shown in Fig. 18(c) was devised to conduct short-transverse tests on sections that are too thin for tensile specimens or C-rings to be obtained (Ref 21). As with other tuning fork specimens, the relationship between strain on the grooved surface and the deflection at the ends of the legs can be determined through the use of strain gages.

Plastic Strain Specimens

Many accelerated SCC tests are performed with plastically deformed specimens, because these specimens are simple and economical to manufacture and use. These specimens are convenient for multiple replication tests of self-stressed (fixed-deflection) specimens in all environments. Because they usually contain large amounts of elastic and plastic strain, they provide one of the most severe tests available for smooth SCC test specimens.

Generally, the stress conditions are not known precisely, and the anticipated high level of stress can be obtained consistently only if the precautions described later for each type of specimen are observed. Another consideration is that the cold work required to form the test specimen may change the metallurgical condition and the SCC behavior of certain alloys.

Tests of this type are primarily used as screening tests to detect large differences between the SCC resistance of one alloy in several environments, one alloy in several metallurgical conditions in a given environment, and different alloys in the same environment. These tests are sometimes claimed to be too severe and therefore unsuitable for many applications, but the stress conditions are nevertheless representative of the high locked-in fabrication and assembly stresses frequently responsible for SCC in service.

U-bend specimens are rectangular strips bent approximately 180° around a predetermined radius and maintained in this plastically (and elastically) deformed condition during the test. Standardized test methods for this type of specimen are described in ASTM G 30 (Ref 8) and in Part 3 of Ref 6. Bends slightly less than or greater than 180° are also used, but the term U-bend is generally applied to test specimens that are bent beyond their elastic limits. Figure 19 illustrates typical U-bend configurations showing several different methods of maintaining the applied stress.

U-bend specimens can be used for all materials sufficiently ductile to be formed into a U-con-

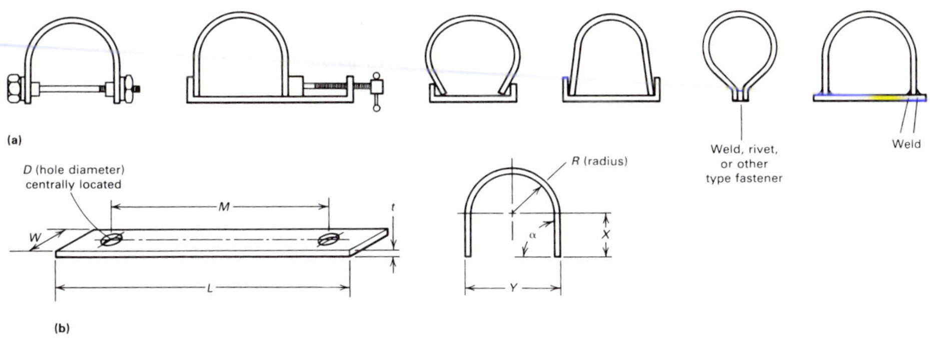

Alternative size	L mm	L in.	M mm	M in.	W mm	W in.	t mm	t in.	D mm	D in.	X mm	X in.	Y mm	Y in.	R mm	R in.
A	80	3.2	50	2.0	20	0.8	2.5	0.098	10	0.4	32	1.26	14	0.55	5	0.2
B	100	4.0	90	3.5	9	0.35	3.0	0.12	7	0.28	25	0.98	38	1.50	16	0.6
C	120	4.7	90	3.5	20	0.8	1.5	0.06	8	0.31	35	1.4	35	1.4	16	0.6
D	130	5.1	100	4.0	15	0.6	3.0	0.12	6	0.24	45	1.77	32	1.26	13	0.51
E	150	5.9	140	5.5	15	0.6	0.8	0.03	3	0.12	61	2.40	20	0.8	9	0.35
F	310	12.2	250	9.8	25	0.98	13.0	0.51	13	0.51	105	4.13	90	3.5	32	1.26
G	510	20.1	460	18.1	25	0.98	6.5	0.26	13	0.51	136	5.35	165	6.5	76	3.0

Note: α = 1.57 rad

Fig. 19 Typical U-bend specimens. (a) Various methods of stressing U-bends. (b) Typical U-bend specimen dimensions. Source: Ref 8 (G 30)

figuration without cracking. A U-bend specimen is most easily made from strips of sheet, but specimens can be machined from plate, bar, wire, tubes, castings, and weldments. Of primary interest in U-bend specimens is circumferential stress, which is not uniform, as discussed previously in the section on "Bent-Beam Specimens" in this chapter. Stress distribution in the U-bend specimen is discussed in detail in Ref 22.

A good approximation of applied strain, ε, can be obtained by:

$$\varepsilon = \frac{t}{2R} \text{ when } t < R \qquad \text{(Eq 4)}$$

where t is the specimen thickness and R is the radius of curvature at the point of interest. Knowledge of the stress-strain curve is necessary to determine the stress. When a U-bend specimen is formed, the material in the outer fibers of the bend is strained into the plastic portion of the true stress/true strain curve, such as in section AB in Fig. 20(a). Several other stress-strain relationships that can exist in the outer fibers of a stressed U-bend test specimen are shown in Fig. 20(b) through (e). The actual relationship obtained depends on the method of stressing used.

Stressing is usually achieved by a one- or two-stage operation, as illustrated in ASTM G 30 (Ref 8). Single-stage stressing is accomplished by bending the specimen into shape and maintaining it in that shape. The two types of stress conditions that can be obtained by single-stage stressing are defined by point X in Fig. 20(b) and (c). In Fig. 20(c), some elastic strain relaxation has occurred by allowing the U-bend legs to spring back slightly at the end of the stressing sequence.

Two-stage stressing involves forming the approximate U-shape and then allowing the elastic strain to relax completely before the second stage of applying the test stress. The applied test strain, as defined by point X, can be a percentage (from 0 to 100%) of the tensile elastic strain that occurred during preforming (Fig. 20d) or can involve additional plastic strain (Fig. 20e). The convex specimen surface is stressed in tension in the region 0NM (Fig. 20d), and the concave surface is in compression. In the region MP, the situation is reversed; that is, compression is on the convex surface, and tension is on the concave surface.

The slope MN of the curve shown in Fig. 20(d) is steep. Therefore, it is often difficult to apply reproducibly a constant percentage of the to-

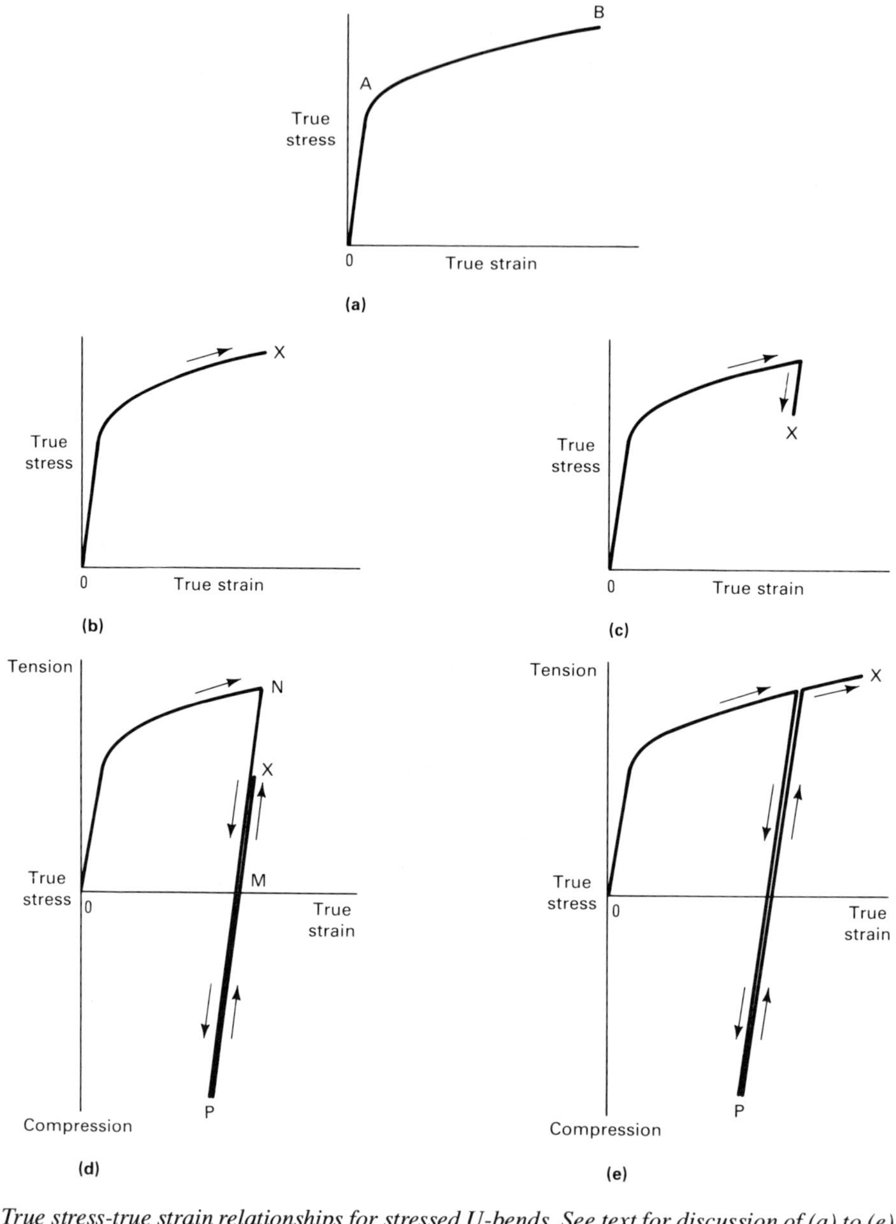

Fig. 20 True stress-true strain relationships for stressed U-bends. See text for discussion of (a) to (e). Source: Ref 8 (G 30)

tal elastic prestrain, and the specimen surface may remain under compressive stress. Therefore, because they result in a more severe test (that is, higher applied stress), the stress conditions in Fig. 20(b) and (e) are recommended. Thus, the final applied strain prior to testing consists of plastic and elastic strain. To achieve the conditions illustrated in Fig. 20(b) and (e), springback of the U-bend legs after achieving the final plastic strain must be avoided. For materials with relatively low creep resistance, there will be some strain relaxation.

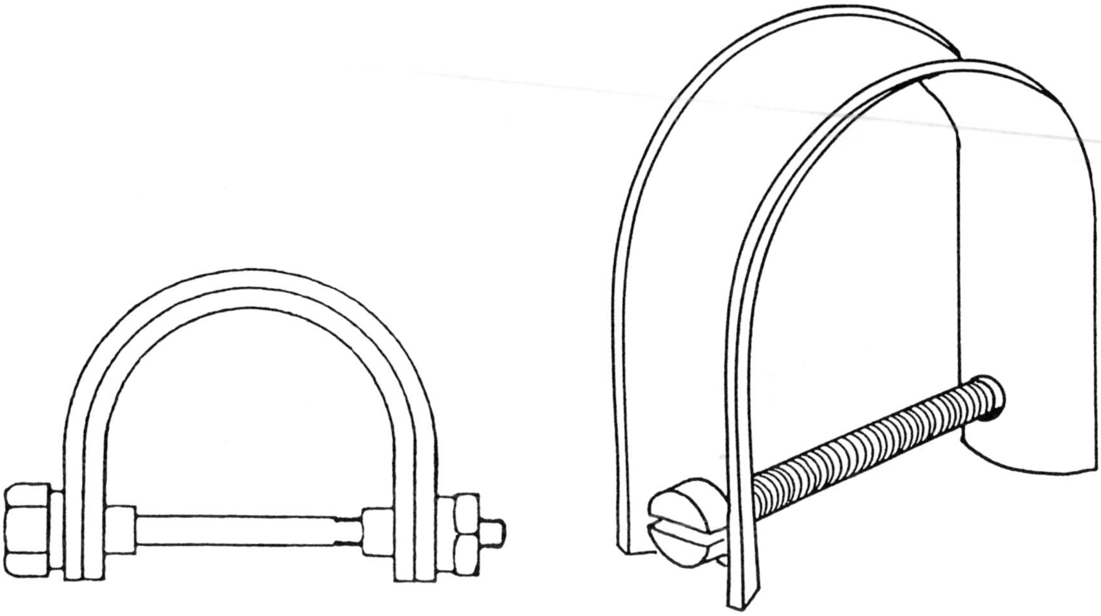

Fig. 21 Two special-purpose U-bend specimens. Source: Ref 23

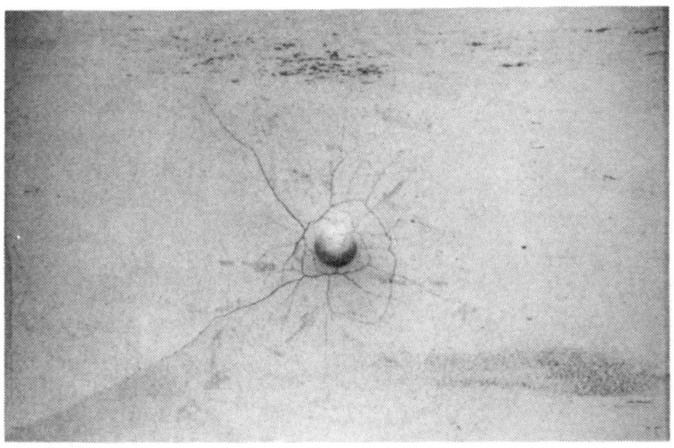

(a)

(b)

Fig. 22 SCC test specimens containing residual stresses from plastic deformation. (a) Cracked cup specimen (Ericksen impression). Source: Ref 4. (b) Joggled extrusion containing SCC in the plastically deformed region. Source: Ref 9

Two special types of U-bend specimens that have been used for simulating the extreme stress and environmental conditions that are likely to be encountered in high-temperature water environments of interest to the nuclear power industry are the double U-bend and the split-tubing U-bend (Fig. 21). To facilitate solution access to the crevice area in the double U-bend specimen, a groove can be machined in the faying surface of the inner bend, allowing a tapered crevice to develop in the area of maximum stress. For a 150 × 15 × 0.8 mm (6 × 0.6 × 0.03 in.) specimen blank, a 25 mm (1 in.) long by 0.125 mm (0.005 in.) deep full-width groove centered on the inner specimen was found to be effective (Ref 24).

Residual Stress Specimens (Self-Stressed)

Most industrial SCC problems are associated with residual stresses developed in the metal during such processes as heat treatment, fabrication, and welding. Therefore, residual stress specimens that simulate anticipated service conditions are useful for assessing the SCC performance of

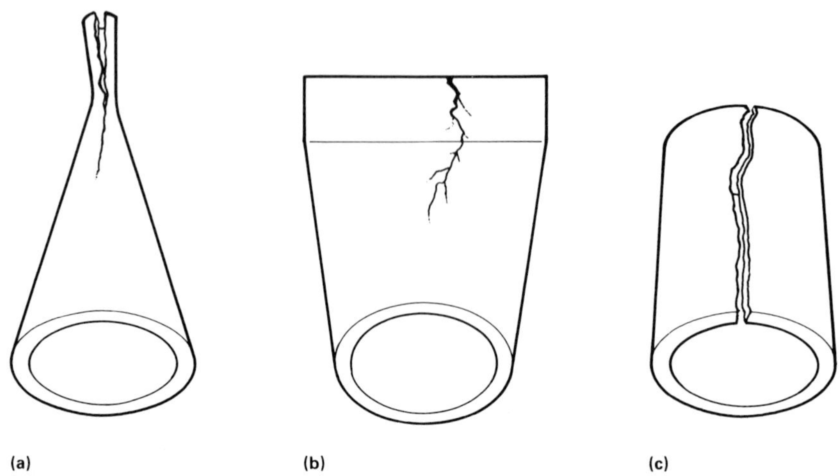

Fig. 23 SCC test specimens containing residual stresses from plastic deformation. Shown are 12.7 mm (0.5 in.) diameter stainless tubular specimens after SCC testing. (a) and (b) Annealed tubing that was cold formed before testing. (c) Cold-worked tubing tested in the as-received condition. Source: Ref 4

Fig. 24 SCC test specimen containing residual stresses from welding. (a) Sandwich specimen simulating rigid structure. Note SCC in edges of center plate. Source: Ref 10. (b) Cracked ring-weld specimen. Source: Ref 4

some materials in particular structures and in specific environments.

Plastic Deformation. Residual stresses resulting from fabricating operations such as forming, straightening, and swaging, which involve localized plastic deformation at room temperature, can exceed the elastic limit of the material. Examples of specimens of this type that have been used are shown in Fig. 22 and 23. Other specimen types used include panels with sheared edges, punched holes, or stamped identification numbers and specimens that represent practical fabricating operations.

Weldments. Residual stresses developed in and adjacent to welds are frequently a source of SCC in service. Longitudinal stresses in the vicinity of a single weld are unlikely to be excessive, because stress in the weld metal is limited by the yield strength of the hot metal that shrinks as it cools. High stresses can be built up, however, when two or more weldments are joined into a more complex structure.

Test specimens containing residual welding stresses are shown in Fig. 24. In fillet welds, residual tensile stress transverse to the weld can be critical, as indicated in Fig. 24(a) for a situation in which the tension stress acts in the short-transverse direction in an aluminum-zinc-magnesium alloy plate.

Surface Preparation of Smooth Specimens

The pronounced effect of surface conditions on the time required to initiate SCC in test specimens is well known (Ref 3). Unless the as-fabricated surface is being studied specifically, the final surface preparation generally preferred is a mechanical process followed by degreasing. However, chemical etches or electrochemical polishes can be used to remove heat-treating films or thin layers of surface metal that may have become distorted during machining.

Care should be exercised to select an etchant that will not selectively attack constituents or phases in the metal and that will not deposit undesirable residues on the surface. Etching or pickling should not be used with alloys that are susceptible to hydrogen embrittlement.

Precautions should be taken when machining specimens to avoid overheating, plastic deformation, or the development of residual stress in the metal surface. Machining should be performed in stages so that the final cut leaves the principal surface with a clean finish of 0.7 μm (30 μin.) rms or smoother. The required machining sequences, types of tools, and feed rate depend on the alloy and metallurgical condition of the test piece. Lapping, mechanical polishing, and similar operations that produce flow of the metal should be avoided.

Static Loading of Precracked (Fracture Mechanics) Specimens

The use of mechanically precracked (fracture mechanics) specimens is based on an engineering design concept that recognizes that crack-like flaws introduced during manufacture or subsequent service are generally present in thick structural components. Moreover, the presence of such flaws can contribute to a susceptibility to SCC, which in some materials, such as titanium, may not be evident from tests of smooth specimens. In the presence of such flaws at the surface of test specimens, the applied nominal stress is inadequate for quantifying the SCC response of the test material. It is instead necessary to invoke fracture mechanics principles, enabling the stress situation to be expressed in terms of the crack-tip stress-intensity factor, K (or J). Above a certain threshold level of K, SCC in a susceptible material will initiate and grow; below that level, however, no measurable propagation is observed. An experimentally determined threshold stress-intensity factor is designated K_{ISCC} (Ref 2, 25-28).

A particular advantage of precracked specimens is that they allow data to be acquired from which critical flaw sizes, above which SCC may occur, can be estimated for components of known geometry subjected to known stresses and environments. They also enable rates of SCC propagation to be determined.

A disadvantage of this method of SCC testing involves the evaluation of thin products, such as sheet. Because of the need to maintain elastically constrained conditions at the crack-tip, precracked specimens are not suitable for the evaluation of thin products and are generally restricted to thicker products, such as plate, forgings, and so on.

Although precracking may shorten the initiation period for SCC of some materials, the initiation stage is not generally circumvented. Therefore, this method of testing may sometimes require long exposure periods.

Test Specimens

Almost all standard plane-strain fracture toughness test specimens can be adapted to SCC testing. Standard configurations should be used to ensure valid fracture analyses. Several examples are illustrated schematically in Fig. 25, where they are classified with respect to loading methods and the relationship with the stress-intensity factor as SCC propagates. Proportional dimensions and tolerances per ASTM E 399 (Ref

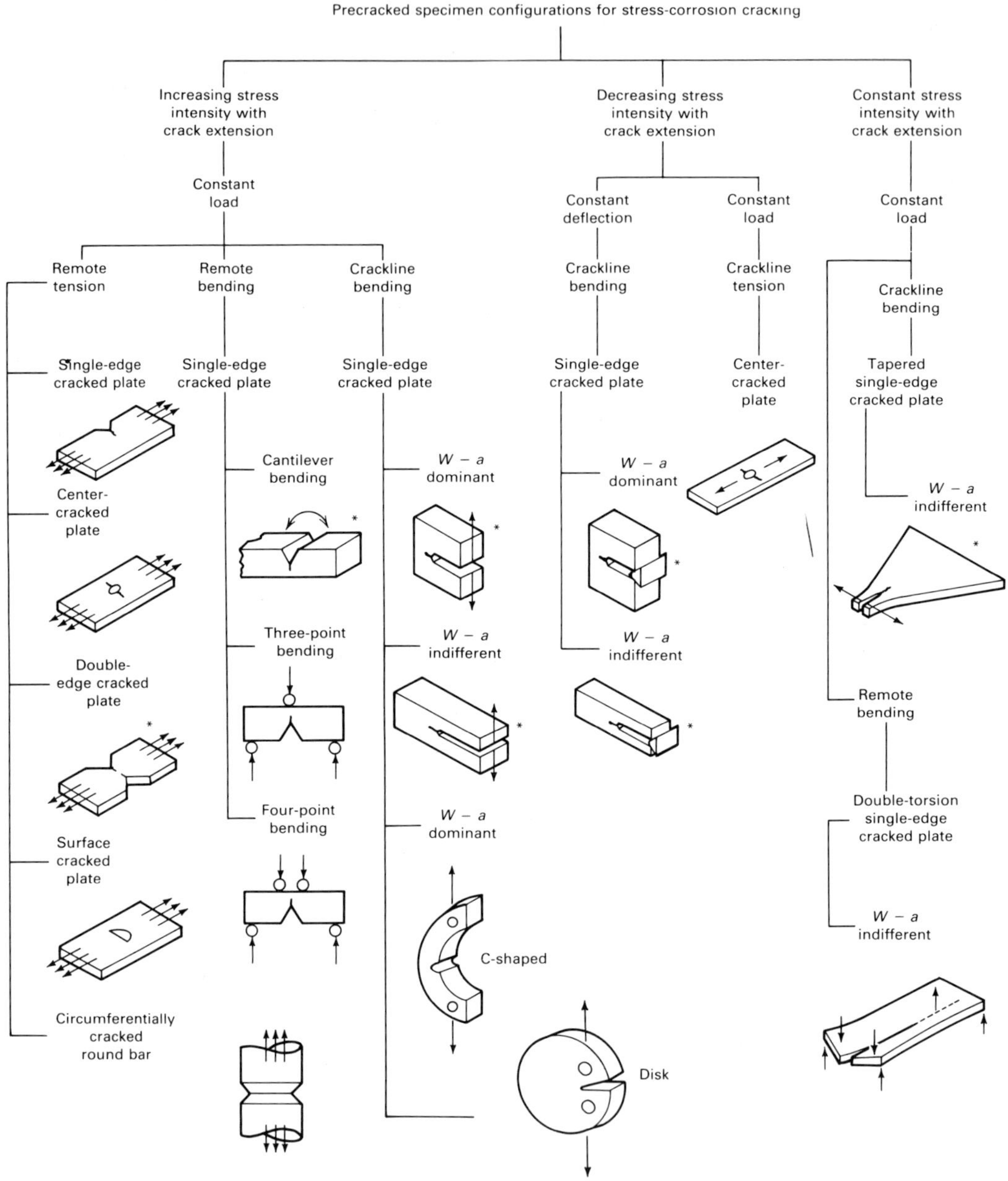

Fig. 25 Classification of precracked specimens for SCC testing. Asterisks denote commonly used configurations. Source: Ref 26

29) for the more commonly used specimens are given in Fig. 26(a) to (c). Minor modifications can be made to these configurations to accommodate different loading arrangements and to facilitate mechanical precracking without invalidating the plane-strain constraints on the specimens. Examples of alternative chevron-notch and face-groove designs are shown in Fig. 27.

Although precracked specimens have been used for SCC testing since about 1966 and many articles have been published on this type of testing (Ref 2, 25-48), it was not until 1989 that the first standard, ISO 7539-6, was published (Ref 6, Part 6). The procedures presented in that standard are also discussed in Ref 23. Practices for the ma-

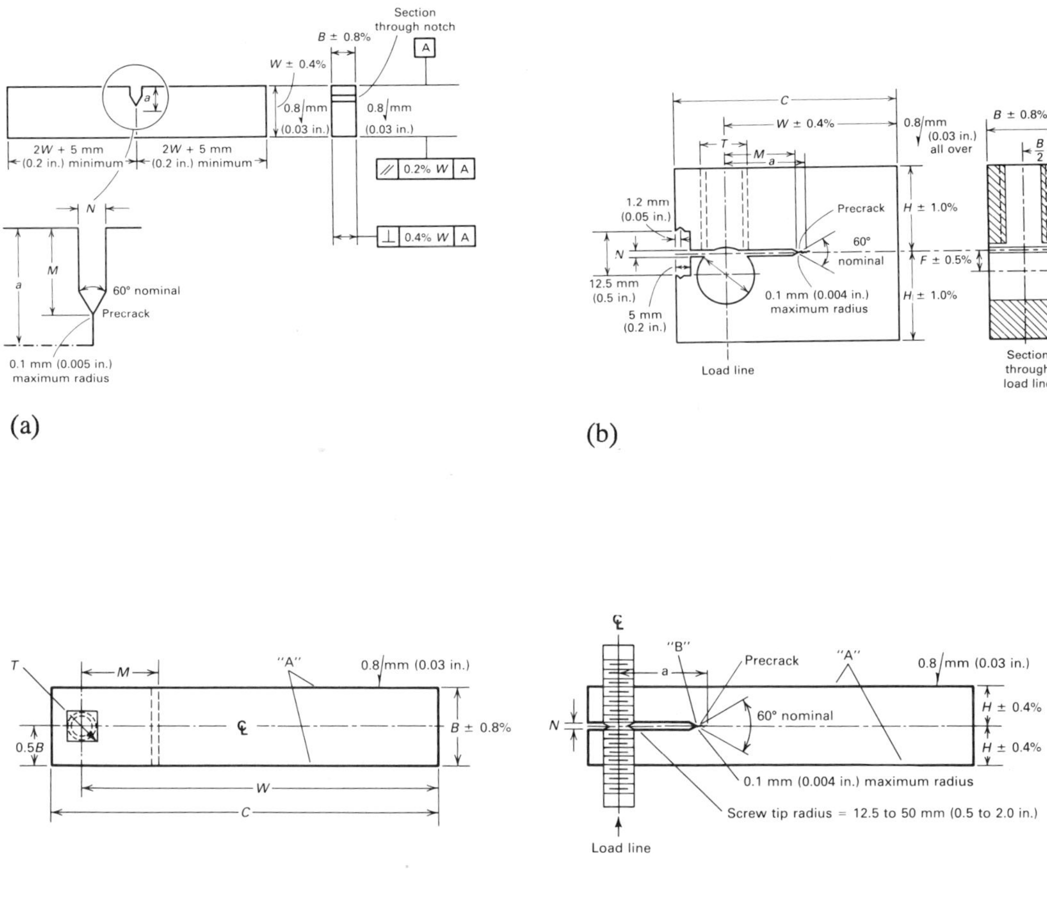

(c)

Fig. 26 (a) Proportional dimensions and tolerances for cantilever bend test specimens. Width = W; thickness (B) = 0.5W; half loading span (L) = 2W; notch width (N) = 0.065W maximum if W > 25 mm (1.0 in.); N = 1.5 mm (0.06 in.) maximum if W ≤ 25 mm (1.0 in.); effective notch length (M) = 0.25 to 0.45W; effective crack depth (a) = 0.45 to 0.55W. (b) Proportional dimensions and tolerances for modified compact specimens. Surfaces should be perpendicular and parallel as applicable to within 0.002H TIR. The bolt centerline should be perpendicular to the specimen centerline within 1°. Bolt of material similar to specimen where practical; fine threaded, square or Allen head. Thickness = B; net width (W) = 2.55B; total width (C) = 3.20B; half height (H) = 1.24B; hole diameter (D) = 0.718B + 0.003B; effective notch length (M) = 0.77B; notch width (N) = 0.06B; thread diameter (T) = 0.625B. (c) Proportional dimensions and tolerances for double-beam specimens. "A" surfaces should be perpendicular and parallel as applicable to within 0.002H TIR. At each side, point "B" should be equidistant from the top and bottom surfaces to within 0.001H. The bolt centerline (load line) should be perpendicular to the specimen centerline to within 1°. Bolt of material similar to specimen where practical; fine threaded, square or Allen head. Half height = H; thickness (B) = 2H; net width (W) = 10H minimum; total width (C) = W + T; thread diameter (T) = 0.75H minimum; notch width (N) = 0.14H maximum; effective notch length (M) = 2H chining and fatigue precracking of the various specimens are covered by ASTM E 399 (Ref 29).

Cantilever bend specimens (Fig. 26a), sometimes referred to as single-edge-notched cantilever bend specimens, have been used in constant-load tests (K-increasing) for characterizing high-strength steels and titanium alloys (Ref 30). Equations 5 to 7 are recommended (Ref 31, 32):

$$K = \frac{6M}{BW^2}\left(\frac{2W}{\pi a}\tan\frac{\pi a}{2W}\right)^{1/2} \times \left[\frac{0.923 + 0.199\left(1 - \sin\frac{\pi a}{2W}\right)^4}{\cos\frac{\pi a}{2W}}\right] \quad \text{(Eq 5)}$$

$$\frac{EBW\,(2V_0)}{M} = e^x \quad \text{(Eq 6)}$$

$$\frac{EBW\,V_{LL}}{M} = e^y \quad \text{(Eq 7)}$$

where e is the base of natural logarithm (2.718); $x = [0.1426 + 11.92(a/W) - 17.42(a/W)^2 + 15.84(a/W)^3 - 2.235(a/W)^4$; $y = [6.188 + 12.98(a/W) - 41.19(a/W)^2 + 54.98(a/W)^3 - 22.28(a/W)^4$; M is the applied bending moment; B is the specimen thickness (face grooves, when present, may be accounted for by replacing B with $\sqrt{BB_n}$, where B_n is the net thickness as the base of the face grooves; see Fig. 27b); W is the depth of the specimen; a is the depth of the notch plus crack; E is the modulus of elasticity; $2V_0$ is the total crack mouth opening displacement at the top face of the specimen; and V_{LL} is the total crack mouth opening displacement measured at the point of load application, which will vary depending on the load arm length.

Equation 5 is an expression for the stress intensity of a rectangular beam in pure bending and is valid over a wide range of a/W values. It applies to mode I loading only, however, and the usual tests include a mode II component from resulting shear stresses.

Equations 6 and 7 were determined by fitting experimental compliance data for cantilever bend specimens with a polynomial equation expressing the natural log of the normalized compliance as a function of a/W. These experimental values are in excellent agreement with those determined from Eq 5 for pure bending, even though the stress state at the crack tip will differ for cantilever bending. It has been suggested that analyses using pure bending expressions related to compliance measurement are suitable for testing with the cantilever bend configuration (Ref 31). Therefore, the simpler Eq 8 is also considered suitable (Ref 6, Part 6):

$$K = \frac{4.12\,Mo\sqrt{(1/\alpha^3) - \alpha^3}}{B\sqrt{W^3}} \quad \text{(Eq 8)}$$

where Mo is the applied bending moment in inch-pounds and $\alpha = 1 - a/W$.

Crack growth measurements can be made with clip gage readings in conjunction with the

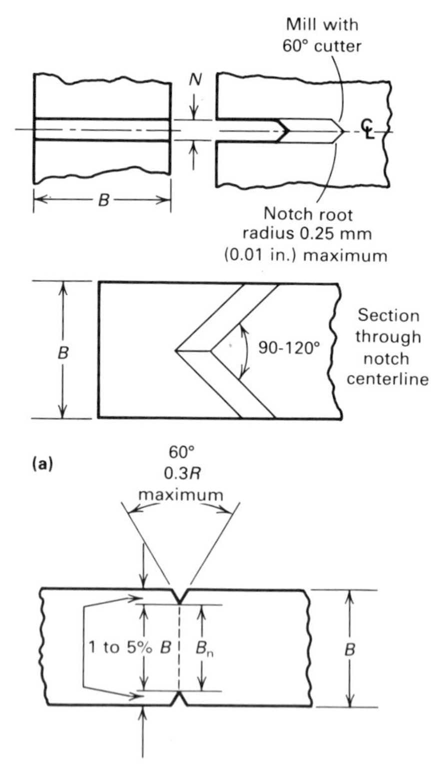

Fig. 27 Alternative chevron notch (a) and face groove (b) designs for single-edge cracked specimens

crack opening displacement calibrations given above or by any other method that can be verified within ±0.127 mm (±0.005 in.). Examples of various methods are given in Ref 31 and 33.

Modified compact specimens (K-decreasing), as shown in Fig. 26(b), are frequently referred to as 1T-WOL (wedge-opening loaded) or modified WOL specimens. Although most frequently used with constant-displacement (bolt) loading (Ref 30, 34), these specimens have also been used with constant load (Ref 1, 35, 36). The specimen configuration shown in Fig. 26(b) is similar to that adopted by the U.S. Navy (Ref 30), except that it does not incorporate face grooves.

Equations 9 to 12 can be used to calculate stress-intensity levels and normalized crack opening displacements for fatigue precracking, for initiation of SCC, and for subsequent intervals during the test. These equations are based on boundary collocation values determined for this type of specimen configuration with face grooves

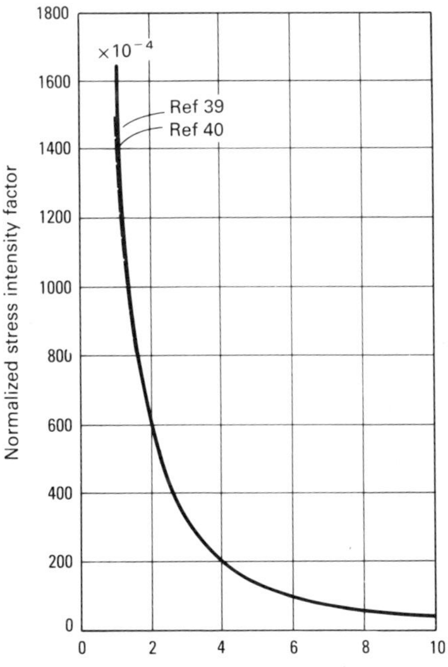

Fig. 28 Configuration and K_I calibration of a double-beam plate specimen. Normalized stress intensity K_I plotted against a/H ratio. (W - a) indifferent, crack-line-loaded, single-edge cracked specimen. Source: Ref 26

and bolt loading (threaded bolt against a rigid loading tip) (Ref 31). The polynominal regression equation agrees with experimentally determined colocation values within 1% for $0.2 \le a/W \le 0.95$:

$$K_{Io} = \frac{P\left(2 + \frac{a_0}{W}\right)}{B\sqrt{W}\left(1 - \frac{a_o}{W}\right)^{3/2}} \times \left[1.308 + 5.278\left(\frac{a_0}{W}\right) - 19.67\left(\frac{a_0}{W}\right)^2 + 24.57\left(\frac{a_0}{W}\right)^3 - 10.27\left(\frac{a_0}{W}\right)^4\right] \quad \text{(Eq 9)}$$

$$2V_0 = \frac{P}{EB} \times e^{x} \quad \text{(Eq 10)}$$

where $x = [1.830 + 4.307(a_0/W) + 5.871(a_0/W)^2 - 17.53(a_0/W)^3 + 14.57(a_0/W)^4]$

$$2V_{LL} = 2V_0 \times \frac{e^{y}}{e^{x}} \quad \text{(Eq 11)}$$

where $y = [1.623 + 3.352)a_0/W + 8.205(a_0/W)^2 - 19.59(a_0/W)^3 + 15.23(a_0/W)^4]$

$$K_{I_i} = \frac{E(2V_{LL})\left(2 + \frac{a_i}{W}\right)}{\sqrt{W}\left(1 - \frac{a_i}{W}\right)^{3/2}} \times \left\{\left[1.308 + 5.278\left(\frac{a_i}{W}\right) - 19.67\left(\frac{a_i}{W}\right)^2 + 24.57\left(\frac{a_i}{W}\right)^3 - 10.27\left(\frac{a_i}{W}\right)^4\right] \div e^{z}\right\} \quad \text{(Eq 12)}$$

where $z = [1.623 + 3.352(a_i/W) + 8.205(a_i/W)^2 - 19.59(a_i/W)^3 + 15.23(a_i/W)^4]$. In Eq 9 to 12, K_{Io} is the desired starting stress intensity, a_0 is the starting crack length, P is the load calculated to develop K_{Io} with measured a_0, W is the net width of the specimen measured from the load line, K_{Ii} is the stress intensity after time interval i, a_i is the crack length after time interval i, and $2V_{LL}$ is the total crack mouth opening displacement at the load line, and 2Vo is the total crack mouth opening displacement at the location of the clip gap. All other quantities are as defined previously.

Standard compact tension specimens have been used for K-decreasing SCC tests (Ref 26, 35, 36). Equations for the stress-intensity relationships are given in ASTM E 399 (Ref 29) and in ISO 7539-6 (Ref 6, Part 6).

Double-beam specimens (K-decreasing or K-increasing), also referred to as double-cantilever beam specimens, are similar to modified compact specimens, but because of their greater width (or length), they are well suited for studying SCC growth rates over a greater range of K_I values. The smaller height of these specimens (Fig. 26c) allows more versatility in performing short-transverse tests from moderate thicknesses of material. Like compact specimens, double-beam specimens are generally used with constant-displacement (bolt) loading for convenience, but they can also be designed for use with constant load.

Equations 13 to 16 are recommended for general use with double-beam specimens:

$$K_{I_i} = \frac{E\,V_{LL}\,H\sqrt{3H(a_i+0.6H)^2+H^3}}{4\,[(a_i+0.6H)^3+H^2a_i]} \qquad \text{(Eq 13)}$$

$$K_{I_i} = \frac{E\,V_{LL}\,H\sqrt{3H(a_i+\alpha H)^2+H^3}}{4\,[(a_i+\alpha H)^3+H^2a_i]} \qquad \text{(Eq 14)}$$

$$K_{I_i} = \frac{\sqrt{3}EV_{LL}}{4\sqrt{H}\left(\frac{a_i}{H}+0.6728\right)^2} \qquad \text{(Eq 15)}$$

$$K_I = \frac{\sqrt{12}Pa\,(1+0.6728\,H/a)}{\sqrt{BH^3}} \qquad \text{(Eq 16)}$$

Equation 13 was suggested by Mostovoy *et al.* (Ref 43) and has been used extensively by Hyatt and others for short-transverse tests of thick high-strength aluminum alloy products (Ref 1, 36-40, 44). This equation is also included in ISO Standard 7539-6. Equation 14 includes a slight refinement of the empirical rotation correction factor, denoted α rather than 0.6 as in Eq 13. Experimental calibration tests by Micheletti and Buratti (Ref 36) indicated a value of $(0.4 - 0.01a)$ for α when $a \leq 40$ mm (1.5 in.) and a value of 0 when $a > 40$ mm (1.5 in.).

Equation 15 is derived from the work of Fichter (Ref 41) and was used in an interlaboratory testing program conducted by ASTM Subcommittee G.01.06 (Ref 45). This simplified equation is believed to provide more versatility with high accuracy for a wider range of configurations and K-values (crack growth) than previously published equations. Equation 16, also from Fichter (Ref 41), is for fatigue precracking of double-beam specimens.

An advantage of the double-beam specimen for testing aluminum alloys is that tests can be started from a "pop-in" mechanical crack, eliminating the need for fatigue precracking. A disadvantage is that tests with these specimens give K_{th} values that, although allowing the selection of materials with regard to this property, cannot be considered valid values of K_{ISCC} because they are not homologous with K_{Ic} values.

Two early K_I calibrations based on stress analysis (Ref 42) and compliance (Ref 43) are illustrated in Fig. 28 and are in excellent agreement. The shape of these curves can also be used as a design guide for preparing specimens. If the test must be completed in the shortest possible time, a_0 should be short to capitalize on the fact that the rate of decrease of K_I with crack extension is maximum for shallow cracks. However, if maximum accuracy is desired, a deeper crack (effective notch length M, Fig. 26c) should be chosen so that errors in crack length measurement do not cause significant errors in K_I.

Although in early work with aluminum alloys (Ref 37, 38) a relatively short effective notch length was used ($a_0/H \cong 0.9$), deeper notches have been used recently ($a_0/H \cong 1.5$ to 2.2), all with a $2H$ value of 25.4 mm (1.0 in.) (Ref 36, 39, 40, 45). The recommended starting a/H value shown in Fig. 26(c) is about 2 to 2.2, depending on the length of the precrack. Limited tests of a smaller beam height of $2H$ = 12.7 mm (0.5 in.) have shown little effect on the amount and rate of crack growth in aluminum alloy 7075 plate (Ref 44); however, additional study is needed in this area.

An alternative double-cantilever beam specimen has been developed for testing relatively thin sections (typically 6.4 mm, or $\frac{1}{4}$ in., thick) of low-alloy steels (Ref 46). The specimen is stressed by forcing an appropriate dimensioned wedge into the slot. These specimens have been used to determine the effect of hardness of low-allow steels on their resistance to SCC in environments containing hydrogen sulfide.

Constant K_I specimens are well suited for studying the mechanisms of SCC, because the stress intensity, K_I, is not dependent on crack depth and can be neglected in kinetic studies. Other attractive features are the relatively simple expressions for stress intensity and compliance and the apparent retention of plane-strain conditions in thin plate and sheet specimens. The cost of specimen preparation and instrumentation, however, prohibits its use for extensive SCC characterizations.

Reference 26 provides equations for the analysis of two types of constant K_I specimens: the tapered double-beam specimen and the double-torsion-loaded single-edge cracked specimen. A recent evaluation of the double-torsion method (Ref 47) used Al-Zn-Mg alloy sheet 3.2 mm (0.125 in.) thick. By using the double-torsion specimen, *V-K* curves were produced for aluminum alloy 7075-T651 sheet with conventional two-stage growth and plateau velocities that were only slightly higher than those for conventional double-cantilever beam tests of plate.

Other precracked specimen configurations, such as those shown in Fig. 25, can be used for special testing conditions. Information on the preparation and use of these specimens and the related fracture mechanics equations are given in Ref 26 and 48 to 50.

Preparation of Precracked Specimens

When using precracked SCC test specimens, the investigator must consider the dimensional (size) requirements of the specimen, its crack configuration and orientation, and machining and precracking of the specimen. These considerations are discussed below. Additional guidelines and recommendations on specimen preparation in conjunction with fracture toughness testing are given in Ref 26, 29, and 49 to 52.

Dimensional Requirements. A basic requirement of all precracked specimen configurations is that the dimensions be sufficient to maintain predominantly triaxial stress (plane-strain) conditions, in which plastic deformation is limited to a very small region in the vicinity of the crack tip. Experience with fracture toughness testing has shown that for a valid K_{Ic} measurement neither the crack depth, *a*, nor the thickness, *B*, should be less than $2.5(K_{Ic}/YS)^2$, where YS is the yield strength of the material (Ref 30). Because of the uncertainty regarding a minimum thickness for which an invariant value of K_{ISCC} can be obtained, guidelines for designing fracture mechanics test specimens should be tentatively followed for SCC test specimens. The threshold stress-intensity value should be substituted for K_{Ic} in the above expression as a test of its validity.

If specimens are to be used for determination of K_{ISCC}, the initial specimen size should be based on an estimate of the K_{ISCC} of the material. Overestimation of the K_{ISCC} value is recommended; therefore, a larger specimen should be

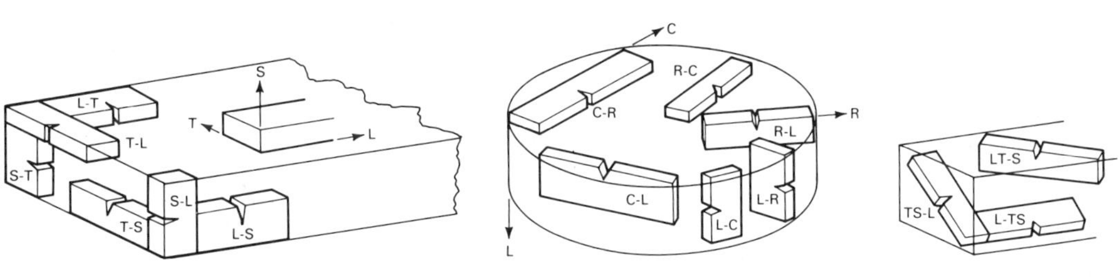

Fig. 29 Specimen orientation and fracture plane identification. L, length, longitudinal, principal direction of metal working (rolling, extrusion, axis of forging); T, width, long-transverse grain direction; S, thickness, short-transverse grain direction; C, chord of cylindrical cross section; R, radius of cylindrical cross section. First letter: normal to the fracture plane (loading direction); second letter: direction of crack propagation in fracture plane. Source: Ref 29

used that may eventually be necessary. When determining stress-corrosion crack growth behavior as a function of stress intensity, specimen size should be based on the highest stress intensity at which crack growth rates are to be measured (substitute K_{Io} in the $2.5(K_{Ic}/YS)^2$ expression).

Notch Configuration and Orientation. For SCC testing, the depth of the initial crack-starter notch—that is, the machined slot with a fatigue or mechanical pop-in crack at its apex— can be as short as 0.2*W*. Guidelines for the depth of the notch depend on the limits of accurate K_I calibration with respect to the range of *a*/*W* or *a*/*H* and the considerations discussed previously for double-beam specimens.

Several designs of crack-starter notches are available for most plate specimens. The machined slot is used to simulate a crack, because it is impractical to produce plane cracks of sufficient size and accuracy in plate specimens. ASTM E 399 (Ref 29) recommends that the notch root radius should not be greater than 0.127 mm (0.005 in.), unless the chevron form is used, in which case it may be 0.25 mm (0.01 in.) or less (Fig. 27). This tolerance can be easily achieved with conventional milling and grinding equipment.

A significant factor in the SCC testing of thick sections of some metals, such as aluminum and titanium, is the direction of applied stress relative to the grain structure. A standardized plan for identifying the loading direction, the fracture plane, and the direction of crack propagation is shown in Fig. 29.

Machining. Specimens of the required orientation should be machined from products in the fully heat-treated and stress-relieved condition to avoid complications due to residual stresses in the finished specimens. Safeguards against the presence of residual stresses are especially important for precracked specimens, because these specimens are usually bulky and contain notches that are machined deep into the metal. For specimens of material that cannot easily be completely machined in the fully heat-treated condition, the final thermal treatment can be given before the notching and finishing operations. However, fully machined specimens should be heat treated only when the heat treatment will not result in distortion, residual stress, quench cracking, or detrimental surface conditions.

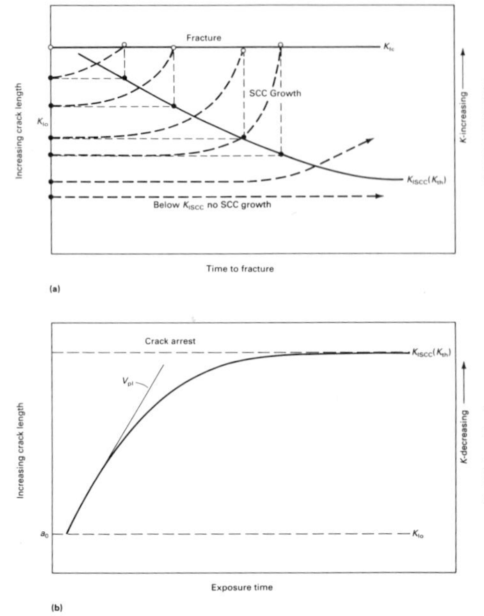

Fig. 30 Schematic comparison of determination of K_{ISCC} or K_{th}. (a) Constant-load (K-increasing) test. (b) Constant crack opening displacement (K-decreasing) test

Precracking. Fatigue precracking should be done in accordance with ASTM E 399 (Ref 29). The *K* level used for precracking each specimen should not exceed about two-thirds of the intended starting *K*-value for the environmental exposure. This prevents fatigue damage or residual compressive stress at the crack tip, which may alter the SCC behavior, particularly when testing at a *K* level near the threshold stress intensity for the specimen.

Aluminum alloy specimens can also be precracked by pop-in methods (wedge-opening loaded to the point of tensile overload), but steel and titanium alloys are usually too strong and tough to pop in without breaking off one of the specimen arms. Chevron notches are usually used to facilitate starting such mechanical precracks, and face grooves are sometimes necessary to produce straight precracks in tough alloys (Fig. 27). These modifications may also be necessary to control fatigue precracking of some materials.

When a specimen is mechanically precracked by pop-in, the load should be maintained and should not be reduced for testing at a lower initial K-value. Reducing the load (crack mouth opening displacement) required for pop-in will result in residual compressive stress at the crack tip, which could interfere with SCC initiation. When testing specimens at a relatively low fraction of K_{Ic}, fatigue precracking is recommended.

Testing Procedures

For all methods using precracked specimens, the primary objective is usually to determine K_{ISCC} or K_{th}, a threshold stress intensity for SCC for the alloy and environment combination. One procedure, similar to that used with smooth specimens, depends on the initiation of SCC at various levels of applied K_{Io} values. Both constant load (K-increasing) and constant-displacement (K-decreasing) tests can be used. The latter procedure, which is unique to precracked specimens, involves crack arrest. This technique requires a K-decreasing constant-displacement test. These methods are compared in Fig. 30, which illustrates the opposite shift in the stress-intensity factor as SCC growth occurs.

K-Increasing Versus K-Decreasing Tests. In constant-load specimens (K-increasing tests), stress parameters can be quantified with confidence. Because crack growth results in an increasing crack opening, there is less likelihood that corrosion products will block the crack or wedge it open. Crack-length measurements can be made readily with several continuous-monitoring methods.

A wide selection of constant-load specimen geometries are available to suit the test material, experimental facilities, and test objective. Therefore, crack growth can be studied under either bend or tension loading conditions. Specimens can be used to determine K_{ISCC} by the initiation of a stress-corrosion crack from a preexisting fatigue crack using a series of specimens or to measure crack growth rates.

The principal disadvantages of constant-load specimens are the expense and bulk associated with the need for an external loading system. Bend specimens can be tested in relatively simple cantilever beam equipment, but specimens subjected to tension loading require constant-load creep-rupture equipment or similar testing machines. In this case, expense can be minimized by testing chains of specimens connected by loading links that are designed to prevent unloading upon failure of individual specimens. Because of the size of these loading systems, it is difficult to test constant-load specimens under operating conditions, but they can be tested in environments obtained from operating systems.

Constant-displacement specimens (K-decreasing tests) are self loaded; therefore, external stressing equipment is not required. Their compact dimensions also facilitate exposure to operating service environments. They can be used to determine K_{ISCC} by the initiation of stress-corrosion cracks from the fatigue precrack, in which case a series of specimens must be used to bracket the threshold value. This can also be achieved by the arrest of a propagating crack, because under constant-displacement testing conditions stress intensity decreases progressively as crack propagation occurs. In this case, a single specimen suffices in principle; in practice, the use of several replicate specimens is recommended to assess variability in test results.

Constant-displacement specimens are subject to several inherent disadvantages. Oxide formation or corrosion products can wedge the crack surfaces open, thus changing the applied displacement and load. Oxide formation or corrosion products can also block the crack mouth, thus preventing the entry of corrodent, and can impair the accuracy of crack length measurements by electrical resistance methods. Applied loads can be measured only indirectly by displacement changes or by other sophisticated instrumentation. Crack arrest must be defined by an arbitrary crack growth rate below which it is impractical to measure cracks accurately (commonly about 10^{-10}m/s, or 1.5×10^{-5} in./h).

Loading Arrangements and Crack Measurement. To monitor crack propagation rate as a function of decreasing stress intensity when testing constant-displacement loaded specimens, two of the three testing variables must be measured: crack depth (a_i) or load (P_i), and crack opening displacement at the load line (V_{LL}). Although crack initiation and growth can be de-

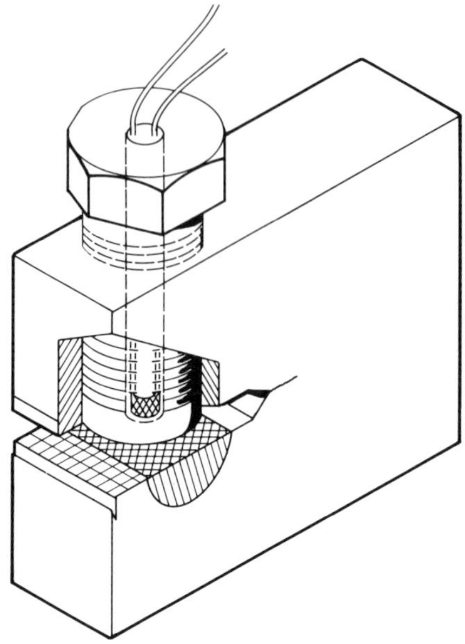

(a)

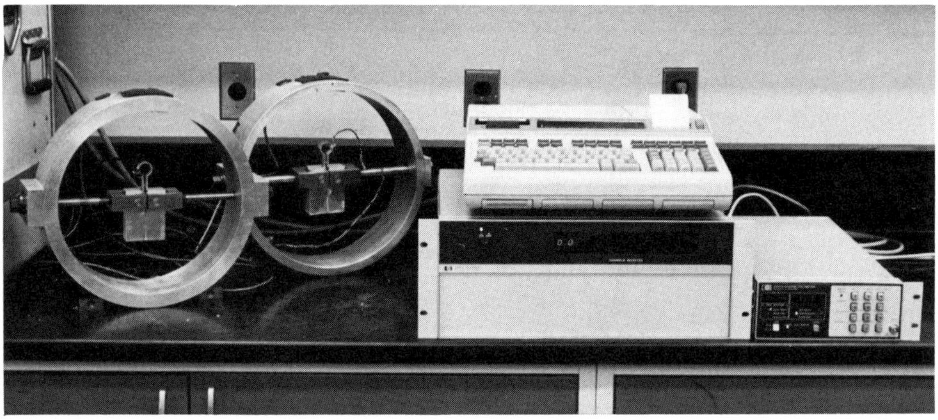

(b)

Fig. 31 (a) Wedge-opening load specimen loaded with instrumented bolt. Source: Ref 52. (b) Ring-loaded wedge-opening load specimen test setup. Box to the left of loading rings contains analog signal conditioning for load and displacement signals. The digital data acquisition system consists of a scanner connected to the analog load and displacement signals, a digital voltmeter, and a portable computer used to read and store data and to control the other instruments. Source: Ref 1

tected from change in either load or crack length, load change is usually more sensitive to these conditions. Therefore, crack advance is easier to detect in specimens loaded in a testing machine, an elastic loading ring, or an instrumented bolt than in specimens loaded with a rigid bolt or wedge. Figures 31(a) and (b) illustrate typical loading arrangements for which load changes can be automatically monitored (Ref 1, 35, 52).

Figure 32 illustrates an ultrasonic method of measuring crack length at the interior (midwidth and quarter widths) of a bolt-loaded double-beam specimen. This method provides a more accurate measure of crack depth than visual measurements

Fig. 32 Ultrasonic crack measurement system for double-beam specimens. Bolt-loaded specimen is mounted on translation stage at center. Ultrasonic transducer is located above specimen, and the oscilloscope at left indicates (left to right) the top of the specimen, the crack plane, and the bottom face reflection. Digital readouts of stage position and peak height for the crack front measurement used to make consistent positioning measurements are shown (right). This system has a crack growth resolution of approximately 0.127 mm (0.005 in.). Source: Ref 1

made on the specimen surfaces. Various other techniques have been used, such as measurement of beam deflection for cantilever beam specimens (Ref 31) and changes in electrical resistance. Such arrangements, however, require calibration. It is feasible and desirable to obtain crack length measurements with a precision of at least ± 0.127 mm (0.005 in.).

Exposure to environment. When practical for laboratory accelerated testing, the test environment should be brought into contact with the specimen before it is stressed or immediately afterward; this enhances access of the corrodent to the crack tip to promote earlier initiation of SCC and to decrease variability in test results. Similarly, in certain cases, it may also be beneficial to introduce the corrodent even earlier, that is, during precracking. However, unless facilities are available to begin environmental exposure immediately after precracking, corrodent remaining at the crack tip may promote blunting due to corrosive attack. In addition, corrosion of the specimen surfaces in the small volume of the precrack or the advancing stress-corrosion crack will change the composition of the environment that is in contact with the crack tip and can significantly affect the test results. Therefore, hydrolysis reactions can drastically reduce the pH of the aqueous test environment (Ref 53) and can induce embrittlement of some steels by corrosion product hydrogen.

Selection of an appropriate test duration presents problems that vary with the testing system; this includes the alloy and metallurgical condition, the test environment, and the loading method. Errors in interpretation of the test results can be caused by test durations that are either too short or too long. The optimum length of exposure can be best approached through recognition of meaningful crack-propagation rates. What is considered meaningful depends on the available precision of measurement of crack lengths and an acceptably low rate for the criterion of a stress-intensity threshold (Fig. 3). A problem also exists with the correlation of SCC crack growth rates in the laboratory test and in an anticipated service environment. The question leads ultimately to the intended application and a determination of what is a tolerable amount of SCC growth for a given length of time.

Calculation of SCC Propagation Rates. There is no generally accepted procedure for calculating crack growth rate, da/dt, as a function of stress intensity from crack growth curves. Various approaches exist; the simplest is a graphical $\Delta a/\Delta t$ technique that may incorporate smoothing of the a versus t curve (Ref 37-40). Another widely used approach is smoothing of the crack growth curve by computer techniques for curve-fitting the entire a versus t curve by a multiple-term polynomial function (Ref 31).

Other techniques include a secant method and an incremental polynomial method, in which derivatives of the smoothed crack growth curve are calculated at various points to determine instantaneous crack growth rates. Instantaneous growth rates are then plotted against the instantaneous stress intensities, K_{Ii}, at corresponding time intervals to obtain graphs similar to that shown in Fig. 3. Additional information on the secant and incremental methods, which are often used in fatigue studies, can be found in the article "Fatigue Crack Growth Data Analysis" in Volume 8 of the *ASM Handbook*.

A limited study of the above four methods of treating crack growth data is presented for a high-strength aluminum alloy in Ref 1. All of the methods used to calculate crack growth rates produced the same general results, which were difficult to interpret because of large amounts of scatter resulting from the use of small crack growth increments. Moreover, the significance of such graphs is dubious when the corrosivity of the environment and the length of exposure can invalidate the estimate of K by causing gross corrosion product wedging effects and/or crack branching.

Reduction of crack length data becomes useless without prior subjective interpretation of crack length versus time curves. Allowances should be made for extraneous effects caused by erratic or apparent initiation of stress-corrosion crack growth, scatter in the measurement data due to excessive crack front curvature, multiple crack planes, crack-tip branching, and gross wedging caused by corrosion products.

A simple method of comparing materials by using crack growth curves is based on average growth rates taken from an exposure time of zero to an arbitrary time that is sufficient to achieve significant crack extension in the most SCC-susceptible materials being compared (Ref 44). This method not only rapidly identifies materials with relatively low resistance to SCC, but also provides numerical test results for highly resistant materials that may not develop a K_I versus da/dt curve with a definite plateau. Additional information is available in Ref 1, 23, 36, 38, 40 and 44.

Dynamic Loading: Slow-Strain-Rate Testing

The most recently developed method for accelerating the SCC process in laboratory testing involves relatively slow-strain-rate tension testing of a specimen during exposure to appropriate environmental conditions. The application of slow dynamic strain exceeding the elastic limit assists in the SCC initiation. This accelerating

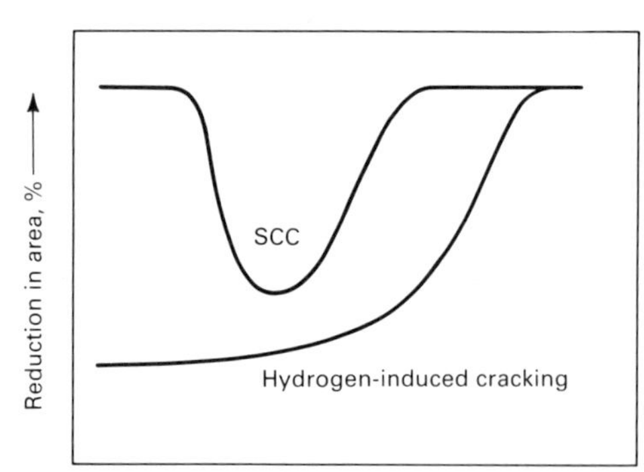

Fig. 33 Schematic showing the effect of strain rate on SCC and hydrogen-induced cracking. Source: Ref 58

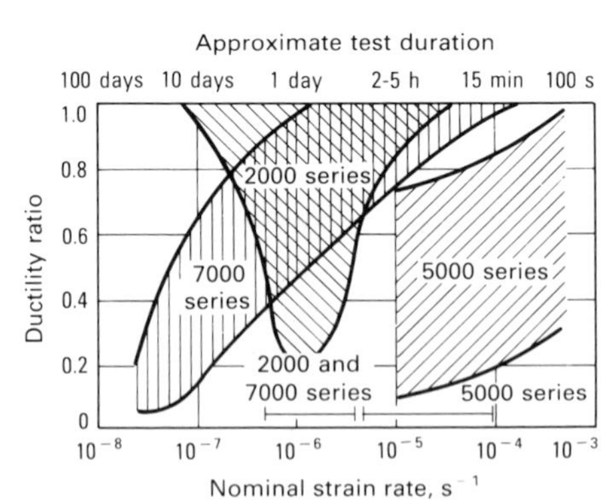

Fig. 34 Strain-rate regimes for studying SCC of various aluminum alloys. Corrodent: 3% NaCl plus 0.3% H_2O_2. Source: Ref 57

Table 2 Critical Strain-Rate Regimes Promoting SCC in Various Metal/Environment Systems

System	Applied strain rate, s^{-1}
Aluminum alloys in chloride solutions	10^{-4} and 10^{-7}
Copper alloys in ammoniacal and nitrite solutions	10^{-6}
Steels in carbonate, hydroxide, or nitrate solutions and liquefied ammonia	10^{-6}
Magnesium alloys in chromate/chloride solutions	10^{-5}
Stainless steels in chloride solutions	10^{-6}
Stainless steels in high-temperature solutions	10^{-7}
Titanium alloys in chloride solutions	10^{-5}

technique is consistent with the various proposed general mechanisms of SCC, most of which involve plastic microstrain and film rupture.

Slow-strain-rate tests can be used to test a wide variety of product forms, including parts joined by welding. Tests can be conducted in tension, in bending, or with plain, notched, or precracked specimens. The principal advantage of slow-strain-rate testing is the rapidity with which the SCC susceptibility of a particular alloy and environment can be assessed.

Slow-strain-rate testing is not terminated after an arbitrary period of time. Testing always ends in specimen fracture, and the mode of fracture is then compared with the criteria of SCC susceptibility for the test material. In addition to its time-saving benefits, less scatter occurs in the test results. Comprehensive discussions on the slow-strain-rate testing technique can be found in Ref 23 and 54 to 57, and in ISO 7539-7 (Ref 6, Part 7).

Critical Strain Rate. The most significant variable in slow-strain-rate testing is the magnitude of strain rate. If the strain rate is too high, ductile fracture will occur before the necessary corrosion reactions can take place. Therefore, relatively low strain rates must be used. However, at too low a strain rate, corrosion may be prevented because of repassivation or film repair so that the necessary reactions of bare metal cannot be sustained, and SCC may not occur. Although typical critical strain rates range from 10^{-5} to 10^{-7} s^{-1}, depending on the alloy and environment system, the most severe strain rate must be determined in each case.

The repassivation reaction that is observed at very low strain rates and that prevents the formation of anodic SCC does not occur when cracking is the result of embrittlement by corrosion product hydrogen. This mechanistic difference can be used to distinguish between anodic SCC (active-path corrosion) and cathodic SCC (hydrogen embrittlement), as illustrated in Fig. 33.

The fastest strain rate that will promote SCC in a given system depends on crack velocity. Generally, the lower the SCC velocity, the slower the strain rate required. Applied strain rates known to have promoted SCC in metal/environment systems are listed in Table 2.

The most relevant strain rates for various aluminum alloys are illustrated in Fig. 34. These trends illustrate that slow-strain-rate tests should be performed in a strain-rate regime that is appropriate for the given alloy and environment system.

Test Specimens. Standard tension specimens (ASTM E 8) (Ref 59) are generally recommended for use with the specified conditions of gage lengths, radii, and so on, unless specialized studies are being conducted. For initially smooth specimens, the strain rate at the onset of the test is clearly defined; however, once cracks have initiated and grown, straining is likely to concentrate in the vicinity of the crack tip, and the effective strain rate is unknown. Rigorous solutions for determining the strain rate at crack tips or notches are not available, but effective strain rates are likely to be higher than for the same deflection rate applied to plain specimens.

Notched or precracked specimens can be used to restrict cracking to a given location—for example, when testing the heat-affected zone associated with a weld. Notched or precracked specimens can also be used to restrict load requirements where bending, as opposed to tensile loading, may offer an added benefit. The section thickness or diameter of such specimens is usually relatively small, so the testing duration is short.

Test Equipment. Constant-strain-rate apparatus requirements include sufficient stiffness to resist significant deformation under the loads

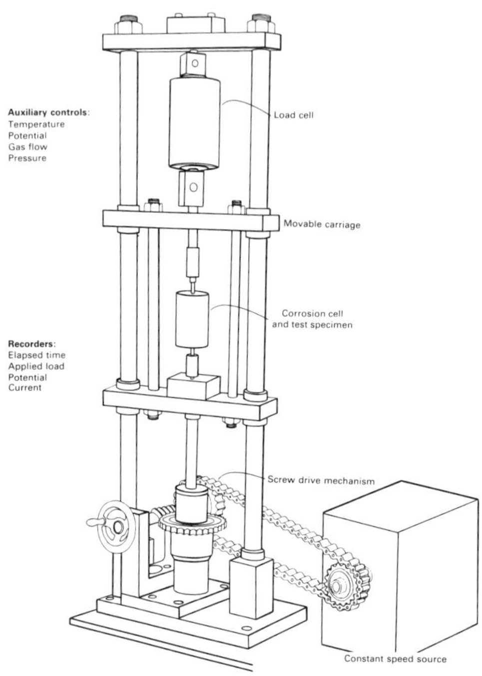

Fig. 35 Typical slow-strain-rate test apparatus. Source: Ref 56

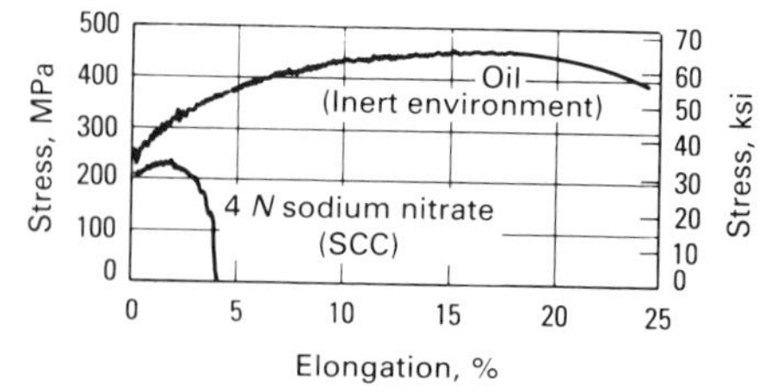

Fig. 36 Nominal stress versus elongation curves for carbon-manganese steel in slow-strain-rate test in boiling 4 N $NaNO_3$ and in oil at the same temperature. Source: Ref 55

necessary to fracture the test specimens, a system to provide reproducible, constant strain rates over the range of 10^{-4} to 10^{-8} s^{-1}, and a cell to contain the test solution. Auxiliary equipment is used to control environmental conditions and to record test data. The testing equipment can also be instrumented to record load-elongation curves, which is convenient when testing at various strain rates. A typical constant-strain-rate unit is illustrated schematically in Fig. 35. Various types of corrosion cells may be required to control the test conditions for specific studies.

In addition to uniaxial tensile units, cantilever constant-strain-rate apparatus has been used in which an extension arm attached to a cantilever beam specimen is lowered at a constant rate. This technique has been successfully used to study SCC of low-carbon steel in carbonate-bicarbonate environments to determine crack velocity, critical strain rates, and inhibitor effectiveness (Ref 60). Additional information on slow-strain-rate testing equipment and procedures is available in Ref 49 and 54.

Fig. 37 Photomacrographs of two carbon steel specimens after slow-strain-rate tests conducted at a strain rate of $2.5 \times 10^{-6} s^{-1}$ and 80 °C (180 °F). The ductility ratio in this example was 0.74 (original diameter: 2.54 mm, or 0.10 in.). (Left) ductile fracture in oil. (Right) SCC in carbonate solution

Assessment of Results. Historically, the principal methods of SCC assessment derived from slow-strain-rate tension testing were based on time to failure, maximum gross section stress de-

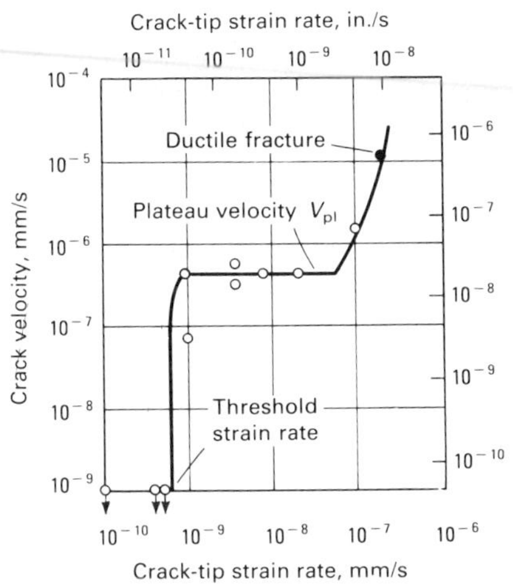

Fig. 38 Effects of beam deflection rate on stress-corrosion crack velocity in precracked cantilever bend specimens of a carbon-manganese steel. Tested in a carbonate-bicarbonate solution at 75 °C (165 °F) and at a potential of – 650 mV (SCE). Source: Ref 55

veloped during the tension test, percent elongation, area bounded by the load-elongation curve, and reduction in area. Figure 36 depicts stress-elongation curves that illustrate how stress-corrosion cracks influence the elongation to fracture as well as the maximum load.

To eliminate non-SCC effects, parallel tests are conducted in an inert environment, and a ratio of the result obtained in the corrodent divided by the result obtained in the inert environment is commonly used as an index of SCC susceptibility. For example, in Fig. 34, higher SCC resistance is denoted by higher ductility ratios.

Figure 37 shows a stress-corroded specimen containing many secondary stress-corrosion cracks and reduced ductility at fracture. Some alloys experience rapid deterioration of mechanical properties on contact with certain corrosive environments; any additional effect of applied straining can best be assessed by comparison with the behavior of unstrained specimens. Therefore, it is essential that the cause of environmental degradation to be verified as SCC.

Slow-strain-rate testing is very efficient in comparing environments in terms of their capability to produce SCC—for example, in steels having similar metallurgical characteristics. However, such comparisons are difficult and not very reliable when applied to groups of steels with different mechanical characteristics (Ref 58).

Slow-strain-rate testing generally does not provide data that can be used for design purposes. Recent work, however, has shown that average SCC velocities, threshold stresses, and threshold strain rates can be obtained with modified techniques combined with microscopy (Ref 55, 60, 61). For example, average SCC crack velocities can be determined from the depth of the largest crack measured on the fracture surfaces of specimens that have failed completely, or in longitudinal sections on the diameter of specimens that have not experienced total failure, divided by the time of testing. With this procedure, SCC is assumed to initiate at the start of the test, which is not always true.

With precracked specimens, other methods can be used to monitor crack growth and thus allow determination of crack velocities. The SCC behavior of a pipeline steel (Fig. 38) has been studied by using a precracked cantilever bend specimen in terms of threshold strain rate for crack growth and also in terms of crack growth rates analogous to the stage II plateau velocity illustrated in Fig. 3. Material properties, such as strength and toughness, that influence SCC performance when measured by tension testing are eliminated as factors; therefore, valid comparisons can be made of alloys with widely different structures and mechanical properties. Additional information on this method of assessment and the effects of strain rate can be found in Ref 62 to 64.

Test Environments

The primary environmental factors in SCC testing are the nature and concentration of anions and cations in aqueous solutions, electrochemical potential, solution pH, the partial pressure and nature of species in gaseous mixtures, and temperature. Separately or in combination, environmental variables can have a profound effect on the thermodynamics and kinetics of the electrochemical processes that control environmentally assisted fracture. Therefore, the choice of environmental

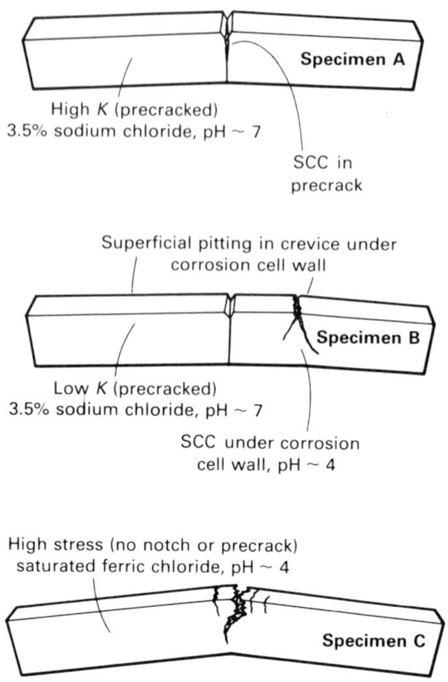

Fig. 39 Cantilever beam specimens of PH13-8Mo stainless steel after testing. Experiments demonstrate that electrochemical factors can override mechanical factors in determining initiation sites of SCC. See text for details. Source: Ref 65

conditions provides an important basis for developing accelerated SCC test methods.

The environmental requirements for SCC vary with different alloys. Although a mechanical precrack or a critical strain rate provides a worse case for SCC from a mechanical standpoint, there does not appear to be a generally applicable worst case from an environmental standpoint. However, because the presence of moisture and salt water is almost universal, the SCC characteristics of alloys in these environments—as well as in any special environment a given engineering structure may experience—are always of interest.

Figure 39 illustrates that electrochemical factors can override mechanical factors in determining SCC initiation sites. Three cantilever beam specimens of PH13-8Mo stainless steel were tested in salt water. Specimen A was tested at a high K level. With the participation of the chloride ions, the protective oxide film ruptured at the bottom of the precrack and initiated SCC, which was halted before the beam fractured completely. Specimen B was loaded at a lower K level. After 1300 h, a stress-corrosion crack initiated, but not in the precrack. Crack initiation occurred under the wall of the cell that surrounded the central portion of the specimen and contained the salt water.

Careful examination of this specimen and replicate specimens revealed small crevice corrosion pits under the wall that initiated SCC in an almost smooth surface. Even if these small pits had been as sharp as a fatigue crack, the K level would have been much lower than at the machined and fatigued notch. In the stagnant situation under the cell wall, the stainless steel reacted with the salt water to form hydrochloric acid and other corrosion products from the metal. Therefore, the low pH in a crevice, due to the hydrolysis of chromium corrosion product, overcame the mechanical disadvantage of the lack of a precrack.

Specimen C was then tested to verify the effectiveness of electrochemical conditions in crack initiation. Saturated ferric chloride was selected to lower the pH to the range inside an active corrosion pit in the stainless steel; application of the solution to the unnotched beam resulted in the immediate initiation of many cracks in the smooth surface. Hydrochloric acid was found to be equally effective.

Stress-corrosion tests can be divided into two broad environmental classes: those conducted in actual service environments and those conducted under laboratory conditions.

Service Environments and Field Testing. The following examples illustrate the value, and in some cases the necessity, of exposure tests performed in actual service environments as an adjunct to laboratory evaluation. The standard 3.5% NaCl alternate immersion test data for aluminum alloys 2024 and 7075 proved useless in predicting the serviceability of these alloys for handling rocket propellant oxidizers such as N_2O_4 and inhibited red fuming nitric acid (Ref 66). The alternate immersion test showed 2024-T351 and 7075-T651 to be susceptible to SCC at low short-transverse stresses, but 2024-T851 and 7075-T7351 were quite resistant even at high stresses. These data were supported by outdoor field tests in seacoast and industrial atmospheres.

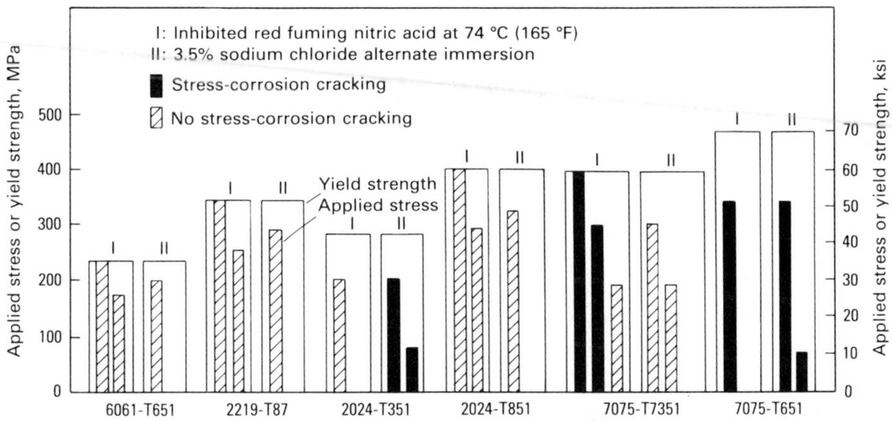

Fig. 40 SCC resistance of various aluminum alloys in inhibited red fuming nitric acid versus alternate immersion in 3.5% NaCl solution. Each bar graph represents an individual short-transverse C-ring test specimen machined from rolled plate and stressed at the indicated level. Source: Ref 66

However, in proof tests consisting of exposure to the actual service environment—inhibited red fuming nitric acid at 74 °C (165 °F)—SCC occurred in both tempers of 7075 alloy and did not occur in either temper of 2024 alloy (Fig. 40). There were no unexpected failures with the 2219-T87 and 6061-T651 materials, however.

Simulated-service tests should be conducted under conditions duplicating the service environment exactly, as illustrated by the following example with Ti-6Al-4V alloy pressure tanks for propellant-grade N_2O_4 (<0.20 wt% moisture) (Ref 67). Preliminary laboratory SCC tests using specimens of Ti-6Al-4V demonstrated satisfactory compatibility with the N_2O_4 and gave no indication of SCC.

In subsequent tests of actual pressurized tanks, however, SCC occurred rapidly, and the tanks failed. It was subsequently shown that the N_2O_4 used in the tanks was of a higher-purity grade than that used in the laboratory tests. When test specimens were exposed to this grade of N_2O_4, they also failed. The small quantities of water and nitric oxide (impurities) present in the N_2O_4 used in previous studies were sufficient to inhibit SCC of Ti-6Al-4V in N_2O_4.

Field testing consists of placing a metal specimen in an environment in which conditions represent those anticipated in service. Typical examples include immersion in seawater, exposure to the atmosphere at marine or industrial sites, chemical plant streams, and so on. Field tests can be performed with test coupons or with actual or simulated structural components.

Laboratory Accelerated Tests. Accelerated testing is conducted extensively in controlled laboratory environments for developmental studies in which it is not considered necessary that the test medium duplicate a particular service environment. Reliance should not be placed on such test results, however, unless they can be correlated with practical service experience or with the results of appropriate field tests.

Ideally, the results of a short exposure in a laboratory test will enable a reliable prediction of the SCC performance of an alloy over a long service period under particular environmental conditions. To fulfill this function, test conditions should be selected with regard to the anticipated service conditions. However, accelerated test media that yield reliable results for one alloy may not yield dependable results for another alloy, even though both alloys are of the same basic metal (Ref 9, 63). Therefore, caution must be exercised in the development and use of standardized test media.

Mercury, other liquid-metal systems, and molten salts can also cause failure by cracking of metal under stress. In most cases of this type, oxidation-reduction reactions (corrosion) are not involved with the alloy, and the mechanisms of the cracking under such conditions are quite different from the mechanism of SCC.

Cracking caused by hydrogen embrittlement has been found to occur in titanium-base alloys and many high-strength ferrous alloys in various service environments. Hydrogen, under some conditions, enters the stressed metal without the presence of a corrosion reaction, and cracking ensues. Hydrogen can also enter stressed metal as a corrosion reaction product, and cracking may be stimulated or may ensue. Therefore, to characterize adequately the environmental cracking of such alloys under sustained tensile stress, testing should be performed both with and without conditions that could involve hydrogen damage.

Test Duration. The question of appropriate test durations for the different material-environment combinations and the apparent need to establish them experimentally on a case-by-case basis has been a major stumbling block in the adoption of standard SCC test methods. The situation becomes acute when the results of an SCC test must be evaluated before all replicate specimens have cracked. In these cases, when a statistical probability evaluation must be performed, it should be recognized that time to failure is often distributed normally with respect to the logarithm of test time (see the section "Interpretation of Test Results" in this chapter). Exposure periods that are too short can result in a test that is trifling; when too long, they can introduce extraneous effects. An example is the excessive pitting of smooth test specimens that leads to failure by a different (and possibly unsuspected) mode of SCC, or by mechanical overload rather than by SCC (Ref 18). Problems in the case of precracked specimens can result from test periods being too short to develop a proper estimate of an apparent threshold stress intensity, especially in K-increasing tests; on the other hand, periods that are too long may result in corrosion product wedging or crack tip blunting, both of which can invalidate calculations of the effective stress intensity and SCC growth rates in K-decreasing tests (Ref 28, 36, 39, 40). Therefore, it is necessary to scrutinize test results carefully so that extraneous effects can be excluded from the analysis of the test results.

Electrochemical Tests. Recognition of the importance of electrical potential as one of the controlling parameters in SCC has resulted in increasing use of tests with potentiostatic control or impressed electrical current. No standards exist for such electrochemcial tests, although several methods are routinely used. These tests involve specific conditions that are applicable only to given alloy and environment systems under consideration. These types of tests offer greater rapidity and precision than free corrosion tests.

Sophisticated approaches for studying systems involving active-path corrosion use poten-

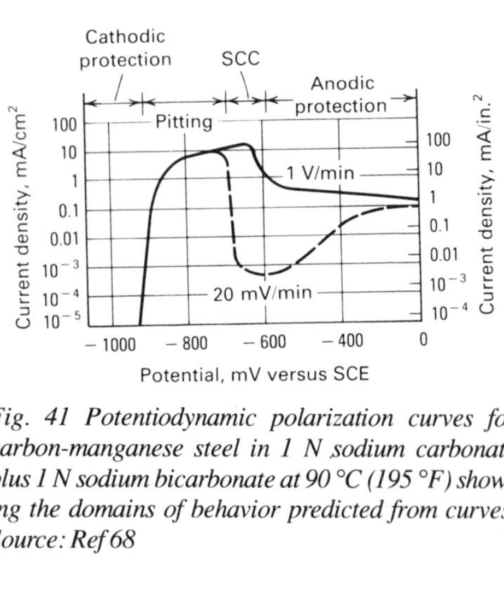

Fig. 41 Potentiodynamic polarization curves for carbon-manganese steel in 1 N sodium carbonate plus 1 N sodium bicarbonate at 90 °C (195 °F) showing the domains of behavior predicted from curves. Source: Ref 68

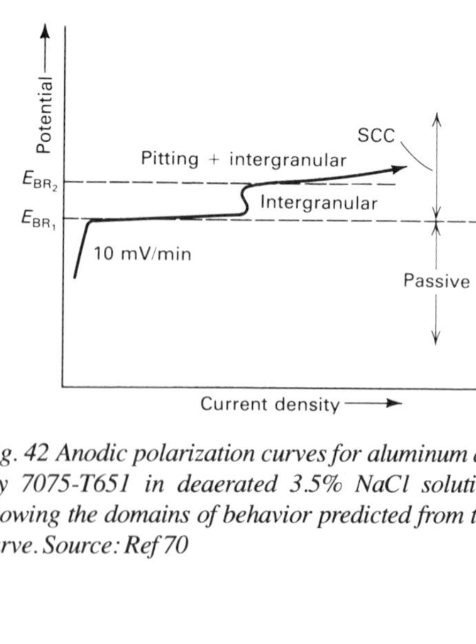

Fig. 42 Anodic polarization curves for aluminum alloy 7075-T651 in deaerated 3.5% NaCl solution showing the domains of behavior predicted from the curve. Source: Ref 70

tiodynamic methods. Anodic polarization curves can provide reasonably accurate predictions of the critical potential ranges and kinetic factors controlling SCC in a given system. One procedure (Ref 68) involves first scanning a range of potentials in the anodic direction at a relatively high scan rate (~1000 mV/min) to determine regions with high current density, in which intense anodic activity is likely. This is followed by a relatively low scan rate (~20 mV/min), which indicates regions where relative inactivity is likely.

Scans should be started at a potential at which the surface is film free. The rapid scan minimizes film formation so that the observed currents related to relatively film-free conditions. The slow scan allows time for filming to occur. Comparison of the two curves reveals any ranges of potential within which high anodic activity in the film-free region reduces to insignificant activity when the time requirements for film formation are met, thus indicating the critical potential range in which SCC is likely.

Figure 41 illustrates this type of experiment for a low-carbon steel in a carbonate-bicarbonate solution with the predicted domains of behavior. Establishing a precise range of critical potentials for SCC requires subjective determination of boundary conditions relating to the current densities.

A different approach with the use of anodic polarization curves has been used to characterize the SCC behavior of aluminum alloys (Ref 69). Corrosion of a susceptible microstructure in aqueous chloride solutions was exclusively intergranular only for a limited range of potentials between the first and second breakdown potentials, E_{BR}, determined from the anodic polarization curves, where E_{BR_1} approximates the critical pitting potential of the active corrosion path at the grain boundaries and E_{BR_2} approximates the critical pitting potential of the grain bodies. Figure 42 depicts a schematic anodic polarization curve for aluminum alloy 7075-T651 that illustrates the predicted domains of behavior.

The use of anodic polarization curves and related measurements is discussed in ASTM G 5 (Ref 8). Information obtained from the anodic polarization curve can be useful for screening environmental test media to reduce the number of actual SCC tests required. However, such electrochemical tests are concerned with testing a bulk environment in contact with an exposed surface, and it should not be assumed that such measurements represent the crack tip under conditions of crack growth. Therefore, even when an environment is identified as innocuous, a potent local environment may exist, such as in a pit when the pitting potential is exceeded or in a crevice or precrack.

Moreover, when laboratory SCC tests involve cracks or crevices, the potential at the crack tip may differ appreciably from that measured at the surface where the crack or crevice emerges. In laboratory tests, it must be considered whether or not environmental changes in composition or potential are involved and how they may relate to service conditions (Ref 63).

Special Considerations for Testing of Weldments

ASTM G 58 (Ref 8) covers test specimens in which stresses are developed by the welding process only (that is, residual stress, Fig. 24), an externally applied load in addition to the stresses due to welding (Fig. 7e), and an externally applied load only, with residual welding stresses removed by annealing.

The National Materials Advisory Board Committee on Environmentally Assisted Cracking Test Methods for High-Strength Weldments recently published the following guidelines on SCC testing of weldments (Ref 28). Fracture mechanics of cracked bodies was found to be a valid and useful approach for designing against environmentally assisted cracking, although several limitations and difficulties must be taken into consideration. For static loading, K_{ISCC} and da/dt versus K_I are useful parameters. They are specific to a material, temperature, and metal/environment system and are functions of local chemical composition, microstructure, and so on.

Superimposed minor load fluctuations and infrequent changes in load can alter environmental cracking response. This effect, which cannot be predicted from K_{ISCC} and da/dt values, may be significant and detrimental. Reexamination of static loading as a design premise may be required. Existing test methodology or environmentally assisted cracking tendency is applicable to the evaluation of weldments. As in other structural components, residual stress must be treated in a quantitative and realistic manner.

The National Materials Advisory Board report supports current design emphasis based on the presumption of preexisting cracklike flaws in the structure and covers testing with precracked (fracture mechanics) specimens only. It contains a critical assessment of the problems associated with environmentally assisted cracking in high-strength alloys and of state-of-the-art design and test methodology.

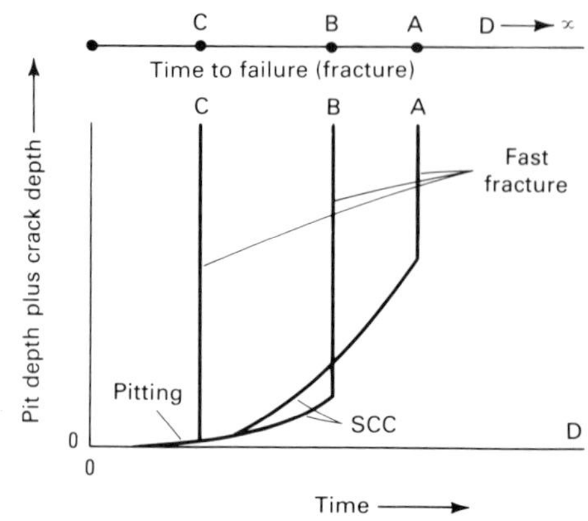

Fig. 43 Various processes in SCC as influenced by the fracture toughness of the metal. Kinetics for pitting (or, in material D, nonpitting), SCC (materials A and B only), and fast fracture. Line at top illustrates how time to failure data can be misleading. Source: Ref 73

Interpretation of Test Results

Interpretation of results is the most fallible part of SCC testing and evaluation; it includes the analysis that leads to conclusions and recommendations. Stress-corrosion test data are at best imprecise and test dependent, and they must be qualified with the testing conditions. It is important to verify the mode of environmental cracking (Fig. 1) and then to review the data to exclude all extraneous results, as discussed previously for the individual test procedures. Following are some comments on the nature of the test dependency of the most commonly used criteria of SCC behavior.

Criteria of SCC Behavior

Specimen Life (Time to Failure). Stress-corrosion testing frequently involves determining the lives of specimens under specific test conditions. This includes the initiation (or incubation) of a stress-corrosion crack and its propagation to the point of fracture (Fig. 2). Such a determination is easily accomplished when only a single crack forms and the specimen fractures within the chosen test period. However, it often happens that SCC occurs but the specimen does not fracture. This is especially likely when testing relatively low-strength material by constant strain loading (Fig. 5b) and when testing at applied stress or stress-intensity levels only slightly above the threshold (Fig. 30a). Cracks may initiate at multiple sites in constant-strain-loaded smooth specimens with relatively low applied stress, and a difficult problem arises in deciding when to consider a specimen failed if it does not fracture visibly.

It is often found that a majority of specimens in a set of replicates in a test fail rapidly; this leaves a few specimens that fail at much longer times or do not fail at all before the test is discontinued. Such behavior presents difficulties, both theoretical and practical, in deciding when to terminate a test, choosing a satisfactory representative value, and comparing such values.

The arithmetic mean specimen life is widely used for smooth specimens because it can be manipulated algebraically and can be used in many standard statistical tests of significance. It should be remembered, however, that extremely large or extremely small values may cause the mean to be atypical of the true distribution. Moreover, in using the arithmetic mean, it is assumed that the population is normally or very nearly normally distributed, which may not be true. The median, on the other hand, has the advantages that it is influenced less by extreme values, requires no assumption about the population distribution, and can be obtained much faster than arithmetic mean values because only about half the number of replicates exposed need to be tested to failure. The median is used in a German specification (Ref 19) that also provides for the use of the geometric mean if the replication is small.

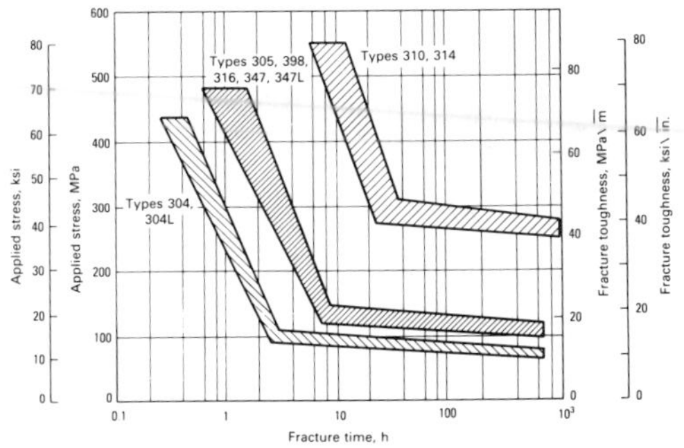

Fig. 44 Relative SCC behavior of austenitic stainless steels in boiling $MgCl_2$. Source: Ref 74

References 3 and 71 contain examples for highly susceptible steel and aluminum alloys, respectively; these examples demonstrate normal distributions for the logarithms of the specimen lives. With such distributions, a geometric mean would be the best representative value of the specimen life. It has also been shown that a Weibull distribution can be appropriate for the non-normally distributed test data for a relatively resistant aluminum alloy (Ref 72). Thus, it should not be assumed that any one distribution is applicable for all testing situations.

Comparisons of alloys with differing strength and fracture toughness based on time to failure can be completely misleading. For example, SCC growth curves are illustrated schematically in Fig. 43 for alloys with different fracture toughnesses. Curves A, B, and C represent materials with decreasing toughness, with curve C showing fast fracture initiated by corrosion pits or fissures with no SCC.

The behavior of a material that does not develop localized pitting or intergranular attack is represented by a line coincident with the abscissa, designated D in Fig. 43. The time-to-failure ranking above the graph indicates D as best and A, B, and C as poorest. Actually, the SCC responses of C and D were not measured, and the true SCC ranking of A and B (indicated by depth of SCC at the time of fracture) exhibits a trend opposite to that inferred from the time-to-failure data above.

Further difficulties may arise, because the total time to fracture is also influenced by non-SCC factors, such as specimen type and size, method of loading, initial stress level, and initiation behavior of the alloy. Consequently, the SCC ranking of materials may vary among investigators using different testing techniques. Nevertheless, comparisons of specimen lives derived from smooth specimens can be useful in certain mechanistic studies and in tests for comparing environmental variations if the mechanical aspects of the investigation are held constant.

Threshold Stress (Stress-Time Curve). More information about the resistance to SCC of a material can be obtained when testing smooth specimens by using a range of applied stresses. Such data are usually presented graphically with the applied gross section stress plotted against specimen life (Fig. 44). The primary interest is generally in the long-life portion of the curve to obtain an estimate of the threshold stress.

A common method of estimating threshold stress involves the experimental determination of the lowest stress at which cracking occurs in at least one specimen and the highest stress at which cracking does not occur in several specimens (for example, three or more, depending on variability). An average of the lowest failure and highest no-failure stresses is usually taken as the threshold stress. Such determinations of critical stresses are carried out at specified test times that are known through experience or preliminary tests to be sufficient to produce SCC in the alloy/environment system of interest. Statistical methods are available for determining threshold stresses more precisely (for example, the Probit method or staircase method) as are commonly used for the determination of fatigue limits. However, the additional testing involved can be quite extensive.

Apparent threshold stresses determined in laboratory tests of coupon specimens are useful for ranking the SCC susceptibility of various materials, but, because such data are dependent on test conditions, they are not realistic for the purpose of engineering design (Ref 2). Also, when using such data as an aid in selecting the material for a specific structure, caution should be exercised in trying to relate the laboratory test conditions to the anticipated service conditions. Not only the environmental condition but also the geometry and size of the test specimens and the

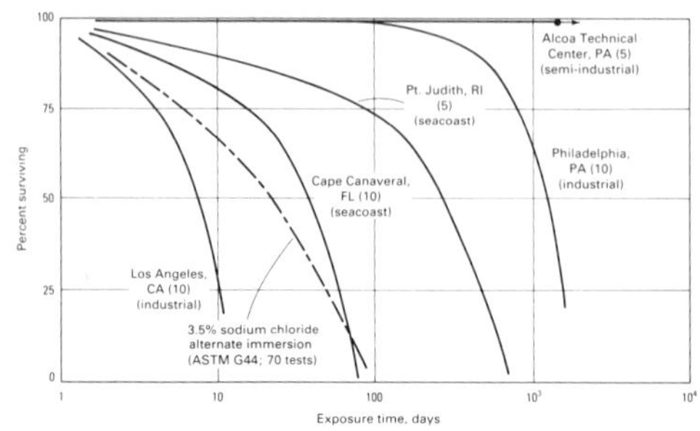

Fig. 45 Effect of variations in atmospheric environment on the probability and time to failure by SCC of a material with an intermediate susceptibility. Tests were made on short-transverse 3.2 mm (0.125 in.) diameter tension specimens from 7075-T7651 type plate stressed 310 MPa (45 ksi). Parenthetical values indicate replication of tests. Source: Ref 1

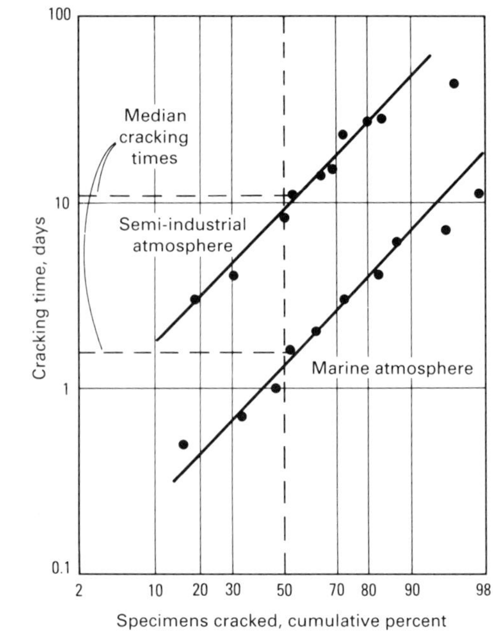

Fig. 46 Distribution of SCC test results for stainless steel. Source: Ref 3

method of stressing should be compared (Fig. 6, 17). Furthermore, threshold stresses obtained with statically loaded smooth specimens are likely to be nonconservative; it has been shown in tests on carbon-manganese steel that the threshold stress obtained by static loads is reduced by applied constant slow strain rates and that it can be reduced even further with cyclic loading (Ref 54). With appropriate frequency, load change, and temperature, average creep rates can be sustained over extended periods, but with static loading, the creep rate may fall below the level needed to promote SCC. In general, experimentally determined threshold stresses for materials with only limited susceptibility to SCC are more sensitive to variations in testing conditions that in the case of highly susceptible materials.

Percent Survival. This method of analyzing stress-corrosion test results is especially useful when some of the specimens in a group survive the duration of the test. Examples of various comparisons by this technique are shown in Fig. 17 and 45. Although the curves can be drawn on regular coordinate graph paper, the percent survival values will often life along a straight line when plotted on normal probability paper, as illustrated in Fig. 46. The linearity of this plot indicates that the statistical distribution of the test results is logarithmic-normal. The vertical positions of the lines indicate the cracking ability of the environment, which can be represented by median cracking times. The slopes of the lines correspond to the variance, which can be used to calculate confidence limits.

Threshold Stress Intensity (K_{ISCC}, K_{th}). Linear elastic fracture mechanics is well established as a basis for materials characterization, including environmental cracking (Ref 28, 75, 76). In practice, it is most practical to define K_{ISCC} as the K_I level associated with some generally acceptable and definably low rate of crack growth that is commensurate with the design service life. When K_{ISCC} values are reported, the criterion for their assessment and the exposure time in the environment must accompany the threshold values. A rational approach to the development of useful data for design is to establish an operational definition of K_{ISCC} that is appropriate for the structure under consideration.

Such characterization requires that linear elastic fracture mechanics and plane-strain conditions be satisfied. However, for certain low-strength steels and aluminum alloys, existing data show that SCC can occur under conditions that deviate substantially from plane-strain, and that

Table 3 Influence of Cutoff Time on Apparent K_{ISCC} Using the SCC Initiation Method

Exposure time, h	Apparent K_{ISCC}, $MPa\sqrt{m}$	Apparent K_{ISCC}, $ksi\sqrt{in.}$
100	187	170
1000	127	110
10,000	28	25

Note: The initiation method was used on a constant-load cantilever bend specimen (K-increasing) of alloy steel with a yield strength of 1240 MPa (180 ksi). Test environment was synthetic seawater at room temperature. Source: Ref 28

Table 4 Comparison of K_{ISCC} Values Determined by Initiation and Arrest Methods

Steel alloy	K_{ISCC}, $MPa\sqrt{m}$ ($ksi\sqrt{in.}$), Initiation	K_{ISCC}, $MPa\sqrt{m}$ ($ksi\sqrt{in.}$), Arrest
10Ni, normal purity	24 (22)	26 (24)
10Ni, high purity	59 (54)	57 (52)
18Ni, normal purity	22-33 (20-30)	28 (25)
18Ni, high purity	<33 (<30)	<33 (<30)

Note: Based on a crack growth rate of 2.5×10^{-4} mm/h (10^{-5} in./h). Modified compact specimens: constant load for initiation and wedge-loaded with a bolt for arrest. Test environment: salt water at room temperature. Source: Ref 28

SCC is by no means limited to or is most severe under plane-strain loading conditions (Ref 28, 39, 77, 78). In these cases, the application of linear elastic fracture mechanics is no longer valid, and the parameter K_{ISCC} is no longer meaningful. Similarly, when testing materials with a high resistance to SCC, loading to high percentages of K_{Ic} may cause a relaxation of stress due to creep. In this case also, the apparent K_{ISCC} values are meaningless. Constant-load tests, therefore, are preferred for lower-strength materials (Ref 79).

The symbol K_{th} has been used to identify threshold stress-intensity factors developed under test conditions that do not satisfy all the requirements for plane-strain stress. Design calculations using such values should not be employed unless it is clear that the laboratory tests exhibit the same stress state as that for the intended application. Nevertheless, properly determined K_{th} values can be useful for ranking materials.

In principle, experimentally determined K_{ISCC} values should be the same whether they are determined by the initiation or the arrest test method (Fig. 30). In both tests, there are dimensional requirements for ensuring that the test results are independent of geometrical effects (see the section "Preparation of Precracked Specimens" in this chapter). However, precautions must be exercised during testing to avoid the potential problems involved with the environmental exposure (incubation, corrosion product wedging, crack branching, crack-tip blunting, and so on). Comparisons of K_{ISCC} values determined for selected steels by both methods, along with examples of overestimated values resulting from insufficient length of exposure, are shown in Tables 3 and 4. It is advisable, when practicable, to use a test that matches the type of loading encountered in the anticipated service.

From the parameter K_{ISCC}, a value of a_{cr} can be calculated using the relation $a_{cr} = 0.2 (K_{ISCC}/TYS)^2$ (see the section "Static Loading of Precracked (Fracture Mechanics) Specimens" in this chapter). This is the shallowest crack (surface length is long compared to its depth) that will propagate as a stress-corrosion crack at a yield strength level of gross stress under the given environmental conditions. This can be a very useful parameter for comparing materials, especially when the measured a_{cr} values can be related to the capability of the flaw inspection system used for a given engineering structure (Fig. 47). Straight lines representing assumed values of a_{cr} in Fig.

47, illustrate how K_{ISCC} values for the various steel alloys related to the maximum depth of long surface flaws that can be tolerated without growth of SCC.

Such a plot can be used as follows. If the inspection system to be used can detect all long surface flaws deeper than 0.25 mm (0.01 in.), then the materials engineer would select an alloy with a K_{ISCC} above the 0.25 mm (0.01 in.) line. Conversely, if the K_{ISCC} and tensile yield strength of a material are known, the equation can be used to estimate the maximum tolerable flaw size. Substitution of an anticipated design stress in terms of percentage of tensile yield strength in the formula for a_{cr} will generate a new series of a_{cr} lines of lower slope.

Alternatively, the following method, which is specific for a given loading method, can be used. Inasmuch as the flaw depth and applied stress are uniquely related for a specific loading situation, a family of curves for constant K_{th} values can be developed within these parameters. Figure 48 shows such curves for long, shallow flaws; the curves were generated according to the following equations (Ref 82-84):

$$K_I = M_B\, \sigma \sqrt{\pi \frac{a}{Q}} \qquad \text{(Eq 17)}$$

$$M_B = \frac{Y}{6}\sqrt{\frac{t}{\pi a}} \qquad \text{(Eq 18)}$$

$$Y = \sqrt{139\left(\frac{a}{t}\right) - 221\left(\frac{a}{t}\right)^2 + 783\left(\frac{a}{t}\right)^3} \qquad \text{(Eq 19)}$$

where K_I is stress intensity, σ is applied stress, a is flaw depth, Q is a shape parameter, and t is thickness. When K_{th} is substituted into Eq 17 and a flaw shape is assumed ($Q = 0.8$), this parameter is related to the flaw size and the applied stress. In this representation, SCC growth will not occur below the curve for the appropriate K_{th}. For example, for a steel with Kth of 60.5 MPa $\sqrt{m}$ (55 ksi$\sqrt{in.}$), a very deep flaw (6 mm, or 0.25 in.) would be required to cause crack propagation in hydrogen gas of a steel component stressed to 360 MPa (52 ksi) in bending. Such representations are useful for relating test data and K_{th} in design and in the development of crack inspection requirements, as well as for ranking alloys.

The requirements for K_{ISCC} tests, as well as for other fracture mechanics tests, include very explicit criteria regarding the minimum crack length for ensuring that the test results can be ana-

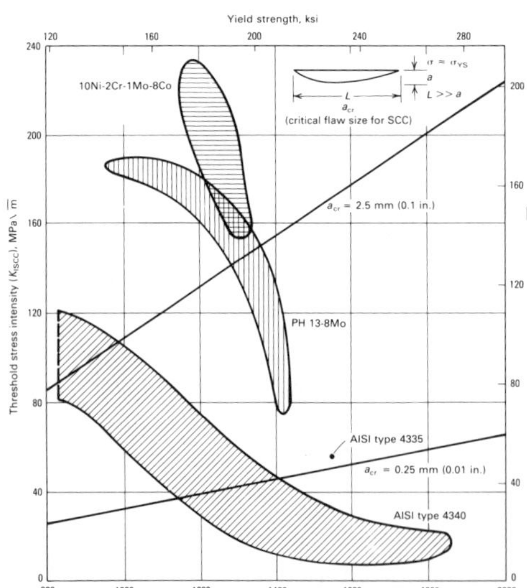

Fig. 47 Comparison of SCC behavior of several high-strength steels based on threshold stress-intensity (K_{ISCC}) values in salt water. Source: Ref 80

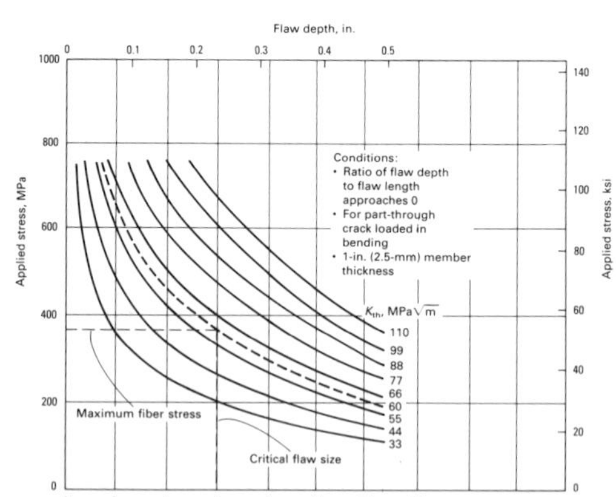

Fig. 48 Relationship of applied stress and flaw depth to crack propagation in hydrogen gas. Dashed lines show an example of the use of such a chart for a steel with K_{th} of 60.5 MPa $\sqrt{m}$ (55 ksi$\sqrt{in.}$) at an operating stress of 359 MPa (52 ksi). Source: Ref 81

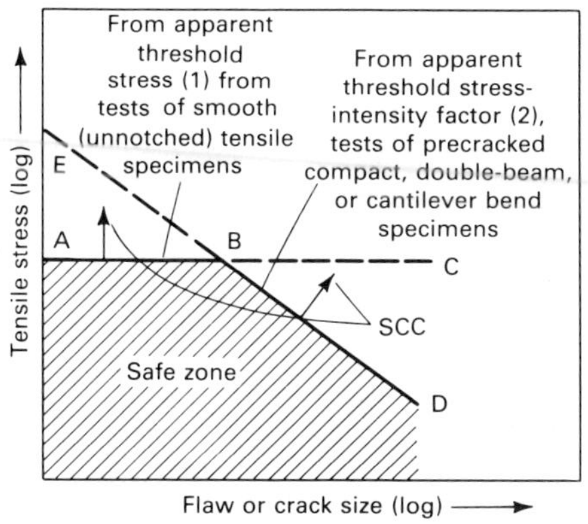

Fig. 49 Concept for combining SCC thresholds obtained on smooth and linear elastic fracture mechanics specimens to yield a conservative assessment of materials. (1) Minimum stress at which small tensile specimens fail by SCC when stressed in environment of interest. (2) Minimum stress intensity at which significant stress-corrosion crack growth occurs in environment of interest. Source: Ref 86

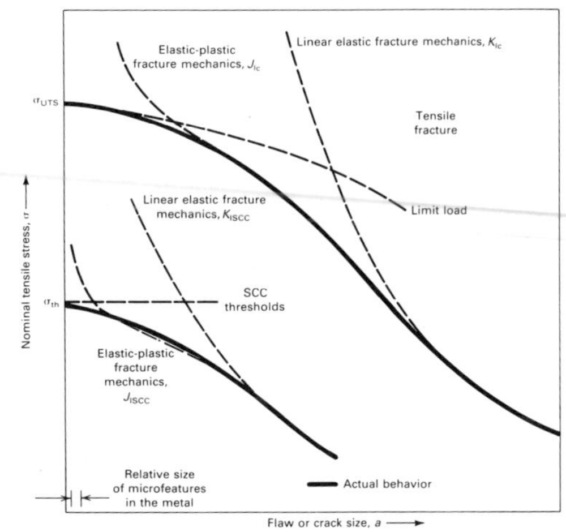

Fig. 50 Proposed linear elastic and elastic-plastic models for describing critical combinations of stress and flaw size at SCC thresholds and at the onset of rapid tensile fracture. Source: Ref 1

lyzed properly using existing linear fracture mechanics concepts. Therefore, a typical K_{ISCC} test uses a relatively large starting crack, on the order of 25 mm (1 in.) long in a 25 mm (1 in.) thick specimen. In many types of service, however, initial defects of this size are rare. For example, damage-tolerant design criteria for military aircraft specify a flaw size on the order of 1.27 mm (0.05 in.) as the initial worse-case damage assumption upon introduction of a new part into service (Ref 75). Experimental work on high-strength steels exposed to hydrogen sulfide gas indicates that for a given combination of materials and applied stress there may indeed be a defect size below which the direct applicability of linear elastic fracture mechanics is questionable (Ref 85). Because of their susceptibility to SCC, however, high-strength steels should not be contemplated for service in the presence of hydrogen sulfide.

For example, Fig. 49 represents a concept of combining SCC thresholds based on smooth specimen and linear elastic fracture mechanics tests of aluminum alloy plate to give a conservative estimate of materials for design. As shown in Fig. 49, the threshold stress-intensity analysis breaks down in the small flaw region (ABE) when the smooth specimen threshold stress is exceeded. Therefore, the definition of a safe zone requires results from both types of tests; the exclusive use of either one of the test methods can yield nonconservative conclusions. It is anticipated that application and further development of elastic-plastic fracture mechanics theory will lead to improved estimates of critical stress/flaw size combinations for the onset of SCC and tensile fracture, as proposed in Fig. 50.

SCC Propagation Rate (*V*-K_I Curves). The issues of crack initiation and crack growth must be addressed for proper consideration of SCC response (Fig. 2). The use of mechanically precracked specimens provides a convenient approach for kinetic measurements where the crack growth rate can be determined as a function of the crack-tip stress-intensity factor (Fig. 3). Current emphasis, however, is on the identification of a steady-state response (plateau velocity) for the ranking of materials (Fig. 51). A disadvantage of this approach is that in some testing situations a plateau velocity is not observed (Fig. 52). However, for design purposes, the phenomena of incubation and the non-steady-state response (stage I, Fig. 3) must also be taken into account.

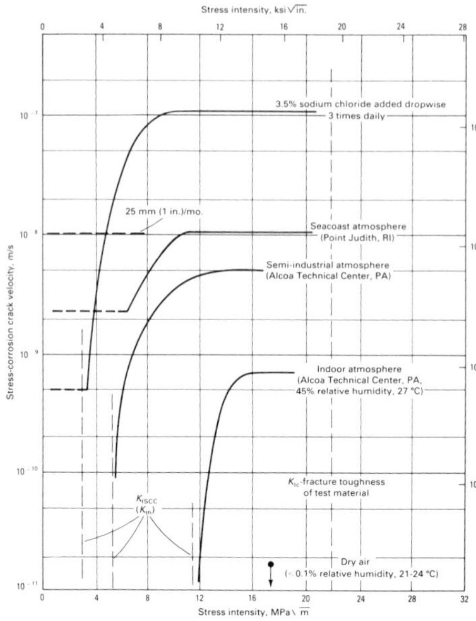

Fig. 51 Effect of corrosive environment on SCC velocity and threshold stress intensity for 7079-T651 plate (64 mm, or 2.5 in., thick) stressed in the short-transverse direction (S-L; see Fig. 29). Double-beam specimens bolt-loaded to pop-in. No SCC occurred during 3 years of exposure to dry air in a desiccator; however, the plateau velocity (horizontal part of each curve) and the apparent threshold stress intensity (K_{ISCC} or K_{th}) vary with the environment. Dashed portions of the curves represent the effect of corrosion product wedging. Source: Ref 44

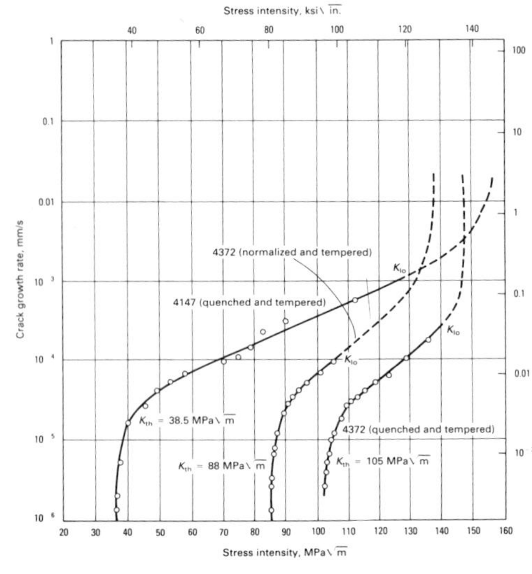

Fig. 52 Crack growth kinetics of three steels in hydrogen at 21 MPa (3000 psi). Source: Ref 81

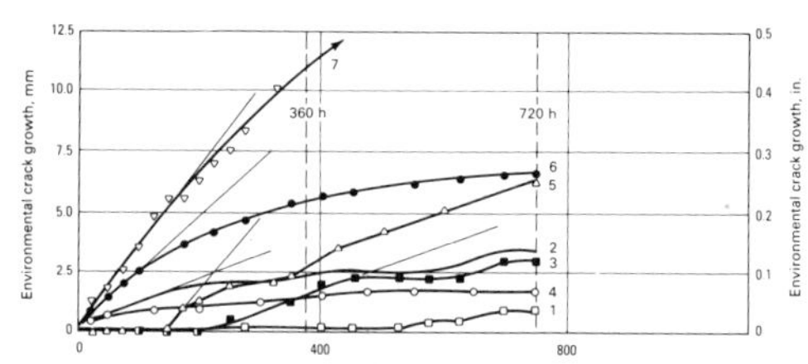

Fig. 53 Examples of SCC crack growth in various aluminum alloys with relatively high resistance to SCC. S-L (see Fig. 29) double-beam specimens bolt-loaded to pop-in and wetted with 3.5% NaCl three times daily; relative humidity 45%. The numbers 1 to 7 indicate different test materials listed in order of decreasing resistance to SCC (see Table 5); thin sloping lines indicate plateau velocities for first growth.. The alloy 2 specimen failed by mechanical fracture rather than intergranular SCC. Source: Ref 39, 44

These latter requirements are not well understood.

An approximate approach to this problem of involving both the time of incubation (initiation) and the maximum crack growth rate of SCC at high stress intensities (plateau velocity) was proposed earlier in this chapter for use with testing of aluminum alloys using constant crack opening displacement tests (Fig. 53 and Table 5). This involves a simple average rate taken from time zero to a judiciously chosen test duration. Useful ranking of materials can be displayed on V-K_I diagrams such as that shown in Fig. 54.

Another approach to the average velocity concept is available through an elastic-plastic fracture mechanics interpretation of breaking-load test data obtained from stress-corrosion tests of smooth tension specimens (Ref 17). The stress-corrosion crack growth rates estimated from the slopes of the curves in Fig. 16 from 0 to 4 or 6 days are shown in Table 6 to agree reasonably well with the plateau (K-independent) growth rates determined from conventional fracture mechanics tests using bolt-loaded double-beam specimens. It is also noteworthy that a distinction between the intermediate resistance T7X1 temper and the nearly immune T7X2 temper was made more readily using either the smooth specimen test or the average crack growth over 42 days with the double-beam specimen. Fractographic examination confirmed that this was a correct distinction.

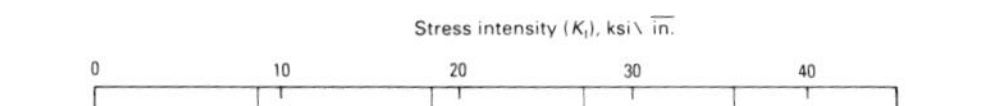

Fig. 54 SCC propagation rates for various aluminum alloy 7050 products. Double-beam specimens (S-L; see Fig. 29) bolt-loaded to pop-in and wetted three times daily with 3.5% NaCl. Plateau velocity averaged over 15 days. The right-hand end of the band for each product indicates the pop-in starting stress intensity (K_{Io}) for the tests of that material. Data for alloys 7075-T651 and 7079-T651 are from Ref 37. Source: Ref 87

The implication here, which needs further investigation, is that from a single test method (that is, the breaking-load test) it is possible to compare materials (1) on the basis of their probabilities of initiating and propagating SCC flaws to an arbitrary depth or (2) by their respective crack growth rates, both being meaningful engineering descriptors of SCC damage. An additional advantage to this approach is that the effects of specimen size and alloy strength and toughness can be normalized (Fig. 16).

Ductility Ratio (DR Versus Strain-Rate Curves). Various ratio criteria, such as reduction of area, elongation, fracture stress, fracture energy, and time to failure, have been found to be useful in environmental studies with the slow-strain-rate test method. However, such criteria have limited use in comparing various materials because of their dependence on the strength and toughness of the alloy and the specificity of the critical strain rate and the environmental species (Fig. 34 and Table 2). Recent work has shown that average SCC growth rates, threshold stresses, and threshold strain rates can be obtained with modified techniques combined with microscopy (Ref 55, 60, 61).

Other Criteria. Several other more specialized criteria can be found in the literature, such as:

- Jones Stress Corrosion Index (Ref 88, 89)
- Critical strain (Ref 90)
- Mean critical stress (Ref 91)

Table 5 Correlation of SCC Plateau Crack Velocities With Smooth Specimen SCC Ratings

		Plateau crack velocity			
	Smooth specimen	First growth		Average (0 to 15 days)	
Alloy	rating(a)	m/s	in. × 10^{-5}/h	m/s	in. × 10^{-5}/h
1	A	6×10^{-10}	10	2×10^{-10}	3
2	A	7×10^{-9}	100(c)	1.8×10^{-9}	27(c)
3	B	2.1×10^{-9}	30	1.2×10^{-9}	19
4	B	4.2×10^{-9}	60	1.3×10^{-9}	20
5	B	7×10^{-9}	100	2.1×10^{-9}	30
6	C	6.3×10^{-9}	90	4.2×10^{-9}	60
7	D	1.1×10^{-8}	160	8.4×10^{-9}	120

(a) Short-transverse ratings per ASTM G 64 (Ref 8). **(b)** S-L (see Fig. 29) double-beam specimens bolt-loaded to pop-in and wetted three times daily with 3.5% NaCl. **(c)** Fractographic examination revealed mechanical fracture rather than the intergranular SCC verified in all other materials. Source: Ref 44

Table 6 Stress-Corrosion Crack Growth Rates in Aluminum Alloy 7075 Obtained by Different Test Methods

Test method	Stress-corrosion crack growth rate 7075-T651 m/s	7075-T651 in. × 10^{-5}/h	7075-T7X1 m/s	7075-T7X1 in. × 10^{-5}/h	7075-T7X2 m/s	7075-T7X2 in. × 10^{-5}/h
Breaking load test using smooth tensile bar stressed 207 MPa (30 ksi); 4 or 6 days	$> 3.8 \times 10^{-9}$	> 54(a)	7×10^{-10}	10(b)	1.3×10^{-10}	2(b)
Bolt-loaded double-beam, pop-in stress; plateau velocity obtained from V-K_I curves	3.1×10^{-9}	44	7.5×10^{-10}	11	(c)	
Bolt-loaded double-beam, pop-in stress; average growth 0-15 days	3.4×10^{-9}	48	4×10^{-10}	6	4×10^{-10}	6
Bolt-loaded double beam, pop-in stress; average growth 0-42 days	2.8×10^{-9}	40	4×10^{-10}	6	2×10^{-10}	3

(a) Over 4 days. **(b)** Over 6 days. **(c)** No plausible estimate could be made, because of the slight crack growth. Source: Ref 1

Precision of SCC Data

SCC test data are at best relatively imprecise. Variability in test results can arise from several sources: errors in determination of stress (or stress intensity), variations in the test environment, inhomogeneity of the test material, and duration of the test. An awareness of each of these sources, reviewed in perspective with the test objective, is necessary for a rational and productive interpretation of test results.

In the production of sophisticated high-strength and high-toughness alloys, close metallurgical control of the fabricating and thermal treatments is necessary to ensure that the required mechanical properties satisfy specifications. It is equally important from the standpoint of SCC that the metallurgical condition of the alloy be properly controlled. Just as there is a range of applicable mechanical properties for a given alloy and metallurgical condition (temper), a range in the SCC behavior can be expected from one heat (or lot) to another. Also, there can be an appreciable variation in the behavior of different-sized mill products of the same material. An example of the variations in the SCC behavior of different lots and various mill products is shown in Fig. 54.

When selecting a material for a particular structural component, SCC tests should be made on the specific mill products and sizes that will be required. Costly mistakes have been made in the past in the evaluation of prototype structures by testing parts fabricated from substitute products for economic expediency.

It is unfortunate that the precision of SCC test results is generally lower for materials with an intermediate resistance to SCC than for materials with a very low or a very high resistance. This presents a special challenge in the comparison of competitive materials with "improved" resistance to SCC. For example, there is no way to determine what portion of the scatterbands shown in Fig. 54 is due to material variations and how much is due to the precision of the test measurements.

Variability can occur even in carefully controlled investigations. Table 7 contains a summary of K_{ISCC} values obtained in an interlaboratory program of tests on a sample of AISI 4340 alloy steel plate with a yield strength of 1254 MPa (182 ksi). Two types of test specimens, K-increasing and K-decreasing, were exposed to 3.5% NaCl at room temperature (Ref 92).

In another investigation, the precision in the measurement of plateau velocities was determined for a number of replicate bolt-loaded double-beam specimens of short-transverse orientation from a sample of 25 mm (1 in.) thick 7075-T651 aluminum alloy plate (Ref 93). Tension pop-in specimens were exposed by continuous immersion in aqueous solutions of 1 M NaCl and 1 M $NaClO_4$. For six tests in the chloride solution, the mean plateau velocity was 1.4×10^{-8}

Table 7 Estimates of K_{ISCC} in AISI 4340 Steel Alloy Obtained by Different Test Methods in an Interlaboratory Test Program

Test method	Average K_{ISCC} MPa$\sqrt{m}$	ksi$\sqrt{in.}$
K-increasing test		
Cantilever bend [SE(B)] specimen; 1000 h; 8 labs(a)	34.5±5.8	31.1±5.2
K-decreasing test		
Modified compact (WOL) specimen; bolt loaded [MC(W_b)]; 4000 h; 14 labs; K_{Io} = 55 MPa$\sqrt{m}$ (49.5 ksi$\sqrt{in.}$)	21.7	19.5
Standard deviation within labs	2.54	2.29
Standard deviation between labs	2.86	2.57

(a) Because K_{ISCC} is subjectively determined with the cantilever bend test, no meaningful statistical information could be obtained.

m/s (1×10^{-3} in./h) with a standard deviation of 0.2×10^{-8} m/s (0.1×10^{-3} in./h, and the range was 1 to 2×10^{-8} m/s (7×10^{-4} to 1.4×10^{-3} in./h) (±35%). For nine tests in the perchlorate solution, the mean plateau velocity was 8×10^{-9} m/s (6×10^{-4} in./h) with a standard deviation of 0.2×10^{-9} m/s (1.4×10^{-4} in./h), and the range was 5×10^{-9} to 1×10^{-8} m/s (4 to 8×10^{-4} in./h) (±31%).

Normalizing SCC Data

It is often necessary to normalize test results with respect to one of the mechanical properties of critical interest for a given engineering structure in order to place the SCC response in proper perspective. This is commonly done by expressing the exposure stress (or stress intensity) and the apparent threshold values in terms of percent yield strength (or percent critical stress intensity) rather than in "absolute" units. This process can sometimes result in a different ranking order of alloys, and controversy can arise over which ranking is more pertinent. The use of normalized data is usually most appropriate when the utmost resistance to SCC is required; that is, the material should be resistant even when stressed to nearly 100% of the normalizing property. For alloy development screening tests, consideration should be given to evaluating materials in both ways.

Predicting Service Life

The ultimate customers for SCC evaluations are the materials users who must select the alloys, designs of structural components and the amount of protection that may be required to avoid the hazard of SCC. One of the toughest problems for the SCC investigator is that of convincing the decision makers that service life cannot be predicted in hard numbers because materials traditionally have been evaluated by comparisons. Hopefully, the guess-work will have been minimized by use of the most appropriate test procedures and interpretation of the test results.

Life predictions are difficult and should be approached with caution because there are no workable mathematical models. The field is plagued with confusion created by :

- The complex, multifaceted nature of the phenomenon, which involves metallurgy, mechanics, chemistry, and time—all of which influence the SCC behavior of an alloy
- Extensive data scatter (imprecision)
- Lack of correlations between laboratory test results and service experience
- Difficulty in assessing precisely the service conditions that the part must withstand

Designing to avoid SCC has traditionally taken the approach of preventing the initiation of SCC (safe-life concept). There is considerable service experience to justify this; the materials that have given the most service problems are capable of developing very high SCC propagation rates (on the order of 25 mm, or 1 in., per month, as shown in Fig. 51). Growth rates of this order, combined with relatively low thresholds of stress or stress intensity, make the fail-safe concept rather impractical. However, with the availability of advanced materials with higher thresholds for initiation of SCC and lower propagation rates, the damage-tolerant concept of design may become more practical for both SCC and corrosion fatigue.

References

1. D.O. Sprowls *et al.*, "A Study of Environmental Characterization of Conventional and Advanced Aluminum Alloys for Selection and Design: Phase II—The Breaking Load Test Method," Contract NASI-16424, NASA Contractor Report 172387, Aug 1984
2. B.F. Brown, *Stress Corrosion Cracking Control Measures*, National Bureau of Standards Monograph 156, U.S. Department of Commerce, June 1977
3. R.N. Parkins *et al.*, Report Prepared for the European Federation of Corrosion Working Party on Stress Corrosion Test Methods. *Br. Corros. J.*, Vol 7, July 1982, p 154-167
4. A.W. Loginow, Stress Corrosion Testing of Alloys, *Mater. Prot.*, Vol 5 (No. 5), 1966, p 33-39
5. E.H. Dix, Jr., Aluminum-Zinc-Magnesium Alloys, Their Development and Commercial Production, *Trans. ASM*, Vol 42, 1950, p 1057-1127
6. "Corrosion of Metals and Alloys—Stress Corrosion Testing," ISO 7539, International Organization for Standardization
7. G. Vogt, Comparative Survey of Type of Loading and Specimen Shape for Stress Corrosion Tests, *Werkst. Korros.*, Vol 29, 1978, p 721-725
8. *Metal Corrosion, Erosion, and Wear*, Vol 03.02, Section 3, *Annual Book of ASTM Standards*, ASTM
9. H.L. Craig, Jr., D.O. Sprowls, and D.E. Piper, Stress-Corrosion Cracking, *Handbook on Corrosion Testing and Evaluation*, W.H. Ailor, Ed., John Wiley and Sons, 1971, p 231-290
10. M.B. Shumaker *et al.*, Evaluation of Various Techniques for Stress Corrosion Testing Welded Aluminum Alloys, *Stress Corrosion Testing*, STP 425, ASTM, 1967, p 317-341
11. R.A. Davis, Stress Corrosion Cracking Investigation of Two Low Alloy, High Strength Steels, *Corrosion*, Vol 19 (No. 2), 1963, p 45t-55t
12. S.O. Fernandez and G.F. Tisinai, Stress Analysis of Un-notched C-rings Used for Stress Cracking Studies, *J. Eng. Ind.*, Feb 1968, p 147-152
13. F.S. Williams, W. Beck, and E.J. Jankowsky, A Notched Ring Specimen for Hydrogen Embrittlement Studies, *Proc. ASTM*, Vol 60, 1960, p 1192
14. D.O. Sprowls *et al.*, "Investigation of the Stress Corrosion Cracking of High Strength Aluminum Alloys," Final Technical Report for U.S. Government NASA Contract NAS-8-5340, Control No. 1-4-50-001167-01(lf), CPB-02-1215-64, 1967
15. Report of Task Group 1, of ASTM Subcommittee B-3/X, Stress Corrosion Testing Methods, *Stress Corrosion Testing*, STP 425, ASTM, 1967, p 3-20; *Proc. ASTM*, Vol 65, 1965, p 182-197
16. B.W. Lifka and D.O. Sprowls, Stress Corrosion Testing of 7079-T6 Aluminum Alloy in Various Environments, *Stress Corrosion Testing*, STP 425, ASTM, 1967, p 342-362
17. R.J. Bucci *et al.*, The Breaking Load Method: A New Approach for Assessing Resistance to Growth of Early Stage Stress Corrosion Cracks, *Corrosion Cracking*, V.S. Goel, Ed., American Society for Metals, 1986, p 267-277
18. D.O. Sprowls *et al.*, Evaluation of a Proposed Standard Method of Testing for Susceptibility of SCC of High Strength 7XXX Series Aluminum Alloy Products, *Stress Corrosion—New Approaches*, STP 610, H.L. Craig, Jr., Ed., ASTM, 1976, p 3-31
19. "Testing of Light Metals, Stress Corrosion Test," German Standard DIN 50908, 1964
20. P. Brenner, Realistic Stress Corrosion Testing, *Metallurgy*, Vol 23 (No. 9), 1969, p 879-886
21. F.H. Haynie *et al.*, "A Fundamental Investigtion of the Nature of Stress Corrosion Cracking in Aluminum Alloys," Technical Report AFML 66-267, USAF Contract No. AF 33(615)-1710 Air Force Materials Laboratory, June 1966
22. H. Nathorst, Stress Corrosion Cracking in Stainless Steels, Part II—An Investigation of the Suitability of the U-bend Specimen, *Weld. Res. Counc. Bull.*, Series No. 6, Oct 1950
23. A.J. Sedricks, Stress Corrosion Cracking Test Methods, Corrosion Testing Made Easy, Vol 1, National Association of Corrosion Engineers, 1990
24. H.R. Copson and S.W. Dean, Effect of Contaminants on Resistance to Stress Corrosion Cracking of Ni-Cr Alloy 600 in Pressurized Water, *Corrosion*, Vol 21 (No. 1), 1965, p 8
25. B.F. Brown, The Application of Fracture Mechanics to Stress Corrosion Cracking, *Met. Rev.*, Vol 13, 1968, p 171-183
26. H.R. Smith and D.E. Piper, Stress Corrosion Testing with Precracked Specimens, *Stress Corrosion Cracking in High Strength Steels and in Titanium and Alunimum Alloys*, B.F. Brown, Ed., Naval Research Laboratory, 1972, p 17-78
27. R.P. Wei, S.R. Novak, and D.P. Williams, Some Important Considerations in the Development of Stress Corrosion Cracking Test Methods, *Mater. Res. Stand.*, Vol 12, 1972, p 25
28. "Characteriation of Environmentally Assisted Cracking for Design—State of the Art," National Materials Advisory Board Report No. NMAB-386, National Academy of Sciences, 1982
29. "Standard Method Test for Plane-Strain Fracture Toughness Testing of Metallic Materials, E 399, *Annual Book of ASTM Standards*, Vol 03.01, ASTM
30. J.A. Hauser, II, R.W. Judy, Jr., and T.W. Crooker, "Draft Standard Method of Test for Plane-Strain Stress-Corrosion-Cracking Resistance of Metallic

Materials in Marine Environments," NRL Memorandum Report 5295, Naval Research Laboratory, March 1984

31. W.B. Lisagor, Influence of Precracked Specimen Configuration and Starting Stress Intensity on the Stress Corrosion Cracking of 4340 Steel, *Environment-Sensitive Fracture: Evaluation and Comparison of Test Methods*, STP 821, S.W. Dean, E.N. Pugh, and G.M. Ugiansky, Ed., ASTM, 1984, p 80-97
32. H. Tada, P. Paris and G. Irwin, *The Stress Analysis of Cracks Handbook*, Del Research Corp., 1973
33. J.A. Joyce, D.F. Hasson, and C.R. Crowe, Computer Data Acquisition Monitoring of the Stress Corrosion Cracking of Deplated Uranium Cantilever Beam Specimens, *J. Test Eval.*, Vol 8 (No. 6), 1980, p 293-300
34. S.R. Novak and S.T. Rolfe, Modified WOL Specimen for K_{ISCC} Environmental Testing, *J. Met.*, Vol 4 (No. 3), 1969, p 701-728
35. J.G. Kaufman, J.W. Coursen, and D.O. Sprowls, An Automated Method for Evaluating Resistance to Stress-Corrosion Cracking with Ring-Loaded Precracked Specimens, *Stress Corrosion—New Approaches*, STP 610, H.L. Craig, Jr., Ed., ASTM, 1976, p 94-107
36. C. Micheletti and M. Buratti, New Testing Methods for the Evaluation of the Stress-Corrosion Behavior of High-Strength Aluminum Alloys by the Use of Precracked Specimens, *Symposium Proceedings, Aluminum Alloys in the Aircraft Industry* (Turin, Italy), Oct 1976, Technicopy Ltd., 1978, p 149-159
37. M.V. Hyatt, Use of Precracked Specimens in Stress Corrosion Testing of High Strength Aluminum Alloys, *Corrosion*, Vol 26 (No. 11), 1970, p 487-503
38. M.O. Speidel and M.V. Hyatt, Stress-Corrosion Cracking of High Strength Aluminum Alloys, *Advances in Corrosion Science and Technology,* Vol 2, M.G. Fontana and R.W. Staehle, Ed., Plenum Press, 1972, p 115-335
39. D.O. Sprowls *et al.*, "Evaluation of Stress Corrosion Cracking Susceptibility Using Fracture Mechanics Techniques," Contract NAS 8-21487, Contractor Report NASA CR-124469, May 1973
40. R.C. Dorward and K.R. Hasse, "Flaw Growth of 7075, 7475, 7050, and 7049 Aluminum Plate in Stress Corrosion Environments," Final Technical Report for U.S. Government Contract NAS 8-30890, Oct 1976; *Corrosion*, Vol 34 (No. 11), 1978, p 386-395
41. W.B. Fichter, The Stress Intensity Factor for the Double Cantilever Beam, *Int. J. Fract.*, Vol 22, 1983, p 133-143
42. J.E. Srawley and B. Gross, Stress Intensity Factors for Crackline-Loaded Edge-Crack Specimens, *Mater. Res. Stand.*, Vol 7 (No. 4), 1967, p 155-162
43. S. Mostovoy, R.B. Crosley, and E.J. Ripling, Use of Crackline Loaded Specimens for Measuring Plane Strain Fracture Toughness, *J. Mater.*, Vol 2 (No. 3), 1967, p 661-681
44. D.O. Sprowls and J.D. Walsh, Evaluating Stress-Corrosion Crack Propagation Rates in High Strength Aluminum Alloys with Bolt Loaded Precracked Double Cantilever Beam Specimens, *Stress Corrosion—New Approaches*, STP 610, H.L. Craig, Jr., Ed., ASTM, 1976, p 143-156
45. M.S. Domack, Evaluation of K_{ISCC} and da/dt Measurements for Aluminum Alloys Using Precracked Specimens, *Environment-Assisted Cracking: Science and Engineering*, STP 1049, W.B. Lisagor, T.W. Crooker, and B.N. Leis, Ed., ASTM, 1990, p 391-409
46. R.B. Heady, Evaluation of Sulfide Corrosion Cracking Resistance in Low Alloy Steels, *Corrosion*, Vol 33 (No. 3), 1977, p 98-107
47. T.L. Bond, R.A. Yeske, and E.N. Pugh, Studies of Stress Corrosion Crack Growth in Al-Zn-Mg Alloys by the Double Torsion Method, *Environment-Sensitive Fracture: Evaluation and Comparison of Test Methods,* STP 821, S.W. Dean, E.N. Pugh, and G.M. Ugiansky, Ed., ASTM, 1984, p 128-149
48. J.C. Lewis and G. Sines, Ed., *Fracture Mechanics: 14th Symposium, Vol II: Testing and Application,* STP 791, ASTM, 1983
49. S.W. Dean, E.N. Pugh, and G.M. Ugiansky, Ed., *Environment-Sensitive Fracture: Evaluation and Comparison of Test Methods,* STP 821, ASTM, 1984
50. J.H. Underwood, S.W. Freiman, and F.I. Baratta, Ed., *Chevron-Notched Specimens: Testing and Stress Analysis,* STP 855, ASTM, 1984
51. D.O. Sprowls, Evaluation of Stress-Corrosion Cracking, *Metals Handbook,* Vol 13, 9th ed., ASM International, 1987, p 245-282
52. W.B. Gilbreath and M.J. Adamson, Aqueous Stress Corrosion Cracking of High Toughness D6AC Steel, *Stress-Corrosion*—New Approaches, STP 610, H.L. Craig, Jr., Ed., ASTM, 1976, p 176-187
53. B.F. Brown, Concept of Occluded Corrosion Cells, *Corrosion*, Vol 26 (No. 8), 1970, p 249
54. G.M. Ugiansky and J.H. Payer, Ed., *Stress Corrosion Cracking—The Slow Strain-Rate Technique,* STP 665, ASTM, 1979
55. R.N. Parkins, Development of Strain-Rate Testing and Its Implications, *Stress Corrosion Cracking—The Slow Strain-Rate Technique*, STP 665, G.M. Ugiansky and J.H. Payer, Ed., ASTM, 1979, p 5-25
56. J.H. Payer, W.E. Berry, and W.K. Boyd, Constant Strain Rate Technique for Assessing Stress-Corrosion Susceptibility, *Stress Corrosion—New Approaches*, STP 610, H.L. Craig, Jr., Ed., ASTM, 1976, p 82-93

57. N.J.H. Holroyd and G.M. Scamans, Slow-Strain-Rate Stress Corrosion Testing of Aluminum Alloys, *Environment-Sensitive Fracture: Evaluation and Comparison of Test Methods*, STP 821, S.W. Dean, E.N. Pugh, and G.M. Ugiansky, Ed., ASTM, 1984, p 202-241
58. C.D. Kim and B.E. Wilde, A Review of Constant Strain-Rate Stress Corrosion Cracking Test, *Stress Corrosion Cracking—The Slow Strain-Rate Technique*, STP 665, G.M. Ugiansky and J.H. Payer, Ed., ASTM, 1979, p 97-112
59. "Standard Methods of Tension Testing of Metallic Materials," E 8, *Annual Book of ASTM Standards*, Vol 03.01, ASTM
60. R.N. Parkins, Fifth Symposium on Line Pipe Research, Catalog No. L30174, UI-40, American Gas Association
61. W.R. Wearmouth, G.P. Dean, and R.N. Parkins, Role of Stress in the Stress Corrosion Cracking of a Mg-Al Alloy, *Corrosion,* Vol 29 (No. 6), 1973, p 251-258
62. R.N. Parkins and Y. Suzuki, Environment Sensitive Cracking of a Nickel-Aluminum Bronze Under Monotonic and Cyclic Loading Conditions, *Corros. Sci.*, Vol 23 (No. 6), 1983, p 577-599
63. R.N. Parkins, A Critical Evaluation of Current Environment-Sensitive Fracture Test Methods, *Environment-Sensitive Fracture: Evaluation and Comparison of Test Methods*, STP 821, S.W. Dean, E.N. Pugh, and G.M. Ugiansky, Ed., ASTM, 1984, p 5-31
64. J. Yu, N.J.H. Holroyd, and R.N. Parkins, Application of Slow-Strain Rate Tests to Defining the Stress for Stress Corrosion Crack Initiation in 70/30 Brass, *Environment-Sensitive Fracture: Evaluation and Comparison of Tests Methods*, STP 821, S.W. Dean, E.N. Pugh, and G.M. Ugiansky, Ed., ASTM, 1984, p 288-309
65. B.F. Brown, The Contributions of Physical Metallurgy and of Fracture Mechanics to Containing the Problem of Stress Corrosion Cracking, *Philos. Trans. R. Soc. (London) A*, Vol 282 (No. 1307), 1976, p 235-245
66. D.O. Sprowls and R.H. Brown, Stress Corrosion Mechanisms for Aluminum Alloys *Fundamental Aspects of Stress-Corrosion Cracking*, R.W. Staehle, A.J. Forty, and D. van Rooyen, Ed., National Association of Corrosion Engineers, 1969, p 466-512
67. J.D. Jackson, W.K. Boyd, and R.W. Staehle, "Corrosion of Titanium, " DMIC Memorandum 218, Defense Metals Information Center, Battelle Memorial Institute, Sept 1966
68. R.N. Parkins, Predictive Approaches to Stress Corrosion Cracking Failure, *Corros. Sci.*, Vol 20, 1980, p 147-166
69. J.R. Galvele and S.M. DeMichelli, Mechanism of Intergranular Corrosion of Al-Cu Alloys, *Corros. Sci.*, Vol 10, 1970, p 795-807
70. S. Maitra and G.C. English, Mechanism of Localized Corrosion of 7075 Alloy Plate, *Metall. Trans. A*, Vol 12, March 1981, p 535-541
71. F.F. Booth and G.E. Tucker, Statistical Distribution of Endurance in Electrochemical Stress-Corrosion Tests, *Corrosion*, Vol 21 (No. 5), 1965, p 173-177
72. J.H. Harshbarger, A.I. Kemppinen, and B.W. Strum, Statistical Treatment of Non-normally Distributed Stress-Corrosion Data, *Handbook on Corrosion Testing and Evaluation*, W.H. Ailor, Ed., John Wiley and Sons, 1971, p 87-100
73. B.F. Brown, Chapter 1, *Stress Corrosion Cracking in High Strength Steels and in Titanium and in Aluminum Alloys*, B.F. Brown, Ed., Naval Research Laboratory, 1972, p 13
74. E. Denhard, Effect of Composition and Heat Treatment on the Stress Corrosion Cracking of Austenitic Stainless Steels, *Corrosion,* Vol 16 (No. 7), 1960, p 131-141
75. "Military Specification for Airplane Damage Tolerance Requirements," MIL-A-83444, U.S. Air Force, 2 July 1974
76. *Damage Tolerant Handbook, A Compilation of Fracture and Crack-Growth Data for High-Strength Alloys*, Vol 1, MCIC-HB-O1R, Battelle Columbus Laboratories, Dec 1983
77. S.R. Novak, Effect of Prior Uniform Plastic Strain on the K_{ISCC} of High-Strength Steels in Sea Water, *J. Eng. Fract. Mech.*, Vol 5, 1973, p 727-763
78. W.G. Clark, Jr., Some Problems in the Application of Fracture Mechanics, *Fracture Mechanics—13th Conference*, STP 743, R. Roberts, Ed., ASTM, 1981, p 269
79. M.O. Speidel, Fracture Mechanics and Stress Corrosion, *Blech, Rohre, Profile*, Vol 25 (No. 1), 1978, p 14-18 (in German)
80. G. Sandoz, High Strength Steels, *Stress Corrosion Cracking in High Strength Steels and in Titanium and in Aluminum Alloys,* B.F. Brown, Ed., Naval Research Laboratory, Washington, DC, 1972, p 79-145
81. A.W. Loginow and E.H. Phelps, Steels for Seamless Hydrogen Pressure Vessels, *Corrosion*, Vol 31 (No. 11), 1975, p 404-412
82. C.F. Tiffany, J.N. Masters, and F. Pall, Some Fracture Considerations in the Design of Spacecraft Pressure Vessels, *Proceedings of the ASM Metals Congress*, American Society for Metals, 1966
83. G.M. Sinclair and S.T. Rolfe, Analytical Procedure for Relating Subcritical Crack Growth to Inspection Requirements, *Proceedings of the ASME Metals Engineering Division Conference*, American Society of Mechanical Engineers, 1969

84. F.W. Smith, "Stress-Intensity Factors for a Semi-Elliptical Surface Flaw," Structural Development Research Memorandum 17, Boeing Company, 1966

85. W.G. Clark, Jr., Applicability of the K_{ISCC} Concept to Very Small Defects, *Cracks and Fracture (Ninth Conference)*, STP 601, ASTM, 1976, p 138-153

86. J.G. Kaufman, Stress Corrosion—Traditional Versus Fractured Mechanicians, *Corrosion*, Vol 35 (No. 4), April 1979, p i

87. R.E. Davies, G.E. Nordmark, and J.D. Walsh, "Design Mechanical Properties, Fracture Toughness, Fatigue Properties, Exfoliation and Stress Corrosion Resistance of 7050 Sheet, Plate, Extrusions, Hand Forgings and Die Forgings," Final Report, Naval Air Systems, Contract N00019-72-C-0512, July 1975

88. F.F. Booth and H.P. Godard, An Anodic Stress-Corrosion Test for Aluminum-Magnesium Alloys, *First International Congress on Metallic Corrosion*, Butterworths, 1962, p 703-712

89. E.L. Jones, Stress Corrosion of Aluminum Magnesium Alloys II—Method for Expressing Stress Corrosion Susceptibility on a Comparative Basis, *J. Appl. Chem.*, Jan 1954, p 7-10

90. J.P. Fraser, G.G. Eldredge, and R.S. Treseder, Laboratory and Field Methods for Quantitative Study of Sulfide Corrosion Cracking, *Corrosion*, Vol 14, 1958, p 517t-523t

91. J.T. Staley, Evaluating New Aluminum Alloy Forging Alloys, *Met. Eng. Quart.*, Vol 14 (No. 4), 1974, p 50-55

92. R.P. Wei and S.R. Novak, Interlaboratory Evaluation of K_{ISCC} Measurement Procedures for Steels, *Environment-Sensitive Fracture: Evaluation and Comparison of Test Methods*, STP 821, S.W. Dean, E.N. Pugh, and G.M. Ugiansky, Ed., ASTM, 1984, p 75-79

93. A.H. Le and R.T. Foley, Stress Corrosion Cracking of AA7075-T651 in Various Electrolytes—Statistical Treatment of Data Obtained Using DCB Precracked Specimens, *Corrosion*, Vol 39 (No. 10), 1983, p 379-383

CHAPTER 18

Failure Analysis of Stress-Corrosion Cracking

S.W. Stafford, El Paso Natural Gas Company
W.H. Mueller, Texas Operations, Dow Chemical

Stress-corrosion cracking (SCC) is a term used to describe service failures that occur by a process of environmentally induced crack initiation and propagation. It represents a major form of corrosion failure and has a complex dependence on various mechanical, electrochemical, and metallurgical factors. Stress-corrosion cracks can be transgranular, intergranular, or both and may grow gradually over a period of time until they reach a critical size, resulting in a stress state of sufficient magnitude to induce brittle fracture.

Analysis of SCC service failures is quite involved and requires prescribed and thorough examination procedures. Appropriate methodologies for evaluating such service failures are the subject of this chapter. Other processes that cause failure of structural components by the conjoint action of mechanical stress and corrosion include hydrogen damage, corrosion fatigue, liquid-metal embrittlement, and fretting. However, the crack-initiation sites, crack-growth patterns, crack surface topographies, and evidence of corrosion all provide clues as to the causes, origins, and mechanisms unique to SCC.

Nondestructive Evaluation (NDE)

Many conventional NDE methods are quite useful for the detection of SCC, even in its early stages. In advanced stages, SCC is often visible to the unaided eye, and suspect areas such as pits and rust discoloration on austenitic stainless steels are readily visible. The methods, applications, and effectiveness of various NDE techniques are shown in Table 1.

Visual, dye-penetrant, magnetic-particle, and eddy-current inspection are widely used for the detection of SCC. However, their use is limited to periods of plant shutdown, or to times when systems can be made available for inspection (Ref 1). Obviously, these techniques can also be used to locate cracks on samples removed from service. Liquid-penetrant examination is useful in locating SCC, but it may not reveal fine cracks. Because the substances found in liquid penetrants are chemically similar to those that may cause SCC, the use of penetrants may interfere with identification of fracture-surface deposits. Consequently, their undesirable side effects on the overall analysis must be considered.

Ultrasonic inspection can be used for on-line monitoring, provided surface contact is maintained. Unfortunately, this technique is time intensive and requires highly skilled personnel. Therefore, it is often impractical for thoroughly inspecting large vessels and extensive piping systems (Ref 1).

Acoustic emission allows detection of growing stress-corrosion cracks on-line in particularly large vessels or systems that are generally too complex to inspect by other NDE techniques. However, operational or extraneous noise is often difficult to distinguish from crack growth. To compensate for the low amplitude of emission stemming from SCC, a considerably greater number of transducers than normal must often be

Table 1 NDE Techniques for Evaluating Suspected SCC Damage

Material	Ultrasonic, shear wave	Ultrasonic, longitudinal wave	Ultrasonic, surface wave	Magnetic particle, dry	Magnetic particle, wet fluorescent	Penetrant, visible	Penetrant, fluorescent	Radiographic, gamma ray	Radiographic, x-ray	Eddy current, standard	Eddy current, remote field	Acoustic emmission	Visual	In situ metallography
	NDE method (a)													
Austenitic stainless steel	P	P	F	NA	NA	G	E	F	G	E	F	F	F	G
Martensitic stainless steel	P	P	F	P	E	G	E	F	G	NA	F	G	P	G
Ferritic stainless steel	P	P	F	P	E	G	E	F	G	NA	F	G	P	G
Nickel and nickel-base alloys	P	P	F	E(b)	E(b)	G	E	F	G	E(c)	F	F	P	G
Copper and copper-base alloys	P	P	F	NA	NA	G	E	F	G	E	F	F	P	F
Aluminum and aluminum alloys	P	P	F	NA	NA	G	E	P	G	E	F	F	P	F
Titanium and titanium alloys	P	P	F	NA	NA	G	E	P	G	E	F	F	P	F
Carbon and low-alloy steels	F	P	F	P	E	G	E	F	G	NA	F	G	F	E

(a) Detectability: NA, not applicable to this material; E, excellent; G, good; F, fair; P, poor. **(b)** For magnetic alloys only. **(c)** For nonmagnetic alloys only

used. Once the general location of the stress-corrosion crack is located using acoustic emission, features of orientation and dimensions can be provided by dye-penetrant or ultrasonic methods (Ref 1).

Analysis

Procedures for analyzing SCC failures are generally the same as those for analyzing any service-induced failure. Ideally, the failed component should be brought to the laboratory shortly after failure occurs, with minimal disturbance to the physical state of the component. Attention must also be given to the chemical environments to which the component was exposed during manufacturing, shipment, storage, and service, particularly to such factors as concentration, temperature, pressure, acidity or alkalinity, contaminants, and any accidental or deliberate changes in the control of these variables. In effect, a complete service history is required.

Inspection of the actual service site allows observation and evaluation of the environmental conditions and mechanical processes involved. The relationship of the failed part to the overall operation can be established, as well as possible sources of stress. Photographs or other forms of visual documentation are critical for recording the location of the failure(s), the extent of cracking, the presence of corrosion products, evidence of mechanical damage, and other pertinent details. The application of dye penetrants to prepared surfaces will greatly enhance the determination of crack length, orientation, branching, and colonization.

Removal of service failures and selection of samples must be performed carefully. Suitable precautions must be taken to avoid (1) destruction of evidence, such as corrosion products or other deposits, which could prove valuable to the analysis, (2) overheating, contaminating, or damaging the component and adjacent components and structures, and (3) contaminating samples of the environment.

When sampling the environment to which the failed component was exposed, the investigator should be guided by information already collected about the operating conditions and history of the failure. The general environment should be sampled, as should the environment in immediate contact with the failure. General information on the chemical substances and environments likely to cause SCC in specific metal alloys is provided in Table 2 (Ref 2).

After initial visual examination of the general conditional state of the component and contained fracture surfaces, corrosion products should be removed to facilitate further examination and chemical analysis. The fracture surface may be stripped with an acetate film, which not only allows removal of significant portions of the corrosion products but also serves to contain the residue. The acetate film can then be used for microanalysis, using an electron microprobe or a scanning electron microscope (SEM) with an energy-dispersive analysis system or other suitable analytical equipment. After stripping, the fracture surface should be ultrasonically cleaned in acetone. Additional information on fracture surface cleaning procedures can be found in Ref 3.

The following criteria provide a general characterization of SCC features:

- Stress-corrosion cracking represents a brittle rupture. Brittle fracture and an absence of plastic deformation should generally be present.
- The stresses required for SCC are usually small (below the macroscopic yield stress) and are static and tensile in nature. Applied stresses often add to a residual stress condition that exists within the structure. These residual stresses result from fabrication processes such as deep drawing, punching, rolling, mismatch in riveting, welding, and so on. Unfavorable residual stresses also arise from differential thermal expansion and from installation.
- Various alloy systems have characteristic preferred crack paths (see Table 2).
- Stress-corrosion cracking usually follows branched paths; this applies to both transgranular and intergranular cracking. However, this criterion is not universally applicable, because stress-corrosion cracks are not always branched (see Fig. 1).
- Stress-corrosion cracks initiate on the surface, and numerous initiation sites are possible (in contrast with mechanical failures, which usually have a single preferred site). However, the initiation sites may be submicroscopic and are largely determined by local chemical/microstructural differences, protective film features, corrodent concentration, and stress concentrations. A preexisting mechanical crack or other surface discontinuity, including chemically and/or mechanically induced damage, may act as a stress raiser and thus initiate SCC.
- As is characteristic of brittle failures, the general crack path is usually aligned normal to the tensile axis.
- Surface cracks may follow a regular, random, or aggregated pattern (Ref 4).
- There usually is minimal evidence of corrosion damage. However, surface staining on exposed surfaces may be present.
- Stress-corrosion cracking is a progressive type of fracture, somewhat similar to fatigue. The crack or cracks grow gradually over a period of time until a critical size is achieved, creating a diminished cross-sectional area that may cause the remaining metal to suddenly fail.
- Microfractography of transgranular cracks may reveal cleavage or quasi-cleavage fracture surfaces. These surfaces may also exhibit characteristic chevron (feathery) markings in the direction of crack propagation (see Fig. 2 and 3). Intergranular cracks usually show grain-boundary facets, free from signs of ductility (Ref 5). Figure 4 shows the general characteristics of "classic" intergranular fracture.

Macroscopic Examination

Those portions of a fracture directly attributable to SCC are brittle in nature. If a material/environment combination leads to susceptibility, then often the tensile stress required for SCC may be very low or take the form of residual stress.

Cracking usually initiates only after an incubation period, which may involve incipient corrosion damage to the material through pitting (Ref 5).

Even in highly ductile metals, SCC results in fractures that exhibit virtually no plastic deformation, although neighboring regions of the material reveal an absence of embrittlement. Multiple cracks are often formed running normal to the axis of the principal tensile stress in the component and away from the surface. Preferential crack formation in the region of welds can be

Table 2 Environments that Cause SCC in Numerous Metal Alloy Systems

FERROUS ALLOYS				AUSTENITIC STAINLESS STEELS			Martensitic	COPPER BASE ALLOYS			
CAST IRON			Mild	302, 304	316, 317	ACI CN-20	Stainless 405–410	Copper	Brass 70–80Cu	Brass 59–93Cu	Cupro-Nickel
Gray	Nickel	Silicon	Steel	321, 347		20 Cr–30Ni		85–99.9	+Zn, Sn or Pb	+Al, Zn or As	66–88:11–33
1	2	3	4	5	6	7	8	9	10	11	12

NICKEL BASE ALLOYS						Al	Ti
Nickel	Ni–Cu	Ni–Cr–Fe	Ni–Fe–Cr	Ni–Mo 62–28	Ni–Cr–Mo 54–15–16	Aluminum	Titanium
99	66–32	76–16–7	32–47–20	+Fe, V	+Fe, W		
13	14	15	16	17	18	19	20

ENVIRONMENTS CAUSING STRESS CORROSION CRACKING

Corrosive	1	2	3	4	5	6	7	8	9	10	11	12	13	14	15	16	17	18	19	20
Acetic acid + mercury salts									•	•	•								•	
Alkyl aryl sulfonates					•															
Aluminum chloride					•	•														
Aluminum sulfate					•															
Ammonia				•						•	•									
Ammonium biophosphate					•	•														
Ammonium chloride					•	•														
Ammonium fluosilicate														•						
Ammonium hydroxide									•	•	•									
Ammonium nitrate				•																
Aniline											•									
Barium chloride					•	•														
Barium nitrate				•																
Calcium bromide					•	•														
Calcium chloride						•														
Calcium nitrate				•																
Carbon tetrachloride					•	•														
Chlorine					•	•														
Chloranilines											•									
Hydrochloric acid (aerated)																			•	
Hydrocyanic acid + hydrogen cyanide					•		•													
Hydrofluoric acid vapors														•						
Hydrofluoric acid (aerated)					•									•						
Hydrofluoric acid (no air)										•	•		•	•	•					
Hydrogen					•															
Hydrogen chloride				•	•															
Hydrogen sulfide (wet)					•														•	
Lead bromide					•															
Levulinic acid					•	•														
Lithium chloride					•	•														
Magnesium chloride				•	•	•	•			•	•			•						
Magnesium chloride + calcium chloride					•															
Magnesium fluosilicate													•	•						
Magnesium sulfate																	•	•		
Manganese chloride					•	•														
Mercuric chloride					•	•			•	•	•	•		•					•	
Mercuric cyanide														•						
Mercuric nitrate														•						
Mercurous nitrate					•				•	•	•			•						
Methallylamine											•									
Methanol																				•
Monoethanolamine				•																
Naphtha feedstock					•															
Naphthenic acids					•															
Nickel chloride					•															
Nickel nitrate				•																
Nitric acid				•							•									
Pentachloroethane					•	•														
Perchloroethylene																				•
Phosphoric acid						•														
Phosphoric acid (aerated)					•															
Chlorobenzene														•						

Corrosive	1	2	3	4	5	6	7	8	9	10	11	12	13	14	15	16	17	18	19	20
Chloroform					•	•														
Chlorotoluidines											•									
Chromic acid				•							•			•						
Copper tetramine											•									
Cresol				•																
Cresylic acid vapors				•																
Cyanogen				•																
Dichlorophenol					•	•														
Epichlorhydrin					•	•														
Ethanol																				•
Ethylamine				•																
Ethyl chloride					•	•														
Ferric chloride				•							•									
Ferrous chloride					•	•														
Fluosilicic acid													•	•			•	•		
Gasoline vapor					•															
Glutamic acid					•	•														
Glycerol					•															
Hexachloroethane					•	•														
Polythionic acid					•															
Potassium carbonate				•																
Potassium chloride					•	•			•	•	•									
Potassium chromate											•									
Potassium hydroxide				•	•	•								•						
Potassium permanganate											•									
Silver nitrate				•																
Sodium aluminate				•																
Sodium bisulfate					•	•														
Sodium bisulfite					•	•														
Sodium carbonate					•	•														
Sodium chloride					•	•														•
Sodium fluoride				•	•															
Sodium hydroxide	•	•		•	•	•	•						•	•	•		•			
Sodium nitrate				•									•	•						
Sodium phosphate (tribasic)					•	•														
Sodium sulfate					•															
Sodium sulfide															•					
Steam															•				•	
Steam, geothermal (aerated)					•	•														
Strontium nitrate				•																
Sulfate liquor				•	•															
Sulfonated oil														•	•					
Sulfuric acid				•																
Sulfuric acid + chlorine					•	•														
Sulfurous acid						•														
Toluene										•	•									
Triethanolamine											•									
Trimethylamine										•										
Uranyl sulfate																				•
Water + hydrogen sulfide + light hydrocarbons						•														
Zinc chloride					•	•														
Zinc fluorosilicate														•						

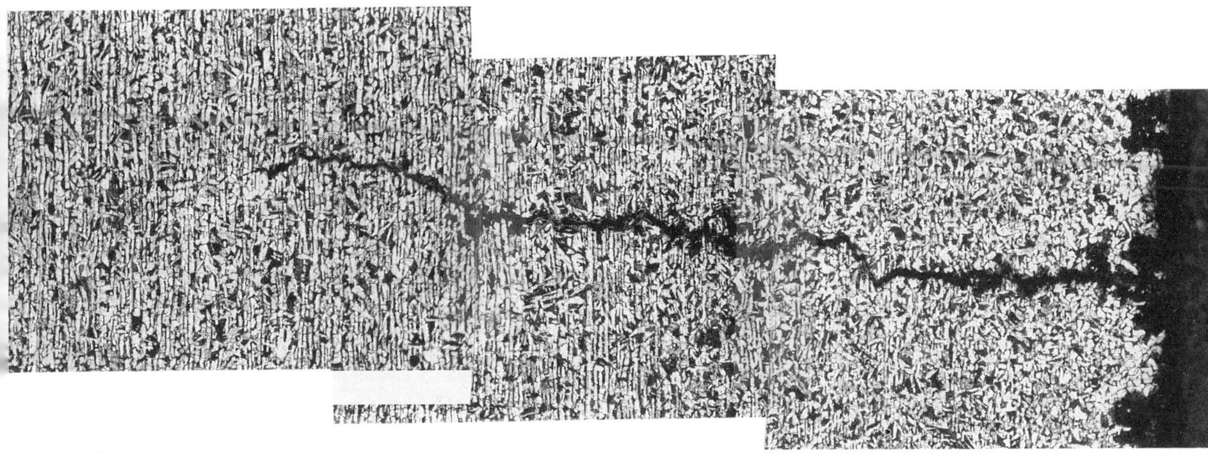

Fig. 1 Metallographic view of an intergranular crack penetrating a ferrite/pearlite matrix. 2% nital etchant, 100×

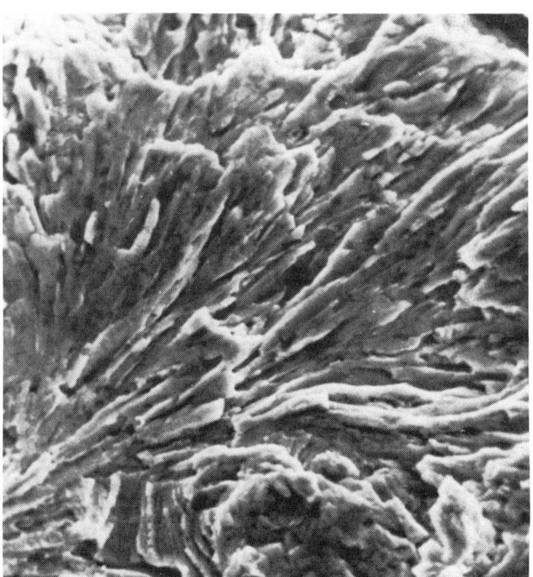

Fig. 2 Typical evidence of feathery transgranular SCC rupture in an 18Cr-9Ni austenitic stainless steel in a hot $MgCl_2$ solution. 2100×. Source: Ref 6

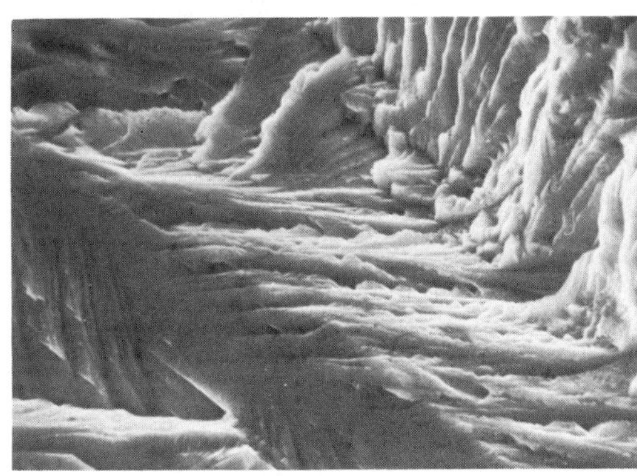

Fig. 3 SCC in an austenitic stainless steel (0.1% C, 1% Si, 2% Mn, 18% Cr, 10% Ni, 2% Mo, 0.6% Ti) caused by $MgCl_2$ solution at approximately 100 °C (212 °F). Transgranular rupture reveals feathery pattern. 5500×. Source: Ref 6

caused both by the associated residual stress and by local changes in material characteristics — for example, composition of the weld metal, microstructure in the heat-affected zone (HAZ), etc.

Many transgranular stress-corrosion cracks characteristically change fracture planes as they propagate, producing flat facets. These faceted features, as well as those associated with intergranular stress-corrosion cracks, are ordinarily observed at low magnifications. The fracture surfaces usually contain distinguishable regions of crack initiation, crack propagation, and final rupture. The portion of the fracture surface developed by SCC propagation may be discolored or covered with corrosion products, or it may be only slightly darkened compared with the final separation zone. Many of these macroscopic features can be seen in Fig. 5. However, the presence

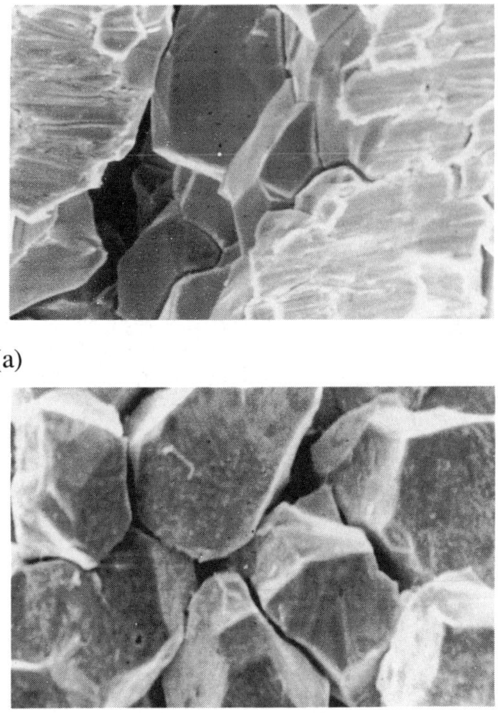

(a)

(b)

Fig. 4 Intergranular fracture of type 304 austenitic stainless steel following exposure to an aqueous $CuSO_4$ + H_2SO_4 solution. (a) Primary rupture plane is shown intersecting the surface. Note the secondary intergranular cracks. 650×. (b) Classic intergranular fracture. 650×

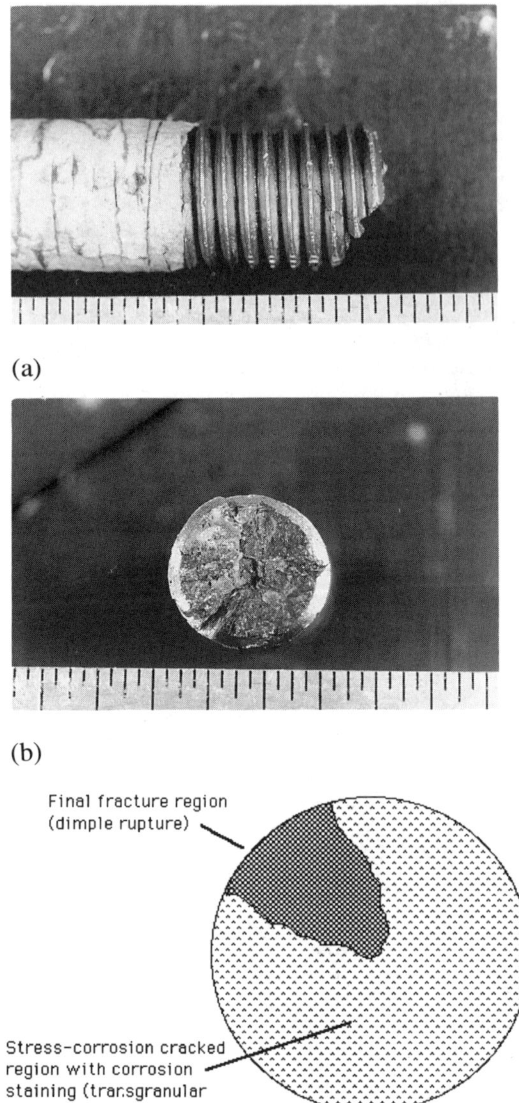

(a)

(b)

(c)

Fig. 5 SCC failure of a type 304 austenitic stainless steel bolt exposed to seacoast environment. (a) Lateral view demonstrating absence of gross ductility. (b) Brittle fracture surface with corrosion staining on facets and distinct overload rupture region. (c) Schematic of (b) representing progressive cracking region and overload region

of staining or corrosion products on the fracture surface is by no means positive proof of SCC. Many fractures become corroded before a failure investigation can commence. Also, some SCC failures are not discolored, especially in highly corrosion-resistant materials. The final fracture zone is usually caused by tensile overload and may show macro/microscopic ductility.

Fatigue failures also occur without evidence of ductility, but the fractured surfaces are usually smoother than those associated with SCC. Both types of failures can initiate at surface stress raisers. However, if the environment is sufficiently corrosive and the tensile stresses are sufficiently high, SCC can initiate even on smooth surfaces. Fracture caused by either of these mechanisms always initiate at surfaces or at some site where the aggressive chemical species can contact the metal.

Microscopic Examination

As in all fractographic studies, careful correlation of microscopic fracture topography with macroscopic fracture surface features is essential. The crack initiation site(s) and the directions of crack propagation must be identified accurately so that information concerning the sequence of events and the micromechanisms of fracture can be correlated with the circumstances of crack initiation and the mechanism of crack propagation. It is not uncommon for several different fracture mechanisms to be observed on a single fracture surface. Accordingly, correct identification of the initiating fracture mechanism and any changes in the mechanism during crack propagation is of vital importance in arriving at a factual understanding of the failure.

The characteristic features of SCC are transgranular and/or intergranular fractures that show little deformation in otherwise ductile materials, without the formation of easily visible corrosion products. Although the processes leading to crack initiation and propagation are still unclear, localized deformation constantly produces discontinuities on the metal surface that serve as initiation sites for SCC. These localized discontinuities may be formed from slip-band penetration of protective surface films or from anodic dissolution at surface-intersecting grain boundaries. The result is the spontaneous formation of a crack that is devoid of deformation and cleavage-like or intergranular in nature. In either case, the microfractographic features are indicative of classic brittle failure. Cracking can either traverse the entire cross section or propagate incrementally. It is not unusual in cases of transgranular SCC for the fracture surfaces to exhibit evidence of progressive crack movement similar to that associated with fatigue. Intermittent crack growth is characterized by the presence of crack-arrest lines or striations on the fracture surfaces (see Fig. 2 and 3).

Chemical Analysis

The service environment, be it atmospheric or process fluid or gas, is a major factor in SCC. Stress-corrosion cracking is caused by specific ions under given conditions of temperature, stress state, and alloy composition. It has been widely documented that certain ions are detrimental to specific alloy systems. In most cases, increasing the specific ion concentration will increase the cracking tendency and decrease the time to failure. It should also be noted that the responsible ion may be present in very low concentrations (<10 ppm) in the environment, yet be highly concentrated in crevices, pits, under deposits, and so on. For instance, when type 304 stainless steel cracks in a solution containing only a few parts per million of chloride, the chloride concentration is usually found to be significantly higher at the site of crack initiation. Figure 6 shows a case of chloride SCC for a type 304 austenitic stainless steel used as a storage vessel exposed to a seacoast environment. Furthermore, these areas of localized concentration can be quite acidic, even though the environment remains basic in nature. It is thus apparent that minor changes in the environment can cause drastic changes in the threshold and propagation of SCC. These minor changes are often difficult to determine.

Chemical analysis should include analysis of the environment, the subject metal, surface de-

(a)

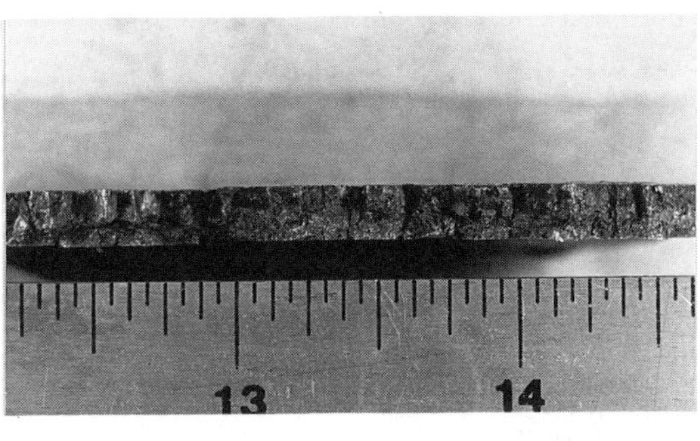

(b)

Fig. 6 Transgranular SCC of type 304 austenitic stainless steel following prolonged seacoast service. (a) Painted exterior side of tank exposed to environment. (b) View of representative macroscopic fracture surface. Note the brittle, faceted appearance.

posits, and substances contained within or on the fracture surfaces. The investigation should determine the exact roles (if any) these various elements played in the failure. When information on the environment and service history of the component is incomplete, or when transient conditions may have existed, results of the investigation may be inconclusive. Some circumstances warrant tests that simulate service and investigations of the effects of variations in environment composition and operating conditions. Table 2 presents a partial listing of documented SCC environments for various metal alloy systems (Ref 5).

Impurities and substances present at low concentrations or in trace amounts are of special interest in SCC investigations. The identification of chemicals contained in surface deposits on the metal and/or within the cracks is important in determining the identity of the corrodents and the nature of the damaging corrosion processes. However, care should be taken to ensure that the samples are representative of the service environment and are not merely residues of extraneous materials introduced during operations such as nondestructive testing or sectioning.

The entire operating cycle, including start-up and shutdown of a unit or operation, must be considered in an analysis. Particular chemical species, temperatures, and stresses acting on a component at any time during operation can promote SCC. Numerous failure analysis case studies have documented occurrences of SCC during unusual conditions experienced in start-up or shutdown phases or during breaks in routine operation (Ref 7).

Metallographic Analysis

The examination of polished metallographic specimens is highly recommended for determining the nature of SCC. Although it is possible to observe cracks on the surfaces of specimens or plant equipment, it is usually good practice to supplement visual inspection with nondestructive inspection. Visual observation usually indicates only that some form of cracking is occurring and does not identify the type of cracking or the depth of penetration (Ref 8). Crack depth and

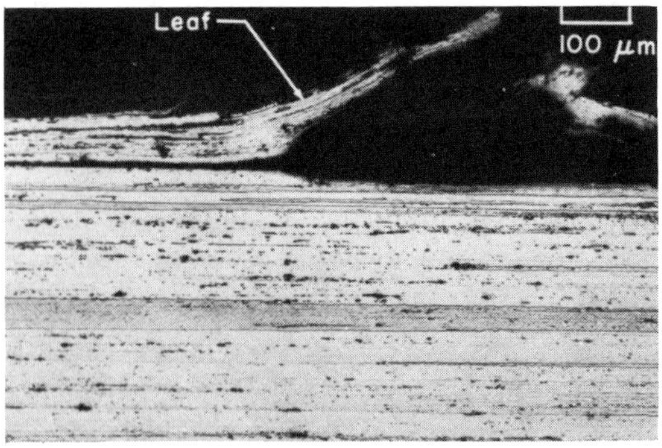

Fig. 7 Metallographic section through a SCC fractured aluminum bolthead cap showing "leaves" along the fracture surface. The primary and secondary cracks are intergranular. Keller's reagent, 100×. Source: Ref 9

crack length-to-depth ratios are particularly helpful in evaluating crack-growth rates and residual-strength issues. Even after metallographic examination, it is sometimes difficult to diagnose cracks as SCC.

Metallographic analysis should be used to examine the features associated with crack advancement. Metallography will reveal the transgranular and/or intergranular nature of the primary rupture interface and associated secondary crack segments. Cracks emanating from pits and corrosion of crack walls can also be documented more thoroughly by metallographic examination. Stress-corrosion cracking is usually intergranular in aluminum alloys, α-brasses, and low-carbon steels. Cracking is primarily transgranular in chloride SCC of austenitic stainless steels. There are always exceptions; one should look for trends and expect to find occasional mixed-mode failures.

In addition to observation of microstructural features associated with the crack paths, special attention should be paid to the following possibilities:

- Preferred grain orientation and associated crack path(s) (see Fig. 7). Some precipitation-hardenable aluminum alloys are susceptible to SCC in the short-transverse direction
- The possible existence of surface alteration (for example, hardening or softening result-

ing from processing or service), which could promote crack-initiation sites

- The presence of chemically/microstructurally altered surface zones due to contamination
- The possibility that processing has introduced hydrogen into the metal, which can promote hydrogen-induced delayed fracture. Hydrogen manifests itself microstructurally in the form of precipitated hydrides in a limited number of metal alloy systems (titanium, for example). Otherwise, the presence of hydrogen is difficult to detect
- Detailed analysis of grain-boundary regions to determine whether composition or structure was not controlled or was altered such that the grain boundaries became preferential sites for attack, as in the sensitization of austenitic stainless steel
- The possibility that an unspecified impurity may be present that would render an alloy particularly susceptible to SCC. In stainless steel, such impurities include arsenic and antimony (tramp elements for which chemical analysis is usually not conducted)

Carbon and Low-Alloy Steels

The susceptibilities of carbon and low-alloy steels are largely controlled by carbon content and achievable strength level, respectively. Carbon steels with carbon contents lower than 0.10 wt% appear to be the most susceptible. The behavior of high-strength low-alloy steels is less alloy dependent and more a function of strength. Steels with yield strengths of about 1379 MPa (200 ksi) or higher are especially susceptible to this type of failure (Ref 10).

The environments that most commonly produce SCC in mild steels are sodium hydroxide and nitrate solutions. Cracking in sodium hydroxide, formerly known as caustic embrittlement, once was a serious problem in riveted steam boilers (Ref 11). Fortunately, chemical inhibition and proper heat treatment are now widely used to eliminate most occurrences. Cracking failures of this type generally originate near faying surfaces, where localized surface sites serve to allow the concentration of soluble salts to accumulate high local concentrations of caustic soda (NaOH). The preferred crack-propagation path is intergranular. Although failures of this type have been produced at caustic concentrations as low as 5%, concentrations higher than 15% generally are required, with some level of oxygen. Failures typically occur when the operating temperature is in the range of 200 to 250 °C (390 to 480 °F). The concentration of NaOH necessary to produce cracking increases as the temperature decreases.

Case Study 1: HCN Cracking of Steel Pipe. A 25 mm (1 in.) OD drain line made from Schedule 80 ASTM A53 black pipe steel cracked in a weld. The drain line connected a small wash water tank containing mainly water, with 1 to 2% organics and 20 to 100 ppm HCN.

Intergranular SCC originating on the inside diameter was contained in the weld deposit. Hardness testing indicated an average base-metal hardness of 173 HB, while the weld averaged 245 HB. It was apparent that the weld had not been postheat treated. Figure 8 reveals the multiple intergranular crack patterns associated with this state of SCC.

Stress-corrosion cracking in nitrate solutions has been cited in tubing and couplings used in high-pressure condensate wells and ion storage tanks containing radioactive wastes. Cracking in nitrate solutions follows an intergranular path. Boiling solutions of several nitrates, including NH_4NO_3, $Ca(NO_3)_2$, $LiNO_3$, KNO_3, and $NaNO_3$, have been found to produce cracking (Ref 10). In general, more acidic solutions have more potent effects. The threshold stress necessary to produce cracking decreases with increasing nitrate solution concentration.

Crack initiation and propagation in nitrate solutions appear to be closely related to carbon and, possibly, to nitrogen in the grain boundaries. It has been suggested that cracking proceeds by a mechanism in which carbon-rich areas serve as cathodes, leading to anodic dissolution of neighboring ferrite grains. Thus, any processing/service that alters the carbon distribution appreciably raises the susceptibility of the steel to SCC (Ref 10). Strong carbide-formers alloyed in the steel are generally helpful in improving cracking resistance. As in caustic cracking, stress-relief annealing after welding or cold working can assist in preventing cracking of assemblies.

It has been shown that both plain and low-alloy steels are susceptible to SCC in ammonia. The higher-strength steels are more susceptible. Gen-

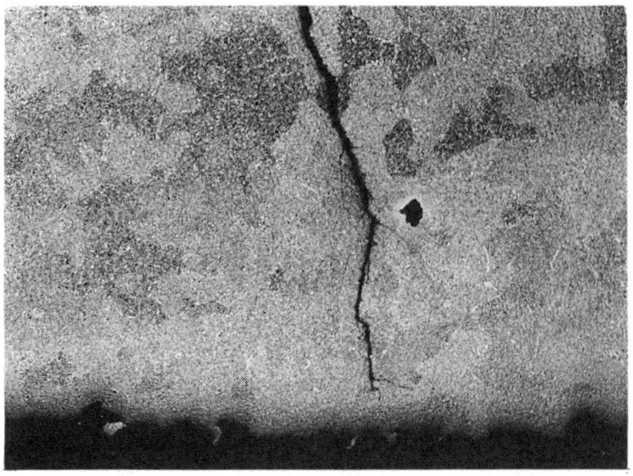

(a)

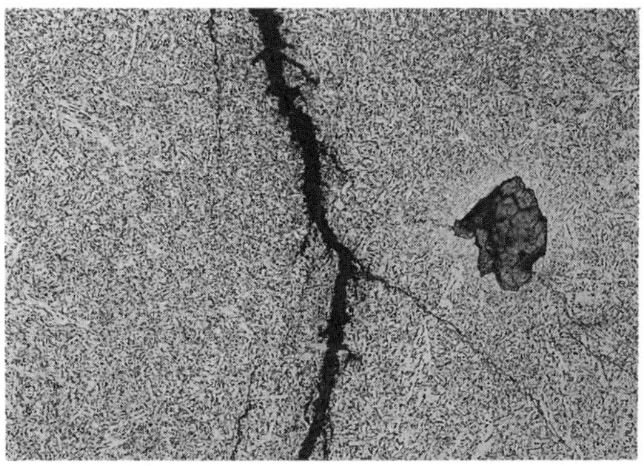

(b)

Fig. 8 HCN cracking in a Schedule 80 ASTM A53 black pipe steel drain line. (a) Primary crack penetration has advanced toward the outside diameter 2% nital etch. 50×. (b) The primary crack is intergranular and contains numerous intergranular branches. 2% nital, 200×

erally, these failures do not develop in pure states of ammonia, but in environments that also contain oxygen and carbon dioxide. Oxygen is the key element, with nitrogen serving only to aggravate the cracking problem. The addition of water to liquid ammonia completely eliminates any susceptibility to cracking. The Department of Transportation mandates water additions for interstate transportation of ammonia in high-strength steel vessels.

Case Study 2: Caustic Cracking Reactor Wall. A jacketed batch reactor failed by SCC cracking along the outside diameter of the vessel on the inside of the jacket (see Fig. 9). The reactor operated with a heat cycle produced from 1.4 MPa (200 psi) steam pressure and cooling from a

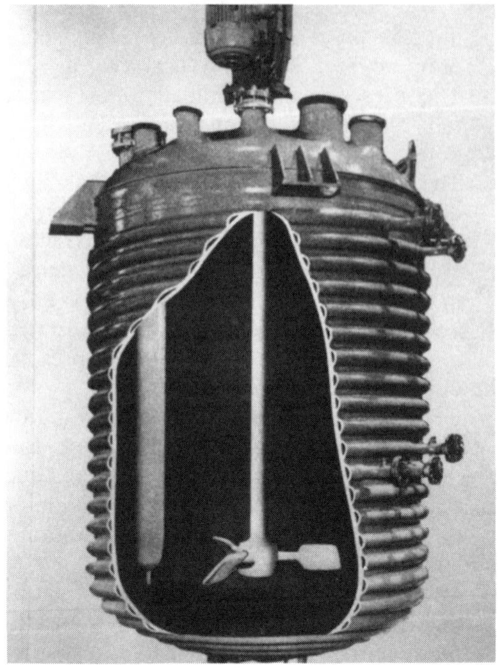

(a)

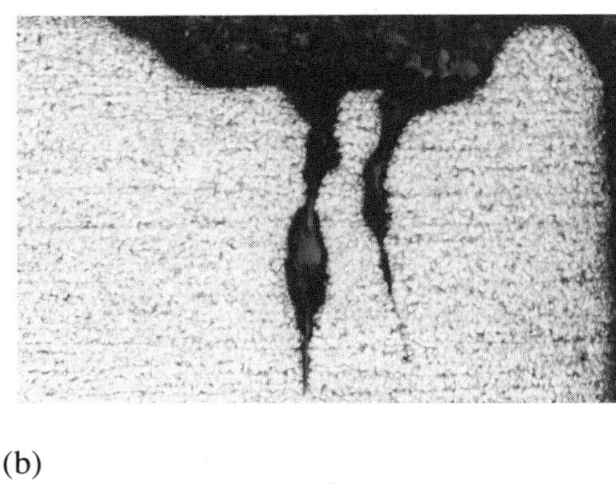

(b)

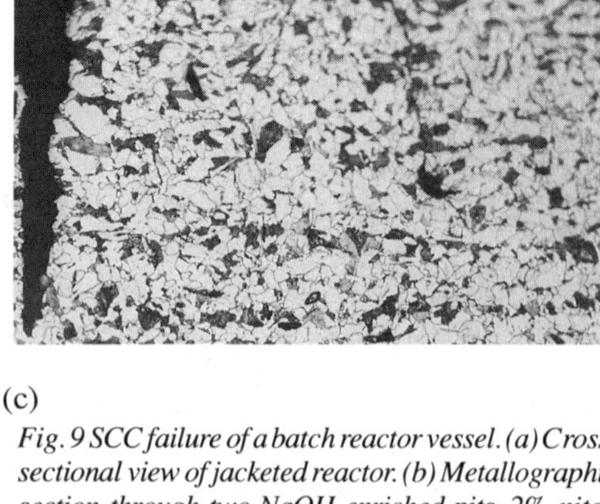

(c)

Fig. 9 SCC failure of a batch reactor vessel. (a) Cross-sectional view of jacketed reactor. (b) Metallographic section through two NaOH-enriched pits. 2% nital, 50×. (c) Intergranular cracking initiated from pit penetrations. 2% nital, 100×

Fig. 10 Pipe section immersed in liquid nitrogen and then ruptured to reveal the intersecting stress-corrosion cracks (dark portions on the fracture surfaces). Crack initiation was at the outside diameter.

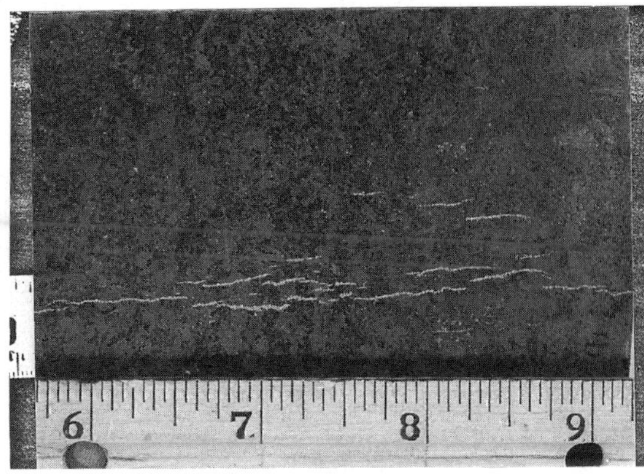

Fig. 11 Fluorescent-dye/magnetic-particle indication of a SCC colony on the bottom of a natural-gas pipe

cooling-tower water supply. Thus, the system was operating in a classic wet-dry cycle. Intergranular cracking emanated from pits along the jacket wall. The pits were found to contain from 50 to 100 ppm NaOH. Figure 9 illustrates the pit developments along the inside jacket surface, with SCC originating in the more deeply penetrating pits.

Stress-corrosion cracking of natural-gas pipelines has been well documented in varying soil environments. Slightly alkaline water solutions containing sodium bicarbonate ($NaHCO_3$) and sodium carbonate (Na_2CO_3) have been identified in the vicinity of stress-corrosion crack colonies. These failures have been characterized by multiple intergranular cracks, often branched, aligned in the axial direction (see Fig. 1). When exposed, the crack surfaces appear darkly stained due to the presence of magnetite (Fe_3O_4) and iron carbonate ($FeCO_3$) (Ref 12) (see Fig. 10).

Case Study 3: SCC of a Natural-Gas Pipeline. A 0.76 m (30 in.) OD API 5LX52 steel began leaking after 20 years in buried service. Upon removal of the protective surface coating, fluorescent-dye/magnetic-particle testing was performed to locate the SCC site(s). Surface cracks of varying lengths were located on the bottommost portion of the pipe (see Fig. 11).

Representative samples were obtained following replacement of the damaged sections to analyze the nature of the cracking. Figure 12 reveals intergranular cracking penetrating approximately one-third of the wall thickness. Multiple cracks are illustrated along with the characteristic staining on a selected sample that was ruptured following immersion in liquid nitrogen (–196 °C, or –320 °F) (see Fig. 10). X-ray diffraction analysis confirmed the presence of $FeCO_3$ on the stained regions. The characteristics of this field failure are consistent with bicarbonate-carbonate SCC.

Stress-corrosion cracking can take the form of hydrogen-assisted or induced cracking in specific material/environment systems (Ref 13). Hydrogen-assisted crack growth has been suggested for ferritic steels, nickel-base alloys, titanium alloys, and aluminum alloys (Ref 13). This form of hydrogen damage has been a problem in many structural steels for years. This form of SCC is generally more common in higher-strength steels with accompanying lower-temperature transformation products. Thus, hydrogen-assisted cracking is not generally observed in steels with yield strengths below 504 MPa (72 ksi). In many cases, pure water, steam condensate, or even atmospheric moisture acts as a sufficiently aggressive corrosive medium. However, it is much more common for hydrogen-assisted cracking to occur in acid solutions or in environments containing sulfides or arsenides, which accentuate hydrogen entering the structural metal.

Case Study 4: Hydrogen-Induced Cracking in a Pipeline. A 0.66 m (26 in.) OD natural gas pipeline with a wall thickness of 7.14 mm (0.28 in.) was found to have a leak site consisting

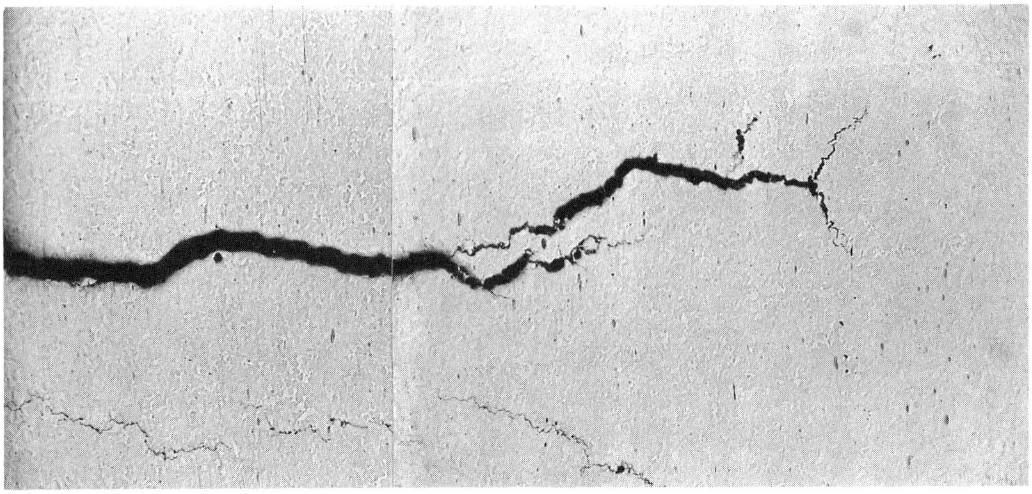

Fig. 12 Circumferential cross section of two stress-corrosion cracks following a branched, intergranular path. The major crack penetrates to a depth of approximately 2.5 mm (0.1 in.).

Fig. 13 Axially oriented crack in a pipe wall that led to gas leakage. Source: Ref 13

of a 0.18 m (7 in.) crack (see Fig. 13). The API 5LX52 steel was in service for 10 years under cathodic protection, with an average pipe-to-soil potential of –1.27 V.

Preliminary findings suggested that the crack occurred in a hard spot. Thus, hardness traverses were made along the pipe section following the longitudinal and circumferential orientations. The hardness evaluation verified that the crack was contained in a zone whose average hardness was 415 HB, while the average hardness of the rest of the pipe was about 175 HB.

Additional sectioning allowed the fracture surfaces to be examined. Figure 14 demonstrates brittle rupture, with an apparent initiation site located on the outside surface of the pipe just above the "3" on the ruler in the photograph.

The microstructure of the steel near the failure origin consisted of martensite, as illustrated in Fig. 15. Small secondary, intergranular cracks, branching from the primary rupture interface, were also visible. The microstructural state away from this zone typically consisted of ferrite and pearlite.

The preponderance of physical evidence in this case suggests that failure was caused by hydrogen-assisted cracking. In the absence of any corrosion-related damage, it appears that hydrogen, generated by the cathodic protection system at the outer pipe surface, was absorbed into the metal. The combination of the presence of hydrogen, the crack-sensitive microstructure, and the sustained stress from the internal gas pressure led to hydrogen-assisted cracking (Ref 14).

Hardenable Stainless Steels

The hardenable stainless steels, including martensitic chromium stainless steels such as type 410 and chromium-nickel precipitation-hardenable steels such as 17-4 PH, have been known to crack in hot chloride solutions and in high-temperature, high-purity water. Beyond these, the environments that cause failure are not specific; almost any corrosive environment capable of sus-

Fig. 14 Sectioning allowed partial exposure of the fracture surfaces of the crack in Fig. 13. Note the radial line pattern converging at the surface site of crack initiation just above the "3" on the ruler. Source: Ref 13

taining hydrogen evolution can cause cracking. However, the hardenable stainless steels provide maximum resistance after tempering at 600 °C (1110 °F), and the precipitation-hardenable stainless steels are most resistant after overaging at approximately 620 °C (1150 °F) (Ref 11). Thus, susceptibility to failure is determined primarily by the metallurgical structural state.

Case Study 5: SCC of a 17-4 PH Shroud Band. Turbine blades were mounted in a shroud band, with the blade tendons set through holes in the shroud (see Fig. 16). Each blade tendon was then cold formed as a riveted head. The turbine assembly was normally exposed to steam generated from a geothermal well operating in the temperature range of 110 to 135 °C (230 to 275 °F). Analysis of the steam varied, but generally contained 30 to 50 ppm H_2S, 500 to 1000 ppm chlorides, and varying amounts of potassium and vanadium.

Both the turbine blade tendons and shroud assembly stress-corrosion cracked as shown in Fig. 17. However, the back of each tendon revealed more extensive cracking, probably caused by cold working sustained in the areas associated with fretting damage.

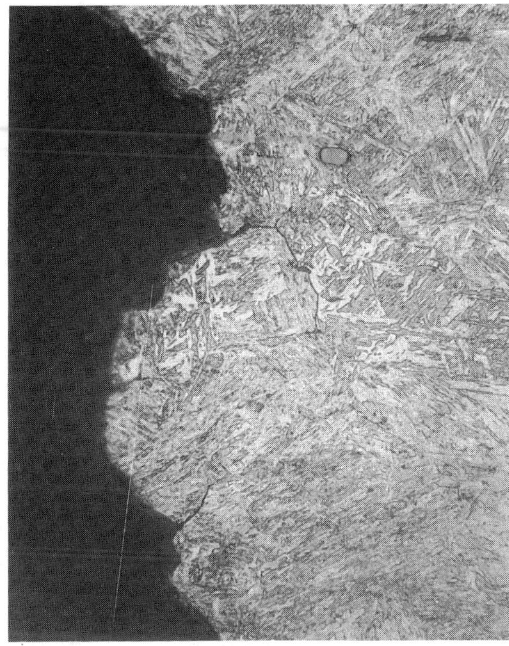

Fig. 15 Cross section of the primary fracture surface of the pipe in Fig. 13, showing secondary cracks contained within the martensite matrix. 2% nital, 500×. Source: Ref 13

Fig. 16 Steam turbine assembly after removal from service. SCC was found in the shroud bands

Austenitic Stainless Steels

Although the corrosion resistance of austenitic stainless steels is usually quite outstanding, they are susceptible to both transgranular and intergranular SCC. Intergranular cracking most often occurs in alloys that have been sensitized, whereas transgranular cracking is independent of sensitization. The most infamous agent for cracking austenitic stainless steels is the chloride ion,

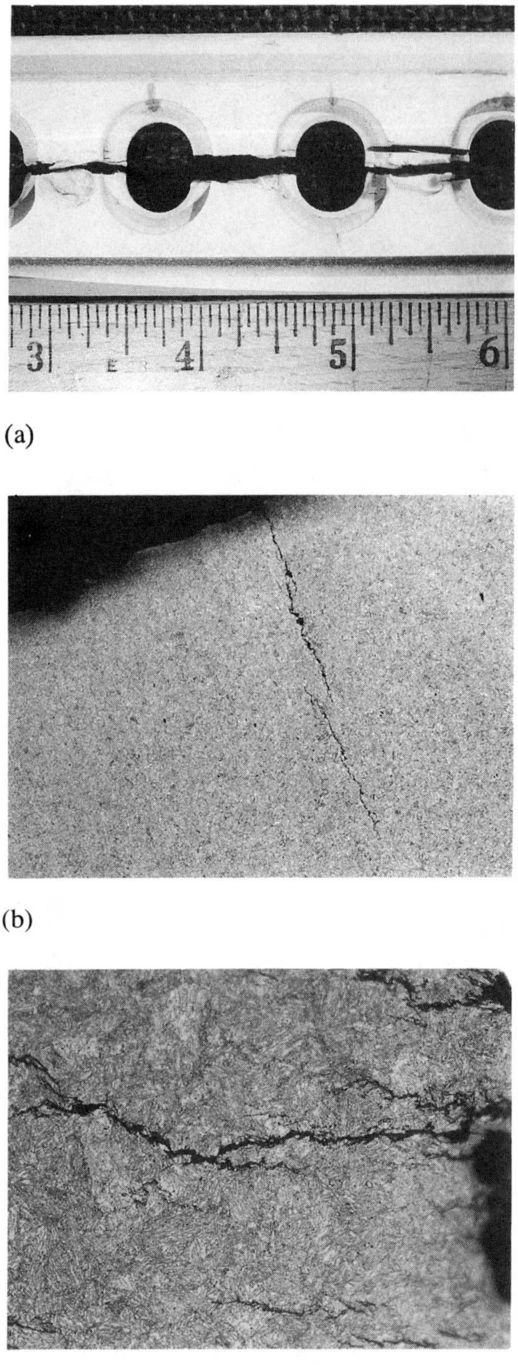

(a)

(b)

(c)

Fig. 17 SCC of a 17-4 PH stainless steel turbine shroud band. (a) Numerous circumferential cracks were observed in the shroud-band holes. (b) Partially penetrating stress-corrosion cracks emanating from a tendon opening. 50×. (c) Intergranular cracking with numerous branches. 400×

and chloride attack constitutes the most important drawback to the use of these steels. These alloys are most susceptible to hot concentrated chloride solutions, but failures have been reported in solutions with chloride contents as low as 5 ppm (Ref 11). Another source of transgranular cracking is hot caustic solutions. Austenitic stainless steels will resist attack by caustic at low temperatures and concentrations, but may crack in solutions above 50% caustic at temperatures above 150 °C (300 °F).

If a steel is solution heat treated above 1000 °C (1830 °F) and slowly cooled from this temperature or reheated in the range of 550 to 750 °C (1020 to 1380 °F), its resistance to intergranular corrosion is greatly decreased. During these detrimental heat treatments, chromium-rich carbides, usually in the form of $Cr_{23}C_6$, precipitate along the grain boundaries. This process temporarily produces a chromium-depleted zone adjacent to the carbide precipitates. Because the carbides preferentially precipitate at grain boundaries, and because chromium is the element that promotes corrosion resistance, this depletion leads to intergranular corrosion (Ref 15).

The corrosive attack in sensitized alloys such as these is almost always intergranular. Figure 4 illustrates an austenitic stainless steel previously sensitized and then exposed to corrosion attack. As sensitization progresses, the intergranular attack increases to the point where the sample ruptures with minimal resistance. Whether sensitizing is damaging in a given instance depends on the requirements of the specific application, the environment and the applied and/or residual stress state, and the prior history of the steel.

For nonsensitized stainless steels in solutions containing chlorides, SCC takes the form of multibranched, transgranular cracking. Chloride SCC has been reported in such diverse media as natural waters, geothermal steam, industrial steam condensate, acetic acid, and caustic (Ref 1). Higher temperatures intensify the stress-corrosion state, probably by affecting the concentration of chlorides at the metal surface. Even very low chloride concentrations are sufficient to cause cracking if the temperature is raised.

(a)

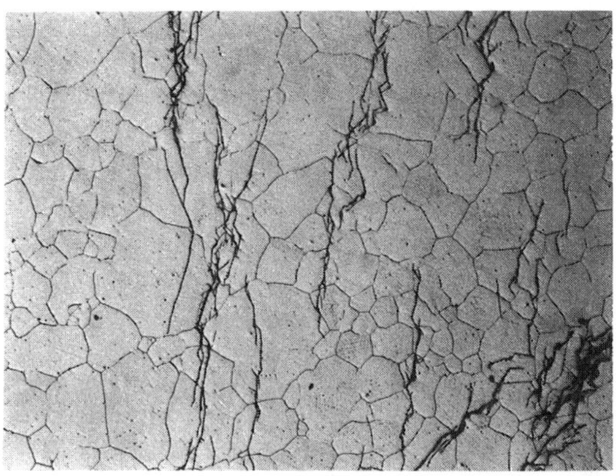

(b)

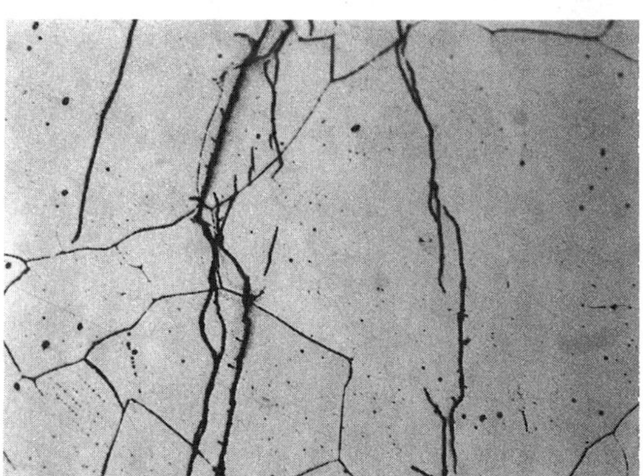

(c)

Fig. 18 Chloride cracking in a sensitized steel thermowell pipe cap weld. (a) Cracking was contained in the region of the circumferential weld. (b) Carbide enrichment is observed in the austenitic grain boundaries. Multiple transgranular crack segments are also visible. Marble's reagent, 50×. (c) Higher-magnification view of sensitized structure and transgranular SCC. Marble's reagent, 200×

Intergranular cracking also develops in systems that handle hydrogen sulfide. The cause can be traced to the formation of polythionic acids ($H_2S_xO_6$, where $x = 3, 4, 5$) formed by the reaction of residual metallic sulfides on oil-refinery equipment with moist air at room temperature. Sensitized stainless steels are particularly susceptible to polythionic acid attack (Ref 16, 17).

Duplex stainless steels demonstrate enhanced resistance to sensitization and intergranular corrosion compared with austenitic grades of similar carbon content (Ref 18). In addition, their structure, which is a mixture of austenite and ferrite, provides substantially improved strength and toughness with no loss of corrosion resistance. The cracking characteristics of these materials are very dependent on the stress state. At low to moderate stresses, the austenite grains crack transgranularly, but continuation of the transgranular crack path through the ferrite grains is blocked, resulting in intermittent transgranular and intergranular crack segments in a zigzag pattern. At higher stresses, the crack path through both phases is transgranular (Ref 18).

Case Study 6: Chloride SCC in a Sensitized Type 304 Stainless Steel Thermowell Pipe. A 2.8 MPa (400 psi) pressure steam service line of type 304 austenitic stainless steel (0.087 wt% C) cracked from the outer surface. Operating in a marine/industrial environment with intermittent wet-dry cycles, chlorides were concentrated on the outside pipe surfaces. Sensitization was also observed around the circumferential weld (see Fig. 18), along with transgranular SCC.

Copper-Base Alloys

Copper alloys, even with minor amounts of phosphorus, zinc, antimony, nickel, silicon, and aluminum, are susceptible to intergranular SCC in ammonia and ammoniacal solutions. In some cases, depending on the pH of the solution, both intergranular cracking and transgranular cracking are observed. Transgranular cracking more commonly occurs under more acidic ($pH < 4$) and more alkaline ($pH > 8$) conditions. Ammoniacal SCC of brasses shows intergranular cracking and a minimum time to failure at a pH of 7.

The fact that SCC requires water and oxygen is confirmed by the observation that anhydrous

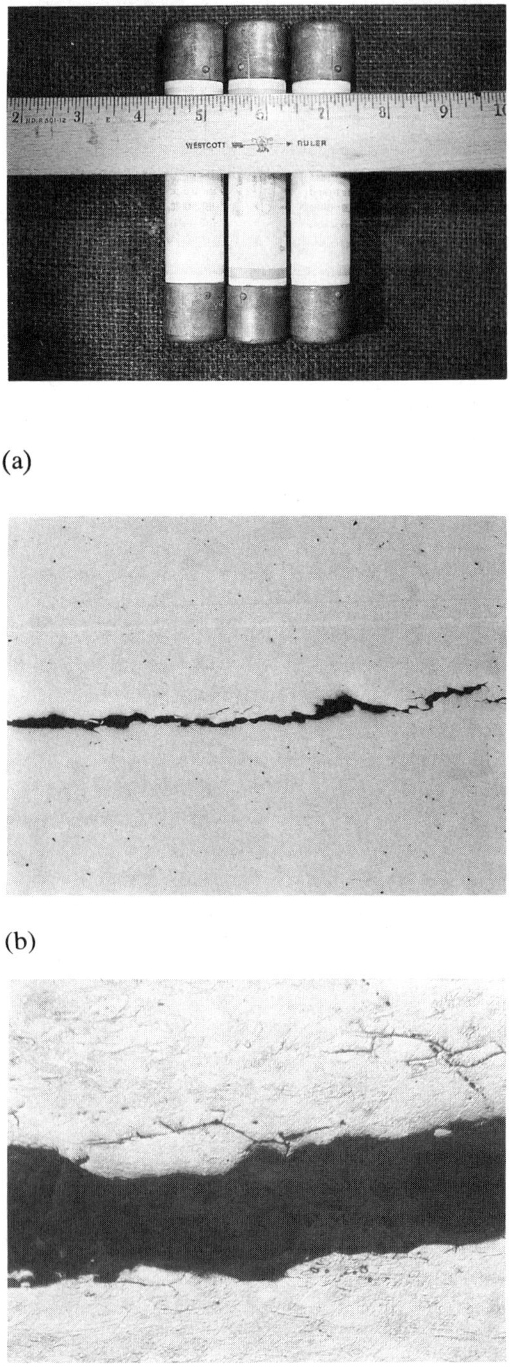

(a)

(b)

(c)

Fig. 19 SCC in cold-formed brass fuse caps. (a) Numerous longitudinal cracks are visible in the brass caps. (b) Unetched metallographic section showing the primary crack opening with limited branching. 50×. (c) Etched metallographic section detailing limited intergranular branching. 50% HNO_3 + 50% distilled water by volume. 250×

ammonia does not cause SCC of copper or any of its alloys (Ref 1). However, using copper in such a service environment would be inappropriate unless the ammonia was kept free of air and water under all conditions.

The sensitivity of some copper alloys to ammonia SCC also accounts for failures where the presence of ammonia was at first unknown. Brass heat-exchanger tubes stored in industrial atmospheres have been known to crack from ammonia absorbed by accumulated moisture on the tubes. Nitrogenous compounds such as nitrates, amines, and oxides of nitrogen, as well as sulfates, sulfur dioxide, and steam, have caused similar failures. There is little doubt that ammonia is the most worrisome agent for copper alloys. However, not all cracking failures should be attributed solely to ammonia (Ref 1).

Case Study 7: Cracking of Brass Fuse Caps. Figure 19 illustrates longitudinal cracking along unplated 70% Cu cartridge brass fuse caps. A high-residual-stress state stemming from cold forming and exposure to a marine/industrial atmosphere led to cracking. Although the fuses were in service in a high-humidity environment, no ammonia was used in production or in the service areas. Some of the fuses actually cracked in un-air-conditioned storage prior to service exposure. Mercury nitrate testing also indicated that the fuses were highly susceptible to SCC. Figures 19(a) and (b) reveal crack segments in the unetched and etched states.

Titanium and Titanium Alloys

Over the years, titanium and many of its alloys have been shown to be susceptible to SCC in certain environments — for example, distilled water, seawater, chlorinated solvents, red fuming nitric acid, and molten salts. The two environments (Ref 15) that have received the greatest attention are (1) aqueous solutions of NaCl at both neutral and acidic pH values and (2) organic liquids, such as carbon tetrachloride and especially methanol.

Generally, α and $\alpha + \beta$ alloys are susceptible, while β and pure titanium are not. In α alloys, a critical concentration of tin or aluminum is usually required before significant SCC is observed. Precipitation of the ordered phase Ti_3Al and an increase in grain size also increase susceptibility.

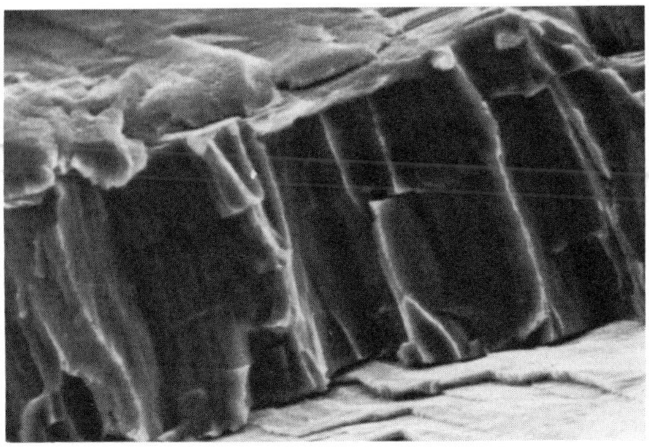

Fig. 20 Flutes and intergranular cleavage resulting from SCC of β-annealed Ti-8Al-1Mo-1V in methanol. Source: Ref 9

In α + β alloys, the susceptibility decreases as the amount of β increases. The cracks are usually transgranular cleavage and are restricted to the α phase. Mild stress-corrosion attack can exhibit fracture that cannot be easily distinguished from normal overload, while more severe attack results in cleavage or quasicleavage fracture. Stress-corrosion fractures exhibit cleavage along with fluting and quasicleavage (Ref 9). Figure 20 illustrates elongated grooves or voids connecting widely separated cleavage planes that resemble pleats in window draperies.

Titanium of various purity levels along with the α alloys will corrode intergranularly if left unstressed in a methanol/halogen solution. Alloys containing the β phase usually do not corrode in the unstressed condition, but do so readily when stressed. In unnotched samples, the fracture usually initiates along grain boundaries and then begins to propagate transgranularly. In notched samples, the fracture tends to be more completely transgranular, although at low stress intensities intergranular cracking will be observed (Ref 15).

The primary difference between cracking in methanol/halogen solutions and that in aqueous NaCl solutions arises from the fact that titanium will dissolve in methanol to produce methoxide; therefore, when a stress or halogens are present to break down the passive film, corrosion will occur. Whether the corrosion is transgranular or intergranular depends on the slip mode and stress level. When corrosion occurs in unstressed samples, grain boundaries are the preferred paths, perhaps because grain boundaries are more difficult to passivate due to structural asperities or because segregation of certain species make them more reactive. At low stresses in all samples, independent of notches, the cracking mode is intergranular (Ref 15).

Aluminum and Aluminum Alloys

Stress-corrosion cracking in aluminum alloys is usually confined to higher-strength alloys, such as the 2000 and 7000 series, whose strength properties stem from heat treatment rather than cold working. Corrosion and SCC in these alloys is almost invariably intergranular. When heat treated for strength, these alloys sometimes contain precipitates in the grain boundaries that render them susceptible to intergranular SCC. Cracking also may occur in aluminum-magnesium alloys that contain more than 3.5% Mg, and failures have occurred in certain high-strength casting alloys. Most studies involving aluminum alloys have focused on chloride-containing solutions. Additional testing has been performed in water vapor and in aqueous solutions containing other anions.

As a general rule, the SCC susceptibility of aluminum alloys increases after shorter aging times (underaged condition) and then decreases after longer times. Peak susceptibility may occur at peak hardness. Overaging invariably reduces susceptibility to SCC. In all of these alloys, the pathway in typically intergranular. Sensitivity to SCC is strongly dependent on grain orientation with respect to the principal axis, the short transverse direction being the most sensitive to intergranular cracking (Ref 9). Service failures of extrusions, rolled plate, and forgings have been caused chiefly by tensile stresses acting in the short transverse direction. Figure 7 shows intergranular cracking in a fractured bolthead cap. In this case, SCC was favored by simultaneous application of tensile stress and exposure to a marine environment in service (Ref 19). Specific thermal treatments, such as those that produce T7 tempers in Al-Zn-Mg-Cu alloys and T8 in Al-Cu alloys, have been developed to provide improved resistance to SCC in the more critical short transverse direction. Specifically, the yield strength of alloy 7075-T6, tested under normal conditions, is reported to decrease from 552 MPa (80 ksi) in the longitudinal orientation to 414 MPa (60 ksi) in the short transverse. However, the maximum stress that the material can withstand without failure when subjected to cyclic alternate immersion in a 3.5% NaCl solution for 10 min and then in air for 50 min decreased from approximately 414 MPa (60 ksi) for testing in the longitudinal direction to 83 MPa (12 ksi) for the short transverse direction.

Case Study 8: SCC of Aircraft Hinge Brackets. Forged aluminum alloy 2014-T6 hinge brackets in naval aircraft rudder and aileron linkages were found cracked in service. The cracks were in the hinge lugs.

Examination of the microstructure of a lug from a typical failed hinge bracket revealed intergranular corrosion and cracking in the region of the cracks discovered in service. The fracture was not located along the flash line nor was it oriented in the short transverse direction, both of which are characteristic of most SCC failures of aluminum forgings.

It was evident from the interference fit of the bushing in the hole of the lug that a hoop stress of sustained tension existed in the corroded area. The overall condition of the paint film was good, but chipping had occurred on the lug. Figure 21 shows the crack in the lug and the branched secondary cracking adjacent and parallel to the fracture surface.

Failure of the hinge brackets occurred by SCC. The corrosion was caused by exposure to a marine environment in the absence of an adherent paint in the stressed area. The stress resulted from the interference fit of the bushing in the lug hole (Ref 10).

Case Study 9: Intergranular Corrosion and SCC in an Aluminum Pressure-Switch Housing. Figure 22 shows excessive metal degradation of a cast 308 aluminum pressure-switch housing. Cracking was localized in the highly stressed region of the treaded connections exposed to atmospheric chlorides from the seacoast. Deterioration included intergranular corrosion and SCC.

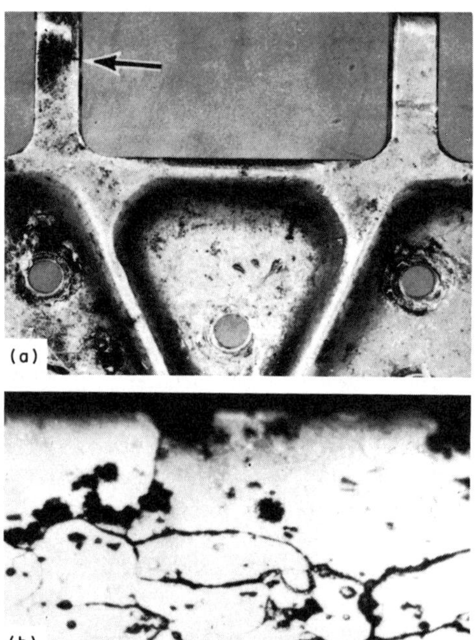

Fig. 21 Aluminum alloy 2014-T6 hinge bracket that failed by SCC in service. (a) Hinge bracket. Arrow indicates crack. (b) Micrograph showing secondary cracking adjacent and parallel to the fracture surface. Keller's reagent, 250×. Source: Ref 10

Nickel and Nickel Alloys

Commercially pure nickel is quite resistant to SCC in various chloride salts and has excellent resistance to all nonoxidizing halides (Ref 16). Oxidizing acid chlorides are extremely corrosive and should be used only in low concentrations with wrought nickel. The most serious cases of SCC failures of nickel alloys involve contact with hydrofluoric acid vapor and hot concentrated sodium hydroxide solutions (Ref 16). Failures of highly stressed nickel and nickel alloys exposed to hot concentrated caustics have been reported, although nickel, nickel-clad steel, and Monel 400 generally are considered to be satisfactory in hot caustics. For temperatures above 450 °C (840 °F), a low-carbon nickel is recommended. Nickel with higher carbon contents may become brittle, and, if stressed in fused caustic and steam, may fail, (Ref 20).

The susceptibility of austenitic nickel-chromium alloys to transgranular cracking in chlorides generally decreases as the nickel content

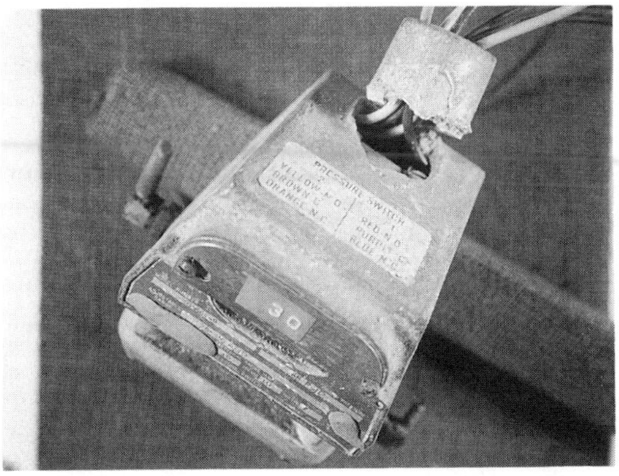

(a)

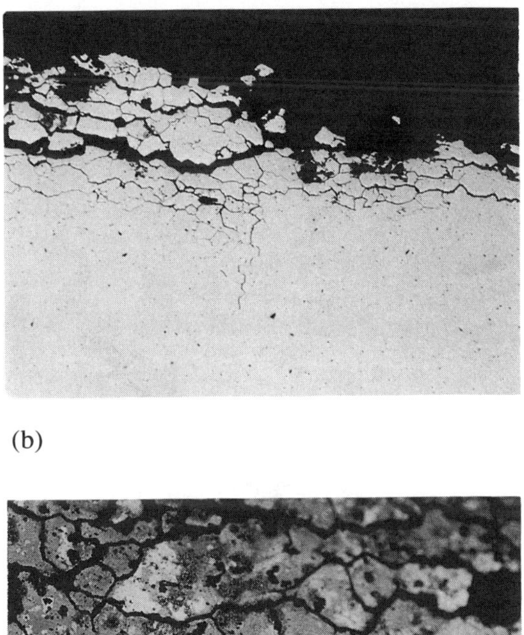

(b)

(c)

Fig. 22 SCC failure of a cast aluminum pressure-switch housing. (a) Failed assembly. (b) Unetched metallographic section along the fracture interface, revealing extensive intergranular separation. 50×. (c) Etched metallographic section identifies extensive intergranular separation. 4 mL HBF_4 + 100 mL distilled water. 200×

increases. Inconel 600, with a minimum nickel content of 72%, is virtually immune to chloride SCC. This alloy is subject to SCC in high-temperature caustic alkalis in high concentrations. When Inconel 600 is used in this type of environment, it should be fully stress relieved prior to service, and the operating stress should be kept to a minimum. The recommendation given for alkali-caustic service should be followed if the alloy is used in an application that involves contact with mercury at elevated temperatures. Nickel alloys also may be susceptible to cracking in H_2S streams when sensitized or galvanically coupled to steel. Figure 23 illustrates the combined damage of cyanides and sulfides on sensitized Inconel 600.

Inconel 601 exhibits good resistance to SCC in various corrosive media, including 45% $MgCl_2$ and 10 to 98% NaOH. Nickel-chromium-molybdenum alloys (alloys C-276 and 625) are often employed for their resistance to SCC in many industrial chemicals (Ref 10).

References

1. D.R. McIntyre and C.P. Dillon, *Guidelines for Preventing Stress Corrosion Cracking in the Chemical Process Industries*, MTI Publ. No. 15, Materials Technology Institute of the Chemical Process Industries, 1985, p 108
2. *Corrosion Data Survey*, 5th ed., National Association of Corrosion Engineers, 1981, p 268-269
3. D.A. Ryder, T.J. Davies, I. Brough, and F.R. Hutchings, General Practice in Failure Analysis, in *Metals Handbook*, Vol 11, 9th ed., *Failure Analysis and Prevention*, American Society for Metals, 1986, p 19
4. R.N. Parkins, B.N. Leis, and T.K. Christman, "Spatial Densities of Stress-Corrosion Cracks in Line-Pipe Steels," NG-18 Report No. 195, Battelle-Columbus Laboratories, Aug 1991, p 35
5. J. Hickling, Failures Due to the Conjoint Action of Corrosion and Mechanical Loading, in *Systematic Analysis of Technical Failures*, DGM Informationsgesellschaft, 1986, p 229-243
6. L. Engel and H. Klingele, *An Atlas of Metal Damage*, Prentice-Hall, 1981, p 105
7. J.H. Payer, Environmental Cracking, in *Forms of Corrosion — Recognition and Prevention*, C.P. Dillon, Ed., National Association of Corrosion Engineers, 1982, p 55
8. E.H. Phelps and M.E. Komp, Techniques for Diagnosis of Corrosion Failures, in *Source Book on Failure Analysis*, American Society for Metals, 1974, p 346

Fig. 23 Intergranular cracking of sensitized nickel alloy 600 piping in water service containing H_2S and cyanides. 15×. Source: Ref 1

9. *Metals Handbook*, Vol 12, 9th ed., *Fractography*, ASM International, 1987, p 27, 433
10. B.E. Wilde, Stress-Corrosion Cracking, in *Metals Handbook*, Vol 11, 9th ed., *Failure Analysis and Prevention*, American Society for Metals, 1986, p 203
11. M.E. Holmberg and T.V. Bruno, Metallurgy and the New Corrosion Engineer: What He Should Know About Corrosion Induced Brittle Failures, *Mater. Protect.*, May 1966, p 8
12. R.R. Fessler and T.J. Barlo, Many Causes Possible for Stress-Corrosion Cracking, *Pipeline Gas J.*, March 1979, p 25
13. R.H. Jones and R.E. Ricker, Environmentally Induced Cracking, in *Metals Handbook*, Vol 13, 9th ed., *Corrosion*, ASM International, 1987, p 145
14. T.P. Groeneveld and A.R. Elsea, "Investigation of a Failure in a 26 Inch Diameter Pipeline," Battelle Research Report, Jan 1967
15. C.L. Brient, Stress Corrosion Cracking, in *Metallurgical Aspects of Environmental Failures*, Elsevier Science, 1985, p 31
16. H.L. Logan, *The Stress-Corrosion Cracking of Metals*, John Wiley and Sons, 1966, p 264
17. H.H. Uhlig and R.W. Revie, *Corrosion and Corrosion Control*, 3rd ed., Wiley Interscience, 1985, p 323
18. J.E. Truman, Stress-Corrosion Cracking of Martensitic and Ferritic Stainless Steels, *Int. Met. Rev.*, Vol 6, 1981, p 301
19. J.C. Scully, *The Fundamentals of Corrosion*, 2nd ed., Pergamon, 1975, p 179
20. H.L. Logan, Stress Corrosion, in *NACE Basic Corrosion Course*, National Association of Corrosion Engineers, 1969, p 10.1-10.19

Index